Communications in Computer and Information Science 443

T0213769

Anne Laurent   Olivier Strauss
Bernadette Bouchon-Meunier
Ronald R. Yager (Eds.)

Information Processing and Management of Uncertainty in Knowledge-Based Systems

15th International Conference, IPMU 2014
Montpellier, France, July 15-19, 2014
Proceedings, Part II

 Springer

Volume Editors

Anne Laurent
Université Montpellier 2, LIRMM, France
E-mail: laurent@lirmm.fr

Olivier Strauss
Université Montpellier 2, LIRMM, France
E-mail: olivier.strauss@lirmm.fr

Bernadette Bouchon-Meunier
Sorbonne Universités, UPMC Paris 6, France
E-mail: bernadette.bouchon-meunier@lip6.fr

Ronald R. Yager
Iona College, New Rochelle, NY, USA
E-mail: ryager@iona.edu

ISSN 1865-0929   e-ISSN 1865-0937
ISBN 978-3-319-08854-9   e-ISBN 978-3-319-08855-6
DOI 10.1007/978-3-319-08855-6
Springer Cham Heidelberg New York Dordrecht London

Library of Congress Control Number: 2014942459

Typesetting: Camera-ready by author, data conversion by Scientific Publishing Services, Chennai, India

Printed on acid-free paper

Springer is part of Springer Science+Business Media (www.springer.com)

Preface

Here we provide the proceedings of the 15th International Conference on Information Processing and Management of Uncertainty in Knowledge-based Systems, IPMU 2014, held in Montpellier, France, during July 15–19, 2014. The IPMU conference is organized every two years with the focus of bringing together scientists working on methods for the management of uncertainty and aggregation of information in intelligent systems.

This conference provides a medium for the exchange of ideas between theoreticians and practitioners working on the latest developments in these and other related areas. This was the 15th edition of the IPMU conference, which started in 1986 and has been held every two years in the following locations in Europe: Paris (1986), Urbino (1988), Paris (1990), Palma de Mallorca (1992), Paris (1994), Granada (1996), Paris (1998), Madrid (2000), Annecy (2002), Perugia (2004), Malaga (2008), Dortmund (2010) and Catania (2012).

Among the plenary speakers at past IPMU conferences, there have been three Nobel Prize winners: Kenneth Arrow, Daniel Kahneman, and Ilya Prigogine. An important feature of the IPMU Conference is the presentation of the Kampé de Fériet Award for outstanding contributions to the field of uncertainty. This year, the recipient was Vladimir N. Vapnik. Past winners of this prestigious award were Lotfi A. Zadeh (1992), Ilya Prigogine (1994), Toshiro Terano (1996), Kenneth Arrow (1998), Richard Jeffrey (2000), Arthur Dempster (2002), Janos Aczel (2004), Daniel Kahneman (2006), Enric Trillas (2008), James Bezdek (2010), Michio Sugeno (2012).

The program of the IPMU 2014 conference consisted of 5 invited academic talks together with 180 contributed papers, authored by researchers from 46 countries, including the regular track and 19 special sessions. The invited academic talks were given by the following distinguished researchers: Vladimir N. Vapnik (NEC Laboratories, USA), Stuart Russell (University of California, Berkeley, USA and University Pierre et Marie Curie, Paris, France), Inés Couso (University of Oviedo, Spain), Nadia Berthouze (University College London, United Kingdom) and Marcin Detyniecki (University Pierre and Marie Curie, Paris, France).

Industrial talks were given in complement of academic talks and highlighted the necessary collaboration we all have to foster in order to deal with current challenges from the real world such as Big Data for dealing with massive and complex data.

The success of IPMU 2014 was due to the hard work and dedication of a large number of people, and the collaboration of several institutions. We want to acknowledge the industrial sponsors, the help of the members of the International Program Committee, the reviewers of papers, the organizers of special sessions, the Local Organizing Committee, and the volunteer students. Most of all, we appreciate the work and effort of those who contributed papers to the conference. All of them deserve many thanks for having helped to attain the goal of providing a high quality conference in a pleasant environment.

May 2014   Bernadette Bouchon-Meunier
                                                                Anne Laurent
                                                             Olivier Strauss
                                                           Ronald R. Yager

Conference Committee

General Chairs

Anne Laurent   Université Montpellier 2, France
Olivier Strauss   Université Montpellier 2, France

Executive Directors

Bernadette Bouchon-Meunier   CNRS-UPMC, France
Ronald R. Yager   Iona College, USA

Web Chair

Yuan Lin   INRA, SupAgro, France

Proceedings Chair

Jérôme Fortin   Université Montpellier 2, France

International Advisory Board

Giulianella Coletti (Italy)   Benedetto Matarazzo (Italy)
Miguel Delgado (Spain)   Manuel Ojeda-Aciego (Spain)
Mario Fedrizzi (Italy)   Maria Rifqi (France)
Laurent Foulloy (France)   Lorenza Saitta (Italy)
Salvatore Greco (Italy)   Enric Trillas (Spain)
Julio Gutierrez-Rios (Spain)   Llorenç Valverde (Spain)
Eyke Hüllermeier(Germany)   José Luis Verdegay (Spain)
Luis Magdalena (Spain)   Maria-Amparo Vila (Spain)
Christophe Marsala (France)   Lotfi A. Zadeh (USA)

Special Session Organizers

Jose M. Alonso   European Centre for Soft Computing, Spain
Michal Baczynski   University of Silesia, Poland
Edurne Barrenechea   Universidad Pública de Navarra, Spain
Zohra Bellahsene   University Montpellier 2, France
Patrice Buche   INRA, France
Thomas Burger   CEA, France
Tomasa Calvo   Universidad de Alcalá, Spain

Brigitte Charnomordic	INRA, France
Didier Coquin	University of Savoie, France
Cécile Coulon-Leroy	École Supérieure d'Agriculture d'Angers, France
Sébastien Destercke	CNRS, Heudiasyc Lab., France
Susana Irene Diaz Rodriguez	University of Oviedo, Spain
Luka Eciolaza	European Centre for Soft Computing, Spain
Francesc Esteva	IIIA - CSIC, Spain
Tommaso Flaminio	University of Insubria, Italy
Brunella Gerla	University of Insubria, Italy
Manuel González Hidalgo	University of the Balearic Islands, Spain
Michel Grabisch	University of Paris Sorbonne, France
Serge Guillaume	Irstea, France
Anne Laurent	University Montpellier 2, France
Christophe Labreuche	Thales Research and Technology, France
Kevin Loquin	University Montpellier 2, France
Nicolás Marín	University of Granada, Spain
Trevor Martin	University of Bristol, UK
Sebastia Massanet	University of the Balearic Islands, Spain
Juan Miguel Medina	University of Granada, Spain
Enrique Miranda	University of Oviedo, Spain
Javier Montero	Complutense University of Madrid, Spain
Jesús Medina Moreno	University of Cádiz, Spain
Manuel Ojeda Aciego	University of Málaga, Spain
Martin Pereira Farina	Universidad de Santiago de Compostela, Spain
Olga Pons	University of Granada, Spain
Dragan Radojevic	Serbia
Anca L. Ralescu	University of Cincinnati, USA
François Scharffe	University Montpellier 2, France
Rudolf Seising	European Centre for Soft Computing, Spain
Marco Elio Tabacchi	Università degli Studi di Palermo, Italy
Tadanari Taniguchi	Tokai University, Japan
Charles Tijus	LUTIN-CHArt, Université Paris 8, France
Konstantin Todorov	University Montpellier 2, France
Lionel Valet	University of Savoie, France

Program Committee

Michal Baczynski	Ulrich Bodenhofer	Rita Casadio
Gleb Beliakov	P. Bonissone	Yurilev Chalco-Cano
Radim Belohlavek	Bernadette	Brigitte Charnomordic
Salem Benferhat	Bouchon-Meunier	Guoqing Chen
H. Berenji	Patrice Buche	Carlos A. Coello Coello
Isabelle Bloch	Humberto Bustince	Giulianella Coletti

Additional Members of the Reviewing Committee

Moulin Bernard
Christophe Billiet
Fernando Bobillo
Gloria Bordogna
Christian Borgelt
Stefan Borgwardt
Carole Bouchard
Antoon Bronselaer
Thomas Burger
Ana Burusco
Manuela Busaniche
Silvia Calegari
Tomasa Calvo
Jesus Campaña
Andrés Cano
Philippe Capet
Arnaud Castelltort
Laëtitia Chapel
Yi-Ting Chiang
Francisco Chiclana
Laurence Cholvy
Davide Ciucci
Chloé Clavel
Ana Colubi
Arianna Consiglio
Didier Coquin
Pablo Cordero
María Eugenia Cornejo
Chris Cornelis
Cécile Coulon-Leroy
Fabio Cozman
Madalina Croitoru
Jc Cubero
Fabio Cuzzolin
Inmaculada de Las Peñas
  Cabrera
Gert Decooman
Afef Denguir
Thierry Denœux
Lynn D'Eer
Guy De Tré
Irene Diaz Rodriguez
Juliette
  Dibie-Barthélémy
Pawel Drygaś

Jozo Dujmovic
Fabrizio Durante
Luka Eciolaza
Nicola Fanizzi
Christian Fermüller
Javier Fernandez
Tommaso Flaminio
Carl Frélicot
Ramón
  Fuentes-González
Louis Gacogne
Jose Galindo
José Luis
  García-Lapresta
Brunella Gerla
Maria Angeles Gil
Lluis Godo
Manuel Gonzalez
Michel Grabisch
Salvatore Greco
Przemyslaw
  Grzegorzewski
Kevin Guelton
Thierry Marie Guerra
Robert Hable
Allel Hadjali
Pascal Held
Sascha Henzgen
Hykel Hosni
Celine Hudelot
Julien Hué
Eyke Hüllermeier
Atsushi Inoue
Uzay Kaymak
Anna Kolesarova
Jan Konecny
Stanislav Krajci
Ondrej Kridlo
Rudolf Kruse
Naoyuki Kubota
Maria Teresa Lamata
Cosmin Lazar
Florence Le Ber
Fabien Lehuede
Ludovic Lietard

Berrahou Lilia
Chin-Teng Lin
Xinwang Liu
Nicolas Madrid
Ricardo A. Marques
  Pereira
Luis Martinez
Carmen Martinez-Cruz
Sebastià Massanet
Brice Mayag
Gaspar Mayor
Jesús Medina Moreno
Corrado Mencar
David Mercier
Radko Mesiar
Andrea Mesiarova
Arnau Mir
Franco Montagna
Susana Montes
Tommaso Moraschini
María Moreno García
Olivier Naud
Manuel Ojeda-Aciego
Daniel Paternain
Andrea Pedrini
Carmen Peláez
Martín Pereira-Farina
David Perez
Irina Perfilieva
Nathalie Perrot
Valerio Perticone
Davide Petturiti
Pascal Poncelet
Olga Pons
Thierry Pun
Erik Quaeghebeur
Benjamin Quost
Dragan Radojevic
Eloisa Ramírez
Jordi Recasens
Juan Vicente Riera
Antoine Rolland
Daniel Ruiz
M. Dolores Ruiz
Nobusumi Sagara

Table of Contents – Part II

Fuzzy Logic in Boolean Framework

Management of Uncertainty in Social Networks

From Different to Same, from Imitation to Analogy

Soft Computing and Sensory Analysis

Database Systems

Fuzzy Set Theory

Measurement and Sensory Information

Aggregation

Formal Methods for Vagueness and Uncertainty in a Many-Valued Realm

Graduality

Preferences

Uncertainty Management in Machine Learning

Philosophy and History of Soft Computing

Soft Computing and Sensory Analysis

Supplier Selection Using Interpolative Boolean Algebra and Logic Aggregation

Ksenija Mandic[*] and Boris Delibasic

University of Belgrade, Faculty of Organizational Sciences, Jove Ilica 154, Belgrade, Serbia
ksenija.mandic@crony.rs, boris.delibasic@fon.bg.ac.rs

Abstract. The interest of the decision makers in the selection process of suppliers is constantly growing as a reliable supplier reduces costs and improves the quality of products/services. This process is essentially reducible to the problem of multi-attribute decision-making. Namely, the large number of quantitative and qualitative attributes is considered. This paper presents a model of supplier selection. Weighted approach for solving this model was used combined with logical interactions between attributes. Setting logical conditions between attributes was carried out by using the Boolean Interpolative Algebra. Then the logical conditions are transformed into generalized Boolean polynomial that is through logical aggregation translated into a single value. In this way, the ranking of the suppliers is provided. Using this model managers will be able to clearly express their demands through logical conditions, i.e. will be able to conduct a comprehensive analysis of the problem and to make an informed decision.

Keywords: Fuzzy logic, Interpolative Boolean algebra, Generalized Boolean polynomial, Logical aggregation, Supplier selection problem.

1 Introduction

Selection of the most favorable supplier is a strategic decision that ensures profitability and long-term survival of the company. The company's goal is to carefully choose the right suppliers who will provide the requested product at the right time. In most cases, the strengths and weaknesses of suppliers vary over time, so that managers are in a position to have to make complex decisions in the selection.

In real situations, managers often want to set up mutual relationships between the attributes in order to bring the best possible decision. As conventional fuzzy methods of multi-attribute decision-making do not allow setting of logical interactions between attributes, i.e. they are not in the Boolean frame, the consistent fuzzy logic is introduced. The basis of this approach is interpolative realization of Boolean algebra that transforms logical conditions between attributes into a generalized Boolean polynomial, then the set logical conditions merge into a single value by using the logic

---

[*] Corresponding author.

A. Laurent et al. (Eds.): IPMU 2014, Part II, CCIS 443, pp. 1–9, 2014.
© Springer International Publishing Switzerland 2014

aggregation. In this way, a suitable tool is developed for mapping linguistic requirements of decision-makers in the appropriate Boolean polynomial.

The paper is organized as follows: Section 2 provides an introduction to Boolean consistent fuzzy logic and transformation of logic functions in generalized Boolean polynomial and applying Boolean aggregation. Section 3 analyzes the problem of selecting suppliers and using Boolean consistent fuzzy logic is presented. The paper concludes with Section 4 where the conclusive considerations are presented.

2   Boolean Consistent Fuzzy Logic

Classical Boolean algebra [1] is based on the statements that are true/false, yes/no, white/black. However, there are situations in which classical two-valued realization of Boolean algebra is not adequate. Often it is impossible to express in the absolutely precise way, but we are forced to use vague constellations. In this regard, the necessity of gradation in relations is recognized, so that fuzzy logic is introduced [2], which in its implementation uses the principle of Many-valued logic [3]. The main advantage of fuzzy logic is that it is very close to human perception and does not require completely exact data. Indicating that it is not precisely defined by an element belonging to a certain set, but elements can take values from the interval [0,1]. However, the main disadvantage of fuzzy logic is that it is not in the Boolean frame.

Extension of fuzzy logic by introducing logical interactions is enabled by using Interpolative Boolean Algebra - IBA [4,5], which is a consistent generalization of fuzzy logic. IBA is a real valued, and/or, [0,1] value realization of Boolean algebra [6]. Under the IBA all Boolean axioms and theorems apply.

IBA consists of two levels: a) symbolic or qualitative - at this level the elements structure is defined, and is the matter of final Boolean framework, b) semantic or value - at this level the values are introduced in this way to preserve all the laws set symbolically, in the general case it is a matter of interpolation [7,8].

In fact, the IBA represents an atomic algebra (as a result of a finite number of elements) and is substantially a different gradation approach in comparison to fuzzy approach. Atoms as the simplest elements of algebra play a fundamental role. One of the basic concepts of symbolic levels is the structure of IBA elements. The structure of any IBA element and/or the principle of structural functionality [9] is a bridge between these two levels and basis of generalization, as long as the elements are value-independent. The structure of the analyzed elements determines which atom is (of the final set of elements IBA) included and/or not included. The principle of structural functionality indicates that the structure of any element of IBA may be directly calculated based on the structure of its components. The structure is an independent value and that is the key to preserving Boolean laws both at the symbolic and at the level of values [10]. This principle requires that the IBA transformations are performed at the symbolic level before the introduction of value. Indicating that the negation is treated in a different way, at a structural level, rather than negated variable immediately transforms into value. Thus the observation of negation allows preservation of all Boolean laws. Also, within the IBA applies the law of excluded middle, the axiom of

Boolean logic, where $a \vee \neg a = 1$, which is not respected in the conventional fuzzy logic [11]. Based on all the foregoing, we conclude that fuzzy logic is not in the Boolean frame.

2.1 Generalized Boolean Polynomial and Logical Aggregation

IBA is an algebraic structure with elements which is represented by the Eq. (1) [12]:

$$\langle BA, \wedge, \vee, \neg \rangle \tag{1}$$

where BA is the set of finite elements, binary operators of conjunction $\wedge$ and disjunction $\vee$ and unary negation operator $\neg$, for which all Boolean axioms and theorems are valid.

Under the IBA every Boolean function can be uniquely transformed into the corresponding generalized Boolean polynomial (GBP) [8]. Technically, if any element of Boolean algebra can be represented in a canonical disjunction way, it can be represented also by appropriate GBP. And thus, it allows for the processing of the corresponding element of Boolean algebra into the value on the real interval [0, 1] using operators such as classical (+), classical (-) and generalized product ($\otimes$) [13]. Generalized product (GP) can be any function ($\otimes$): $[0,1] \times [0,1] \rightarrow [0,1]$ which meets all the requirements that one function be a t-norm (commutativity, associativity, monotonicity and limitation Eq. (2,3,4,5)), as well as additional non negativity condition, which is defined as Eq. (6) [7]:

1. $A_i \otimes A_j = A_j \otimes A_i$ $\tag{2}$
2. $A_i \otimes (A_j \otimes A_k) = (A_i \otimes A_j) \otimes A_k$ $\tag{3}$
3. $A_i \leq A_j \implies A_i \otimes A_k \leq A_j \otimes A_k$ $\tag{4}$
4. $A_i \otimes 1 = A_j$ $\tag{5}$
5. $\sum_{K \in P(\Omega/S)} (-1)^{|K|} \otimes A_i(x) \geq 0, \ S \in P(\Omega), \ A_i(x) \in [0,1], \ A_i \in \Omega$ $\tag{6}$

Within the IBA, the method enabling unification of factors is referred to as Logical Aggregation (LA). The main task of LA is a merger of the primary attributes $\Omega = \{a_1, ..., a_n\}$ into a single value, which represents a given set, by using the logical/pseudological function. If we consider the problem of multi-attribute decision making, which is the subject of this paper, LA can be realized in two steps [14]:

1. The normalization of attributes values, which is represented by the Eq. (7):

$$||\cdot||: \Omega \rightarrow [0,1] \tag{7}$$

2. Aggregation of normalized attributes values into one, by using a logical aggregation or pseudological functions as LA operator, defined by Eq. (8) [15]:

$$Aggr\ [0,1]^n \rightarrow [0,1] \tag{8}$$

Boolean function enables the aggregation of factors, i.e. it is an expression that transforms into GBP. Pseudological function is a convex combination of GBP. LA is a technique that gives the user the most options in modeling and treating negation in the right way.

3   The Method of Solving the Problem of Supplier Selection by Using IBA

The problem which is analyzed in the paper is the selection of the best suppliers within a telecommunications company. The company specializes in the manufacture of the equipment necessary for building, monitoring and maintenance of telecommunication systems and wants to choose the best company for the supply of repeater transmission frequencies that allow coverage area without GSM signal or a very weak signal. Three suppliers' companies were considered that are ranked based on four basic attributes and nine sub-attributes (Table 1).

Table 1. Presentation of attributes and sub-attributes

Attributes	Sub-attributes	Attribute type	Unit	Max/Min
Production characteristics (K_1)	Technical performances (k_{11})	Quantitative	Excellent, Very good, Good, Satisfactory, Unsatisfaction	Max
	Product quality (k_{12})	Qualitative	Excellent, Very good, Good, Satisfactory, Unsatisfaction	Max
	Delivery time (k_{13})	Quantitative	Day	Min
Supplier profile (K_2)	Reference (k_{21})	Qualitative	Excellent, Very good, Good, Satisfactory, Unsatisfaction	Max
	Brand position (k_{22})	Qualitative	Excellent, Very good, Good, Satisfactory, Unsatisfaction	Max
Financial aspect (K_3)	Product price (k_{31})	Quantitative	Eur	Min
	Product costs (k_{32})	Quantitative	Eur	Min
Support and services (K_4)	Service (k_{41})	Qualitative	Excellent, Very good, Good, Satisfactory, Unsatisfaction	Max
	Technical support (k_{42})	Qualitative	Excellent, Very good, Good, Satisfactory, Unsatisfaction	Max

Within this paper it will be displayed how IBA can help managers include their preferences in a more sophisticated way compared to weights approach. In many techniques of decision making (conventional fuzzy) a weighted approach is used which allows exclusively linear relationship between the attributes. However, when solving problems with multi-attribute decision-making method, such as the problem of selection of suppliers, often attributes are interdependent and it is needed to establish between them the logical interactions. Logical interactions are based on the introduction of Boolean algebra operators ∧,∨, ¬, by which managers can more clearly show dependence and comparisons between attributes. In this way, a large number of real problems can be expressed by the Boolean algebra.

As part of Table 2 the quantitative and qualitative values of the sub-attributes are presented.

Table 2. The values of sub-attributes

	Production characteristics			Supplier profile		Financial aspect		Support and services	
	Tech perform.	Quality	Delivery Time	References	Brand	Price	Costs	Service	Tech. support
S₁	good	very good	30	good	satisfactory	250	120	excellent	excellent
S₂	very good	very good	45	excellent	very good	345	85	excellent	very good
S₃	satisfactory	good	30	good	good	275	110	good	excellent

The problem was analyzed in the initial interval $[1,5]$, the values of Table 2 are converted into the quantitative values presented in Table 3.

Table 3. Quantitative values of the sub-attributes

	Production characteristics			Supplier profile		Financial aspect		Support and services	
	Technical perform.	Quality	Delivery Time	References	Brand	Price	Costs	Service	Technical support
S₁	3	4	3	3	2	3	4	5	5
S₂	4	4	2	5	4	2	5	5	4
S₃	2	3	3	3	3	3	4	3	5

As mentioned above, fuzzy logic takes values from the interval $[0,1]$, it indicates that it is necessary to convert the value of the sub-attributes from the initial val $[1,5]$ to interval $[0,1]$, i.e. it is necessary to perform a normalization. After the normalization, the values of the sub-attributes are presented in Table 4.

Table 4. Normalized values of sub-attributes

	Production characteristics			Supplier profile		Financial aspect		Support and services	
	Technical perform.	Quality	Delivery Time	References	Brand	Price	Costs	Service	Technical support
S₁	0,6	0,8	0,6	0,6	0,4	0,6	0,8	1	1
S₂	0,8	0,8	0,4	1	0,8	0,4	1	1	0,8
S₃	0,4	0,6	0,6	0,6	0,6	0,6	0,8	0,6	1

In order to select the best supplier it is necessary to introduce a weighted sum of the attributes/sub-attributes, which is represented by Eq. (9):

$$w_1 * k_1 + w_2 * k_2 = p, \qquad (9)$$

where w_1 and w_2 represent the weight in this model, k_1 and k_2 are the values of attributes/sub-attributes and p represents supplier's total point in interval $[0,1]$.

Managers believe that for the selection of suppliers, in this case, it is important to take into consideration sub-attributes Technical performances and Quality, as an sub-attribute does not exclude other. Hence, the logical relation was established by using the Boolean operator $\wedge$ and the sub-attribute function have the following form Eq. (10):

$$0,7 * (k_{11} \wedge k_{12}) + 0,3 * k_{13} = 0,7 * (k_{11} \otimes k_{12}) + 0,3 * k_{13} = p, \quad (10)$$

where weights w_1 and w_2 have following values 0.7 and 0.3. Also, within the equation was used standard product as appropriate operator of GP.

Weight sum for sub-attributes Reference and Brand position have values 0.6 and 0.4 respectively, shown in Eq. (11):

$$0,6 * k_{21} + 0,4 * k_{22} = p, \qquad (11)$$

for sub-attributes Price and Costs weight sum are 0.7 and 0.3 Eq. (12):

$$0,7 * k_{31} + 0,3 * k_{32} = p, \qquad (12)$$

and for sub-attributes Service and Technical support are 0.6 and 0.4 Eq. (13):

$$0,6 * k_{41} + 0,4 * k_{42} = p, \qquad (13)$$

By the inclusion of normalized k-values from Table 4 sub-attributes functions were set and by the application of LA we obtain the values of alternatives (suppliers) for the four basic attributes as shown in the Table 5.

Table 5. Values of suppliers for the four basic attributes

	Production characte-ristics (K_1)	Supplier Profile (K_2)	Financial aspect (K_3)	Support and Services (K_4)
S_1	0,516	0,52	0,66	1
S_2	0,568	0,92	0,58	0,92
S_3	0,348	0,6	0,66	0,76

However, in real situations, managers often want to set the mutual relationships between the attributes in order to bring the best possible decision. This was enabled by using the logical conditions, presented hereinafter:

Condition 1: "If the production characteristics are at a high level, then the product is acceptable, if it is not at high level pay attention to the profile of the supplier, the financial aspect and the support and services." (Eq. (14))

$$k_1 \lor (\neg\, k_1 \land k_2 \land k_3 \land k_4) \tag{14}$$

Condition 2: "If a supplier profile is satisfying he should also have good production characteristics, if the profile of the supplier is not satisfactory attention should be paid to the financial aspect and the support and services." (Eq. (15))

$$(k_2 \land k_1) \lor (\neg\, k_2 \land k_3 \land k_4) \tag{15}$$

Condition 3: "If the financial aspect is high, attention should be paid to the manufacturing characteristics, if not high, attention should be paid to profile of supplier." (Eq. (16))

$$(k_3 \land k_1) \lor (\neg\, k_3 \land k_2) \tag{16}$$

Each of these logical conditions is transformed to the GBP, by using standard product as appropriate operator of GP. Transformation is given in the following steps Eq. (17):

$$
\begin{aligned}
k_1 \lor (\neg\, k_1 \land k_2 \land k_3 \land k_4) \\
= \; & k_1 + (\neg\, k_1 \land k_2 \land k_3 \land k_4) - k_1 \otimes (\neg\, k_1 \land k_2 \land k_3 \land k_4) \\
= \; & k_1 + \big((1 - k_1) \otimes k_2 \otimes k_3 \otimes k_4\big) - k_1 \\
& \otimes \big((1 - k_1) \otimes k_2 \otimes k_3 \otimes k_4\big) \\
= \; & k_1 + k_2 \otimes k_3 \otimes k_4 - k_1 \otimes k_2 \otimes k_3 \\
& \otimes k_4
\end{aligned} \tag{17}
$$

In the same way the remaining two logical conditions are transformed, which is represented by the Eq. (18,19):

$$(k_2 \land k_1) \lor (\neg\, k_2 \land k_3 \land k_4) = k_2 \otimes k_1 + k_3 \otimes k_4 - k_2 \otimes k_3 \otimes k_4 \tag{18}$$

$$(k_3 \land k_1) \lor (\neg\, k_3 \land k_2) = k_2 - k_2 \otimes k_3 + k_3 \otimes k_1 \tag{19}$$

In the presented GBP equations we will introduce the attributes values from Table 5 based on which by using LA we obtain the values in Table 6.

Table 6. The values of logical conditions for three suppliers (S_1, S_2, S_3)

	Condition 1	Condition 2	Condition 3
S_1	0,682	0,585	0,517
S_2	0,78	0,565	0,715
S_3	0,544	0,409	0,433

The final ranking of suppliers is obtained by placing another of the weighted function, where instead of individual values of sub-attributes we introduce previously mentioned logical conditions, shown in Eq. (20):

$$0,5 * (k_1 \lor (\neg k_1 \land k_2 \land k_3 \land k_4)) + 0,2 * ((k_2 \land k_1) \lor (\neg k_2 \land k_3 \land k_4) + 0,3 \\ * (k_3 \land k_1) \lor (\neg k_3 \land k_2) = p \qquad (20)$$

Weights for conditions are 0.5, 0.2 and 0.3 respectively, and p represents the final supplier rank.

By entering the obtained values of logical conditions of Table 6 in the expression and by application of pseudo logical aggregation we obtain values of Table 7.

Table 7. The final ranking of suppliers

S_1	0,613
S_2	0,717
S_3	0,484

From Table 7 we can see that the order of suppliers is as follows: $S2>S1>S3$.

4   Conclusion

The reason of analysis of the presented model is primarily to provide practical support to decision-makers i.e. managers when choosing suppliers in the telecommunications sector. Also, with the use of logical conditions managers are able to present more clearly their requirements. In this way, they can make a more comprehensive and better decision than would be the case with conventional fuzzy methods which are not in the Boolean framework and that in a different way treat negation.

In addition to solving the observed problems in this paper is used the weighted approach combined with the Boolean consistent fuzzy logic. IBA logic enabled the transformation of logic functions to a generalized Boolean polynomial. While by the use of Logical/pseudological aggregation GBP is reduced to values. In this way we achieved the ranking of suppliers ($S2>S1>S3$). What makes this logic more suitable way to solve these types of problems compared to conventional fuzzy logic is that the structural transformations are performed before the introduction of values.

Further research will be directed towards the inclusion of logical conditions into the multi-attribute decision-making method AHP, TOPSIS, ELECTRE and comparison of the obtained results.

References

1. Boole, G.: The calculus of logic. The Cambridge and Dublin Mathematical Journal 3, 183–198 (1848)
2. Zadeh, A.L.: Fuzzy sets. Information and Control 8(3), 338–353 (1965)
3. Lukasiewicz, J.: Selected Works. In: Brokowski, L. (ed.). North-Holland Publ. Comp., Amsterdam and PWN (1970)
4. Radojevic, D.: Logical measure of continual logical function. In: 8th International Conference IPMU – Information Processing and Management of Uncertainty in Knowledge-Based Systems, Madrid, pp. 574–578 (2000)
5. Radojevic, D.: New [0, 1] – valued logic: A natural generalization of Boolean logic. Yugoslav Journal of Operational Research – YUJOR 10(2), 185–216 (2000)
6. Dragovic, I., Turajlic, N., Radojevic, D., Petrovic, B.: Combining Boolean consistent fuzzy logic and AHP illustrated on the web service selection problem. International Journal of Computational Intelligence Systems 7(1), 84–93 (2013)
7. Radojevic, D.: Fuzzy Set Theory in Boolean Frame. International Journal of Computers, Communications & Control 3, 121–131 (2008)
8. Radojevic, D.: Interpolative Realization of Boolean Algebra as a Consistent Frame for Gradation and/or Fuzziness. In: Nikravesh, M., Kacprzyk, J., Zadeh, L.A. (eds.) Forging New Frontiers: Fuzzy Pioneers II. STUDFUZZ, vol. 218, pp. 295–317. Springer, Heidelberg (2008)
9. Radojevic, D.: Logical measure – structure of logical formula. In: Bouchon-Meunier, B., Gutiérrez-Ríos, J., Magdalena, L., Yager, R.R. (eds.) Technologies for Constructing Intelligent System 2: Tools. STUDFUZZ, vol. 90, pp. 417–430. Springer, Heidelberg (2002)
10. Radojevic, D.: Interpolative realization of Boolean algebra. In: Proceedings of the NEUREL 2006, The 8th Neural Network Applications in Electrical Engineering, pp. 201–206 (2006)
11. Radojevic, D.: Interpolative Relations and Interpolative Preference Structures. Yugoslav Journal of Operations Research 15(2), 171–189 (2005)
12. Radojevic, D.: Real probability (R-probability): fuzzy probability idea in Boolean frame. In: 28th Linz Seminar on Fuzzy Set Theory (2007)
13. Milošević, P., Nešić, I., Poledica, A., Radojević, D.G., Petrović, B.: Models for Ranking Students: Selecting Applicants for a Master of Science Studies. In: Balas, V.E., Fodor, J., Várkonyi-Kóczy, A.R., Dombi, J., Jain, L.C. (eds.) Soft Computing Applications. AISC, vol. 195, pp. 93–103. Springer, Heidelberg (2012)
14. Mirkovic, M., Hodolic, J., Radojevic, D.: Aggregation for Quality Management. Yugoslav Journal for Operational Research 16(2), 177–188 (2006)
15. Radojevic, D.: Logical Aggregation Based on Interpolative Boolean Algebra. Mathware & Soft Computing 15, 125–141 (2008)

Finitely Additive Probability Measures in Automated Medical Diagnostics

Milica Knežević[1], Zoran Ognjanović[1], and Aleksandar Perović[2]

[1] Mathematical Institute of the Serbian Academy of Sciences and Arts,
Kneza Mihaila 36, 11000 Belgrade, Serbia
knezevic.milica@gmail.com, zorano@mi.sanu.ac.rs
[2] University of Belgrade – Faculty of Transport and Traffic Engineering,
Vojvode Stepe 305, 11000 Belgrade, Serbia
pera@sf.bg.ac.rs

Abstract. We describe one probabilistic approach to classification of a set of objects when a classification criterion can be represented as a propositional formula. It is well known that probability measures are not truth functional. However, if μ is any probability measure and α is any propositional formula, $\mu(\alpha)$ is uniquely determined by the μ-values of conjunctions of pairwise distinct propositional letters appearing in α. In order to infuse truth functionality in the generation of finitely additive probability measures, we need to find adequate binary operations on $[0, 1]$ that will be truth functions for finite conjunctions of pairwise distinct propositional letters. The natural candidates for such truth functions are t-norms. However, not all t-norms will generate a finitely additive probability measure. We show that Gödel's t-norm and product t-norm, as well as their linear convex combinations, can be used for the extension of any evaluation of propositional letters to finitely additive probability measure on formulas. We also present a software for classification of patients with suspected systemic erythematosus lupus (SLE), which implements the proposed probabilistic approach.

Keywords: probability measures, classification, fuzzy logic, soft computing, medical diagnosis.

1 Introduction

Reasoning under uncertainty is one of the most prominent research themes in theoretical computer science for more than four decades. Many-valued logics, i.e. logics with more than 2 truth values (usually the set of truth values is the real or the hyper-real unit interval $[0,1]$) have been extensively studied and developed in order to formalize various phenomena related to artificial intelligence. The most significant many-valued logics are fuzzy logics [4, 6, 7, 9, 10, 13], possibilistic logics [1–3] and probability logics [5, 8, 11, 12, 14–21, 27–31].

Since the late sixties, probability theory has found application in development of various medical expert systems. Bayesian analysis, which is essentially an optimal path finding through a graph called Bayesian network, has been (and

A. Laurent et al. (Eds.): IPMU 2014, Part II, CCIS 443, pp. 10–19, 2014.

still is) successfully applied in so called sequential diagnostics, when the large amount of reliable relevant data is available. The graph (network) represents our knowledge about connections between studied medical entities (symptoms, signs, diseases); the Bayes formula is applied in order to find the path (connection) with maximal conditional probability. Moreover, a priori and conditional probabilities were used to define a number of measures designed specifically to handle uncertainty, vague notions and imprecise knowledge. Some of those measures were implemented in MYCIN in the early seventies. The success of MYCIN has initiated construction of rule based expert systems in various fields.

However, designing expert systems with the large number of rules (some of them like CADIAG-2 have more than 10000) has an unfortunate consequence, i.e., some of them are turned to be inconsistent. On the other hand, the emergence of theoretical computer science as a new scientific discipline has lead to discovery that the completeness techniques from mathematical logic are well found (and sometimes the only ones) methods for proving correctness of hardware and software. Consequently, mathematical logic has become a theoretical foundation of artificial intelligence.

In this paper we will present the mathematical background of the medical decision support system for the systemic erythematosus lupus (SLE) that we have been lately working on with our colleagues from the Institute of Allergology and Immunology in Belgrade. The methodology is based on the work of Dragan Radojević [22–25] and our joint work [21].

2 Probabilities and Truth Functionality

It is well known that probability measures are not truth functional. However, if μ is any probability measure and α is any propositional formula, then $\mu(\alpha)$ is uniquely determined by the μ-values of conjunctions of pairwise distinct propositional variables appearing in α.

Example 1. Let $\alpha = (p_0 \rightarrow p_1) \wedge p_2$, $\mu(\alpha)$ is calculated as:

$$
\begin{aligned}
\mu((p_0 \rightarrow p_1) \wedge p_2) &= \mu((\neg p_0 \vee p_1) \wedge p_2) \\
&= \mu((\neg p_0 \wedge p_2) \vee (p_1 \wedge p_2)) \\
&= \mu(\neg p_0 \wedge p_2) + \mu(p_1 \wedge p_2) - \mu(\neg p_0 \wedge p_1 \wedge p_2) \\
&= \mu(p_2) - \mu(p_0 \wedge p_2) + \mu(p_1 \wedge p_2) - \mu(p_1 \wedge p_2) \\
&\quad + \mu(p_0 \wedge p_1 \wedge p_2) \\
&= \mu(p_2) - \mu(p_0 \wedge p_2) + \mu(p_0 \wedge p_1 \wedge p_2).
\end{aligned}
$$

It turns out that the previous example is not a singularity but rather a pattern that indicates how to fuse truth functionality and finitely additive probability measures. Namely, if μ is a finitely additive probability measure on the set of propositional formulas, using the complete disjunctive normal form and finite additivity, it is easy to see that $\mu(\alpha)$ is a finite sum of terms of the form

$$
\mu(\pm p_1 \wedge \cdots \wedge \pm p_n),
$$

where $p_1, \ldots, p_n$ are all propositional letters appearing in α, and $+p_i$ and $-p_i$ denote p_i and $\neg p_i$, respectively. Since

$$\mu(\neg\beta) = 1 - \mu(\beta) \tag{1}$$

and

$$\mu(\beta \wedge \neg\gamma) = \mu(\beta) - \mu(\beta \wedge \gamma), \tag{2}$$

negation can be eliminated from each $\mu(\pm p_1 \wedge \cdots \wedge \pm p_n)$, so $\mu(\alpha)$ depends only on the logical structure of α and μ-values of finite conjunctions of pairwise distinct propositional letters.

Hence, in order to infuse truth functionality in the generation of finitely additive probability measures, we need to find adequate binary operations on $[0,1]$ that will be truth functions for finite conjunctions of pairwise distinct propositional letters.

The natural candidates for such truth functions are t-norms. However, not all t-norms will generate a finitely additive probability measure, which will be explicitly shown in this section. We will show that Gödel's t-norm $T_G(x,y) = \min(x,y)$ and product t-norm $T_\Pi(x,y) = xy$ can always be used for the extension of any evaluation $e : Var \longrightarrow [0,1]$ to finitely additive probability measures e^G and e^Π on formulas. In particular, for pairwise distinct propositional letters $p_1, \ldots, p_n$,

$$e^\Pi(p_1 \wedge \cdots \wedge p_n) = e(p_1) \cdots e(p_n) \tag{3}$$

and

$$e^G(p_1 \wedge \cdots \wedge p_n) = \min(p_1, \ldots, p_n). \tag{4}$$

Product measures e^Π correspond to one extreme situation: stochastic or probability independence of propositional letters. Gödel's measures e^G correspond to another kind of extreme situation: logical dependence of propositional letters. While stochastic independence is a measure-theoretic property and cannot be forced by some nontrivial logical conditions (see [8]), logical dependence is expressible in classical propositional calculus. For instance, logical condition

$$p \to q$$

clearly entails that

$$\bar{e}(p \wedge q) = \min(e(p), e(q)) \ ,$$

where $\bar{e}$ is a finitely additive probability measure on propositional formulas that extends e.

Linear convex combinations of finitely additive probability measures are finitely additive probability measures as well, so using e^G and e^Π we can construct an infinite scale of probability measures

$$e^{(s)} = se^\Pi + (1-s)e^G, \quad s \in (0,1) \cap \mathbb{Q} \ . \tag{5}$$

From the uncertainty point of view, the measures $e^{(s)}$ correspond to various degrees of dependence between propositional letters. From the fuzzyness point of view, the measures $e^{(s)}$ provide countably many ways to extend any initial evaluation e of propositional letters, which enables probability evaluations of fuzzy quantities. As before, we will try to illustrate our intended meaning with the following simple example:

Example 2. Suppose that we have to classify compounds C_1, C_2, and C_3 of the substances p and q according to the criteria of minimal harmfulness of a compound. It is known that both p and q are harmful, but they neutralize each other. Concentrations of substances p and q in compounds C_1 , C_2, and C_3 are given in Table 1.

Table 1. Concentrations of substances in compounds

Compound	Concentration of p	Concentration of q
C_1	0.95	0.05
C_2	0.15	0.85
C_3	0.65	0.35

Syntactically, we consider p and q as propositional letters. Since substances p and q neutralize each other, minimal harmfulness criteria is adequately represented by the formula $p \leftrightarrow q$. If e is any $[0, 1]$-evaluation of p and q, then

$$
\begin{aligned}
e^{\Pi}(p \leftrightarrow q) &= e^{\Pi}((p \wedge q) \vee (\neg p \wedge \neg q)) \\
&= e^{\Pi}(p \wedge q) + e^{\Pi}(\neg p \wedge \neg q) \\
&= e(p) \cdot e(q) + e^{\Pi}(\neg p) - e^{\Pi}(\neg p \wedge q) \\
&= e(p) \cdot e(q) + 1 - e(p) - e(q) + e(p) \cdot e(q) \\
&= 1 - e(p) - e(q) + 2e(p) \cdot e(q) \ .
\end{aligned}
$$

Similarly,

$$
e^{G}(p \leftrightarrow q) = 1 - e(p) - e(q) + 2 \min(e(p), e(q))
$$

and

$$
e^{(0.25)}(p \leftrightarrow q) = 0.25\, e^{\Pi}(p \leftrightarrow q) + 0.75\, e^{G}(p \leftrightarrow q) \ .
$$

If we interpret C_1, C_2, and C_3 as $[0, 1]$-evaluations of p and q, then we can easily evaluate $p \leftrightarrow q$ in e^{G}, e^{Π}, and $e^{(0.25)}$. The results are displayed in Table 2.

All three columns $e^{G}(p \leftrightarrow q)$, $e^{\Pi}(p \leftrightarrow q)$, and $e^{(0.25)}(p \leftrightarrow q)$ induce the obviously correct classification: the least harmful compound is C_3, then follows C_2, and the most harmful compound is C_1.

Arguably, a mathematical form of classification is an ordering of some finite set of evaluations according to certain criteria. More precisely, classification of the finite set of evaluations $\{e_0, \ldots, e_k\}$, $e_i : Var \longrightarrow [0, 1]$ with respect to the

Table 2. Evaluation of formula $p \leftrightarrow q$ in e^G, e^Π, and $e^{(0.25)}$

Evaluation	$e(p)$	$e(q)$	$e^G(p \leftrightarrow q)$	$e^\Pi(p \leftrightarrow q)$	$e^{(0.25)}(p \leftrightarrow q)$
C_1	0.95	0.05	0.1	0.095	0.099
C_2	0.15	0.85	0.3	0.255	0.289
C_3	0.65	0.35	0.7	0.455	0.639

set of attributes (propositional letters) $\{p_1, \ldots, p_n\}$ and the criterium function $f : [0,1]^n \longrightarrow [0,1]$ is the partial ordering $<$ defined by

$$e_i < e_j \text{ iff } f(e_i(p_1), \ldots, e_i(p_n)) < f(e_j(p_1), \ldots, e_j(p_n)) \ .$$

In practice, we can apply the measures $e_i^{(s)}$ in any classification problem where at least one part of the computation of the criterion function f involves computation of the truth value of certain formula $\alpha(p_1, \ldots, p_n)$.

For example, suppose that we want to develop a fuzzy relational database for automated trade of furniture, where database entries are evaluations of predefined quality attributes. User's queries should be stated in the form of propositional formulas over the quality attributes. The resolution process we will illustrate on the example of the query "find me a sturdy but light wooden chair that is not too expensive":

- Prompt the user to chose a rational $s \in [0,1]$. Here s represents user's estimation of dependence between quality attributes in the query. 1 represents stochastic independence, while 0 represents logical dependence;
- Compute

$$e^{(s)}(\text{sturdy} \wedge \text{light} \wedge \text{wooden} \wedge \neg(\text{too expenssive}))$$

for all relevant database entries e;
- Return to the user all relevant database entries e with maximal $e^{(s)}$-values.

Of course, one might ask the adequacy of such approach, especially in the evaluation of the queries such as "find me a sturdy but light wooden chair that is not too expensive", which, at least at the first sight, have no natural probabilistic interpretation.

We can offer several justification arguments. The most obvious is the capricious one "why not", which is rooted in the fact that any classification criterion expressible as a classical propositional formula can be effectively evaluated from the initial evaluation e of primary attributes (propositional letters) using e^G, e^Π and $e^{(s)}$ measures. Substantiality (with respect to the question of adequacy), this is not too much different from any other evaluation method involving truth functionality principle. The only difference is in the computation of the value of the underlying formula.

Deeper arguments are connected with the choice of the adequate parameter s, as well as with the construction of the initial fuzzy sets that represent underlying fuzzy attributes. For instance, attributes such as "expensive" are more psychological than anything else, so it is natural to construct their mathematical representation - fuzzy set using relevant statistical data.

As we have mentioned before, the product measure corresponds to the one extreme - stochastic or probability independence of attributes (propositional letters), while Gödel's measure corresponds to the other extreme - logical dependence of attributes. The standard statistical techniques, such as linear or nonlinear regression, can be applied to measure stochastic independence of fuzzy attributes.

Intermediate measures $e^{(s)}$ are particularly useful in the cases where both e^G and e^{Π} do not classify observed objects. Namely, it is easy to construct an example with the measurement results given in Table 3.

Table 3. Ordering of objects according to classification criteria α

object	$e^{\Pi}(\alpha)$	$e^G(\alpha)$	$e^{(0.5)}$
A	0.3	0.4	0.35
B	0.3	0.6	0.45
C	0.5	0.6	0.55

As a consequence, neither the product nor Gödel's measure provide classification of objects A, B and C according to the classification criteria α, while the arithmetic mean $e^{(0.5)}$ provides a classification - linear ordering $A < B < C$ that is sound with both partial orderings induced by e^{Π} and e^G. In this example, e^{Π} induces partial ordering $A < C$ and $B < C$, while e^G induces partial ordering $A < C$ and $A < C$.

What we want to say is that, in cases where we disregard independence issues and only evaluate formulas with both e^{Π} and e^G, intermediate measures $e^{(s)}$ might give additional information that is sound with the partial orderings induced by e^{Π} and e^G, and provide a finer classification.

3   Implementation

Based on the previously presented theoretical framework, a software component was developed as a utility to measure similarity among patients, with suspected SLE, based on patients' medical test results and symptoms. Classifying patients in this way and comparison of new patients with those with already given diagnosis would help and support medical decision making process. The software component is implemented as a desktop application written in Java and Prolog. The graphical user interface (GUI) is written in Java, while Prolog was a more natural choice for implementation of the logic component that lies in the background of the algorithm.

Patients' data are stored in JavaScript Object Notation (JSON) format[32]. It is an open-standard format for representing data as human readable text. JSON format is language independent and parsers for JSON data are readily available in nearly all modern programming languages. Patients' data are organized in two JSON files. First file is *parameterList* and it contains descriptions of 80 medical symptoms and tests related and relevant for diagnosing SLE. We will use the term parameter to denote both medical symptoms and tests. Second file is *patientList* and it contains actual patients' results for parameters defined in *parameterList*. Description of a parameter in the file *parameterList* is given as a JSON data object of the following structure:

```
parameterCode:{
    "name":name,
    "possibleValues":{
        value : valueName,
        ...
        value : valueName
    }
}
```

Parameter code is an integer that is used for easy identification of the parameter itself. Name is a string that represents medical term that describes the parameter. Possible values is a JSON object that represents a discrete set of values that a given parameter can have. Value name is a human readable description of the associated value. For better understanding of the described object structure we give examples that show representation of Antinuclear Antibodies (ANAs) and Coombs test as JSON objects:

```
"0":{
    "name":"ANA",
    "possibleValues":{
        "0":"negative",
        "0.5":"positive ≤ 1:80",
        "1":"positive > 1:80"
    }
},

"1":{
    "name":"Coombs test",
    "possibleValues":{
        "0":"negative",
        "1":"positive"
    }
}
```

A user enters a formula φ with names of the symptoms and medical tests from the file *parameterList* as propositional letters. We give an example of such formula, which can be used to express level of dependency of leukopenia, lymphopenia, malar rash and ANA:

$$\varphi = ((leukopenia \lor lymphopenia) \land malar\ rash) \to ANA$$

Similarity of the patients is then computed with regard to φ. In order to compute $\mu(\varphi)$ for each patient, formula φ is transformed into a logically equivalent formula φ' for which it is simpler to compute $\mu(\varphi')$ by applying rules from Sect. 2. Disjunctive normal form is selected as a suitable logical equivalent of the entered formula φ. φ_{dnf} will denote the disjunctive normal form of φ. Formula φ is passed to the Prolog module of the software component and this module is responsible for the transformation $\varphi \leadsto \varphi_{dnf}$. Formula φ_{dnf} is in the form $Clause_1 \lor ... \lor Clause_n$, where each $Clause_i$ is in the form $\pm p_{i_1} \land ... \land \pm p_{i_k}$. Internally, in Prolog, φ_{dnf} is represented as a list of lists: $[[\pm p_{1_1}, ..., \pm p_{1_k}], ..., [\pm p_{n_1}, ..., \pm p_{n_k}]]$. This representation enables computation of $\mu(\varphi)$ using dynamic programming and applying inclusion-exclusion principle.

$$\mu(\varphi) = \mu(\varphi_{dnf}) = \sum_{l=1}^{n} (-1)^l \sum_{1 \leq i_1 < ... < i_l \leq n} \mu(Clause_{i_1} \land ... \land Clause_{i_l})$$

where each $\mu(Clause_{i_1} \land ... \land Clause_{i_l})$ is of the form $\mu(\pm p_{j_1} \land ... \land \pm p_{j_m})$ and thus can be computed easily following the rules from Sect. 2. For each patient, $\mu(\varphi)$ is then computed by assigning actual values to propositional letters in accordance with data stored in *patientList* file. Value of $\mu(\varphi)$ for a given patient, expresses the level of accuracy of the formula φ in accordance with patient's data.

Acknowledgments. This work is supported by the Ministry of Science, Technology and Development, Republic of Serbia (project III 044006).

References

1. Dubois, D., Lang, J., Prade, H.: Possibilistic logic. In: Gabbay, D.M., Hogger, C.J., Robinson, J.A., Seikmann, J.H. (eds.) Handbook of Logic in Artificial Intelligence and Logic Programming, vol. 3, pp. 439–513. Oxford University Press, Inc., New York (1994)
2. Dubois, D., Prade, H.: Possiblistic logic: a retrospective and prospective view. Fuzzy Sets and Systems 144, 3–23 (2004)
3. Dubois, D., Godo, L., Prade, H.: Weighted logic for artificial intelligence: an introductory discussion. In: Godo, L., Prade, H. (eds.) ECAI 2012 Workshop, Weighted Logic for Artificial Intelligence, pp. 1–7 (2012)
4. Esteva, F., Godo, L., Montagna, F.: The $L\Pi$ and $L\Pi\frac{1}{2}$ logics: two complete fuzzy logics joining Łukasiewicz and product logic. Archive for Mathematical Logic 40, 39–67 (2001)
5. Fagin, R., Halpern, J., Megiddo, N.: A logic for reasoning about probabilities. Information and Computation 87(1-2), 78–128 (1990)

6. Flaminio, T.: Strong non-standard completeness for fuzzy logic. Soft Computing 12(4), 321–333 (2008)
7. Godo, L., Marchioni, E.: Coherent conditional probability in a fuzzy logic setting. Logic Journal of the IGPL 14(3), 457–481 (2006)
8. Hailperin, T.: Sentential Probability Logic. Associated University Presses, Inc., London (1996)
9. Hajek, P., Esteva, F., Godo, L.: Fuzzy logic and probability. In: Proceedings of the 11th Conference on Uncertainty in Artificial Inteligence, Montreal, Canada, pp. 237–244 (1995)
10. Hájek, P.: Methemathematics of Fuzzy Logic. Kluwer Academic Publishers, Dordrecht (1998)
11. Halpern, J.: Reasoning about Uncertainty. The MIT Press, Cambridge (2003)
12. Lehmann, D.: Generalized qualitative probability: savage revisited. In: Proceedings of the 12th Conference on Uncertainty in Artificial Intelligence (UAI 1996), pp. 381–388 (1996)
13. Marchioni, E., Montagna, F.: On triangular norms and uninorms definable in $L\Pi\frac{1}{2}$. International Journal of Approximate Reasoning 47(2), 179–201 (2008)
14. Narens, L.: On qualitative axiomatizations for probability theory. Journal of Philosophical Logic 9(2), 143–151 (1980)
15. Nilsson, N.: Probabilistic logic. Artificial Intelligence 28(1), 71–87 (1986)
16. Ognjanović, Z., Rašković, M.: A logic with higher order probabilities. Publications de l'Institut Mathematique, Nouvelle série 60(74), 1–4 (1996)
17. Ognjanović, Z., Rašković, M.: Some probability logics with new types of probability operators. Journal of Logic and Computation 9(2), 181–195 (1999)
18. Ognjanović, Z., Rašković, M.: Some first-order probability logics. Theoretical Computer Science 247(1-2), 191–212 (2000)
19. Ognjanović, Z., Perović, A., Rašković, M.: Logic with the qualitative probability operator. Logic Journal of IGPL 16(2), 105–120 (2008)
20. Ognjanović, Z., Rašković, M., Marković, Z.: Probability logics. In: Ognjanović, Z. (ed.) Logic in Computer Science, vol. 12(20), pp. 35–111. Mathematical Institute of Serbian Academy of Sciences and Arts (2009)
21. Ognjanović, Z., Perović, A., Rašković, M., Radojević, D.: Finitely additive probability measures on classical propositional formulas definable by Gödel's t-norm and product t-norm. Fuzzy Sets and Systems 169(1), 65–90 (2011)
22. Radojević, D.: [0, 1]-valued logic: a natural generalization of Boolean logic. Yugoslav Journal on Operations Research 10(2), 185–216 (2000)
23. Radojević, D.: Interpolative realization of Boolean algebra. In: 8th Seminar on Neural Network Applications in Electrical Engineering (NEUREL 2006), pp. 201–206 (2006)
24. Radojević, D.: Interpolative realization of Boolean algebra as a consistent frame for gradation and/or fuzziness. In: Nikravesh, M., Kacprzyk, J., Zadeh, L.A. (eds.) Forging New Frontiers: Fuzzy Pioneers II. STUDFUZZ, vol. 218, pp. 295–317. Springer, Heidelberg (2008)
25. Radojević, D.: Logical aggregation based on interpolative realization of Boolean algebra. Mathware and Soft Computing 15(1), 125–141 (2008)
26. Radojević, D., Perović, A., Ognjanović, Z., Rašković, M.: Interpolative Boolean logic. In: Dochev, D., Pistore, M., Traverso, P. (eds.) AIMSA 2008. LNCS (LNAI), vol. 5253, pp. 209–219. Springer, Heidelberg (2008)
27. Rašković, M.: Classical logic with some probability operators. Publications de l'Institut Mathematique, Nouvelle Série 53(67), 1–3 (1993)

28. Rasković, M., Ognjanović, Z.: A first order probability logic LP_Q. Publications de l'Institut Mathematique, Nouvelle Série 65(79), 1–7 (1999)
29. Rašković, M., Marković, Z., Ognjanović, Z.: A logic with approximate conditional probabilities that can model default reasoning. International Journal of Approximate Reasoning 49(1), 52–66 (2008)
30. van der Hoek, W.: Some considerations on the logic $P_F D$: a logic combining modality and probability. Journal of Applied Non-Classical Logics 7(3), 287–307 (1997)
31. Wellman, M.P.: Some varieties of qualitative probability. In: Bouchon-Meunier, B., Yager, R.R., Zadeh, L.A. (eds.) IPMU 1994. LNCS, vol. 945, pp. 171–179. Springer, Heidelberg (1995)
32. JSON (JavaScript Object Notation), http://json.org

Demonstrative Implications of a New Logical Aggregation Paradigm on a Classical Fuzzy Evaluation Model of «Green» Buildings

Milan Mrkalj

Faculty of Organizational Sciences, Belgrade, Serbia
mrki2000@mail.bg

Abstract. This paper demonstrates the application possibilities of the generalised model for Logical Aggregation (LA) in performance and quality evaluation of Green buildings through establishing a linear order of them by a (scalar) general aggregated quality parameter and sorting them into categories. An existing classical fuzzy model is experimentally modified in order to demonstrate the advanced capabilities of adequate articulation of the partial demands for quality via the new generalised model for logical aggregation.

Keywords: Logical Aggregation, Quality management, Green buildings.

1   Introduction

The advanced abilities of generalised model for Logical Aggregation presented in [1,4,5,6,7] expand the boundaries of existing fuzzy model presented in [2]. The existing model is in essence an aggregation model since the inputs are multidimensional values and the output is a scalar quantity which provides base for simple linear sorting. In existing fuzzy model only trivial quality attributes are used, without interaction between them through logical functions which should improve the articulation of partial demands for quality. Also, the aggregation operator (explained in [3,6]) is only dot product.

The model for Logical Aggregation (*LA*) introduces the use of logical functions between trivial attributes which emerge from the finite set of possible functions – Boolean algebra generated over the set of attributes. The new complex aggregation operators are introduced with the use of different generalised products which belong to a subclass of T-norms with additional axiomatic term of nonnegativity, as shown in [1,4,6].

Implementation of the model for Logical Aggregation [1] is possible in both phases of aggregation in the existing model [2]. The *LA* paradigm emanates an algebraic structure which provides the basis for arithmetic calculation of the output results.

A. Laurent et al. (Eds.): IPMU 2014, Part II, CCIS 443, pp. 20–27, 2014.

2   Implementaton of Logical Aggregation in the First Phase of Aggregation of the Existing Fuzzy Model

Implementation of *LA* on the first phase of aggregation of existing model is demonstrated, and the values of the output are compared. The aim is to show the ability to create complex attributes which have more power of expressing the partial demands for quality in quality management. The term *attributes* from [1,3,6] are named as "factors" in [2] and are organized as two "factor sets".

Vector of weights of secondary attributes of the primary attribute *energy efficient and energy utilization* from [2] is: $A_2 =$ [0,35   0,35   0,2   0,1].

Weights of secondary attributes are respectively $w_a = 0,35$ $w_b = 0,35$ $w_c = 0,2$ and $w_d = 0,1$. In this case, generalised pseudo-Boolean polynomial is trivial (contains only single basic attributes).

Now we shall unify attributes *architecture design* (*a*) and *envelope structure* (*b*) into one single new complex attribute on the basis of the fact that these two single attributes are significantly positive correlated in qualitative sense (*architecture design and envelope structure which assures the rational heat transfer coefficient of external surfaces*) [2].

Single attributes *a* and *b* will be dismissed (their respective weights become zero instead of 0,35), and new complex attribute *a∧b* (1) will be introduced, with weight $w_{a∧b} = w_a + w_b = 0.7$.

$$\varphi(a,b) = (a \cap b) \tag{1}$$

Generalised Boolean polynomial of this logical expression is given in (2).

$$\varphi^{\otimes}(a,b) = ((a \cap b))^{\otimes} = a \otimes b \tag{2}$$

New Logical Aggregation operator based on [3] is given in (3).

$$\text{Agg}^{\otimes}(a,b,c,d) = \frac{7}{10}\varphi^{\otimes}(a,b) + \frac{1}{5}c + \frac{1}{10}d = \frac{7}{10}(a \otimes b) + \frac{1}{5}c + \frac{1}{10}d \tag{3}$$

Corresponding aggregation measure [3] is given in (4).

$$\mu = \frac{7}{10}(\sigma_a \wedge \sigma_b) + \frac{1}{5}\sigma_c + \frac{1}{10}\sigma_d \tag{4}$$

Aggregation measures (4) are shown in Table 1:

Table 1.

S	a	b	0,2 c	0,1 d	0,7 $a∧b$	$\mu(S)$
{0}	0	0	0	0	0	**0**
{a}	1	0	0	0	0	**0**
{b}	0	1	0	0	0	**0**

Table 1. (*continued*)

{c}	0	0	1	0	0	**0,2**
{d}	0	0	0	1	0	**0,1**
{a,b}	1	1	0	0	1	**0,7**
{a,c}	1	0	1	0	0	**0,2**
{a,d}	1	0	0	1	0	**0,1**
{b,c}	0	1	1	0	0	**0,2**
{b,d}	0	1	0	1	0	**0,1**
{c,d}	0	0	1	1	0	**0,3**
{a,b,c}	1	1	1	0	1	**0,9**
{a,b,d}	1	1	0	1	1	**0,8**
{a,c,d}	1	0	1	1	0	**0,3**
{b,c,d}	0	1	1	1	0	**0,3**
{a,b,c,d}	1	1	1	1	1	**1**

On value level input parameters from R_2 matrix from [2] are processed through operator given in (3).

Table 2.

«Energy efficient and energy utilization» - values from R_2	CATEGORIES		
	one star *	two stars **	three stars ***
Architecture design **a**	0,1	0,2	0,7
Envelope structure **b**	0,1	0,2	0,7
HVAC **c**	0,15	0,25	0,6
Lighting system **d**	0,1	0,25	0,65

In qualitative sense, attributes given in Table 2 are highly positive correlated, so the most suitable *generalised product* [3,4,6] should be the *min* function ($\otimes := min$). Aggregation operator (3) processes one by one column of R_2 matrix and outputs into vector B_2'.

$$Agg^{\otimes}(a,b,c,d):\ R_2 \rightarrow B_2'$$

$$B_2' = [0{,}11 \quad 0{,}215 \quad 0{,}675]$$

This result of Logical Aggregation coincides with the result of first phase aggregation of the existing fuzzy model developed in [2], so $B_2' = B_2$.

Now we shall use dot product as generalised product for *LA* ($\otimes: = *$).

$$B_2' = [0{,}047 \quad 0{,}115 \quad 0{,}675]$$

Different choice of generalised product obviously induces changes on the level of values, $B_2' \neq B_2$. After first phase of aggregation, the resulting values of quality are different from the values given by existing model.

3   Simulation of the Use of Aggregation Operator of Existing Fuzzy Model through Logical Aggregation Operator

New model of *LA* [1,4,6] will be applied again on the first phase of aggregation of the existing model [2], and the results will be compared. The aim is to show the possibilities of manipulation on the level of internal structure of the attributes which is fragmented via the new *LA* paradigm and to achieve the same results as with usage of the existing model.

Vector of weights of the secondary attributes of the primary attribute *operation management* [2] is: A_6 = [0,3   0,3   0,4]. Weights of secondary attributes *a* (*garbage classification and biologic treatment*),*b* (*intellectualized system*) and *c* (*all kinds of management system*) are respectively w_a = 0,3 w_b = 0,3   and w_c = 0,4.

Generalised pseudo-Boolean polynomial [1] of the single attributes is equal to the attributes themselves:

$$\varphi^{\otimes}(a) = a^{v} \qquad \varphi^{\otimes}(b) = b^{v} \qquad \varphi^{\otimes}(c) = c^{v}$$

From the aspect of immanent structure, single attributes are complex elements (contain more than one *atomic attribute* [1]). In this particular case, every attribute *a*, *b*, or *c* includes 4 atomic attributes. Some of the atomic attributes (which present internal algebraic structure) are shared among the attributes.

Now aggregation measures $\mu(S)$ are assigned to the atomic attributes (S) according to weights given in vector A_6.

Table 3.

Atoms (S)	$w\,\vec{\sigma}_\varphi = 0.3$	$w\,\vec{\sigma}_\varphi = 0.1$	$w\,\vec{\sigma}_\varphi = 0.6$	$\mu(S)$
{0}	0	0	0	0
{a}	1	0	0	0.3
{b}	1	0	0	0.3
{c}	1	1	0	0.4
{a,b}	0	0	1	0.6
{a,c}	0	1	1	0,7
{b,c}	0	1	1	0,7
{a,b,c}	1	1	1	1

In Table 3 one of the possible weights assignment and structures is presented, and it defines the immanent structure $\vec{\sigma}_\varphi$ (as explained in [1,4,6,7]) of the new attributes which are used to simulate the aggregation operator of existing fuzzy model [2].

The corresponding generalised pseudo-Boolean polynomial is generated by summation of atomic pseudo-Boolean polynomials included into structure defined by structure vectors $\vec{\sigma}_{\varphi}$. This procedure gives the aggregation operator shown in equation (5):

$$\text{Agg}^{\otimes}(a,b,c) = \frac{3}{10}[\,a - 2(a \otimes c) - 2(a \otimes b) + 4(a \otimes b \otimes c) + b - 2(b \otimes c) + c\,] + \frac{1}{10}$$

$$c + \frac{3}{5}[\,a \otimes b - 2(a \otimes b \otimes c) + a \otimes c + b \otimes c\,] \tag{5}$$

On the level of values (arithmetic level), one by one column of matrix of input parameters R_6 from [2] is processed through the operator given in (5) and put into vector B_6'. For generalised product, dot product is used.

$$B_6' = [0,17 \quad 0,285 \quad 0,545], \quad B_6' = B_6$$

If the *min* function is used as the generalised product, the yield is:

$$B_6' = [0,17 \quad 0,285 \quad 0,545], \quad B_6' = B_6$$

As shown above, using the Logical Aggregation operator (5) and generalised product ($\otimes := *$) and ($\otimes := min$) respectively, the yield is the same as the one in the first phase of the existing model presented in [2], and the aim of simulating the model from [2] is successfully achieved.

4   Implementation of Logical Aggregation in the Second Phase of Aggregation of the Existing Fuzzy Model

The second phase of aggregation in [2] will be modified for the purpose of demonstration of two interesting abilities of the *LA* model.

- If there is a need to satisfy only one of the two or more quality demands and not all of them at the same time
- If there is a quality demand which should "swap" the relative importance of other quality demands

Table 4.

PRIMARY ATTRIBUTES (A)	WEIGHT VECTOR OF PRIMARY ATTRIBUTES A=[$w(a_i)$]
Land-saving and outdoor environment (*a*)	0,1
Energy efficient and energy utilization (*b*)	0,5
Water-saving and water utilization (*c*)	0,2
Material-saving and material utilization (*d*)	0,1
Indoor environment quality (*e*)	0,05
Operation management (*f*)	0,05
	$\sum w(a_i) = 1$

In Table 4 the primary attributes with their respective weights are shown. The weight vector is A = [0,1   0,5   0,2   0,1   0,05   0,05].

If we wish to have satisfied only one of the two quality demands *material-saving and material utilization* and *indoor environment quality*, and not both of them at the same time, this can be achieved by removing single attributes (their respective weights become zero), and introducing new complex attribute *d∨e* with weight $w_{d\vee e}$ = $w_d + w_e = 0,15$.

$$\varphi(d,e) = (d \cup e) \tag{6}$$

Generalised Boolean polynomial [1] of this logical expression is given in (7).

$$\varphi^{\otimes}(d,e) = ((d \cup e))^{\otimes} = d + e - d \otimes e \tag{7}$$

If we wish to have a quality demand which can "swap" the relative importance of other quality demands, it can be done through a logical function of these attributes.

For example, if *Energy efficient and energy utilization* **is not** highly satisfied, then a complex attribute can be created in a way that it gives less importance to it (overall quality will be less affected by this unsatisfaction), and more importance to *Water-saving and water utilization, Material-saving and material utilization* and *Land-saving and outdoor environment*.

But if *Energy efficient and energy utilization* **is** highly satisfied, then the importance will be brought back to itself (so the overall quality will be more affected by *Energy efficient and energy utilization* attribute), and importance of other attributes is reduced. Logical expression of such complex attribute is given in (8).

$$\varphi(a,b,c,d) = b \cup (Cb \cap a \cap c \cap d) \tag{8}$$

Generalised pseudo-Boolean polynomial [1] of this logical expression is given in (9).

$$\varphi^{\otimes}(a,b,c,d) = b + [(1 - b) \otimes a \otimes c \otimes d]$$

$$\varphi^{\otimes}(a,b,c,d) = b + a \otimes c \otimes d - a \otimes b \otimes c \otimes d \tag{9}$$

Weight of this complex attribute will be taken from the single attribute *Energy efficient and energy utilization*.

$$w_{b\vee(\neg b \wedge a \wedge c \wedge d)} = 0,25 \qquad w_b = 0,5 - 0.25 = 0.25$$

After both modifications of the model shown in chapter 4, instead of second phase operator from [2], the new operator of Logical Aggregation [3] is given in (10).

$$Agg^{\otimes}(a,b,c,d) = a + \frac{1}{4}b + \frac{1}{5}c + \frac{1}{20}f + \frac{3}{20}(d + e - d \otimes e) + (b + a \otimes c \otimes d - a \otimes b \otimes c \otimes d) \tag{10}$$

On the arithmetic level, the input parameters matrix R from [2] will be processed through operator given in (10).

$$\text{Agg}^{\otimes}(a,b,c,d): \quad R \rightarrow B_* \quad \text{or} \quad B_{min}$$

$$R = \begin{bmatrix} 0,16875 & 0,285 & 0,54625 \\ 0,11 & 0,215 & 0,675 \\ 0,2025 & 0,2925 & 0,505 \\ 0,22 & 0,285 & 0,495 \\ 0,265 & 0,265 & 0,47 \\ 0,17 & 0,285 & 0,545 \end{bmatrix}$$

If generalised product is dot product ($\otimes := *$):

$$B_* = [0,186553 \quad 0,284584 \quad 0,641322]$$

If generalised product is *min* function ($\otimes := min$):

$$B_{min} = [0,175312 \quad 0,269 \quad 0,594625]$$

Final performance index T from [2] is calculated as a product of vector [1 2 3] with transposed vector B, and it is also valid for our modified model. It is shown in (11).

$$T = [1 \quad 2 \quad 3] * B^{T \, (\text{transposed})} \tag{11}$$

$$T = \begin{bmatrix} 1 & 2 & 3 \end{bmatrix} * B.^{T} = \begin{bmatrix} 1 & 2 & 3 \end{bmatrix} * \begin{bmatrix} 0,186553 \\ 0,284584 \\ 0,641322 \end{bmatrix} = 2,6796$$

$$T = \begin{bmatrix} 1 & 2 & 3 \end{bmatrix} * B_{min}^{T} = \begin{bmatrix} 1 & 2 & 3 \end{bmatrix} * \begin{bmatrix} 0,175312 \\ 0,269 \\ 0,594625 \end{bmatrix} = 2,4971$$

Depending on performance index, the building belongs to one of the three categories (one, two or three stars) [2].

$T \in [1,0 \quad 1,7]$ - one star building
$T \in [1,7 \quad 2,4]$ - two stars building
$T \in [2,4 \quad 3,0]$ - three stars building.

When modified model presented in chapter 4 of this paper is applied, the building from the example given in [2] stays in the interval $T \in [2,4 \quad 3,0]$, which implies that it belongs to the category of three stars building.

5   Conclusion

Models for performance evaluation of Green buildings enable good quality and objective evaluation, minimizing the subjectivity bias. Experts in the subject area contributed as the weight coefficients providers, while the final result (performance index) is achieved through processing on arithmetic level, through the fixed mathematic model.

Conventional tools for aggregation are often inadequate due to limitations in the sense of disability of using logical interactions between quality attributes. Improvements of the model's "articulation abilities" by using advanced techniques of Logical Aggregation significantly expand the possibilities of customization of the model to more specific needs. This is especially noticeable in the ability to include complex logical functions in which nontrivial attributes emerge. New paradigm treats logical functions – partial aggregation demands as a generalised Boolean polynomial, which processes values from the unitary real interval [0, 1]. Aggregation in general is a generalised pseudo-logical function.

Comparative review of the results of modified improved model presented in this paper, and existing fuzzy model [2] is shown, and the advanced abilities of the new approach to aggregation of quality parameters and performance of Green buildings are demonstrated.

References

1. Mirković, M., Hodolič, J., Radojević, D.: Aggregation for Quality Management. Yugoslav Journal of Operations Research 16(2), 177–188 (2006)
2. Sun, J., Wu, Y., Hao, Y., Dai, Z.: Fuzzy Comprehensive Evaluation Model and Influence Factors Analysis on Comprehensive Performance of Green Buildings. In: ICEBO 2006, Shenzhen, vol. VIII-4-2. China Renewable Energy Resources and a Greener Future (2006)
3. Radojević, D.: Logical Aggregation Based on Interpolative Realization of Boolean Algebra. In: Eusflat Conf., vol. (1), pp. 119–126 (2007)
4. Radojevic, D.: Fuzzy Set Theory in Boolean Frame. Int. J. of Computers, Communications & Control III (2008)
5. Radojevic, D.: New [0,1]-valued logic: A natural generalization of Boolean logic. Yugoslav Journal of Operational Research - YUJOR 10(2) (2000)
6. Radojevic, D.: Logical Aggregation Based on Interpolative Boolean Algebra. Mathware & Soft Computing 15 (2008)
7. Radojević, D.: Interpolative relations and interpolative preference structures. Yugo-slav Journal of Operational Research – YUJOR 15(2) (2005)

Structural Functionality as a Fundamental Property of Boolean Algebra and Base for Its Real-Valued Realizations

Dragan G. Radojević

University of Belgrade, Institute Mihajlo Pupin, Belgrade
dragan.radojevic@pupin.rs

Abstract. The value of the complex Boolean function can be calculated directly on the basis of its components value. It is a principle known as the truth functionality. Properties of the Boolean algebra have indifferent values. The truth functional principle is taken as a valid principle in general case in the conventional generalization: multi-valued and/or real-valued realizations (fuzzy logic in the broad sense). This paper presents that truth functionality is not valued indifferent property of the Boolean algebra and it is valid only in two-valued realization, and thus it cannot be the basic of the value generalization. The value generalization (real-valued realizations) enables incomparably more descriptiveness than the two-valued classical Boolean algebra, so that the finite Boolean algebra is enough for any real application. Each finite Boolean algebra is atomic. Every Boolean function (the element of the analyzed finite Boolean algebra) can be presented uniquely as disjunction of the relevant atoms – disjunctive canonical form. Which atoms are and which are not included in the analyzed Boolean function is defined by its structure: 0-1 vector which dimension matches the number of atoms (in the case of n independent variables, the number of atoms is 2^n). Atom corresponds uniquely to each vector structure position and value 0 means that the adequate atom is not included in the analyzed function, and 1 means that it is included. The principle of the structural functionality is: the structure of the complex Boolean function is defined directly on the basis of its components structure. The truth functionality is a value image of the structural functionality only in the case of two-valued realization. Each insisting on the truth functionality, such as in the case of conventional multi-valued logic and fuzzy logic in general sense, is unjustified from the point of the Boolean consistency.

Keywords: Boolean algebra, atomic Boolean functions, disjunctive canonical form, Boolean function structure, structural functionality, truth functionality, generalized value realization of the Boolean functions.

1   Introduction

The truth functionality is a property of the two-valued realization of Boolean algebra. The value of complex Boolean function can be calculated directly based on its

A. Laurent et al. (Eds.): IPMU 2014, Part II, CCIS 443, pp. 28–36, 2014.
© Springer International Publishing Switzerland 2014

components values. The principle of the truth functionality is taken as a basis of generalization in regular (conventional) value generalization of the theories based on the Boolean algebra: multi-valued and/or real-valued logic (fuzzy logic in general sense). Practically, there is no explanation why the truth functionality is taken as a basis of the valued generalization except ``because it is usual and technically useful`` (obviously not mathematical motivation!).

This paper presents that truth functionality is valid only in two-valued realization of the Boolean algebra. Every finite Boolean algebra is atomic. The Boolean algebra generated with n independent variables contains 2n atoms. Atoms are the simplest elements from the point of the valued realization (in the classical case for the free set of 0-1 independent variables values , only one atom has value 1 and all the others 0). The important atom property of the analyzed Boolean algebra is the fact that they do not have anything in common and/or that conjunction of two different atoms is identical to 0 element of the Boolean algebra i.e. it has value realization identical to 0. Each Boolean function (element of the analyzed Boolean algebra) can uniquely be presented as a union of relevant atoms – disjunctive canonical form. Which atoms are and which are not included is defined with 0-1 vector dimension 2^n – by the structural function. Atom uniquely corresponds to each vector position and value 0 means that the appropriate atom is not included in the analyzed function, and 1 means that it is included.

The structure of the complex Boolean function is defined directly on the basis of the structure of its components – the principle of the structural functionality. This principle is value indifferent – algebra and it must be saved in all Boolean consistent value realizations. The truth functionality is a value image of the structural functionality and only in the case of two-valued realization – known truth table, for example. Each insisting on the truth functionality in general case (multi-valued and/or real-valued realization) has as a result impossibility of simultaneous keeping all properties of Boolean algebra. So, the answer on the question stated in the paper title is that the structural functionality is the basic of the Boolean consistent generalization of the finite Boolean algebra valued realization in general case.

The Boolean consistent generalization enables direct generalization of all theories based on the classical finite Boolean algebra. From the point of the Boolean consistency, it is completely unjustified to insist on the truth functionality such as in the case of the conventional multi-valued logic and fuzzy logic in general sense.

2   Fuzzy Logic in Boolean Frame

Fuzzy logic, realized in the Boolean frame, means that all Boolean axioms and theorems are valid in the most general case i.e. in the real-valued case. Since the appropriate classical techniques are based on the Boolean algebra, precisely, on its two-valued realization, the consistent generalization should be based on the real-valued realization of the Boolean algebra. The real-valued realization of the finite or atomic Boolean algebra [4, 5 and 6] is described here in detail.

The main problem of the conventional approaches (the main stream of the usual realization) is fact that they are based on the principle of the truth functionality, which is taken from the classical logic based on the two-valued realization. From the point of the Boolean algebra, this principle is adequate or correct only in two-valued case. The reason is simple: the Boolean function has the vector nature in general case, but in the classical case the attention is drawn only to one component (which is defined with 0-1 values of the independent variables). In general case, when the values of the independent variables are not only 0 or 1 but also include everything in between, it is necessary to include more components in calculation and in the most general case all components of the vector immanent to the analyzed Boolean function in the analyzed Boolean algebra. In order to illustrate the main idea, we will use the Boolean function of two independent variables x and y, from the famous Boolean paper from 1848 [2]:

$$\phi(x,y) = \phi(1,1)xy + \phi(1,0)x(1-y) + \phi(0,1)(1-x)y + \phi(0,0)(1-x)(1-y). \quad (1)$$

Actually, this equation can be treated also as a special case of the Boolean polynomial [4]:

$$\phi^{\otimes}(x,y) = \phi(1,1)x \otimes y + \phi(1,0)(x - x \otimes y) +$$
$$\phi(0,1)(y - x \otimes y) + \phi(0,0)(1 - x - y + x \otimes y). \quad (2)$$

The independent variables in general case take the value from the unit real interval $x, y \in [0,1]$.

$\otimes$ is generalized product [4] or t-norm with the following property:

$$\max(x + y - 1, 0) \leq x \otimes y \leq \min(x, y).$$

2.1 Boolean Polynomial

The real-valued realization of the finite (atomic) Boolean algebra is based on the Boolean polynomials. The free Boolean function can be uniquely transformed into the appropriate Boolean [6].

Example 1: *Using the equation (2) of the equivalence relation, exclusive disjunction and implication, respectively*

   a. $\phi(x,y) =_{def} x \Leftrightarrow y$

   $\phi(1,1) = 1; \quad \phi(1,0) = 0; \quad \phi(0,1) = 0; \quad \phi(0,0) = 1;$

   $\boxed{x \Leftrightarrow y = 1 - x - y + 2x \otimes y}.$

b. $\phi(x,y) =_{def} x \veebar y$

   $\phi(1,1) = 0; \quad \phi(1,0) = 1; \quad \phi(0,1) = 1; \quad \phi(0,0) = 0;$

   $\boxed{x \veebar y = x + y - 2x \otimes y}.$

c. $\phi(x,y) =_{def} x \Rightarrow y$

   $\phi(1,1) = 1; \quad \phi(1,0) = 0; \quad \phi(0,1) = 1; \quad \phi(0,0) = 1;$

   $\boxed{x \Rightarrow y = 1 - x + x \otimes y}.$

The finite (atomic) Boolean algebra generated by the set of independent variables $\Omega = \{x_1, ..., x_n\}$, is $BA(\Omega) = P(P(\Omega))$, where: $P(\Omega)$ is a set of all subsets Ω. The atomic elements of the analyzed Boolean algebra $BA(\Omega)$ are [6]:

$$\alpha(S)(x_1, ..., x_n) = \bigwedge_{x_i \in S} x_i \bigwedge_{x_j \in \Omega \setminus S} \bar{x}_j, \qquad S \in P(\Omega). \tag{3}$$

The atomic Boolean polynomial $\alpha^{\otimes}(S)(x_1, ..., x_n)$ uniquely corresponds to the atomic element $\alpha(S)(x_1, ..., x_n)$, and it is defined by the following equation [6]:

$$\alpha^{\otimes}(S)(x_1, ..., x_n) = \sum_{C \in P(\Omega \setminus S)} (-1)^{|C|} \bigotimes_{x_i \in C \cup S} x_i; \quad S \in P(\Omega). \tag{4}$$

Example 2: *The atomic Boolean polynomials for the Boolean algebra generated with* $\Omega = \{x, y\}$ *are:*

$\alpha^{\otimes}(\{x, y\}) = x \otimes y;$

$\alpha^{\otimes}(\{x\}) = x - x \otimes y;$

$\alpha^{\otimes}(\{y\}) = y - x \otimes y;$

$\alpha^{\otimes}(\{\ \}) = 1 - x - y + x \otimes y.$

The values of the atomic polynomials in the real-valued case are negative $\alpha^{\otimes}(S)(x_1, ..., x_n) \in [0,1]$ $S \in P(\Omega)$, and their sum is identically equal to 1. In the case of the example described by the equation (2) for $x, y \in [0,1]$:

$$x \otimes y + (x - x \otimes y) + (y - x \otimes y) + (1 - x - y + x \otimes y) \equiv 1.$$

The classical two-valued case is just a special case which satisfies this fundamental equity, since the value of only one atom is equal to 1 and all others are identically equal to 0.

The free Boolean function, the appropriate element of the analyzed Boolean algebra $\phi(x_1, ..., x_n) \in BA(\Omega)$, can uniquely be presented in disjunctive canonical form as a disjunction of relative atomic elements:

$$\phi(x_1, ..., x_n) = \bigvee_{S \in P(\Omega)} \sigma_\phi(S) \alpha(S)(x_1, ..., x_n). \tag{5}$$

Where: $\sigma_\phi(S)$, $(S \in P(\Omega))$ the relation of inclusion of corresponding atom $\alpha(S)(x_1, ..., x_n)$ in the analyzed Boolean function $\phi(x_1, ..., x_n)$, is defined in the following way:

$$\sigma_\phi(S) =_{def} \phi(\chi_S(x_i) | i = 1, ..., n),$$
$$\left(\chi_S(x_i) =_{def} \begin{cases} 1, & x_i \in S \\ 0, & x_i \notin S \end{cases} \right), \quad (S \in P(\Omega)). \tag{6}$$

The relation of inclusion determines which atoms $(S \in P(\Omega))$ are included in the analyzed Boolean function:

$$\sigma_\phi(S) = \begin{cases} 1, & \alpha(S)(x_1, ..., x_n) \subset \phi(x_1, ..., x_n) \\ 0, & \alpha(S)(x_1, ..., x_n) \not\subset \phi(x_1, ..., x_n) \end{cases}, \tag{6.1}$$

(The values of the corresponding relation of inclusion are equal to 1) and which are not included (the values of the relation of inclusion are equal to 0).

The Boolean polynomial uniquely corresponds to the analyzed Boolean function as Figure (5):

$$\phi^\otimes(x_1, ..., x_n) = \sum_{S \in P(\Omega)} \sigma_\phi(S) \alpha^\otimes(S)(x_1, ..., x_n). \tag{7}$$

The Boolean polynomial (7) can be presented as a scalar product for two vectors:

$$\boxed{\phi^\otimes(x_1, ..., x_n) = \vec{\sigma}_\phi \vec{\alpha}^\otimes(x_1, ..., x_n)}, \quad x_1, ..., x_n \in [0, 1] \tag{8}$$

$\vec{\sigma}_{\phi} = \left[\sigma_{\phi}(S) \middle| S \in P(\Omega)\right]$ is a structure of the analyzed Boolean func-

tion $\phi(x_1,...,x_n) \in BA(\Omega)$, i.e. a vector of the relation of inclusion of atomic functions in the analyzed function.

$\vec{\alpha}^{\otimes}(x_1,...,x_n) = \left[\alpha(S)(x_1,...,x_n) \middle| S \in P(\Omega)\right]^T$ is a vector of atomic polynomials of the analyzed finite (atomic) Boolean algebra $BA(\Omega)$.

Example 2: *Structures of the analyzed Boolean functions from the example 1:*

$$\vec{\sigma}_{x \Leftrightarrow y} = \begin{bmatrix} 1 & 0 & 0 & 1 \end{bmatrix}, \quad \vec{\sigma}_{x \underline{\vee} y} = \begin{bmatrix} 0 & 1 & 1 & 0 \end{bmatrix},$$
$$\vec{\sigma}_{x \Rightarrow y} = \begin{bmatrix} 1 & 0 & 1 & 1 \end{bmatrix}.$$

and atomic polynomials vectors for two independent variables:

$$\vec{\alpha}^{\otimes}(x_1, x_2) = \begin{bmatrix} x_1 \otimes x_2 \\ x_1 - x_1 \otimes x_2 \\ x_2 - x_1 \otimes x_2 \\ 1 - x_1 - x_2 + x_1 \otimes x_2 \end{bmatrix}.$$

3   The Structural Functionality

The principle of the structural functionality: The structure of the free complex Boolean function can be calculated directly on the basis of its components using the following identities:

$$\vec{\sigma}_{\phi \wedge \psi} = \vec{\sigma}_{\phi} \wedge \vec{\sigma}_{\psi}$$
$$\vec{\sigma}_{\phi \vee \psi} = \vec{\sigma}_{\phi} \vee \vec{\sigma}_{\psi}$$
$$\vec{\sigma}_{\neg \phi} = \neg \vec{\sigma}_{\phi} = \vec{1} - \vec{\sigma}_{\phi}.$$

The known principle of the truth functionality is just an image of the structural functionality at the value level and thus only in the case of two-valued realization. In general case (multi-valued and/or real-valued realizations), the principle of the truth functionality is not able to keep all Boolean algebra properties. That is the reason why fuzzy approaches, based on the principle of the truth functionality (conventional fuzzy logic in wider sense), cannot be in the Boolean frame and/or they are not the Boolean consistent generalizations.

Structures as algebra properties of the Boolean functions keep all Boolean laws [6]:

3.1   Laws of Monotonicity

Associativity

$$\vec{\sigma}_\phi \vee \left(\vec{\sigma}_\psi \vee \vec{\sigma}_\xi\right) = \left(\vec{\sigma}_\phi \vee \vec{\sigma}_\psi\right) \vee \vec{\sigma}_\xi,$$
$$\vec{\sigma}_\phi \wedge \left(\vec{\sigma}_\psi \wedge \vec{\sigma}_\xi\right) = \left(\vec{\sigma}_\phi \wedge \vec{\sigma}_\psi\right) \wedge \vec{\sigma}_\xi. \tag{9}$$

Commutatively

$$\vec{\sigma}_{\phi\vee\psi} = \vec{\sigma}_{\psi\vee\phi}, \qquad \vec{\sigma}_\phi \vee \vec{\sigma}_\psi = \vec{\sigma}_\psi \vee \vec{\sigma}_\phi;$$
$$\vec{\sigma}_{\phi\wedge\psi} = \vec{\sigma}_{\psi\wedge\phi}, \qquad \vec{\sigma}_\phi \wedge \vec{\sigma}_\psi = \vec{\sigma}_\psi \wedge \vec{\sigma}_\phi. \tag{10}$$

Distributive

$$\vec{\sigma}_\phi \wedge \left(\vec{\sigma}_\psi \vee \vec{\sigma}_\xi\right) = \left(\vec{\sigma}_\phi \wedge \vec{\sigma}_\psi\right) \vee \left(\vec{\sigma}_\phi \wedge \vec{\sigma}_\xi\right),$$
$$\vec{\sigma}_\phi \vee \left(\vec{\sigma}_\psi \wedge \vec{\sigma}_\xi\right) = \left(\vec{\sigma}_\phi \vee \vec{\sigma}_\psi\right) \wedge \left(\vec{\sigma}_\phi \vee \vec{\sigma}_\xi\right). \tag{11}$$

Identity

$$\vec{\sigma}_\phi \vee \vec{0} = \vec{\sigma}_\phi, \qquad \vec{\sigma}_\phi \vee \vec{1} = \vec{1};$$
$$\vec{\sigma}_\phi \wedge \vec{0} = \vec{0}, \qquad \vec{\sigma}_\phi \wedge \vec{1} = \vec{\sigma}_\phi. \tag{12}$$

Idempotent

$$\vec{\sigma}_\phi \vee \vec{\sigma}_\phi = \vec{\sigma}_\phi, \qquad \vec{\sigma}_\phi \wedge \vec{\sigma}_\phi = \vec{\sigma}_\phi. \tag{13}$$

Absorption

$$\vec{\sigma}_\phi \wedge \left(\vec{\sigma}_\phi \vee \vec{\sigma}_\xi\right) = \vec{\sigma}_\phi, \qquad \vec{\sigma}_\phi \vee \left(\vec{\sigma}_\phi \wedge \vec{\sigma}_\xi\right) = \vec{\sigma}_\phi. \tag{14}$$

3.2   No Monotonicity Laws

Complementarity

$$\vec{\sigma}_\phi \wedge \vec{\sigma}_{\neg\phi} = \vec{0}, \qquad \vec{\sigma}_\phi \vee \vec{\sigma}_{\neg\phi} = \vec{1}. \tag{15}$$

De Morgan Laws

$$\neg\left(\vec{\sigma}_\phi \wedge \vec{\sigma}_\psi\right) = \vec{\sigma}_{\neg\phi} \vee \vec{\sigma}_{\neg\psi}, \qquad \neg\left(\vec{\sigma}_\phi \vee \vec{\sigma}_\psi\right) = \vec{\sigma}_{\neg\phi} \wedge \vec{\sigma}_{\neg\psi}. \tag{16}$$

4   Conclusion

This approach based on the application of the structural functionality keeps all Boolean algebra laws in all possible value realizations (from the classical two-valued to the most general real-valued realization) independently from the selected operators of the generalized product.

If you interpret the general case concretely (real-valued realization), then even the special case (two-valued realization) is treated in a different way.

Excluded middle and non-contradiction are defined for two-valued realization case (in logic, for example, the statement is either truth or untruth and it cannot be both truth and untruth) and thus they are not adequate for general case. It seems that in the case of the conventional approaches to the generalization, the excluded middle may even not be valid. In general case one statement can be partially truth but then it is untruth with complementary intensity. From the point of the structure, the complementary function corresponds to each Boolean function so that it contains all atoms which the analyzed function does not contain and whereat they do not have a single common atom. The consequence is that the disjunction of the free Boolean function and its complementary functions contain all atoms and /or it is identical to the Boolean constant 1, i.e. it has a value identically equal to 1 – the excluded middle. Also, its conjunction does not include a single atom and/or it is identical to the Boolean constant 0, i.e. it has a value identically equal to 0 – non-contradiction. However, these fundamental laws are valid in all consistent value realizations and known definitions are valid only in classical two-valued case.

Actually, excluded middle and non-contradiction uniquely define the complementary property of the analyzed property and thus these two laws are fundamental and important for cognition in general.

This can be illustrated in a simple example, such as a glass filled with water. The classical two-valued case treats only full or empty glass. Empty is complement of full and vice versa. In general or real case, the glass can be partially full and, at the same time, with the rest it is empty with complementary intensity, so that the sum of the intensity "full" and intensity "empty" is identically equal to 1.

It is obvious that, besides the fact that properties "full" and "empty" do not have anything in common, they are both simultaneously in the same glass and thus the union intensity is equal to 1 and intersection intensity to 0.

The free classical theory, based on the finite Boolean algebra using the real-valued realization of the Boolean algebra, can be generalized directly [6]. This is very important for many interesting examples which are logically complex such as: artificial intelligence, mathematical cognition, prototype theory in psychology, concept theory, etc.

The structural functionality sheds a new light on the relationship between syntax and semantics in the classical logic which will be presented in the next paper.

References

1. Zadeh, L.A.: From Circuit Theory to System Theory. In: Proc. of Institute of Ratio Engineering, vol. 50, pp. 856–865 (1962)
2. Boole, G.: The Calculus of Logic. The Cambridge and Dublin Mathematical Journal 3, 183–198 (1848)
3. Radojevic, D.: New [0,1]-valued logic: A natural generalization of Boolean logic. Yugoslav Journal of Operational Research - YUJOR 10(2), 185–216 (2000)
4. Radojevic, D.: There is Enough Room for Zadeh's Ideas, Besides Aristotle's in a Boolean Frame. In: 2nd International Workshop on Soft Computing Applications, SOFA 2007 (2007)
5. Radojevic, D.: Interpolative Realization of Boolean algebra as a Consistent Frame for Gradation and/or Fuzziness. In: Nikravesh, M., Kacprzyk, J., Zadeh, L.A. (eds.) Forging New Frontiers: Fuzzy Pioneers II. STUDFUZZ, vol. 218, pp. 295–317. Springer, Heidelberg (2008)
6. Radojevic, D.: Real-valued realization of Boolean algebra is natural frame for consistent fuzzy logic. In: Seising, R., Trillas, E., Moraga, C., Termini, S. (eds.) On Fuzziness, A Homage to Lotfi Zadeh. STUDFUZZ, vol. 2, pp. 559–566. Springer, Heidelberg (2013)

Fuzzy Concepts in Small Worlds and the Identification of Leaders in Social Networks[*]

Trinidad Casasús-Estellés[1] and Ronald R. Yager[2]

[1] Faculty of Economics-IEI, University of Valencia, Spain
casasus@uv.es
[2] Machine Intelligence Institute, Iona College, New Rochelle, NY 10801 USA
yager@panix.com

Abstract. In the study of the Social Networks, the Small World phenomenon appears frequently. We apply some techniques of graph theory and fuzzy sets to characterize the Small World features as well as the existence of the figure of leader in Social Networks. These techniques help to the conceptual formalization in relational networks analysis, by transforming linguistic and human-focused manner concepts related to social networks in some formal representation. These techniques are also applied when the similarity among nodes wants to be measured in order to study the current homophily present in a Network.

Keywords: Fuzzy Sets, Small Worlds, Social Network, Leader.

1   Introduction

Social Networks spread through almost all the aspects of our economic and social lives and they have been widely studied from several approaches given their presence in a huge range of fields [1]. The study of Random Graphs has proved that these graphs can exhibit some of the features of observed social networks but they lack some others [1]. Some of these social networks present a feature of "high local clustering and short global separation" [2], known in the literature as the small world phenomenon. In this sense Watts & Strogatz [3] developed a variation of a random network in order to generate networks that simultaneously exhibit high clustering and low diameter, a combination which they called a small-world network referring to the Stanley Milgram's "Small Worlds" experiment [4]: Many social networks tend to have small diameter and small average path length, where small is on the order of the log of the number of nodes or less.

In [5] the natural connection between graph theory and granular computing, particularly fuzzy set theory is pointed out, and fuzzy sets are used to formalize some concepts associated with social networks. In [6] the idea of fuzzy relationships and their role in modeling weighted social relational networks as well as the idea of vector-valued nodes are discussed.

---

[*] This work has been partially supported by the project TIN2008-06872-C04-02.

A. Laurent et al. (Eds.): IPMU 2014, Part II, CCIS 443, pp. 37–45, 2014.

Following this reasoning, we are going to use fuzzy set methodology to define Small Worlds as a part of the structure of Social Networks. Let's start by introducing some notation we will use in the following chapters. (See [5] and [7] for more notation and definitions).

1.1   Some Definitions and Notation

Let $G = \langle V, E \rangle$ be a graph, where V is the set of vertices and E is the collection of edges. We consider associated with G a relationship R: $VxV \rightarrow \{0,1\}$, such that R(x, y) = 1 = R(y, x) when (x, y) is an edge of G, and R(x, y) = 0 = R(y, x) if (x, y) is not an edge of G. By the term graph we will understand undirected graph, this is, the graph with the associated relationship R being symmetric R(x, y) = R(y, x).

Definition 1. Let define the *neighbor* of a node x, $Neigh_x(y) = R(x, y)$, the *Neighborhood* of x as $N_x = \{y\ /\ Neigh_x\ (y) = 1\}$, and the *size of Neighborhood* of x, $d_x = Card(N_x)$.

Given two vertices x and y of V, we will understand by an *x-y path* in G a sequence of distinct vertices $\{x_1, x_2,..., x_n\}$ beginning with x, $x_1 = x$, and ending with y, $x_n = y$, so that there is an edge between any two adjacent vertices in the sequence. We will say that two nodes are *connected* if it is possible to find a path from one node to the other. A subset $S \subseteq V$ is *completely connected* if each pair of nodes belonging to S are connected.

The *length* of a path is defined as the number of edges the path contains. A *cycle* is an x-y path in which x = y and it must contain at least three distinct vertices. We refer to a shortest path as a *geodesic* and denote the shortest path between x and y as Geo (x, y). If Len (p) indicates the length of a path then the *distance* (x, y) = Len (Geo (x, y)). Given two relationships R_1 and R_2, over the same vertex set V, their composition, over the set V, is defined as

$$R_1* R_2(x, z) = Max_{y \in V}[Min(R_1(x, y), R_2(y, z))]. \tag{1}$$

It will be denote by $R^k = R*R*...*R$, k-times, and the distance (x, y) can be understood as the smallest k for which $R^k(x, y) = 1$.

Given $G = \langle V, E \rangle$ an undirected graph representing a social network, with R the associated relationship, a subset $S \subseteq V$ is called a *cluster of order k* ([5]) if the following conditions hold:

a. for all $x, y \in S,$ $Len(Geo(x, y)) \leq k$
b. for all $z \notin S$, $Len(Geo(x, z)) > k$, for some $x \in S$.

S is a *clique* in the special case when S is a cluster of order k =1, that is, a maximal completely connected sub-network of a given network.

2   Small Worlds

To define the Small-World behavior in a Network $G = \langle V, E \rangle$, $V = \{x_i\}$, following [3], two concepts are considered, the Characteristic Path Length L, a global property of the graph, and the Clustering Coefficient C, a local property that quantifies the neighborhood 'cliquishness'. In a small world the coincidence of high local clustering and short global separation are given. Thus, the Small World (SW from now on) behavior is characterized by [8]:

A. Low Characteristic Path Length, where the characteristic path length L (average connectivity of the network), as defined by [3], is the average distance between all pair of vertices, given by the formula

$$L = \frac{1}{n(n-1)} \sum_{i \neq j} d_{ij} \tag{2}$$

being d_{ij} the distance between the nodes x_i and x_j. It is a global property of the graph that describes the average number of the shortest path inside the network. In case of not-fully connected network, there exist another ways to give a measure of the average path length of the network (See for instance [9])

B. High Clustering Coefficient, it is the proportion of pairs of neighbors of a node that are also neighbors of each other; it quantifies how close neighbors are to being a clique. Given the node x_i, the clustering coefficient of x_i, C_i, denotes the fraction of the links actually present among its neighbors

$$C_i = \frac{pairs\ of\ neighbors\ of\ x_i\ connected}{possible\ connections\ between\ neighbors}$$

$$C_i = \frac{2}{d_i(d_i-1)} \sum_{k=j+1}^{n} \sum_{j=1, j \neq i}^{n} Neigh_i(x_k) Neigh_i(x_j) Neigh_j(x_k), \tag{3}$$

where $d_i = Card(N_i)$ was defined in Paragraph 1.1., the Neighborhood size of node x_i. The Clustering Coefficient of whole network is given by

$$\bar{C} = \frac{1}{n} \sum_{i=1}^{n} C_i \tag{4}$$

Roughly speaking, a small-world graph with n vertex is a large, sparsely connected, decentralized graph, which exhibits a characteristic path length close to that of an equivalent random graph, yet with a clustering coefficient much greater.

In the spirit of human perception, we could say that "small" in the environment of Social Networks means that almost every element of the network is somehow "close" to almost every other element, even those that are perceived as likely to be far away. The Small World implies the claim that even when two people do not have a friend in common, only a short chain of intermediaries separates them [5] & [6]). In the next paragraph we are giving some soft definitions by using fuzzy tools, following some techniques introduced by those authors. This method will help us to approach some ideas, as for instance those of "small" and "close".

2.1   Soft Clustering Coefficient

We can give the softer fuzzy definition of the Clustering Coefficient as the proportion of pairs of close neighbors of a node that are close neighbors of each other.

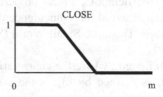

Fig. 1. Prototypical definition of Close

Let $Q(m)$ be the degree m links satisfies the idea of "close", $Q: Z^+ \rightarrow [0,1]$ like in Figure 1. One way to define the degree to which node x_j is close neighbor of x_i can be given by

$$QNeigh_i(x_j) = Max_m[Q(m) \wedge R^m(x_i, x_j)]$$

Then, the Soft Clustering Coefficient of Node x_i can be defined as

$$SC_i = \frac{\sum_{k=j+1}^{n} \sum_{j=1, i \neq j}^{n} QNeigh_i(x_k) QNeigh_i(x_j) QNeigh_j(x_k)}{\sum_{k=j+1}^{n} \sum_{j=1, i \neq j}^{n} QNeigh_i(x_k) QNeigh_i(x_j)} \tag{5}$$

2.2   Soft Characteristic Path Length

In a similar way to the previous Paragraph, it is possible to give a definition of Soft Path Length SL. In communal vocabulary a short path is expressed as a fuzzy subset S defined on the set of positive integers. Here S must be such that $S(1) = 1$ and $S(l_1) \geq S(l_2)$, when $l_1 < l_2$, where $S(k)$ denotes the degree k links is short distance, similar to that of close defined in Paragraph 2.1.

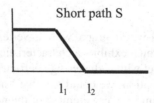

Fig. 2. Prototypical definition of Short Path

The satisfaction of the criteria "there is a Short Distance from x_i to x_j" is given by

$$SP_i(x_j) = Max_k[S(k) \wedge R^k(x_i, x_j)]$$

Then, we could define the Soft Characteristic Path Length as

$$SL = \frac{1}{n(n-1)} \sum_{i \neq j} SP_i(x_j) \in [0, 1] \tag{6}$$

And SL will give us a value that reflects how "short" is the average length of our network according to our criteria of Short Path S.

Note that given S(k) is decreasing with respect to k, we see $SP_i(x_j)$ will be equal to S(k) for the first k for which there is a path from x_i to x_j. If we denote this k_j then $SP_i(x_j) = S_i(k_j) \in [0,1]$

3   Leadership Inside the Network

To study the position of a given node inside the network is an interesting question that can be formulated as the analysis of the connections of that node in the overall network. [1] writes about many different measures of centrality that tend to capture different aspects of the position that a node has, been one of them the Betweenness introduced by [10], together with the Degree – how connected a node is, the Closeness – how easily a node can reach other nodes, and the Neighbors' Characteristics C- how important, central, or influential a node's neighbors are. Newman [11] , refers to the Betweenness of a node x, as the total number of shortest paths between pairs of nodes that pass through x. This quantity would be an indicator of the most influential people in a network, the one who control the flow of information between the others.

Following with our proposal, let us define in a human-meaningful way what we will understand by leader of a SW and provide a procedure for evaluating when a node satisfies our definition. To do this, we will take into account the number of edges arriving to a node (idea of congestion and centrality, [5]) as well as the betweenness of it.

Definition 2. A node x of G = ⟨V, E⟩ will be considered a *leader* inside a small world of a social network if the following conditions are given:

1. There is a high congestion on this node.
2. Most of the elements, in the SW x belongs to, are connected with it. (Most of the nodes close to x are connected directly to it)
3. High betweenness of the node x.
4. The node x is connected with most of other nodes that verify the same conditions 1, 2 and 3.

Let's go in depth in each point of the above criteria:

1. We shall say that a node is a congested node if it has many incident nodes, that is, it has a high density. Given the set of nodes V, it is possible to define a fuzzy set C over the set V of nodes in the network such that C(x) ∈ [0,1] is going to indicate the degree to which x is a congested node.

Following [5], we consider MANY as a fuzzy set over the set N = {1,2,..., n}, with n the number of nodes in the network G, such that for each y ∈ N the value

MANY(y) $\in$ [0,1], MANY: N $\rightarrow$ [0,1]   a monotonically increasing function of y that indicates the degree to which the quantity y satisfies the concept many.

The degree to which a node x_i is a congested node is defined by

$$C(x_i) = MANY(\sum_{j=1,i\neq j}^{n} R(x_i, x_j))$$

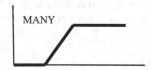

Fig. 3. Prototypical definition of MANY

Other definitions of MANY (see [5]) may be considered, as for example, in terms of proportion:

$$CG(x_i) = MANY_p(\frac{\sum_{j=1,i\neq j}^{n} R(x_i,x_j)}{N}) \tag{7}$$

where by $MANY_p(y) \in$ [0,1] indicates the degree to what y (as proportion) satisfies the concept many.

2. For most nodes of the small world x belongs to, the path from x to each one of them is short and highly centered (relatively inside the small world):

The idea of Most is given, as in point 1, by a fuzzy subset Q: [0,1] $\rightarrow$ [0,1], where for any proportion p the value Q(p) indicates the degree to it satisfies the concept Most, Q(p) $\in$ [0,1] is a monotonically increasing function of p, such that Q(0) = 0, Q(1) = 1.

We have seen in Paragraph 2.2. the definition of Soft Characteristic Path between x_i and x_j as

$$SP_i(x_j) = Max_k[S(k) \wedge R^k(x_i, x_j)] = S_i(k_j)$$

To check that x satisfies Condition 2, one way would be (see [5] for other alternative procedures, using OWA operators):

We first order the set of values of $SP_x(x_j)$, the degree to which there is a short path from x to x_j, $x \neq x_j$, and let d_k be the value of the kth largest of the $SP_x(x_j)$. Using this and the fuzzy subset Q representing our concept of Most, we calculate

$$CT(x) = Max_{k=1...n-1}[Q(\frac{k}{n-1}) \wedge d_k] \tag{8}$$

which will give us the degree of centrality of the considered node x.

3. High betweenness of the node x would mean that the leader x lies on most of the Soft Short paths between pairs of nodes being in the small world considered.

Given two nodes x and x_j, if we consider the soft short path between them defined previously, $SP_x(x_j)$, the degree to the node x satisfies the Soft Betweenness can be calculated as follows:

Let's denote by B the set of paths z-y such that x is one of the nodes included in the path, and let D = Card (B) be. If we consider the corresponding set of values Soft Short Path $\{SP_z(y)\}_{z-y\in B}$ we can assume, as before, these paths ordered being d_k the value of the kth largest of the set. Then, the Soft Betweenness of x can be defined as:

$$SB(x) = Max_{k=1\ldots n-1}[Q(\frac{k}{D}) \wedge d_k] \qquad (9)$$

and this value will give us a criteria about the betweenness of node x.

The last condition we consider for leaders has to be with the relationship among nodes with similar characteristics. Are there any correlation patterns in the degrees of connected nodes? (see [1] and [9]). For instance, do relatively high degree nodes have a higher tendency to be connected to other high degree nodes? This is termed as Positive Assortativity. While there is little systematic study of assortativity, there is a hypothesis that positive assortativity is a property of many socially generated networks, and it contrasts with the opposite relationship that is more prevalent in technological and biological networks. In our concept of leader we can include one more criteria related to the connectivity among the leaders of SW in a given network:

4. Most of the leaders are connected each other (there exists positive assortativity among them). Most of $x \in V$ that verify simultaneously conditions 1, 2 and 3 are connected each other. Let $L = \{y_i\}_{i=1}^r$ be the set of nodes that verifies conditions 1, 2, and 3. As we have seen in the Condition 2, given that the Soft Short Path between y_i and y_j is obtained from

$$SP_i(y_j) = Max_k[S(k) \wedge R^k(y_i, y_j)]$$

we proceed in a similar way for a given y_i, and we order the set of $SP_i(y_j)$, $j \neq i$, such that let d_k be the value of the kth largest of the $SP_i(y_j)$. Using this and the fuzzy subset Q representing our concept of Most, we calculate

$$PA(y_i) = Max_{k=1\ldots n-1}[Q(\frac{k}{n-1}) \wedge d_k]$$

(also in this point it is possible to consider other alternative definitions, see [5]). This value will give us the degree of positive assortatitivity corresponding to the given node y_i with the others nodes considered as leaders of small worlds.

Once these criteria have been valuated for some nodes, one way to analyze the degree of leadership of a given node x would be, for instance, to calculate the minimum value

$$SL(x) = min[CG(x), CT(x), SB(x), PA(x)]$$

4   About Homophily

A feature observed frequently in many social networks, that we have considered in Condition 4 of Paragraph 3, is that many leaders use to be connected each other. The tendency of individuals to associate and bond with 'similar' others is known as Homophily [12]. In our study, the homophily may be understood as the tendency of nodes to be attached to other nodes with similar characteristics. But, if we speak about similarity among nodes, this is because some features do exist on each node suscepti-ble to be compared.

We keep on working with the fuzzy set based technologies to introduce and study the similarity among the nodes of a social network and the presence of homophily:

Let's consider a Network $G = \langle V,G \rangle$ and let's assume attached to each node $v_i \in V$ a vector of attributes that take their values in the space (see [6]) $U = U_1 x U_2 x \ldots x U_q$, such that $v_i(u_1,u_2,\ldots u_q) := (u_{1i},u_{2i},\ldots u_{qi}) \in U$ where each $u_{ri} \in U_r$ is the value of the rth attribute for the node v_i. If we are able to compare the attributes of the nodes, it is because some kind of 'metric' is defined on the set of these attributes. We are going to consider each attribute belonging to the Subset U_r, $u_{ri} \in U_r$, a set where we have defined a metric $(U_r, \|\cdot\|_r)^1$ (to be defined in each case).

Definition 3. We define the Soft Similarity on $G = \langle V,G \rangle$ by means of the following criteria:

Given two nodes, v_i, $v_j \in V$, with attributes $v_i(u_1,u_2,\ldots u_q) := (u_{1i},u_{2i},\ldots u_{qi}) \in U$ and $v_j(u_1,u_2,\ldots u_q) := (u_{1j},u_{2j},\ldots u_{qj}) \in U$, the similarity of v_i to v_j is calculated as

$$SSIM_i(v_j) = SIM((u_{1i}, u_{2i}, \ldots u_{qi}), (u_{1j}, u_{2j}, \ldots u_{qj})) = SIM \left(\frac{\sum_{r=1}^{q} \|u_{ri} - u_{rj}\|_r}{qMAX_{i,j,r}\|u_{ri} - u_{rj}\|_r} \right) \quad (9)$$

$SSIM_i(v_j) \in [0,1]$, where by SIM we understand a decreasing function defined (as seen in previous concepts)

$\quad$ SIM: $[0,1] \rightarrow [0,1]$ $\quad$ such that $\quad$ SIM(0)=1 $\quad$ and SIM(1)=0.

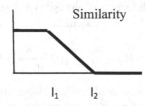

Fig. 4. Prototypical definition of Similar

In the study of the small worlds, it would be interesting to analyze the *similarity* among the members of a given small world $S \subseteq V$.

We can speak about Soft Homophily in a given subset of a network when most of the elements of the subset are similar in the sense given above, this is, we can calcu-late the degree to which v_i is similar to most of the nodes v_j in the subset. We are going to follow a similar technique to that seen previously, but we may also follow a way based on OWA operators (see [5] and [6]).

---

[1] It would be also possible to consider a soft similarity defined on each U_r and proceed later in a similar way to define the Global Soft Similarity among the nodes.

We order the values $SIM_i(v_j)$ and let d_k be the value of the kth largest of the $SIM_i(v_j)$. Using this and the fuzzy subset Q representing our concept of Most, we calculate

$$H_i = H(v_i) = Max_{k=1\ldots n-1}[Q(\frac{k}{n-1}) \wedge d_k], \text{ for each node } v_i$$

where the idea of Most is given by a fuzzy subset $Q: [0,1] \to [0,1]$ as previously.

A measure of the Soft Homophily of the world S will be given by

$$H(S) = Max_{i=1\ldots n}\left[Q(\frac{i}{n}) \wedge H_i\right]$$

We must stress that we have considered undirected edges. Obviously, if we consider the above metrics verifying the symmetric property, it will happen that $SSIM_i(v_j) = SSIM_j(v_i)$.

5   Conclusions

We follow the ideas started in [6] and [5] to formalize concepts related to the Small Worlds in Social Networks, by using Fuzzy techniques and the theory of graphs. We give also a proposal for the identification of a node as a leader in the Small World and to identify some common features among the elements belonging to a given Social Network. The fuzzy techniques are useful to represent linguistic concepts related to the theory of Social Networks.

References

1. Jackson, M.: Social and Economic Networks. Princeton University Press (2008)
2. Watts, D.: Networks, Dynamics, and the Small-World Phenomenon. American Journal of Sociology 105(2), 493–527 (1999)
3. Watts, D., Strogatz, S.: Collective Dynamics of ´Small World´ Networks. Nature 393, 440–442 (1998)
4. Milgram, S.: The Small-World Problem. Psychology Today 2, 60–67 (1967)
5. Yager, R.R.: Intelligent Social Network Analysis using Granular Computing 23, 1197–1220 (2008)
6. Yager, R.R.: Concept Representation and Data base Structures in Fuzzy Social Relational Networks. IEEE Transactions on Systems. Man & Cib. 40(2), 413–419 (2010)
7. Jungnickel, D.: Graphs, networks and algorithms, 4th edn. Springer (2013)
8. Mathias, N., Gopal, V.: Small - worlds: How and why.arXiv:cond-mat/0002076v1, 1-20 (2000)
9. Newman, M.: The structure and function of complex networks. SIAM Review 45, 167–256 (2003)
10. Freeman, L.C.: A set of measures of centrality based on betweenness. Sociometry 40, 35–41 (1977)
11. Newman, M.: Scientific Collaboration Networks II. Shortest paths, weighted networks, and centrality. Phys. Rev. E 64, 016132 (2001)
12. Lazarsfeld, P., Merton, R.: Friendship as a social process: a sustantive and methodological analysis. In: Freedom and Control in Modern Society. Van Nostrad, NewYork (1954)

Generating Events for Dynamic Social Network Simulations

Pascal Held, Alexander Dockhorn, and Rudolf Kruse

Department of Knowledge Processing and Language Engineering
Faculty of Computer Science
Otto von Guericke University of Magdeburg, Magdeburg, Germany
{pascal.held,rudolf.kruse}@ovgu.de, alexander.dockhorn@st.ovgu.de
http://fuzzy.cs.uni-magdeburg.de

Abstract. Social Network Analysis in the last decade has gained remarkable attention. The current analysis focuses more and more on the dynamic behavior of them. The underlying structure from Social Networks, like facebook, or twitter, can change over time. Groups can be merged or single nodes can move from one group to another. But these phenomenas do not only occur in social networks but also in human brains. The research in neural spike trains also focuses on finding functional communities. These communities can change over time by switching the stimuli presented to the subject. In this paper we introduce a data generator to create such dynamic behavior, with effects in the interactions between nodes. We generate time stamps for events for one-to-one, one-to-many, and many-to-all relations. This data could be used to demonstrate the functionality of algorithms on such data, e.g. clustering or visualization algorithms. We demonstrated that the generated data fulfills common properties of social networks.

1 Introduction

Social Network Analysis is a hot topic since some years. The first investigations where done in a static analysis of the networks. Examples for this are networks representing friendship, or co-author relationships, between people. But social networks, like facebook, or Twitter, are very dynamic. So a static view is too simple. In [8] we described some algorithms which are based on events between nodes. This could be e.g. an interaction between users of a social network, or interactions between neurons in human brains.

The human brain consists of different regions. In each region there are a lot of neurons which are connected. These connections cause that stimuli from outside are passed through this network of neurons by fire events of neurons. A group of neurons which handle the same stimuli are called functional groups or communities. The discovery of such functional communities is very similar to social networks analysis. [14] The history of fire events of a single neuron is called spike train.

A. Laurent et al. (Eds.): IPMU 2014, Part II, CCIS 443, pp. 46–55, 2014.
© Springer International Publishing Switzerland 2014

There are some data sets available, e.g. the Enron Dataset, extracts from twitter, or recordings from human brains. These are all real data sets, where we know nothing about the base structure of the underlying data.

In this paper we present a data generator to generate events for social networks or fire events in spike trains. This data represents interaction between different nodes in the network. We do not only represent static behavior, but also dynamics in the underlying structure. So we can configure the generator in a way, that clusters of nodes are changing, like in real social structures.

The generated data is based on clusters of nodes in a graph. Nodes within the same cluster have a higher communication frequency than nodes in different clusters. During the simulation it is possible to generate new clusters, or modify them by adding nodes, moving nodes from one to another cluster, merging with other clusters, split up the clusters.

The rest of the paper is organized as follows. In the second section we present some fundamentals in the field of social networks and human brain spike trains. The third section describes the generator in detail, followed by experiments with the generated data in Section 4. In the last section we discuss our results.

2 Fundamentals

The generation of graphs, especially in social network analysis is nothing new. There are a lot of generators, like the Kronecker graph generators [10] or the generator from McGlohon et al. [12]. The most algorithms generate graphs which hold some typical properties or have problems with a growing number of nodes. Akoglu and Faloutsos developed the realistic graph generator [1] which holds most typical properties and also enables the evolving of graphs. A good survey of different generators is given in [3].

The main drawback of these algorithms is that evolution of the graph in most cases is a static growing, where new nodes and edges are created. In some cases there are also changes in the connections, but the main structure is constant.

Our focus is to create a graph, where we know the main structure of the underlying communities, so we have a ground truth to check cluster algorithms on dynamic graphs. Also we want to be able to change this ground structure.

Another point is that we are primary not interested in the graph itself, but on the events between the nodes, which could be used to create such a graph.

In the following we present same requirements for spike train simulation as well as for social network generation.

2.1 Simulating Spike Trains

A neuron in the human brain can be simulated as a list of events. An event occurs when the potential of the concerning neuron increases. This means that the neuron fires. Such an event list is commonly named spike train.

The easiest way to simulate such spike trains is a point process simulation based on a Poisson process model [15]. This model is based on the assumption that the probability of a neuron firing in a given time frame $[t, t + \delta t]$ is simply given by $R\delta t$ where R is the firing rate, for sufficient small δt. The probability is absolutely independent from the position of the last firing.

One interpretation of interacting neurons are ensembles in parallel spike trains [11]. The main idea is, that neurons that fire together are wired together. Borgelt et al. generate data parallel spike trains with multiple Poisson processes. Every spike train has one generator process. Additionally one process for every ensemble is present. The events from the ensemble process will be copied into the individual spike trains with a given probability. These probabilities describes how close the neurons are interacting with each other. The individual point processes of neurons within an ensemble have a lower fire rate, so the combined fire rate together with the common fire events is the same as from other neurons.

2.2 Social Networks

One aspect of our data generator is the simulation of interactions in social networks. For this simulation the generated data should have similar characteristics as real social networks. In this section we would like to present some of these characteristics and how to show them. These are the small-world property, scale-free characteristics, and the structure of communities and clusters.

The Small-World Property was introduced by Watts and Strogatz [16] for graphs with social network characteristics in 1998. It is based on the six degree of separation from Guare [5]. They focused on the average shortest path between two nodes (global connectivity measure) L and the average clustering coefficient (local neighborhood connectivity measure) C. In their experiments they started with a regular graph where the nodes are placed in a ring. Every node is connected to the k following and previous nodes. With a probability of p they replaced a given edge by a new random one. For $p = 0.0$ this yield to the original graph and $p = 1.0$ yield to a total random graph, with the same amount of nodes and edges. They compared three graphs with social network character with random graphs of the same complexity (same number of nodes and edges) and proofed that they all follow the small-world model, which means that $L \gtrsim L_{random}$ and $C \gg C_{random}$.

The Scale-Free Characteristic is a another property of social networks. Typically is that not every node has the same connectivity degree. There are some nodes with a strong connectivity to others and other nodes with much fewer connections. This is based on the fact, that such networks grow over time. New nodes connect themselves more probable with nodes with a strong connectivity. For example, a new person in a friendship network will connect first to high connected friends then to other new nodes, or a new website will link to a known common website then to another new one. Barabasi and Albert investigate in 1999 [2] this phenomenon. They found out that the degree distribution of the nodes from social networks follow a power-law distribution.

To proof this property in the generated data, we will run the same experiments and compare the L and C values with equivalent random graphs.

Communities and Clusters are the core phenomenon in social network. People organize in groups and these groups could be recognized in communities in social networks. Such a group could be a group of people from the same university or a clique of friends.

The main concept of this groups is that the connectivity within a group is much denser then the connectivity to elements outside of the group. In the field of data mining this is called the intra-cluster-density and the inter-cluster-sparseness. In our work we get this property by construction. The data generator itself supports the modeling of such clusters.

Social Networks are Dynamic and not static. They evolve over time. New elements join the network, clusters are emerging, splitting, growing, or shrinking. Nodes change from one cluster to another. In [7] we describe how to analyze this dynamics in clusters.

With our data generator we can generate communication events for such dynamics in social networks.

3   Data Generator

We already introduced a broad area of applications for graphs. However, all variants require the graph structure or the type of output to comply special constraints. For that reason the generator needs to be easily adjustable for a multitude of network characteristics. The implemented data generator is able to produce different types of static and dynamic graphs. It consists of methods for defining the structure of the start graph and allows the import of scripts, which describe changes to specified points in time. The basic behavior of the generator and script functionalities will be outlined in the following sections.

The graph is divided into several clusters. Each network structure of a cluster is generated using the model proposed by Barabási in [2]. This model proposes a growing network starting with m_0 nodes. Further nodes are sequentially added to the graph and connected using $m(\leq m_0)$ edges. The probability of a node having k edges follows a power-law with an exponent of $y_{model} = 2.9 \pm 0.1$. By definition the network structure of each cluster will organize itself into a scale-free stationary state. We use the same model to decide which clusters are connected. Connecting nodes will be drawn at random.

The firing rates and intervals in a cluster can be configured by entering constant values or a distribution where values are drawn of. The generator therefore supports the creation of poisson processes by using gamma distributions to determine time intervals between two events as it was modeled in [15]. Nodes and clusters store corresponding parameters and can be manipulated through script functions as it will be explained later.

Generated events will be stored and exported in a *.csv-file*. The respective format can be set in the command line and be of one of the following types:

Communication with one source and single/multiple targets: Every entry will contain the time of the event, the id of the source node and targets of the communication. *"$time$; id_{source}; $id_{target(s)}$"*

Groups of nodes active at the same time: This format does not include information about the direction of any communication and stores simultaneously active nodes per time frame. *"$time$; $id_{source(s)}$"*

Activity of single nodes: The third format records undirected potentials per node and can be used to record spike trains. *"$time$; id_{node}"*

The main focus of our designed program lies on enabling the user to preconfigure changes of the network in a script file. This way analysis techniques can be brought to the test for already known features of the dynamic graph. Currently provided script-functions and their possible applications are listed below:

Delete/add nodes/clusters: Nodes and clusters can be added/deleted at specified time points. E.g. new people are joining a network, a neuron dies

Change behavior of individual nodes: Adjust firing rates, activity or cluster assignments per node. E.g. a person starts to communicate more frequently, communication partners change because of a new job, a neuron fires more often because a stimulus changed

Change behavior of clusters: Adjust activity of the whole cluster. E.g. a group of people starts to communicate more often to organize an event, a brain region changes activity level while sleeping, the visual part of the brain adapts to changes in visual stimuli

Divide or merge clusters: Multiple clusters are merged to form a new one, a cluster will be divided in parts. E.g. circle of friends splits after finishing school, departments are joined to minimize communication costs, special stimuli only activate specific parts of the brain

4   Experiments

The generator was tested on the simulation of spike trains and social networks. The former was compared to another generator used by Borgelt el al. [11] which is based on a model from Nawrot et al. [13]. We evaluated different measures relating to spike trains and social networks. Our test cases will be presented in the following sections. Our evaluation platform was an HP Z400 with 6GB RAM and a 3.45 GHz 6-core Intel Xeon processor. However the program only used one core for the generation process. Event generation for our biggest test-case (1000 Cluster, 100 000 Nodes and 25 000 000 Events) took about 7 minutes.

4.1   Spike Trains

In spike trains the activity potentials per node are recorded. Therefore we chose the third output type for our generator. Both tools were used to generate

50 nodes, 10 forming a single ensemble and 40 nodes for noise. Events were recorded over ten seconds. The nodes firing rates were set to an average of 20 events per second. The ensemble had a copy rate of 50 percent, therefore an average of half of the ensemble nodes participated on events of the whole ensemble.

First we checked for similarity of individual spike trains of both tools by using a Kolmogorow-Smirnow test for the distributions of time intervals between consecutive events per node. The average of recorded p-values was ~ 0.55. For this reason we can accept the hypothesis that both distributions were drawn from the same continuous distribution.

$$d_{Correlation} = \frac{1}{2} - \frac{n_{11}n_{00} - n_{01}n_{10}}{2\sqrt{(n_{10} + n_{11})(n_{01} + n_{00})(n_{11} + n_{01})(n_{00} + n_{10})}} \qquad (1)$$

$$d_{Yule} = \frac{n_{01} + n_{10}}{n_{11}n_{00} - n_{01}n_{10}} \qquad (2)$$

$$d_{Jaccard} = \frac{n_{10} + n_{01}}{n_{11} + n_{10} + n_{01}} \qquad (3)$$

$$d_{Hamming} = \frac{n_{01} + n_{10}}{n_{**}} \qquad (4)$$

We continued the analysis of our generator by binning resulting spike trains into 1000 bins of size 0.01 second to transform them into a binary matrix. This was used to calculated distance matrices for the four distance measures Correlation (1), Yule (2) [17], Jaccard (3) [9], and Hamming (4) [6]. The results are shown in Figure 1. The distance matrices for hamming and yule distance show clearly the similarity of the first ten nodes forming an ensemble. Distances of noise-noise and noise-ensemble combinations were much higher.

4.2   Social Network

Our second evaluation involves the generation of social networks. We created four graphs with the following configurations.

- Graph 1 = 5 clusters with 20 nodes each
- Graph 2 = 3 clusters with 50 nodes each
- Graph 3 = 5 clusters with 10, 20, 30, 40 and 50 nodes
- Graph 4 = 25 clusters with 25 nodes each

Noise was excluded in all three generation processes. The activity and communication intervals per node were gamma-distributed with $\alpha = 1$ and $\beta = 0.35$.

We compared the average path length and the average cluster coefficient of the generated graphs to random graphs with same number of total edges. Recorded values are shown in Table 1.

It has been shown that for all three generated graphs the small-world model ($L \gtrsim L_{random}$ and $C \gg C_{random}$) is applicable. We used the same set of graphs to test for scale-free characteristics. We used the procedure, described in [4]. It uses a bootstrapping hypothesis test to maximize the Kolmogorow-Smirnow

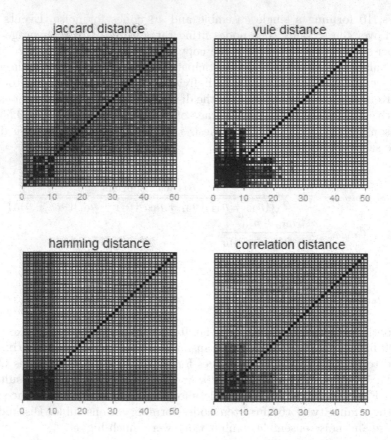

Fig. 1. Distance matrices for jaccard, correlation, yule and hamming distance

statistic to test for a goodness-of-fit between the data and the power law. The resulting p-values need to be ≥ 0.1 to accept the power-law distribution as a plausible hypothesis. Responding p-values per graph are shown in Table 1. The results were all significant (≥ 0.1) and therefore a power-law distribution can be accepted as a hypothesis.

Forming communities and clusters is directly inferred by the generation process. Each cluster will be generated as a separate graph using the model proposed by Barabási in [2]. Noise can be added by defining a base probability for inter-cluster-events for a whole cluster or by adding specified noise nodes. Dependent on the chosen cluster definition it is possible to design clusters of nodes which are active at the same time, in equal distributed time intervals or are more likely to talk to each other than to nodes from different clusters.

To show the dynamic capabilities of our generator we created graphs with two clusters of 20 nodes each and used a simple script, which included the following commands:

- t_1, move five nodes from cluster one to cluster two
- t_3, add five new nodes to cluster one
- t_5, remove five nodes from cluster two
- t_7, lower the cluster activity of cluster two to 50%
- t_9, set cluster activity of cluster two back to 100%

Table 1. Measured average shortest path length and clustering coefficient per graph compared to a random graph with same number of edges, p-values for accepting a power-law distribution as plausible hypothesis

	Graph 1	Graph 2	Graph 3	Graph 4
$L_{generator}$	4.30	4.73	4.38	6.45
L_{random}	2.83	2.97	2.96	3.28
$C_{generator}$	0.40	0.27	0.30	0.42
C_{random}	0.07	0.05	0.35	0.14
p	0.5	0.5	0.5	0.9

We recorded the number of inter- and intra-cluster events per second and averaged those over ten generation processes. Recorded values are binned per separate graph configuration and presented in Table 2 and Figure 2.

Table 2. Evolution of event frequency for dynamic graph, values represent average count of events per second, time-intervals are based on different graph configurations

	t_{0-1}	t_{1-3}	t_{3-5}	t_{5-7}	t_{7-9}	t_{9-10}
Intra-cluster events one	133	67	139	137	134	133
Intra-cluster events two	138	200	197	161	75	164
Inter-cluster events	13	12	16	15	11	12

The influence of the script can be seen in the changes of events. Both clusters start with nearly the same number of events per second. The intra-cluster communication of cluster one is reduced in time frame two, because the number of nodes changed from 20 to 15. Simultaneously an increase for intra-cluster communication of cluster two was recorded. After reinserting five nodes to cluster one, the number of events went back to the initial value. No further significant changes of cluster ones communication level were observed. In contrast records for cluster two include a decrease of events in time frame four, which is correlated with the decrease of nodes. In the following time frames the communication level was first set to 50% and in time frame six set back to 100%, which is observable in the number of events of cluster two, as well. The number of inter-cluster events was constant on a lower level than intra-cluster communication. This can be explained by the set base communication probability of $p_{base} = 0.01$. However, the number of expected events can be changed by adjusting this parameter.

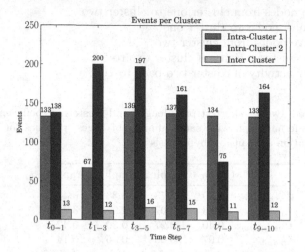

Fig. 2. Evolution of event frequency for dynamic graph, values represent average count of events per second, time-intervals are based on different graph configurations

5 Results and Outlook

We presented a generator for the creation of static and dynamic graphs. The generator in its current version was able to produce reasonable data as a base for multiple graph related problems. The comparison with a previous available spike train generator showed that spike trains with similar distributions of event-intervals can be created. Furthermore predefined ensembles could be detected in calculated distance matrices. Social network experiments demonstrated the ability to create graphs with small world properties and scale-free characteristics. All these processes can be combined with a script for planning changes in the graph structure. Therefore researchers will be able to test analysis techniques for dynamic graphs on event data containing predefined features.

Until the tool gets released we will concentrate on refurbishing the current generation process. Configurations with a high number of events per second (> 100) can still be too much afflicted with noise in the resulting distributions. Additionally further script functions will be added to insert more dynamic characteristics to the graph.

The implementation of our generator could be downloaded from http://iws.cs.uni-magdeburg.de/~pheld/publications/IPMU2014/.

References

1. Akoglu, L., Faloutsos, C.: Rtg: a recursive realistic graph generator using random typing. Data Mining and Knowledge Discovery 19(2), 194–209 (2009)
2. Barabási, A.L., Albert, R.: Emergence of scaling in random networks. Science 286(5439), 509–512 (1999)

3. Chakrabarti, D., Faloutsos, C.: Graph mining: Laws, generators, and algorithms. ACM Comput. Surv. 38(1) (June 2006)
4. Clauset, A., Shalizi, C.R., Newman, M.E.: Power-law distributions in empirical data. SIAM Review 51(4), 661–703 (2009)
5. Guare, J.: Six Degrees of Separation: A Play. Vintage Books, New York (1990)
6. Hamming, R.W.: Error detecting and error correcting codes. Bell System Technical Journal 29(2), 147–160 (1950)
7. Held, P., Kruse, R.: Analysis and visualization of dynamic clusterings. In: 2013 46th Hawaii International Conference on System Sciences (HICSS), pp. 1385–1393. IEEE (2013)
8. Held, P., Moewes, C., Braune, C., Kruse, R., Sabel, B.A.: Advanced analysis of dynamic graphs in social and neural networks. In: Borgelt, C., Gil, M.Á., Sousa, J.M.C., Verleysen, M. (eds.) Towards Advanced Data Analysis. STUD-FUZZ, vol. 285, pp. 205–222. Springer, Heidelberg (2012)
9. Jaccard, P.: Étude comparative de la distribution florale dans une portion des alpes et des jura. Bulletin del la Société Vaudoise des Sciences Naturelles 37, 547–579 (1901)
10. Leskovec, J., Chakrabarti, D., Kleinberg, J., Faloutsos, C.: Realistic, mathematically tractable graph generation and evolution, using kronecker multiplication. In: Jorge, A.M., Torgo, L., Brazdil, P.B., Camacho, R., Gama, J. (eds.) PKDD 2005. LNCS (LNAI), vol. 3721, pp. 133–145. Springer, Heidelberg (2005)
11. Louis, S., Borgelt, C., Grün, S.: Generation and selection of surrogate methods for correlation analysis. In: Grün, S., Rotter, S. (eds.) Analysis of Parallel Spike Trains. Springer Series in Computational Neuroscience, vol. 7, pp. 359–382. Springer US (2010)
12. McGlohon, M., Akoglu, L., Faloutsos, C.: Weighted graphs and disconnected components: Patterns and a generator. In: Proceedings of the 14th ACM SIGKDD International Conference on Knowledge Discovery and Data Mining, KDD 2008, pp. 524–532. ACM, New York (2008)
13. Nawrot, M., Aertsen, A., Rotter, S.: Single-trial estimation of neuronal firing rates: from single-neuron spike trains to population activity. Journal of Neuroscience Methods 94(1), 81–92 (1999)
14. Shalizi, C., Camperi, M., Klinkner, K.: Discovering functional communities in dynamical networks. In: Airoldi, E.M., Blei, D.M., Fienberg, S.E., Goldenberg, A., Xing, E.P., Zheng, A.X. (eds.) ICML 2006. LNCS, vol. 4503, pp. 140–157. Springer, Heidelberg (2007)
15. Vreeswijk, C.: Stochastic models of spike trains. In: Grün, S., Rotter, S. (eds.) Analysis of Parallel Spike Trains. Springer Series in Computational Neuroscience, vol. 7, pp. 3–20. Springer US (2010)
16. Watts, D.J., Strogatz, S.H.: Collective dynamics of 'small-world' networks. Nature 393(6684), 440–442 (1998)
17. Yule, G.U.: On the association of attributes in statistics: with illustrations from the material of the childhood society, &c. Philosophical Transactions of the Royal Society of London. Series A, Containing Papers of a Mathematical or Physical Character 194, 257–319 (1900)

A Model for Preserving Privacy
in Recommendation Systems*

Luigi Troiano[1] and Irene Díaz[2]

[1] University of Sannio, Italy
[2] University of Oviedo, Spain

Abstract. The problem of preserving privacy in recommendation systems is faced in this work. The approach presented reduces the study of privacy threats to the study of frequent property set obtained from the characteristics of the objects the recommendation system provides to a target user. This study is made by defining a prominence index for each item and by using efficient methods to explore the lattice of item characteristics.

Keywords: Recommendation System, Privacy, Prominence Index, Frequent Item Sets.

1 Introduction

Research in Recommendation Systems (RS) has been growing with the development of e-commerce mainly because many web sites and e-shops produce recommendations to users in order to retain them. RS provides a rating or a preference for a user using certain information about his/her preferences. This information can be acquired either explicitly from the rates provided by users or implicitly from monitoring users' behavior (booked hotels or heard songs). In addition to that, RS can benefit of other kinds of information such as location or demographic features.

The research related to RS has been focused on movie, music and book recommendation [3], with music recommendation being the most studied topic. Recently it has been applied to other domains such as e-commerce. Many of the works about recommender systems are focused on the understanding a users browsing and purchase history via implicit and explicit actions and how to recommend them products that they have a higher probability of being preferred [18].

Privacy issues have been largely studied for data bases [20] and also for Social Networks. Brief reviews to this topic are presented in [7,25]). There are other works presenting different approaches related to the study of privacy concerns in Social Networks [5,8,9,19,24]).

* Author acknowledges financial support by Grant TEC2012-38142-C04-04 from Ministry of Education and Science, Government of Spain and by Grant UNOV-13-EMERG-GIJON-10 from University of Oviedo, Spain.

A. Laurent et al. (Eds.): IPMU 2014, Part II, CCIS 443, pp. 56–65, 2014.

However, privacy issues involved in recommender systems have been studied only in few works. For example in [13] the problem of producing recommendations from collective user behavior while simultaneously providing guarantees of privacy for these users is studied.

In [16] it is presented a privacy preserving recommendation framework based on groups with the aim of protecting users from unreliable service providers. The starting point of this work is to find a way out of the opposition between anonymity and personalization, that is, how a certain level of anonymitycan be maintained without sacrificing useful and accurate recommendations. Other works have been focused on some concern related to both privacy issues or recommendation systems [19,9].

In this paper we face the problem of privacy concerns in a recommendation system based on frequent item sets. The basic approach is to consider all the recommendations provided by a given web site or e-shop and to study the set of frequent properties of these recommended objects. The remainder of this contribution is organized as follows: Section 2 provides some preliminaries regarding recommender systems and frequent itemsets theory, Section 3 outlines the model we adopted, Section 4 describes the model application by means of an example, Section 5 draws conclusions and future directions.

2 Preliminaries

2.1 Recommendation Systems

Recommendation systems were created out of the user needs to handle the increasing volume of information that is available in the world wide web [3]. The definition of a RS depends on several issues such as the type of data available to provide the recommendation. For example, it is possible to use the ratings or the information provided by the user when registration or social relationships. The data can be provided both explicitily and implicitily. Explicit data is given by a customer (for example a rate) while implicit data is obtained from the user's behavior (for example if he looks at the description of an object, marks the object or refers to it). On the other hand, a recommendation algorithm is essentially a filtering algorithm. Therefore, according to the filtering approach, the RS engine can be classified as *Content-based recommenders* if recommendations are based on the past user preferences ([15]), *Collaborative recommenders* if recommendations to each user are based on the information provided by similar users ([6]), *Demographic recommenders* if they categorize the user based on personal attributes assuming that common personal attributes such as age or sex derive on some common preferences ([14]) and *Hybrid methods* which combine the aforementioned methods ([4]).

Many data mining techniques have been used to provide recommendations [11]. In addition, other issues such as scarcity, scalability or privacy should be addressed when working with RS (see for example [17,12]). This last property, privacy, is faced in this work.

2.2   Frequent Item Set Theory

Frequent item sets theory represents one of the major families of techniques for characterizing data. This problem is often viewed as the discovery of association rules. The reason is that they are strongly related because the discovery of frequent item sets is a prior step to discover association rules. Frequent Item Set mining is based on the Market-Basket Model of data. Therefore it looks for sets that frequently co-occur together.

Let $\mathcal{I} = \{i_1, \ldots, i_m\}$ be a set of items and let $\mathcal{D} = \{d_1, \ldots, d_n\}$ be the set of transactions. Each d_k, with $1 \leq k \leq n$ is represented by a vector $d_k = (i_{\sigma(1)}, \ldots, i_{\sigma(r)})$ with $i_{\sigma(j)} \subset \mathcal{I}$ and $1 \leq r \leq m$.

A transaction $d \in D$ covers an item set $\mathcal{J}$ if and only if $\mathcal{J} \subset d$. The support of an item set $\mathcal{J}$ is defined as the number of transactions covering it. In other terms, the support provides a measure of how often a combination of values is presented.

The problem of mining frequent item set is defined as searching all item sets that have support greater than the user-specified minimum support (called *minsup*). The most well-known algorithm to produce frequent item sets is Apriori. This algorithm in addition computes association rules from the obtained frequent item sets by selecting those association rules according to a minimum confidence (called *minconf*). Confidence measures how often the association between values occur. For a detailed description, authors refer to [2].

3   The Model

Let $\mathcal{I} \equiv \{I_1, I_2, \ldots, I_n\}$ be the collection of items considered by the recommender system as suggestion to user. Each item I_h is scored by $\rho(I_h)$, that is the relevance assigned by the recommender system to the item with respect to the user preferences. This score is assigned over a ratio scale, and we can assume $\rho(I_h) \in [0, 1]$ without any loss of generality. In addition we assume that each itemset I_k is tagged by a set of keywords, $K(I_h) = \{k_{h,1}, \ldots, k_{h,m}\} \subseteq \mathcal{K}$. Table 1 outlines an example of result set obtained by the recommender system. In this case, the system replied with 10 suggested items, each with a relevance scoring between 0.9 (the highest) and 0.1 (the lowest). Each itemset is tagged with a subset of the keywords $\mathcal{K} \equiv \{A, B, C, D\}$.

We assume keywords are the sensitive information the user would like to protect against association a recommender systems could infer for his/her profile. So the question we aim to answer is the following: what are the keyword subsets that are most prominently linked to the user profile? To answer this question it is considered the inclusion lattice, as that depicted in Figure 1.

An *index of prominence* is a function

$$\pi : 2^{\mathcal{K}} \rightarrow [0, 1] \tag{1}$$

It is a function of the relevence assigned to each item. In particular

$$\pi(K) = p_K(\rho(I_1), \ldots, \rho(I_q)) \quad K \subseteq K(I_h), h = 1 \ldots q \tag{2}$$

Table 1. An example of result set

Item ($\mathcal{I}$)	Relevance (ρ)	Keywords (K)
1	0.9	B, C
2	0.9	C, D
3	0.8	A, B, C
4	0.8	B, C
5	0.7	A, B, C
6	0.7	C, D
7	0.6	C, D
8	0.6	A
9	0.2	A, D
10	0.1	A, D

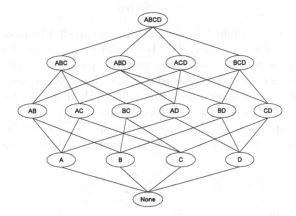

Fig. 1. The keyword powerset lattice

where

$$p_K : [0,1]^q \rightarrow [0,1] \ \ K \subseteq \mathcal{K} \tag{3}$$

is a function of relevance assigned to each item, thus to the associated keyword set.

We expect π to be anti-monotone w.r.t. inclusion and monotone w.r.t. relevance, that is

$$\pi(K_1) = p_{K_1}(\rho_1, \ldots, \rho_{q_1}) \geq p_{K_2}(\rho_1, \ldots, \rho_{q_2}) = \pi(K_2) \ \ \forall K_1 \subset K_2 \subseteq \mathcal{K} \tag{4}$$

and

$$p_K(\rho_1, \ldots, \rho_q) \leq p_K(\rho'_1, \ldots, \rho'_q) \ \ \forall \rho_i \leq \rho'_i, i = 1..n, \forall K \subseteq \mathcal{K} \tag{5}$$

Condition expressed by Eq.(4) states that the index associated to a combination of keywords is lower than the index associated to all its parts, while Eq.(5) states that a result set with higher relevance will provide a higher prominence index.

Other properties are related to the *symmetry* of information, that is

$$\pi(K_1) = \pi(K_2) \ \ \forall K1, K2 \subseteq \mathcal{K}, |K_1| = |K_2| \tag{6}$$

and

$$p(\rho_1, \ldots, \rho_q) = p(\rho_{(1)}, \ldots, \rho_{(q)}) \tag{7}$$

where $(\cdot)$ is a generic permutation of arguments.

Finally, we expect that considering 0-relevant supersets should not affect the result, that is

$$p_K(\rho_1, \ldots, \rho_q) = p_K(\rho_1, \ldots, \rho_{q-1}) \quad \rho_q = 0, \forall K \subseteq \mathcal{K} \tag{8}$$

An index of prominence satisfying all the previous properties is the *relevance average* over the whole result set.

$$\pi(K) = \frac{1}{n} \sum_{I_h \in \mathcal{I}, K \subseteq K(I_h)} \rho(I_h) \tag{9}$$

Once the index is determined, it is established a threshold τ in order to decide which subset is prominent or not. The threshold τ can be fixed depending on problem characteristics and solution conservativeness. Higher threshold values will reduce the number of risky keyword subsets, thus relaxing the solution conservativeness. Differently, lower values will enlarge the number of risky keyword subset, thus enforcing the solution conservativeness.

The method proposed to set the threshold consists in computing the expected index $E[\bar{\pi}]$ assuming for each subset K the average relevance value $\bar{\rho} = 0.5$. Formally,

$$\tau = E[\pi] = \frac{1}{2^m} \sum_{K \subseteq \mathcal{K}} \bar{\pi}(K) \tag{10}$$

where

$$\bar{\pi}(K) = p_K(\overbrace{\bar{\rho}, \ldots, \bar{\rho}}^{\frac{qn}{m}-\text{times}}), \quad |K| = q \tag{11}$$

It could be objected that computing τ is computationally prohibitive, as it requires to move over the whole power set. However, as the average value $\bar{\rho}$ is used, this is not necessary, since

$$\bar{\pi}(K_1) = \bar{\pi}(K_2) = p_K(\overbrace{\bar{\rho}, \ldots, \bar{\rho}}^{\frac{qn}{m}-\text{times}}) = \bar{\pi}_q$$
$$\forall K_1, K_2, K \subseteq \mathcal{K}, |K_1| = |K_2| = |K| = q \tag{12}$$

Therefore,

$$\tau = E[\pi] = \frac{1}{2^m} \sum_{K \subseteq \mathcal{K}} \bar{\pi}(K) = \frac{1}{2^m} \sum_{q=0}^{m} \binom{m}{q} \bar{\pi}_q \tag{13}$$

In the case of relevance average we have

$$\tau = \frac{1}{2^m} \sum_{q=0}^{m} \binom{m}{q} \bar{\pi}_q = \frac{1}{2^m} \sum_{q=0}^{m} \binom{m}{q} \frac{q}{m} \bar{\rho} = \frac{1}{2} \bar{\rho} = 0.25 \tag{14}$$

In this case, the lattice is splitted in two parts as depicted in Fig. 2. The upper part of the lattice is made of non-prominent subsets, while the bottom is filled by prominent subsets. We can observe the following:

- if a keyword subset is prominent, all subsets will be prominent as well (e.g., subset BC)
- a subset can be prominent due the joint contribution of its supersets (e.g., subset D)
- a subset, although having relevance above $\bar{\rho}$, can be not prominent (e.g., subset A)
- the empty set is subset of any keyword set, thus $\pi(\emptyset)$ provides the prominence index for the whole lattice and the index upper limit (e.g., subset None)

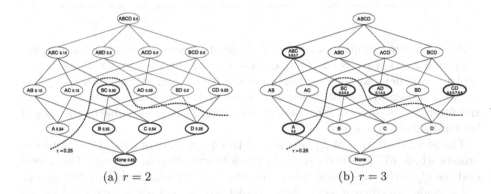

(a) $r = 2$ (b) $r = 3$

Fig. 2. Prominent (left) and Generator (right) subsets

The prominence index is computed from recommendations given by the result set. They represent the source of information. For this reason, we call *generators* the keyword subsets that are associated to items suggested by the recommender system. Their mapping to lattice is given by Fig. 2. Generators are circled by a bold line; inside the relvance of related recommendations is given. Obviously, when a generator is prominent, all its subsets are prominent as well (e.g., subset BC). Differently, non-prominent generators might contain prominent keyword subsets (e.g., subsets AD and CD), or not (e.g. subset ABC).

Since prominence of keyword subset depends on contribution provided by supersets, it might be necessary to extend the analysis to the whole keyword powerset lattice. This requires efficient methods to explore the lattice. Anti-monotonicity w.r.t. inclusion opens to the possibility of adopting data mining algorithm developed for the search of itemsets [1,10,23,22,21].

The first and most noticeable algorithm for mining frequent itemsets is known as *Apriori* [1]. Apriori is a levelwise, breadth-first, bottom-up algorithm, as outlined by Algorithm 1.

Algorithm 1. Apriori Prominence

1: L_q: prominent keyword subsets of size q
2: G_q: generated keyword subset of size q
3: ml: size of the largest keyword subset
4: τ: the prominence threshold
5:
6: find L_1
7: $q = 1$
8: **while** $L_q \neq \emptyset$ and $q < ml$ **do**
9:     $G_{q+1} = L_q$ join L_q
10:     $L_{q+1} = \{K \in G_{q+1} | \pi(K) \geq \tau\}$
11:     $q = q + 1$
12: **end while**
13: **return** $\bigcup_k L_k$

After selecting prominent keywords, pairs of keywords are generated by joining prominent keywords and filtered according to the threshold τ. So, keyword triplets are generated from prominent keywords and filtered. At each step, $(q+1)$-keyword subsets (candidates) are generated from q-keyword prominent subsets and filtered. The process is iterated ultil no candidate is available or the largest keyword subset is reached.

The main limitation of Apriori is that the generation of candidates produces subsets which will not be cosidered at the following step as too long. This might lead the algorithm to consider the same subset at different iterations. FP-growth [10] provides an efficient search based on FP-tree structure, an extended prefix-tree structure able to store minimal information regarding freuent itemsets. Indeed FP-trees allow to pack the representation of dataset transactions, avoiding the generation of candidate itemsets and employing a partitioning-based method to decompose the mining task into a set of smaller tasks.

Connections to data mining are not only limited to the discovery of frequent itemset. For instance, it is possible to transpose the concept of *closed* keyword sets to

So far, we considered the average relevance as prominence index. Other possibilities are feasible. For instance we might consider the *maximum relevance*, defined as

$$\pi(K) = \max_{I_h \in \mathcal{I}, K \subseteq K(I_h)} \rho(I_h) \tag{15}$$

In this case, according to Eq.(13), the prominence threshold can be computed as

$$\tau = \frac{1}{2^m} \sum_{q=0}^{m} \binom{m}{q} \bar{\rho} = \bar{\rho} = 0.5 \tag{16}$$

and the prominence area is depicted if Fig. 3.

We can observe how the prominence region enlarged. Indeed, if on one side we increased the threshold, we generally provide higher prominence indexes to

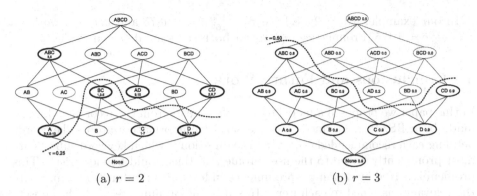

(a) $r = 2$ (b) $r = 3$

Fig. 3. Closed (left) and Prominence-max (right) subsets

keyword subsets: a keyword subset is prominent if there is at least one recommendation with a relevance over the threshold. We can generalize this criterion and look for keyword subsets with at least r recommendations with a relevance over the threshold. This is can be obtained by considering

$$\pi(K) = \underset{I_h \in \mathcal{I}, K \subseteq K(I_h)}{r^{th} - ordstat} \rho(I_h) \tag{17}$$

Threshold can be computed considering that τ for the maximum, assuming $i = qn$, can be written as

$$\tau = \frac{1}{2^m} \sum_{q=0}^{m} \binom{m}{q} \bar{\rho} = \frac{\bar{\rho}}{2^m} \left(1 + \frac{1}{n} \sum_{i=0}^{nm-1} \binom{m}{\lfloor \frac{i}{n} \rfloor} \right) \tag{18}$$

This equation can be generalized to statistic order r as

$$\tau = \frac{\bar{\rho}}{2^m} \left(1 + \frac{1}{n} \sum_{i=(r-1)m}^{nm-1} \binom{m}{\lfloor \frac{i}{n} \rfloor} \right) \tag{19}$$

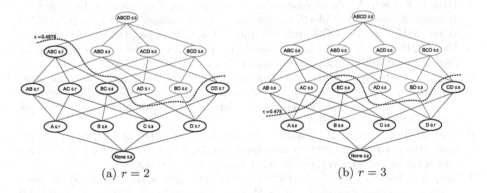

(a) $r = 2$ (b) $r = 3$

Fig. 4. Prominence regions (order statistic)

In our example, if $r = 2$ we have $\tau = \frac{156}{160}\bar{\rho} = 0.4875$ and if $r = 3$ we have $\tau = \frac{152}{160}\bar{\rho} = 0.475$. Prominence regions for both cases are depicted in Fig. 4.

4   Conclusions and Future Work

In this paper we proposed a model to identify threats from the recommendations made by a RS. It is assumed that the sensitive information are the keywords defining each recommended object. So the question is which keyword subsets are most prominently linked to the user profile and thus, could be more risky. This problem is solved by defining a prominence index which is a function to measure the relevance assigned to each item. To study the prominence of all the subsets we propose the use of efficient methods to explore the lattice of keywords. In the future we plan to test these approach using standard data bases. In addition, it is neccessary a deep study about prominence indexes that could be used as well as aggregation procedures.

References

1. Agrawal, R., Mannila, H., Srikant, R., Toivonen, H., Verkamo, A.I.: Fast discovery of association rules. In: Advances in Knowledge Discovery and Data Mining (1996)
2. Agrawal, R., Srikant, R.: Fast algorithms for mining association rules in large databases. In: Proceedings of the 20th International Conference on Very Large Data Bases, VLDB 1994, pp. 487–499. Morgan Kaufmann Publishers Inc., San Francisco (1994)
3. Bobadilla, J., Ortega, F., Hernando, A., Gutiérrez, A.: Recommender systems survey. Knowledge-Based Systems 46, 109–132 (2013)
4. Burke, R.: Knowledge-based recommender systems (2000)
5. Campan, A., Truta, T.M.: Data and structural k-anonymity in social networks. In: Bonchi, F., Ferrari, E., Jiang, W., Malin, B. (eds.) PinKDD 2008. LNCS, vol. 5456, pp. 33–54. Springer, Heidelberg (2009)
6. Candillier, L., Meyer, F., Boullé, M.: Comparing state-of-the-art collaborative filtering systems
7. Díaz, I., Ralescu, A.: Privacy issues in social networks: A brief survey. In: Greco, S., Bouchon-Meunier, B., Coletti, G., Fedrizzi, M., Matarazzo, B., Yager, R.R. (eds.) IPMU 2012, Part IV. CCIS, vol. 300, pp. 509–518. Springer, Heidelberg (2012)
8. Díaz, I., Rodríguez-Muñiz, L.J., Troiano, L.: Fuzzy sets in data protection: strategies and cardinalities. Logic Journal of IGPL 20(4), 657–666 (2012)
9. Díaz, I., Rodríguez-Muñiz, L.J., Troiano, L.: On mining sensitive rules to identify privacy threats. In: Pan, J.-S., Polycarpou, M.M., Woźniak, M., de Carvalho, A.C.P.L.F., Quintián, H., Corchado, E. (eds.) HAIS 2013. LNCS, vol. 8073, pp. 232–241. Springer, Heidelberg (2013)
10. Han, J., Pei, J., Yin, Y., Mao, R.: Mining frequent patterns without candidate generation: A frequent-pattern tree approach. In: Mannila, H. (ed.) Data Mining and Knowledge Discovery, pp. 53–87. Kluwer, New York (2004)
11. Lee, M., Choi, P., Woo, Y.: A hybrid recommender system combining collaborative filtering with neural network. In: De Bra, P., Brusilovsky, P., Conejo, R. (eds.) AH 2002. LNCS, vol. 2347, pp. 531–534. Springer, Heidelberg (2002)

12. Luo, X., Xia, Y., Zhu, Q.: Incremental collaborative filtering recommender based on regularized matrix factorization. Know.-Based Syst. 27, 271–280 (2012)
13. McSherry, F., Mironov, I.: Differentially private recommender systems: Building privacy into the net. In: Proceedings of the 15th ACM SIGKDD International Conference on Knowledge Discovery and Data Mining, KDD 2009, pp. 627–636. ACM, New York (2009)
14. Pazzani, M.J.: A framework for collaborative, content-based and demographic filtering. Artif. Intell. Rev. 13(5-6), 393–408 (1999)
15. Salter, J., Antonopoulos, N.: Cinemascreen recommender agent: Combining collaborative and content-based filtering. IEEE Intelligent Systems 21(1), 35–41 (2006)
16. Shang, S., Hui, Y., Hui, P., Cuff, P.W., Kulkarni, S.R.: Privacy preserving recommendation system based on groups. CoRR, abs/1305.0540 (2013)
17. Takács, G., Pilászy, I., Németh, B., Tikk, D.: Scalable collaborative filtering approaches for large recommender systems. J. Mach. Learn. Res. 10, 623–656 (2009)
18. Troiano, L., Díaz, I., Kriplani, A.: A recommender system based on dempster-shafer theory. In: Eurofuse, pp. 232–241 (2013)
19. Troiano, L., Díaz, I., Rodríguez-Muñiz, L.J.: A model for assessing the risk of revealing shared secrets in social networks. In: Greco, S., Bouchon-Meunier, B., Coletti, G., Fedrizzi, M., Matarazzo, B., Yager, R.R. (eds.) IPMU 2012, Part IV. CCIS, vol. 300, pp. 499–508. Springer, Heidelberg (2012)
20. Troiano, L., Rodríguez-Muñiz, L.J., Ranilla, J., Díaz, I.: Interpretability of fuzzy association rules as means of discovering threats to privacy. Int. J. Comput. Math. 89(3), 325–333 (2012)
21. Troiano, L., Scibelli, G.: Mining frequent itemsets in data streams within a time horizon. Data & Knowledge Engineering 89, 21–37 (2014)
22. Troiano, L., Scibelli, G.: A time-efficient breadth-first level-wise lattice-traversal algorithm to discover rare itemsets. Data Mining and Knowledge Discovery 28(3), 773–807 (2014)
23. Troiano, L., Scibelli, G., Birtolo, C.: A fast algorithm for mining rare itemsets. In: ISDA 2009, pp. 1149–1155 (2009)
24. Zhou, B., Pei, J.: The k-anonymity and l-diversity approaches for privacy preservation in social networks against neighborhood attacks. Knowledge and Information Systems 28(1), 1–38 (2010)
25. Zhou, B., Pei, J., Luk, W.: A brief survey on anonymization techniques for privacy preserving publishing of social network data. SIGKDD Explor. Newsl. 10, 12–22 (2008)

Classification of Message Spreading in a Heterogeneous Social Network*

Siwar Jendoubi[1,2], Arnaud Martin[2],
Ludovic Liétard[2], and Boutheina Ben Yaghlane[1]

[1] LARODEC, University of Tunis, Avenue de la liberté, 2000 Le Bardo, Tunisie
[2] IRISA, University of Rennes 1, Rue E. Branly, 22300 Lannion, France
jendoubi.siwar@yahoo.fr, {Arnaud.Martin,ludovic.lietard}@univ-rennes1.fr,
boutheina.yaghlane@ihec.rnu.tn

Abstract. Nowadays, social networks such as Twitter, Facebook and LinkedIn become increasingly popular. In fact, they introduced new habits, new ways of communication and they collect every day several information that have different sources. Most existing research works focus on the analysis of homogeneous social networks, *i.e.* we have a single type of node and link in the network. However, in the real world, social networks offer several types of nodes and links. Hence, with a view to preserve as much information as possible, it is important to consider social networks as heterogeneous and uncertain. The goal of our paper is to classify the social message based on its spreading in the network and the theory of belief functions. The proposed classifier interprets the spread of messages on the network, crossed paths and types of links. We tested our classifier on a real word network that we collected from Twitter, and our experiments show the performance of our belief classifier.

Keywords: Information propagation, heterogeneous social network, classification, evidence theory.

1 Introduction

Nowadays, social networks such as Twitter, Facebook and LinkedIn become increasingly popular. In fact, they introduced new habits and new ways of communication. Besides, one of the distinguishing features of on-line social networks is the information spreading through social links. This is due to the "word-of mouth" exchanges, *i.e.* user-to-user exchanges, which makes the information more accessible and it spreads and reaches a large scale in few minutes. The volume and the dynamic of the exchange has attracted the attention of research communities. This research is motivated by the fact that the study of the diffusion of information is useful for understanding the dynamic behind social networks and the evolution of human relationships. Thus, they have focused on

---

* These research works and innovation are carried out within the framework of the device MOBIDOC financed by the European Union under the PASRI program and administrated by the ANPR.

A. Laurent et al. (Eds.): IPMU 2014, Part II, CCIS 443, pp. 66–75, 2014.

the processing of such data to extract high quality information, this information may be an important event, it can also be useful for optimizing business performance, or even for preventing terrorist attacks, etc.

The processing of a social network, always, starts by studying its structural properties, in fact the simple visualization of the network cannot give us a clear analysis about it. In the literature, we found a lot of structural properties measures like the degree, the betweenness, the closeness, the eigenvector centrality, etc. Quantifying structural properties and interpreting them will be essential to characterize the behavior of social actors, their position in the network, their interactions and how do they diffuse the information. Hence, the analysis of the network structural properties is an essential step when we study and model information propagation.

In our work we are interested in the classification of the spreading of the information in a heterogeneous social network. We assume that each type of content has some specific behavior when it propagates in the network. Hence, we propose a new algorithm of information propagation in a heterogeneous social network that takes into account the behavior of the content to be propagated. Therefore, we introduce an evidential algorithm to classify the propagation of the information through the network.

In the next section, we outline the literature review of the information propagation in social networks, the social message classification and the theory of belief functions. In section three, we introduce our algorithm of information propagation in a heterogeneous social network. In section four, we present our classification algorithm. Finally, we present our experiments in the fifth section.

2   Literature Review

2.1   Information Propagation in Social Networks

Information dissemination is a wide research domain that attracted the attention of researchers from various field such as physics and biology. We find the family of epidemiological models that are used to understand how diseases spread through populations. The simplest version is SI (*Suspected-Infected*), in this model, an individual is suspected if he has not the disease yet but he can catch it and become infected. This model was extended and many other version appeared to model specific diseases. Hence, we find SIS model (*Suspected-Infected-Suspected*), SIR model (*Suspected-Infected-Recovered*), SIRS model (*Suspected-Infected-Recovered-Suspected*), etc. The reader can refer to [1,17] for further details.

Computer scientists are generally interested in studying information propagation in on-line social networks. Mainly, their goal is to develop a model that simulates the diffusion process. Basic models are *Linear Threshold Model* (LTM) [7] and *Independent Cascade Model* (ICM) [6]. They assume the existence of a structure of a directed graph where each node can be activated or not knowing that you can not inactivate already activated nodes. The ICM model requires a probability distribution which must be associated with each link and LTM requires a degree of influence that must be set on each link and a threshold of influence for each node [12]. These two models were reused and improved in a lot of works like [5,18].

In this paper, we focus on information propagation in a heterogeneous social network, *i.e.* on which we find several types of links and/or nodes. In fact, in real word social networks we find many types of objects (users, groups, applications, etc) that are connected *via* many types of social links (friendship, membership, colleague, etc). Information dissemination in homogeneous social networks has been widely studied and the reader can refer to [8] for a recent survey. Now, research works start focusing on the processing of heterogeneous social networks. We find the work of [19] that simulates the propagation of the information in heterogeneous social networks based on the configuration model approach. In [13], authors propose to consider the behavior of individuals to model the influence propagation, their model is based on a heterogeneous social network.

2.2   Social Message Classification

Social message classification approaches, presented in the literature, are generally based on the content of the information and text mining techniques. They search to classify the user generated content to positive or negative about a some specific product. This task is so called sentiment classification and it is used to mine opinions. It starts by an item and/or feature extraction step, then it compares the extracted items and/or features to an existing corpus, finally comes the sentiment classification that can be based on items, features or both of them [14]. We find the work of [15] in which the author used a random sample of 3516 tweets to classify the feelings of consumers with respect to well-known brands. He classified the opinions (tweets) into positive and negative to see what is the most dominant opinion. In [10], a detailed case study that applies text mining to analyze unstructured textual content published on Twitter and Facebook and that talks about three chains of pizza. The reader can refer to the work of [16] for a recent study of the state of the art of social networks data mining.

2.3   Theory of Belief Functions

Upper and Lower probabilities [4] was the first ancestor of the theory of belief functions. Then comes the *Mathematical theory of evidence* [20] which defines the basic framework of information management and processing in the evidence theory, often called *Shafer model*. The main purpose of the theory of belief functions is to achieve more reliable, precise and coherent information. Here we present a short introduction of this theory, for more details the reader can refer to [20].

Let $\Omega = \{\omega_1, \omega_2, \ldots, \omega_n\}$ be a set of all possible decisions that can be made in a particular problem, it is called frame of discernment. The basic belief assignment (BBA), m^Ω, represents the agent belief on Ω, and it must respect $\sum_{A \subseteq \Omega} m^\Omega(A) = 1$. In the case where we have $m^\Omega(A) > 0$, A is called focal set of m^Ω. The basic belief assignment can be converted into other functions defined

---

Algorithm 1. Information propagation algorithm

Inputs:
- **N:** number of iteration
- **S:** source of the message
- **Str:** propagation strategy
- **Network:** the heterogeneous social network

Output:
- **PrNet:** propagation network

Algorithm:
1. ReadyNodes.add(S);
2. **For** $i = 1$ **to** N **do**
   (a) **for** $j = 1$ **to** ReadyNodes.size() **do**
       i. Node ←ReadyNodes.get(j);
       ii. if(Node.propagate()=True)
           foreach LinkType **do**
               $x \leftarrow$ Node.outdegree() * Node.propagationTendancy()
               *Str.LinkTypeProportion();
               R← (Node.randomSelection(x, $LinkType$));
   (b) Pr.refine(R);
   (c) R1.addAll(R);
   (d) ReadyNodes.addAll($R1$);
   (e) R1.clear;

---

from 2^{Ω} to $[0, 1]$. This theory presents a rich framework for information fusion and combining pieces of information (evidence). We find the Dempster's rule [4], the conjunctive and disjunctive combination rule [21], etc.

3 Propagation Algorithm in a Heterogeneous Social Network

In this section we introduce an algorithm of information propagation in a heterogeneous social network. This new algorithm takes four different inputs which are the number of iterations (stopping condition), the source of the message, the propagation strategy and the heterogeneous social network. As output, we have the propagation network that preserves the traversed paths. Algorithm 1 shows outlines of our propagation process. It starts by the source node. First, we verify if the current node is ready (wants) to propagate the message. Then, for each type of link in the network we compute the number of neighbors that will receive the message.

We assume that each type of message has some special characteristics of propagation in the network that is related to the types of links, so we define a propagation strategy for each type of message. Moreover, we consider the tendency of a particular node to propagate the message as a propagation parameter. Indeed, this parameter models the fact that a node can choose to distribute the message to a subset of its contacts (that he selects) or to retain it. The novelty of this algorithm is that we consider the type of the message while propagating it. Moreover our algorithm works with heterogeneous social networks where we have different types of links.

4   Classification of Information Propagation

The main purpose of this paper is to classify the spreading of the information through the network in order to characterize its content. In this section, we introduce our classification process that is composed of two steps; parameter learning step and the classification step. As mentioned in the algorithm 2, to learn the parameters of the model we need a set of propagation networks. First of all, we compute the number of nodes that have received the message *via* each type of link. We do this computation for each propagation level, *i.e.* we call propagation level the number of links between the source of the message and the target node. Second, we calculate the accrued effective by summing the effective of each level with the effective of the one before, this computation is done in order to preserve the propagation history at each propagation level. After that we transform the effective set of each level to a probability distribution defined on types of links, this transformation is done for two reasons; the first one, we need a probability distribution for the probabilistic classifier and the second one, it is an essential step to get the basic belief assignment distribution. Finally we transform each probability distribution to a BBA distribution using the consonant transformation [2,3].

Algorithm 2. Parameter learning algorithm

Input:
 − **PrNetSet:** a set of propagation networks
Output:
 − **ProbaSet:** a set of probabilities distributions (a probability distribution by propagation level).
 − **BbaSet:** a set of BBA distributions (a BBA distribution by propagation level).
Algorithm:
//effective computation
Foreach PrNet **in** PrNetSet **do**
 1. **Foreach** Level **in** PrNet **do**
  (a) **Foreach** TypeLink **do**
          N (TypeLink, Level) $\leftarrow N$ (TypeLink, Level)
             $+$ComputeNodes($TypeLink$);
//Accrued effective calculation
For Level$=$ 2 **to** NbrLevels **do**
 1. **Foreach** TypeLink **do**
  (a) N (TypeLink, Level) $\leftarrow N$ (TypeLink, Level)
       $+N$ (TypeLink, Level $-$ 1);
//ProbaSet and BbaSet computation
ProbaSet$\leftarrow$ProbabilitiesCalculation(N);
BbaSet$\leftarrow$ConsonantTransformation(ProbaSet);

Once model's parameters are learned, we can use it to classify new coming message (propagation network of the message) as shown in algorithm 3. Our classification algorithm starts by applying the same parameter learning process (algorithm 2) on the propagation network to be classified. Then for each level in the network we compute the distance between its probability distribution and the probability distribution of each propagation strategy, then we choose the class of the nearest propagation strategy (with the shortest distance) to be the class of the message in the current level. The same process is done with BBA distributions as mentioned in the algorithm.

Algorithm 3. Classification algorithm

Input:
- **ProbaSets:** a set of probabilities distributions for each strategy of propagation.
- **BbaSets:** a set of BBA distributions for each strategy of propagation.
- **PrNet:** The propagation network to be classified

Output:
- In order to see the impact of the level of propagation on the classification results, in our output we have a class by level.

Algorithm:
1. (ProbaPr, BbaPr) $\leftarrow$ ParameterLearning $(PrNet)$;
2. **For** $i = 1$ **to** NbrStrategies **do**
    (a) **Foreach** Level **do**
        i. ProbaDist$(i, Level)$ $\leftarrow$ Distance (ProbaPr, ProbaSets (i));
        ii. BbaDist$(i, Level)$ $\leftarrow$ Distance (BbaPr, BbaSets (i));
3. **Foreach** Level **do**
    (a) ProbaClasses$(Level)$ $\leftarrow$ StrategyMinDistance (ProbaDist $(:, Level)$);
    (b) BbaClasses$(Level)$ $\leftarrow$ StrategyMinDistance (BbaDist $(:, Level)$);

---

5 Experiments and Results

In this section, we present some experiments to show the power of the proposed evidential classification algorithm.

5.1 Data Description

We used NodeXL V 1.0.1.245 [9] to collect social network data from Twitter. We collected the network shown in figure 1. It is a directed network in which nodes are Twitter users. Table 1 shows the characteristics of our network data.

Table 1. Data characteristics

Vertices	Edges	Geodesics distance	Betweenness	Closeness	Eigenvector
97	350	6	184.99	0.004	0.01

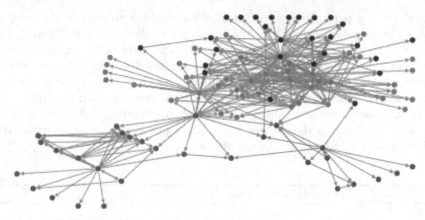

Created with NodeXL (http://nodexl.codeplex.com)

Fig. 1. Network visualization

As mentioned above, we need a heterogeneous social network to test proposed algorithms. Therefore, we used the structure of the network collected from Twitter and we generated, randomly, the types of links. We assume four types of link in the network which are "Professional", "Familial", "Friendly" and "Undefined". Then we obtained a heterogeneous social network that is used as input for our propagation algorithm.

5.2   Experiment Configuration

In the following experiments, we defined three different propagation strategies for three types of messages which are: "Spam", "Professional" and "Familial". Each strategy is defined as the proportion of the nodes that will receive the message from each type of links. Hence, we have to define four proportions for each propagation strategy. To be as near as possible to the reality, we added a noise rate to the strategy. We note that the noise value can be added or removed from the proportions of kind of messages. We used the euclidean distance for the probabilistic classifier:

$$d_E\left(Pr_1, Pr_2\right) = \sqrt{\sum_{i=1}^{card}\left(Pr_1\left(i\right) - Pr_2\left(i\right)\right)^2} \tag{1}$$

and the Jousselme distance [11] for the evidential one:

$$d_J\left(m_1, m_2\right) = \sqrt{\frac{1}{2}\left(m_1 - m_2\right)^T \underline{\underline{D}}\left(m_1 - m_2\right)} \tag{2}$$

such that $\underline{\underline{D}}$ is an $2^n \times 2^n$ matrix and $D\left(A, B\right) = \frac{|A \cap B|}{A \cup B}$. We fixed the number of levels in the network to three (three iterations in the propagation algorithm).

Then we run the proposed propagation algorithm to create a training set for each propagation strategy, we fixed the size of the strategy training set to 100 propagation networks. Also, we created a testing set of size 100.

5.3   Results and Discussion

In this section we present our results and a comparison between the probabilistic and the evidential classifier. To obtain accurate results we turned the experimental process ten times and we take the mean of the percentage of correctly classified (PCC) propagation networks. Figure 2 shows the impact of the propagation level on the PCC of the probabilistic results (figure 2a) and the evidential results (figure 2b). Figures 2a and 2b illustrate that the PCC increases when the propagation level increases and we observe this fact starting from the noise level 20%. In figure 2a we observe that the curve of the second level coincides with the curve of third level and practically there is no improvement in the PCC. However, in figure 2b (evidential results), we note that the PCC increases with the propagation level, this fact is observed starting from the noise rate 20%. Hence, we have the PCC of the third level greater than the PCC of the first and the second levels, and the PCC of the second level is higher than the PCC of the first one. Therefore, more the message propagates in the network, more we can characterize it.

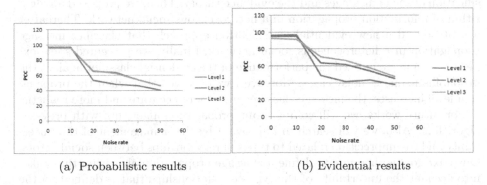

(a) Probabilistic results   (b) Evidential results

Fig. 2. The impact of the propagation level on the PCC

In figure 3, we compare the probabilistic and the evidential results of the third propagation level. We note that without noise (0%) the probabilistic PCC is about 96% (with a 95% confidence interval of ± 1.27) and the evidential PCC is equal to 93% (with a 95% confidence interval of ± 1.60), but in real world social networks the absence of the noise is an ideal fact and cannot be realistic. When the noise rate increases, the curve shows that the percentage of correctly classified propagation networks (messages) decreases. However, we see that the evidential (Belief) PCC starts to be greater than the probabilistic (Proba) one.

We observe this fact from the noise rate 20% where we have an evidential PCC equals to 70.7% (±4.33) and a probabilistic PCC equals to 65.8% (±4.18). Thus, we can conclude that the evidential classifier is more robust against the noise and gives better classification rates than the probabilistic classifier.

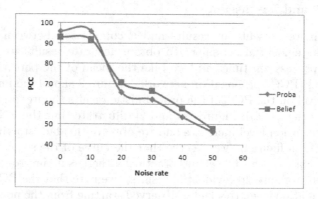

Fig. 3. Comparison between probabilistic results and evidential results (level three)

6   Conclusion

To conclude, we presented a state of the art of the information propagation, classification of social messages and the evidence theory. Then, we proposed an algorithm of information propagation in a heterogeneous social network. Thereafter we introduced a new evidential classification approach that classifies message propagation in a heterogeneous social network. Finally, we presented some experiments and we noticed the performance of the evidential classifier against the probabilistic one in noisy cases. Moreover, we observed that when the propagation level increases, the message class becomes more accurate and more realistic.

For future works, we will compare our propagation algorithm with previous algorithms. Also, we will search to improve it by the management of the uncertainty and the imprecision related to types of relationships between social actors. Our next goal is therefore to define a message propagation algorithm that takes into account the uncertainty of the types of relationships that is defined on the links, also we will search to consider the heterogeneity of nodes in the network. Second, we will run our classification algorithm with a more complex heterogeneous social network in order to prove its applicability.

References

1. Anderson, R.M., May, R.M.: Infectious Diseases of Humans. Oxford University Press (1991)
2. Aregui, A., Denœux, T.: Consonant belief function induced by a confidence set of pignistic probabilities. In: Mellouli, K. (ed.) ECSQARU 2007. LNCS (LNAI), vol. 4724, pp. 344–355. Springer, Heidelberg (2007)

3. Aregui, A., Denoeux, T.: Constructing consonant belief functions from sample data using confidence sets of pignistic probabilities. Int. J. Approx. Reasoning 49(3), 575–594 (2008)
4. Dempster, A.P.: Upper and Lower probabilities induced by a multivalued mapping. Annals of Mathematical Statistics 38, 325–339 (1967)
5. Galuba, W., Aberer, K., Chakraborty, D., Despotovic, Z., Kellerer, W.: Outtweeting the twitterers - predicting information cascades in microblogs. In: WOSN 2010, pp. 3–11 (2010)
6. Goldenberg, J., Libai, B., Muller, E.: Talk of the network: A complex systems look at the underlying process of word-of-mouth. Marketing Letters 12(3), 211–223 (2001)
7. Granovetter, M.: Threshold models of collective behavior. American Journal of Sociology, 1420–1443 (1978)
8. Guille, A., Hacid, H., Favre, C., Zighed, D.A.: Information diffusion in online social networks: a survey. SIGMOD Rec. 42(1), 17–28 (2013)
9. Hansen, D.L., Shneiderman, B., Smith, M.A.: Analysing social media network with nodeXL insights from a connected world. Elsevier Inc. (2011)
10. He, W., Zhab, S., Li, L.: Social media competitive analysis and text mining: A case study in the pizza industry. International Journal of Information Management 33, 464–472 (2013)
11. Jousselme, A.L., Grenier, D., Bossé, E.: A new distance between two bodies of evidence. Information Fusion 2, 91–101 (2001)
12. Kempe, D., Kleinberg, J., Tardos, E.: Maximizing the spread of influence through a social network. In: Proceedings of the Ninth ACM SIGKDD International Conference on Knowledge Discovery and Data Mining, KDD 2003, pp. 137–146. ACM Press (2003)
13. Li, C.T., Lin, S.D., Shan, M.K.: Influence propagation and maximization for heterogeneous social networks. In: WWW 2012-Poster Presentation, pp. 559–560 (April 2012)
14. Lo, Y.W., Potdar, V.: A review of opinion mining and sentiment classification framework in social networks. In: 3rd IEEE International Conference on Digital Ecosystems and Technologies, DEST 2009 (June 2009)
15. Mostafa, M.M.: More than words: Social networks text mining for consumer brand sentiments. Expert Systems with Applications 40, 4241–4251 (2013)
16. Nettleton, D.F.: Survey data mining of social networks represented as graphs. Computer Sciences Review 7, 1–34 (2013)
17. Newman, M.E.J.: Networks: An introduction. Oxford University Press (2010)
18. Saito, K., Ohara, K., Yamagishi, Y., Kimura, M., Motoda, H.: Learning diffusion probability based on node attributes in social networks. In: Kryszkiewicz, M., Rybinski, H., Skowron, A., Raś, Z.W. (eds.) ISMIS 2011. LNCS, vol. 6804, pp. 153–162. Springer, Heidelberg (2011)
19. Sermpezis, P., Spyropoulos, T.: Information diffusion in heterogeneous networks: The configuration model approach. In: 2013 Proceedings IEEE INFOCOM, pp. 3261–3266 (April 2013)
20. Shafer, G.: A mathematical theory of evidence. Princeton University Press (1976)
21. Smets, P.: Belief Functions: the Disjunctive Rule of Combination and the Generalized Bayesian Theorem. International Journal of Approximate Reasoning 9, 1–35 (1993)

Measures of Semantic Similarity of Nodes in a Social Network

Ahmad Rawashdeh[1], Mohammad Rawashdeh[1], Irene Díaz[2], and Anca Ralescu[1]

[1] EECS Department, University of Cincinnati, Cincinnati, OH 45221-0030, USA
[2] Computer Science Department, University of Oviedo, Spain
{rawashay,rawashmy}@mail.uc.edu, sirene@uniovi.es, Anca.Ralescu@uc.edu

Abstract. Assessing the similarity between node profiles in a social network is an important tool in its analysis. Several approaches exist to study profile similarity, including semantic approaches and natural language processing. However, to date there is no research combining these aspects into a unified measure of profile similarity. Traditionally, semantic similarity is assessed using keywords, that is, formatted text information, with no natural language processing component. This study proposes an alternative approach, whereby the similarity assessment based on keywords is applied to the output of natural language processing of profiles. A *unified similarity measure* results from this approach. The approach is illustrated on a real data set extracted from Facebook and compared with other similarity measures for the same data.

1 Introduction

Social networks allow people to connect and share their personal details. Many social networking websites have been created and they vary in the services which they provide. Mainly, they allow users to comment and post pictures or video and share.

Facebook is a social networking website that has over one billion users. It allows the user to connect to friends, create personal profiles by specifying their interest –TV, movies, sports, and books – and by posting images and videos of their activities. The website also allows anyone to create pages for their business or favorite personality. Users can even create pages for special interest groups which are open on a restricted basis to group members.[6]

People tend to form relationships with people who are similar to them. Alternatively, it can be said that if a relationship is formed between two people, then there must be some similarity between them. Indeed, it has been found that 80% of social network users form relationships with the contact of their friends [3].

Analysis of similarity between Facebook profiles can be assessed from the study of keyword similarity [3]. To find the relationship between the keywords, these are arranged in a hierarchical structure to form trees of different heights. In the forest model more than one tree is generated for each profile. Related words

A. Laurent et al. (Eds.): IPMU 2014, Part II, CCIS 443, pp. 76–85, 2014.

are retrieved by search in these profile trees, implemented as heuristic search. Semantic relationships between the words can be assessed by using Wordnet. [10]

This study proposes to find the semantic relationship between attribute entries in the social network, not only between keywords. Therefore the category of the words which appear in these entries must be found. This can be accomplished by using a tagger, a program which tags a word by its semantic category. These categories are used to extract the words suitable to assess profile similarity [1]. The (semantic) distance between profiles is very important to this process, as it has been shown that the similarity between profiles deteriorates as the distance between them increases [4].

2   Finding Similar Profiles

The approach to profile similarity proposed here, semantic distance based similarity, combines Wordnet [8] with the cosine similarity, which is a very common device to assess document similarity [9]. The results obtained are compared with other similarity measures, including, (i) occurrence frequency, (ii) set similarity (Jaccard index), (iii) syntactic similarity, (iv) word frequency vector similarity. Before defining each of these measures of similarity, Wordnet and the semantic tagger are briefly described.

2.1   Wordnet

Wordnet is a free lexical database that organizes English words into concepts and relations, well-known for assessing semantic similarity. English nouns, verbs, adjectives, and adverbs form hierarchies of *synset* where relations exist that connect them. The relations are Synonymy, Antonymy, Hypernymy, Meronymy, Troponymy, Entailment.

Hypernym of a Word. *Hypernym* of a word conveys its place in a hierarchy of concepts/words and can be retrieved using Wordnet. Consider for example, the two senses of word "comedy":

– comedy as a "humorous drama"
– comedy as "comic incident"

Taking the first sense, since comedy is kind of drama, drama is a hypernym of comedy. Similarly, since drama is kind of literary work, literary work is a hypernym of drama [8]. The hierarchy determined by the hypernym relationship is a *synset*. Therefore, based on the above, the synset for comedy (with respect to the first meaning) is

Synset 1: [entity] ← [abstract entity] ← [abstraction] ← [communication]
    ← [expressive style,style] ← [writing style,literary genre,genre]
    ← [drama] ← [comedy] -

$$\text{light and humorous drama with a happy ending} \tag{1}$$

while the Synset with respect to the second meaning is:

Synset 2: [entity] ← [abstract entity] ← [abstraction]
    ← [communication] ← [message,content,subject matter,substance]
    ← [wit, humor, humor, witticism, wittiness] ← [fun, play,sport]
    ← [drollery, clowning, comedy, funniness] -

$$\text{a comic incident or series of incidents} \tag{2}$$

2.2 Semantic Tagger

A semantic tagger takes as input text and outputs a collection of word-tags pairs (w, t_w) [7]. Tags denote semantic categories. The innovative aspect of the current study is the use of tags instead of words to assess profile similarity. The tags used in this study are shown in Table 1.

Table 1. Word tags and their descriptions [1]

Tag	Description
NN	noun, proper, singular or mass
NNP	noun, proper, singular
NNS	noun, common, plural
$NNPS$	noun, proper, plural

3 Semantic Similarity Measures

The literature of intelligent data processing, including information retrieval, document processing, etc. is rich in similarity measures. The similarity measures considered here are all adapted to deal with word tags /semantic categories rather than (key) words. For each of the measures described below D_i, $i = 1, 2$ denote two documents or profiles.

3.1 Cosine Similarity for Vectors

Cosine similarity for vectors [9] has been successfully used as measure of similarity between documents. A vocabulary of size N of words of interest is defined and each document is described by an N *dimensional vector of word frequencies*. The similarity of two documents is then based on the cosine of the angle made between their corresponding vectors. More precisely, given the documents D_i, $i = 1, 2$, with corresponding word frequency vectors v_1 and v_2, the *cosine similarity* between D_1 and D_2 is defined as

$$CS(D_1, D_2) = \frac{v_1 \cdot v_2}{\|v_1\|\|v_2\|} \tag{3}$$

where $\cdot$ is the dot product between two vectors, and $\|v\|$ denotes the norm of a vector v. Since frequencies are positive quantities, it can be easily seen that this similarity is consistent with the distance between vectors.

3.2   Wordnet Cosine Similarity

Adapting to the encoding using Wordnet and tags, the cosine similarity is applied to the vectors of the frequencies of tags (rather than vectors of frequencies of word) used to encode a profile. This means that two profiles that contain different words can be evaluated as similar (even identical) if these words fall into the same semantic categories.

3.3   Set-Based Similarity

A frequently used, set-based similarity is the well-known Jaccard Index. Given two subsets A and B of the same universe of discourse, their similarity is $J(A, B)$ is defined by

$$J(A, B) = \frac{|A \cap B|}{|A \cup B|} \tag{4}$$

where $|A|$ denotes the size of the set A. When Jaccard index is used in conjunction with Wordnet, given two documents, D_1 and D_2, their corresponding Jaccard index is computed by equation (5).

$$J(D_1, D_2) = \frac{|P_1 \cap P_2|}{|P_1 \cup P_2|} \tag{5}$$

where P_i denotes the collection of semantic categories from the parent sets for each word in the document D_i, $i = 1, 2$.

3.4   The Occurrence Frequency Similarity(OF)

Let D and D' denote two documents/profiles, each having the multiple valued attribute i. Following the work in [2] the *occurrence frequency* similarity measure between D and D' is defined by equation (6).

$$OF(i_D, i_{D'}) = \begin{cases} 1 & \text{if } i_D.n = i_{D'}.n \\ \frac{1}{B}\sum_{k=1}^{B}(1 + A \times B)^{-1} & \text{if } i_D.n \neq i_{D'}.n \end{cases} \tag{6}$$

where i_D denotes the value of attribute i in the profile D, $i_D.n$ denotes the value of the nth subfield for i_D, N is the total number of item values, and $f(\cdot)$ is the number of records; $A = \log(\frac{N}{1+f(i_u.n)})$, and $B = \log(\frac{N}{f(i_x.k)})$.

4   A Unified Similarity Measure: Wordnet-Cosine Similarity

A unified similarity measure takes into account the semantics of the input, encodes it into some numerical representation, and computes the similarity based on this representation. The general algorithm to compute such a measure for two documents/profiles is as shown below:

1. Extract the text in the feature field (movies, title) if the data-set is not formatted well.
2. Natural Language Processing: Parse the sentence to obtain its structure.
3. Get the first synset of the word using Wordnet.
4. Encode the word
   − Get all hypernym of the synset of the word.
   − Find the distance from the word to the root of the synset.
5. Each feature field of a profile is encoded as a vector of such distances.
6. Apply cosine similarity between vectors of such distances.

Each profile is represented as a collection of word-tag pairs (w, t_w). Given a word-tag pair, (w, t_w), w is considered for inclusion in the similarity evaluation if and only if $t_w \in Tags$ of Table 1. Each selected word, w is input to Wordnet which returns the list of hypernyms, in the hierarchical synset representation of w. As already mentioned, in this study, only the first sysnset is used for similarity assessment. The encoding of w is the distance to it from the top hypernym ('entity') in the synset. For example, the encoding of the word "comedy" based on the first synset 1 is equal to 7. If a word has no hypernym (e.g., it is not in Wordnet) then its encoding is 0. Therefore, for each word w_i, use Wordnet to extract its first synset and encode it as $d_i = d(w_i)$ where, for a given word w,

$$d(w) = \begin{cases} dist(w, [\text{entity}]) & \text{if } w \text{ is in Wordnet} \\ 0 & \text{otherwise} \end{cases} \qquad (7)$$

where $dist$ is the distance to [entity], the top hypernym of w in its first synset, output by Wordnet. The encoding of the profile D is a mapping $e : D \mapsto \Re_+^k$ such that

$$e(D) = (d_1, \ldots, d_k)$$

Given two profiles, D, and D' and their corresponding encoding $e(D) = (d_1, \ldots, d_k)$ and $e(D') = (d'_1, \ldots, d'_k)$ the similarity between D and D' is defined as the cosine similarity of $e(D)$ and $e(D')$, as shown in equation (8)

$$Sim(D, D') = CS(e(D), e(D')) \qquad (8)$$

where CS is defined as in equation (3). The process described above converts the problem of similarity assessment between unstructured data into a more rigorously defined problem of similarity between real valued vectors. In principle, it is possible, for a given word w (and hence for a profile), to obtain more than one encoding, by using all the synsets to encode a line of text using several synsets. However, this case is beyond the scope of the current study. Figure 1 illustrates the approach proposed in this study and described above.

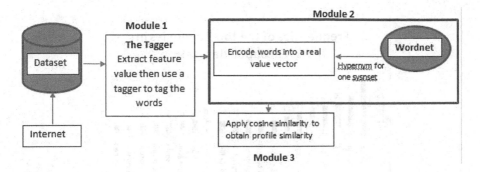

Fig. 1. Diagram for computing the unified similarity measure using Wordnet and cosine similarity

5 Experimental Results

The approach described in the preceding section is applied to a Facebook data set as shown next.

5.1 Facebook Profiles Data-Set

The Facebook data-set considered in experiments contains 2013 profile pages from Facebook (raw data before the introduction of the Facebook time-line). Skull security has a list of publicly available Facebook URLs which is used to download this data-set that consists of 2013 profiles [5]. More specifically, *Dataset.txt* (Facebook Data-set) contains all the movies interest for different Facebook profile numbers. The format of the data-set is as follows: *Profile_id* followed by the Movies interest entered by the user identified by the *Profile_id*. Furthermore, various characteristics are extracted from the Facebook Data-set, as shown in Table 2. Figure 2 shows the frequency of the top 20 movies in the Facebook data-set.

Table 3 illustrates the encoding the Movie Attribute for three Facebook profiles.

Table 2. Characteristics of the Facebook profile data

Number of Facebook profiles	2013
Average movies entries per profile	2.9
Number of movies entries for all profiles	1744
Maximum movies entries	8
Most Common Genre type [1]	which is the genre type "unknown"
Minimum movies entries	0
Different movies count	1089

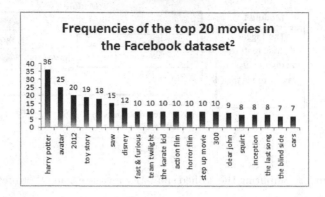

Fig. 2. Frequency of the top 20 movies from the Facebook data-set

Table 3. Illustration of Movie Attribute of Facebook profiles: their tags and Hypernyms

Profile 1: Movie Attribute	Harry Potter, Transformers, Mr. & Mrs. Smith						
Words	Harry	Potter	Transformers	Mr.	&	Mrs.	Smith
Tags	NNP	NNP	NNPS	NNP	CC	NNP	NNP
dist to root in synset	0	7	8	8	ignored	8	0
Profile 2: Movie Attribute	Sherina's Adventure						
Words	Sherina	's	Adventure				
Tags	NNP	POS	NNP				
dist to root in synset	0	ignored	8				
Profile 3: Movie Attribute	Love mein Gum, Maqsood Jutt Dog Fighter						
Words	Love	mein	Gum	Maqsood	Jutt	Fog	Fighter
Tags	NNP	NNP	NNP	NNP	NNP	NNP	NNP
dist to root in synset	7	0	7	0	0	6	4

5.2   Results

All the similarity measures described here, including the algorithm of [2], were implemented in Java. In the first set of experiments, the similarity was calculated between each adjacent nodes' row in the data-set using both the OF measure and Wordnet-cosine approach. Table 4 illustrates these similarity results for two profiles.

Figure 3 shows the result of applying the OF similarity and the Wordnet-cosine similarity for all the node pairs connected by an edge in the data set. Using OF, most of the data are similar, with similarity value equal to 1. By contrast, using Wordnet, the similarity values are distributed over all the data having a peak value at 0.2.

In the second experiment, for the same data, the similarities obtained by using Wordnet cosine similarity, Jaccard index, semantic similarity and vector cosine

Table 4. OF and Wordnet Similarity of two Facebook profiles along their Movie Attribute. Profile IDs are partially masked for privacy.

Data Set	Facebook
Profile-1 ID	10000006XXXXXX.html
Movies Interests	Captain Jack Sparrow, Meet The Spartans, Ice Age Movie, Spider-Man
Profile-2 ID	100000067XXXXXX.html
Movies Interests	Clash of the Titans, Ratatouille, Independence Day, Mr. Nice Guy, The Lord of the Rings Trilogy (Official Page)
OF Similarity	0.9472
Wordnet based similarity	0.1892

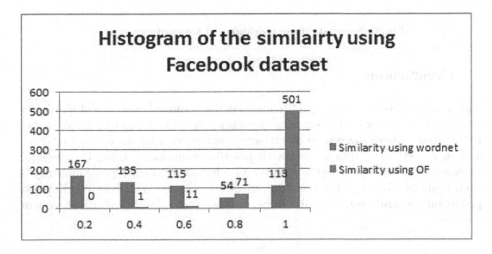

Fig. 3. OF and Wordnet similarity results for the Facebook data-set

Table 5. Results for the similarity of two profiles based on the four similarities measures used in this study. Profile IDs are partially masked.

Profile ID	Profile
132XXXXXXX.html	*Comedy, Action films, American, El El*
774XXXXXXX.html	*Haunted 3D, Saw, Transformers, Pirates of the Caribbean, Mind Hunter*

SIMILARITY MEASURES			
Wordnet cosine	Set similarity	Semantic similarity	Word frequency vector similarity
0.862795963	0.0659340066	0.877526909	0.74900588

similarity were compared. Table 5 shows the difference in the similarity results for two profiles, using four similarity measures. The results for all four similarity measures on the Facebook data set, are shown in Figure 4.

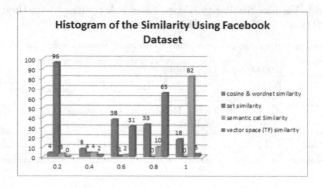

Fig. 4. Similarity using four different similarity measures

6   Conclusions

This study introduces a new approach towards a unified measure of similarity between node profiles, and in general, between pieces of unstructured text. Natural language processing is used to extract speech parts from the texts of interest, and to encode them into vectors with positive components using the distance between the words extracted to the root of a hierarchy of concepts. Similarity is then evaluated between the resultant encoding vectors. While the results seem promising, several issues remain to be discussed and developed in subsequent studies.

References

1. The university of pennsylvania (penn) treebank tag-set,
   http://www.comp.leeds.ac.uk/ccalas/tagsets/upenn.html
   (accessed October 1, 2013)
2. Akcora, C.G., Carminati, B., Ferrari, E.: Network and profile based measures for user similarities on social networks. In: 2011 IEEE International Conference on Information Reuse and Integration (IRI), pp. 292–298. IEEE (2011)
3. Bhattacharyya, P., Garg, A., Wu, S.F.: Analysis of user keyword similarity in online social networks. Social Network Analysis and Mining 1(3), 143–158 (2011)
4. Boriah, S., Chandola, V., Kumar, V.: Similarity measures for categorical data: A comparative evaluation. Red 30(2), 3 (2008)
5. Bowes, R.: Return of the Facebook Snatchers (2012),
   http://www.skullsecurity.org/blog/2010/
   return-of-the-facebook-snatchers (accessed July 19, 2012)
6. Ellison, N.B., et al.: Social network sites: Definition, history, and scholarship. Journal of Computer-Mediated Communication 13(1), 210–230 (2007)

7. The Stanford Natural Language Processing Group. Pos Tagger FAQ,
   http://nlp.stanford.edu/software/pos-tagger-faq.shtml
   (accessed July 19, 2012)
8. Miller, G.A.: Wordnet: a lexical database for english. Communications of the
   ACM 38(11), 39–41 (1995)
9. Peat, H.J., Willett, P.: The limitations of term co-occurrence data for query expansion in document retrieval systems. JASIS 42(5), 378–383 (1991)
10. Spear, M., Lu, X., Matloff, N.S., Wu, S.F.: Inter-profile similarity (IPS): A method
    for semantic analysis of online social networks. In: Zhou, J. (ed.) Complex 2009.
    LNICST, vol. 4, pp. 320–333. Springer, Heidelberg (2009)

Imitation and the Generative Mind

Jacqueline Nadel

CNRS Centre Emotion, Pierre & Marie Curie University, ICM,
La Salpêtrière Hospital, Paris, France
jacqueline.nadel@upmc.fr

Abstract. In its perpetual capacity to imagine, create and revisit artifacts and representations, human mind is the perfect example of generativity. Yet if we agree with Epstein (1996)'s theory of generativity, new ideas result from interconnections among old ones. That is, cultural knowledge heavily influences our individual minds. In this line, our minds need meeting other minds to generate innovation. I will argue in this article that the basis of a generative meeting between minds is imitation. This proposal is developed against the well-established reputation of imitation as an idiotic behaviour stifling creativity.

Keywords: imitation, generativity, flexibility, development, brain dynamics.

1   Introduction

An ancient tradition hinders the reputation of imitation. This tradition comes from the great philosopher Plato. Plato described imitation as dangerous because it stifles creativity, hampers the development of personal identity and disrupts the perception of other people as unique beings. Girard recalls in his book, 'Things hidden since the foundation of the world' (Girard, 1987), that in certain cultures, one child out of every set of twins would be killed, as would a son who looked too much like his father. Who exactly was at risk, in a world where such little importance was given to the concept of individuality? Surely the danger was not for the imitator but rather for the social group, where too close a physical resemblance might have caused confusion about roles in the community (Vernant, 1983). It remains that for centuries and centuries, imitation has been an object of contempt. For instance, Piaget (1945) called "intelligent imitation," a reproduction that is not stuck in the present (i.e., imitating an absent model), nor is it stuck with what the infant already knows how to do; thanks to representation, an action can be performed without requiring a direct perception of it. In sum, according to Piaget's theory of intelligence, simply doing what the other does is not 'generative'. The aim of this paper is to demonstrate that this view does not take into account the capacities required in order to imitate, and the generativity it allows to brain, behavior and mind.

A. Laurent et al. (Eds.): IPMU 2014, Part II, CCIS 443, pp. 86–92, 2014.

2   Generativity and the Meeting of Minds

Generativity is described via Wikipedia as 'a self-contained system from which its user draws an independent ability to create, generate, or produce new content unique to that system without additional help or input from the system's original creators'. In its perpetual capacity to imagine, create and revisit artifacts and representations, human mind is the perfect example of generativity. As a consequence, human mind inspires generative models in computer modelling. Linguistic theories such as the famous Chomsky's theory have emphasized the unique role of language in the expression of our generative structure of mind (Chomsky, 1985). Language offers us the means to express our thoughts through unique pieces of linguistic creation. These pieces are built thanks to a generative and transformational grammar that possesses compositionality (Dennett, 1971) and provides us the capacity to construct complex messages. Chomsky's model of generative syntax contributes to the theory of mind's perspective. It meets probabilistic generative models that aim to infer invisible variables of the investigated phenomenon on the basis of visible variables. Indeed it is what we do each time we infer the unobservable mental states of others on the basis of probabilistic computation on observed events Now suppose we adopt Ziffrain (2008)'s definition of web generativity as " a system's capacity to produce unanticipated change through unfiltered contributions from broad and varied audiences." Then, we have to broaden our definition of a system to the assembly of two or more persons. In this view, our mind possesses means to produce generativity proviso it works in concert with other minds.

3   Social Cognition and Social Interaction in Cognitive Sciences

This way of thinking is in line with a burgeoning field in cognitive and neurocognitive sciences. After a long focus on mentalizing processes studied in subjects in isolation, cognitive sciences are now turning to analyze the role of social cognition in online social interaction. Yet the need for a clear-cut distinction, at the theoretical and methodological levels, between the generic term of social cognition and the specific phenomenon described as social interaction is not shared by all specialists in the field. A traditional cognitive interpretation holds that the brain is simply entering another mode of functioning when immerged in a social interactive context. Of course social interaction mostly involves social cognition as an underlying process by which humans understand, anticipate, or infer the intentional behavior of others. Moreover, it is the place where social cognition most frequently occurs in everyday life. Yet social interaction is a specific online phenomenon which cannot be considered merely as a category of inputs to be processed by individual mechanisms (De Jaeger & Di Paolo, 2012; Dumas, Martinerie, Soussignan & Nadel, 2012). The reason is that social interaction is a co-regulated coupling between at least two agents who are mutually influencing each other. This definition and the underlying dynamical theory are currently gaining ground in social neuroscience against a solipsistic view of the generative mind.

4   Developmental Psychology, Imitation and the 'Two-Person' Stance

Inspiration for an alternative approach to social interaction comes from developmental psychology, in which, beginning in the 1970s, attention turned to studying real-time dynamic interactions involving two or more partners. The emphasis was then put on dyadic variables (Nadel & Camaioni, 1993) such as imitation, joint attention, turn-taking and coregulation, various aspects of which have been referred to as co-regulation (Fogel, 1993), synchrony (Trevarthen, 1977), or harmonization (Stern, 1977). The concept of coregulation suggests a dynamically changing individual during the process of transaction with others. In the same line, cognition is considered to be constantly evolving in dynamic interactions (Varela, Thompson & Rosch, 1991).

Immediate imitation is often defined as a social behaviour leading to the individual benefit of learning. There is however another function of imitation which fits well the two-person perspective. Studying spontaneous imitation in an online meeting of peers aged 12, 15, 18, 24, 30, 36 and 42 months, we have shown that young children take advantage of the two facets of imitation to get two roles (imitator and model) that they switch to take turns (Nadel & Butterworth, 1999). Indeed the dynamics of imitation makes it a genuine communicative system which presents the three parameters of any interactive system: synchrony, joint attention and turn-taking. Like in conversation, roles are exchanged smoothly on the basis of a coregulation. Prepin and Revel (2007) have shown that two oscillators facing each other loose progressively thei specific tempo and adopt a common tempo different from their own. Similarly, young children imitating each other form a system which generates novel common actions differing from the repertory of action of each partner: it is literally a generative two-person system. Thus, it appears that imitation serves both the traditionally-recognized function of promoting skill acquisition and a previously unacknowledged interactive function (Andry, Gaussier, Moga, Banquet & Nadel, 2001).. Where does this double function of imitation emerge from?

5   Neonatal Imitation and the Foundation of Social Interaction

Social interaction in its basic foundation is well represented by imitation from birth on. Literally from birth, typical neonates are able to imitate a tongue protrusion (Meltzoff & Moore, 1977). They are even able to imitate a tongue protrusion presented on a screen (Soussignan, Courtial, Canet, Danon-Apter & Nadel, 2011). It is not a prowess: Protruding tongue is already in the motor repertory of foetuses of gestational age 25 weeks (Piontelli, 2010). The prowess is that they are able to use their motor repertory according to their perception. So doing, they relate their motor patterns to the others' motor patterns. Moreover, the newborns match more and more exactly the perceived stimulus after repeated attempts, which shows that they are able to modulate their motor repertoire (Soussignan et al, 2011). Thus neonatal imitation, though experience- dependent, adapts action to perception with great plasticity. For Lepage and Théoret (2007), this plasticity renders plausible the hypothesis of a gradual development of the Mirror neuron System (MNS) through repeated motor activity and related sensory feed-backs of the foetus. Similarly, the adult MNS is experience-dependent and plastic: our mirror neurons resonate to the observation of actions that are not part of our motor repertoire only after repeated exposure (Calvo-Merino et al., 2005).This demonstration of a flexible repertoire is of paramount importance. Indeed the individual deprived of social encounters would not have the opportunity to enrich their repertory according to the observed actions of the others. It is the beginning of a perception-action coupling that will take many different forms, from acts to thoughts, but will never stop. A few months later, having enriched their motor repertoire thanks to their matching of others' actions, toddlers will start being able to store representations of actions they have never done: How? The storage originates from somatotopic and proprioceptive recalls of past experience (Raos, Evangeliou &Savaki, 2007) involving elements of the observed actions. This mental recombination is a powerful multiplier of experiences as it prints in our memory of actions those actions performed by others that we have observed but never done. This is possible, proviso we have elements of the observed action in our repertory. Then we can build new possibilities with old ones -a basic illustration of Epstein (1996)'s theory which asserts that new ideas result from interconnections among old ones. Notice that though this novel repertory is built thanks to the actions of the others, it is different from the others' repertory just because it is issued from our own history of actions, and our own gestural procedures. From acts to thoughts, the process is similar, as shown by Fadiga's team (Fadiga, Craighero & D'Ausilio, 2009). The benefit of innovation is individual here at first but through the process of interactive imitation, it will be revisited as a common and innovative by-product of the interaction.

6   Imitation and Social Neuroscience

Inspired by our developmental research, we have built an innovative fMRI platform which allows synchronizing behavioural and brain recordings during online imitative interaction. Our results replicated previous findings demonstrating the existence of an imitative neural network (Iacoboni et al., 1999), and most importantly revealed the involvement of the dorsolateral prefrontal cortex and other regions involved in social anticipation and adjustment, thus verifying that reciprocal imitation is a prototype of two-person coregulation (Guionnet, Nadel, Bertasi, Delaveau, Sperduti, & Fossati, 2011). Our fMRI work thus supported the notion that imitation is a useful model for two-person neuroscience.

Two-person neuroscience aimed at investigating the simultaneous activity of two brains recorded simultaneously during a dyadic encounter. The novel technique known as 'hyperscanning' allows for simultaneous recording (through fMRI or EEG) of brain activity in multiple participants, facilitating both within- and between-brain analyses. We used hyperscanning in a dyadic context of free interaction and it was the first experiment of this kind, to our own knowledge. The dyads were composed of two unacquainted subjects seated in separate experimental cabins and viewing each other's hand gestures. Dyads engaged in imitation (i.e., made hand gestures of similar morphology) roughly about 65% of the time and synchronized hand movements (i.e., gestures began and ended at the same time, but did not necessarily share the same morphology) about 78% of the time. Within each dyad, we observed a spontaneous emergence of a balanced turn taking between the role of model and imitator; EEG data showed emergent synchronization of brainwaves in subjects who were engaged in spontaneous imitation with interactional synchrony (Dumas, Nadel, Soussignan, Martinerie & Garnero,2010). This inter-brain relationship was strongly present in the alpha-mu frequency band where it symmetrically linked the right parietal regions of the two subjects (Figure 5B). Inter-brain synchronization of right parietal regions in this range of rhythmic activity suggests a link between inter-individual coordination and the intra-individual temporal estimation and anticipation necessary for an effective alternation of roles (Wilson and Wilson, 2005). Interbrain synchronization was also observed in higher frequency bands, though not between homologous brain regions according to the role of imitator or model. (Dumas, Martinerie, Soussignan & Nadel,. 2012).

Besides an understanding of the other's action, turn-taking requires anticipation of other's intention and active co-regulation of complementary action on the part of the two partners. Our PsychoPhysical Interaction results suggest that these sophisticated aspects of an ongoing social interaction involve both the mirror and the mentalizing systems (Sperduti, Guionnet, Delaveau, Fossati & Nadel, 2014). The mirror system allows understand and anticipate action schemes leading to synchronized actions and the mentalizing system accounts for the novelty emerging from the imitative interaction.

7   Conclusion

Cognition involving others, or social cognition, is often conceptualized as the solitary third person computation of mental states. Relatively little attention has been paid to how individuals use their cognitive capacities at the behavioral and brain levels in social exchanges. We introduced imitation as a valuable model of dynamic social interactive phenomenon, and described laboratory procedures for studying it in behavioral and neuroimaging contexts. From birth on, imitation allows us to continuously revisit our resources thanks to the observation of others. Interacting with others multiplies the effect of observation. Indeed it generates novelty emerging from the dynamic coregulation of two different repertories that couple perception and action and anticipate each other's responses.

We reviewed research that reveals behavioural and neural synchronization of individuals engaged in imitation. In the latter case, brain activity is correlated in imitative partners but the pattern expressed by an individual depends on the individual's role (i.e., model or imitator). We linked these findings to theoretical notions about mirroring and mentalizing brain systems, and then described how mirroring and mentalizing support the notion of generative cognition, even in basic forms of communication such as reciprocal imitation. And finally we showed that the traditional view of imitation does not take into account the exceptional potential of generativity that it allows.

References

Andry, P., Gaussier, P., Moga, S., Banquet, J.P., Nadel, J.: Learning and communication in imitation: an autonomous robot perspective. IEEE Transactions on Systems, Man and Cybernetics 31, 431–444 (2001)

Calvo-Merino, B., Glaser, D.E., Grèzes, J., Passingham, R.E., Haggard, P.: Action observation and acquiredmotor skills: an fMRI study with expert dancers. Cerebral Cortex 15, 1243–1249 (2005)

Chomsky, N.: Aspects of the theory of syntax. MIT Press, Cambridge (1965)

Dennett, D.C.: Intentional systems. Journal of Philosophy 68, 87–106 (1971)

De Jaegher, H., Di Paolo, E.: Enactivism is not interactionism. Frontiers in Human Neuroscience 6 (2012), doi:10.3389/fnhum.00128

Di Paolo, E., De Jaegher, H.: The interactive brain hypothesis. Frontiers in Human Neuroscience 6 (2012), doi:10.3389/fnhum.2012.00128

Dumas, G., Nadel, J., Soussignan, R., Martinerie, J., Garnero, L.: Interbrain synchronization durting social interaction. PlosOne 5, e12166 (2010)

Dumas, G., Martinerie, J., Soussignan, R., Nadel, J.: Does the brain know who is at the origin of what in an imitative interaction? Frontiers in Human Neuroscience 6 (2012), doi:10.3389/fnhum.2012.00128

Epstein, R.: Cognition, Creativity, and Behavior: Selected Essays. Praeger, München (1996)

Fadiga, L., Craighero, L., D'Audilio, A.: Broca's area in language, action and music. Ann. N.Y. Acad; Sci. 119, 448–458 (2009)

Fogel, A.: Two principles of communication: Co-regulation and framing. In: Nadel, J., Camaioni, L. (eds.) New Perspectives in Communicative Development, pp. 9–22

Girard, R.: Things hidden since the foundation of the world (1987). Stanford University Press, Stanford (2005) (French edition: 1978)

Guionnet, S., Nadel, J., Bertasi, E., Sperduti, M., Delaveau, P., Fossati, P.: Reciprocal imitation: toward a neural basis of social interaction. Cerebral Cortex 22(4), 971–978 (2012)

Iacoboni, M., Woods, R.P., Brass, M., Bekkering, H., Mazziota, J., Rizzolatti, G.: Cortical mechanisms of human imitation. Science 286, 2526–2528 (1999)

Lepage, J.F., Théoret, H.: The mirror neuron system: grasping others' actions from birth? Developmental Science 10, 513–523 (2007)

Meltzoff, A.N., Moore, M.: Newborn infants imitate adult facial gestures. Child Development 54, 702–709 (1983)

Nadel, J., Butterworth, G.: Imitation in infancy. Cambridge University Press, Cambridge (1999)

Nadel, J., Camaioni, L.: New perspectives in communicative development. Routledge, London (1993)

Piaget, J.: Play, dreams and imitation in childhood. Norton (translated from: La formation du symbole chez l'enfant), New York (1945/1962)

Piontelli, A.: Development of normal fetal movements. Springer, Milano (2010)

Prepin, K., Revel, A.: Human-machine interaction as a model of machine-machine interaction: how to make machines interact as humans do. Advanced Robotics 21, 1709–1723 (2007)

Raos, V., Evangeliou, M.N., Savaki, H.E.: Mental simulation of action in the service of action perception. Journal of Neuroscience 27, 12675–12683 (2007)

Soussignan, R., Courtial, A., Canet, P., Danon-Apter, G., Nadel, J.: Human newborns match tongue protrusion of disembodied human and robotic mouths. Developmental Science (2010), doi:10.1111/j.1467-7687.2010.00984

Sperduti, M., Guionnet, S., Fossati, P., Nadel, J.: Mirror Neuron System and Mentalizing System connect during online social interaction. Cognitive Processing (2014)

Stern, D.: The first relationship: infant and mother. Harvard University Press, Harvard (1977)

Trevarthen, C.: Descriptive analyses of infant communicative behaviour. In: Schaffer, H.R. (ed.) Studies of Infant-Mother Interaction. The Loch Lomond Symposium, pp. 227–270. Academic Press, London (1977)

Varela, F.J., Thompson, E.T., Rosch, E.: The embodied mind: Cognitive science and human experience. MIT Press, Cambridge (1991)

Vernant, J.-P.: Myth and thought among the Greeks. Routledge & Kegan Paul, London (1983)

Wilson, M., Wilson, T.: An oscillator model of the timing of turn-taking. Psychon. Bull. Rev. 12, 957–968 (2005)

Zittrain, J.: The future of the internet and how to stop it. Yale University Press, Yale (2008)

Conditions for Cognitive Plausibility of Computational Models of Category Induction

Daniel Devatman Hromada

Slovenská Technická Univerzita, Fakulta Elektrotechniky a Informatiky
Ilkovičova 3, 812 19 Bratislava 1
hromi@giver.eu

Abstract. We present two axiomatic and three conjectural conditions which a model inducing natural language categories should dispose of, if ever it aims to be considered as "cognitively plausible". 1st axiomatic condition is that the model should involve a bootstrapping component. 2nd axiomatic condition is that it should be data-driven. 1st conjectural condition demands that the model integrates the surface features – related to prosody, phonology and morphology – somewhat more intensively than is the case in existing Markov-inspired models. 2nd conjectural condition demands that asides integrating symbolic and connectionist aspects, the model under question should exploit the global geometric and topologic properties of vector-spaces upon which it operates. At last we shall argue that model should facilitate qualitative evaluation, for example in form of a POS-i oriented Turing Test. In order to support our claims, we shall present a POS-induction model based on trivial k-way clustering of vectors representing suffixal and co-occurrence information present in parts of Multext-East corpus. Even in very initial stages of its development, the model succeeds to outperform some more complex probabilistic POS-induction models for lesser computational cost.

Keywords: categorization, part-of-speech induction, surface features, vector spaces, categorization-oriented Turing Test, clustering of formal syntactic analogies, cognitive plausibility.

1 Introduction

The notion of "cognitive plausibility" and "part-of-speech induction" shall be defined in subsection 1.1. Subsection 1.2 shall clarify the position of syntactic category induction within the field of Natural Language Processing (NLP). The last subsection (1.3) shall offer a brief overview of the history of the problem, arguing that the current paradigm is probabilistic and English-centered one.

1.1 Cognitive Plausibility

This article enumerates some basic conditions which should be fulfilled, we believe, by engineers aiming to transform their computational models into "cognitively

A. Laurent et al. (Eds.): IPMU 2014, Part II, CCIS 443, pp. 93–105, 2014.

plausible" artificial agents. We label as "cognitively plausible" a model which tends to address some basic function of human cognitive system not only by simulating, in a sort of "black-box apparatus", the mapping of inputs (stimuli, corpus data etc.) upon outputs (results), but also tends to faithfully represent the way how the respective function/skill is accomplished by a human mind and its material substrate – the brain.

In other terms, we believe that a cognitively plausible model should not only aim to attain the most quantitatively accurate results, but also to do so by processing the information similarly to the way mind does it.

The aim of this article is to elucidate the notion of "cognitive plausibility" (CP) by relating it to one particular problem, that of construction of grammatical categories present in natural languages. More concretely, we shall try to illustrate our point on the problem of construction of part-of-speech (POS) classes. We precise that the term POS-induction (POS-i) designates the process which endows the human or an artificial agent with the competence to attribute the POS-labels (like "verb", "noun", "adjective") to any token observable in agent's linguistic environment. For the simplicity of the argument, only parts of textual corpora like Multext-East (Erjavec, 2012) shall be considered as such "linguistic environment" of the computational agent introduced below.

1.2   Part-of-Speech Induction in NLP and Language Acquisition Studies

POS-i is often considered to be "one of the most popular tasks in research on unsupervised NLP" (Christodoulopoulos et al., 2010). The problem of construction of grammatical categories is closely related to problem of "grammar induction" and language acquisition. Since "syntactic category information is part of the basic knowledge about language that children must learn before they can acquire more complicated structures" (Schütze, 1993), it is hard to imagine any computational model of grammar induction - aiming to discover the set of rules of the grammar of the language under study- without it being able to construct, in the first place, the equivalence classes upon which the rules-to-discover shall be applied (Elman, 1989; Solan et al., 2005).

Acquisition of formal grammatical categories, be it parts-of-speech or others, is thoroughly studied in psycholinguistic literature – for introductory overview c.f. Levy et al. (1988). For few decades the main motivation in the field was the question **"whether grammatical categories are innate, or induced through interaction with environment by means of imitation and analogy?"**. Whole spectrum of answers were proposed during the deba, generating as a by-product vase amount of both empiric and theoretic knowledge. Such knowledge can be further exploited by engineers aiming to bring together disparate disciplines of artificial intelligence and developmental psychology.

1.3   POS-i Paradigm(s)

While already latent in worthy POS-i models, like that of (Elman, 1989) existed before, or were published more or less in parallel (Schütze, 1993), the paradigm currently dominating the POS-i domain was fully born with article published by Brown et al. in 1992. Without going into detail, we precise that the model was successful because of its ability to apply both Markovian probabilistic concepts and those coming from information theory (Shannon & Weaver, 1949) upon the information contained in the co-occurrences of the words in the sequences, thus becoming the flagship of what we label hereby as "co-occurrence distribution" or "contextual distribution" (CD) paradigm. In decades to follow, the CD paradigm has clearly dominated the POS-i field. Be it hidden Markov Models tweaked with variational Bayes (Johnson, 2007) , Gibbs sampling (Goldwater & Griffiths, 2007), morphological features (Berg-Kirkpatrick, Bouchard-Côté, DeNero, & Klein, 2010; Clark, 2003) or graph-oriented methods (Biemann, 2006) – all such approaches and many others consider contextual co-occurrence to be the primary source of POS-irelevant information.

But as comparative study of (Christodoulopoulos et al., 2010) indicates when demonstrating that models integrating morphological features tend to better than those who do not, it seems plausible that the uncontested primary role of CD in POS should be revised. While it is evident that the CD indeed must furnish relevant information if ever distributional hypothesis is valid (Harris, 1954) and it is axiomatic that distributional hypothesis applies in case of any agent creating categories consistently with Hebb's law (Hebb, 1964) we shall argue in subsection 3.1 that pertinent POS-I clues can be extracted not only from word's "external" contextual properties but also from word's very "internal" Μορφε.

2   Axiomatic Conditions of Cognitive Plausibility

This section deals with what we believe are necessary (i.e. sine qua non) conditions of cognitive plausibility of a computational model. Subsection 2.1 deals with the "bootstrapping" condition stating that categories which are being built are based on categories which have already been built. Emergence of bootstrapping effect shall be illustrated on a trivial multi-iterative re-clustering of clusters pre-clustered according to CD features. Subsection 2.2 discusses the assumption that in order to be cognitively plausible, the model should be data and/or oracle-driven.

2.1   Bootstrapping the Bootstrapping

From biochemistry to social sciences it is a well known fact that structuring structures are the structures structured. Computational Linguistics and NLP in particular is not an exception. The most general definition of the term bootstrapping (B) – i.e. that B is a multi-iterative process whereby outputs of the previous iteration modify the very execution of the next iteration – could be indeed apply upon so many computational "recurrent", "self-feeding" (Riloff & Jones, 1999), "auto-organizing" (Nowak et al., 1999) approaches that have been already applied in so many NLP studies, that to state about a NLP algorithm X that "X bootstraps" may sometimes seem to be plain tautology.

In certain sense almost any POS-i model based on CD paradigm are, ex vi termini, bootstrapping ones because even in the most simplistic models, the information about the membership of the target word WT in the candidate class C is inferred from the probabilities of membership of WL (WT's left context) and WR (WT's right context) to their respective candidate POS classes. Given the fact that the WT plays the role of right context for WL and the role of left context for WR, whole problem is circular and as such often calls for a bootstrapping solution.

Solan et al. (2005) refer to a crucial 4th component of their automatic distillation of structure (ADIOS) algorithm as "generalized bootstrapping". Differently from the "geometric approach" which shall be presented in our experiment below, ADIOS implements graph-like structures in order to attain its aim of construction of equivalence classes useful in subsequent grammar induction. But in its very essence, the approach of Solan et al., i.e. that one should substitute the vertices "subsumed" by a "subsuming" non-terminal class-denoting vertex is analogical, mutatis mutandi, to the approach presented in the following paragraphs.

1st Experiment: Bootstrapping k-way POS Clustering Seeded by Token Co-occurrence Features.

Experiment was performed with data contained in English (en), Czech (cs) and Slovak (sk), corpora contained in 4th version of Multext-East corpus (Erjavec, 2012).

Table 1. Overall statistics of analyzed corpora

Corpus	Word Types	Tokens	Tags$_{POS}$	Feat$_{COOC}$
Cs	19283	100368	13	70426
En	10511	134832	12	36774
Sk	20588	103452	13	74912

Table 1 presents summary statistics concerning the quantities of distinct word tokens, word types (i.e. tokens without context) and the most coarse-grained "gold standard" POS-tags is presented along with total number of distinct co-occurrence features which is equivalent to the number of columns (dimensions) in the resulting co-occurrence matrix.

Every word WT type was characterized by a (row) vector of values [W1L, W2L ...WNL, W1R, W2R ... WNR], W1L referring to cases when the word W1 occurred to the left of WT, W2L to cases when W2L was to the left, W3R to cases when W3 was to the right from the target word. What results is a simple co-occurrence matrix with N rows and maximum of FeatCOOC==2*N columns. Given that in the experiment we were actually looking two words to the left and two words to the right from WT, the maximum possible number of columns was FeatCOOC =4*N. But since not all word couples do occur asides each other, the final number FeatCOOC was always below the theoretical limit.

The matrix has been clustered in C={2 ... 50} clusters by the fast & frugal repeated bisection k-way clustering algorithm as implemented in the clustering tool CLUTO (Karypis, 2002). Columns were scaled according to IDF principle and the clustering was done according to cosine metrics. Once finished, comparison with "gold standard" yielded V-measure (Rosenberg & Hirschberg, 2007) values presented in Table 2 for cluster sizes {10, 20, 30, 40, 50}. which are also illustrated as NO curves on Figure 1.

We have implemented the bootstrapping component in a following manner: After each clustering, the information about the proposed cluster is added as a new feature to target's word vector description. Thus, if matrix with 20 columns entered the first iteration which clustered the vectors into 5 clusters, the matrix entering the second iteration shall have 20+5 columns. If second iteration yields 6 clusters, a matrix with 25+6 columns will become the input for the third iteration etc. Figure 1 shows that in case of all 3 studied corpora, the bootstrapping BO method always attains higher scores than the static NO approach.[1]

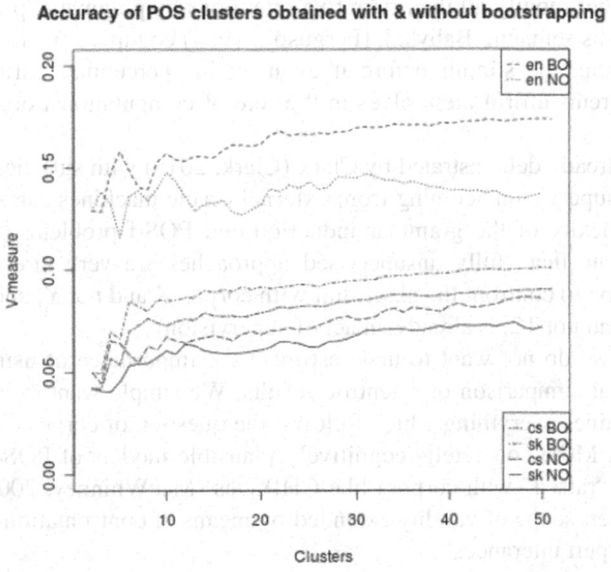

Fig. 1. Bootstrapping of contextual co-occurrence statistics

2.2   Data and Oracle-Driven Learning

Computational models unable to analyze what they have previously synthesized and synthesize what they have previously analyzed could be hardly labeled as

---

[1] Note that the V-measure of NO-bootstrap curves seem to be relatively stable in regards to increase of number of clusters. Contrary to many-to-one accuracy (purity) which increases with number of clusters, V-measure thus seems to be better evaluation measure for cases when solutions containing different numbers of clusters have to be compared.

"cognitively plausible". But even the presence of such "dialectic" component cannot be the guarantee of absolute success, if ever the model's initial prima materia – the data with which the whole bootstrapping is initiated – are not adapted to model's prewired "innate" state.

It is unfortunately often the case in computational linguistics that whenever the model does not attain the expected performance, huge amount of effort is invested into tuning the model by diverse ad hoc modifications. After hours of exhaustive search, both intellectual as well as automatic, diverse parameters, meta-parameters and hyper-parameters are finally discovered which allow the model to attain somewhat superior performances when confronted, for example, with Wall Street Journal (WSJ) corpus But human categorization faculties – POS-i included – do not develop in such a way. While it seems plausible that same sort of "tuning of parameters" indeed takes place during initial period of language acquisition, it seems to be so efficient because the data itself is well adapted to ever-evolving state of baby's neuro-linguistic structures. Said more concretely, parents do not recite to its children the WSJ or Eulex corpora in order to adjust the synaptic weights in the brains of their children, they rather modify all their narrative intentions by pragmatic, prosodic, phonological as well as semantic Babytalk (Ferguson, 1964) cognitive filters. In doing so – by pre-processing the stimuli before it even attains perceptual buffers of child agent's ears – parents affirm themselves in the role of computational oracle (Turing, 1939).

Since it was already demonstrated by Clark (Clark, 2010) with sufficient analytical clarity that the "supervision" coming from external oracle machines can significantly reduce the complexity of the grammar induction and POS-i problems, we found it worthwhile to state that "fully unsupervised approaches are very rare because the engineer's decision to confront the algorithm with corpus X and not Y, and to do so in the moment T1 and not T2, is already an act of supervision".

By saying so we do not want to underestimate the importance of using the same corpora for mutual comparison of scientific results. We simple want to indicate that, because it determines everything which follows, the question of corpus choice should not be neglected. More concretely, cognitively plausible models of POS-i should be firstly tuned and "raised" with corpora like CHILDes (MacWhinney, 2000) and only later should be their scope of validity extended by means of confrontation with corpora of adult and expert utterances.

3   Conjectural Conditions of Model's Cognitive Plausibility

Subsection 3.1 discuss the role of non-distributional "surface" features for POS-induction. Discussion is followed by results of an experiment suggesting that features like suffix can indeed offer quite strong clues for the creation of syntactic categories. Subsection 3.2 introduces a conjectural condition for model's CP by proposing to base it principally on geometric grounds. It is followed by subsection 3.3 arguing that CP model should facilitate evaluation by means of qualitative inspection. In general, these sections deal with CP's conjectural conditions, meaning that while they may seem less self-evident that the axiomatic ones, we nonetheless consider them as valid.

3.1   Integration of Surface Features

Natural languages are very redundant communication channels (de Saussure., 1922; Shannon & Weaver, 1949). Three facets of the word – its morpho-phonological signifiant, its invisible signifiée and its its syntactic function – are not independent from one another and more often than not do they significantly overlap (Jackendoff, 2003; Lakoff, 1990). Thus it is not surprising that especially in morphologically rich languages, token's very syntactic function is encoded by morphemes present in the surface, i.e. objectively perceivable form, of the token itself. And results obtained by Clark (Clark, 2003) or (Berg-Kirkpatrick et al., 2010) indeed point in this direction – it may be no coincidence that approaches which exploit the morphological features turned out, in (Christodoulopoulos et al., 2010) comparative study, to perform better than models which do not.

2nd Experiment: Assessing the Impact of Sufixal Features on Part-of-Speech Categorisation.

We used the same three Multext-East corpora as in the first experiment. Ultimate character trigram was extracted from every word type and considered to be a feature. Word types are subsequently clustered in C clusters according these FeatSUFFIX orthogonal dimensions. The comparison with Mutext-East gold standard subsequently yields V-measures (V), entropies (H) and purities (P) presented in Table 2.

Table 2. Performance of model's inducing C categories solely according to suffixal features

	C=10	C=30	C=50
Cs **534**	V=0.178 H=0.487 P=0.582	V=0.24 H=0.392 P=0.642	**V=0.26** H=0.34 P=0.69
En **286**	**V=0.248** H=0.428 P=0.639	V=0.215 H=0.4 P=0.652	V=0.2 H=0.39 P=0.66
Sk **523**	V=0.17 H=0.5 P=0.504	V=0.272 H=0.373 P=0.685	**V=0.274** H=0.339 P=0.714

Amount below the corpus name in the above table denotes the length of the FeatSUFFIX vector, i.e. the number of distinct suffixal trigrams observed in their respective corpora.

FeatSUFFIX-driven model attains lesser V-measures as had obtained (Christodoulopoulos et al., 2010) when evaluating models of (Clark, 2003) or (Berg-Kirkpatrick et al., 2010) within their 2013 comparative study. The very same study however also indicates that even the simplistic FEATSUFFIX-driven model can be worth of certain interest since it seems to be quite fast – in comparison to models harnessing the power of more than dozen computational cores to attain comparable or even better

V-measures than FEATSUFFIX-driven method , we are glad to state that in order to attain results presented above, our dual-core Pentium needed in average TEN=1.8, TSK=3.2, TCS=3.6 seconds per simulation.

3.2   Knowledge Is Geometric

After the Turing machine symbol-operating paradigm started to put more importance upon ever-still more & more fine-grained modular to probabilistic and connectionist models. But in recent years, a "geometric" paradigm starts to gain momentum in diverse fields of cognitive sciences including computational linguistics and NLP. In experiments described above such paradigm was harnessed in a sense that instead of modulating weights along different dimensions, geometers often modulate the number of dimensions itself. It could be possibly reproached to such a geometric approach that associating every plausible feature with a new dimension can induce some serious matrix-sparsity problems and|or that such an approach would be, sooner or later, confronted with insurmountable computational and memory limits. It is true that methods by means of which some older approaches deal with the problem of huge co-occurrency matrices can be very costly, as is the case, for example, in singular value decomposition within LSA (Landauer & Dumais, 1997). But since very elegant, simple and concise representations of sparse matrices can be very easily generated (Karypis, 2002) and since lemma of Johnson-Lindenstrauss (W. B. Johnson & Lindenstrauss, 1984) indicates that sparse high-dimensional matrices can be easily projected into low-dimensional as is often done in random-indexing (Sahlgren, 2005), it seems to be plausible to state that construction of vector spaces which are 1) dense but 2) transformable for low computational cost 3) encode huge amount of features attributed to huge amount of objects is not so problematic as it used to be in time when HMM-mastered POS-i paradigm was born.

Series of articles by Sahlgren (2002; 2005), Cohen (2010), Widdows (2004) and their colleagues offer valuable initiation into advantages of random-projection based semantic models. For more general discussion of "geometrization of thought" in diverse fields of cognitive sciences, see (Gärdenfors, 2004). Within all such geometric models, categories can be considered as local subspaces of a global space derived from the data.

3.3   Mix of Quantitative and Qualitative Evaluation

Performance of early grammatical category induction models was evaluated manually by introspection into induced equivalence classes and articles published in the period of "golden age" of POS-i often used to enumerate members of at least one particularly pleasing class or presenting their dendograms. Such an approach was later critiqued by Clark (2003) as "inadequate" and attention of POS-I community turned towards more quantitative measures like perplexity, conditional entropy, cross-validation

(Gao & Johnson, 2008), one-to-one (Haghighi & Klein, 2006) or many-to-1 accuracy (purity); variation of information (Meila, 2003) , substituable F-score (Frank et al., 2009) etc.

For the purposes of this article we had decided to present our simulations principally in terns of V-measure. Given its elegance, stability in regards to growing number of clusters but also certain "strictness" (note that even the best performing models present in comparative study (Christodoulopoulos et al., 2010) rarely surpass the V>0.6 limit), we consider the V-measure to be very valuable quantitative measure of performance of clustering POS-i algorithms.

But we also believe that the "old school" many-to-1 purity measure can be of certain interest, especially for those aiming to create a "semi-supervised bridge" between POS-induction and POS-tagging models; or by those aiming not to evaluate the performance of the model by rather to gain insights of correct annotations of analyzed corpora. In other terms, asides to "global" statistic measures informing the researcher about the overall performance of the model, more "local" measures can still offer interesting and useful information about individual induced classes themselves. Values presented in Table 3 represent the number C of clusters into which the corpus has to be partitioned in order to obtain at least Φ absolutely pure (i.e. Purity=1) classes.

Table 3. Distillation of absolutely pure categories

	SFFX	CD	CD+BO	SFFX+CD+BO
$\Phi=1$	72	168	107	69
$\Phi=2$	92	194	142	71
$\Phi=3$	105	196	180	80
$\Phi=4$	126	248	189	90
$\Phi=5$	131	281	194	96
$\Phi=10$	160	377	256	116

For example, in order to obtain an absolutely pure cluster on the basis of contextual distribution (CD) features, one would have to partition the English part of Multext-East corpus into 168 clusters among which shall emerge following noun-only cluster:

authority, character, frontispiece, judgements, levels, listlessness, popularity, sharpness, stead, successors, translucency, virtuosity.

Interesting insights can also be attained by inspection of some exact points of the clustering procedure. Let's inspect, as an example, the case when one clusters the English corpus into 7 clusters according to features both internal to the word – i.e. suffixes – and external – i.e. co-occurrence with other words . Such an inspection indicates that the model somehow succeeds to distinguish verbs from nouns. As is shown on Table 4, whose columns represent the "gold standard" tags and rows denote the artificially induced clusters, our naïve computational model tends to put nouns in clusters 4 and 6 while putting verbs into clusters 2, 3 and 5.

Table 4. Formal origins of Noun-Verb distinction

	N	V	M	D	R	A	S	C	I	P	X	G
0	10	3	0	0	413	30	0	0	0	0	1	0
1	568	67	0	0	1	0	1	2	0	1	0	0
2	97	668	0	0	1	137	3	2	0	0	0	0
3	13	1011	1	0	275	0	2	0	0	0	0	0
4	1173	67	4	0	6	133	0	0	0	4	3	0
5	608	958	72	67	252	321	99	72	7	106	3	12
6	1977	97	22	0	42	1091	3	0	3	0	2	0

The objective of our ongoing work is to align as much as possible such "seeding" states like that presented on Table 4. with data consistent with psycholinguistic knowledge about diverse stages of language acquisition process.

At last but not least, we believe that the temporal aspects of model's performance, i.e. the answer to the question "How long does the model need to run in order to furnish reasonable results?" should be always seriously considered. One way how to evaluate such temporal aspects of categorization could be a simplistic Turing-Test (TT) like POS-i oriented scenario where the evaluator asks the model (or an agent) to attribute the POS-label to word posed by evaluator, or at least to return a set of members of the same category. In such a real-life scenario, an absolute perfection of possible future answer could be possibly traded off for less perfect (yet still locally optimal) answer given in reasonable time.

But because with this TTPOS proposal we already depart from the domain of unsupervised induction towards semi-supervised "learning with oracle" or fully supervised POS-tagger, we conclude that we consider the condition "cognitively plausible model of part of speech induction should be evaluated by both quantitative and qualitative means" to be the weakest among all proposals concerning the development of an agent inducing the categories of natural language in a "cognitively plausible" way.

4   Conclusion

Model should be labeled as "cognitively plausible" model of certain human faculty if and only if it not only accurately emulates the input (problem) $\rightarrow$ output (solution) mapping executed by the faculty, but also emulates the basic "essential" characteristics associated to such mapping operation in case of human cognitive systems, i.e. emulates not only WHAT but also HOW the problem $\rightarrow$ solution mapping is done.

In relation to the problem of how part-of-speech induction is effectuated by human agents, two characteristic conditions have been defined as axiomatic (necessary). First postulates that POS-i should involve a "bootstrapping" multi-iterative process able to subsume terminals sharing common features under a new non-terminal and to subsequently exploit the information related to occurrence of the new non-terminal to extend the (vectorial) definition terminals represented in the memory. Ideally the process should converge to partitions "optimally" corresponding to the gold standard. First experiment has shown for three distinct corpora that even a very simple model based on clustering of the most trivial co-occurrence information can attain higher accuracies if such a bootstrapping component is involved. The second necessary condition of POS-i's CP is that it should be data or oracle-driven. It should perform better when first confronted with simple corpora like CHILDes (MacWhinney, 2000) and only latter with more complex ones than if it would be first confronted with complex corpora.

Another condition of POS-i's CP proposed that morphological and surface features should not be neglected and instead of playing a secondary "performance increasing role", they should possibly "seed" whole bootstrapping process which shall follow. This condition is considered to be conjectural (i.e. "weaker") just because it points to somewhat orthogonal direction than does a traditionally acclaimed distributional hypothesis (Harris, 1954). It may be the case, however, that especially native speakers of some morphologically rich languages shall consider the "syntax-is-also-IN-the-word" paradigm not only as conjectural but also axiomatic.

Another "weak" condition of cognitive plausibility postulates that many phenomena related to mental representations and thinking, POS-i included, can be not only described but also explained and represented in geometric and topologic terms. Ideally, the geometric paradigm (Gärdenfors, 2004) should not be contradictory but rather complementary to symbolic and connectionist paradigms. The last and weakest condition of CP proposed that computational models of part-of-speech induction should be not only easily quantitatively analyzed but should be also transparent for researcher's or supervisor's qualitative analyses. They should facilitate and not complicate posing of all sorts of "Why?" questions and the results should be easily interpretable. A sort of categorization-faculty Turing Test was proposed which could be potentially embedded into the linguistic component of the hierarchy of Turing Tests which we propose elsewhere (Hromada, 2012).

It may be the case that the list of conditions of cognitive plausibility presented in this article is not sufficient one and should be extended with other terms like "modularity", "self-referentiality" or notions coming from complex systems and evolutionary computing. Regarding the problem of elucidation of how could a machine induce, from the environment-representing corpus, the categories in a way analogical to that of a child learning by imitating its parents, we consider even the list of 2 strong precepts and 3 weak precepts hereby presented as quite useful and possibly necessary.

References

1. Berg-Kirkpatrick, T., Bouchard-Côté, A., DeNero, J., Klein, D.: Painless unsupervised learning with features. In: Human LanguageTechnologies: The 2010 Annual Conference of the North American Chapter of the Association for Computational Linguistics, pp. 582–590 (2010)
2. Biemann, C.: Unsupervised part-of-speech tagging employing efficient graph clustering. In: Proceedings of the 21st International Conference on Computational Linguistics and 44th Annual Meeting of the Associationfor Computational Linguistics: Student Research Workshop, pp. 7–12 (2006)

3. Brown, P.F., Desouza, P.V., Mercer, R.L., Pietra, V.J.D., Lai, J.C.: Class-based ngram models of natural language. Computational Linguistics 18(4), 467–479 (1992)
4. Christodoulopoulos, C., Goldwater, S., Steedman, M.: Two Decades of Unsupervised POS induction: How far have we come? In: Proceedings of the 2010 Conference on Empirical Methods in Natural Language Processing, pp. 575–584 (2010)
5. Clark, A.: Combining distributional and morphological information for part of speech induction. In: Proceedings of the Tenth Conference on European Chapter of the Association for Computational Linguistics, vol. 1, pp. 59–66 (2003)
6. Clark, A., de Jong, J.: Towards general algorithms for grammatical inference. In: Hutter, M., Stephan, F., Vovk, V., Zeugmann, T. (eds.) ALT 2010. LNCS, vol. 6331, pp. 11–30. Springer, Heidelberg (2010)
7. Cohen, T., Schvaneveldt, R., Widdows, D.: Reflective Random Indexing and indirect inference: A scalable method for discovery of implicit connections. Journal of Biomedical Informatics 43(2), 240–256 (2010)
8. Elman, J.L.: Representation and structure in connectionist models. DTIC Document (1989)
9. Erjavec, T.: MULTEXT-East: morphosyntactic resources for Central and Eastern European languages. Language Resources and Evaluation 46(1), 131–142 (2012)
10. Ferguson, C.A.: Baby talk in six languages. American Anthropologist 66(6_PART2), 103–114 (1964)
11. Frank, S., Goldwater, S., Keller, F.: Evaluating models of syntactic category acquisition without using a gold standard. In: Proc. 31st Annual Conf. of the Cognitive Science Society, pp. 2576–2581 (2009)
12. Gao, J., Johnson, M.: A comparison of Bayesian estimators for unsupervised Hidden Markov Model POS taggers. In: Proceedings of the Conference on Empirical Methods in Natural Language Processing, pp. 344–352 (2008)
13. Gärdenfors, P.: Conceptual spaces: The geometry of thought. MIT Press (2004)
14. Goldwater, S., Griffiths, T.: A fully Bayesian approach to unsupervised part-of-speech tagging. In: Annual Meeting Association for Computational Linguistics, vol. 45, p. 744 (2007)
15. Haghighi, A., Klein, D.: Prototype-driven learning for sequence models. In: Proceedings of the Main Conference on Human LanguageTechnology Conference of the North American Chapter of the Association of Computational Linguistics, pp. 320–327 (2006)
16. Harris, Z.S.: Distributional structure. Word (1954)
17. Hebb, D.O.: The Organization of Behavior: A Neuropsychlogical Theory. John Wiley & Sons (1964)
18. Hromada, D.D.: Taxonomy of Turing Test Scenarios. In: Proceedings of AISB/IACAP 2012 Symposium, Birmingham, United Kingdom (2012)
19. Jackendoff, R.: Foundations of language: Brain, meaning, grammar, evolution. OxfordUniversity Press, USA (2003)
20. Johnson, M.: Why doesn't EM find good HMM POS-taggers. In: Proceedings of the 2007 Joint Conference on Empirical Methods in Natural Language Processing and Computational Natural Language Learning (EMNLP-CoNLL), pp. 296–305 (2007)
21. Johnson, W.B., Lindenstrauss, J.: Extensions of Lipschitz mappings into a Hilbert space. Contemporary Mathematics 26, 1 (1984)
22. Karypis, G.: CLUTO-a clustering toolkit. DTIC Document (2002)
23. Lakoff, G.: Women, fire, and dangerous things. Univ. of Chicago Press (1990)
24. Landauer, T.K., Dumais, S.T.: A solution to Plato's problem: The latent semantic analysis theory of acquisition, induction, and representation of knowledge. Psychological Review 104(2), 211–240 (1997)

25. Levy, Y., Schlesinger, I.M., Braine, M.D.S.: Categories and Processes in Language Acquisition. Lawrence Erlbaum (1988)
26. MacWhinney, B.: The CHILDES Project: Tools for Analyzing Talk. Transcription, format and programs, vol. 1. Lawrence Erlbaum (2000)
27. Meilă, M.: Comparing clusterings by the variation of information. In: Schölkopf, B., Warmuth, M.K. (eds.) COLT/Kernel 2003. LNCS (LNAI), vol. 2777, pp. 173–187. Springer, Heidelberg (2003)
28. Nowak, M.A., Plotkin, J.B., Krakauer, D.C.: The evolutionary language game. Journal of Theoretical Biology 200(2), 147–162 (1999)
29. Riloff, E., Jones, R.: Learning dictionaries for information extraction by multi-level bootstrapping. In: Proceedings of the National Conference on Artificial Intelligence, pp. 474–479 (1999)
30. Rosenberg, A., Hirschberg, J.: V-measure: A conditional entropy-based external cluster evaluation measure. In: Proceedings of the 2007 Joint Conference on Empirical Methods in Natural Language Processing and Computational Natural Language Learning (EMNLP-CoNLL), vol. 410, p. 420 (2007)
31. Sahlgren, M.: An introduction to random indexing. In: Methods and Applications of Semantic Indexing Workshop at the 7th International Conference on Terminologyand Knowledge Engineering, TKE, vol. 5 (2005)
32. Sahlgren, M., Karlgren, J.: Vector-based semantic analysis using random indexing for cross-lingual query expansion. In: Peters, C., Braschler, M., Gonzalo, J., Kluck, M. (eds.) CLEF 2001. LNCS, vol. 2406, p. 169. Springer, Heidelberg (2002)
33. De Saussure, F., Bally, C., Séchehaye, A., Riedlinger, A., Calvet, L.J., De Mauro, T.: Cours de linguistique générale. Payot, Paris (1922)
34. Schütze, H.: Part-of-speech induction from scratch. In: Proceedings of the 31st Annual Meeting on Association for Computational Linguistics, pp. 251–258 (1993)
35. Shannon, C.E., Weaver, W.: The mathematical theory of information, vol. 97. University of Illinois Press, Urbana (1949)
36. Solan, Z., Horn, D., Ruppin, E., Edelman, S.: Unsupervised learning of natural languages. Proceedings of the National Academy of Sciences 102(33), 11629 (2005)
37. Turing, A.M.: Systems of logic based on ordinals. Proceedings of the London Mathematical Society 2(1), 161–228 (1939), Language and Speech 40(1), 47–62
38. Vlachos, A., Korhonen, A., Ghahramani, Z.: Unsupervised and constrained Dirichlet process mixture models for verb clustering. In: Proceedings of the Workshop on Geometrical Models of Natural Language Semantics, pp. 74–82 (2009)
39. Widdows, D., Kanerva, P.: Geometry and meaning. CSLI Publications Stanford (2004)

3D-Posture Recognition Using Joint Angle Representation

Adnan Al Alwani[1], Youssef Chahir[1], Djamal E. Goumidi[2],
Michèle Molina[2], and François Jouen[3]

[1] GREYC CNRS (UMR 6072)
[2] PALM, EA 4649
Université de Caen, Basse-Normandie
Caen, France
[3] CHArt EA 4004
Ecole Pratique des Hauts Etudes
{firstname.secondname}@unicaen.fr,
gmdjml@yahoo.com,
francois.jouen@ephe.sorbonne.fr

Abstract. This paper presents an approach for action recognition performed by human using the joint angles from skeleton information. Unlike classical approaches that focus on the body silhouette, our approach uses body joint angles estimated directly from time-series skeleton sequences captured by depth sensor. In this context, 3D joint locations of skeletal data are initially processed. Furthermore, the 3D locations computed from the sequences of actions are described as the angles features. In order to generate prototypes of actions poses, joint features are quantized into posture visual words. The temporal transitions of the visual words are encoded as symbols for a Hidden Markov Model (HMM). Each action is trained through the HMM using the visual words symbols, following, all the trained HMM are used for action recognition.

Keywords: 3D-joint locations, Action recognition, Hidden Markov Model, Skeleton angle.

1 Introduction

Action recognition from video is considered as one of the most active research area in the field of computer vision, especially in the field of video analysis, surveillance system, and human-computer interaction. There is a rich literature in action recognition in a wide range of applications, including computer vision, machine learning, and pattern recognition [22,23]. In the past years, efforts have focused on recognizing actions from video sequences with single camera. Among the different approaches, spatial-temporal interest points (STIP) and 2D binary silhouettes are the most popular representations of the human activity and action [8,14,17,18]. In the past decade, several silhouette-based methods for action recognition were mainly categorized into two

A. Laurent et al. (Eds.): IPMU 2014, Part II, CCIS 443, pp. 106–115, 2014.
© Springer International Publishing Switzerland 2014

subsets. One is designed to handle the sequences of silhouettes in order to extract action descriptors. Then conventional classification strategies are frequently used for recognition [1,2,3,4]. The other category models the dynamics of the action explicitly based on the features extracted from each silhouette [5,6,7,16].

However, particular challenges in the human action recognition can alter the performance of actions descriptor from 2D image sequences: intra-class variation, inter-class dependence of action, different contexts of the same action and occlusions are the major challenges in action recognition. The use of several cameras significantly alleviates the challenges such as, occlusion, cluttered background, and viewpoints changes, which are the major low-level difficulties that reduce the recognition performance from traditional 2D imagery. Furthermore, using multiple cameras provided stable information of actions from certain viewpoints. For example, taking account a direction of the camera makes possible to distinguish object pointing from reaching from depth map rather than in RGB space. However, earlier range sensors were either difficult to use on human subjects, or may provide poor measurement. To overcome the limitations of range sensors, depth has to be inferred from stereoscopic using low-cost visible light cameras. Furthermore, 3D body configurations can be captured by multiple cameras in a predefined environment [25].

Skeleton is an articulated system, which consists of limbs segments and the joints between segments. Joints connect rigid segments and articulated motion can be considered as a continuous evolution of local poses configuration [10]. Therefore, for a given sequences of 3D maps, if we can get the stream of the 3D joints location, then reliably action recognition can be achieved by using the tracked joints locations, which significantly improves human action recognition that is under-recognized by traditional techniques.

Recently 3D information has been interpreted using special release of the Microsoft Kinect®, which provides both depth and RGB image streams. Although mainly targeted for commercial purpose, this device has brought considerable interest to the research in computer vision, and hand gesture control.

In this article, we recommend a method for posture-based human action recognition. In the proposed work, 3D locations of joints from skeleton configurations are considered as inputs. Skeletal joints positions are extracted and simple relation between coordinates vector is used to describe the 3D human poses. We perform first the representation of human postures by selecting 7 primitive joints positions. The collection of joint-angle features is quantized through unsupervised clustering into k pose vocabularies. Then encoding temporal joint-angle features into discrete symbols is performed to generate Hidden Markov Model HMM (HMM) for each action. We recognize individual human action using generated HMM. The proposed method is evaluated with public 3D dataset.

The Contribution Parts in This Work Consist of Two Parts: First, we use joint-angle positions to describe posture representation as human action recognition system. Second, our method presents low computational cost since only 7 joints are adopted, and includes representation of poses that is view-invariant.

The organization of this paper is as follows. We introduce the related works in section 2. Section3 describes the method we used to elaborate the architecture of proposed system from postures representation to features extraction. Section 4 addresses action recognition by an HMM. Section 5 explains experimental results. Section 6 concludes the paper.

2   Related Work

Efforts have been reported for the problem of human action recognition, by exploring different kind of visual information. Review on the categories of visual features can be found in [8,22,25]. However, only few attempts on action recognition using depth maps have been recently proposed. Therefore, we present a review of works based on 3D poses action recognition since they are related to our work.

The recent trends in the field of action recognition that use depth maps have induced further progress. Uddin et al. [13] reported a novel method of action recognition using body joint angles estimated from a pair of stereo images from stereo cameras. Thang et al. [21] developed a method for estimating body joint angles from time-series of paired stereo images recreded with a single stereo camera. Yu and Aggarwal [11] adopted an approach for action recognition where body parts are considered as a semantic representation of postures. Weinland et al. [26] proposed a model action involving 3D sequences of prototypes, which are represented as visual structures captured by a system of 5 cameras. The work proposed by Li et al.[9] suggested to map the dynamic actions as a graph, and sample a set of 3D points from the depth maps to describe a set of salient postures, that correspond to the nodes in the graph. However, the challenge in the sampling technique is view dependent. Xai et al. [12] presented a method of action recognition based on 3D skeleton joints location. They proposed a compact representation of postures by characterizing human poses as histogram of 3D joints locations sampled inside spherical coordinates system.

3   Body Parts Representation

In this section we describe the human poses representation and joints position estimation from skeleton model. This kind of representation involves 3D joints coordinates to describe a basic body structure reduced to 20 skeletal joints. Recent release of Kinect® system offers better solution for the estimation of the 3D joint positions. Figure 1 demonstrates the result of the application of depth map and the 3D skeletal joints extraction according to algorithm of Shotton et al [24] who proposed to extract 3D body joint locations from a depth map.

This algorithm is used to estimate pose locations of skeletal joints. Starting with a set of 20 joints coordinates in a 3D space, we compute a set of features to form the representation of postures. Among the 20 joints, 7 primitive joints coordinates are selected to describe geometrical relations between body parts. The category of primitive joints offers redundancy reduction to the resulting representation. Most importantly, primitive joints achieve view invariance to the resulting pose representation, by aligning the cartesian coordinates with the reference direction of the person. Moreover, we propose an efficient and view-invariant representation of postures using 7 skeletal joints, including L/R hand, L/R feet, L/R hip, and hip center.

Fig. 1. (a) Depth map image. (b) Skeletal joints positions proposed by [15].

The hip center is considered as the center of coordinate system, and the horizontal direction is defined according to left hip and right hip junction. The remaining 4 skeletal locations are used for poses joint angles descriptor.

3.1   Action Coordinates Description for Skeletal Joints

View invariance is a challenging problem in action recognition. With the use of 3D body skeleton, we can capture the 3D positions of the human body. We propose a viewpoint-invariant representation of body poses by using 3D joint angles from skeletal data. In our approach of poses features inference, we achieve the view-invariant by aligning the Kinect® cartesian system with the direction of human body as shown in the Fig 2. We consider the hip center joint as the center of the new orthogonal coordinates. We define the horizontal offset vector γ to represent the vector from left to right of the hip center, the reference vertical vector ρ as the vector that is perpendicular to the horizontal reference vector computed by rotating the vector γ by 90°. The depth reference vector β is obtained by cross product operation between γ and ρ. The next steps demonstrate the procedure of aligning the orthogonal coordinates with the specific reference direction of the body.

Let the system Landmark be defined as Rs (O,i,j,k), and the actions landmark as $Ra(\acute{O}, \gamma, \rho, \beta)$. If we define the hip center as origin of the action coordinates, then the action horizontal direction γ is written as:

$$\vec{\gamma} = \begin{pmatrix} hipcenter_x & + & \lambda_x \\ hipcenter_y & + & \lambda_y \\ hipcenter_z & + & \lambda_z \end{pmatrix} = \begin{pmatrix} \gamma_x \\ \gamma_y \\ \gamma_z \end{pmatrix}. \tag{1}$$

Where $\lambda = \dfrac{\vec{u}}{|u|}$, is the normal unit vector, and u is defined as:

$$\vec{u} = \begin{pmatrix} lh_x & - & rh_x \\ lh_y & - & rh_y \\ lh_z & - & rh_z \end{pmatrix} = \begin{pmatrix} u_x \\ u_y \\ u_z \end{pmatrix}. \tag{2}$$

where l_h, r_h are left hip and right hip.

By performing some vector manipulations, the reference vector ρ is defined as the vertical vector that is perpendicular to the horizontal plane, and the vector β is calculated from the cross product operation between γ and ρ vectors.

For the point in the 3D coordinate system M(x,y,z), the unit vector translation from OM to ÓM is defined as:

$$\overrightarrow{O'M} = \overrightarrow{O'O} + \overrightarrow{OM}$$

$$= \overrightarrow{OM} + \overrightarrow{O'O}$$

$$= M_x\vec{i} + M_y\vec{j} + M_z\vec{k} - hc_x\vec{i} - hc_y\vec{j} - hc_z\vec{k}. \tag{3}$$

where h_c, is the hip center, and i,j,k are the unit direction vectors of coordinates system.

In order to express the ÓM as a function of skeletal landmarks, we first specify the system unit vectors i,j,k in terms of the action system coordinates as:

$$\vec{i} = i_\gamma\vec{\gamma} + \; j_\rho\vec{\rho} + \; k_\beta\vec{\beta}$$

$$\vec{j} = i_\gamma\vec{\gamma} + j_\rho\vec{\rho} + k_\beta\vec{\beta}$$

$$\vec{k} = i_\gamma\vec{\gamma} + j_\rho\vec{\rho} + k_\beta\vec{\beta}. \tag{4}$$

Substituting eq. 4 into eq. 3, we get the final formula for the vector ÓM as:

$$\overrightarrow{OM} = M_\gamma\vec{\gamma} + M_\rho\vec{\rho} + M_\beta\vec{\beta}. \tag{5}$$

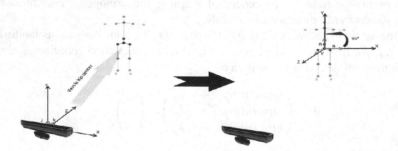

Fig. 2. Coordinates system description of skeletal joints

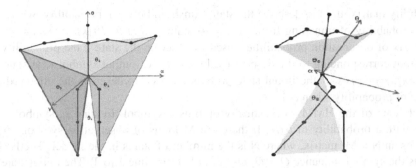

Fig. 3. Joints angles features.(a) XY plane. (b) ZY plane.

3.2   Features Description

In our approach human poses are distinguished by the idea of angles groups estimated from the four junctions, which are mentioned above. The joints angles groups are sampled from two planes, XY plane and ZY plane. In the XY plane, four angles represent the angles between hand left-foot left, hand left-hand right, hand right-foot right, and foot right-foot left respectively. Same joints angles are also defined from the plane ZY. The final features vector includes eight joints angles $Ft=\{\theta_1,\theta_2,......\theta_8\}$ at each pose instant t.   Fig. 3 shows the joint angles of two planes, where each angle is defined according to the corresponding four junctions which were illustrated in section 3.

4   HMM for Action Recognition

To apply HMMs to the problem of human action recognition, video frames $V=\{I_1,I_2,...I_T\}$ are transformed into symbols sequences O. The transformation is done during the learning and recognition phases. From each video frames, a feature vector $f_i \in R,\{ (i=1,2,.....T)$,T is the number of the frames} is extracted, and f_i is assigned to a symbol v_j chosen from the set of symbols V. In order to specify observation symbols, we perform the clustering of features vector into k clusters using K-means. Then each posture is represented as a single number of a visual word. In this way, each action is a time series of visual words. The obtained symbol sequences are used to train HMMs to learn the correct model for each action. For the recognition of a test action, the obtained observation symbol sequence $O = \{O_1,O_2,....O_N\}$ is used to   determine across all trained HMMs which is the most accurate for the tested human action.

HMMs, which have been recently applied with particular success to speech recognition, are a kind of stochastic state transit model [20]. HMMs use observation sequence to determine the hidden states. We suppose $O = \{O_1,O_2,....O_N\}$ as the observation of the stochastic sequence. HMM with N state is specified by three groups of parameters: $\beta=\{A,B,\pi\}$,where $A=\{a_{ij},=pr(q_t=s_j/q_{t-1}=s_i)\}$ denotes the state transition probability matrix, used to describe the state transition between probability, where, a_{ij} is the probability of transiting from state q_i to state q_j, and $B=\{b_j(k)=pr(v_k/q_t=s_j)\}$, is the matrix of observation probabilities, used to describe the state j, the probability of the output corresponding to the observed values $b_j(k)$ of output symbol v_k at state q_j, and $\pi=\pi\{\pi_i=pr(q_1=s_i)\}$ the initial state probability used to describe the observed sequence of probability when t=1.

Each state of the HMM stochastically outputs a symbol. In state s_i, symbol v_k is output with a probability of $b_i(k)$. If there are M kinds of observation symbols, $b_j(k)$ becomes an N x M matrix, where N is the number of states in the model. The HMM outputs the symbol sequence $O = O_1, O_2, \ldots, O_Y$ from time 1 to T. The initial state of the HMM is also stochastically determined by the initial state probability π.

To recognize the observed symbol sequences, we create a single HMM for each action. For a classifier of C actions, we choose the model which best matches the observations from C HMMs $(\beta i=\{Ai, Bj, \pi i\})$, i = 1 . . . C. This means that when a sequence of unknown category is given, we calculate $Pr(\beta_i /O)$ for each HMM βi, and select βc˜. For instance, given the observation sequence $O = O_1, \ldots O_T$ and the HMM β_i, according to the Bayes's rule, the problem is how to evaluate $Pr(O|\beta_i)$, the probability that the sequence was generated by HMM β_i, which can be solved using the forward algorithm. Then we classify the action as the one that presents the largest posterior probability

$$c˜=\text{argmax}_i(Pr(\beta_i|O)). \tag{6}$$

where i indicates the likelihood of test sequence for the ith HMM.

5   Experiments

We evaluate the performance of our algorithm with the public G 3D dataset collected by Bloom et al.[19]. In addition, we evaluated the algorithm with the MSR Action 3D dataset collected by Li et al.[9] and we compared our results with results reported in [9].

Table 1. The subsets of actions used with the MSR Action 3D dataset

Action Set 1 (AS1)	Action Set2 (AS2)	Action set3 (AS3)
Horizontal arm wave	High arm wave	High throw
Hammer	Hand catch	Forward kick
Forward punch	Draw x	Side kick
High throw	Draw tick	Jogging
Hand clap	Draw circle	Tennis swing
Bend	Two hand wave	Tennis serve
Tennis serve	Forward kick	Golf swing
Pickup & throw	Side boxing	Pickup & throw

Table 2. Recognition rates of our method on the G3D action dataset. Results are compared with Bloom et al. [19].

Action Category	Bloom et al.	Ours method
Fighting	70.46%	79.84%
Golf	83.37%	100%
Tennis	56.44%	78.66%
FPS	53.57%	54.10%
Driving a car	84.24	81.34%
Misc.	78.21%	89.40%
Overall	71.04%	80.55%

Table 3. Recognition rates of our method on the MSR Action 3D dataset. Results are compared result with Li et al. [9].

Action subset	Li et al.	Ours method
AS1	72.9%	86.30%
AS2	71.9%	65.40%
AS3	79.2%	77.70%
Overall	74.7%	76.46%

5.1   Experimental Results

The results of our approach with the G3D dataset collected by Bloom et al.[19], containing 22 types of human actions are summarized in table 2. Each action was performed by 10 individuals for 3 times. Note that we only used the information from the skeleton for action recognition in our algorithm. The experiment was repeated 20 times, and the averaged performance is reported in Table 2. The set of clusters was fixed to K=80, and the number of states to N=6. Half of the subjects were used for training and the rest of the subjects were used for testing. Across experiments, the overall mean accuracy is 80.55% demonstrating that our method performs better recognition than Bloom et al [19].

We also tested our algorithm on the public MSR Action3D database that contains 20 actions. We divided the actions into three subsets (similar to [9]), each comprising 8 actions (see table 1). We used the same parameter settings as previously described. In this test, half of the subjects were used for training and the rest of the subjects were used for testing. Each test was repeated 20 times, and the averaged performance is given in Table3. We compared our performance with Li et al[9]: our algorithm achieves considerably better recognition rates than Li et al.

6   Conclusion

This paper presents a framework to recognize human action from sequences of skeleton data. We use 3D joints positions inferred from skeleton data as input. We propose a method for postures representation that involves joint angles in xy and zy planes within a modified action coordinates system as description of postures. In order to classify action types, we model sequential postures with HMMs. Experimental results illustrate the performance of the proposed method, and also refer to a promising approach to perform recognition tasks using 3D points.

References

1. Bobick, A., Davis, J.: The recognition of human movement using temporal templates. IEEE Trans. PAMI 23(3), 257–267 (2001)
2. Meng, H., Pears, N., Bailey, C.: A human action recognition system for embedded computer vision application. In: Proc. CVPR (2007)
3. Davis, J.W., Tyagi, A.: Minimal-latency human action recognition using reliable-inference. Image and Vision Computing 24(5), 455–473 (2006)
4. Chen, D.-Y., Liao, H.-Y.M., Shih, S.-W.: Human action recognition using 2-D spatio-temporal templates. In: Proc. ICME, pp. 667–670 (2007)
5. Kellokumpu, V., Pietikainen, M., Heikkila, J.: Human activity recognition using sequences of postures. In: Proc. IAPR Conf. Machine Vision Applications, pp. 570–573 (2005)
6. Sminchisescu, C., Kanaujia, A., Li, Z., Metaxas, D.: Conditional models for contextual human motion recognition. In: Proc. ICCV, vol. 2, pp. 808–815 (2005)
7. Zhang, J., Gong, S.: Action categorization with modified hidden conditional random field. Pattern Recognitoin 43, 197–203 (2010)
8. Li, W., Zhang, Z., Liu, Z.: Expandable data-driven graphical modeling of human actions based on salient postures. IEEE Transactions on Circuits and Systems for Video Technology 18(11), 1499–1510 (2008)
9. Li, W., Zhang, Z., Liu, Z.: Action recognition based on a bag of 3D points. In: CVPRW (2010)
10. Zatisiorsky, V.M.: Kinematics of Human motion. Human Kinetics Publisher
11. Yu, E., Aggarawal, J.K.: Human action recognition with extremities as semantic posture representation. In: Proc. CVPR (2009)
12. Xai, L., Chen, C.C., Aggarwal, J.K.: View invariant human action recognition using histogram of 3D Joints. In: 2nd International Workshop on Human Action Understanding from 3D Data in Conjunction with IEEE CVPR, pp. 20–27 (2012)
13. Uddin, M.Z., Thang, N.D., Kim, J.T., Kim, T.S.: Human Activity Recognition Using Body Joint-Angle Features and Hidden Markov Model. ETRI Journal 33(4), 569–579 (2011)
14. Yamato, J., Ohya, J., Ishii, K.: Recognition Human action in time-sequential images using hidden markov model. In: IEEE Int. Conf. Computer Vision Pattern Recognition, pp. 379–385 (1992)
15. http://dipresec.king.ac.uk/G3D
16. Niu, F., Abdel-Mottaleb, M.: View-invariant human activity recognition based on shape and motion features. In: IEEE 6th Int. Symp. Multimedia Software Eng., pp. 546–556 (2004)
17. Uddin, M.Z., et al.: Human activity recognition using independent components features from depth images. In: 5th Int. Conf. Ubiquitous Healthcare, pp. 181–183 (2008)
18. Uddin, M.Z., Lee, J., Kim, T.-S.: independent shape component-based human activity recognition via hidden markov model. Appl. Intellig. 33(2), 193–206 (2009)

19. Bloom, V., Makris, D., Argyriou, V.: G3D: a Gaming action dataset and real time action recognition evaluation framework. In: 3rd IEEE Inter. Workshop on Computer Vision for Computer Games, CVCG (2012)
20. Rabiner, L.R.: A Tutorial on Hidden Markov Models and Selected Applications in Speech Recognition. Proc. of the IEEE 77(2), 257–285 (1989)
21. Thang, N.D., et al.: Estimation of 3D Human Body Posture via Co-registration of 3D Human Model and sequential stereo information. Applied Intell (2010), doi:10.1007/s10489-009-0209-4
22. Turaga, P., Chellapa, R., Subrahmanian, V.S., Udrea, O.: Machine recognition of human activities: A survey. IEEETransactions on Circuits and Systems for Video Technology 18(11), 1473–1488 (2008)
23. Aggarwal, J.K., Ryoo, M.S.: Human activity analysis: A review. ACM Computing Surveys (2011)
24. Shotton, J., Fitzgibbon, A., Cook, M., Sharp, T., Finocchio, M., Moore, R., Kipman, A., Blake, A.: Real-Time Human Pose Recognition in Parts from a Single Depth Image. In: CVPR IEEE (June 2011)
25. Moeslund, T.B., Granum, E.: A survey of computervision-based human motion capture. Computer Vision and Image Understanding 81, 231–268 (2001)
26. Weinland, D., Boyer, E., Ronfard, R.: Action recognition from arbitrary views using 3D exemplars. In: ICCV 2007 (2007)

Gesture Trajectories Modeling Using Quasipseudometrics and Pre-topology for Its Evaluation

Marc Bui[1], Soufian Ben Amor[2], Michel Lamure[3], and Cynthia Basileu[3]

[1] CHaRt-EA4004, Université paris 8 & EPHE
marc.bui@univ-paris8.fr
[2] Laboratoire PRISM, Université de Versailles-Saint-Quentin-en-Yvelines, 45,
Versailles, France
soufian.ben-amor@uvsq.fr
[3] Laboratoire SIS EA 4128, Université Lyon 1-UFR d'odontologie-11 rue Guillaume
Paradin 69372
lamure@universite-lyon1.fr

Abstract. The main question addressed in this work deals with the difficulty to compare different data trajectories patterns in particular due to the non symmetry properties. In order to tackle this fundamental issue in its generality from a theoretical point of view, we introduce the quasipseudo-metrics concepts, and with the induced pre-topological space on the datasets we can identify proximity between trajectories. We will illustrate the ideas by discussing the application of the theoretical framework on gestual analysis.

1 Introduction

Complex systems present characteristic features linked to their capability to be constituted from strongly interconnected heterogeneous components, to be characterized by complex hierarchical network structure and by to be highly open to their environment. This is the case with human centric-sensing systems such as location-based services, ubiquitous computing facilities or smart wearable devices. In order to understand, predict and manage such systems we need to define an adequate mathematical framework to model their structural dynamics. In fact, the main characteristics of complex systems are not totally handled by graph theory. It is the case when a structure appears, based on a family of intricate relationships between individuals. Complex systems understanding is mainly based on the analysis of the associated complex data. Given the importance and specificity of the data associated with complex systems both in qualitative and quantitative terms, we looked into the adapted techniques, namely quasipseudometrics, in analyzing and classification of complex data in order to extract valuable information concerning the systems behavior.

Knowledge extraction from massive amount of tracking data need news mathematical theoretical models and tool. Clustering techniques are based, as a rule,

A. Laurent et al. (Eds.): IPMU 2014, Part II, CCIS 443, pp. 116–134, 2014.

on the concept of distance that supposes a precise axiomatic. This axiomatic sometimes proves to be very coercive, in particular in the applications in social sciences. In various cases, a problem that arises is the property of symmetry which expresses that the distance from x to y is equal to the distance from y to x. The question is therefore to know what it happens when this hypothesis is relaxed.

The concept of similarity between objects is very well modeled when the working data can be immersed in a metric space where the notion of distance can precisely quantify the similarity between two objects. It is quite different when the data, by their nature, can not be immersed in such a metric space without being denatured. The purpose of this paper is to address this question of measuring the similarity between objects in the latter case. For this, we propose a model based on a generalization of the topology, called pretopology ([5], [6], [7], [4],[3],[2],[1]), associated with the definition of quasipseudometric , extension of the concept of distance or metric.

The paper is divided into five sections. The first is a reminder of the basics on pretopological spaces, the second presents the concept of quasipseudometric, the third explores the link with a specific type of pretopological space, the fourth sets an example and the fifth concludes with future prospects .

2 Pretopological Spaces

In this section, we introduce the pretopology theory with its basic definitions. We define *pretopology* by mean of its two fundamental set functions: the *pseudoclosure* and *interior* maps instead of introducing a definition based on the family of open sets (or closed sets) as in *topology*. There is a theoretical reason for this: in pretopological spaces, family of open subsets or closed subsets do not characterize the pretopological structure. According to properties fulfilled by pseudoclosure and interior maps, we get different pretopological spaces, which will be detailed in a subsequent section. We will then be able to present the two different concepts of pretopological subset and pretopological subspace of a pretopological space.

2.1 Definitions

In this subsection, we define the *pseudoclosure* and *interior* maps and the pretopological space this defined.

From now, we consider a non empty set E and we define two functions from $\mathcal{P}(E)$ into $\mathcal{P}(E)$.

Definition 1. *We call pseudoclosure defined on E, any function $a(.)$ from $\mathcal{P}(E)$ into $\mathcal{P}(E)$ such as:*

- $a(\emptyset) = \emptyset$
- $\forall A \subset E, A \subset a(A)$

We can note that pseudoclosure fulfills two of properties of a topological closure.

Definition 2. *We call interior defined on a set E, any function $i(.)$ from $\mathcal{P}(E)$ into $\mathcal{P}(E)$ such as:*

- $i(E) = E$
- $\forall A, A \subset E, i(A) \subset A$

Very often, $i(.)$ is defined by c-duality. If we denote A^c the set $E - A$, then, given a pseudoclosure $a(.)$, we can define $i(.)$ by putting:
$\forall A, A \subset E, i(A) = (a(A^c))^c$. And conversely $\forall A, A \subset E, a(A) = (i(A^c))^c$.

Definition 3. *Given, on E, $a(.)$ and $i(.)$, the couple $s = ((a(.), i(.))$ is called pretopological structure on E an the 3-uple $(E, a(.), i(.))$ is called a pretopological space.*

Thus, $a(.)$ and $i(.)$ are two means to transform subsets of E in such a way that we have $\forall A \subset E, i(A) \subset A \subset a(A)$. As in topology, we can define define closed and open subsets of a pretopological space $(E, a(.), i(.))$.

Definition 4. *Given a pretopological space $(E, a(.), i(.))$, any subset A of E is said to be a closed subset of E if and only if $A = a(A)$.*

Definition 5. *Given a pretopological space $(E, a(.), i(.))$, any subset A of E is said to be an open subset of E if and only if $A = i(A)$.*

If a subset A of a pretopological space $(E, a(.), i(.))$ is both a closed and an open subset, we call it is an *oc* of the pretopological space.

In the same way as in topology, we obviously obtain the following result.

Proposition 1. *Given a pretopological space $(E, a(.), i(.))$ where $a(.)$ and $i(.)$ are defined by c-duality, then, for any $A, A \subset E$, we have:*
A closed subset of $E \Leftrightarrow A^c$ open subset of E.

Then, it is possible to generalize the two concepts of closure and opening of any subset of a pretopological space.

Definition 6. *Given a pretopological space $(E, a(.), i(.))$, we call closure of any subset A of E, when exists, the smallest closed subset of $(E, a(.), i(.))$ which contains A.*
It is also the intersect of all closed subsets of $(E, a(.), i(.))$ which contains A. The closure of A is denoted by $F(A)$.

And, in a same way:

Definition 7. *Given a pretopological space $(E, a(.), i(.))$, we call opening of any subset A of E, when exists, the biggest open subset of $(E, a(.), i(.))$ which is included in A.*
It is also the union of all open subsets of $(E, a(.), i(.))$ which are included A. The opening of A is denoted by $O(A)$.

Here, we can note a fundamental difference between topology and pretopology: as in topology, the family of closed subsets characterizes the topological structure, we do not get the same result in pretopology. Two different pseudoclosure $a(.)$ and $a'(.)$ can give the same family of closed subsets on a set E.

2.2   Different Types of Pretopological Spaces

In this subsection, we give the various types of pretopological before restricting to the most common one, and then, focus on how to elaborate the proximity concept tools with the fundamentals set operations available.

Given a pretopological space $(E, a(.), i(.))$, we can define the following family of subsets of E.

$$\forall x \in E, \mathcal{U}(x) = \{B \subset E | x \in i(B)\}$$

Then:

(i) $\forall B \in \mathcal{U}(x), x \in B$

(ii) $\forall B \in \mathcal{U}(x), \forall B' \in \mathcal{U}(x), B \cap B' \neq \emptyset$

Conversely, let us suppose we consider for any x in E, a family $\mathcal{U}(x)$ which satisfies (i) and (ii), and let us define $i(.)$ from $\mathcal{P}(E)$ into itself such as:

$$\forall A \subset E, i(A) = \{y \in E | A \in \mathcal{U}(y)\}$$

then, obviously $i(\emptyset) = \emptyset$ and $\forall A \subset E, i(A) \subset A$.

Thus, $i(.)$ fulfills the two properties of an interior. By duality, we can define the related pseudoclosure $a(.)$:

$$\forall A \subset E, a(A) = (\{y \in E | A^c \in \mathcal{U}(y)\})^c$$

So:

Definition 8. *Given a pretopological space* $(E, a(.), i(.))$, *the family defined by* $\forall x \in E, \mathcal{U}(x) = \{B \subset E | x \in i(B)\}$ *is called the family of neighborhoods of* x.

Thus, as in topology, we get a concept of neighborhoods. However, up to now, pretopological neighborhoods do not verify the same properties than topological neighborhoods. For example, it is very easy to see that if U is a neighborhood of a given x in E and if U is included in a subset V of E, that does not mean that V is a neighborhood of x. So, we were led to define different types of pretopological spaces which are less general than the basic ones but for which, good properties are fulfilled by neighborhoods. In this section, we propose the different types of pretopological spaces which have been defined:

1. $\mathcal{V}$-type spaces: $a(.)$ also verifies $A \subset B \Rightarrow a(A) \subset a(B)$
2. $\mathcal{V}_D$-type spaces: $a(.)$ also verifies $a(A \cup B) = a(A) \cup a(B)$
3. $\mathcal{V}_S$-type spaces: $a(.)$ also verifies $a(A) = \bigcup_{x \in A} a(\{x\})$
4. and at last Topological spaces: $a(.)$ also verifies $a(A \cup B) = a(A) \cup a(B)$ and $a(a(A))$

Among these different types of pretopological spaces, we further only consider the first one for for the interest it has.

2.3   A Useful Type: The $\mathcal{V}$-type Pretopological Space

As previously suggested, a $\mathcal{V}$-type pretopological space is defined as follows:

Definition 9. *Let a pretopological space* $(E, a(.), i(.))$, *we say that it is a* $\mathcal{V}$ *type space if and only if:*

$$\forall A \subset E, \forall B \subset E, (A \subset B \Rightarrow a(A) \subset a(B))$$

Equivalently, we can put:

Definition 10. *Let a pretopological space* $(E, a(.), i(.))$, *we say that it is a* $\mathcal{V}$ *type space if and only if:*

$$\forall A \subset E, \forall B \subset E, (A \subset B \Rightarrow i(A) \subset i(B))$$

In this case, we can use the concept of neighborhood to characterize a $\mathcal{V}$-type space $(E, a(.), i(.))$.
For that, given such a space, let us consider the following family:
$\forall x \in E, \mathcal{V}(x) = \{V \subset E | x \in i(V)\}$ which is the family of neighborhoods of x as already defined. Then:

Proposition 2. *Given a* $\mathcal{V}$ *type space* $(E, a(.), i(.))$, *the family* $\mathcal{V}(x)$ *is a prefilter of subsets of* E.

Proof. Given a family $\mathcal{B}$ of subsets of a set E, we say that $\mathcal{B}$ is a prefilter of subsets of E if and only if:
(i) $\emptyset \notin \mathcal{B}$
(ii) $\forall A \in \mathcal{B}, (A \subset B \Rightarrow B \in \mathcal{B})$
Let $x \in E$,
$V \in \mathcal{V}(x) \Rightarrow x \in V$. So it is impossible that $\emptyset$ belongs to $\mathcal{V}(x)$.
Let $V \in \mathcal{V}(x)$ and $V \subset W$.
$V \in \mathcal{V}(x) \Rightarrow x \in i(V)$. But $(E, a(.), i(.))$ is a $\mathcal{V}$ type space. So $V \subset W \Rightarrow i(V) \subset i(W)$ then $x \in i(W)$ which implies that $W \in \mathcal{V}(x)$ by definition of $\mathcal{V}(x)$.

By duality, for any $x \in E$, we can consider the family $\mathcal{A}(x)$ defined by $\mathcal{A}(x) = \{A \subset E | x \notin a(A)\}$. It is easy to prove that $\mathcal{A}(x)$ is a preideal of subsets of E, i.e.
(i) $E \notin \mathcal{A}(x)$
(ii) $\forall A \in \mathcal{A}(x), (B \subset A \Rightarrow B \in \mathcal{A}(x))$
$\mathcal{A}(x)$ is said the family of points of E which are "far away" from x.

Proposition 3. *Let* $\mathcal{V}(x)$ *a prefilter of subsets of* E *for any* x *in* E.
Let $i(.)$ *and* $a(.)$ *the functions from* $\mathcal{P}(E)$ *into itself defined as:*
(i) $\forall A \subset E, i(A) = \{x \in E | \exists V \in \mathcal{V}(x), V \subset A\}$
(ii) $\forall A \subset E, a(A) = \{x \in E | \forall V \in \mathcal{V}(x), V \cap A \neq \emptyset\}$
Then $(E, a(.), i(.))$ *is a* $\mathcal{V}$ *type space. We say it is generated by the family* $\{\mathcal{V}(x), x \in E\}$

Proof. First, we can note that functions $a(.)$ et $i(.)$ are c-dual.

$$x \in (a(A^c))^c \Leftrightarrow x \notin a(A^c)$$
$$\Leftrightarrow \neg(\forall V \in \mathcal{V}(x), V \cap A^c \neq \emptyset)$$
$$\Leftrightarrow \exists V, V \in \mathcal{V}, \neg(V \cap A^c \neq \emptyset)$$
$$\Leftrightarrow \exists V, V \in \mathcal{V}, V \subset A$$
$$\Leftrightarrow x \in i(A)$$

$a(\emptyset) = \{x \in E | \forall V \in \mathcal{V}(x), V \cap \emptyset \neq \emptyset\} = \emptyset$
Obviously, $\forall A \subset E, A \subset a(A)$ and let A and B such as $A \subset B$, then:
$\forall V \in \mathcal{V}(x), V \cap A \neq \emptyset \Rightarrow V \cap B \neq \emptyset$
so $a(A) \subset a(B)$

At this point, given a $\mathcal{V}$-type space $(E, a(.), i(.))$, we are able to determine the family of neighborhoods $\mathcal{V}(x)$ for any $x \in E$. And if for any $x \in E$, we have a *prefilter* of subsets of E, we are able to determine a pseudoclosure $a(.)$ (and so an interior function $i(.)$) such as we get a $\mathcal{V}$-type space $(E, a(.), i(.))$. The problem then is to answer the following question: given the initial $\mathcal{V}$-type space $(E, a(.), i(.))$, from the family $\mathcal{V}(x), \forall x, x \in E$, we define a new pseudoclosure function and a new interior function, are they the same as the functions of the initial space?

The following proposition gives this answer.

Proposition 4. *The $\mathcal{V}$-type pretopological space $(E, a(.), i(.))$ is generated by an unique family of prefilters $\mathcal{V}(x)$ and conversely any family of prefilters $\mathcal{V}(x)$ generates an unique pretopological structure on E.*

Proof. In a first step, let us prove the family $\mathcal{V}(x), \forall x \in E$ is the only one family of prefilters which generates $(E, a(.), i(.))$. For that, we suppose there exists another family of prefilters $\mathcal{W}(x)$ which contains x and which generates $(E, a(.), i(.))$. Let us suppose we can find an element V such as $V \in \mathcal{V}(x) - \mathcal{W}(x)$. This implies that $x \in i(V)$ as V is an element of $\mathcal{V}(x)$ and that $x \notin i(V)$ as V is not an element of $\mathcal{W}(x)$, which leads to a contradiction. The same conclusion holds if $V \in \mathcal{W}(x) - \mathcal{V}(x)$.

Now, second step, let us suppose the family $\mathcal{V}(x)$ generates another space $(E, a^*(.), i^*(.))$.

In this case, $i(.) \neq i^*(.)$, so $\exists A \subset E$ such as $i(A) \neq i^*(A)$. Let $y \in i^*(A) - i(A)$, then $\exists V \in \mathcal{V}(y)$ such as $V \subset A$, which leads to $y \in i(A)$.

An advantage of $\mathcal{V}$ type spaces is the fact that the family of neighborhoods of elements is a prefilter which characterizes the space. This gives a practical way to build spaces. However, it can be difficult to specify this family of neighborhoods.

A practical concept is the following, by introducing the basis of neighborhood.

Definition 11. *Given a $\mathcal{V}$ type pretopological space $(E, a(.), i(.))$, for any x in E, the family $\mathcal{B}(x)$ is called a basis of neighborhoods of x if and only if $\forall x \in E, \forall V \in \mathcal{V}(x), \exists B \in \mathcal{B}(x)$ such as $B \subset V$.*

The question then is to know how $\mathcal{B}(x)$ works in the definition of the pseudo-closure $a(.)$ and the interior $i(.)$.

Proposition 5. *Given a $\mathcal{V}$ type pretopological space $(E, a(.), i(.))$ defined by the family $\mathcal{V}(x), \forall x \in E$ and the basis $\mathcal{B}(x)$, then:*
(i) $\forall A \subset E$, $a(A) = \{x \in E | \forall B \in \mathcal{B}(x), B \cap A \neq \emptyset\}$
(ii) $\forall A \subset E$, $i(A) = \{x \in E | \exists B \in \mathcal{B}(x), B \subset A\}$

Proof. (i) $x \in \{x \in E | \forall B \in \mathcal{B}(x), B \cap A \neq \emptyset\}$
$\Rightarrow x \in \{x \in E | \forall V \in \mathcal{V}(x), V \cap A \neq \emptyset\}$
Conversely, let us suppose $x \in \{x \in E | \forall V \in \mathcal{V}(x), V \cap A \neq \emptyset\}$ and
$x \notin \{x \in E | \forall B \in \mathcal{B}(x), B \cap A \neq \emptyset\}$
$\Rightarrow \exists B \in \mathcal{B}(x), B \cap A = \emptyset$
$\Rightarrow \exists V \in \mathcal{V}(x), B \cap A = \emptyset$
$\Rightarrow x \notin a(A)$
(ii) is obvious from (i) by c-duality.

We have defined open and closed subsets in a previous paragraph, which leads to closure and opening of a subset A. In the most general case, opening and closure does not necessarily exist.

In the case of $\mathcal{V}$-type pretopological spaces, we get the following results which lead us to a specific result about opening and closure.

Proposition 6. *Given a $\mathcal{V}$-type pretopological space $(E, a(.), i(.))$*
(i) $A \subset E$ is open if and only if A is a neighborhood for each of its elements
(ii) Let $x \in E$ and $V \subset E$, if there exists an open subset A such as $\{x\} \subset A \subset V$, then V is a neighborhood of x. The converse generally is not true.
(iii) Any union of open sets is an open set
(iv) Any intersection of closed sets is a closed set.

Proof. (i) A is open so $A = i(A)$. Then, for any $x \in A$, $x \in i(A)$ which implies A is a neighborhood of any x in A.
(ii) $\{x\} \subset A \subset V$, so as A is open, $x \in i(A)$ which implies $x \in i(V)$. Therefore V is a neighborhood of x.
(iii) Let $A_j, j \in J$ a family of open sets and let us consider $\bigcup_{j \in J} A_j$. $x \in \bigcup_{j \in J} A_j$, then
$\exists j_0$ such as $x \in A_{j_0}$. A_{j_0} is an open set, then it is a neighborhood of x which is included in $\bigcup_{j \in J} A_j$. Therefore $\bigcup_{j \in J} A_j$ is a neighborhood of x.
(iv) is proven by c-duality

This last result leads to the following which establish existence of opening and closure of any subset in a $\mathcal{V}$-type space.

Proposition 7. *In a $\mathcal{V}$-type pretopological space $(E, a(.), i(.))$, opening and closure of any subset A of E always exist.*

Proof. Given $A \subset E$, let us consider $\mathcal{F}_A$ the family of all closed subsets which contain A. $\mathcal{F}_A \neq \emptyset$ because $E \in \mathcal{F}_A$

Let us consider $H(A) = \bigcap_{G \in \mathcal{F}_A} G$. $H(A)$ is a closed subset from property (iv) of the previous proposition. Obviously, it is the closest regarding inclusion, so $H(A) = F(A)$.

We have settle down the theoretical part concerning pre topological spaces. These spaces corresponds to datasets in which we have complex objects linked by various relations, possibly valued, and for which we want to identify similarities according to a proximity measure. This is why we turn now to the concept of quasipseudo-metrics.

3   Concept of Quasipseudometrics

Clustering techniques are based, as a rule, on the concept of distance that supposes a precise axiomatic. This axiomatic sometimes proves to be very coercive, in particular in the applications in social sciences. In various cases, a problem that arises is the property of symmetry which expresses that the distance from x to y is equal to the distance from y to x. The question is therefore to know what it happens when this hypothesis is relaxed.

Obviously, the first questions, in this case, are relevant of theoretical points:

- Is it possible to define a concept of "metric" without the symmetry axiom, but with sufficient properties to build get theoretical results?
- What are the basic properties of such a "metric" ?
- What kind of topological space can we endow the metric space with ?

In this work, we propose to give some answers these questions related to such a measure of distance fulfilling neither the property of symmetry nor the triangular inequality property. We derive from this definition the first properties relative to the structures which the set could be endowed with. In particular, we show that it is not more possible to get a topology and that the structures that we can drift of such a distance measure are only pretopological structures ([5] [6] [7] [1]), however endowed with interesting properties ([4]).

3.1   Definitions and Basic Properties

In this section, we shall focus ourselves on defining more general spaces than metric ones and we shall study properties of those spaces. This extension of metric spaces is obtained by way of keeping only the first axiom of a metric. We present all definitions and basic results about quasipseudometrics hereafter.

Definition 12. *(Quasipseudometric) Let be E a non empty set, we call quasipseudometric on E, any mapping from $E \times E$ into $P(E)$ such as* $\forall (x,y), (x,y) \in E \times E, d(x,y) = 0 \Leftrightarrow x = y.$

Example 1. Let $E = \mathbb{R}^2$.

For any $x = (x_1, x_2)$ and $y = (y_1, y_2)$ in E, we set:

$d(x,y) = 2(x_1 - y_1) + 2(x_2 - y_2)$ if $y_1 \leq x_1$ and $y_2 \leq x_2$

$\quad d(x,y) = (y_1 - x_1) + (y_2 - x_2)$ if $y_1 \geq x_1$ and $y_2 \geq x_2$

$\quad d(x,y) = 2(x_1 - y_1) + (y_2 - x_2)$ if $y_1 < x_1$ and $x_2 < y_2$

$\quad d(x,y) = 2(x_2 - y_2) + (y_1 - x_1)$ if $y_1 > x_1$ and $x_2 < y_2$

It is obvious to see that $d(x,y) = 0 \Leftrightarrow x = y$ and furthermore, if we take $x = (0,0)$ and $y = (0,1)$ we get $d(x,y) = 1$ and $d(y,x) = 2$. Thus, d is a quasipseudometric on E.

Example 2. Let E be the set $E = \{x, y, z, t\}$ and $\Re$ a binary relationship on E characterized by the following table:

	x	y	z	t
x	0	1	0	1
y	0	0	1	1
z	1	1	0	0
t	0	0	1	0

We define d as:

$\quad d : E \times E \to \mathbb{R}^+$

$\quad \forall a \in E, \ d(a, a) = 0$

$\quad \forall a \in E, \forall b \in E, \ d(a, b) = n(a, b)$ where $n(a, b)$ is the length of the shortest path from a to b. So, we get the following *distance* table:

d	x	y	z	t
x	0	1	2	1
y	2	0	1	1
z	1	1	0	2
t	2	2	1	0

It is not a symmetric table, which shows that d is a quasipseudometric, not a metric in the usual sense.

Example 3. Let us consider $E = \{a, b, c, d, e\}$ and a distance table on E given by:

d	a	b	c	d	e
a	0	1	$\sqrt{2}$	$\sqrt{17}$	4
b	1	0	1	4	$\sqrt{17}$
c	$\sqrt{2}$	1	0	3	$\sqrt{10}$
d	$\sqrt{17}$	4	3	0	1
e	4	$\sqrt{17}$	$\sqrt{10}$	1	0

d is a classical metric on E. Let us now consider δ defined on E by: $\delta(x, y) = k \Leftrightarrow$ y is the kth nearest neighbor of x. Then, we get the following table for δ:

δ	a	b	c	d	e
a	0	1	2	4	3
b	1.5	0	1.5	3	4
c	2	1	0	3	4
d	4	3	2	0	1
e	3	4	2	1	0

We can see that δ fulfills axioms of a quasipseudometric, not of a metric.

Definition 13. *(Quasipseudometric space) The couple (E, d) where E is a non empty set and d is a quasipseudometric on E, is called quasipseudometric space*

In clustering techniques, we are used to work with metrics which do not fulfill the third axiom, the triangle axiom. We know it is without consequences on existing methods of clustering. However, when the axiom of symmetry is not fulfilled, we cannot use the classical methods of clustering and we have to design new ones. We also can imagine an intermediate case between the case where d is symmetric and the case where d is not symmetric.

Definition 14. *Let (E, d) be a quasipseudometric space. If there exists a mapping t from E into itself such that $\forall (x, y), (x, y) \in E^2$, $d(x, y) = d(t(y), t(x))$, then we say that (E, d) is t-pseudosymmetric*

We can note that, if t is identity mapping, we get the symmetry property. From now, we shall work with a quasipseudometric d which is not a symmetric one.

Definition 15. *Let (E, d) a quasipseudometric space, we call surface of symmetry the set S defined by $S = \{(x, y) \in E^2 | d(x, y) = d(y, x)\}$*

Then, it is obvious to note that:

Proposition 8. *Let (E, d) a quasipseudometric space with a surface of symmetry S*

- $S \neq \emptyset$
- $\forall (x, y) \in E^2, (x, y) \in S \Leftrightarrow (y, x) \in S$

Moreover, we can note that S is a neighborhood of the diagonal Δ of E ($\Delta = (x, x) \mid x \in E$). It is also obvious that d is a symmetric quasipseudometric if and only if $S = E \times E$. By analogy with metric spaces, we can study the concept of open or closed ball, with center x and radius r, but in the case of a quasipseudometric space, we have to distinguish between right and left balls.

Definition 16. *Let (E, d) be a quasipseudometric space, r a positive real number and x a point of E*

- *We call half open right ball with center x and radius r, the set, noted $\dot{B}_d(x, r)$, defined by*

$$\dot{B}_d = \{y \in E | d(x, y) < r\}$$

- *We call half closed right ball with center x and radius r, the set, noted $\bar{B}_d(x, r)$, defined by*

$$\bar{B}_d = \{y \in E | d(x, y) \le r\}$$

- *We call half open left ball with center x and radius r, the set, noted $\dot{B}_g(x, r)$, defined by*

$$\dot{B}_g = \{y \in E | d(y, x) < r\}$$

- *We call half closed left ball with center x and radius r, the set, noted $\bar{B}_g(x, r)$, defined by*

$$\bar{B}_g = \{y \in E | d(y, x) \le r\}$$

Example 4. If we consider the previous example 1 with $x = (0, 0)$ and $r = 1$, we get the following half right and left balls (see Figures 1, 2):

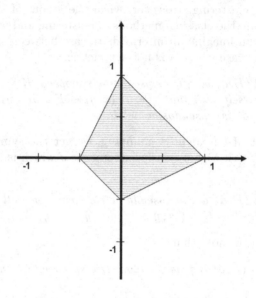

Fig. 1. Half right ball

Definition 17. *Let us consider a quasipseudometric space (E, d). For any x in E and any r positive real number:*

- *We call lower open ball (resp. lower closed ball), with center x and radius r, the set, noted $\dot{B}_{inf}(x, r)$ (resp. $\bar{B}_{inf}(x, r)$), defined by $\dot{B}_{inf}(x, r) = \dot{B}_d(x, r) \cap \dot{B}_g(x, r)$*
  (resp. $\bar{B}_{inf}(x, r) = \bar{B}_d(x, r) \cap \bar{B}_g(x, r)$).
- *We call upper open ball (resp. lower closed ball), with center x and radius r, the set, noted $\dot{B}_{sup}(x, r)$ (resp. $\bar{B}_{sup}(x, r)$), defined by $\dot{B}_{sup}(x, r) = \dot{B}_d(x, r) \cup \dot{B}_g(x, r)$*
  (resp. $\bar{B}_{sup}(x, r) = \bar{B}_d(x, r) \cup \bar{B}_g(x, r)$)

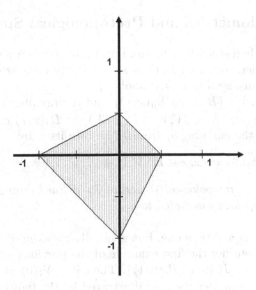

Fig. 2. Half left ball

We then get the obvious following result:

Proposition 9. *Let (E, d) be a quasipseudometric space.*
(i) $\forall r, r > 0, \forall x, x \in E, \forall y, y \in E, y \in \dot{B}_d(x, r) \Leftrightarrow x \in \dot{B}_g(y, r)$
(ii) $\forall r, r > 0, \forall x, x \in E, \forall y, y \in E, y \in \dot{B}_{inf}(x, r) \Leftrightarrow x \in \dot{B}_{inf}(y, r)$
(iii) $\forall r, r > 0, \forall x, x \in E, \forall y, y \in E, y \in \dot{B}_{sup}(x, r) \Leftrightarrow x \in \dot{B}_{sup}(y, r)$
(iv) $\forall r, r > 0, \forall x, x \in E, x \in \dot{B}_d(x, r) \cap \dot{B}_g(x, r)$

The last proposition obviously holds for closed balls. Let us suppose that the quasipseudometric d is t-pseudo symmetric, then

$$\forall (x, y) \in E^2, d(x, y) = d(t(y), t(x)).$$

So

$$\begin{aligned}
y \in \dot{B}_d(x, r) &\Leftrightarrow d(x, y) < r \\
&\Leftrightarrow d(t(y), t(x)) < r \\
&\Leftrightarrow t(y) \in \dot{B}_g(x, r).
\end{aligned}$$

If $t^{-1}(A)$ denotes the set defined by
$t^{-1}(A) = \{x, x \in E | t(x) \in A\}$, we can write:

Proposition 10. *Let (E, d) be a quasipseudometric space, if d is t-pseudo symmetric, then :*

$$\begin{aligned}
&(i) \ \dot{B}_d(x, r) = t^{-1}(\dot{B}_g(t(x), r)) \\
&(ii) \ B_d(x, r) = t^{-1}(B_g(t(x), r))
\end{aligned}$$

4   Quasipseudometrics and Pretopological Spaces

In this section, we shall study how to build pretopological structures related to a given quasipseudometric on a set E, then we will study what are their properties.

First, we shall consider a given threshold $r, r > 0$.

Let us consider $\mathcal{B}_r(x) = \{\dot{B}_d(x,r), \dot{B}_g(x,r)\}$ and the prefilter $\mathcal{V}_r$ of subsets of E generated by $\mathcal{B}_r(x)$, i.e. $\mathcal{V}_r = \{V, V \in \mathcal{P}(E) \mid V \supset \dot{B}_d(x,r) \text{ or } V \supset \dot{B}_g(x,r)\}$. Then, let us define the mapping a_r from $\mathcal{P}(E)$ into itself by

$$a_r(A) = \{x, x \in E \mid \forall V, V \in \mathcal{V}_r, V \cap A \neq \emptyset\}$$

Definition 18. *The pretopological structure $\mathcal{P}_r$ defined onto E by families $\mathcal{V}_r$ is called the r-pretopology associated to d.*

This pretopology is a $\mathcal{V}$-type one, but generally speaking it is not a $\mathcal{V}_D$-type one (see [1]). If we consider the first example of the previous section, in the case where $r = 1$, $\mathcal{B}_1(0) = \{\dot{B}_d(0,1), \dot{B}_g(0,1)\}$. Then $V \in \mathcal{V}_1(0) \Leftrightarrow V \supset \dot{B}_d(0,1)$ or $V \supset \dot{B}_d(0,1)$. Let us consider the case illustrated by the figure 3:

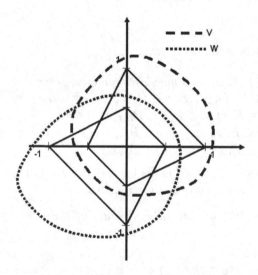

Fig. 3. Neighbourhood

V is a neighbourhood of 0, W is also a neighbourhood of 0, but $V \cap W$ is not a neighbourhood of 0. This implies that $\mathcal{V}_1(0)$ is not a filter of subsets of E and then the pretopology is not a $\mathcal{V}_D$ one.

The pseudoclosure map can then be expressed as follows:

Proposition 11. $\forall A, A \in \mathcal{P}(E)$
$a_r(A) = \{x \in E \mid \dot{B}_d(x,r) \cap A \neq \emptyset \text{ and } \dot{B}_g(x,r) \cap A \neq \emptyset\}$

Thus, we know the main properties of the pretopology $\mathcal{P}_r$ with a fixed r.

Let us examine what it can be said when we consider two different pretopologies generated by two distinct thresholds r_1 and r_2. Let us consider the case where $r_1 \leq r_2$. Then, it is immediate to note that :

$\forall x, x \in E, \dot{B}_d(x, r_1) \subset \dot{B}_d(x, r_2)$

$\forall x, x \in E, \dot{B}_g(x, r_1) \subset \dot{B}_g(x, r_2)$. Let us note by $\mathcal{B}_{r_1}(x)$ and $\mathcal{B}_{r_2}(x)$ the corresponding neighborhoods basis.

$$\forall V, V \in \mathcal{V}_{r_2}(x), V \supset \dot{B}_d(x, r_2) \vee V \supset \dot{B}_d(x, r_2)$$

$$\Rightarrow \forall V, V \in \mathcal{V}_{r_2}(x), V \supset \dot{B}_d(x, r_1) \vee V \supset \dot{B}_d(x, r_1)$$

$$\Rightarrow \forall V, V \in \mathcal{V}_{r_2}(x) \Rightarrow V \in \mathcal{V}_{r_1}(x)$$

Then:

Proposition 12. *If* $r_1 \leq r_2$, *the pretopology* $\mathcal{P}_{r_1}$ *is coarser than the pretopology* $\mathcal{P}_{r_2}$ *(we denote* $\mathcal{P}_{r_1} \prec \mathcal{P}_{r_2}$)

These pretopologies, generated by a given threshold, rely upon the concepts of half right balls and half left balls. But, we also have the concepts of sup and inf balls.It is possible to generate two new pretopologies from them by setting:

$$\underline{\mathcal{P}}_r = \{\underline{\mathcal{V}}(x), x \in E\}$$

where

$$\underline{\mathcal{V}}(x) = \{V \subset E / V \supset \dot{B}_{inf}(x, r)\}$$
$$\overline{\mathcal{P}}_r = \{\overline{\mathcal{V}}(x), x \in E\}$$

where

$$\overline{\mathcal{V}}(x) = \{V \subset E / V \supset \dot{B}_{sup}(x, r)\}$$

These two pretopologies are obviously $\mathcal{V}_D$ pretopologies and by the definitions of $\dot{B}_{inf}(x, r)$ and $\dot{B}_{sup}(x, r)$, we get:

Proposition 13. *The pretopologies* $\underline{\mathcal{P}}_r$ *and* $\overline{\mathcal{P}}_r$ *are* $\mathcal{V}_D$ *pretopologies and we have* $\underline{\mathcal{P}}_r \prec \mathcal{P}_r \prec \overline{\mathcal{P}}_r$

Up to now, we worked with pretopologies generated by giving a threshold r. The selection of that threshold may introduce a bias, so it would be important to be able to define associated pretopologies to a quasipseudometric without considering such a threshold. For that, let us consider

$$\forall x, x \in E, \mathcal{B}(x) = \{\dot{B}_d(x, r), \dot{B}_g(x, r), r > 0\}$$

and let denote by $\mathcal{V}(x)$ the prefilter generated by $\mathcal{B}(x)$, i.e. $V \in \mathcal{V}(x)$
$\Leftrightarrow$
$(\exists r, r > 0, V \supset \dot{B}_d(x, r)) \vee (\exists r', r' > 0, V \supset \dot{B}_g(x, r'))$ So we can put

$$\forall A, A \subset E, a(A) = \{x \in E / \forall V, V \in \mathcal{V}(x), V \cap A \neq \emptyset\}$$

It is obvious to see that :
$a(\emptyset) = \emptyset$
$\forall A, A \in E, A \subset a(A)$
It follows the definition of the induced pretopology on E by the quasipseudo-metric:

Definition 19. *(Induced pretopology) The pretopology which is defined by the family* $\mathcal{V}(x)$ *given above and the pseudo closure a of which is the function defined above, is called the pretopology induced on E by the quasipseudometric d*

Proposition 14. $\forall A, A \subset E$
$a(A) = \{x \in E \mid \forall r, r > 0, (\dot{B}_d(x,r) \cap A \neq \emptyset) \wedge (\dot{B}_g(x,r) \cap A \neq \emptyset)\}$

Proof.
$$\forall A, a(A) = \{x \in E/\forall V, V \in \mathcal{V}(x), V \cap A \neq \emptyset\}$$

$\Rightarrow$
$$x \in A \Rightarrow \forall r, r > 0, (\dot{B}_d(x,r) \cap A \neq \emptyset) \wedge (\dot{B}_g(x,r) \cap A \neq \emptyset)$$

Conversely, let us suppose that : $\forall r, r > 0$
$\dot{B}_d(x,r) \cap A \neq \emptyset \wedge \dot{B}_g(x,r) \cap A \neq \emptyset) \vee (\exists V^0, V^0 \in \mathcal{V}(x), V^0 \cap A = \emptyset)$
As $\exists r_0$ such that $V^0 \supset \dot{B}_d(x,r_0)$ or $V^0 \supset \dot{B}_g(x,r_0)$
This leads us to $\dot{B}_d(x,r_0) \cap A = \emptyset$ or $\dot{B}_g(x,r_0) \cap A = \emptyset$ which is contradictory.

Proposition 15. *The pretopology induced by a quasipseudometric d is a* $\mathcal{V}$ *one.*

Proof. Let us suppose that $A \subset B$
$x \in a(A) \Leftrightarrow \forall r, r > 0, \dot{B}_d(x,r) \cap A \neq \emptyset \wedge \dot{B}_g(x,r) \cap A \neq \emptyset$
$\Rightarrow$
$x \in a(A) \Leftrightarrow \forall r, r > 0, \dot{B}_d(x,r) \cap B \neq \emptyset \wedge \dot{B}_g(x,r) \cap B \neq \emptyset$
Then $A \subset B \Rightarrow a(A) \subset a(B)$

Remark. This pretopology is not a $\mathcal{V}_D$-one because, generally speaking, $\mathcal{V}(x)$ is not a filter of subsets of E.

As in the case of pretopologies using a given threshold for r, we can use the concepts of upper balls and lower balls to define two other pretopological structures on E. Let us consider $\underline{\mathcal{B}}(x) = \{\dot{B}_{inf}(x,r), r > 0\}$ and $\overline{\mathcal{B}}(x) = \{\dot{B}_{sup}(x,r), r > 0\}$. We can put:

$$\underline{a}(A) = \{x \in E/\dot{B}_{inf}(x,r) \cap A \neq \emptyset, \forall r, r > 0\}$$

and

$$\overline{a}(A) = \{x \in E/\dot{B}_{sup}(x,r) \cap A \neq \emptyset, \forall r, r > 0\}$$

Obviously, we define two pseudoclosures of two pretopologies respectively noted $\underline{\mathcal{P}}$ and $\overline{\mathcal{P}}$. We straightforwardly get the following result:

Proposition 16. *If* $\mathcal{P}$ *denotes the pretopology induced by the quasipseudometric d, we have :* $\underline{\mathcal{P}} \prec \mathcal{P} \prec \overline{\mathcal{P}}$

Remark. The two pretopologies $\underline{P}$ and $\overline{P}$ are $\mathcal{V}_D$ ones.

An interesting particular case is the case when d fulfills the triangle axiom although being non symmetric. In that case, what happens to the pretopology induced by d ?

Proposition 17. *If d fulfills the triangle axiom, the pseudoclosure function a of the pretopology $\mathcal{P}$ induced by d is an idempotent function.*

Proof. We have to show that
$\forall A, A \subset E, a(a(A)) = a(A)$.

In fact, it is sufficient to prove that
$\forall A, A \subset E, a(a(A)) \subset a(A)$.

Let us consider $x \in a(a(A))$.
Then, $\forall r, r > 0, \exists y_r, y_r \in a(A)$,
$d(x, y_r) < r \wedge \exists y'_r, y'_r \in a(A), d(y'_r, x) < r$
But $y_r \in a(A) \wedge y'_r \in a(A)$, then: $\forall s, s > 0, \exists z_s, z_s \in A, d(y_r, z_s) < s \wedge \exists z'_s, z'_s \in A, d(z'_s, y_r) < s$. By the triangle axiom, we can say: $\forall r, \forall s, \exists z_s, z_s \in A$, $d(x, z_s) < r + s \wedge \exists z'_s, z'_s \in A, d(z'_s, x) < r + s$ It is sufficient to prove that $x \in a(A)$ and then $\forall A, A \subset E, a(a(A)) \subset a(A)$.

Now, we have defined a family of pretopologies on E : the pretopologies associated to the quasipseudometric d for a given value of r. We also have the pretopology induced on E by the quasipseudometric d.

What are the links between all these pretopologies?

The answer is given by the following result.

Proposition 18. *$\forall A, A \subset E, a(A) = \bigcap_{r>0} a_r(A)$.*
The pretopology $\mathcal{P}$ induced by the quasipseudometric d is the lower bound of the pretopologies $\mathcal{P}_r$.

Proof. Let us consider $x, x \in a(A)$
$\Rightarrow \forall r, r > 0, \dot{B}_d(x, r) \cap A \neq \emptyset$ and $\dot{B}_g(x, r) \cap A \neq \emptyset$
$\Rightarrow \forall r, r > 0, x \in a_r(A)$

The following definition and result allows to characterize elements of $a(A)$.

Definition 20. *Let (E, d) a quasipseudometric space, A a subset of E. For every x in E, we put:*
$d(x, A) = \inf\{d(x, y), y \in A\}$
$d(A, x) = \inf\{d(y, x), y \in A\}$

We can then write:

Proposition 19. *Let $A \in \mathcal{P}(E)$, the two following assertions are equivalents*
(i) $x \in a(A)$
(ii) $d(x, A) = 0$ and $d(A, x) = 0$

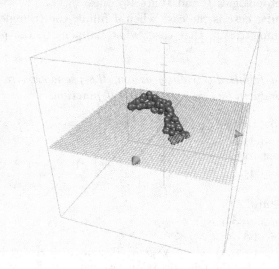

Fig. 4. Expert gesture

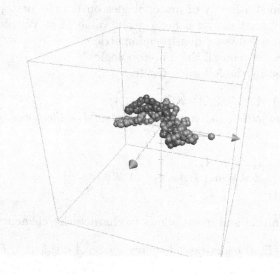

Fig. 5. Expert gesture VS Newbie test 1

Proof. $x \in a(A)$
$\Leftrightarrow \forall r, r > 0, \dot{B}_d(x, r) \cap A \neq \emptyset$ and $\dot{B}_g(x, r) \cap A \neq \emptyset$
$\Leftrightarrow \forall r, r > 0, \exists y_r, y_r \in A, 0 \leq d(x, y_r) < r$ and $\forall r, r > 0, \exists y'_r, y'_r \in A, 0 \leq d(y'_r, x) < r$
$\Leftrightarrow \inf\{d(x, y), y \in A\} = 0$ and $\inf\{d(y, x), y \in A\} = 0$
$\Leftrightarrow d(x, A) = 0$ and $d(A, x) = 0$

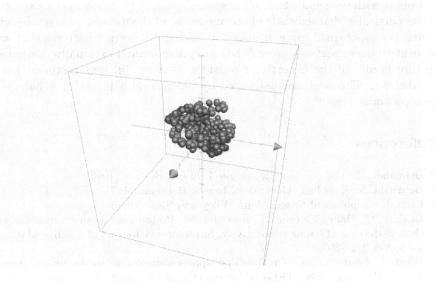

Fig. 6. Expert gesture VS Newbie test 2

A consequence of this result is that it is possible to characterize the neighborhoods of x by means of the quasipseudometric d:

Proposition 20. *Let $V \in \mathcal{P}(E)$, a necessary and sufficient condition for V to be a neighborhood of x is that $d(x, V^c) \neq 0$ or $d(V^c, x) \neq 0$*

Proof. It is sufficient to note that if V is a neighborhood of x, that means that x does not belong to $a(V^c)$.

5   Discussion

The results presented in this paper are the first results of a complete analysis on quasipseudometrics and on related pretopological spaces. Other theoretical questions are still subject to further investigations, in particular the analysis of conditions allowing a $\mathcal{V}$-pretopological space to be associated to a quasipseudometric space. Otherwise, concerning the applications in the field of data analysis, it remains to define clustering methods founded on that concept of quasipseudometric.

In our illustrating example, we intend to use our modeling approach to identify the proximity between various trajectories corresponding to the gesture of an expert serving a glass of wine versus the gesture of a newbie learning the technique. The question is thus to decide, given a threshold, the trajectories which look alike in order to identify when a newbie has been trained enough to master the gesture — which should then be analog to the one of the expert .

Coping with the symmetry (in the semi-circular trajectories in our case study) is the particular characteristic of our modeling of the dataset with quasipseudometric pre topological space. It allows to distinguish gesture patterns that would be next to the expert one even if they are going around externally compared to gesture border of the expert and consider as more different gestures that are too direct ... This is clearly addressed by the general half right/left balls of the quasipseudometrics.

References

1. Belmandt, Z.: Manuel de prétopologie. Editions Hermés (1993)
2. Bourbaki, N.: Topologie Générale. 2 tomes, Hermann (1971)
3. Cech, E.: Topological Spaces. John Wiley and Sons (1966)
4. Lamure, M., Dalud-Vincent, M., Brissaud, M.: Pretopology as an extension of graph theory: the case of strong connectivity. International Journal of Applied Mathematics 5, 455–472 (2001)
5. Duru, G.: Contribution à l'analyse des espaces abstraits, le cas des images digitales. Technical report, Thèse d'Etat, Université Claude Bernard - Lyon (November 1987)
6. Duru, G., Auray, J.P., Brissaud, M.: Connexité des espaces pré férenciés. In: Colloque Mathématiques Discrètes: Codes et hypergraphes (1978)
7. Lamure, M.: Contribution à l'analyse des espaces abstraits, le cas des images digitales. Technical report, Thèse d'Etat, Université Claude Bernard - Lyon (November 1987)

Analogy and Metaphors in Images

Charles Candau, Geoffrey Ventalon, Javier Barcenilla, and Charles Tijus

Laboratory Cognitions Humaine et Artificielle, University Paris 8, France

Abstract. Museums have large databases of images. The librarians that are using these databases are doing two types of images search: either they know what they are looking for in the database (a specific image or a specific set of well defined images such as kings of France), or they do not know precisely what they are looking for (e.g., when they are required to build images portfolios about some concepts such as "decency"). As each image is having a number of metadata, searching for a well-defined image, or for set of images, is easily solved. On the contrary, this is a hard problem when the task is to illustrate a given concept such as "freedom", "decency", "bread", or "transparency" since these concepts are not metadata. How to find images that are somewhat analogs because they illustrate a given concept?

We collected and analyzed the search results of librarians that were given themselves the task of finding images related to a given concept. Seven relations between the concept and the images were found as explanation of the selection of images for any concept: conceptual property, causality, effectivity semantic, anti-logic, metaphorical-vehicle and metaphorical-topic. The inter-rate agreement of independent judges that evaluated the relations was of .78.

Finally, we designed an experiment to evaluate how much metaphor in images can be understandable.

1 Introduction: Prospective Ergonomics about the Design of Search Engine for Images

Everywhere, homes or workplaces, in the streets or the public areas, where we go for business, service, shopping, leisure or travel, they are digital systems with which we interact. Because they have to be adapted to humans, these digital systems include a model of theirs users [1] and are more and more made of Cognitive Technologies that are technologies that process as inputs data provided by their users. These emerging Cognitive Technologies are flourishing areas of multifaceted scientific research and research development, including neuroscience (e.g., brain computing), psycho-physiology (e.g., emotive computing), psychophysics (e.g., actimetry), cognitive psychology (e.g., digital reading and learning), computational linguistics (e.g., texts processing) to be used in association with artificial intelligence, cognitive robotics, distributed Human-Machine systems, cognitive ergonomics, and cognitive engineering.

The framework of the content of this paper is the open innovation process of imaging future things (prospective ergonomics) based on cognitive technologies [2], in the

A. Laurent et al. (Eds.): IPMU 2014, Part II, CCIS 443, pp. 135–142, 2014.

context of Living Labs, more precisely in the context of LUTIN, which is a Living Lab located in Universcience - City of Science and of Industry in Paris, a member of LabEx Smart dedicated to "Smart Human/Machine/Human Interactions In The Digital Society". For an user centered approach of conception of future services and products, these are made for and by the people and then industrially manufactured, and a step further, one considers as a citizen duty to participate in innovation process: "innovation needs you, innovation needs your expertise" [3]. Thus, although we still need the best methods of prospective innovation, the problem at hand is not how to imagine, but what can be ergonomics of things that do not exist: their usability and learnability. In addition, with cognitive technologies, ergonomics studies for prospective ergonomics are not solely how to facilitate interactions with a digital device, but also how to implement the system with a pertinent model of the users (data to collect, computing modes).

The emerging field of prospective ergonomics has potential for many domains, such as everyday life technologies, conception of teaching and learning in the classroom, e-learning, science and technology-related museology, e-government applications, heath, military and intelligence applications, and so on. However, among these fields, because of its contents, searching for images is a special challenge. Although, it appears easy to group pictures according to one object, it is a difficult topic about how to group things that are analogs. For instance, Museums have large databases of images (e.g., one of the database involved in the present project, - which is one of the Réunion des Musées Nationaux, RMN, is about of 700 000 images). The librarians that are using these databases are doing two types of images search: either they know what they are looking for in the database (a specific image or a specific set of well defined images such as kings of France), or they do not know precisely what they are looking for (e.g., when they are required to build images portfolios about some concepts such as "decency"). As each image is having a number of metadata, searching for a well-defined image, or for set of images, is easily solved. On the contrary, this is a hard problem when the task is to illustrate a given concept such as "freedom", "decency", "bread", or "transparency" since these concepts are not metadata. How to find images that are somewhat analogs because they illustrate a given concept?

2   Searching for Images

Searching for images is a challenging cognitive activity for several reasons. When the search is a well-defined search (we know what we want to find), a profitable search engine for images must be able to quickly find what we want by providing a small result set of images that contains what is sought. However, most of the times, the research is often poorly defined (we do not know precisely what we are looking for) and, even more, is also multi-criteria (about contents, but also definition, size...). These criteria are often approximate: we would like a particular criterion being met, but a compromise on this criterion can be done if another criteria are quite satisfied.

When the search is poorly defined, the size of the group of images that correspond to what would satisfy the search is unknown and when you got the results, it is unclear if any others images who would be more satisfactory. It follows that judgment about the search criterion is problematic because we do not know when to stop searching. Another difficulty comes from results that were not been retained (because we sought better): they are difficult to find again because the capacity of working memory (how did I do?). Short-term human cognition capacities are limited.

Finally, there is above all the discovery and the encounter during search of images that were not searched for (serendipity) and for which we would be eventually ready to question the selection criteria and to renew the search.

For prospective ergonomics, we can therefore expect the user of an images search engine find quite useful and enjoyable a system that facilitates his research if it becomes so easy to find, not only what we want, but also to find what was not researched. This, especially, if the interface is the most appropriate way and if users have a good understanding of its operations, in order to maximize them.

Our work was to define specific needs of users in terms of image search, and new services in this area. Users in question are professionals who use the database of about 700,000 images. Activity of librarians was searching for specific images, but also the realization of portfolios. Achieving portfolios carries a significant theoretical interest. To make a portfolio (i.e., a collection of images of works in a given subject), is to find a set of images that are analogs in the sense that they are exemplars of images of a given concept.

Thus, the question at hand is how librarians are searching for images related to a concept, and how to help them finding images that are analogs.

3   Analogy in Images

We have seen that the task is to find a set of images corresponding to a theme (e.g., bread) or to a concept (modesty) in order to get a set of images to be retained in a portfolio.

This type of task is emblematic of a poorly defined search: "How to illustrate abstract concepts?" "How to find what we are unaware ... and we do not even know name?" An example is how to find the illustration of a work of Umberto Eco entitled "To say almost the same thing. " Thus, the task is to find a set of images constituting a portfolio of images corresponding to "almost the same" in order for the author or publisher to finally choose the one he likes the most.

The theory we adopt, is given by Tijus et al. [4]. Searching for a specific image is a well-defined problem solving [5] that requires legal reasoning. In opposite, searching for images that could illustrate a concept is a kind of innovative problem solving that requires finding ideas of images content. The main theory is the analogy-based theory: innovation is the transportation of a solution process from a source domain to the problem domain (if to illustrate "freedom", I was using a photo of the statue of liberty, to illustrate "decency", I will search for a statue). Thus, the making of analogies, through Case-based or analogy-based systems to find analogous cases could be of help [6]. By contrast, we explore a cognitive mechanism that is based on the making of substitutes as sources.

Tijus et al. [4] reported that for Sir Ernst Gombrich [7], a well-known art specialist, a ordinary hobbyhorse "Should we describe it as an 'image of a horse'?" Is it rather "a substitute for a horse", a "horse like" that can be ridden by a child. There is no analogy between the horse and the stick. Gombrich's conclusion was that "substitution may precede portrayal and creation communication".

According to [8], the finding of possible sources for analogy is as follows. First, the goal being defined, a matching process starts by carrying out a large number of comparisons between the components (objects, objects attributes and relations) of the sources and of the goal. Second, source and structure are mapped for "global" identifications. However, [9] discussed the order of these successive phases (encode target, find sources with local matches, match structure) in human analogical processes. With the additional problem of how components are selected for mapping [10], since analogy is based on similarity with already encoded data, it would be difficult to find a never used solution.

We advocate that the solution is a substitute that is inferred while searching how the problem could be solved. Differently of analogy, the process does not involve matching sources to be modified and adapted to target, but supposes target adaptation to find sources. When searching for images related to a concept, the solution might be finding substitutes.

4   Experiment: Observing and Analyzing the Conceptual Search of Images

This task was to analyze the results of research done by librarians, in order to inductively find the reasons for the selection of images. If it is possible to determine the selection criteria, then an automatic search engine could use these criteria to facilitate the search. These criteria are of semantic order, conceptual.

The librarians have an indexed images base (700,000 images). Besides a number of information about each image that might be used for the search (size, definition, author, date...): for example, to provide images about an exhibition on Egypt for the general public, descriptors (eras, collections...) are known and can be used. In opposite, librarians are also required to conduct more conceptual research as to form an online portfolio on a particular theme (the naked, modesty, the taste). For this second type of research, conventional descriptors cannot be used: this is the kind of research during which we do not know what to look for.

4.1   Method

The interviews were made with 15 Participants (8 librarians, 3 Iconographers, 2 illustrators and 2 Art Directors in advertising). The interview was about the use of search engines for finding images. The method was the Critical Incident Technique [11], which is about recollection of facts, which comprises the set of pictures that were found. Some of these groups of analog pictures can be found at

http://www.photo.rmn.fr/cf/htm/
CDocT.aspx?V=CDocT&E=2C6NU0O0Z17S&DT=ALB
http://www.photo.rmn.fr/cf/htm/
CDocT.aspx?V=CDocT&E=2C6NU0YDXTJU&DT=ALB

We collected a number of search results for a number of topics such as "Shame", "transparency", "decency", "bread", "hair" and "prejudices about women", with sub-searches such as "women at work", "parity", "nudity", "parenthood". For bread, there were combinations such as "bread making", "crafts related to bread making", "manufacturing", "baker", "bakery", "consumption", "bread: a daily concern around the world", "the sacred bread", "the symbolic bread".

Thus, our work for PCE was to analyze the results of research librarians in order to find inductively the reasons for the selection of images.

If we could determine the selection criteria librarians were using, then we could use these criteria to help facilitate the search. These criteria, which are of semantic nature, might be added to the search engine in addition to other kinds of image components. The principle is that users' needs stems from their embodiment of the task.

Note that searching for a well-defined image is as a well-defined problem and that looking for images related to a given concept is as an ill-defined problem.

Because we need a specific goal, we reasoned that a concept refers to objects that have this conceptual property (ugliness refers to objects that are ugly; freedom to objects which are free), but also because of causality (which makes it ugly, makes it free), of effectivity (what are the consequences to be ugly; to be free), or semantic: near to be (could be ugly, might be free), anti-logical (a great beauty to designate ugliness; undressed to represent decency by the lack of decency; lack of freedom to designate freedom), metaphorical-vehicle to mean something else (ugliness to designate the absence of moral; freedom to designate democracy), metaphorical-topic (to be represented by another object: the mud to indicate ugliness, bird to indicate freedom).

4.2   Results

Figure 1 shows samples of pictures that were grouped as instances of decency and as instances of transparency, one image can be in both sets. Figure 2 shows the kind of relation that links the image to the concept.

Fig. 1. Some pictures found as analogous for being examples of « decency» (top) and of "transparency" (bottom)

Nom   attribut   cause   effet   sémantique

Fig. 2. Images for freedom, according to the relations between the image and the concept to be illustrated, from left to right: Name, Attribute, Causality, Effectivity, and Semantics

We reasoned that, starting from a set of images corresponding to a concept, it is possible to determine what in each image refers to the concept using non-exclusive relations as criteria.

Thus, we defined six types of relationships between the image and the concept. Then, we define some coding rules to attribute or not an image with each of the six relationships and three judges coded 170 photos of 3 portfolios ("bread", "modesty" and "transparency"), while two other judges coded 196 photos of 2 portfolios ('freedom" and "decency") attributing each of the six relations, assessed independently for each image. The inter-rate agreement was of .78 and of .84 respectively.

Note that the judges as well as feedback from the librarians affected the final 6 types of " concept-to-images" relationships that were identified:

- Name: first relationship between the concept and the presence of his name in the image (e.g. concept of "freedom " with " Statue of Liberty ").
- Cause: the object (or action) in the picture helps to explain the existence of the content of the concept. (e.g., the "revolt" contributes to the "freedom") .
- Effect: the object represented occurs as a consequence of the concept application (e.g. "freedom" by excess may arise as "debauchery.")
- Attribute: the object is represented as a single component or a concept (e.g., "freedom " offers the quality to be free, autonomous, independent, etc.

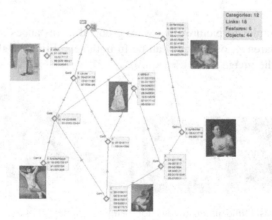

Fig. 3. Galois Lattice describing how the 6 object-to-concept relations distributed over the categories of images (one single image is given to illustrate the category)

- Typicality: the object represented is clearly characteristic and could serve as a model of the concept (e.g., the "bread" can be characterized by a loaf of round, brown, notched at the top, etc.)
- Semantics: this is a relationship that is reported with a higher degree of cognitive flexibility (compared to the relationships mentioned above) on the linkage between the figurative object and the concept.

5   Discussion and Conclusion

The making of portfolios could well be done with intuitive navigation through images, by linking automatic metadata associated with each image corresponding to the selected theme, as well as navigation data, leading to relevant images for this topic. Thus, if a librarian decided to create a portfolio with a search engine related with a thesaurus. The system processes all available information (navigation data, comparison of formal and textual metadata images...), and provides relevant findings taking into account the six types of relationship.

These relationships therefore participate to guide the user in his search by a semantic reformulation, thus contributing to the "serendipity" of images search in the database. The librarians say that their search were very intuitive and did not follow the kind of rules we listed, they found this method impressive and agree that it could be useful to have a search engine that could use them.

Thus, having a detailed description of the future semantic search engine, we started defining how the search could be simple (search in turn for each of the relationships according to the librarian preferences), familiar (using procedures of advanced search), feedbacks (an image should indicate which of the relationships it is exemplified), transparency (display the chosen criteria), presence (good image definition if needed; low definition for first survey), safety (does not be losing preceding results) and affordances (have an image being the prototype of the relationship).

Next step of this current research for future search engines is to test these relationships as research criteria with naïve users.

Acknowledgment. This work was supported by FUI EGONOMY.

References

1. Tijus, C., Cambon de Lavalette, B., Poitrenaud, S., Leproux, C.: L'interaction autorégulatrice entre dispositif et utilisateur: une modélisation des inférences sur les durées du parcours routier. Le Travail Humain 66(1), 23–44 (2003)
2. Tijus, C., Barcenilla, J., Jouen, F., Rougeaux, M.: Open innovation and prospective ergonomics for smart clothes. In: 2nd International Conference on Ergonomics in Design (2014)
3. Barcenilla, J., Tijus, C.: Ethical issues raised by the new orientations in ergonomics and living labs. Work 41, 5259–5265 (2012)

4. Tijus, C., Poitrenaud, S., Léger, L., Brézillon, P.: Counterfactual Based Innovation: A Galois Lattice Approach of Creative Thinking. In: International Conference on Computing and Communication Technologies, RIVF 2009, pp. 1–4. IEEE (2009)
5. Newell, A., Simon, H.A.: Human problem solving. Prentice Hall, Englewood Cliffs (1972)
6. Roth-Berghofer, T.R.: Explanations and Case-Based Reasoning. Foundational Issues. In: Funk, P., González Calero, P.A. (eds.) ECCBR 2004. LNCS (LNAI), vol. 3155, pp. 389–403. Springer, Heidelberg (2004)
7. Gombrich, E.H.: Mediations on a hobby horse. In: Meditations on a Hobby Horse and Other Essays on the Theory of Art, London (1963)
8. Gentner, D., Toupin, C.: Systematicity and surface similarity in the development of analogy. Cognitive Science 10(3), 277–300 (1986)
9. Ripoll, T., Eynard, J.: A Critical Analysis of Current Models of Analogy. In: Proceedings of the 2002 Information Processing And Management of Uncertainty in Knowledge-Based Systems, IPMU 2002 (2002)
10. Kwon, H., Im, I., Van de Walle, B.: Are you thinking what I am thinking – A comparison of decision makers' cognitive map by means of a new similarity measure. In: Proceedings of the 35th Hawaii International Conference on System Sciences, vol. 78, p (2002)
11. Flanagan, J.C.: The critical incident technique. Psychological Bulletin 5, 327–358 (1954)

Fuzzy Transform Theory in the View of Image Registration Application

Petr Hurtík, Irina Perfilieva, and Petra Hodáková

University of Ostrava, Centre of Excellence IT4Innovations,
Institute for Research and Applications of Fuzzy Modeling,
30. dubna 22, 701 03 Ostrava 1, Czech Republic
{petr.hurtik,irina.perfilieva,petra.hodakova}@osu.cz

Abstract. In this paper, the application of the fuzzy transforms of the zero degree (F^0-transform) and of the first degree (F^1-transform) to the image registration is demonstrated. The main idea is to use only one technique (F-transform generally) to solve various tasks of the image registration. The F^1-transform is used for an extraction of feature points in edge detection step. The correspondence between the feature points in two images is obtained by the image similarity algorithm based on the F^0-transform. Then, the shift vector for corresponding corners is computed, and by the image fusion algorithm, the final image is created.

Keywords: image registration, feature detection, edge detection, image similarity, image fusion.

1 Introduction

In computer graphics, interactions between the machine and the real worlds are basically ensured by the image processing. One of the tasks is to represent data for computer processing to be similar to the humen eye vision as much as possible. Therefore, this task is very popular in developing soft-computing methods. It became a common practice that soft computing methods work with uncertain information and can achieve better result than methods based on crisps information.

One of the effective soft computing methods is fuzzy transform (F-transform for short) developed by Irina Perfilieva. The main theoretical preliminaries were described in [1][2]. The F-transform is a technique that performs a transformation of an original universe of functions into a universe of their "skeleton models". Each component of the resulting skeleton model is a weighted local mean of the original function over an area covered by a corresponding basic function. The F-transform consists of two steps: direct and inverse transform. This method proved to be very general and powerful in many applications. Particularly, image compression [3][4], where the user can control the strength and the quality of compression by choosing the number of components used in F-transform. Another application is image fusion [7][8], where several damaged images are fused in one image which then has better quality than all the particular images. Image reduction and interpolation [5] is another application where

A. Laurent et al. (Eds.): IPMU 2014, Part II, CCIS 443, pp. 143–152, 2014.

the direct F-transform can reduce (shrink) the original image and the inverse F-transform can be used as an interpolation method. The F-transform of a higher degree (F^s-transform, $s \geq 1$) [10] can approximate the original function even better. Moreover, the F^1-transform can approximate the partial derivatives of the original function and therefore, it can be used in edge detection to compute the image gradient [9].

The task of the image registration is to match up two or more images. There are several examples where the image registration is used - images taken by different sensors, in different time, from different positions, with different size, etc. One of the most natural applications is to match several images of landscape which are partially overlapped into one large image. There exists a lot of methods how to register images [6], most of them consist of four basic steps: detect important features in each image; match the features from all images; find a suitable mapping function which describes image shift, rotation, etc.; interpolate images and fuse their overlaps.

This contribution, we demonstrate the use of the F-transform technique for all those steps: the F^1-transform for the gradient detection and for the feature points extraction; the F^0-transform for image similarity measures and for the image fusion.

2   Fuzzy Transform

2.1   Generalized Fuzzy Partitions

A *generalized fuzzy partition* appeared in [10] in connection with the notion of the higher-degree F-transform. Its even weaker version was implicitly introduced in [3] for the purpose of meeting the requirements of image compression. We summarize both these notions and propose the following definition.

Definition 1. *Let $[a, b]$ be an interval on the real line $\mathbb{R}$, $n > 2$, and let $x_1, \ldots, x_n$ be nodes such that $a \leq x_1 < \ldots < x_n \leq b$. Let $[a, b]$ be covered by the intervals $[x_k - h'_k, x_k + h''_k] \subseteq [a, b]$, $k = 1, \ldots, n$, such that their left and right margins $h'_k, h''_k \geq 0$ fulfill $h'_k + h''_k > 0$.*

We say that fuzzy sets $A_1, \ldots, A_n : [a, b] \to [0, 1]$ constitute a generalized fuzzy partition of $[a, b]$ (with nodes $x_1, \ldots, x_n$ and margins h'_k, h''_k, $k = 1, \ldots, n$), if for every $k = 1, \ldots, n$, the following three conditions are fulfilled:

1. *(locality) — $A_k(x) > 0$ if $x \in (x_k - h'_k, x_k + h''_k)$, and $A_k(x) = 0$ if $x \in [a, b] \setminus (x_k - h'_k, x_k + h''_k)$;*
2. *(continuity) — A_k is continuous on $[x_k - h'_k, x_k + h''_k]$;*
3. *(covering) — for $x \in [a, b]$, $\sum_{k=1}^{n} A_k(x) > 0$.*
4. *(monotonicity) — $A_k(x)$, for $k = 2, \ldots, n$, strictly increases on $[x_k - h'_k, x_k]$ and $A_k(x)$, for $k = 1, \ldots, n - 1$, strictly decreases on $[x_k, x_k + h''_k]$;*

An (h, h', h'')-*uniform* generalized fuzzy partition of $[a, b]$ is defined for equidistant nodes $x_k = a + h(k-1)$, $k = 1, \ldots, n$, where $h = (b-a)/(n-1)$; $h', h'' > h/2$ and two additional properties are satisfied:

4. $A_k(x) = A_{k-1}(x - h)$ for all $k = 2, \ldots, n - 1$ and $x \in [x_k, x_{k+1}]$, and
   $A_{k+1}(x) = A_k(x - h)$ for all $k = 2, \ldots, n - 1$ and $x \in [x_k, x_{k+1}]$.
5. $h_1' = h_n'' = 0$, $h_1'' = h_2' = \ldots = h_{n-1}'' = h_n' = h'$ and for all $k = 2, \ldots, n - 1$
   and all $x \in [0, h']$, $A_k(x_k - x) = A_k(x_k + x)$.

An (h, h')-uniform generalized fuzzy partition of $[a, b]$ can also be defined using the *generating function* $A_0 : [-1, 1] \to [0, 1]$, which is assumed to be *even*[1], continuous and positive everywhere except for on boundaries, where it vanishes. Then, basic functions A_k of an (h, h')-uniform generalized fuzzy partition are shifted copies of A_0 in the sense that

$$A_1(x) = \begin{cases} A_0\left(\frac{x - x_1}{h'}\right), & x \in [x_1, x_1 + h'], \\ 0, & \textit{otherwise}, \end{cases}$$

and for $k = 2, \ldots, n - 1$,

$$A_k(x) = \begin{cases} A_0\left(\frac{x - x_k}{h'}\right), & x \in [x_k - h', x_k + h'], \\ 0, & \textit{otherwise}. \end{cases} \tag{1}$$

$$A_n(x) = \begin{cases} A_0\left(\frac{x - x_n}{h'}\right), & x \in [x_n - h', x_n], \\ 0, & \textit{otherwise}, \end{cases}$$

2.2 F^0-transform

The direct and inverse F^0transform (originally just as F-transform) of a function of two (and more) variables is a direct generalization of the case of one variable. We introduce the discrete version only, because it is used in our applications below. Let us refer to [2] for more details.

Suppose that the universe is a rectangle $[a, b] \times [c, d] \subseteq \mathbb{R} \times \mathbb{R}$ and that $x_1 < \ldots < x_n$ are fixed nodes of $[a, b]$ and $y_1 < \ldots < y_m$ are fixed nodes of $[c, d]$ such that $x_1 = a$, $x_n = b$, $y_1 = c$, $y_m = d$ and $n, m \geq 2$. Assume that $A_1, \ldots, A_n$ are basic functions that form a generalized fuzzy partition of $[a, b]$ and $B_1, \ldots, B_m$ are basic functions that form a generalized fuzzy partition of $[c, d]$. Then, the rectangle $[a, b] \times [c, d]$ is partitioned into fuzzy sets $A_k \times B_l$ with the membership functions $(A_k \times B_l)(x, y) = A_k(x)B_l(y)$, $k = 1, \ldots, n$, $l = 1, \ldots, m$.

In the discrete case, an original function f is assumed to be known only at points $(p_i, q_j) \in [a, b] \times [c, d]$, where $i = 1, \ldots, N$ and $j = 1, \ldots, M$. In this case, the (discrete) F^0-transform of f can be introduced in a manner analogous to the case of a function of one variable.

Definition 2. *Let a function f be given at points $(p_i, q_j) \in [a, b] \times [c, d]$, for which $i = 1, \ldots, N$ and $j = 1, \ldots, M$, and $A_1, \ldots, A_n$ and $B_1, \ldots, B_m$, where $n < N$ and $m < M$, be basic functions that form generalized fuzzy partitions*

---

[1] The function $A_0 : [-1, 1] \to \mathbb{R}$ is even if for all $x \in [0, 1]$, $A_0(-x) = A_0(x)$.

of $[a, b]$ and $[c, d]$ respectively. We say that the $n \times m$-matrix of real numbers $\mathbf{F}[f] = (F_{kl})_{nm}$ is the discrete F^0-transform of f with respect to $A_1, \ldots, A_n$ and $B_1, \ldots, B_m$ if

$$F_{kl} = \frac{\sum_{j=1}^{M} \sum_{i=1}^{N} f(p_i, q_j) A_k(p_i) B_l(q_j)}{\sum_{j=1}^{M} \sum_{i=1}^{N} A_k(p_i) B_l(q_j)} \tag{2}$$

holds for all $k = 1, \ldots, n$, $l = 1, \ldots, m$.

The inverse F^0-transform of a discrete function f of two variables is defined as follows.

Definition 3. Let $A_1, \ldots, A_n$ and $B_1, \ldots, B_m$ be basic functions that form generalized fuzzy partitions of $[a, b]$ and $[c, d]$, respectively. Let function f be defined on the set of points $(p_i, q_j) \in P \times Q$ where $P = \{p_1, \ldots, p_N\} \subseteq [a, b]$, $Q = \{q_1, \ldots, q_M\} \subseteq [c, d]$ and both sets P and Q are sufficiently dense with respect to corresponding partitions, i.e $\forall k, l \; \exists i, j; \; A_k(p_i) B_l(p_j) > 0$. Moreover, let $\mathbf{F}[f] = (F_{kl})_{nm}$ be the discrete F^0-transform of f w.r.t. $A_1, \ldots, A_n$ and $B_1, \ldots, B_m$. Then, the function $\hat{f} : P \times Q \to \mathbb{R}$ represented by

$$\hat{f}(p_i, q_j) = \frac{\sum_{k=1}^{n} \sum_{l=1}^{m} F_{kl} A_k(p_i) B_l(q_j)}{\sum_{k=1}^{n} \sum_{l=1}^{m} A_k(p_i) B_l(q_j)} \tag{3}$$

is called the inverse F^0-transform of f.

2.3 F^1-transform

We can generalize the F-transform with constant components to the F^1-transform with linear components. The latter are orthogonal projections of an original function f onto a linear subspace of functions with the basis of polynomials $P_k^0 = 1, P_k^1 = (x - x_k)$. We say that the n-tuple

$$F^1[f] = [F_1^1, \ldots, F_n^1] \tag{4}$$

is the F^1-transform of f w.r.t. $A_1, \ldots, A_n$ where the $k-$th component F_k^1 is defined by

$$F_k^1 = c_{k,0} P_k^0 + c_{k,1} P_k^1, \; k = 1, \ldots, n. \tag{5}$$

For the h-uniform fuzzy partition and the triangular-shaped basic functions we can compute the coefficients $c_{k,0}, c_{k,1}$ for each $k = 1, \ldots, n$ as follows

$$c_{k,0} = \frac{1}{h} \sum_{i=1}^{N} f(p_i) A_k(p_i), \tag{6}$$

$$c_{k,1} = \frac{12}{h^3} \sum_{i=1}^{N} f(p_i)(p_i - x_k) A_k(p_i). \tag{7}$$

It can be shown that the coefficient $c_{k,0}$ is equal to the F-transform component F_k, $k = 1, \ldots, n$. The next theorem shows the important property of the coefficient $c_{k,1}$ which will be useful for the proposed edge detection technique. The theorem is formulated for the continuous version of the F^1-transform.

Theorem 1. *Let $A_1, \ldots, A_n$, be an h-uniform partition of $[a, b]$, let functions f and A_k, $k = 1, \ldots, n$ be four times continuously differentiable on $[a, b]$, and let $F^1[f] = [F_1^1, \ldots, F_n^1])$ be the F^1-transform of f with respect to $A_1, \ldots, A_n,$. Then, for every $k = 1, \ldots, n$, the following estimation holds true:*

$$c_{k,1} = f'(x_k) + O(h). \tag{8}$$

We refer to [10] for a proof of **Theorem 1** and for a detailed description of the F^1-transform.

3   Image Registration

This section focuses on applications of the F-transforms theory into image registration. The developed method is divided into four steps: feature extraction; feature matching; image mapping; image fusion.

3.1   Feature Extraction

Let us remark that there exist many algorithms for feature extraction, the most used are FAST, ORB or SIFT [11]. In this contribution, we understand the problem of feature extraction as a procedure that selects small corner areas in the image. According to the accepted terminology, we call the latter *point features*. Extracted point features in a reference and sensed images should be detected on the similar places even if the sensed image is rotated, resized or has different intensity. We propose an original technique of point features detection using the first degree F-transform (F^1-transform) adopted from [9].

By **Theorem 1**, coefficients $c_{k,1}$ of the F^1-transform give us a vector whose components approximate the first derivative of the original function at certain nodes. We use these coefficients as components of the inverse F-transform and we get the approximation of the first derivative of the original image function in each pixel.

Let triangular fuzzy sets $A_1, \ldots, A_n$ establish a fuzzy partition of $[1, N]$ and triangular $B_1, \ldots, B_m$ do the same for $[1, M]$. Let $x_1, \ldots, x_n \in [1, N]$, $h_x = x_{k+1} - x_k$, $k = 1, \ldots, n$ and $y_1, \ldots, y_m \in [1, M]$, $h_y = y_{l+1} - y_l$, $l = 1, \ldots, m$ be nodes on $[1, N]$, $[1, M]$ respectively. Then we can determine the approximation of the first derivative for each $(p_i, p_j) \in D$ in the horizontal direction

$$G_x(p_i, p_j) \approx \sum_{k=1}^{n} \sum_{l=1}^{m} c_{k,1}(y_l) A_k(p_i) B_l(p_j) \tag{9}$$

and in the vertical direction

$$G_y(p_i, p_j) \approx \sum_{k=1}^{n} \sum_{l=1}^{m} c_{l,1}(x_k) A_k(p_i) B_l(p_j) \tag{10}$$

as the inverse F-transform of the image function u where the coefficients $c_{k,1}(y_l)$, $c_{l,1}(x_k)$, $k = 1, \ldots, n$, $l = 1, \ldots, m$ are given by the F^1-transform

$$c_{k,1}(y_l) = \frac{12}{h_x^3} \sum_{i=1}^{N} f(p_i, y_l)(p_i - x_k)A_k(p_i), \tag{11}$$

$$c_{l,1}(x_k) = \frac{12}{h_y^3} \sum_{j=1}^{M} f(x_k, p_j)(p_j - y_l)B_l(p_j). \tag{12}$$

Then, the gradient magnitude G of an edge at point (p_i, p_j) is computed as

$$G(p_i, p_j) = \sqrt{G_x(p_i, p_j)^2 + G_y(p_i, p_j)^2} \tag{13}$$

and the gradient angle Θ is determined by

$$\Theta(p_i, p_j) = \arctan \frac{G_y(p_i, p_j)}{G_x(p_i, p_j)} \tag{14}$$

where for simplicity in according to [9] the gradient angle will be quantized by: $\Theta^Q : \Theta \to \{0, 45, 90, 135\}$.

Definition 4. *We say that a corner is a set of neighboring pixels (we call them corner points) where at least three different quantized angles show up. The center of gravity of a corner is called a feature point.*

Many corner points can be found in an image. It may happen that corner points are close to each other, and in this case, we have to choose only one of them. We modify computer graphic *flood fill* algorithm to detect clusters of close corner points and then compute centers of gravity of each cluster. These centers constitute the set of point features.

Image 1 simply demonstrate comparison of proposed algorithm with SIFT[11]. Two top images are original; two middle images were firstly vertically flipped, processed by both algorithms and then flipped back for comparison. Two bottom images were lighten and then processed by both algorithms. The result show that both algorithms works correctly for this simplest image modification. The feature points detection should hold invariance for difficult cases such as scale transformation or rotation. The research of rotation and scale invariance of the proposed algorithm deserve future work.

3.2   Feature Matching

In this step, a correspondence between the point features detected in the reference and sensed images is established. As a main technique (among various similarity measures or spatial relationships) we propose to measure similarity by a (inverse) distance between F^0-transform components of various levels.

In more details, the lowest (first) level is comprised by the F^0-transform components of image f and corresponds to the discretization given by the respective

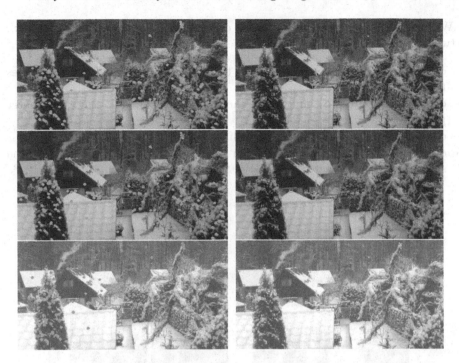

Fig. 1. Left: inpainted circles by SIFT. Right: inpainted squares by the own algorithm.

fuzzy partition of the domain. This first level $F^{(1)}[f]$ is given by the F^0-transform of f so that

$$F^{(1)}[f] = F[f] = (F_{11}, ..., F_{nm}).$$ (15)

The vector of the F^0-transform components $(F_{11}, ..., F_{nm})$ is a linear representation of a respective matrix of components. This first level serves as a new image for the F^0-transform components of the second level and so on. For a higher level ℓ we propose the following recursive formula:

$$F^{(\ell)}[f] = F[F^{(\ell-1)}] = (F_{11}^{(\ell-1)}, ..., F_{n_{(\ell-1)}m_{(\ell-1)}}^{(\ell-1)}).$$ (16)

The top (last) level $F^{(t)}[f]$ consists of only one final component F^{fin}.

The F^0-transform based similarity S of two image functions $f, g \in \mathcal{I}$ is proposed to be as follows:

$$S(f, g) = 1 - |F^{fin} - G^{fin}| \cdot \frac{\sum_{k=1}^{n} \sum_{l=1}^{m} |F_{kl}^{(1)} - G_{kl}^{(1)}|}{nm}$$ (17)

where F^{fin}, G^{fin} are the top F^0-transform components of f and g, and $F_{kl}^{(1)}, G_{kl}^{(1)}$, $k = 1, ..., n$, $l = 1, ..., m$ are the first level F^0-transform components of f and g, respectively. The justification that S is a similarity measure with respect to the product t-norm was given in [4].

We propose the following procedure in order to establishes a feature matching between two point features:

- choose two point features P and Q from the reference and sensed images respectively,
- create square areas S_P and S_Q of the same size around P and Q as center points,
- compute similarity $S(S_P, S_Q)$ according to (17),
- establish a feature matching between P and Q, if there is no point R in the sensed image such that $S(S_P, S_Q) < S(S_P, S_R)$.

Fig. 2. Green inpainted square illustrates an area around a point feature. Every couple of image fragments illustrates detected matching between corresponding point features.

The Figure 2 demonstrates feature matching in different images. Because of images can be obtained in different time, different light conditions or some noise can be in images, it is necessary to create similarity measure which will be robust enough to these changes. The used F^0-transform based similarity approximate image function with user defined accuracy defined by value of h parameter - see approximation theorem 2 in [2]. This property allow to compare images even if there are changes between them.

3.3   Image Mapping

The goal of the image mapping step is to find a shift between two images in x and y axis. The image mapping should determine a perspective distortion and then transform images in such a way that they would fit each other. For simplicity of demonstration, we show on Fig. 3 the result of registration of seven images where detected points features were shifted, but not interpolated. The shift between every two images is computed as an average shift between all pairs of matched points features. Figure 4 demonstrates result of existing software called AutoStitch published in [12]. It is obvious that the proposed approach is on the right way, but should be improved.

3.4   Image Fusion

Image fusion is the last step of our registration algorithm. It is applied to each overlapping part of input images with the purpose to extract the best representative pixels. The detailed description of used image fusion that has been applied in the proposed registration algorithm is in [8]. Figure 3 shows the final result of the proposed algorithm of image registration that uses seven input images. The result image is not perfect, there are lot of artifacts inside. The reason why the artifacts are there is because of mapping function without flexible grid cannot work with perspective distortion properly. Therefore some existing algorithm for an landscape composition can achieve better result.

Fig. 3. Registered and fused image of seven images

Fig. 4. Same result obtained by AutoStitch [12]

4   Conclusion

In the paper we apply the techniques of the F^0-transform and the F^1-transform to the problem of image registration. The technique of F^0-transform is used in computation of image similarity and that of the F^1-transform is a part of the gradient detection algorithm. The proposed theory is applied to image registration problem where F^1-transform edge detection is a base of detecting important areas (corners) in image. In order to find a correspondence between corner areas we used a newly proposed image similarity algorithm that helps in computation of shift vectors between images. Finally, overlapped part of images are inputs of image fusion algorithm that is based on F^0-transform. As a final result panorama image is created. Measuring of computation time show that the process of detection of feature points and map them is lower than one second without any perfect programming skill. The paper demonstrate how the unique technique of F-transform can be successfully used in various algorithms that comprise the multi-step problem of registration. The future work will be focused on the last step: to find image mapping function where the F-transform can be used (again) for operation of an image interpolation. Without these missing part the result is now worse than result obtained by existing approach called AutoStitch.

Acknowledgement. This work was supported by the European Regional Development Fund in the IT4Innovations Centre of Excellence project (CZ.1.05/1.1.00/02.0070). This work was also supported by SGS14/PrF/2013 project and "SGS/PrF/2014 – Vyzkum a aplikace technik soft-computingu ve zpracovani obrazu" project.

References

1. Perfilieva, I.: Fuzzy transforms: Theory and applications. Fuzzy Sets and Systems 157, 993–1023 (2006)
2. Perfilieva, I., De Baets, B.: Fuzzy Transform of Monotonous Functions with Applications to Image Processing. Information Sciences 180, 3304–3315 (2010)
3. Hurtik, P., Perfilieva, I.: Image compression methodology based on fuzzy transform. In: Herrero, Á., Snášel, V., Abraham, A., Zelinka, I., Baruque, B., Quintián, H., Calvo, J.L., Sedano, J., Corchado, E. (eds.) Int. Joint Conf. CISIS'12-ICEUTE'12-SOCO'12. AISC, vol. 189, pp. 525–532. Springer, Heidelberg (2013)
4. Hurtik, P., Perfilieva, I.: Image compression methodology based on fuzzy transform using block similarity. In: 8th Conference of the European Society for Fuzzy Logic and Technology (EUSFLAT 2013). Atlantis Press (2013)
5. Hurtik, P., Perfilieva, I.: Image Reduction/Enlargement Methods Based on the F-Transform, pp. 3–10. European Centre for Soft Computing, Asturias (2013)
6. Zitova, B., Flusser, J.: Image registration methods: a survey. Image and Vision Computing 21(11), 977–1000 (2003)
7. Perfilieva, I., Daňková, M.: Image fusion on the basis of fuzzy transforms. In: Proc. 8th Int. FLINS Conf., Madrid, pp. 471–476 (2008)
8. Vajgl, M., Perfilieva, I., Hod'kov, P.: Advanced F-Transform-Based Image Fusion. Advances in Fuzzy Systems 2012 (2012)
9. Perfilieva, I., Hodáková, P., Hurtík, P.: F^1-transform edge detector inspired by Canny's algorithm. In: Greco, S., Bouchon-Meunier, B., Coletti, G., Fedrizzi, M., Matarazzo, B., Yager, R.R. (eds.) IPMU 2012, Part I. CCIS, vol. 297, pp. 230–239. Springer, Heidelberg (2012)
10. Perfilieva, I., Daňková, M., Bede, B.: Towards F-transform of a higher degree. Fuzzy Sets and Systems 180, 3–19 (2011)
11. Lowe, D.G.: Distinctive Image Features from Scale-Invariant Keypoints. International Journal of Computer Vision 60 (2004)
12. Brown, M., Lowe, D.: Automatic Panoramic Image Stitching using Invariant Features. International Journal of Computer Vision 74(1), 59–73 (2007)

Improved F-transform Based Image Fusion*

Marek Vajgl and Irina Perfilieva

Institute for Research and Applications of Fuzzy Modeling
Centre of Excellence IT4Innovations
University of Ostrava, Czech Republic
{marek.vajgl,irina.perfilieva}@osu.cz

Abstract. The article summarizes current approaches to image fusion problem using fuzzy transformation (F-transform) with their weak points and proposes improved version of the algorithm which suppress them. The first part of this contribution brings brief theoretical introduction into problem domain. Next part analyses weak points of current implementations. Last part introduces improved algorithm and compares it with the previous ones.

Keywords: Image processing, image fusion, F-transform.

1 Introduction

The contribution focus on the problem of image fusion, what is one of the important research areas in image processing field. The main aim of image fusion is the comparison and integration of the distorted scenes into result image, which contains the best part extracted from the each input scene. The problem is how to define, select and extract the "best" pixel from all input scenes. The definition of the "best" pixel depends on the requested fusion result. In this contribution we aim at the *multi focus* images, which differ by focused area (e.g. see Fig. 1 (a),(b)). Moreover, if each pixel is "best" in the some input image, we are speaking about *mosaic multi focus*.

A measurement of the local focus is the main task, typically based on evaluation of the high frequencies in the image. The idea comes from the fact that blurry parts (which are unfocused) suppress high frequencies, so focus measure decreases with scene blur.

There already exist many approaches used to solve this problem. They differs by mathematical fields – statistical methods(e.g., usage of aggregation operators ([1]), estimation theory ([2]), fuzzy methods ([3], [4]), approaches based on the optimization using genetic algorithms ([5]), etc.).

In this article approaches based on the fuzzy transform (called F-transform for short) will be discussed. F-transform is integral transformation, for which

---

* This work was supported by the European Regional Development Fund in the IT4Innovations Centre of Excellence project (CZ.1.05/1.1.00/02.0070). This work was also supported by SGS14/PF/2013 project and "SGS/PF/2014 – Výzkum a aplikace technik soft-computingu ve zpracování obrazu" project.

A. Laurent et al. (Eds.): IPMU 2014, Part II, CCIS 443, pp. 153–162, 2014.
© Springer International Publishing Switzerland 2014

motivation came from fuzzy modeling ([6], [7]) and which was successfully applied in multiple areas of image processing, like image compression ([11]), edge detection ([12]) or image reconstruction ([13]).

The contribution presents image fusion solutions using F-transform as the main tool for image fusion([9], [10], [14]). After the experiments, further research was done to obtain further improvements. Those improvements are presented in this paper.

2   F-transform

As presented in the introduction, all methods explained in this contribution refers to the F-transform technique, which will be explained only briefly. For full description see [6].

F-transform performs linear mapping from the set of ordinary continuous or discrete functions over domain P into a set of discrete functions defined on a fuzzy partition of P using direct and inverse F-transform.

Let image function u is discrete function $P \to \mathbb{R}$ of two variables, defined over the set of pixels $P = \{(i,j)|i = 1, ..., N; j = 1, ..., M\}$. Let fuzzy sets $A_k \times B_l$, $k = 1, ..., n$, $l = 1, ..., m$, (in the meaning defined in [6]) establish a fuzzy partition of $[1, N] \times [1, M]$. The (direct) *F-transform* of u (with respect to the chosen partition) is an image of the map $F[u] : \{A_1, ..., A_n\} \times \{B_1, ..., B_m\} \to \mathbb{R}$ defined by

$$F[u](A_k \times B_l) = \frac{\sum_{i=1}^{N} \sum_{j=1}^{M} u(i,j) A_k(i) B_l(j)}{\sum_{i=1}^{N} \sum_{j=1}^{M} A_k(i) B_l(j)}, \tag{1}$$

where $k = 1, ..., n, l = 1, ..., m$. The value $F[u](A_k \times B_l)$ is called an *F-transform component* of u and is denoted by $F[u]_{kl}$. The components $F[u]_{kl}$ can be arranged into the matrix $\mathbf{F}_{nm}[u]$.

The *inverse F-transform* of u is a function on P representing inversed image u_{nm} obtained by the following inversion formula, where $i = 1, ..., N, j = 1, ..., M$:

$$u_{nm}(i,j) = \sum_{k=1}^{n} \sum_{l=1}^{m} F[u]_{kl} A_k(i) B_l(j). \tag{2}$$

It can be shown that the inverse F-transform u_{nm} approximates the original function u on the domain P. The proof can be found in [6,7].

Processing of direct and inverse F-transform applied sequentially is called *filter F-transform*. Values N, M are referred as *number-of-components – noc*, minimal value is 2. Number of pixels covered by one basis function A_k or B_l is called as *points-per-base – ppb* and minimal value is 3.

2.1   F-transform Application in Image Processing

The main approach used in almost all of the image processing application areas using F-transform is calculation of *residuals*, which can be obtained as a subtraction between original input and input processed by direct and inverse F-transform.

As the main property of the F-transform is removal of high frequencies from the input image, the reconstructed image do not contain high frequency artifacts, like noise, or sharp edges. The amount of removed frequencies depends on the *points–per–base/number-of-components* values.

2.2 Image Fusion Using F-transform

The main aim of the image fusion is integration of multiple input images into new one, where new image is in some way "better" than original input ones. In case of the multifocus area, "better" image means the image containing the most focused parts from the all input images.

Generally, for multifocus analysis, let u represents ideal image and $c_1, \ldots, c_K$ are acquired (input) images. Then the relation between each c_i and u can be expressed by

$$c_i(x,y) = d_i(u(x,y)) + e_i(x,y), \ i = 1, \ldots, K$$

where d_i is an unknown operator describing the image degradation, and e_i is some random noise. The aim of the fusion is to obtain fused image $\hat{u}$ such that it is closer to u (and therefore "better") than any of $c_1, \ldots, c_K$.

The basic idea of usage of F-transform for image decomposition. We assume that the image u is a discrete real function $u = u(x,y)$ defined on the $N \times M$ array of pixels $P = \{(i,j) \mid i = 1, \ldots, N, j = 1, \ldots, M\}$ so that $u : P \to \mathbb{R}$. Moreover, let fuzzy sets $A_k \times B_l$, $k = 1, \ldots, n$, $l = 1, \ldots, m$, where $2 \leq n \leq N, 2 \leq m \leq M$ establish a fuzzy partition of $[1, N] \times [1, M]$.

We begin with the following representation of u on P:

$$u(x,y) = u_{nm}(x,y) + e(x,y), \tag{3}$$

$$e(x,y) = u(x,y) - u_{nm}(x,y), \tag{4}$$

where $0 < n \leq N, 0 < m \leq M$, and u_{nm} is the inverse F-transform of u and e is the respective first difference. Value e represents *residuals* of the image u. If we replace e in (3) by its inverse F-transform e_{NM} with respect to the finest partition of $[1, N] \times [1, M]$, the above representation can then be rewritten as follows:

$$u(x,y) = u_{nm}(x,y) + e_{NM}(x,y), \forall (x,y) \in P. \tag{5}$$

We call (5) a *one-level decomposition* of u on P.

If function u is smooth, then the function e_{NM} is small, and the one-level decomposition (5) is sufficient for our fusion algorithm (see Section 3.2). However, for complex images there may be need to process next level decomposition of first difference e in (3) in following manner: we decompose e into its inverse F-transform $e_{n'm'}$ (with respect to a finer fuzzy partition of $[1, N] \times [1, M]$ with $n' : n < n' \leq N$ and $m' : m < m' \leq M$ basic functions, respectively) and the second difference e'. Thus, we obtain the *second-level decomposition* of u on P:

$$u(x, y) = u_{nm}(x, y) + e_{n'm'}(x, y) + e'(x, y),$$
$$e'(x, y) = e(x, y) - e_{n'm'}(x, y).$$

In the same manner, we can obtain a *higher-level decomposition* of u on P (see [14]).

In recursion, the whole iteration can be repeated to achieve such e decomposition fulfilling algorithm requirements.

3   Previous Implementations

3.1   Original (CA) Algorithm

Complete algorithm (CA for short) was the first approach researched in the image fusion based on the F-transform. The algorithm processes iteratively the input images where each iteration increases number-of-components value (and takes into account higher frequencies) and processes residuals from the previous iteration until stopping condition (too high number-of-components value or residuals are not significant).

As CA algorithm was described precisely in [14] with its behavior, and is replaced by ESA algorithm due to its memory and time consumption, it will not be explained.

3.2   Simple (SA, ESA) Algorithm

The main idea of *simple algorithm* (SA) is to process input image via direct and inverse F-transform to obtain residuals and then create final image by taking pixels from the image with the higher residual value (further referred as *greatest*). This algorithm is based on the idea of the minimization time and memory consumption.

Enhanced version (called ESA) of this algorithm (presented at [14]) did improve behavior. The main improvement is that result pixel is calculated as weighted sum of all input image pixel and weights represented by first level decomposition residuals (further referred as *weighted*) instead of picking one input image (referred as *greatest*), what suppress disturbing effect (compare Fig 2a vs. Fig. 2b).

The main advantage is lowered time and memory consumption. The main disadvantage of this approach is requirement to set input variable points-per-base by some expert, in wide range 3 – 50, where invalid selection has negative effect on the result. Another disadvantages are occurrences of artifacts as can be seen in 2. Some of them may be avoided by correct ppb selection, but some will remain.

3.3   Results Achieved by Presented Algorithms

Objective comparison of CA and ESA algorithm can be seen in Table 1 and Table 2 (only best result combinations are presented).

(a) Input image, background focus

(b) Input image, foreground focus

(c) Non-mosaic, "greatest" residuals

Fig. 1. ESA algorithm - fusion results, ppb=25

(a) ESA "greatest" ppb=5 – low ppp

(b) ESA "weighted" ppb=25 – "lake" artifact

(c) ESA "greatest" ppb=25 – "lake" artifact

(d) CA – "ghost" artifact

Fig. 2. Fusion results – artifacts

From the subjective point of view (except enormously complex scenes) ESA algorithm provides comparable results to CA algorithm in shorter time. However invalid ppb selection will produces low quality results.

Moreover, both implementations still create artifacts called as a "ghost/lakes", where blur edge is propagated from the blurred image into the final result (see Fig. 2).

4 Improved Implementation Based on ESA Algorithm - IESA

The main aim of improved ESA algorithm is to produce better results than ESA algorithm within acceptable time/memory consumption and to remove dependency on expert decision about points-per-base.

4.1   Ghost Issue Artifact Description

"Ghost" artifact occurs near the significant edges in the result image. They are caused by very blurry edges, which affect the "focus" detection in the image (see Fig. 2), because:

- Exactly at the edge, the more "focused" image has sharp intensity change between both sides of the edge (e.g. between plane wing and sky). Less focused image has only slight color change (due to blur over the edge) and therefore more focused image seems to be more representative.
- Near the edge, the more "focused" image may have no color change (e.g. near the plane wing is only sky). However, less focused image contains "fade" between the edge colors (e.g. sky near plane contains fade between sky and wing) and therefore the less focused image seems to be more representative.

As a result invalid image is preferred to obtain result. Therefore there is need to suppress this behavior by spreading influence of the edge out.

4.2   Improved Fusion

Improved algorithm is built on the idea that edge is taken as significant only if there is no more significant edge in the neighbor on different source scene. Therefore the first step is to influence edge to its neighbor by blurring residuals using F-transform. Second step is to fuse images together using one of the following techniques:

- *Greatest* - the image with the highest difference at the processed pixel is the winner and result image will be taken from the winning image only;
- *Greatest soften* - the image with the highest difference at the processed pixel is the winner (see Fig. 3). However, winning indices (weights) are blurred (using F-transform) to achieve slight fades between pixels from different images.
- *Weighted* - same as in ESA, but with threshold noise reduction.
- *Square-weighted* - same as previous, but with second power of the weights.

In implementation the "greatest" and "greatest-soften" approaches are calculated as weighted approaches with adequate weights set to zero.

4.3   Low Residual Values Reduction

Second part of the algorithm is the low residual values reduction - this occur when residual values are to low, so there is no significant edge. This is easily achieved by setting threshold value for residuals - if residual value is lower than threshold, it is set to zero. Otherwise there will be no change (see Fig. 3). Threshold value will be referred as *residual-threshold* with typical value set to 0.9.

(a) "Evolution" of weights during IESA

Fig. 3. Converting residuals to weights for "greatest soften" option

4.4   Final Algorithm

Final algorithm combines updated SA algorithm together with new features described above. It can be simply described as (numerical values are set according to experimental results):

1. Calculate F-transform filter and residual for each input image (see section 2.2).
2. Blur residuals using F-transform filter (optimal ppb value to do this is 25).
3. Remove low residual values (typical threshold value should be 0.9).
4. For "greatest"/"greatest-soften" approaches convert residuals ($\langle 0; 1 \rangle$) into weights ($\{0; 1\}$).
5. For "greatest-soften" blur the weights using F-transform filter (that's the "soften", for small ppb=5).
6. Calculate result image as combination of input images and weights.

(a) Non-mosaic, greatest

(b) Non-mosaic, greatest soften

(c) Non-mosaic, weighted

(d) Mosaic image

Fig. 4. IESA algorithm - cut-out fusion results, ppb=5, residual threshold=0.9

Cut-out of fusion result examples are shown in Fig. 4 to demonstrate artifacts elimination.

As shown in the following section, presented IESA algorithm achieves better results than preceding CA and ESA algorithms with comparable time consumption requirements.

IESA algorithm has two initial parameters - points-per-base and residual-threshold. Experiments show that both parameters are in most cases independent on input images and preferred values for ppb=6 and for residual-threshold=0.9.

5   Comparison to the Existing Algorithms

From the subjective point of view, results achieved by IESA algorithm are better than results achieved by ESA algorithm and CA algorithm.

For objective evaluation, two characteristics describing image "quality" are used: mean square error – MSE and peak signal to noise ratio – PSNR.

For fused image u_f and optimal (original) image u, MSE is defined as:

$$MSE_{u,u_f} = \frac{1}{M \times N} \sum_{m \in M, n \in N} (u(m,n) - u_f(m,n))^2 \tag{6}$$

The higher the MSE value is, the bigger is the difference between optimal image u and the image created by fusion u_f, so lower values are better. "Error" of MSE is evaluated against "optimal" image for mosaic multifocus, or against the best-available image created by expert in photo editor software for non-mosaic images.

Peak signal-to-noise ratio is calculated from MSE using formula:

$$PSNR_{u,u_f} = 20 \cdot log_{10} \frac{255}{\sqrt{MSE_{u,u_f}}} \tag{7}$$

Higher value in PSNR means lower noise information, so higher values are better.

Results for the algorithms can be seen in Table 1. IESA algorithm gives objectively much better results than so far best results of ESA algorithm (CA algorithm results were worse than SA (ppb=25) results and therefore were not presented).

It is very important to notice that approaches based on the greatest-residual (that are "greatest" and "greatest-soften") give better results than approaches based on weights. This supports idea that its better to pick pixel from specific image than to calculate pixel as weight of input images.

Computation requirements can be described by measurable characteristics: computation time and consumed memory, however they are very dependent on hardware and software. Moreover *number of direct or inverse F-transform evaluations - NoFT* characteristics is presented, describing number of direct of inverse F-transforms evaluations. Results for the algorithms can be seen in table 2.

Computation time of IESA algorithm is higher than ESA algorithm due to additional F-transform operations and more sophisticate residual processing. However, for testing image (cca 11.2 MP) the time is still at level of seconds.

Table 1. Comparison by characteristics for selected image, two input images 4129 × 2726 pixels (for CA algorithm – PPB column defines how ppb value is increased per iteration)

Algorithm	PPB	Other settings	Non-mosaic image		Mosaic image	
			MSE	PSNR	MSE	PSNR
CA	$2i$	–	–	–	6.036	40.300
CA	i^3	–	–	–	25.036	34.145
ESA	5	weighted	45.071	31.591	23.364	34.445
ESA	5	greatest	59.688	30.371	30.968	33.221
ESA	25	weighted	29.577	33.421	7.217	39.546
ESA	25	greatest	29.113	33.489	4.263	41.832
IESA	5	greatest	5.301	40.886	1.386	46.710
IESA	5	greatest-soften	4.507	41.591	0.921	48.485
IESA	5	weighted	5.243	40.934	2.504	44.143
IESA	5	squared-weights	9.408	38.395	1.600	46.089

Table 2. Comparison by resource consumption for selected image, two input images 4129 × 2726 pixels (for CA algorithm – PPB column defines how ppb value is increased per iteration; for CA algorithm – images are half sized as full size images processing did run out of memory)

Algorithm	PPB	Other settings	NoFT	Time (s)	Memory (MB)
CA	$2i$	–	11	25.504	781
CA	i^3	–	4	5.234	369
ESA	5	weighted	$2 \cdot n$	3.202	193
ESA	5	greatest	$2 \cdot n$	3.151	193
ESA	25	weighted	$2 \cdot n$	2.873	128
ESA	25	greatest	$2 \cdot n$	1.854	128
IESA	5	greatest	$4 \cdot n$	3.939	286
IESA	5	greatest-soften	$6 \cdot n$	4.447	214
IESA	5	weighted	$4 \cdot n$	3.281	286
IESA	5	squared-weights	$4 \cdot n$	3.099	279

6   Conclusion

In this paper we presented results about research of effective fusion algorithms based on previous research ([9] and [14]). Improved method presented in paper brings better results within time and memory resources limits and suppression of artifact occurring in result fused image; moreover dependency on input parameter is removed.

References

1. Blum, R.S.: Robust image fusion using a statistical signal processing approach. Information Fusion 6(2), 119–128 (2005)

2. Loza, A., Bull, D., Canagarajah, N., Achim, A.: Non-gaussian model-based fusion of noisy images in the wavelet domain. Computer Vision and Image Understanding 114(1), 54–65 (2010)
3. Singh, H., Raj, J., Kaur, G., Meitzler, T.: Image fusion using fuzzy logic and applications. In: Proceedings of the 2004 IEEE International Conference on Fuzzy Systems, vol. 1, pp. 337–340 (2004)
4. Ranjan, R., Singh, H., Meitzler, T., Gerhart, G.: Iterative image fusion technique using fuzzy and neuro fuzzy logic and applications. In: Annual Meeting of the North American Fuzzy Information Processing Society, NAFIPS 2005, pp. 706–710 (2005)
5. Mumtaz, A., Majid, A.: Genetic algorithms and its application to image fusion. In: 4th International Conference on Emerging Technologies, ICET 2008, pp. 6–10 (2008)
6. Perfilieva, I.: Fuzzy transforms: Theory and applications. Fuzzy Sets and Systems 157, 993–1023 (2006)
7. Perfilieva, I.: Fuzzy transforms: A challenge to conventional transforms. In: Hawkes, P.W. (ed.) Advances in Images and Electron Physics, vol. 147, pp. 137–196. Elsevier Academic Press, San Diego (2007)
8. Perfilieva, I., Daňková, M.: Image fusion on the basis of fuzzy transforms. In: Proc. 8th Int. FLINS Conf., Madrid, pp. 471–476 (2008)
9. Perfilieva, I., Daňková, M., Hodáková, P., Vajgl, M.: The Use of F-Transform for Image Fusion Algorithms. In: Proc. Intern. Conf. of Soft Computing and Pattern Recognition, SoCPaR 2010, pp. 472–477 (2010)
10. Hodáková, P., Perfilieva, I., Daňková, M., Vajgl, M.: F-transform based image fusion. In: Ukimura, O. (ed.) Image Fusion, pp. 3–22. InTech (2011),
http://www.intechopen.com/articles/show/title/
f-transform-based-image-fusion
11. Perfilieva, I., Pavliska, V., Vajgl, M., De Baets, B.: Advanced image compression on the basis of fuzzy transforms. In: Proc. Conf. IPMU 2008, Torremolinos, Malaga, Spain, pp. 1167–1174 (2008)
12. Perfilieva, I., Hodáková, P., Hurtík, P.: F^1-transform Edge Detector Inspired by Canny's Algorithm. In: Greco, S., Bouchon-Meunier, B., Coletti, G., Fedrizzi, M., Matarazzo, B., Yager, R.R. (eds.) IPMU 2012, Part I. CCIS, vol. 297, pp. 230–239. Springer, Heidelberg (2012)
13. Perfiljeva, I., Vlasanek, P., Wrublova, M.: Fuzzy transform for image reconstruction. In: Uncertainty Modeling in Knowledge Engineering and Decision Making, pp. 615–620. World Scientific, Singapore (2012) ISBN 978-987-4417-73-0
14. Prefiljeva, I., Vajgl, M.: Novel Image Fusion Based on F-transform. In: 2nd World Conference on Soft Computing Proceedings, pp. 165–171. Letterpress Publishing House (2012) ISBN 9789952452372

Visual Taxometric Approach to Image Segmentation Using Fuzzy-Spatial Taxon Cut Yields Contextually Relevant Regions

Lauren Barghout

Berkeley Institute for Soft Computing (BISC),
U.C. Berkeley, California, United States
lauren.barghout@gmail.com
http://www.laurenbarghout.org

Abstract. Images convey multiple meanings that depend on the context in which the viewer perceptually organizes the scene. By assuming a standardized natural-scene-perception-taxonomy comprised of a hierarchy of nested spatial-taxons [17] [6] [5], image segmentation is operationalized into a series of two-class inferences. Each inference determines the optimal spatial-taxon region, partitioning a scene into a foreground, subject and salient objects and/or sub-objects. I demonstrate the results of a fuzzy-logic-natural-vision-processing engine that implements this novel approach. The engine uses fuzzy-logic inference to simulate low-level visual processes and a few rules of figure-ground perceptual organization. Allowed spatial-taxons must conform to a set of "meaningfulness" cues, as specified by a generic scene-type. The engine was tested on 70 real images composed of three "generic scene-types", each of which required a different combination of the perceptual organization rules built into our model. Five human subjects rated image-segmentation quality on a scale from 1 to 5 (5 being the best). The majority of generic-scene-type image segmentations received a score of 4 or 5 (very good, perfect). ROC plots show that this engine performs better than normalized-cut [9] on generic-scene type images.

Keywords: visual taxometrics, natural vision processing, image segmentation, spatial taxon cut, fuzzy filter, spatial taxons, scene architecture, scene perception, fuzzy perceptual inference, fuzzy logic, image processing, graph partitioning.

1 Introduction

Segmenting images into meaningful regions is pre-requisite to solving most computer vision interpretation problems. Yet region relevancy depends less on the numeric information stored at each pixel, then on the computer vision task and corresponding scene architecture required to perceptually organize the constituent visual components necessary for the task. This presents a problem for automated image segmentation, because it adds uncertainty to the process of selecting which pixels to include or not include within a segment.

A. Laurent et al. (Eds.): IPMU 2014, Part II, CCIS 443, pp. 163–173, 2014.
© Springer International Publishing Switzerland 2014

An analogous problem exists for text document interpretation. Segmentation[1] of the document into its relevant components, such as characters, words, sentences or paragraphs, is pre-requisite to interpretation. However unlike images, text-documents have a standardized architecture with components designated by punctuation. Traditional punctuation and modern innovations such as hypertext[2] mark-up language (html), minimize uncertainty in the process of selecting which characters to include or not include within a segment.

Standardized architecture for written documents provide an example of a complex system that has proven to be stable across history, culture and technical innovation. As pointed out by Nobel Laureate Herbert Simon [11], "hierarchy is one of the central structural schemes that the architect of complexity uses". He further observes that "hierarchic systems have some common properties that are independent of their specific content" and he roughly defines a complex system as a system in which "the whole is more than the sum of the parts... in the pragmatic sense that given the properties of the parts and the laws of their interaction, it is not a trivial matter to infer the properties of the whole."

Text-document architecture succeeds because its structure is independent of content semantics. Letters, words, sentences and paragraphs follow the same structure regardless of whether they belong to a document discussing fashion, religion or nature.

The standardized natural-scene-perception-architecture described in this paper mimics text-document architecture in several ways: it's structured as a nested taxonomy, scene segment structure is independent of scene content semantics, standardized structure is used to minimize uncertainty as to which pixels belong within a segment; and architecture enables interpretation by delivering visually relevant components.

1.1 Visual Taxometrics and Spatial Taxons

Visual-taxometrics seeks to distinguish categorical visual percepts -such as figure/ground perception, from continuous visual percepts - such as distance or size. Spatial-taxons, categorical variables of 'whole things' such as foreground, object groups or objects (Barghout 2009), are 'building blocks' of scenes. In essence they serve as a proxy for the figural status of the region. When human subjects are asked to mark the center of the subject of the image, they tend choose the center of a spatial taxon with little variance and rarely choose locations defined solely by continuous visual percepts [17] [6]. Furthermore, evidence suggests that the frequency at which people choose spatial-taxons at a particular abstraction level, follow rank-frequency distribution similar to Zipf's law – independent of image content [6]. This is consistent with the law of least effort found in other cognitive systems and with Simon's observations of complex systems.

---

[1] Usually the literature uses the term 'parsing' instead of 'segmentation' to refer to breaking language into constituent parts. I chose this phrase to illustrate the information-normic (similarity) between image and language parsing.

[2] My collaboration with Roger Gregory, who pioneered hypertext with its inventor Ted Nelson, informed my understanding of this point.

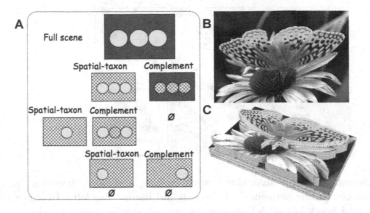

Fig. 1. (*A*) Natural-scene-perception-taxonomy comprised of a hierarchy of nested spatial-taxons. By assuming the taxonomy prior to segmentation, segmentation becomes a series of two-class fuzzy inferences. The full scene, top row, is at the highest level of abstraction. Each subsequent row is at lower level of abstraction within the taxonomy. (*B*) An image of a butterfly on a daisy. (*C*) A 3-dimensional version of image B where the third dimension (height) designates the abstraction level of the spatial taxon as shown in C.

The spatial-taxon view of scene perception assumes that humans parse scenes not between regions of similar features that vary continuously, but instead via discrete spatial 'jumps' biased toward taxometric scene configurations. Theories of visual attention make a similar distinction. The "spotlight theory" [12] assumes that attention regions vary continuously. Theories of "object based" attention assume that attended spatial regions vary in discrete location jumps as it accommodates attended objects.

If humans are parsing scenes by inferring categories, then quantifying pixel-region as to their aggregate "trueness" relative to the category prototype is prerequisite to human inspired computerized image segmentation. Humans assign meaning to visual percepts that they use to infer categories. Fuzzy-logic, which provides tools for handling partial or relative truth of meaning [15], enables inference based on visual percepts [4]. I've coined the phrase "natural-vision-processing" to refer to the parsing of images into psychological variables whose relative truth (fuzzy membership) corresponds to human phenomenological interpretation. Gestalt psychological variables such as similarity, good continuation, symmetry and proximity as first introduced by Wertheimer [13] provide the basis for fitting membership functions. A more detailed discussion on fitting Gestalt variables with fuzzy membership functions can be found in Barghout (2003) [4]. This paper focuses on fuzzy methods for optimizing spatial-taxon inference after a hypothetical set has been posited from Gestalt variables.

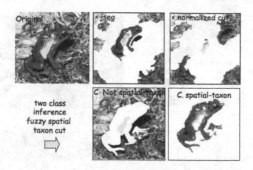

Fig. 2. Image segmentation algorithms, such as jseg and normalized-cut, produce what I call "jig-saw puzzle segments" of the original image (top left). In other words, though they do a good job of delineating regions of similar percepts, they are not meaningful to people. (*A*) Output of jseg algorithm [16]. From UCSB and downloaded 2010 ,version 6b. (*B*) Output of normalized-cut algorithm [18]. 2010 version. (*C*) The natural image processing engine output for not-spatial taxon inference and (*D*) spatial taxon inference.

1.2   Prior Work on Image segmentation

Most image segmentation algorithms stem from the school of thought that attention varies continuously over retinotopic location. Thus it makes sense to view images as a graph, image segmentation as a graph partitioning problem and precise high-dimension descriptive data at each graph node as pre-requisite to solving computer vision problems. For these approaches the criterion for graph partition is vital. They tend choose criterion of maximal contrast, where contrast is defined between summary statistics aggregated over candidate regions [9], [16]. Shi and Malik [9]) provide an excellent review of these methods. Though these methods succeed in parsing dissimilar regions, the regions in and of themselves are not meaningful. For example, figure 2 shows regions parsed to maximize differences between regions. The segments look like jigsaw puzzle pieces. Each jigsaw segment is not relevant to the visual understanding of the context and content or scene organization.

Fuzzy logic, however, provides an alternative school of thought where it makes sense to view images as spatially overlapping universe of discourses, image segmentation as a fuzzy set classification inference problem and the relative truth of the meaning of underlying a segmentation query [14] as pre-requisite to solving computer vision problems. In this way, image segmentation becomes a series of fuzzy two-class inference problems.

2   Fuzzy Natural Vision Processing and Spatial-Taxon Cut

By assuming the taxonomy prior to segmentation, parsing an image becomes a series of two-class fuzzy inferences. In this section, I will describe a system that implements image segmentation as a nested two-class fuzzy inference system. Figure 3 provides an overview of the whole system. The sub-system (box A), shown on the left is similar to other fuzzy systems. It contains a fuzzification phase where the crisp values contained in the original image are reparameterized into fuzzy cognitively relevant variables (CV). CVs are designed to fit human data or mimic human psychophysical and perceptual variables. A discussion along with detailed examples of calculations of fuzzy CVs can be found in Barghout (2003). Meaningfulness cues are composition styles with known CV spatial-taxon configurations. Its inference system, uses CV premises and meaningfulness rules to posit hypothetical spatial-taxons. Thus far, the fuzzy logic system is pretty standard in its design.

The next process (box C on the right), decides on the hypothetical spatial-taxon set and appropriate weighting. It is novel to this system. The system iterates through various combinations of hypothetical spatial-taxons, to infer the defuzzifed spatial-taxon that would result from each combination, and scoring the output for each combination. This enables posits to 'abstain'[3]. The score is a combination of spatial-taxon utility and the attentional resource requirement of the hypothetical spatial-taxon combination. The optimal set is chosen such that it maximizes utility and minimizes attentional resources.

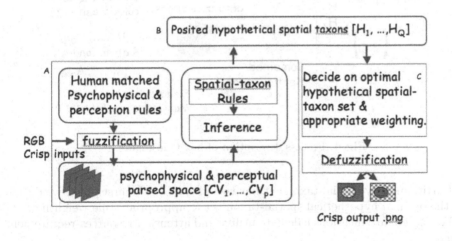

Fig. 3. Fuzzy natural-scene-perception system

The utility function I use to score the posited spatial-taxon was inspired by a seminal study of pictorial object naming [10] that found that objects were identified first at an "entry point" level of abstraction. Curious as to the whether the scene-architecture had an 'entry level' region, I undertook a multi-year study

---

[3] The idea to allow psychological detectors to abstain from contributing information to the system was suggested to me by Lotfi Zadeh in 2006, personal communication.

(2007-2011) surveying participants at the Burningman Arts Festival in NV, the Macworld conference in CA and the department of motor vehicles in Raleigh, N.C. The results suggest that images do indeed have an entry-level spatial taxon. Furthermore the spatial-taxon rank-frequency distribution measured in these studies suggest a law of least effort similar to that found in other cognitive processes [7]. Thus the utility function is inspired by the law of least effort. I define it operationally over an ordinal scale such that entry-level had the most utility, super-ordinate the next highest utility and all sub-ordinate decrease utility as a function of abstraction. This is a soft restriction, with granularity at abstraction levels. Use of attentional resources was also defined on an ordinal scale with granularity at the number of hypothetical spatial-taxons possible in the natural-vision processing engine. It's constrained to be inversely related to the number of significant spatial-taxon combination sets above threshold, where threshold was defined in terms of sub-population variance verses variance of the sub-population with the lowest within-group variance. This process is described in figure 4. Figure 5 provides a pictorial illustration using the image marked as "original" in figure 6.

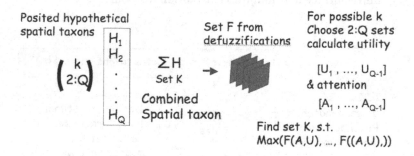

Fig. 4. Process description of box C in Figure 3

Partitioning by spatial-taxon cut has two phases. In the first phase we decide on the optimal hypothetical spatial-taxon set & appropriate rule weighting.

For $\left[\begin{smallmatrix} k \\ 2:Q \end{smallmatrix}\right]$ defuzzifications calculate utility and attention-resources-requirement where

$$Utility(\Phi) = \int \int_{\Phi} hypothetical - spatial - taxon - utility(\Phi)d\Phi \qquad (1)$$

$$Attenional_{r}esources(\Phi) = \int \int_{\Phi} Attentional - inference - load(\Gamma)d\Phi \qquad (2)$$

Let A be a fuzzy set defined on a universe of Φ discrete meaningfulness cues $\Phi = [\Phi_1, \Phi_2, ..., \Phi_a]$ defined on the universe of discourse of two discrete scene architecture states $S = [s_1, s_2]$ where s_1 is a spatial taxon and s_2 is the background[4] Set Φ represents the hypothetical spatial-taxons organizing constraints.

In the second phase, we "cut" the spatial taxon by defuzzifying the fuzzy conclusion. The crisp conclusion is normalized between zero and one. Spatial-taxon threshold is chosen according to use-case. In this system the threshold was set to 0.5.

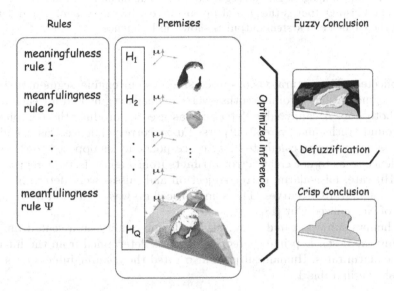

Fig. 5. Pictorial example of spatial-taxon inference as described in Figure 4. The mu axis designates the fuzzy firing power of each spatial taxon.

Figure 5, a woman wearing a hooded poncho in a field of yellow flowers, is used as an example. A series of meaningfulness cues are used to posit hypothetical spatial taxon. To make it easier to follow, I show cut-outs from the original image, next to the meaningfulness cues. An original is shown if the meaningfulness firing power of that pixel exceeds threshold. Note that because the poncho is orange, the intersection of yellow and red, it has membership in spatial-taxons and complement. These conflicting cues abstain because including them in the set drains attentional resources and provides little utility.

3   Performance Test Methods

70 real images composed of four "generic scene types", each of which required a different combination of the perceptual organization rules built into our model,

---

[4] Though the human perceptual state of "ground" extends beyond the subject and thus has fuzzy borders, it's digital image counterpart exists in a defined pixel set such that the "ground" is the complement of the spatial taxon.

Fig. 6. The Golden image was hand segmented by a human and is considered "ground truth". The Crisp Output is the spatial taxon inferred by the system. The difference between the Golden and system output is shown in Difference.

were collected. The natural-vision-processing system engine segmented them. Golden segmentations (ground truths) were manually segmented for each image using photoshop. A canny edge detector was used to produce the contours for both ground truths and system outputs. Fuzzy correspondence was calculated [3]. It was important to calculate fuzzy correspondence as opposed to crisp correspondence so as to not create error artifacts from slight offsets or registration errors. Hit-rate, false-alarm, correct-rejection and misses were determined and used to calculate ROC curves. The same procedure was used on a downloaded version of Normalized Cut [18].

Five human subjects rated image-segmentation quality on a scale from 1 to 5 (5 being the best). D-prime (detectability) was determined from the hit-rates and false alarm rates. Human subjects also rated the meaningfulness cues with results shown in Table 1.

Table 1. shows the four K spatial-taxon sets. An ANOVA was used to extract the relative proportion of meaningfulness cue for each corpus type - shown as linguistic hedges - as scored by 5 human subjects.

Cluster 1		Cluster 2	
Linguistic Hedge	Meaningfulness Cue	Linguistic Hedge	Meaningfulness Cue
abstain	Blurry	some	Blurry
some	Color Surround	abstain	Color Surround
very	Connected Taxon Color	very	Connected Taxon Color
abstain	Wall-like Background	some	Wall-like Background
Cluster 3		**Cluster 4**	
Linguistic Hedge	Meaningfulness Cue	Linguistic Hedge	Meaningfulness Cue
very	Blurry	low	Blurry
very	Color Surround	some	Color Surround
abstain	Connected Taxon Color	some	Connected Taxon Color
some	Wall-like Background	not	Wall-like Background

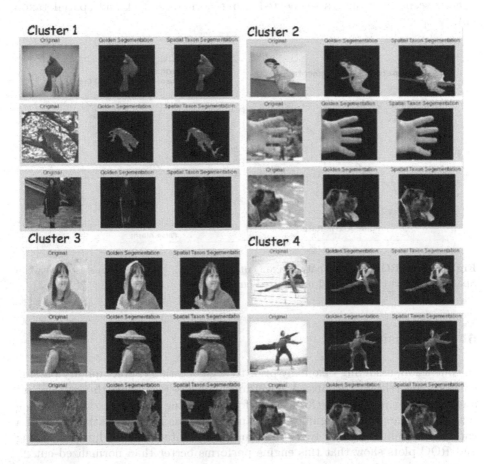

Fig. 7. Example engine outputs organized by meaningfulness cue combination cluster. In each example, the original image is on the left, the golden (hand segmented ground truth) in the middle and spatial taxon segmentation on the right.

4   Segmentation Results

ROC curves for all 70 images, figure 8a, show that the majority of images are well segmented. This is confirmed by humans scoring (5 subjects) that show that the majority of generic-scene-type images segmented via spatial taxon method received a score of 4 or 5 (very good, perfect). Figure 8b, ROC plots for 20 generic-scene-type images segmented using normalized cut and spatial taxon cut.

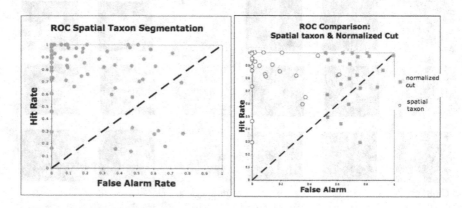

Fig. 8. (Left) ROC plot for spatial-taxon cut (70 images). (Right) Comparison between Spatial-taxon cut (circle) and normalized cut (square).

5   Conclusion

In conclusion, assuming taxonomy prior to segmentation enables quality parsing contextually relevant regions. A novel methodology that finds the optimal set and weight of premises performs well for optimizing spatial-taxon cut. Using fuzzy inference provides significant advantage for quantifying relative truth of a category, enabling cognitively relevant image segmentation. Both human grading and ROC plots show that this engine performs better than normalized-cut [9] on generic-scene type images.

Acknowledgments.   Roger Gregory, Eyegorithm's co-founder and I co-wrote the code base on which the results were obtained. Special thanks to Dr. Christopher Tyler, Dr. Steve Palmer, Dr. Lotfi Zadeh and Dr. Lora Likova provide valuable feedback, suggestions and advice. Other collaborators include Haley Winter, Analucia DaSivla, Yurik Riegal, Colin Rhodes, Eric Rabinowitz, and Shawn Silverman. BurningEyeDeas LLC, an organization that does research at the Burningman art festival. Data posted at www.burningeyedeas.com.

References

1. Ruscio, J., Haslam, N., Ruscio, A.: Introduction To Taxometric Method. Lawrence Eelbaum Associates (2006)
2. Barghout, L.: Linguistic Image Label Incorporating Decision Relevant Perceptual, Semantic, and Relationships Data. USPTO. patent application 20080015843 (2007)
3. Barghout, L.: System and Method for Edge Detection in Image Processing and Recognition. WIPO Patent Application. WO/2007/044828 (2006)
4. Barghout, L., Lee, L.: Perceptual information processing system. USPTO patent application number: 20040059754 (2003)
5. Barghout, L., Sheynin, J.: Real-world scene perception and perceptual organization: Lessons from Computer Vision. Journal of Vision 13(9) (July 24, 2013)
6. Barghout, Winter, Riegal: Empirical Data on the Configural Architecture of Human Scene Perception and Linguistic Labels using Natural Images and Ambiguous figures. In: VSS 2011 (2011)
7. Cancho, Sole: Zipf's law and random texts. Advances in Complex Systems 5(1), 1–6 (2002)
8. James, W.: Principles of psychology, p. 403. Holt, New York (1890)
9. Shi, J., Malik, J.: Normalized Cuts and Image Segmentation. IEEE TPAMI 22(8) (2000)
10. Jolicoeur, Gluck, Kosslyn: Pictures and names: making the connection. Cognitive Psychology 16, 243–275 (1984)
11. Simon, H.: The Architecture of Complexity. Proceedings of the American Philosophical Society 106(6), 467–482 (1962)
12. Treisman, A.M.: Strategies and models of selective attention. Psychological Review 76(3), 282–299 (1969)
13. Wertheimer, M.: Laws of Organization in Perceptual Forms (partial translation). In: Ellis, W.B. (ed.) A Sourcebook of Gestalt Psychology, pp. 71–88. Harcourt Brace (1938)
14. Zadeh, L.: Outline of a new approach to the analysis of complex systems and decision processes. IEEE Trans. Syst. Man & Cybern. SMC-3 (1973)
15. Zadeh, L.: Toward a Restriction-centered Theory of Truth and Meaning (RCT). Information Sciences 248 (2013)
16. Deng, Y., Manjunath, B., Shin, H.: Color image segmentation. In: IEEE Computer Society Conference on Computer Vision and Pattern Recognition, vol. 2 (1999)
17. Barghout, L.: Empirical Data on the Configural Architecture of Human Scene Perception using Natural Image. J. Vis. 9(8), 964 (2009), doi:10.1167/9.8.964
18. Berkeley Segmentation Database, http://www.eecs.berkeley.edu/Research/Projects/CS/vision/bsds

Multi-valued Fuzzy Spaces for Color Representation

Vasile Patrascu

Tarom Information Technology, Bucharest, Romania
patrascu.v@gmail.com

Abstract. This paper proposes two complementary color systems: *red-green-blue-white-black* and *cyan-magenta-yellow-black-white*. Both systems belong to the five-valued category and they represent some particular case of neutrosophic information representation. The proposed multi-valued fuzzy spaces are obtained by constructing fuzzy partitions in the unit cube. In the structure of these five-valued representations, the negation, the union and the intersection operators were defined. Next, using the proposed multi-valued representation in the framework of fuzzy clustering algorithm, it results some color image clustering procedure.

Keywords: fuzzy color space, five-valued representation, intuitionistic fuzzy sets, neutrosophic set.

1 Introduction

A color image generally contains tens of thousands of colors. Therefore, most color image processing applications first need to apply a color reduction method before performing further sophisticated analysis operations such as segmentation. The use of color clustering algorithm could be a good alternative for color reduction method construction. In the framework of color clustering procedure, we are faced with two color comparison subject. We want to know how similar or how different two colors are. In order to do this comparison, we need to have a good coordinate system for color representation and also, we need to define an efficient color similarity measure in the considered system. The color space is a three-dimensional one and because of that for a unique description there are necessary only three parameters. Among of the most important color systems there are the following: *RGB, HSV, HSI, HSL, Luv, Lab, I1I2I3*. This paper presents two systems for color representation called *rgbwk* respectively *cmykw* and they belong to the multi-valued color representation [14]. The presented systems are obtained by constructing a five-valued fuzzy partition of the unit cube. The sum of the parameters r,g,b,w,k and c,m,y,k,w verifies the condition of partition of unity and we can apply some similarities related to this property. Thus, one obtains new formulas for color similarity/dissimilarity. The paper has the following structure: Section 2 presents the construction modality for obtaining of the five-valued color representation *rgbwk*, the inverse transform from the *rgbwk* color system to *RGB* one, and the definition in the framework of the proposed color representation for the negation, the union and the intersection operators. Section 3 presents the using

A. Laurent et al. (Eds.): IPMU 2014, Part II, CCIS 443, pp. 174–183, 2014.

of the proposed multi-valued representation in the framework of the k-means cluster-ing algorithm. The presentation is accompanied with some experimental results. Finally, Section 4 outlines some conclusions.

2   The Construction of a Five-Valued Color Representation

For representing colors, several color spaces can be defined. A color space is a defini-tion of a coordinate system where each color is represented by a single vector. The most commonly used color space is *RGB* [6]. It is based on a Cartesian coordinate system, where each color consists of three components corresponding to the primary colors *red*, *green* and *blue*. Other color spaces are also used in the image processing area: linear combination of *RGB* (similar to *I1I2I3* [9]), color spaces based on human color terms like *hue*, *saturation* and *luminosity* (similar to *HIS* [4], *HSV* [18], *HSL* [8]), or perceptually uniform color spaces (similar to *Lab* [5], *Luv* [3].

2.1   The Fuzzy Color Space rgbwk

We will construct this new representation starting from the *RGB* (*red*, *green*, *blue*) color system. We will suppose that the three parameters take value in the interval [0,1]. We will define the maximum V, the minimum v, the hue H, the luminosity L [12] and the saturation S [10]:

$$V = \max(R,G,B), \quad v = \min(R,G,B) \tag{1}$$

$$H = atan2 \left(\frac{(G-B)\sqrt{3}}{2}, R - \frac{B+G}{2} \right) \tag{2}$$

$$L = \frac{V}{1+V-v} \tag{3}$$

$$S = \frac{2(V-v)}{1+|v-0.5|+|V-0.5|} \tag{4}$$

Firstly, we will define a fuzzy partition with two sets: the fuzzy set of chromatic col-ors and the fuzzy set of achromatic colors. These two fuzzy sets will be defined by the following two membership functions:

$$\rho_C = S \tag{5}$$

$$\rho_A = 1 - S \tag{6}$$

We obtained the first fuzzy partition for the color space:

$$\rho_C + \rho_A = 1 \tag{7}$$

The parameter ρ_C is related to the color chromaticity while ρ_A is related to the color achromaticity.

Next in the framework of the chromatic colors, we will define the reddish, bluish and greenish color sets by the following formulae:

$$r = \frac{S}{\sigma}\left(\gamma_R - \frac{\min(\gamma_R, \gamma_G) + \min(\gamma_R, \gamma_B)}{2}\right) \tag{8}$$

$$g = \frac{S}{\sigma}\left(\gamma_G - \frac{\min(\gamma_R, \gamma_G) + \min(\gamma_G, \gamma_B)}{2}\right) \tag{9}$$

$$b = \frac{S}{\sigma}\left(\gamma_B - \frac{\min(\gamma_R, \gamma_B) + \min(\gamma_G, \gamma_B)}{2}\right) \tag{10}$$

where

$$\gamma_R = \cos(H), \quad \gamma_G = \cos\left(H - \frac{2\pi}{3}\right), \quad \gamma_B = \cos\left(H + \frac{2\pi}{3}\right) \quad \text{and}$$

$$\sigma = \max(\gamma_R, \gamma_G, \gamma_B) - \min(\gamma_R, \gamma_G, \gamma_B)$$

There exists the following equality:

$$r + g + b = \rho_C \tag{11}$$

After that, in the framework of the achromatic colors, we define two subsets: one related to the white color and the other related to the black color:

$$w = \rho_A \cdot L \tag{12}$$

$$k = \rho_A \cdot (1 - L) \tag{13}$$

There exists the following equality:

$$w + k = \rho_A \tag{14}$$

From (7), (11) and (14) it results the subsequent formula:

$$r + g + b + w + k = 1 \tag{15}$$

We obtained a five-valued fuzzy partition of unity and in the same time we obtained a five-valued color representation having the following five components: r (*red*), g (*green*), b (*blue*), w (*white*) and k (*black*). We must observe that among the three chromatic components r, g, and b at least one of them is zero, explicitly $\min(r, g, b) = 0$.

2.2   The Inverse Transform from rgbwk to RGB

In this section, we will present the computing formulas for the *RGB* components having as primary information the *rgbwk* components. Firstly, we will compute the *HSL* components and then the *RGB* ones. Thus for the computing of luminosity *L*, we will use the achromatic components *w,k*.

$$L = \frac{w}{w+k} \tag{16}$$

For the computing of saturation S and hue H, we will use the chromatic components *r,g,b*.

$$S = r + g + b \tag{17}$$

$$H = atan2\left(\frac{\sqrt{3}(\omega_G - \omega_B)}{2}, \omega_R - \frac{\omega_G + \omega_B}{2}\right) \tag{18}$$

where

$$\omega_R = r + \min(r, b + g) \tag{19}$$

$$\omega_G = g + \min(g, b + r) \tag{20}$$

$$\omega_B = b + \min(b, r + g) \tag{21}$$

For the *RGB* components, we have the following formulae:

$$R = (V - v)\frac{\omega_R}{S} + v \tag{22}$$

$$G = (V - v)\frac{\omega_G}{S} + v \tag{23}$$

$$B = (V - v)\frac{\omega_B}{S} + v \tag{24}$$

The parameters V and v can be determined solving the system of equations (3) and (4) and taking into account (16) and (17).

2.3   Negation, Union and Intersection for rgbwk Space

In the following, we consider the negation of color $Q = (R,G,B)$, namely $\overline{Q} = (C,M,Y) = (1-R, 1-G, 1-B)$. Using (2), (8), (9), (10), (12) and (13) it results:

$$\bar{r} = c = \frac{S}{\sigma}\left(\frac{\max(\gamma_R,\gamma_G) + \max(\gamma_R,\gamma_B)}{2} - \gamma_R\right) \tag{25}$$

$$\bar{g} = m = \frac{S}{\sigma}\left(\frac{\max(\gamma_R,\gamma_G) + \max(\gamma_G,\gamma_B)}{2} - \gamma_G\right) \tag{26}$$

$$\bar{b} = y = \frac{S}{\sigma}\left(\frac{\max(\gamma_B,\gamma_G) + \max(\gamma_R,\gamma_B)}{2} - \gamma_B\right) \tag{27}$$

$$\bar{w} = k = \rho_A \cdot (1-L) \tag{28}$$

$$\bar{k} = w = \rho_A \cdot L \tag{29}$$

The five components defined by (25), (26), (27), (28) and (29) verify the condition of partition of unity, namely:

$$c + m + y + k + w = 1 \tag{30}$$

and in addition:

$$c + m + y = \rho_C$$

After the negation operation, we obtained a new five-valued fuzzy partition of unity and in the same time we obtained a five-valued color representation having the following components: c (*cyan*), m (*magenta*), y (*yellow*), k (*black*) and w (*white*). We must observe that among the three chromatic components c, m and y at least one of them is zero, explicitly,

$$min(c,m,y) = 0$$

There exist the following equivalent relations between these two complementary systems:

$$r = \min(y,m) + \frac{m+y}{2} - \frac{\min(c,m) + \min(c,y)}{2} \tag{31}$$

$$g = \min(y,c) + \frac{c+y}{2} - \frac{\min(c,m) + \min(m,y)}{2} \tag{32}$$

$$b = \min(c,m) + \frac{m+c}{2} - \frac{\min(c,y) + \min(m,y)}{2} \tag{33}$$

$$c = \min(g,b) + \frac{g+b}{2} - \frac{\min(r,g) + \min(r,b)}{2} \tag{34}$$

$$m = \min(r,b) + \frac{r+b}{2} - \frac{\min(r,g)+\min(g,b)}{2} \qquad (35)$$

$$y = \min(r,g) + \frac{r+g}{2} - \frac{\min(b,r)+\min(b,g)}{2} \qquad (36)$$

We must highlight that the pair *Red-Cyan* (R,C) defines a fuzzy set [19] for the reddish color and it verifies the condition of fuzzy sets, namely $R+C=1$. The pair *red-cyan* (r,c) defines an Atanassov's intuitionistic fuzzy set [1] for the reddish colors and it verifies the condition $r+c \le 1$. Similarly, the pair *blue-yellow* (b,y) defines an Atanassov's intuitionistic fuzzy set for bluish colors, the pair *green-magenta* (g,m) defines an Atanassov's intuitionistic fuzzy set for greenish colors while the pair (w,k) defines an Atanassov's intuitionistic fuzzy set for the white color. Thus for the color $Q = (0.3,0.5,0.8)$, one obtains the fuzzy set $R = 0.3$ and $C = 0.7$, while for intuitionistic fuzzy description one obtains $r=0$, $c=0.53$. The intuitionistic description is better than fuzzy description because the color Q is a bluish one and then reddish membership degree must be zero. More than that for the white color $Q = (1,1,1)$ one obtains for the fuzzy set description $R=1$ and $C=0$ while for intuitionistic description one obtains $r=0$, $c=0$. Again, the intuitionistic description is better than the fuzzy one. Thus, the fuzzy description is identically with that of the red color while in the framework of intuitionist fuzzy description, the intuitionistic index is 1. This value is a correct value for an achromatic color like the white color.

Taking into account the neutrosophic theory proposed by Smarandache [15], [16], [17] we can consider that the vector (r,g,b,w,k) provides a neutrosophic set for reddish colors. From this point of view, r represents the membership function, b,g represent two non-membership functions while w,k represent two neutralities functions. For this neutrosophic set, we define the union and intersection .

The Union
For any two colors $p = (r_p, g_p, b_p, w_p, k_p)$ and $q = (r_q, g_q, b_q, w_q, k_q)$ we define the union by formulae:

$$r_{p \cup q} = \max(r_p, r_q) \qquad (37)$$

$$g_{p \cup q} = \min(g_p, g_q) \qquad (38)$$

$$b_{p \cup q} = \min(b_p, b_q) \qquad (39)$$

$$w_{p \cup q} = \max((w_p + r_p, w_q + r_q) - \max(r_p, r_q) \qquad (40)$$

$$k_{p \cup q} = \min(k_p + g_p + b_p, k_q + g_q + b_q) - \min(g_p, g_q) - \min(b_p, b_q) \qquad (41)$$

The Intersection

For any two colors $p = (r_p, g_p, b_p, w_p, k_p)$, $q = (r_q, g_q, b_q, w_q, k_q)$ we compute the intersection using two steps. Firstly, we compute the intersection for the color negations by formulae:

$$c_{p \cap q} = \max(c_p, c_q) \tag{42}$$

$$m_{p \cap q} = \min(m_p, m_q) \tag{43}$$

$$y_{p \cap q} = \min(y_p, y_q) \tag{44}$$

$$k_{p \cap q} = \max((k_p + c_p, k_q + c_q) - \max(c_p, c_q) \tag{45}$$

$$w_{p \cap q} = \min(w_p + m_p + y_p, w_q + m_q + y_q) - \min(m_p, m_q) - \min(y_p, y_q) \tag{46}$$

Secondly, having $c_{p \cap q}, m_{p \cap q}, y_{p \cap q}$ and using (31),(32) and (33) we compute $r_{p \cap q}$, $g_{p \cap q}$ and $b_{p \cap q}$.

The results of union and intersection verify the condition of partition of unity.

Also, the union and intersection are associative, commutative and verify the De Morgan properties. Similarly, we can define these two operations for blue, green, yellow, magenta or cyan colors.

More than that, we can construct five-valued neutrosophic set for any color hue but this construction will not be subject of this paper.

3   Color Clustering in the rgbwk Color Space

For any two colors $p = (r_p, g_p, b_p, w_p, k_p)$, $q = (r_q, g_q, b_q, w_q, k_q)$ we compute the Bhattacharyya similarity [2]:

$$F(p, q) = \sqrt{r_p r_q} + \sqrt{g_p g_q} + \sqrt{b_p b_q} + \sqrt{w_p w_q} + \sqrt{k_p k_q} \tag{47}$$

and its dissimilarity:

$$D(p, q) = \sqrt{1 - F(p, q)} \tag{48}$$

Using the dissimilarity defined by (48) in the framework of k-means algorithm, one obtains a color clustering algorithm. The algorithm k-means [7], [11] is one of the simplest algorithms that solve the clustering problem. The procedure classifies a given data set through a certain number of clusters fixed a priori. The main idea is to define k centroids, one for each cluster. The next step is to take each point belonging to the data set and associate it to the nearest centroid. After that, the cluster centroids are recalculated and k new centroids are obtained. Then, a new binding has to done between the same data set points and the nearest new centroid. A loop has been generated. This algorithm aims at minimizing an objective function, in this case a squared error function. The objective function is defined by:

$$J = \sum_{j=1}^{k} \sum_{i=1}^{n} D^2(x_i^{(j)}, c_j) \qquad (49)$$

where $D^2(x_i^{(j)}, c_j)$ is a chosen dissimilarity measure between a data point x_i^j and the cluster center c_j. The function J represents an indicator of the dissimilarity of the n data points from their respective cluster centers. Using in (49) the dissimilarity (48), we obtained the experimental results shown in figures 1 and 2.

For the image "bird" shown in figure 1, only in the case (i) for the *rgbwk* system the orange color was separated. For the image "bird", the uniform green background was well separated for the *HIS*, *HSV*, *Lab*, *Luv* and *rgbwk* color systems. For the image "flower" shown in figure 2, the orange color was separated in the case (h) for the *Lab* system and in the case (i) for the *rgbwk* system. For the image "flower", the uniform gray background was well separated using the *Lab*, *Luv* and *rgbwk* color systems.

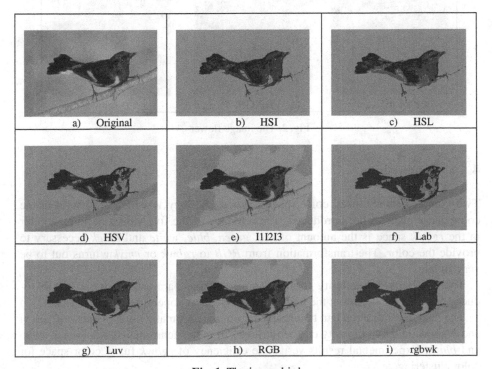

Fig. 1. The image bird

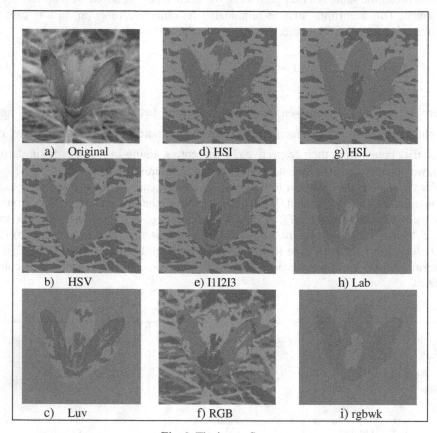

Fig. 2. The image flower

4   Conclusions

Two complementary fuzzy color spaces, *rgbwk* and *cmykw* which are useful in the color image analysis are introduced. The semantic of the five values defining a color in the *rgbwk* space is the amount of *red, green, blue, white* and *black* necessary to provide the color. The transformation from *RGB* to *rgbwk* or *cmykw* turns out to be very simple.

The similarity/dissimilarity formula using the five parameters *r,g,b,w,k* is introduced and also the negation, the union and intersection operators were defined.

The hue and saturation can be retrieved from the chromatic components *red, green* and *blue* while the luminosity can be retrieved from the achromatic components *white* and *black*. Experimental results verify the efficiency of *rgbwk* fuzzy color space for color clustering.

References

1. Atanassov, K.T.: Remark on a Property of the Intuitionistic Fuzzy Interpretation Triangle. Notes on Intuitionistic Fuzzy Sets 8, 34 (2002)
2. Bhattacharyya, A.: On a measure of divergence between two statistical populations defined by their probability distributions. Bulletin of the Calcutta Mathematical Society 35, 99–109 (1943)
3. Fairchild, M.D.: Color Appearance Models. Addison-Wesley, Reading (1998)
4. Gonzales, J.C., Woods, R.E.: Digital Image Processing, 1st edn. Addison-Wesley (1992)
5. Hunter, R.S.: Accuracy, Precision, and Stability of New Photoelectric Color-Difference Meter. JOSA 38(12) (1948), Proceedings of the Thirty Third Annual Meeting of the Optical Society of America
6. Jain, A.K.: Fundamentals of Digital Image Processing. Prentice Hall, New Jersey (1989)
7. MacQueen, J.B.: Some Methods for classification and Analysis of Multivariate Observations. In: Proceedings of 5-th Berkeley Symposium on Mathematical Statistics and Probability, vol. 1, pp. 281–297. University of California Press, Berkeley (1967)
8. Michener, J.C., van Dam, A.: A functional overview of the Core System with glossary. ACM Computing Surveys 10, 381–387 (1978)
9. Ohta, Y., Kanade, T., Sakai, T.: Color information for region segmentation. Computer Graphics and Image Processing 13(3), 222–241 (1980)
10. Patrascu, V.: New fuzzy color clustering algorithm based on *hsl s*imilarity. In: Proceedings of the Joint 2009 International Fuzzy Systems Association World Congress (IFSA 2009), Lisbon, Portugal, pp. 48–52 (2009)
11. Patrascu, V.: Fuzzy Image Segmentation Based on Triangular Function and Its n-dimensional Extension. In: Nachtegael, M., Van der Weken, D., Kerre, E.E., Philips, W. (eds.) Soft Computing in Image Processing. STUDFUZZ, vol. 210, pp. 187–207. Springer, Heidelberg (2007)
12. Patrascu, V.: Fuzzy Membership Function Construction Based on Multi-Valued Evaluation. In: Proceedings of the 10th International FLINS Conference. Uncertainty Modeling in Knowledge Engineering and Decision Making, pp. 756–761. World Scientific Press (2012)
13. Pătraşcu, V.: Cardinality and Entropy for Bifuzzy Sets. In: Hüllermeier, E., Kruse, R., Hoffmann, F. (eds.) IPMU 2010. CCIS, vol. 80, pp. 656–665. Springer, Heidelberg (2010)
14. Patrascu, V.: Multi-valued Color Representation Based on Frank t-norm Properties. In: Proceedings of the 12th Conference on Information Processing and Management of Uncertainty in Knowledge-Based Systems (IPMU 2008), Malaga, Spain, pp. 1215–1222 (2008)
15. Smarandache, F.: Neutrosophy. / Neutrosophic Probability, Set, and Logic. American Research Press, Rehoboth (1998)
16. Smarandache, F.: Definiton of neutrosophic logic - a generalization of the intuitionistic fuzzy logic. In: Proceedings of the Third Conference of the European Society for Fuzzy Logic and Technology, EUSFLAT 2003, Zittau, Germany, pp. 141–146 (2003)
17. Smarandache, F.: Generalization of the Intuitionistic Fuzzy Logic to the Neutrosophic Fuzzy Set. International Journal of Pure and applied Mathematics 24(3), 287–297 (2005)
18. Smith, A.R.: Color Gamut transform pairs. Computer Graphics SIGGRAPH 1978 Proceedings 12(3), 12–19 (1978)
19. Zadeh, L.A.: Fuzy sets. Inf. Control 8, 338–353 (1965)

A New Edge Detector Based on Uninorms

Manuel González-Hidalgo, Sebastia Massanet,
Arnau Mir, and Daniel Ruiz-Aguilera

Department of Mathematics and Computer Science
University of the Balearic Islands, E-07122, Palma, Spain
{manuel.gonzalez,s.massanet,arnau.mir,daniel.ruiz}@uib.es

Abstract. A new fuzzy edge detector based on uninorms is proposed and deeply studied. The behaviour of different classes of uninorms is discussed. The obtained results suggest that the best uninorm in order to improve the edge detection process is the uninorm $\mathcal{U}_{\min}$, with underlying Łukasiewicz operators. This algorithm gets statistically substantial better results than the others obtained by well known edge detectors, as Sobel, Roberts and Prewitt approaches and comparable to the results obtained by Canny.

Keywords: edge detection, uninorm, Canny, Sobel, Roberts, Prewitt.

1 Introduction

One of the main operations in image processing is edge detection. Determining the borders of an image allows to use more complex analysis such as segmentation and it can be used in computer vision and recognition. Its performance is crucial for the final results of the image processing technique. In recent decades, a large number of edge detection algorithms have been developed. These different approaches vary from classical algorithms [18] based on the use of a set of convolution masks, to new techniques based on fuzzy sets and their extensions [3]. From all classical algorithms, the Canny edge detector [4] is one of the most used due to its performance.

A different approach to edge detection is based in mathematical morphology [20]. The classical setting has been generalized using concepts and techniques of fuzzy sets and their extensions [5,6,8,9]. In particular, uninorms as a particular case of conjunction and aggregation operator have been used in image processing and analysis [1,6,8,10,16]. All these applications of uninorms are due to their structure: like a t-norm in the region $[0, e]^2$, like a t-conorm in the region $[e, 1]^2$, and like a compensatory operator on the remaining regions. This structure may be useful in order to magnify the output value whenever all the input values are greater than the neutral element, to reduce the output value whenever all the input values are smaller than the neutral element, and finally to compensate the output value when the input values are some of them greater than the neutral element and some of them are smaller than this element.

Having this in mind, and taking into account the Canny edge detector, we propose a new algorithm based on uninorms. So, after a brief revision about

A. Laurent et al. (Eds.): IPMU 2014, Part II, CCIS 443, pp. 184–193, 2014.

uninorms (Section 2) we describe our edge detection algorithm (Section 3). In Section 4 the experimental environment is specified, followed by the obtained results. We finish the communication with some conclusions and future work to expand our study.

2 Preliminaries

In these preliminaries we introduce the basic definitions and notations of the operators that will be used in the paper. More details can be found in [12].

Definition 1. *A* t-norm (t-conorm) *is a two-place function* $T : [0,1]^2 \to [0,1]$ *(*$S : [0,1]^2 \to [0,1]$*) which is associative, commutative, non-decreasing in each place and such that that* $T(1,x) = x$ *(*$S(0,x) = x$*) for all* $x \in [0,1]$*.*

Well known t-norms are the product t-norm $T_{\mathbf{P}}(x,y) = x \cdot y$, the Łukasiewicz t-norm $T_{\mathbf{L}}(x,y) = \max(x+y-1,0)$, and the nilpotent minimum t-norm

$$T_{\mathbf{nM}}(x,y) = \begin{cases} 0 & \text{if } x+y \leq 1, \\ \min(x,y) & \text{otherwise.} \end{cases}$$

A way to construct t-conorms from t-norms is the duality, by using the formula $S(x,y) = 1 - T(1-x, 1-y)$, for all $x,y \in [0,1]$, where T is any t-norm and S, its dual t-conorm. Therefore, the dual t-conorms of the t-norms presented before are the probabilistic sum $S_{\mathbf{P}}(x,y) = x + y - x \cdot y$, the Łukasiewicz t-conorm $S_{\mathbf{L}}(x,y) = \min(x+y,1)$, and the nilpotent maximum t-conorm

$$S_{\mathbf{nM}}(x,y) = \begin{cases} 1 & \text{if } x+y \geq 1, \\ \max(x,y) & \text{otherwise.} \end{cases}$$

Definition 2 ([21]). *A* uninorm *is a two-place function* $U : [0,1] \times [0,1] \to [0,1]$ *which is associative, commutative, non-decreasing in each place and such that there exists some element* $e \in [0,1]$*, called* neutral element*, such that* $U(e,x) = x$ *for all* $x \in [0,1]$*.*

It is clear that U becomes a t-norm when $e = 1$ and a t-conorm when $e = 0$. Any uninorm satisfies $U(0,1) \in \{0,1\}$ and so U is called *conjunctive* when $U(1,0) = 0$ and *disjunctive* when $U(1,0) = 1$.

Given any uninorm U, a t-norm T and a t-conorm S can be defined from its values in $[0,e]^2$ and $[e,1]^2$, respectively. The values of U in $[0,1]^2 \setminus ([0,e]^2 \cup [e,1]^2)$ are between minimum and maximum. The general structure of a uninorm is depicted in Figure 1. In general, a uninorm U with neutral element e and underlying t-norm T and t-conorm S will be denoted by $U \equiv \langle T, e, S \rangle$.

In the last years some classes of uninorms have been characterized (idempotent, representable, continuous in $]0,1[^2$) but here we will only focus on uninorms in $\mathcal{U}_{\min}$ and $\mathcal{U}_{\max}$, the classes that will be used in the experimental results.

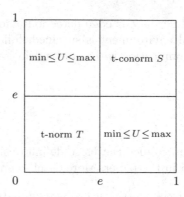

Fig. 1. General structure of a uninorm U with neutral element e

Theorem 1 ([7]). *Let U be a uninorm with neutral element $e \in \;]0,1[$ having functions $x \mapsto U(x,1)$ and $x \mapsto U(x,0)$ $(x \in [0,1])$ continuous except (perhaps) at the point $x = e$. Then U is given by one of the following forms*

(a) If U is conjunctive ($U(0,1) = 0$), then U is given by

$$U(x,y) = \begin{cases} eT\left(\frac{x}{e}, \frac{y}{e}\right) & \text{if } (x,y) \in [0,e]^2 \\ e + (1-e)S\left(\frac{x-e}{1-e}, \frac{y-e}{1-e}\right) & \text{if } (x,y) \in [e,1]^2 \\ \min(x,y) & \text{otherwise.} \end{cases} \quad (1)$$

(b) If U is disjunctive ($U(0,1) = 1$), then U is given by

$$U(x,y) = \begin{cases} eT\left(\frac{x}{e}, \frac{y}{e}\right) & \text{if } (x,y) \in [0,e]^2 \\ e + (1-e)S\left(\frac{x-e}{1-e}, \frac{y-e}{1-e}\right) & \text{if } (x,y) \in [e,1]^2 \\ \max(x,y) & \text{otherwise.} \end{cases} \quad (2)$$

In both formulas T is a t-norm and S is a t-conorm.

The class of all uninorms with expression (1) will be denoted by $\mathcal{U}_{\min}$, and a uninorm in $\mathcal{U}_{\min}$ with associated t-norm T and t-conorm S will be denoted by $U \equiv \langle T, e, S \rangle_{\min}$. Similarly, $\mathcal{U}_{\max}$ will denote all uninorms with expression (2), and a uninorm in this class will be referred as $U \equiv \langle T, e, S \rangle_{\max}$.

3   The Edge Detector Based on Uninorms: Algorithm

Before introducing the main algorithm, we need to explain what we understand by a directional gradient based on uninorms. So, for each pixel location (i,j) in the set of coordinates Ω of a gray level original image X, we set the eight neighbours in a 3×3 windows of the pixel $x_{i,j}$, see Fig. 2.a). Let us now consider the two convolution masks shown in Fig. 2.b) then, the basic directional gradients are defined as the absolute value of the difference of the intensity values

$x_{i-1,j-1}$	$x_{i,j-1}$	$x_{i+1,j-1}$	1	1	1	1	-1
$x_{i-1,j}$	$x_{i,j}$	$x_{i+1,j}$				1	-1
$x_{i-1,j+1}$	$x_{i,j+1}$	$x_{i+1,j+1}$	-1	-1	-1	1	-1

(a)   (b)

Fig. 2. (a) Neighbourhood of a current pixel $x_{i,j}$. (b) Convolution masks in order to compute the basic directional gradients and the directional uninorm based gradients, at left for y direction, at right for x direction.

between its horizontal and vertical neighbours respectively. The basic directional gradients along x-direction are defined as

$$y^l_{i-1,i+1} = |x_{i-1,l} - x_{i+1,l}|, \quad l \in \{j-1, j, j+1\}$$

and, the basic directional gradients along y-direction are defined as

$$x^k_{j-1,j+1} = |x_{k,j-1} - x_{k,j+1}|, \quad k \in \{i-1, i, i+1\}.$$

Let $U \equiv \langle T, e, S \rangle$ be a uninorm, we define the y directional U based gradient for the pixel $x_{i,j}$, as the aggregation of the basic directional gradients along y-direction using the uninorm U. This can be represented by

$$U_y(x_{i,j}) = U(y^{j-1}_{i-1,i+1}, y^j_{i-1,i+1}, y^{j+1}_{i-1,i+1}).$$

In a same way, we define the x directional U based gradient for the pixel $x_{i,j}$, as the aggregation of the basic directional gradients along x-direction using the uninorm U. This can be represented by

$$U_x(x_{i,j}) = U(x^{i-1}_{j-1,j+1}, x^i_{j-1,j+1}, x^{i+1}_{j-1,j+1}).$$

Observe that if the values of the basic directional gradients are high (belonging to the interval $[e, 1]$), the respective directional uninorm based gradient is computed using the underlying t-conorm S, as an indicator of a presence of a remarkable edge point. If these values are low (belonging to the interval $[0, e]$) the respective directional uninorm based gradient is computed using the underlying t-norm T, and we are in the presence of a point which can not be an edge point. In other cases we compute the directional uninorm based gradient as a kind of average. Recall that for all $(x, y) \in [0, 1]^2$, $S(x, y) \geq \max(x, y) \geq \min(x, y) \geq T(x, y)$ and if $(x, y) \in [0, e] \times [e, 1] \cup [e, 1] \times [0, e]$, $\min(x, y) \leq U(x, y) \leq \max(x, y)$.

Once we have defined the bidirectional uninorm based gradients $U_x(x_{i,j})$ and $U_y(x_{i,j})$, if we want to view the result at this point the two components must be combined. The magnitude of the result is computed at each pixel location (i, j) as

$$M_U(i, j) = \sqrt{U_x(x_{i,j})^2 + U_y(x_{i,j})^2}. \tag{3}$$

As we can observe the magnitude is computed in the same manner as it was for the gradient, which is in fact what is being computed. Note that these operators have the desirable property of yielding zeros for uniform regions. Other ways to calculate the magnitude and combine the bidirectional uninorm based gradients are possible, for instance, taking the maximum norm or the 1-norm, or even using another appropriate aggregation function.

After the magnitude given by Eq. 3 has been calculated, it should be normalized in order to obtain the fuzzy edge map. So, our fuzzy edge image is the normalized image of M_U that, without confusion, we denote in the same way.

However, the fuzzy edge map generated by the magnitude is an image where the value of a pixel represents its membership degree to the set of edges. This idea contradicts the restrictions of Canny [4], forcing a representation of the edge images as binary images with edges of one pixel width. Therefore the fuzzy edge image must be thinned and binarized. The fuzzy edge map will contain large values where there is a strong image gradient, but to identify edges the broad regions present in areas where the slope is large must be thinned so that only the magnitudes at those points which are local maxima remain. Non Maxima Suppression (NMS) performs this by suppressing all values along the line of the gradient that are not peak values [4]. NMS has been performed using P. Kovesis' implementation in Matlab [13].

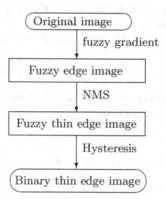

Fig. 3. Block diagram of the proposed edge detector

Finally, to binarize the image, we have implemented an automatic non-supervised hysteresis based on the determination of the instability zone of the histogram to find the threshold values [15]. Hysteresis allows to choose which pixels are relevant in order to be selected as edges, using their membership values. Two threshold values T_1, T_2 with $T_1 \leq T_2$ are used. All the pixels with a membership value greater than T_2 are considered as edges, while those which are lower to T_1 are discarded. Those pixels whose membership value is between the two values are selected if, and only if, they are connected with other pixels above T_2. The method needs some initial set of candidates for the threshold values. In this case, $\{0.01, \ldots, 0.25\}$ has been introduced, the same set used in [15]. In Figure 3, we display the block diagram of the algorithm describing the edge detector proposed in this section.

4   Experimental Environment

In this section, the different configurations of the edge detector based on uninorms which are going to be analysed in Section 5 are introduced. Furthermore, we settle the experimental environment based on an objective performance comparison of several edge detectors in order to determine which one of those considered in the analysis obtains the best results.

4.1   Configurations of the Edge Detector Based on Uninorms

In the previous section, the edge detector based on uninorms has been presented. This edge detector depends on the uninorm chosen to aggregate the six basic directional gradients obtained using the two masks, the vertical one and the horizontal one. Therefore, an unavoidable step consists on determining which uninorm has to be used to obtain the best results. With this aim in mind, we will consider 24 different uninorms of the classes of $\mathcal{U}_{min}$ and $\mathcal{U}_{max}$ with different underlying t-norms and t-conorms and with several neutral elements. Namely, the uninorms $U \equiv \langle T, e, S \rangle_{min}$ and $U \equiv \langle T, e, S \rangle_{max}$ with the underlying t-norms T_{nM}, T_P, T_L, introduced in Section 2, and as underlying t-conorms their dual operations, S_{nM}, S_P and S_L, respectively, have been considered. For each of the previous uninorms, four neutral elements e have been used: 0.02, 0.04, 0.06 and 0.08. Note that these values correspond to the values 5, 10, 15 and 20 if we consider the usual chain $\{0, \dots, 255\}$ with the possible gray levels of a pixel. These values have been considered since if we consider a higher value than 20 there could be some basic gradients higher than 20 that were aggregated using the underlying t-norm of the uninorm and thus, this potential edge would not be stand out. Similarly, if we consider a lower value than 5, there could be some basic gradients lower than 5 that were aggregated using the underlying t-conorm of the uninorm and thus, this potential noise would not be weakened.

Remark 1. From their expressions, it is evident that the fuzzy edge image obtained using the uninorm $U \equiv \langle T, e, S \rangle_{max}$ is going to contain more edges than the obtained using the uninorm $U \equiv \langle T, e, S \rangle_{min}$ for some fixed t-norm T, t-conorm S and neutral element $e \in (0, 1)$. This does not imply that the configuration using the uninorm of the class of $\mathcal{U}_{max}$ obtains always better results than the configuration of the corresponding uninorm of $\mathcal{U}_{min}$ since, in addition to the non-supervised hysteresis which does not satisfy always the monotonicity because the thresholds can differ in both cases, some pixels which are not edges as texture or noise could be detected by the first configuration.

4.2   Objective Performance Comparison Method

Nowadays, it is well-established in the literature that the visual inspection of the edge images obtained by several edge detectors can not be the unique criterion with the aim of proving the superiority of one edge detector with respect to the others. This is because each expert has different criteria and preferences and consequently, the reviews given by two experts can differ substantially. For this reason, the use of objective performance measures on edge detection is growing in popularity to compare the results obtained by different edge detection algorithms. There are several measures of performance for edge detection in the literature, see [14] and [17]. These measures require, in addition to the binary edge image with edges of one pixel width (DE) obtained by the edge detector we want to evaluate, a reference edge image or *ground truth* edge image (GT) which is a binary edge image with edges of one pixel width containing the real edges of the original image. In this work, we will use the following quantitative performance measures

- The ρ-coefficient ([11]), defined as

$$\rho = \frac{card(E_{TP})}{card(E) + card(E_{FN}) + card(E_{FP})},$$

where E_{TP} is the set of well-detected edge pixels, E_{FN} is the set of ground truth edges missed by the edge detector and E_{FP} is the set of edge pixels detected but with no counterpart on the ground truth image.
- The F-measure ([19]), defined as the harmonic mean between precision (PR) and recall (RE) given by

$$PR = \frac{card(E_{TP})}{card(E_{TP}) + card(E_{FP})}, \quad RE = \frac{card(E_{TP})}{card(E_{TP}) + card(E_{FN})}.$$

Larger values of ρ and F ($0 \leq \rho, F \leq 1$) are indicators of better capabilities for edge detection.

Consequently, we need a dataset of images with their ground truth edge images in order to compare the outputs obtained by the different algorithms. So, the first 15 images and their edge specifications from the public dataset of the University of South Florida[1] ([2]) have been used. In [2], the details about the ground truth edge images and their use for the comparison of edge detectors are included.

5   Obtained Results

In Table 1, the mean and the standard deviation of the 15 values of the ρ-coefficient obtained by each configuration, the number of best and worst images for each edge detector and the mean of the rankings of each configuration in every image are collected. The best three uninorms are the following: $\langle T_{\mathbf{P}}, 0.04, S_{\mathbf{P}} \rangle_{\min}$, $\langle T_{\mathbf{P}}, 0.02, S_{\mathbf{P}} \rangle_{\min}$ and $\langle T_{\mathbf{L}}, 0.02, S_{\mathbf{L}} \rangle_{\min}$. This last configuration obtains the higher mean value according to the ρ-coefficient. We have performed a Wilcoxon and a t-test to see if there are significant differences between the

---

[1] This image dataset can be downloaded from
ftp://figment.csee.usf.edu/pub/ROC/edge_comparison_dataset.tar.gz

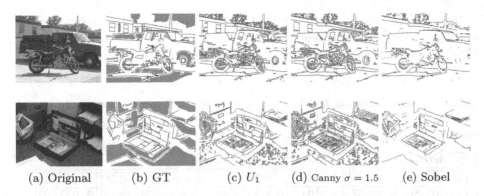

| | | | | |
| (a) Original | (b) GT | (c) U_1 | (d) Canny $\sigma = 1.5$ | (e) Sobel |

Fig. 4. Some original images, ground truth edge images and the results obtained by some of the considered edge detectors, where $U_1 = \langle T_{\mathbf{L}}, 0.02, S_{\mathbf{L}} \rangle_{\min}$

Table 1. Mean, standard deviation, number of best and worst images and mean of rankings of the different uninorm-based configurations according to the ρ-coefficient values

Method	$\overline{x}$	σ	✓	×	$\overline{x}_p$	Method	$\overline{x}$	σ	✓	×	$\overline{x}_p$
$\langle T_{\mathbf{nM}}, 0.02, S_{\mathbf{nM}} \rangle_{\min}$	0.671	0.124	0	0	16.0	$\langle T_{\mathbf{P}}, 0.06, S_{\mathbf{P}} \rangle_{\min}$	0.736	0.126	1	0	9.1
$\langle T_{\mathbf{nM}}, 0.02, S_{\mathbf{nM}} \rangle_{\max}$	0.647	0.124	0	2	20.5	$\langle T_{\mathbf{P}}, 0.06, S_{\mathbf{P}} \rangle_{\max}$	0.743	0.103	0	0	9.5
$\langle T_{\mathbf{nM}}, 0.04, S_{\mathbf{nM}} \rangle_{\min}$	0.658	0.135	0	0	17.7	$\langle T_{\mathbf{P}}, 0.08, S_{\mathbf{P}} \rangle_{\min}$	0.715	0.134	0	0	12.3
$\langle T_{\mathbf{nM}}, 0.04, S_{\mathbf{nM}} \rangle_{\max}$	0.652	0.118	0	1	19.5	$\langle T_{\mathbf{P}}, 0.08, S_{\mathbf{P}} \rangle_{\max}$	0.726	0.112	0	0	11.5
$\langle T_{\mathbf{nM}}, 0.06, S_{\mathbf{nM}} \rangle_{\min}$	0.633	0.143	0	1	20.5	$\langle T_{\mathbf{L}}, 0.02, S_{\mathbf{L}} \rangle_{\min}$	0.765	0.092	3	0	5.4
$\langle T_{\mathbf{nM}}, 0.06, S_{\mathbf{nM}} \rangle_{\max}$	0.652	0.119	0	0	19.5	$\langle T_{\mathbf{L}}, 0.02, S_{\mathbf{L}} \rangle_{\max}$	0.757	0.087	3	1	7.6
$\langle T_{\mathbf{nM}}, 0.08, S_{\mathbf{nM}} \rangle_{\min}$	0.615	0.145	0	8	22.5	$\langle T_{\mathbf{L}}, 0.04, S_{\mathbf{L}} \rangle_{\min}$	0.757	0.105	2	0	6.1
$\langle T_{\mathbf{nM}}, 0.08, S_{\mathbf{nM}} \rangle_{\max}$	0.650	0.124	0	1	19.6	$\langle T_{\mathbf{L}}, 0.04, S_{\mathbf{L}} \rangle_{\max}$	0.751	0.089	0	0	8.5
$\langle T_{\mathbf{P}}, 0.02, S_{\mathbf{P}} \rangle_{\min}$	0.762	0.096	3	0	5.8	$\langle T_{\mathbf{L}}, 0.06, S_{\mathbf{L}} \rangle_{\min}$	0.732	0.120	0	0	10.3
$\langle T_{\mathbf{P}}, 0.02, S_{\mathbf{P}} \rangle_{\max}$	0.751	0.095	1	1	8.5	$\langle T_{\mathbf{L}}, 0.06, S_{\mathbf{L}} \rangle_{\max}$	0.745	0.104	1	0	9.3
$\langle T_{\mathbf{P}}, 0.04, S_{\mathbf{P}} \rangle_{\min}$	0.756	0.107	1	0	5.7	$\langle T_{\mathbf{L}}, 0.08, S_{\mathbf{L}} \rangle_{\min}$	0.705	0.131	0	0	13.7
$\langle T_{\mathbf{P}}, 0.04, S_{\mathbf{P}} \rangle_{\max}$	0.748	0.094	0	0	8.9	$\langle T_{\mathbf{L}}, 0.08, S_{\mathbf{L}} \rangle_{\max}$	0.728	0.113	0	0	11.9

three previous uninorms and the uninorms $\langle T_{\mathbf{nM}}, e, S_{\mathbf{nM}} \rangle_{\min}$ and their counterparts in $\mathcal{U}_{\max}$. The results show that the three best uninorms are statistically better than these other ones.

Next, a comparison of this group of three uninorms and the classical methods of Canny, Sobel, Roberts and Prewitt has been made. The results concerning the ρ-coefficient are shown in Table 2. If we perform a Wilcoxon test and a t-test using the configuration $\langle T_{\mathbf{L}}, 0.02, S_{\mathbf{L}} \rangle_{\min}$ and the classical methods, the results show that there are no significant differences between this uninorm based method and the methods of Canny with $\sigma \in \{0.5, 1, 1.5\}$ but the uninorm is statistically better that the method of Canny with $\sigma \in \{2, 2.5\}$ and the methods of Sobel, Robert and Prewitt. Nevertheless, the results indicate that the uninorm based method is the best one according to the mean value and the mean of the rankings. In Fig. 4 we can observe some of the results obtained by some of the edge detectors considered in this paper.

If we choose the F-measure instead of the ρ-coefficient, the results are the same in the sense that the best and worst images and the means of rankings are identical and the conclusions of the statistical tests are the same.

Table 2. Mean, standard deviation, number of best and worst images and mean of rankings of the best uninorm-based configurations and some classical methods according to the ρ-coefficient values

Method	$\bar{x}$	σ	✓	×	$\bar{x}_p$	Method	$\bar{x}$	σ	✓	×	$\bar{x}_p$
$\langle T_{\mathbf{P}}, 0.04, S_{\mathbf{P}}\rangle_{\min}$	0.756	0.107	3	0	4.267	Canny $\sigma = 0.5$	0.700	0.217	5	4	4.733
$\langle T_{\mathbf{P}}, 0.02, S_{\mathbf{P}}\rangle_{\min}$	0.762	0.096	0	0	3.933	Canny $\sigma = 1$	0.762	0.149	3	0	3.800
$\langle T_{\mathbf{L}}, 0.02, S_{\mathbf{L}}\rangle_{\min}$	0.765	0.092	2	0	3.600	Canny $\sigma = 1.5$	0.738	0.108	2	0	4.667
Sobel	0.604	0.122	0	0	7.867	Canny $\sigma = 2$	0.666	0.096	0	0	6.400
Roberts	0.516	0.107	0	11	10.400	Canny $\sigma = 2.5$	0.586	0.086	0	0	8.600
Prewitt	0.602	0.124	0	0	7.733						

6   Conclusions and Future Work

In this work, we have introduced a new fuzzy edge detector based on uninorms and the performance of several uninorms in the class of $\mathcal{U}_{\min}$ has been studied. From the results, we can conclude that the configuration with the uninorm of $\mathcal{U}_{\min}$, with Łukasiewicz underlying operations gets statistically better results than the the ones obtained by well known edge detectors, as Sobel, Roberts and Prewitt approaches and comparable to the results obtained by Canny.

As future work, it would be interesting to evaluate the proposed method with respect to Canny's algorithm using the same post-processing steps in order to assess the quality of the gradient estimation exclusively, and to evaluate the method with respect to the subsequent steps: object segmentation or recognition. In addition, we want to study other classes of uninorms, such as idempotent and representable uninorms. The use of other aggregation functions and other directions of the basic directional gradients could further improve the results.

Acknowledgments. This paper has been partially supported by the Spanish Grants MTM2009-10320 and TIN2013-42795-P with FEDER support.

References

1. Bede, B., Nobuhara, H., Rudas, I.J., Fodor, J.: Discrete cosine transform based on uninorms and absorbing norms. In: IEEE International Conference on Fuzzy Systems (FUZZ-IEEE), pp. 1982–1986 (2008)
2. Bowyer, K., Kranenburg, C., Dougherty, S.: Edge detector evaluation using empirical ROC curves. Computer Vision and Pattern Recognition 1, 354–359 (1999)

3. Bustince, H., Barrenechea, E., Pagola, M., Fernandez, J.: Interval-valued fuzzy sets constructed from matrices: Application to edge detection. Fuzzy Sets and Systems 160(13), 1819–1840 (2009)
4. Canny, J.: A computational approach to edge detection. IEEE Trans. Pattern Anal. Mach. Intell. 8(6), 679–698 (1986)
5. De Baets, B., Kerre, E., Gupta, M.: The fundamentals of fuzzy mathematical morfologies part I: basics concepts. International J. of General Systems 23, 155–171 (1995)
6. De Baets, B., Kwasnikowska, N., Kerre, E.: Fuzzy morphology based on uninorms. In: Proc. of the 7th IFSA World Congress, Prague, pp. 215–220 (1997)
7. Fodor, J.C., Yager, R.R., Rybalov, A.: Structure of uninorms. Int. J. Uncertainty, Fuzziness, Knowledge-Based Systems 5, 411–427 (1997)
8. González, M., Ruiz-Aguilera, D., Torrens, J.: Algebraic properties of fuzzy morphological operators based on uninorms. In: Artificial Intelligence Research and Development. Frontiers in Artificial Intelligence and Applications, vol. 100, pp. 27–38. IOS Press, Amsterdam (2003)
9. González-Hidalgo, M., Massanet, S.: A fuzzy mathematical morphology based on discrete t-norms: fundamentals and applications to image processing. Soft Computing, 1–15 (December 2013)
10. González-Hidalgo, M., Mir-Torres, A., Ruiz-Aguilera, D., Torrens, J.: Edge-images using a uninorm-based fuzzy mathematical morphology: Opening and closing. In: Tavares, J., Jorge, N. (eds.) Advances in Computational Vision and Medical Image Processing, ch. 8. Computational Methods in Applied Sciences, vol. 13, pp. 137–157. Springer, Netherlands (2009)
11. Grigorescu, C., Petkov, N., Westenberg, M.A.: Contour detection based on non-classical receptive field inhibition. IEEE Transactions on Image Processing 12(7), 729–739 (2003)
12. Klement, E.P., Mesiar, R., Pap, E.: Triangular norms. Kluwer Academic Publishers, London (2000)
13. Kovesi, P.D.: MATLAB and Octave functions for computer vision and image processing. Centre for Exploration Targeting, School of Earth and Environment. The University of Western Australia (2012)
14. Lopez-Molina, C., De Baets, B., Bustince, H.: Quantitative error measures for edge detection. Pattern Recognition 46(4), 1125 (2013)
15. Medina-Carnicer, R., Muñoz-Salinas, R., Yeguas-Bolivar, E., Diaz-Mas, L.: A novel method to look for the hysteresis thresholds for the Canny edge detector. Pattern Recognition 44(6), 1201–1211 (2011)
16. Nagau, J., Regis, S., Henry, J.-L., Doncescu, A.: Study of aggregation operators for scheduling clusters in digital images of plants. In: 26th International Conference on Advanced Information Networking and Applications Workshops, 2012, pp. 1161–1166 (2012)
17. Papari, G., Petkov, N.: Edge and line oriented contour detection: State of the art. Image and Vision Computing 29(2-3), 79 (2011)
18. Pratt, W.K.: Digital Image Processing, 4th edn. Wiley Interscience (2007)
19. Rijsbergen, C.: Information retrieval. Butterworths (1979)
20. Serra, J.: Image analysis and mathematical morphology, vol. 1, 2. Academic Press, London (1982)
21. Yager, R.R., Rybalov, A.: Uninorm aggregation operators. Fuzzy Sets and Systems 80, 111–120 (1996)

Context-Aware Distance Semantics
for Inconsistent Database Systems

Anna Zamansky[1], Ofer Arieli[2], and Kostas Stefanidis[3]

[1] Department of Information Systems, University of Haifa, Israel
annazam@is.haifa.ac.il
[2] School of Computer Science, The Academic College of Tel-Aviv, Israel
oarieli@mta.ac.il
[3] Institute of Computer Science, Foundation for Research and Technology
Hellas (FORTH), Greece
kstef@ics.forth.gr

Abstract. Many approaches for consistency restoration in database systems have to deal with the problem of an exponential blowup in the number of possible repairs. For this reason, recent approaches advocate more flexible and fine grained policies based on the reasoner's preference. In this paper we take a further step towards more personalized inconsistency management by incorporating ideas from context-aware systems. The outcome is a general distance-based approach to inconsistency maintenance in database systems, controlled by context-aware considerations.

1 Introduction

Inconsistency handling in constrained databases is a primary issue in the context of consistent query answering, data integration, and data exchange. The standard approaches to this issue are usually based on the principle of minimal change, aspiring to achieve consistency via a minimal amount of data modifications (see, e.g., [2,7,10]). A key question in this respect is how to *choose* among the different possibilities of restoring the consistency of a database (i.e., 'repairing' it).

Earlier approaches to inconsistency management were based on the assumption that there should be some fixed, pre-determined way of repairing a database. Recently, there has been a paradigm shift towards user-controlled inconsistency management policies. Works taking this approach provide a possibility for the user to express some *preference* over all possible database repairs, preferring certain repairs to others (see [18] for a survey and further references). While such approaches provide the user with flexibility and control over inconsistency management, in reality they entail a considerable technical burden on the user's shoulders of calibrating, updating and maintaining preferences or policies. Moreover, in many cases these preferences may be *dynamic*, changing quickly on the go (e.g., depending on the user's geographical location). In the era of ubiquitous computing, users want *easy* – and sometimes even *fully automatic* – inconsistency management solutions with little cognitive load, while still expecting them

A. Laurent et al. (Eds.): IPMU 2014, Part II, CCIS 443, pp. 194–203, 2014.

to be *personalized* to their particular needs. This leads to the idea of introducing *context-awareness* into inconsistency management.

Context-awareness is defined as the use of contexts to provide task-relevant information and services to a user (see [1]). We believe that inconsistency management has natural relations to the concept of context. To capture this idea, we incorporate notions and techniques that have been studied by the context-aware computing community to consistency management for database systems, by combining the following two ingredients:

- *Distance-based semantics* for restoring the consistency of inconsistent databases according to the principle of minimal change, and
- *Context-awareness considerations* for incorporating user preferences.

Example 1. Let us consider the following simple database instance:

empNum	name	address	salary
1	John	Tower Street 3, London, UK	70K\$
1	John	Herminengasse 8, Wien, AT	80K\$
2	Mary	42 Street 15, New York, US	90K\$

Two functional dependencies that may be violated here are empNum $\to$ address and empNum $\to$ salary. Thus, a database with the above relation and integrity constraints is not consistent. Minimal change considerations (which will be expressed in what follows by distance functions) imply that it is enough to delete either the first or the second tuple for restoring consistency. Now, the decision which tuple to delete may be *context-dependent*. For instance, for tax assessments tuples with higher salaries may be preferred, while tuples with lower salaries may have higher priority when loans or grants are considered. The choice between the first two tuples may also be determined by more dynamic considerations, such as geographic locations, etc.

2   Inconsistent Databases and Distance Semantics

For simplicity of presentation, in this paper we remain on the propositional level and reduce first-order databases to our framework by grounding them. In the sequel, $\mathcal{L}$ denotes a propositional language with a *finite* set of atomic formulas Atoms($\mathcal{L}$). An *$\mathcal{L}$-interpretation* I is an assignment of a truth value in $\{T, F\}$ to every element in Atoms($\mathcal{L}$). Interpretations are extended to complex formulas in $\mathcal{L}$ in the usual way, using the truth tables of the connectives in $\mathcal{L}$. The set of two-valued interpretations for $\mathcal{L}$ is denoted by $\Lambda_{\mathcal{L}}$. An interpretation I is a *model* of an $\mathcal{L}$-formula ψ if $I(\psi) = T$, denoted by $I \models \psi$, and it is a model of a set Γ of $\mathcal{L}$-formulas, denoted by $I \models \Gamma$, if it is a model of every $\mathcal{L}$-formula in Γ. The set of models of Γ is denoted by $mod(\Gamma)$. We say that Γ is *satisfiable* if $mod(\Gamma)$ is not empty.

Definition 1. A *database* $\mathcal{DB}$ in $\mathcal{L}$ is a pair $\langle \mathcal{D}, \mathcal{IC} \rangle$, where $\mathcal{D}$ (the *database instance*) is a finite subset of Atoms($\mathcal{L}$), and $\mathcal{IC}$ (the *integrity constraints*) is a finite and consistent set of $\mathcal{L}$-formulas.

The meaning of $\mathcal{D}$ is determined by the conjunction of its facts, augmented with Reiter's *closed world assumption*, stating that each atomic formula that does not appear in $\mathcal{D}$ is false: $\mathsf{CWA}(\mathcal{D}) = \{\neg p \mid p \notin \mathcal{D}\}$. Henceforth, a database $\mathcal{DB} = \langle \mathcal{D}, \mathcal{IC} \rangle$ will be associated with the theory $\Gamma_{\mathcal{DB}} = \mathcal{IC} \cup \mathcal{D} \cup \mathsf{CWA}(\mathcal{D})$.

Definition 2. A database $\mathcal{DB}$ is *consistent* iff $\Gamma_{\mathcal{DB}}$ is satisfiable.

When a database is not consistent at least one integrity constraint is violated, and so it is usually required to look for "repairs" of the database, that is, changes of the database instance so that its consistency will be restored. There are numerous approaches for doing so (see, e.g., [2,7,10] for some surveys on this subject). Here we follow the distance-based approach described in [3,5], which we find suitable for our purposes since it provides a modular and flexible framework for a variety of methods of repair and consistent query answering. In the context of database systems this approach aims at addressing the problem that when $\mathcal{DB}$ is inconsistent $mod(\Gamma_{\mathcal{DB}})$ is empty, so reasoning with $\mathcal{DB}$ is trivialized. This may be handled by replacing $mod(\Gamma_{\mathcal{DB}})$ with the set $\Delta(\mathcal{DB})$ of interpretations that, intuitively, are 'as close as possible' to (satisfying) $\mathcal{DB}$, while still satisfying the integrity constraints. When $\mathcal{DB}$ is consistent, $\Delta(\mathcal{DB})$ and $mod(\Gamma_{\mathcal{DB}})$ coincide (see Proposition 3 below), which assures that distance-based semantics is a conservative generalization of standard semantics for consistent databases.

In what follows, we recall the relevant definitions for formalizing the intuition above (see also [3,5]).

Definition 3. A *pseudo-distance* on a set U is a total function $d : U \times U \to \mathbb{R}^+$, which is symmetric (for all $\nu, \mu \in U$, $d(\nu, \mu) = d(\mu, \nu)$) and preserves identity (for all $\nu, \mu \in U$, $d(\nu, \mu) = 0$ if and only if $\nu = \mu$). A pseudo-distance d is called a *distance* (*metric*) on U, if it satisfies the triangular inequality: for all $\nu, \mu, \sigma \in U$, $d(\nu, \sigma) \leq d(\nu, \mu) + d(\mu, \sigma)$.

Definition 4. A *(numeric) aggregation function* is a function f, whose domain consists of multisets of real numbers and whose range is the real numbers, satisfying the following properties:
1. f is non-decreasing when a multiset element is replaced by a larger element,
2. $f(\{x_1, \ldots, x_n\}) = 0$ if and only if $x_1 = x_2 = \ldots x_n = 0$, and
3. $f(\{x\}) = x$ for every $x \in \mathbb{R}$.
An aggregation function f is *hereditary*, if $f(\{x_1, \ldots, x_n\}) < f(\{y_1, \ldots, y_n\})$ entails that $f(\{x_1, \ldots, x_n, z_1, \ldots, z_m\}) < f(\{y_1, \ldots, y_n, z_1, \ldots, z_m\})$.

In what follows we shall aggregate distance values. Since distances are non-negative numbers, aggregation functions in this case include the summation and the maximum functions, the former is also hereditary.

Example 2. One may define the following distances on $\Lambda_{\mathcal{L}}$:

$$d_U(I, I') = \begin{cases} 1 & \text{if } I \neq I', \\ 0 & \text{otherwise.} \end{cases} \qquad d_H(I, I') = \mid \{p \in \mathsf{Atoms}(\mathcal{L}) \mid I(p) \neq I'(p)\} \mid .$$

d_U is sometimes called the uniform distance and d_H is known as the Hamming distance. More sophisticated distances are considered, e.g., in [5] and [12].

Definition 5. A *distance setting* (for a language $\mathcal{L}$) is a pair $\mathsf{DS} = \langle d, f \rangle$, where d is a pseudo-distance on $\Lambda_{\mathcal{L}}$ and f is an aggregation function.

The next definition is a common way of using distance functions for maintaining inconsistent data (see, e.g., [15,16]).

Definition 6. For a finite set $\Gamma = \{\psi_1, \ldots, \psi_n\}$ of formulas in $\mathcal{L}$, an interpretation $I \in \Lambda_{\mathcal{L}}$, and a distance setting $\mathsf{DS} = \langle d, f \rangle$ for $\mathcal{L}$, we denote: $d_{\mathsf{DS}}(I, \psi_i) = \min\{d(I, I') \mid I' \models \psi_i\}$ and $\delta_{\mathsf{DS}}(I, \Gamma) = f(\{d_{\mathsf{DS}}(I, \psi_1), \ldots, d_{\mathsf{DS}}(I, \psi_n)\})$.

Proposition 1. [3,16] *For every interpretation $I \in \Lambda_{\mathcal{L}}$ and a distance setting* $\mathsf{DS} = \langle d, f \rangle$, *it holds that* $I \models \psi$ *iff* $d_{\mathsf{DS}}(I, \psi) = 0$ *and* $I \models \Gamma$ *iff* $\delta_{\mathsf{DS}}(I, \Gamma) = 0$.

Definition 7. Given a database $\mathcal{DB} = \langle \mathcal{D}, \mathcal{IC} \rangle$ in $\mathcal{L}$ and a distance setting $\mathsf{DS} = \langle d, f \rangle$ for $\mathcal{L}$, the set of *the most plausible interpretations of* $\mathcal{DB}$ (with respect to DS) is defined as follows:
$$\Delta_{\mathsf{DS}}(\mathcal{DB}) = \{I \in \mathrm{mod}(\mathcal{IC}) \mid I' \in \mathrm{mod}(\mathcal{IC}) \Longrightarrow$$
$$\delta_{\mathsf{DS}}(I, \mathcal{D} \cup \mathrm{CWA}(\mathcal{D})) \leq \delta_{\mathsf{DS}}(I', \mathcal{D} \cup \mathrm{CWA}(\mathcal{D}))\}.$$

Note 1. Since $\mathcal{IC}$ is satisfiable, for every database $\mathcal{DB} = \langle \mathcal{D}, \mathcal{IC} \rangle$ and a distance setting DS for its language, it holds that $\Delta_{\mathsf{DS}}(\mathcal{DB}) \neq \emptyset$.

Definition 8. Let $\mathcal{DB} = \langle \mathcal{D}, \mathcal{IC} \rangle$ be a database and $\mathsf{DS} = \langle d, f \rangle$ a distance setting. We say that $\mathcal{R}$ is a DS-*repair* of $\mathcal{DB}$, if there is an $I \in \Delta_{\mathsf{DS}}(\mathcal{DB})$ such that $\mathcal{R} = \{p \in \mathrm{Atoms}(\mathcal{L}) \mid I(p) = T\}$. We shall sometimes denote this repair by $\mathcal{R}(I)$ and say that it is *induced by* I (or that I is the *characteristic model* of $\mathcal{R}$). The set of all the DS-repairs is denoted by $\mathrm{Repairs}_{\mathsf{DS}}(\mathcal{DB}) = \{\mathcal{R}(I) \mid I \in \Delta_{\mathsf{DS}}(\mathcal{DB})\}$.

An alternative characterization of the DS-repairs of $\mathcal{DB}$ is given next:

Proposition 2. *Let* $\mathcal{DB} = \langle \mathcal{D}, \mathcal{IC} \rangle$ *be a database and* $\mathsf{DS} = \langle d, f \rangle$ *a distance setting. Let* I_S *be the characteristic function of* $S \subseteq \mathrm{Atoms}(\mathcal{L})$ *(that is, $I_S(p) = T$ if $p \in S$ and $I_S(p) = F$ otherwise). The* DS*-inconsistency value of* S *is:*

$$\mathrm{Inc}_{\mathsf{DS}}(S) = \begin{cases} \delta_{\mathsf{DS}}(I_S, \mathcal{D} \cup \mathrm{CWA}(\mathcal{D})) & \text{if } I_S \in \mathrm{mod}(\mathcal{IC}), \\ \infty & \text{otherwise.} \end{cases}$$

Then $\mathcal{R} \subseteq \mathrm{Atoms}(\mathcal{L})$ *is a* DS*-repair of* $\mathcal{DB}$ *iff its* DS*-inconsistency value is minimal among the* DS*-inconsistency values of the subsets of* $\mathrm{Atoms}(\mathcal{L})$.

Proof. Let $\mathcal{R} \subseteq \mathrm{Atoms}(\mathcal{L})$ such that $\mathrm{Inc}_{\mathsf{DS}}(\mathcal{R}) \leq \mathrm{Inc}_{\mathsf{DS}}(S)$ for every $S \subseteq \mathrm{Atoms}(\mathcal{L})$. Since $\mathcal{IC}$ is satisfiable, $\mathrm{Inc}_{\mathsf{DS}}(\mathcal{R}) < \infty$, and so $I_{\mathcal{R}} \in \mathrm{mod}(\mathcal{IC})$. Let now $\mathcal{R}'$ be a DS-repair of $\mathcal{DB}$. Then there is an element $I' \in \Delta_{\mathsf{DS}}(\mathcal{DB})$ such that $\mathcal{R}' = \{p \in \mathrm{Atoms}(\mathcal{L}) \mid I'(p) = T\}$. But $\delta_{\mathsf{DS}}(I_{\mathcal{R}}, \mathcal{D} \cup \mathrm{CWA}(\mathcal{D})) \leq \delta_{\mathsf{DS}}(I', \mathcal{D} \cup \mathrm{CWA}(\mathcal{D}))$, and so $I_{\mathcal{R}} \in \Delta_{\mathsf{DS}}(\mathcal{DB})$ as well, which implies that $\mathcal{R}$ is a DS-repair of $\mathcal{DB}$.

For the converse, let $\mathcal{R}$ be a DS-repair of $\mathcal{DB}$ and let $S \subseteq \mathrm{Atoms}(\mathcal{L})$. We have to show that $\mathrm{Inc}_{\mathsf{DS}}(\mathcal{R}) \leq \mathrm{Inc}_{\mathsf{DS}}(S)$. Indeed, if $I_S \notin \mathrm{mod}(\mathcal{IC})$ then $\mathrm{Inc}_{\mathsf{DS}}(S) = \infty$ and so the claim is obtained. Otherwise, both $I_{\mathcal{R}}$ and I_S are models of $\mathcal{IC}$, and since $\mathcal{R}$ is a DS-repair of $\mathcal{DB}$, $I_{\mathcal{R}} \in \Delta_{\mathsf{DS}}(\mathcal{DB})$. It follows that $\delta_{\mathsf{DS}}(I_{\mathcal{R}}, \mathcal{D} \cup \mathrm{CWA}(\mathcal{D})) \leq \delta_{\mathsf{DS}}(I_S, \mathcal{D} \cup \mathrm{CWA}(\mathcal{D}))$ and so $\mathrm{Inc}_{\mathsf{DS}}(\mathcal{R}) \leq \mathrm{Inc}_{\mathsf{DS}}(S)$. $\square$

By Proposition 1 and Definition 8, we also have the following result:

Proposition 3. *Let $\mathcal{DB} = \langle \mathcal{D}, \mathcal{IC} \rangle$ be a database and* DS *a distance setting. The following conditions are equivalent: (1) $\mathcal{DB}$ is consistent, (2) $\Delta_{DS}(\mathcal{DB}) = mod(\Gamma_{\mathcal{DB}})$, (3)* Repairs$_{DS}(\mathcal{DB}) = \{\mathcal{D}\}$, *(4) The* DS*-inconsistency value of every* DS*-repair of $\mathcal{DB}$ is zero.*

Example 3. Let us return to the database in Example 1. The projection of the database table on the attributes id and salary is: $\{\langle 1, 70\mathsf{K\$} \rangle, \langle 1, 80\mathsf{K\$} \rangle\rangle, \langle 2, 90\mathsf{K\$} \rangle\}$. After grounding the database and representing the tuple $\langle \mathsf{empNum}, \mathsf{salary} \rangle$ by a propositional variable $\mathsf{T}^{\mathsf{empNum}}_{\mathsf{salary}}$, we have:

$$\mathcal{D} \cup \mathsf{CWA}(\mathcal{D}) = \left\{ \mathsf{T}^1_{70K\$}, \mathsf{T}^1_{80K\$}, \neg\mathsf{T}^1_{90K\$}, \neg\mathsf{T}^2_{70K\$}, \neg\mathsf{T}^2_{80K\$}, \mathsf{T}^2_{90K\$} \right\},$$

and the functional dependency $\mathsf{empNum} \to \mathsf{salary}$ is formulated as follows:

$$\mathcal{IC} = \left\{ \mathsf{T}^x_y \to \neg\mathsf{T}^x_z \mid y \neq z,\ y, z \in \{70K\$, 80K\$, 90K\$\},\ x \in \{1, 2\} \right\}.$$

Using the distance d_H from Example 2 and $f = \Sigma$, we compute:

I	$d_H(I, \mathsf{T}^1_{70})$	$d_H(I, \mathsf{T}^1_{80})$	$d_H(I, \neg\mathsf{T}^1_{90})$	$d_H(I, \neg\mathsf{T}^2_{70})$	$d_H(I, \neg\mathsf{T}^2_{80})$	$d_H(I, \mathsf{T}^2_{90})$	$\delta_{d_H, \Sigma}(I, \Gamma_{\mathcal{DB}})$
$\emptyset$	1	1	0	0	0	1	3
$\{\mathsf{T}^1_{70}\}$	0	1	0	0	0	1	2
$\{\mathsf{T}^1_{80}\}$	1	0	0	0	0	1	2
...	...	...	...	...	...	...	...
$\{\mathsf{T}^1_{70}, \mathsf{T}^2_{90}\}$	0	1	0	0	0	0	1
$\{\mathsf{T}^1_{80}, \mathsf{T}^2_{90}\}$	1	0	0	0	0	0	1
...	...	...	...	...	...	...	...

It follows that $\Delta_{\langle d_H, \Sigma \rangle}(\mathcal{DB}) = \{I_1, I_2\}$ and Repairs$_{DS}(\mathcal{DB}) = \{\mathcal{R}(I_1), \mathcal{R}(I_2)\}$, where $\mathcal{R}(I_1) = \{\mathsf{T}^1_{70}, \mathsf{T}^2_{90}\}$ and $\mathcal{R}(I_2) = \{\mathsf{T}^1_{80}, \mathsf{T}^2_{90}\}\}$. Thus, only T^2_{90} holds in all the repairs of $\mathcal{DB}$, that is, only the salary of employee 2 is certain.

3   Context-Aware Inconsistency Management

3.1   Context Modeling

As defined in [1], *"Context is any information that can be used to characterize the situation of an entity. An entity is a person, place or object that is considered relevant to the interaction between a user and an application, including the user and application"*. This notion has been found useful in several domains, such as machine learning and knowledge acquisition (see, e.g., [8,9]). We shall consider as a context any data that can be used to characterize database-related situations, involving database entities, user contexts and preferences, etc. [11]. There is a wide variety of methods for modeling contexts. Here we follow the data-centric approach introduced in [20], and refer to contexts using a finite set of special purpose variables (which may not be part of the database).

Definition 9. A *context environment* (or just a *context*) C is a finite tuple of variables $\langle c_1, \ldots, c_n \rangle$, where each variable c_i ($1 \leq i \leq n$) has a corresponding range Range(c_i) of possible values. A *context state* for C (a C-state, for short) is an assignment S such that $S(c_i) \in$ Range(c_i). The set of context states is denoted by States(C).

Intuitively, a context environment C represents the parameters that may be taken into consideration for the database inconsistency maintenance.

We are now ready to incorporate context-awareness into distance considerations. We do so by making the 'most plausible' interpretations in $\mathcal{DB}$, the elements in $\Delta_{DS}(\mathcal{DB})$, *sensitive to context*, in the sense that more 'relevant' formulas have higher impact on the distance computations than less 'relevant' formulas. Thus, while we still strive to minimize change, the latter will be measured in a more subtle, context-aware way.

Definition 10. A *relevance ranking* for a set Γ of formulas and a context environment C, is a total function $R : \Gamma \times \mathsf{States}(C) \to (0, 1]$.

Given a set Γ and a context environment C, a relevance ranking function for Γ and C assigns to every formula $\psi \in \Gamma$ and every state S of C a (positive) *relevance factor* $R(\psi, S)$ indicating the relevance of ψ according to S. Intuitively, higher values of these factors correspond to higher relevance of their formulas, which makes changes to these formulas in computing database repairs less desirable.[1]

Definition 11. A *context setting* for a set of formulas Γ is a triple $\mathsf{CS}(\Gamma) = \langle C, S, R \rangle$, where C is a context environment, $S \in \mathsf{States}(C)$ is a C-state, and R is a relevance ranking function for Γ and C. In what follows we shall sometimes denote by $\mathsf{CS}(\mathcal{L})$ a context setting $\mathsf{CS}(\Gamma)$ in which Γ is the set of all the well-formed formulas of $\mathcal{L}$.

Consistency restoration for databases can now be defined as before (see Definitions 6 and 7), except that the underlying distance setting $\mathsf{DS} = \langle d, f \rangle$ should now be context-sensitive in the sense that d_{DS} preserves the order induced by ranking in the following way:

Definition 12. Let $\mathsf{CS}(\mathcal{L}) = \langle C, S, R \rangle$ be a context setting for a language $\mathcal{L}$. A distance setting $\mathsf{DS} = \langle d, f \rangle$ is called CS-*sensitive*, if for every two atomic formulas p_1 and p_2 such that $R(p_1, S) > R(p_2, S)$, it holds that $d_{DS}(I_2, p_1) > d_{DS}(I_1, p_2)$ for every $I_1 \in mod(p_1) \setminus mod(p_2)$ and $I_2 \in mod(p_2) \setminus mod(p_1)$.

Clearly, Proposition 3 holds also for context-sensitive distance settings.

Next, we demonstrate the effect of incorporating context sensitive distance settings on inconsistency management.

Proposition 4. *Let* $\mathcal{DB} = \langle \mathcal{D} \sqcup \{p_1, p_2\}, \mathcal{IC} \rangle$ *be a database[2],* $\mathsf{CS} = \langle C, S, R \rangle$ *a context setting and* $\mathsf{DS} = \langle d, f \rangle$ *a* CS-*sensitive distance setting in which* f *is hereditary. If* $R(p_1, S) > R(p_2, S)$, *for every* $\mathcal{D}' \subseteq \mathsf{Atoms}(\mathcal{L}) \setminus \{p_1, p_2\}$ *such that* $\mathcal{D}' \sqcup \{p_1\} \models \mathcal{IC}$, *the* DS-*inconsistency value of* $\mathcal{D}_1 = \mathcal{D}' \sqcup \{p_1\}$ *is smaller than the* DS-*inconsistency value of* $\mathcal{D}_2 = \mathcal{D}' \sqcup \{p_2\}$.

---

[1] Relevance factors may be thought of as a context-dependent interpretation of weights in prioritized theories (see, for example, [4]).

[2] We denote by $\mathcal{D} \sqcup \{p_1, p_2\}$ the disjoint union of $\mathcal{D}$ and $\{p_1, p_2\}$.

Proof. Let $\mathcal{D}' \subseteq \mathsf{Atoms}(\mathcal{L}) \setminus \{p_1, p_2\}$ and $\mathcal{D}_1 = \mathcal{D}' \cup \{p_1\}$. Since $\mathcal{D}_1 \models \mathcal{IC}$, we have that $\mathsf{Inc}_{\mathsf{DS}}(\mathcal{D}_1) < \infty$. Thus, $\mathsf{Inc}_{\mathsf{DS}}(\mathcal{D}_1) < \mathsf{Inc}_{\mathsf{DS}}(\mathcal{D}_2)$ whenever $\mathcal{D}_2 \not\models \mathcal{IC}$. Suppose then that $\mathcal{D}_2 \models \mathcal{IC}$ as well. In this case, in the notations of Proposition 2, we have that $I_{\mathcal{D}_1}$ and $I_{\mathcal{D}_2}$ differ only in the assignments for p_1 and p_2 (I.e., $I_{\mathcal{D}_1}$ satisfies p_1 and falsifies p_2 while $I_{\mathcal{D}_2}$ satisfies p_2 and falsifies p_1. Elsewhere, both interpretations are equal to $I_{\mathcal{D}'}$). Now, since DS is CS-sensitive, by the facts that (i) $R(p_1, S) > R(p_2, S)$, (ii) $I_{\mathcal{D}_1} \in mod(p_1) \setminus mod(p_2)$ and (iii) $I_{\mathcal{D}_2} \in mod(p_2) \setminus mod(p_1)$, we have that $d_{\mathsf{DS}}(I_{\mathcal{D}_1}, p_2) < d_{\mathsf{DS}}(I_{\mathcal{D}_2}, p_1)$. Let $\mathcal{D} \cup \mathsf{CWA}(\mathcal{D} \sqcup \{p_1, p_2\}) = \{\psi_1, \ldots, \psi_n\}$. By the assumption that f is hereditary,

$\mathsf{Inc}_{\mathsf{DS}}(\mathcal{D}_1) = \delta_{\mathsf{DS}}(I_{\mathcal{D}_1}, \mathcal{DB}) =$
$f(\{d_{\mathsf{DS}}(I_{\mathcal{D}_1}, \psi_1), \ldots, d_{\mathsf{DS}}(I_{\mathcal{D}_1}, \psi_n), d_{\mathsf{DS}}(I_{\mathcal{D}_1}, p_1), d_{\mathsf{DS}}(I_{\mathcal{D}_1}, p_2)\}) =$
$f(\{d_{\mathsf{DS}}(I_{\mathcal{D}_1}, \psi_1), \ldots, d_{\mathsf{DS}}(I_{\mathcal{D}_1}, \psi_n), 0, d_{\mathsf{DS}}(I_{\mathcal{D}_1}, p_2)\}) =$
$f(\{d_{\mathsf{DS}}(I_{\mathcal{D}_2}, \psi_1), \ldots, d_{\mathsf{DS}}(I_{\mathcal{D}_2}, \psi_n), 0, d_{\mathsf{DS}}(I_{\mathcal{D}_1}, p_2)\}) <$
$f(\{d_{\mathsf{DS}}(I_{\mathcal{D}_2}, \psi_1), \ldots, d_{\mathsf{DS}}(I_{\mathcal{D}_2}, \psi_n), 0, d_{\mathsf{DS}}(I_{\mathcal{D}_2}, p_1)\}) =$
$f(\{d_{\mathsf{DS}}(I_{\mathcal{D}_2}, \psi_1), \ldots, d_{\mathsf{DS}}(I_{\mathcal{D}_2}, \psi_n), d_{\mathsf{DS}}(I_{\mathcal{D}_2}, p_2), d_{\mathsf{DS}}(I_{\mathcal{D}_2}, p_1)\}) =$
$\delta_{\mathsf{DS}}(I_{\mathcal{D}_2}, \mathcal{DB}) = \mathsf{Inc}_{\mathsf{DS}}(\mathcal{D}_2).$ $\square$

It follows that when context-sensitive distances are incorporated, "more relevant" formulas are preferred in the repairs. This is shown next.

Corollary 1. *Let $\mathcal{DB} = \langle \mathcal{D} \sqcup \{p_1, p_2\}, \mathcal{IC} \rangle$ be a database, $\mathsf{CS} = \langle C, S, R \rangle$ a context setting and $\mathsf{DS} = \langle d, f \rangle$ a CS-sensitive distance setting in which f is hereditary. If $\mathcal{DB}_1 = \langle \mathcal{D} \sqcup \{p_1\}, \mathcal{IC} \rangle$ is a consistent database, $R(p_1, S) > R(p_2, S)$, and $\mathcal{IC} \cup \{p_1, p_2\}$ is (classically) inconsistent, then no DS-repair of $\mathcal{DB}$ contains p_2.*

Corollary 1 can be generalized as follows:

Corollary 2. *Let $\mathcal{DB} = \langle \mathcal{D}, \mathcal{IC} \rangle$ be a database, $\mathsf{CS} = \langle C, S, R \rangle$ a context setting and $\mathsf{DS} = \langle d, f \rangle$ a CS-sensitive distance setting in which f is hereditary. Suppose that $\mathcal{D} = \mathcal{D}' \sqcup \mathcal{D}''$ can be partitioned to two disjoint nonempty subsets $\mathcal{D}'$ and $\mathcal{D}''$ such that (1): $\mathcal{DB}' = \langle \mathcal{D}', \mathcal{IC} \rangle$ is a consistent database, (2): $\forall p'' \in \mathcal{D}'' \, \exists p' \in \mathcal{D}'$ such that $\mathcal{IC} \cup \{p', p''\}$ is not consistent, and (3): $\forall p' \in \mathcal{D}'$ and $\forall p'' \in \mathcal{D}''$ it holds that $R(p', S) > R(p'', S)$. Then for every DS-repair $\mathcal{R}$ of $\mathcal{DB}$, $\mathcal{R} \cap \mathcal{D}'' = \emptyset$.*

3.2   A Simple Construction of Context-Sensitive Distance Settings

Below we provide a concrete method for defining context-sensitive distance settings and exemplify some of the properties of the settings that are obtained.

Definition 13. Let $\mathsf{CS}(\mathcal{L}) = \langle C, S, R \rangle$ be a context setting for $\mathcal{L}$ and let g be an aggregation function. The (pseudo) distance d_g^{CS} on $\Lambda_{\mathcal{L}}$ is defined as follows:

$$d_g^{\mathsf{CS}}(I, I') = g(\{R(p, S) \cdot |I(p) - I'(p)| \mid p \in \mathsf{Atoms}(\mathcal{L})\}).$$

It is easy to verify that for any CS and g, the function d_g^{CS} is a pseudo-distance on $\Lambda_{\mathcal{L}}$. In particular, for any context setting $\mathsf{CS}(\mathcal{L}) = \langle C, S, R \rangle$ where R is uniformly 1, d_Σ^{CS} coincides with the Hamming distance d_H in Example 2. The next proposition provides a general way of constructing context-sensitive distance settings, based on the functions in Definition 13.

Proposition 5. *Let* $\mathsf{CS} = \langle C, S, R \rangle$ *be a context setting and let* $\mathsf{DS} = \langle d_g^{\mathsf{CS}}, f \rangle$ *be a distance setting, where g is a hereditary. Then* DS *is* CS-*sensitive.*

Proof. Let p_1 and p_2 be atomic formulas such that $R(p_1, S) > R(p_2, S)$, and let $I_1 \in mod(p_1) \setminus mod(p_2)$ and $I_2 \in mod(p_2) \setminus mod(p_1)$. Below, we denote $g(\overline{0}, x) = g(\{0, \ldots, 0, x, 0, \ldots, 0\})$. By Definition 6,

$$d_{\mathsf{DS}}(I_1, p_2) = \min\{d_g^{\mathsf{CS}}(I_1, J) \mid J \models p_2\}$$
$$= \min\{g(\{R(p, S) \cdot |I_1(p) - J(p)| \mid p \in \mathsf{Atoms}(\mathcal{L})\}) \mid J \models p_2\}.$$

Since g is hereditary, the minimum above must be obtained for a model J of p_2 that coincides with I_1 on every atom $p \neq p_2$. It follows, then, that $d_{\mathsf{DS}}(I_1, p_2) = g(\overline{0}, R(p_2, S))$. By similar considerations, $d_{\mathsf{DS}}(I_2, p_1) = g(\overline{0}, R(p_1, S))$. Now, since $R(p_1, S) > R(p_2, S)$ and since g is hereditary, $d_{\mathsf{DS}}(I_2, p_1) > d_{\mathsf{DS}}(I_1, p_2)$. $\square$

The next proposition demonstrates how CS-sensitive distance settings of the form defined above give precedence to "more relevant" facts.

Proposition 6. *Let* $\mathsf{CS} = \langle C, S, R \rangle$ *be a context setting and let* $\mathsf{DS} = \langle d_g^{\mathsf{CS}}, f \rangle$ *be a distance setting, where g and f are hereditary aggregation functions. Let* $\mathcal{DB} = \langle \mathcal{D} \sqcup \{p_1, p_2\}, \mathcal{IC} \rangle$ *be a database such that:*

1. $R(p_1, S) > R(p_2, S)$ *(i.e., p_1 is more relevant than p_2), and*
2. $\mathcal{IC} \cup \{p_1, p_2\}$ *is not consistent but* $\mathcal{DB}_1 = \langle \mathcal{D} \sqcup \{p_1\}, \mathcal{IC} \rangle$ *is consistent[3].*

Then $\Delta_{\mathsf{DS}}(\mathcal{DB}) = \{I_1\}$, *where I_1 is the (unique) model of $\mathcal{DB}_1$.*

Proof. Again, we denote: $g(\overline{0}, x) = g(\{0, \ldots, 0, x, 0, \ldots, 0\})$. Then, for all p, I,

$$d_{\mathsf{DS}}(I, p) = \begin{cases} 0 & \text{if } I \models p, \\ g(\overline{0}, R(p, S)) & \text{otherwise.} \end{cases}$$

Suppose that $\mathcal{D} \cup \mathsf{CWA}(\mathcal{D} \sqcup \{p_1, p_2\}) = \{\psi_1, \ldots, \psi_n\}$. Let $I \in \Delta_{\mathsf{DS}}(\mathcal{DB})$ and $I_1 \in mod(\Gamma_{\mathcal{DB}_1})$ (Such a model exists, since $\mathcal{DB}_1$ is consistent). By Corollary 1, since DS is CS-sensitive (Proposition 5), $I \not\models p_2$, and so $d_{\mathsf{DS}}(I, p_2) = g(\overline{0}, R(p_2, S))$. Now,

$$\delta_{\mathsf{DS}}(I, \mathcal{DB}) = f(\{d_{\mathsf{DS}}(I, \psi_1), \ldots, d_{\mathsf{DS}}(I, \psi_n), d_{\mathsf{DS}}(I, p_1), d_{\mathsf{DS}}(I, p_2)\}) =$$
$$f(\{d_{\mathsf{DS}}(I, \psi_1), \ldots, d_{\mathsf{DS}}(I, \psi_n), d_{\mathsf{DS}}(I, p_1), g(\overline{0}, R(p_2, S))\}).$$

and so, since f is non-decreasing,

$$\delta_{\mathsf{DS}}(I, \mathcal{DB}) \geq f(\{0, \ldots, 0, 0, g(\overline{0}, R(p_2, S))\}) =$$
$$f(\{d_{\mathsf{DS}}(I_1, \psi_1), \ldots, d_{\mathsf{DS}}(I_1, \psi_n), d_{\mathsf{DS}}(I_1, p_1), d_{\mathsf{DS}}(I_1, p_2)\}) =$$
$$\delta_{\mathsf{DS}}(I_1, \mathcal{DB}).$$

Thus, $I_1 \in \Delta_{\mathsf{DS}}(\mathcal{DB})$. On the other hand, if there is some $q \in \{\psi_1, \ldots, \psi_n, p_1\}$ for which $d_{\mathsf{DS}}(I, q) \neq 0$, then since f is hereditary the above inequality becomes strict, which contradicts the assumption that $I \in \Delta_{\mathsf{DS}}(\mathcal{DB})$. It follows that for every $q \in \{\psi_1, \ldots, \psi_n, p_1\}$ $d_{\mathsf{DS}}(I, q) = d_{\mathsf{DS}}(I_1, q) = 0$, i.e., $I \models q$. One concludes, then, that I is a model of $\mathcal{DB}_1$, that is, $I = I_1$. $\square$

Proposition 6 may be extended in various ways. Below is one such extension.

---

[3] $\mathcal{DB}_2 = \langle \mathcal{D} \sqcup \{p_2\}, \mathcal{IC} \rangle$ may be consistent as well, but this is not a prerequisite.

Proposition 7. *Let $\mathcal{DB} = \langle \mathcal{D}, \mathcal{IC} \rangle$ $\mathsf{CS} = \langle C, S, R \rangle$ and $\mathsf{DS} = \langle d_g^{\mathsf{CS}}, f \rangle$, where g and f are hereditary aggregation functions. Suppose that $\mathcal{D}$ can be partitioned to two nonempty subsets $\mathcal{D}'$ and $\mathcal{D}''$, such that*

1. *$\mathcal{DB}' = \langle \mathcal{D}', \mathcal{IC} \rangle$ is a consistent database,*
2. *$\forall p'' \in \mathcal{D}''\ \exists p' \in \mathcal{D}'$ s.t. $\mathcal{IC} \cup \{p', p''\}$ is not consistent, and*
3. *$\forall p' \in \mathcal{D}'$ and $\forall p'' \in \mathcal{D}''$, $R(p', S) > R(p'', S)$.*

Then $\Delta_{\mathsf{DS}}(\mathcal{DB}) = \{I'\}$, where I' is the (unique) model of $\mathcal{DB}'$.

Proof. The proof is similar to that of Proposition 6. We omit the details.   □

Example 4. Consider again the database in Example 1. By Example 3, the distance setting $\mathsf{DS} = \langle d_H, \Sigma \rangle$ leads to the following two equally good repairs:

Repair 1 :

eNum	name	address	salary
1	John	..., UK	70K$
2	Mary	..., US	90K$

Repair 2 :

eNum	name	address	salary
1	John	..., AT	80K$
2	Mary	..., US	90K$

Sensitivity to context may differentiate between these repairs, preferring one to another. Let us again denote by $\mathsf{T}_{\mathsf{UK}}^1$, $\mathsf{T}_{\mathsf{AT}}^1$ and $\mathsf{T}_{\mathsf{US}}^2$ the tuple according to which John lives in the UK and is payed 70K$, John lives in Austria and is payed 80K$, and the tuple with the information about Mary.

Now, consider the context setting $\mathsf{CS}(\mathcal{L}) = \langle C, S, R \rangle$ and the distance setting $\mathsf{DS} = \langle d_\Sigma^{\mathsf{CS}}, \Sigma \rangle$, where $C = \{\mathsf{country}\}$, $\mathsf{Range}(\mathsf{country}) = \{\mathsf{US}, \mathsf{UK}, \mathsf{AT}\}$, $S(\mathsf{country}) = \mathsf{UK}$, and the relevance ranking is given by the following functions:

$$R(\mathsf{T}_\mathsf{c}^i, S) = \begin{cases} 1, & \text{if } c = S(\mathsf{country}), \\ 0.5, & \text{otherwise.} \end{cases} \qquad R(\neg \mathsf{T}_\mathsf{c}^i, S) = \begin{cases} 0.5, & \text{if } c = S(\mathsf{country}), \\ 1, & \text{otherwise.} \end{cases}$$

Computation of Δ_{DS} is given below (where we abbreviate $d(\psi, S)$ for $d_\Sigma^{\mathsf{CS}}(\psi, S)$).

I	$d(I, \mathsf{T}_{\mathsf{UK}}^1, S)$	$d(I, \mathsf{T}_{\mathsf{AT}}^1, S)$	$d(I, \neg \mathsf{T}_{\mathsf{US}}^1, S)$	$d(I, \neg \mathsf{T}_{\mathsf{UK}}^1, S)$	$d(I, \neg \mathsf{T}_{\mathsf{AT}}^1, S)$	$d(I, \mathsf{T}_{\mathsf{US}}^2, S)$	$\delta_{\mathsf{DS}}(I, \Gamma, S)$
$\emptyset$	1	0.5	0	0	0	0.5	2
$\{\mathsf{T}_{\mathsf{UK}}^1\}$	0	0.5	0	0	0	0.5	1
$\{\mathsf{T}_{\mathsf{AT}}^1\}$	1	0	0	0	0	0.5	1.5
$\{\mathsf{T}_{\mathsf{US}}^1\}$	1	0.5	1	0	0	0.5	3
...	...	...	...	...	...	...	...
$\{\mathsf{T}_{\mathsf{UK}}^1, \mathsf{T}_{\mathsf{US}}^2\}$	0	0.5	0	0	0	0	**0.5**
$\{\mathsf{T}_{\mathsf{AT}}^1, \mathsf{T}_{\mathsf{US}}^2\}$	1	0	0	0	0	0	1
...	...	...	...	...	...	...	...

According to CS, the single element in $\Delta_{\mathsf{DS}}(\mathcal{DB})$ satisfies $\{\mathsf{T}_{\mathsf{UK}}^1, \mathsf{T}_{\mathsf{US}}^2\}$, and so Repair 1 is preferred. Dually, in a state S' where $S'(\mathsf{country}) = \mathsf{AT}$, Repair 2 is preferred. Thus, context-aware considerations lead us to choose different repairs according to the relevance ranking, as indeed guaranteed by Propositions 6 and 7.

4   Conclusion

As observed in [13], contexts have largely been ignored by the AI community. In the database community context awareness has only recently been addressed in relation to user preference in querying (consistent) databases [17,19]. To the best of our knowledge, the approach presented here is the first one to combine inconsistency management with context-aware considerations. Combined with the extensive work available on personalization and automatically determining user's context and preferences (see, e.g., [6,14]), it may open the door to new inconsistency management solutions and novel database technologies. Implementation and evaluation of the methods in this paper is currently a work in progress.

References

1. Abowd, G.D., Dey, A.K.: Towards a better understanding of context and context-awareness. In: Gellersen, H.-W. (ed.) HUC 1999. LNCS, vol. 1707, pp. 304–307. Springer, Heidelberg (1999)
2. Arenas, M., Bertossi, L., Chomicki, J.: Answer sets for consistent query answering in inconsistent databases. TPLP 3(4-5), 393–424 (2003)
3. Arieli, O.: Distance-based paraconsistent logics. International Journal of Approximate Reasoning 48(3), 766–783 (2008)
4. Arieli, O.: Reasoning with prioritized information by iterative aggregation of distance functions. Journal of Applied Logic 6(4), 589–605 (2008)
5. Arieli, O., Denecker, M., Bruynooghe, M.: Distance semantics for database repair. Annals of Mathematics and Artificial Intelligence 50(3-4), 389–415 (2007)
6. Baldauf, M., Dustdar, S., Rosenberg, F.: A survey on context-aware systems. International Journal of Ad Hoc and Ubiquitous Computing 2(4), 263–277 (2007)
7. Bertossi, L.: Consistent query answering in databases. SIGMOD Record 35(2), 68–76 (2006)
8. Bolchini, C., Curino, C., Orsi, G., Quintarelli, E., Rossato, R., Schreiber, F.A., Tanca, L.: And what can context do for data? Comm. ACM 52(11), 136–140 (2009)
9. Brézillon, P.: Context in artificial intelligence: I. A survey of the literature. Computers and Artificial Intelligence 18(4) (1999)
10. Chomicki, J.: Consistent query answering: Five easy pieces. In: Schwentick, T., Suciu, D. (eds.) ICDT 2007. LNCS, vol. 4353, pp. 1–17. Springer, Heidelberg (2006)
11. Dey, A.: Understanding and using context. Personal and Ubiquitous Computing 5(1), 4–7 (2001)
12. Eiter, T., Mannila, H.: Distance measure for point sets and their computation. Acta Informatica 34, 109–133 (1997)
13. Ekbia, H.R., Maguitman, A.G.: Context and relevance: A pragmatic approach. In: Akman, V., Bouquet, P., Thomason, R.H., Young, R.A. (eds.) CONTEXT 2001. LNCS (LNAI), vol. 2116, pp. 156–169. Springer, Heidelberg (2001)
14. Henricksen, K., Indulska, J.: Developing context-aware pervasive computing applications: Models and approach. Pervasive and Mobile Comput. 2, 37–64 (2006)
15. Konieczny, S., Lang, J., Marquis, P.: DA2 merging operators. Artificial Intelligence 157(1-2), 49–79 (2004)
16. Konieczny, S., Pino Pérez, R.: Merging information under constraints: A logical framework. Logic and Computation 12(5), 773–808 (2002)
17. Pitoura, E., Stefanidis, K., Vassiliadis, P.: Contextual database preferences. IEEE Data Engineering Bulletin 34(2), 19–26 (2011)
18. Staworko, S., Chomicki, J., Marcinkowski, J.: Prioritized repairing and consistent query answering in relational databases. Ann. Math. Artif. Intel. 64, 209–246 (2012)
19. Stefanidis, K., Pitoura, E.: Fast contextual preference scoring of database tuples. In: Proceedings of EDBT 2008, pp. 344–355. ACM (2008)
20. Stefanidis, K., Pitoura, E., Vassiliadis, P.: Managing contextual preferences. Information Systems 36(8), 1158–1180 (2011)

An Analysis of the SUDOC Bibliographic Knowledge Base from a Link Validity Viewpoint

Léa Guizol[1], Olivier Rousseaux[2], Madalina Croitoru[1],
Yann Nicolas[2], and Aline Le Provost[2]

[1] LIRMM (University of Montpellier II & CNRS), INRIA Sophia-Antipolis, France
[2] ABES, France

Abstract. In the aim of evaluating and improving link quality in bibliographical knowledge bases, we develop a decision support system based on partitioning semantics. The novelty of our approach consists in using symbolic values criteria for partitioning and suitable partitioning semantics. In this paper we evaluate and compare the above mentioned semantics on a real qualitative sample. This sample is issued from the catalogue of French university libraries (SUDOC), a bibliographical knowledge base maintained by the University Bibliographic Agency (ABES).

1 Introduction

Real World Context. The SUDOC (catalogue du Système Universitaire de Documentation) is a large bibliographical knowledge base managed by ABES (Agence Bibliographique de l'Enseignement Supérieur). The SUDOC contains **bibliographic notices** (document descriptions $\approx$ 10.000.000), and **authorship notices** (person descriptions $\approx$ 2.000.000). An authorship notice possesses some attributes (ppn[1], appellation set, date of birth...). A bibliographic notice also possesses some attributes (title, ppn[1], language, publication date...) and **link**(s) to authorship notices. A link is labeled by a **role** (as *author*, *illustrator* or *thesis advisor*) and means that the person described by the authorship notice has participated as the labeled role to the document described by the bibliographic notice.

One of the most important tasks for ABES experts is to reference a new book in SUDOC. To this end, the expert has to register the title, number of pages, types of publication domains, language, publication date, and so on, in a new bibliographic notice. This new bibliographic notice represents the physical books in the librarian hands which he/she is registering. He/she also has to register people which participated to the book's creation (namely the **contributors**). In order to do that, for each contributor, he/she selects every authorship notice (named *candidates*) which has an appellation similar to the book contributor. Unfortunately, there is not that much information in authorship notices because the librarian politics is to give minimal information, solely in order to distinguish two authorship notices which have the same appellation, and nothing more (they reference books, not people!). So the librarian has to look at bibliographic notices which are linked to authorship notices candidates (the *bibliography* of candidates) in

---

[1] A ppn identifies a notice.

A. Laurent et al. (Eds.): IPMU 2014, Part II, CCIS 443, pp. 204–213, 2014.

order to see whether the book in his/her hands seems to be a part of the bibliography of a particular candidate. If it is the case, he/she links the new bibliographic notice to this candidate and looks at the next unlinked contributor. If there is no good candidate, he/she creates a new authorship notice to represent the contributor.

This task is fastidious because it is possible to have a lot of candidates for a single contributor (as much as 27 for a contributor named "BERNARD, Alain"). This creates errors, which in turn can create new errors since linking is an incremental process. In order to help experts to repair erroneous links, we proposed two **partitioning semantics** in [11] which enables us to detect erroneous links in bibliographic knowledge bases. A partitioning semantics evaluates and compares **partitions**[2].

Contribution. The contribution of this paper is to practically evaluate the results quality of partitioning semantics [11] on a real SUDOC sample. We recall the semantics in section 3, clearly explain on which objects and with which criteria the semantics have been applied in section 2, and present qualitative results in section 4. We discuss the results and conclude the paper in section 5.

2 Qualitative Experiments

In this section, we first adapt the **entity resolution problem**[3][4] to investigate link quality in SUDOC in section 2.1. This problem is known in literature under very different names (as record linkage [16], data deduplication [2], reference reconciliation [14]...). Then we define (section 2.3) and detail (section 2.4) criteria used in order to detect erroneous links in SUDOC. Those criteria are used on SUDOC subsets defined in section 2.2.

2.1 Contextual entities: From Erroneous Links to Entity Resolution

In order to detect and repair erroneous links, we represent SUDOC links into **contextual entity** (the i contextual entity is denoted Nc_i). A contextual entity represents a bibliographic notice Nb_j from the viewpoint of one of its contributor, named the C contributor of Nc_i and denoted $C(Nc_i)$. The contextual entities are compared together with an entity resolution method, in order to see which ones have a contributor representing a same real-word person. As explained in [8], traditional entity resolution methods cannot be directly applied. This entity resolution method is supposed to group (in a same class of the created partition) the contextual entities such as their C contributor represents a same real-word person, and to separate the other ones (to put them in distinct partition classes). A contextual entity Nc_i has several attributes. Most of them are $Nb(Nc_i)$ attributes (as title, publication date, publication language, publication domain codes list) and others depend on the C contributor:

---

[2] A **partition** P of an object set X is a set of **classes** (X subsets) such as each object of X is in one and only one P class.

[3] The **entity resolution problem** is the problem of identifying as equivalent two objects representing the same real-world entity.

– role of the C contributor (there is a set of typed roles as "thesis_advisor"),
– list of the possible appellations of the C contributor. An appellation is composed of a name and a surname, sometimes abbreviated (as "J." for surname),
– list of contributors which are not C. For each of them, we have the identifier of the authorship notice which represents it, and the role.

The publication language attribute is typed (for example, "eng" for English language, "fre" for French language and so on). The publication date is most of the time the publication year ("1984"). Sometimes information is missing and it only gives the century or decade ("19XX" means that the document has been published last century). A publication domain is not a describing string but a code with 3 digits which represent a domain area.

Example 1 (Contextual entity attributes). The authorship notice of ppn **026788861**, which represents "CHRISTIE, Agatha" is linked as "author" to the bibliographic notice of ppn 121495094, which represents *"Evil under the sun"* book. The contextual entity which represents this links has the following attributes:

– title: *"Evil under the sun"*
– publication date: "2001"
– publication language: "eng"
– publication domain codes list: {} (they have not been given by a librarian)
– list of the possible appellations of the C contributor: {"CHRISTIE, Agatha","WEST-MACOTT, Mary","MALLOWAN, Agatha","MILLER, Agathe Marie Clarissa"}
– role of the C contributor: "author"
– list of contributors which are not C: {} (there is no other contributors in this case)

Let Nc_i be the contextual entity identified by i. As any contextual entity, it has been constructed because of a link between an authorship notice and a bibliographic notice, which are respectively denoted $Na(Nc_i)$ and $Nb(Nc_i)$. We define two particular partitions: the initial one and the human one.

The **initial partition** (denoted Pi) of contextual entities is the one such as two contextual entities Nc_i, Nc_j are in a same class if and only if $Na(Nc_i) = Na(Nc_j)$. This represents the original organization of links in SUDOC.

The **human partition** (denoted Ph) of contextual entities is a partition based on an human expert's advice: two contextual entities Nc_i, Nc_j are in a same class if and only if the expert thinks that their C contributor corresponds to a same real word person.

The goal of this paper's work is to distinguish SUDOC subsets constructed as in the following section 2.2 with or without erroneous links. We make the hypothesis that the human partition has to be a best one (because it is the good one according to expert) and that the initial partition has to not be a best partition except if $Pi = Ph$. So, partitioning semantics are approved if Ph is a best partition according to the semantics, but not Pi. Let us determine what is a SUDOC contextual entities subset to partition.

2.2 Selecting contextual entities on Appellation

A SUDOC subset $\mathbb{O}$ selected for an appellation A contains all contextual entities which represent a link between any SUDOC bibliographic notice and a SUDOC authorship notice which has an appellation close to the appellation A. To select a SUDOC subset for a given appellation (as "BERNARD, Alain") is a way to separate SUDOC in subsets which can be treated separately, This is also a simulation of how experts select a SUDOC subset to work on it, as explained in part 1. In the following, we will only be interested into partitioning SUDOC subsets selected for an appellation. Let us define and describe criteria used in order to compare contextual entities together.

2.3 Symbolic Criteria

In the general case, a **criterion** is a function which compares two objects and returns a comparison value. Let c be a criterion, and o_i, o_j are two objects. We denote $c(o_i, o_j)$, the comparison values according to c between o_i and o_j.

In this work case, we use **symbolic criteria** which can return *always*, *never*, *neutral*, a **closeness value** or a **farness value** as comparison value. *always* (respectively *never*) means that objects have to be in a same (respectively distinct) partition class[2]. Closeness (respectively farness) values are more or less intense and far from the *neutral* value, meaning that objects should be in a same (respectively distinct) partition class. Closeness (respectively farness) values are strictly ordered between themselves, specific to a criterion and less intense than *always* (respectively *never*). Those values are denoted $+, ++$ and so on (respectively $-, --$) such as the more $+$ (respectively $-$) symbols they have, the more intense and the further from *neutral* the value is. For a criterion, *always* is more intense than $+ + + + +$, which is more intense than $++$ which is more intense than $+$. $+$ is only more intense than *neutral*. *neutral* means that the criterion has no advice about whether to put objects in a same class or not.

2.4 Criteria for Detecting Link Issues in SUDOC

In order to simulate human expert behaviour, nine symbolic criteria have been developed. Some are closeness-criteria[4] (*title, otherContributors*), farness-criteria[4] (*thesis, thesisAdvisor, date, appellation, language*) and others are both (*role, domain*). Each of these criteria give the *neutral* comparison value when a required attribute of a compared contextual entity is unknown and by default. Let Nc_i, Nc_j be two contextual entities.

- *appellation* criterion is a particular farness-criterion. Indeed, it compares appellation lists to determine which contextual entities can not have a same contributor C. When it is certain (as when appellations are "CONAN DOYLE, Arthur" and "CHRISTIE, Agatha"), it gives a *never* comparison value, which forbids other criteria to compare the concerned authorship notices together. This is also used to divide SUDOC in subsets which should be evaluated separately.
- *title* criterion is a closeness-criterion. This criterion can give an *always* value and 3 closeness comparison values. It is based on a Levenshtein comparison [13]. It is useful to determine which contextual entities represent a same work, edited several times. This is used by the *thesis* criterion.

---

[4] A closeness-criterion (respectively a farness-criterion) c is a criterion which can give a closeness or *always* (respectively a farness or *never*) comparison value to two objects.

- *otherContributors* criterion is a closeness-criterion. It counts the others contributors in common, by comparing their authorship notices. One (respectively several) other common contributor gives a + (respectively ++) comparison value.
- *thesis* criterion is a farness-criterion. $thesis(Nc_i, Nc_j) = -$ means that Nc_i, Nc_j are contextual entities which represent distinct thesis (recognized thanks to the *title* criterion) from their "author" point of view. $thesis(Nc_i, Nc_j) = --$ means that Nc_i, Nc_j have also been submitted simultaneously.
- *thesisAdvisor* criterion is a farness-criterion. $thesisAdvisor(Nc_i, Nc_j) = --$ (respectively $-$) means that Nc_i and Nc_j have a same contributor C if and only if this contributor has supervised a thesis before (respectively two years after) submitting his/her own thesis.
- *date* criterion is a farness-criterion. For 100 (respectively 60) years at least between publication dates, it gives a $--$ (respectively $-$) comparison value.
- *language* criterion is a farness-criterion. When publication languages are distinct and none of them is English, *language* returns a $-$ value.
- *role* criterion returns + when contributor C roles are the same (except for current roles as "author", "publishing editor" or "collaborator"), or $-$ when they are distinct (except for some pairs of roles as "thesis advisor" and "author").
- *domain* criterion compares list of domain codes. Domain codes are pair-wise compared. $domain(Nc_i, Nc_j)$ gives closeness (respectively farness) comparison values if every Nc_i domain codes is close (respectively far) from a Nc_j domain code and the other way around.

Let us resume global and local semantics before to evaluate their relevance with respect to the above mentioned criteria on real SUDOC subsets.

3   Partitioning Semantics

Let us summarize partitioning semantics detailed in [11]. A partitioning semantics evaluates and compares partitions on a same object set. The following partitioning semantics (in sections 3.1 and 3.2) are based on symbolic criteria.

3.1   Global Semantics

In this section we define what is a a best partition on the object set $\mathbb{O}$ (with respect to the $\mathbb{C}$ criteria set) according to global semantics. A partition has to be **valid**[5][2] in order to be a best one. A partition P has also an **intra value** and an **inter value** per criterion of $\mathbb{C}$. The intra value of a criterion c depends of the most intense (explained in section 2.3) farness or *never* value of c such as it compares two objects in a same class (should not be the case according to c). In the same way, the inter value of c depends of the most intense closeness or *always* value of c such as it compares two

---

[5] A partition P is **valid** if and only if there is no two objects o_i, o_j such as: (i) they are in a same class of P and they *never* have to be together according to a criterion (expressed by *never* comparison value), or (ii) they are in distinct P classes but *always* have to be together according to at least a criterion.

objects in distinct P classes. The inter value measures proximity between classes and the intra value measures distance between objects in a class [10]. We note that the *neutral* comparison value does not influence partition values.

A partition P on an object set $\mathbb{O}$ is a best partition according to a criteria set $\mathbb{C}$ if P is valid and P has a best value, meaning that it is impossible to improve an inter or intra value of any criterion $C \in \mathbb{C}$ without decreasing inter or intra value of a criterion $C' \in \mathbb{C}$ (it is a Pareto equilibrium [15]).

Table 1. Example of objects set

id	title	date	domains [...]	appellations
Nc_1	*"Letter to a Christian nation"*		religion	"HARRIS, Sam"
Nc_2	*"Surat terbuka untuk bangsa kristen"*	2008	religion	"HARRIS, Sam"
Nc_3	*"The philosophical basis of theism"*	1883	religion	"HARRIS, Samuel"
Nc_4	*"Building pathology"*	2001	building	"HARRIS, Samuel Y."
Nc_5	*"Building pathology"*	1936	building	"HARRIS, Samuel Y."
Nc_6	*"Aluminium alloys 2002"*	2002	physics	"HARRIS, Sam J."

Example 2 (Global semantics evaluating a partition on an object set $\mathbb{O}$).

Let us represent an object set $\mathbb{O} = \{Nc_1, Nc_2, Nc_3, Nc_4, Nc_5, Nc_6\}$ in table 1. Each object is a contextual entity and represents a link between a bibliographic notice and an authorship notice (here, an "author" of a book). Id is the object identity. For each of them, title, date of publication, publication domain and appellation of the contributor C are given as attributes.

Nc_1 and Nc_2 represent a same person, as Nc_4, Nc_5 does. The human partition on $\mathbb{O}$ is: $Ph = \{\{Nc_1, Nc_2\}, \{Nc_3\}, \{Nc_4, Nc_5\}, \{Nc_6\}\}$. This partition, according to global semantics and with respect to the criteria set $\mathbb{C} = \{appellation, title, domain, date\}$ (criteria are detailed in section 2.4) is not coherent with some of $\mathbb{C}$ criteria. The Ph value is such that:

– inter classes *domain* value is very bad (*always*) because Nc_1 and Nc_2 are in distinct classes but are both about religion.
– intra classes *date* value is bad $(--)$ because Nc_4 and Nc_5 are in a same class, but with publication dates distant of more than 60 years and less than 100 years.

Ph has a best partition value because increasing an inter or intra criterion value (as inter *domain* value by merging $\{Nc_1, Nc_2\}$ and $\{Nc_3\}$ classes) is not possible without decreasing an other inter or intra criterion value (Nc_2 and Nc_3 have publication dates distant more than 100 years, so put them in a same class will decrease *date* intra value).

3.2 Local Semantics

The local semantics, when evaluating a partition on an object set $\mathbb{O}$ with respect to a criteria set $\mathbb{C}$, gives a partition value per parts of $\mathbb{O}$. Parts of $\mathbb{O}$ can be coherent or incoherent. An **incoherent part** $\mathbb{O}_a$ is a subset of $\mathbb{O}$ such as:

- there is no $c(o_i, o_j)$, an *always* or closeness value with $Nc_i \in \mathbb{O} - \mathbb{O}_a$, $Nc_j \in \mathbb{O}_a$, and $c \in \mathbb{C}$;
- there is no subset of $\mathbb{O}_a$ for which the previous property is true;
- there is $b(o_k, o_l)$, a farness or *never* value such as $o_k, o_l \in \mathbb{O}_a$, and $b \in \mathbb{C}$.

An *incoherent part partition value* is based on every comparison between objects which are in it. The **coherent part** of an object set $\mathbb{O}$ is a $\mathbb{O}$ subset containing every $\mathbb{O}$ object which is not in a incoherent part of $\mathbb{O}$. The *coherent part partition value* of $\mathbb{O}$ is based on every comparison between objects which are not in the same incoherent part of $\mathbb{O}$.

Example 3 (Incoherent and coherent parts).

Let us identify incoherent parts of the object set $\mathbb{O}$ according to $\mathbb{C}$ given in example 2. Nc_1, Nc_2, Nc_3 are close together due to *domain* criterion: they are about religion. Nc_1, Nc_2, Nc_3 are not close to Nc_4, Nc_5 or Nc_6 according to any of $\mathbb{C}$ criteria and Nc_2, Nc_3 are far according to *date* criterion ($date(Nc_2, Nc_3) = --$) so $\{Nc_1, Nc_2, Nc_3\}$ is an incoherent part of $\mathbb{O}$. The same way, Nc_4, Nc_5 are close together according to *title* and *domain* criteria, but not close to Nc_6. Nc_4, Nc_5 are also far according to *date* criterion ($date(Nc_4, Nc_5) = -$) so $\{Nc_4, Nc_5\}$ is also an incoherent part.

So, there are 2 incoherent parts in $\mathbb{O}$: $\{Nc_1, Nc_2, Nc_3\}$ and $\{Nc_4, Nc_5\}$. Nc_6 is not in an incoherent part so Nc_6 is in the coherent part of $\mathbb{O}$.

A partition on $\mathbb{O}$ is a *best partition according to local semantics* if it has best partition values for each incoherent part of $\mathbb{O}$ and for the $\mathbb{O}$ coherent part.

Example 4 (Local semantics evaluating a partition on an object sets $\mathbb{O}$).

In example 3, we identified the incoherent parts of the object set $\mathbb{O} = \{Nc_1, Nc_2, Nc_3, Nc_4, Nc_5, Nc_6\}$ according to the criteria set $\mathbb{C} = \{appellation, title, domain, date\}$.

The partition on $\mathbb{O}$ given in example 2: is $Ph = \{\{Nc_1, Nc_2\}, \{Nc_3\}, \{Nc_4, Nc_5\}, \{Nc_6\}\}$. According to local semantics, Ph has 3 values, one for the coherent part and 2 for incoherent parts (1 per incoherent part):

- a perfect value for the coherent part of $\mathbb{O}$;
- the incoherent part $\{Nc_1, Nc_2, Nc_3\}$ has a very bad inter value for the *domain* criterion (*always*);
- the incoherent part $\{Nc_4, Nc_5\}$ has an bad intra value for the *date* criterion ($--$);

This semantics enables us to split an object set into several parts which can be evaluated separately. We explained local and global semantics in this part, which are a way to solve the entity resolution problem. Let us evaluate them on a real SUDOC sample.

4   Results

ABES experts have selected 537 contextual entity divided into 7 SUDOC subsets selected for an appellation. The table 2 shows for each SUDOC subset selected for an appellation A (please see section 2.2):

1. $|Nc|$ is the number of contextual entities which represent a link between a biblio-graphic notice and an authorship notice which has a close appellation to A,
2. $|Na|$ is the number of authorities notices according to human partitions (corre-sponding to class number of human partition),
3. "Ph best" (respectively "Pi best") shows whether the human partition Ph (respec-tively initial partition Pi) has a best value according to global semantics and with respect to all 9 criteria detailed in part 2.4,
4. $Ph \succ Pi$ is true if and only if Ph has a better value than Pi.

Table 2. Human and initial partitions with respect to 9 criteria and global semantics

Appellation	$|Nc|$	$|Na|$	Ph best	Pi best	$Ph \succ Pi$	Ph' best	Repairs
"BERNARD, Alain"	165	27	no	not valid	yes	yes	
"DUBOIS, Olivier"	27	8	no	no	yes	no	1
"LEROUX, Alain"	59	6	no	not valid	yes	yes	
"ROY, Michel"	52	9	yes	not valid	yes	yes	
"NICOLAS, Maurice"	20	3	yes	no	yes	yes	
"SIMON, Alain"	63	13	no	no	yes	no	1
"SIMON, Daniel"	151	16	no	not valid	yes	yes	

Local semantics, has the same results than global semantics on this sample.

For global semantics, Pi is never a best partition. 5 times out of 7, Ph does not have a best value (each time, it is due to the *domain* and *language* criteria, and two times *thesisAdvisor* is also involved), but it is all the time valid and better than Pi, which is encouraging for erroneous link detection. Erroneous links are particularly obvious when Pi is not even valid (4 times out of 7). It is due to the *title* criterion detailed in part 2.4. We regret that Ph is not all the time a best partition, but the global semantics is able to distinguish Pi from Ph in 5 cases out of 7: when Pi is not valid, or when Ph is a best partition but not Pi.

Because the *domain* and *language* criteria often considers that Ph is not a good enough partition, Ph was also evaluated for global semantics according to all cri-teria without *domain* and *language* (shown in table 2 in column "Ph' best") and that increases the human partition which obtains a best value in 3 more cases (for "BERNARD, Alain", "SIMON, Daniel" and "LEROUX, Alain" appellations). This tells us that *domain* and *language* criteria are not reasonably accurate.

In order to evaluate if Ph is far from having a best partition value, we enumerate the number of repairs to transform Ph' into a partition Ph'' which has a best value according to all criteria except *domain* and *language*. We show this repair number in the "Repairs" column of table 2. An atomic repair could be:

– merging two partition classes (corresponds to merging two contextual entities which represent a same real word person), or
– splitting a partition class in two classes (corresponds to separate books which are attributed to a same real word person but belong to two distinct real word persons).

We can see that only a few repairs are needed compared to the number of classes (corresponding to $|Na|$ column in the table): 1 repair for "DUBOIS, Olivier" and for "BERNARD, Alain" appellations.

Let us highlight that observing human partition values has permitted to *detect and correct an erroneous link* (for "ROY, Michel" appellation) in the human reference set, validated with experts. The global semantics does not always consider that the human partition is a best partition, but in the worst case the human partition is very close to be one according to repairs number, and global semantics allow us to detect that initial partitions are much worse than human partitions. This last point is encouraging. This means that the semantics can also be useful to help in criteria tuning, by showing which criteria are bad according to human partitions, and for which authorship notices comparison. For example, the fact that the human partition value is often bad according to the *domain* criterion shows that this criterion is actually not an accurate criterion. Let us talk about other entity resolution methods and conclude.

5   Discussion

The entity resolution problem [4][16][14][6] is the problem of identifying as equivalent two objects representing the same real-world entity. The causes of such mismatch can be due to homonyms (as in people with the same name), errors that occurred at data entry (like "Léa Guizo" for "Léa Guizol"), missing attributes (e.g publication date = XXXX), abbreviations ("L. Guizol") or attributes having different values for two objects representing the same entity (change of address).

The entity resolution problem can be addressed as a rule based pairwise comparison rule approach. Approaches have been proposed in literature [12] using a training pairs set for learning such rules. Rules can be then be chained using different constraints: transitivity [3], exclusivity [12] and functional dependencies [1] [9].

An alternative method for entity resolution problem is partitioning (hierarchical partitioning [5], closest neighbor-based method [7] or correlation clustering [3]). Our work falls in this last category. Due to the nature of treating criteria values, the closest approach to our semantics are [3] and [2]. We distinguish ourself to [3] and [2] because of (1) the lack of *neutral* values in these approaches, (2) the numerization of symbolic values (numerically aggregated into -1 and $+1$ values), and (3) the use of numerical aggregation methods on these values.

Conclusion.   In this paper we proposed a practical evaluation of the global and local semantics proposed in [11]. The conclusions of this evaluation are:

- For SUDOC subsets selected by appellation, *both semantics* are effective to distinguish a human partition from the initial partition; however it is not perfect with respect to our set of criteria (if the human partition is not a best partition, it has a close value).
- *Both semantics* could be useful to detect meaningless criteria.

As immediate next steps to complete this our work we mention using global or local semantics to improve implemented criteria.

Acknowledgements. This work has been supported by the Agence Nationale de la Recherche (grant ANR-12-CORD-0012). We are thankful to Mickaël Nguyen for his support.

References

1. Ananthakrishna, R., Chaudhuri, S., Ganti, V.: Eliminating fuzzy duplicates in data warehouses. In: Proceedings of the 28th International Conference on Very Large Data Bases, VLDB 2002, pp. 586–597. VLDB Endowment (2002)
2. Arasu, A., Ré, C., Suciu, D.: Large-scale deduplication with constraints using dedupalog. In: Proceedings of the 25th International Conference on Data Engineering (ICDE), pp. 952–963 (2009)
3. Bansal, N., Blum, A., Chawla, S.: Correlation clustering 56, 89–113 (2004)
4. Bhattacharya, I., Getoor, L.: Entity Resolution in Graphs, pp. 311–344. John Wiley & Sons, Inc. (2006)
5. Bilenko, M., Basil, S., Sahami, M.: Adaptive product normalization: Using online learning for record linkage in comparison shopping. In: Fifth IEEE International Conference on Data Mining, p. 8. IEEE (2005)
6. Bouquet, P., Stoermer, H., Bazzanella, B.: An entity name system (ens) for the semantic web. In: Bechhofer, S., Hauswirth, M., Hoffmann, J., Koubarakis, M. (eds.) ESWC 2008. LNCS, vol. 5021, pp. 258–272. Springer, Heidelberg (2008)
7. Chaudhuri, S., Ganti, V., Motwani, R.: Robust identification of fuzzy duplicates. In: Proceedings of the 21st International Conference on Data Engineering, ICDE 2005, pp. 865–876. IEEE (2005)
8. Croitoru, M., Guizol, L., Leclère, M.: On Link Validity in Bibliographic Knowledge Bases. In: Greco, S., Bouchon-Meunier, B., Coletti, G., Fedrizzi, M., Matarazzo, B., Yager, R.R. (eds.) IPMU 2012, Part I. CCIS, vol. 297, pp. 380–389. Springer, Heidelberg (2012)
9. Fan, W.: Dependencies revisited for improving data quality. In: Proceedings of the Twenty-Seventh ACM SIGMOD-SIGACT-SIGART Symposium on Principles of Database Systems, pp. 159–170. ACM (2008)
10. Guénoche, A.: Partitions optimisées selon différents critères: évaluation et comparaison. Mathématiques et Sciences Humaines. Mathematics and Social Sciences (161) (2003)
11. Guizol, L., Croitoru, M., Leclere, M.: Aggregation semantics for link validity. In: AI-2013: Thirty-third SGAI International Conference on Artificial Intelligence (page to appear, 2013)
12. Gupta, R., Sarawagi, S.: Answering table augmentation queries from unstructured lists on the web. Proceedings of the VLDB Endowment 2(1), 289–300 (2009)
13. Levenshtein, V.I.: Binary codes capable of correcting deletions, insertions and reversals. Soviet Physics Doklady 10, 707 (1966)
14. Saïs, F., Pernelle, N., Rousset, M.-C.: Reconciliation de references: une approche logique adaptee aux grands volumes de donnees. In: EGC, pp. 623–634 (2007)
15. Wang, S.: Existence of a pareto equilibrium. Journal of Optimization Theory and Applications 79(2), 373–384 (1993)
16. Winkler, W.E.: Overview of record linkage and current research directions. Technical report, BUREAU OF THE CENSUS (2006)

A Fuzzy Extension of Data Exchange

Jesús Medina[1],[*] and Reinhard Pichler[2],[**]

[1] Department of Mathematics, University of Cádiz, Spain
[2] Faculty of Informatics, Vienna University of Technology, Austria

Abstract. Data exchange is concerned with the transformation of data structured under one schema into a different schema. In practice, this task is usually accomplished in a procedural way. In a landmark paper, Fagin et al. have proposed a declarative, purely logical approach to this task. Since then, data exchange has been intensively studied in the database research community. Recently, it has been extended to probabilistic data exchange. In this paper, we propose an extension to fuzzy data exchange.

1 Introduction

Data exchange is a classical problem in the database world. It is concerned with the transformation of data structured under one schema into a different schema. In practice, this task is usually accomplished in a procedural way by so-called ETL (extract-transform-load) scripts. In a landmark paper [9], Fagin et al. have proposed a declarative, purely logical approach to this task. Since then, data exchange has been intensively studied in the database research community.

The crucial concept in this approach are *schema mappings*. A schema mapping is given by two schemas, called the *source schema* $\mathbf{S}$ and the *target schema* $\mathbf{T}$, as well as a set of *dependencies* describing the relationship between the two schemas. The most fundamental kind of dependencies are *source-to-target tuple generating dependencies* (s-t tgds): these are first-order formulas of the form $\forall \boldsymbol{x}(\phi(\boldsymbol{x}) \rightarrow \exists \boldsymbol{y}\psi(\boldsymbol{x},\boldsymbol{y}))$, where the antecedent $\phi(\boldsymbol{x})$ is a conjunctive query (CQ) over $\mathbf{S}$ and the conclusion $\psi(\boldsymbol{x},\boldsymbol{y})$ is a CQ over $\mathbf{T}$. Note that a CQ refers to a (possibly existentially quantified) conjunction of atomic formulas. Intuitively, such an s-t tgd defines a constraint that the presence of certain tuples in the source database instance $\mathcal{I}$ (namely those in the image of some homomorphism h from $\phi(\boldsymbol{x})$ to $\mathcal{I}$) enforce the presence of certain tuples in the target database instance $\mathcal{J}$ (s.t. h can be extended to a homomorphism from $\psi(\boldsymbol{x},\boldsymbol{y})$ to $\mathcal{J}$).

Two further kinds of dependencies may be used to specify constraints on the target instance $\mathcal{J}$, namely again tuple-generating dependencies (target tgds) or *equality generating dependencies* (target egds). The target tgds are formulas of the same form as s-t tgds – but now both the antecedent and the conclusion are

---

[*] Partially supported by the Spanish Science Ministry project TIN2012-39353-C04-04.
[**] Supported by the Austrian Science Fund (FWF):P25207-N23.

A. Laurent et al. (Eds.): IPMU 2014, Part II, CCIS 443, pp. 214–223, 2014.

CQs over the target schema $\mathbf{T}$. Target egds are formulas of the form $\forall \boldsymbol{x}(\phi(\boldsymbol{x}) \to (x_i = x_j))$, where again $\phi(\boldsymbol{x})$ is a CQ over $\mathbf{T}$. Intuitively, egds allow us to express constraints that the presence of certain tuples in the target instance $\mathcal{J}$ enforces the equality of certain values occurring in the instance.

In summary, a *schema mapping* is given by a triple $\mathcal{M} = (\mathbf{S}, \mathbf{T}, \Sigma)$ where $\mathbf{S}$ is the source schema, $\mathbf{T}$ is the target schema, and Σ is a set of dependencies expressing the relationship between $\mathbf{S}$ and $\mathbf{T}$ and possibly also local constraints on $\mathbf{T}$. The *data exchange problem* associated with $\mathcal{M}$ is the following: Given a (ground) source instance $\mathcal{I}$, find a target instance $\mathcal{J}$, s.t. $\mathcal{I}$ and $\mathcal{J}$ together satisfy all dependencies in Σ, written as $\langle \mathcal{I}, \mathcal{J} \rangle \models \Sigma$. Such a $\mathcal{J}$ is called a *solution for $\mathcal{I}$*. Another important computational problem in the area of data exchange is *query answering*, where queries are posed against the target schema. The answers should reflect the source instance and the mapping; they should not depend on the concrete solution that was materialized. Hence, the generally agreed semantics for query answering in data exchange are *certain answers*, i.e., we aim at those answers which are in the result of the query over *every* solution for a given source instance under the given mapping. In case of (unions of) conjunctive queries, efficient query evaluation is possible via so-called *universal solutions* [9], i.e., solutions which admit a homomorphism into any other solution. Moreover, it was shown in [9] that the so-called *chase* procedure [6] can be used to compute a universal solution – provided that any solution exists.

Over the past decade, many extensions of data exchange have been studied such as more complex queries [1,3], data formats different from relational data [4,2,5], etc. Recently, also an extension to *probabilistic data exchange* has been proposed by Fagin et al. [8]. In the latter approach, a probability space over the source and target instances is considered rather than single (source and target) instances. The authors extend all of the central concepts of classical data exchange (such as solutions, universal solutions, the chase, certain answers, etc.) to a probabilistic setting. Moreover, they provide a detailed complexity analysis of the basic computational problems in data exchange. It turns out that query answering in probabilistic data exchange is intractable (more precisely, it is shown $\mathsf{FP}^{\#\mathsf{P}}$-complete) even in the most restricted cases.

The goal of this paper is to introduce *fuzzy data exchange* as another natural extension of data exchange. Analogously to [8], we shall thus extend all of the central concepts of classical data exchange to a fuzzy setting. It will turn out that such an extension is indeed feasible and allows us to carry over the most fundamental properties from the classical setting.

The paper is organized as follows. In Section 2, we briefly recall some basic notions. Due to space limitations, many concepts will be later directly presented in the fuzzy extension without presenting their original classical form first. In Section 3, we introduce fuzzy data exchange and fuzzy solutions. Fuzzy universal solutions and their computation are dealt with in Section 4. Query answering is discussed in Section 5. An outlook to future work is given in Section 6.

2   Preliminaries

A *schema* $\mathbf{R} = \{R_1, \ldots, R_n\}$ is a finite set of relation symbols R_i – each of a fixed arity r_i. An *instance* $\mathcal{I}$ over a schema $\mathbf{R}$ consists of a relation for each relation symbol in $\mathbf{R}$, s.t. both have the same arity. For a relation symbol R, we write $R^{\mathcal{I}}$ (or simply R) to denote the relation of R in $\mathcal{I}$. Tuples of the relations may contain two types of terms: *constants* and *variables*. The latter are also called *labelled nulls*. For every instance $\mathcal{J}$, we write $\mathrm{Dom}(\mathcal{J})$, $\mathrm{Var}(\mathcal{J})$, and $\mathrm{Const}(\mathcal{J})$ to denote the set of terms (i.e., the domain), variables, and constants, respectively, of $\mathcal{J}$. Clearly, $\mathrm{Dom}(\mathcal{J}) = \mathrm{Var}(\mathcal{J}) \cup \mathrm{Const}(\mathcal{J})$ and $\mathrm{Var}(\mathcal{J}) \cap \mathrm{Const}(\mathcal{J}) = \varnothing$. We simply write Const and Var to denote the set of all possible constants and variables, respectively.

Let $\mathbf{S} = \{S_1, \ldots, S_n\}$ and $\mathbf{T} = \{T_1, \ldots, T_m\}$ be two schemas with no relation symbols in common. Then we denote by $\langle \mathbf{S}, \mathbf{T} \rangle$ the schema obtained by concatenation of $\mathbf{S}$ and $\mathbf{T}$. In the same way, given $\mathcal{I}$ and $\mathcal{J}$, instances of $\mathbf{S}$ and $\mathbf{T}$, respectively, the sequence $\langle \mathcal{I}, \mathcal{J} \rangle$ is the instance $\mathcal{K}$ that satisfies $S_i^{\mathcal{K}} = S_i^{\mathcal{I}}$ and $T_j^{\mathcal{K}} = T_j^{\mathcal{J}}$, for all $i \in \{1, \ldots, n\}$ and $j \in \{1, \ldots, m\}$, that is, $\langle \mathcal{I}, \mathcal{J} \rangle$ is the (disjoint) union of $\mathcal{I}$ and $\mathcal{J}$.

3   The Fuzzy Data Exchange Problem

Fuzzy Instances. For our extension of data exchange to a fuzzy setting, we consider a complete lattice $(L, \preceq)$. A *fuzzy instance* $\mathcal{I}$ (over schema $\mathbf{R}$), *f-instance* for short, is a set $\{R_1^{\mathcal{I}}, \ldots, R_k^{\mathcal{I}}\}$, such that each $R_i^{\mathcal{I}}$ is a finite relation of arity $r_i + 1$ with $R_i^{\mathcal{I}} \subseteq (\mathrm{Const} \cup \mathrm{Var})^{r_i} \times L$ for every i. To simplify the notation, we shall write R_i to denote both, the relation symbol and the relation $R_i^{\mathcal{I}}$. We call $\mathcal{I}$ a *fuzzy ground instance* if $\mathrm{Var}(\mathcal{I}) = \varnothing$. The set of all f-instances over $\mathbf{R}$ is denoted by $\mathrm{FInst}(\mathbf{R})$ and the set of ground instances by $\mathrm{FInst}^c(\mathbf{R})$. As in classical data exchange, we shall always assume that source instances are ground.

A tuple $(t_1, \ldots, t_r, \vartheta)$ in $\mathbf{R}$ is denoted as $(R(t_1, \ldots, t_r), \vartheta)$ and it is called *fact*. Hence, an f-instance can be identified by the set of its facts, i.e., $\mathcal{I} = \{(R_1(\mathbf{t}_1), \vartheta_1), \ldots, (R_k(\mathbf{t}_k), \vartheta_k)\}$. The *confidence value associated with* $\mathcal{I}$ is the value $\vartheta_{\mathcal{I}} = \vartheta_1 \,\&\, \ldots \,\&\, \vartheta_k$.

Fuzzy Homomorphisms. Given two f-instances $\mathcal{K}_1$ and $\mathcal{K}_2$ over the same schema, a *fuzzy homomorphism* $h \colon \mathcal{K}_1 \to \mathcal{K}_2$ is a mapping from $\mathrm{Dom}(\mathcal{K}_1)$ to $\mathrm{Dom}(\mathcal{K}_2)$, such that

– $h(c) = c$, for all constants c in $\mathrm{Dom}(\mathcal{K}_1)$,
– for all facts $(R(\mathbf{t}), \vartheta)$ of $\mathcal{K}_1$, a fact $(R(h(\mathbf{t})), \vartheta')$ exists in $\mathcal{K}_2$ with $\vartheta \preceq \vartheta'$, where $h(\mathbf{t}) = (h(t_1), \ldots, h(t_r))$. In this case, we write $\mathcal{K}_1 \to \mathcal{K}_2$.

Example 1. Consider the f-instances $\mathcal{K}_1 = \{(Q(x), 0.5), (R(y), 0.4)\}$ and $\mathcal{K}_2 = \{(Q(a), 0.6), (R(b), 0.4), (P(x), 0.7)\}$. The mapping $h \colon \mathrm{Dom}(\mathcal{K}_1) \to \mathrm{Dom}(\mathcal{K}_2)$ with $h(x) = a$ and $h(y) = b$ can be easily verified to be an f-homomorphism. For instance, the fact $(Q(h(x)), 0.6) = (Q(a), 0.6)$ is in $\mathcal{K}_2$ with $0.5 \leq 0.6$.   ◇

As in the classical case, the *composition* of f-homomorphisms is also an f-homomorphism. Moreover, the *extension* of an f-homomorphism is naturally defined: let $h\colon \mathcal{K}_1 \to \mathcal{K}_2$ be an f-homomorphism between two f-instances $\mathcal{K}_1$ and $\mathcal{K}_2$, and let $\mathcal{K}$ be an f-instance with $\mathcal{K}_1 \subseteq \mathcal{K}$. We call h' an *extension of* h if $h'\colon \mathcal{K} \to \mathcal{K}_2$ is an f-homomorphism satisfying that h' restricted to $\mathcal{K}_1$ is h.

The notion of (crisp) homomorphisms is naturally extended to f-instances, even between an instance and an f-instance. For example, a mapping h from $\mathcal{K}_1 = \{R_1, \ldots, R_k\}$ to $\mathcal{K}_2 = \{(R_1^*, \vartheta_1), \ldots, (R_l^*, \vartheta_l)\}$ is a homomorphism if it satisfies the above conditions of an f-homomorphism ignoring the condition on the confidence values. Moreover, from a crisp instance $K_1 = \{R_1, \ldots, R_k\}$ and a homomorphism $h\colon K_1 \to \mathcal{K}_2$ we can define an f-instance $\mathcal{K}_1^h$. In this f-instance $\mathcal{K}_1^h$, each fact $R(\mathbf{t})$ of K_1 is augmented by the confidence value $\vartheta_R \in L$ such that $\vartheta_R = \sup\{\vartheta \in L \mid (R(h(\mathbf{t})), \vartheta) \in \mathcal{K}_2\}$. Since h is a homomorphism, we can be sure that for each $R(\mathbf{t})$ in K_1, there exists a fact $(R(h(\mathbf{t})), \vartheta) \in \mathcal{K}_2$ for some $\vartheta \in L$. We shall write $\hat{h}$ to denote the resulting f-homomorphism. Hence, we can say that an f-homomorphism $\hat{h}'$ is an *extension of a homomorphism* h, if it is an extension of the f-homomorphism $\hat{h}$ associated with h.

For example, for the instance $K_1 = \{Q(x)\}$, the f-instance $\mathcal{K}_2 = \{(Q(b), 0.5), (Q(a), 0.6)\}$ and the homomorphism h that maps x to a, we obtain the f-instance $\mathcal{K}_1^h = \{(Q(x), 0.6)\}$.

Fuzzy Dependencies. Analogously to classical data exchange, we consider schema mappings consisting of source-to-target tuple generating dependencies (s-t tgds) and target dependencies in the form of tuple generating dependencies (target tgds) and equality generating dependencies (target egds).

Recall that tgds are first-order formulas of the form $\forall \boldsymbol{x}(\phi(\boldsymbol{x}) \to \exists \boldsymbol{y}\psi(\boldsymbol{x}, \boldsymbol{y}))$, where $\phi(\boldsymbol{x})$ and $\psi(\boldsymbol{x}, \boldsymbol{y})$ are CQs, and egds are first-order formulas of the form $\forall \boldsymbol{x}(\phi(\boldsymbol{x}) \to (x_i = x_j))$, where $\phi(\boldsymbol{x})$ is a CQ. To simplify the notation, the universal quantifiers are usually omitted with the understanding that all variables occurring in the antecedent are universally quantified over the entire formula. We now extend the satisfaction of tgds and egds from crisp instances to f-instances.

Definition 1. *Let $\mathbf{R}$ be a schema and $\mathcal{K}$ an instance over $\mathbf{R}$. Consider a tgd d and an egd e over $\mathbf{R}$, i.e.,*

$$d \equiv (S_1(\mathbf{x}) \wedge \cdots \wedge S_n(\mathbf{x})) \to \exists \mathbf{y}(T_1(\mathbf{x}, \mathbf{y}) \wedge \cdots \wedge T_m(\mathbf{x}, \mathbf{y})) \ and$$
$$e \equiv (S_1(\mathbf{x}) \wedge \cdots \wedge S_n(\mathbf{x})) \to (x_1 = x_2)$$

Tgd d is satisfied by $\mathcal{K}$ (written as $\mathcal{K} \models d$) if every homomorphism $h\colon \mathcal{I}_d \to \mathcal{K}$ can be extended to an f-homomorphism $\hat{h}'\colon \mathcal{J}_d \to \mathcal{K}$, where
   $\mathcal{I}_d = \{S_1(\mathbf{x}), \ldots, S_n(\mathbf{x})\}$ *is an instance over $\mathbf{R}$,*
   $\vartheta_{\mathcal{I}_d^h}$ *is the confidence value associated with $\mathcal{I}_d^h$, and*
   $\mathcal{J}_d = \{(T_1(\mathbf{x}, \mathbf{y}), \vartheta_{\mathcal{I}_d^h}), \ldots, (T_m(\mathbf{x}, \mathbf{y}), \vartheta_{\mathcal{I}_d^h})\}$ *is an f-instance over $\mathbf{R}$.*

Likewise, egd e is satisfied by $\mathcal{K}$ (written as $\mathcal{K} \models e$) if for every homomorphism $h\colon \mathcal{I}_d \to \mathcal{K}$, we have $h(x_i) \neq h(x_j)$.

A set Σ of tgds and egds is satisfied by $\mathcal{K}$ (written as $\mathcal{K} \models \Sigma$) if every dependency of Σ is satisfied by $\mathcal{K}$.   ◇

In other words, for the satisfaction of a tgd, we first compute the confidence value $\vartheta_{\mathcal{I}_d^h}$ of an f-homomorphism h from the antecedent of d into the f-instance $\mathcal{K}$ and then check that h can be extended to an f-homomorphism $\hat{h}'$ from the conclusion of d into $\mathcal{K}$ with at least this confidence value. The satisfaction of an egd is defined as in the classical case by ignoring the confidence values. As for the tgd, the above definition clearly applies to both, s-t tgds and target tgds. In case of s-t tgds over a source schema $\mathbf{S}$ and target schema $\mathbf{T}$, we have to consider instances $\mathcal{K} = \langle \mathcal{I}, \mathcal{J} \rangle$ over the combined schema $\mathbf{R} = \langle \mathbf{S}, \mathbf{T} \rangle$.

Example 2. Given the tgd $d = Q(x) \to \exists y R(y)$ and

1. the instances $\mathcal{I}_1 = \{(P(a), 0.7), (Q(a), 0.5)\}$ and $\mathcal{J}_1 = \{(R(b), 0.6)\}$. Consider the homomorphism h from $\mathcal{I}_d = \{Q(x)\}$ to $\mathcal{I}_1$ and $\mathcal{I}_d^h = \{(Q(a), 0.5)\}$. Clearly, the only extension $\hat{h}'$ of h from $\mathcal{J}_d = \{(R(y), 0.5)\}$ to $\mathcal{J}$ maps y to b and $0.5 \leq 0.6$. Hence, $\hat{h}'$ is also an f-homomorphism. Hence, $\langle \mathcal{I}_1, \mathcal{J}_1 \rangle = \{(P(a), 0.7), (Q(a), 0.5), (R(b), 0.6)\}$ satisfies the tgd d.
2. the instances $\mathcal{I}_2 = \{(Q(a), 0.6)\}$ and $\mathcal{J}_2 = \{(R(b), 0.4)\}$. Given $\mathcal{I}_d = \{Q(x)\}$, the only homomorphism h from $\mathcal{I}_d$ to $\mathcal{I}_2$ maps x to a. However there exists no extension $\hat{h}'$ of h from $\mathcal{J}_d = \{(R(y), 0.6)\}$ to $\mathcal{J}_2$, since $0.6 \nleq 0.4$. Therefore, $\langle \mathcal{I}_2, \mathcal{J}_2 \rangle = \{(Q(a), 0.6), (R(b), 0.4)\}$ does not satisfy the tgd d.   ◇

In summary, we get the following definition of fuzzy data exchange.

Definition 2. *A* fuzzy data exchange setting *is a triple* $(\mathbf{S}, \mathbf{T}, \Sigma)$ *consisting of a source schema* $\mathbf{S}$, *a target schema* $\mathbf{T}$, *a finite set* Σ *of dependencies consisting of s-t tgds, target tgds, and target egds.*

The fuzzy data exchange problem *associated with this setting is the following: given a finite source f-instance* $\mathcal{I}$, *find a finite target f-instance* $\mathcal{J}$ *such that* $\langle \mathcal{I}, \mathcal{J} \rangle$ *satisfies* Σ. *In this case,* $\mathcal{J}$ *is called a* fuzzy solution *for* $\mathcal{I}$, *f-solution for short. The set of all solutions for* $\mathcal{I}$ *is denoted by* $FSol(\mathcal{I})$.   ◇

Example 3. Let $\mathbf{S} = \{P, Q\}$, $\mathbf{T} = \{R\}$ and $\Sigma = \{d\}$, where d is an s-t tgd of the form $d = Q(x) \to \exists y R(y)$. The triple $(\mathbf{S}, \mathbf{T}, \Sigma)$ is a fuzzy data exchange setting. In Example 2, it is easy to verify that $\mathcal{J}_1$ is an f-solution for the source f-instance $\mathcal{I}_1$ while $\mathcal{J}_2$ is not an f-solution for $\mathcal{I}_2$.   ◇

4   Universal Solutions

Fuzzy Universal Solutions. In classical data exchange, *universal solutions* are crucial for the evaluation of conjunctive queries over the target schema. We now define fuzzy universal solutions by making use of the extension from crisp homomorphisms to f-homomorphisms introduced in the previous section.

Definition 3. *Given a fuzzy data exchange setting* $(\mathbf{S}, \mathbf{T}, \Sigma)$ *and a source f-instance* $\mathcal{I}$, *a* fuzzy universal solution *of* $\mathcal{I}$ *(universal f-solution, for short) is an f-solution* $\mathcal{J}$ *of* $\mathcal{I}$ *such that, for every f-solution* $\mathcal{J}'$ *for* $\mathcal{I}$, *there exists an f-homomorphism* $h: \mathcal{J} \to \mathcal{J}'$.   ◇

Example 4. Considering the fuzzy data exchange problem in Examples 2 and 3, it can be easily verified that $\mathcal{J}_u = \{R(Y), 0.5\}$, where Y is a labelled null, is a universal f-solution for $\mathcal{I}_1$.   ◇

Fuzzy Chase. In classical data exchange, the so-called *chase* procedure [6] is used to decide if, for a data exchange setting and a given source instance $\mathcal{I}$, a solution $\mathcal{J}$ of $\mathcal{I}$ exists. We now show how the chase can be extended so as to be applicable to fuzzy data exchange.

Definition 4. *Consider an f-instance $\mathcal{K}$ together with the tgd d and egd e of the following form.*

$$d \equiv (S_1(\mathbf{x}) \wedge \cdots \wedge S_n(\mathbf{x})) \to \exists \mathbf{y}(T_1(\mathbf{x}, \mathbf{y}) \wedge \cdots \wedge T_m(\mathbf{x}, \mathbf{y})) \ and$$
$$e \equiv (S_1(\mathbf{x}) \wedge \cdots \wedge S_n(\mathbf{x})) \to (x_1 = x_2)$$

Then the application of tgd d or of egd e to $\mathcal{K}$ is defined as follows.

(tgd) *Let h be a homomorphism $h \colon \mathcal{I}_d \to \mathcal{K}$ which cannot be extended to an f-homomorphism $\hat{h}' \colon \mathcal{J}_d \to \mathcal{K}$, where*
$\mathcal{I}_d = \{S_1(\mathbf{x}), \ldots, S_n(\mathbf{x})\}$ *is an instance over $\mathbf{R}$,*
$\vartheta_{\mathcal{I}_d^h}$ *is the confidence value associated with $\mathcal{I}_d^h$, and*
$\mathcal{J}_d = \{(T_1(\mathbf{x}, \mathbf{y}), \vartheta_{\mathcal{I}_d^h}), \ldots, (T_m(\mathbf{x}, \mathbf{y}), \vartheta_{\mathcal{I}_d^h})\}$ *is an f-instance over $\mathbf{R}$.*
Then we say that d can be applied to $\mathcal{K}$ with homomorphism h. In this case, we extend $\mathcal{K}$ to $\mathcal{K}'$ as follows:
1. *extending h to h^* over $Var(\mathcal{J}_d)$ such that each variable in $\mathbf{y}$ is assigned a new labelled null, and*
2. *extending $\mathcal{K}$ to $\mathcal{K}'$ by the image of the atoms of the conclusion of the tgd d under h^* together with weight $\vartheta_{\mathcal{I}_d^h}$, that is $(T_j(h^*(\mathbf{x}, \mathbf{y})), \vartheta_{\mathcal{I}_d^h})$, with $j \in \{1, \ldots, m\}$.*
The result of applying d to $\mathcal{K}$ with h is $\mathcal{K}'$ and will be denoted as $\mathcal{K} \overset{d,h}{\to} \mathcal{K}'$

(egd) *Let h be a homomorphism $h \colon \mathcal{I}_d \to \mathcal{K}$ with $\mathcal{I}_d = \{S_1(\mathbf{x}), \ldots, S_n(\mathbf{x})\}$, such that $h(x_i) \neq h(x_j)$ holds. Then we say that e can be applied to $\mathcal{K}$ with homomorphism h and we need to consider two possibilities:*
1. *If $h(x_1)$ and $h(x_2)$ are constants, then we obtain a "failure" denoted as $\mathcal{K} \overset{d,h}{\to} \perp$.*
2. *Otherwise, at least one of $h(x_i)$ and $h(x_j)$ is a labelled null, say $h(x_i)$. Then we obtain the f-instance $\mathcal{K}'$ by replacing every occurrence of $h(x_i)$ in $\mathcal{K}$ by $h(x_j)$. In this case, the result of applying e to $\mathcal{K}$ with h is $\mathcal{K}'$ and will be denoted as $\mathcal{K} \overset{e,h}{\to} \mathcal{K}'$.*

The application of tgd d or egd e to $\mathcal{K}$ is called a fuzzy chase step.   ◇

Definition 5. *Consider an f-instance $\mathcal{K}$ and a set Σ of tgds and egds. If we apply several fuzzy chase steps with dependencies from Σ to $\mathcal{K}$, we obtain a fuzzy chase sequence. When, after a finite chase sequence $\mathcal{K} = \mathcal{K}_0 \overset{d_1,h_1}{\to} \mathcal{K}_1 \to$*

$\cdots \rightarrow \mathcal{K}_{m-1} \overset{d_m, h_m}{\Longrightarrow} \mathcal{K}_m$, *no more fuzzy chase step can be applied to the last f-instance $\mathcal{K}_m$ or we get a "failure", that is, $\mathcal{K}_m = \bot$, then we say that $\mathcal{K}_m$ is the result of the* finite chase. *We distinguish between a* successful finite chase *in the first case and a* failing finite chase *in the second case.*   ◇

In the presence of target tgds, there is no guarantee that the chase terminates. Therefore, Fagin et al. [9] introduced a simple syntactic criterion (so-called "weak acyclicity") on the set of target tgds to ensure termination of the chase.

Example 5. Consider the fuzzy data exchange setting $(\mathbf{S}, \mathbf{T}, \Sigma)$ with $\mathbf{S} = \{P, Q\}$, $\mathbf{T} = \{R\}$ and $\Sigma = \{d_1, d_2\}$, such that $d_1 \equiv P(x) \rightarrow R(x, x)$, $d_2 \equiv Q(x) \rightarrow \exists y R(x, y)$. We compute a solution for $\mathcal{I} = \{(P(a), 0.6), (Q(a), 0.5)\}$ using the fuzzy chase.

Initially, we set $\mathcal{K} = \langle \mathcal{I}, \mathcal{J} \rangle$ with $\mathcal{J} = \varnothing$. Suppose that we begin with tgd d_2. Analogously to the previous example we obtain the f-instance $\mathcal{K}' = \{(P(a), 0.5), (Q(a), 0.5), (R(a, Y), 0.5)\}$.

Next, we consider the other tgd. There is the homomorphism h from $\mathcal{K}_{d_1} = \{P(x)\}$ to $\mathcal{K}'$ which maps x to a. Hence, $\mathcal{K}^h_{d_1} = \{(P(a), 0.6)\}$ and $\vartheta_{\mathcal{K}^h_{d_1}} = 0.6$. There is no extension from $\{(R(x, x), 0.6)\}$ to $\mathcal{K}'$, since $R(h(x), h(x)) = R(a, a)$ is not a fact of $\mathcal{K}$. Therefore, we extend $\mathcal{K}'$ to $\mathcal{K}'' = \{(P(a), 0.5), (Q(a), 0.6), (R(a, Y), 0.5), (R(a, a), 0.6)\}$, which satisfies all dependencies in Σ. Hence, we get the solution $\mathcal{J} = \mathcal{K}'' \setminus \mathcal{I} = \{(R(a, Y), 0.5), (R(a, a), 0.6)\}$ of $\mathcal{I}$.

Alternatively, we could start the fuzzy chase by applying tgd d_1, which yields the instance $\mathcal{K}' = \{(P(a), 0.5), (Q(a), 0.6), (R(a, a), 0.6)\}$. Considering now the tgd d_2, the only homomorphism h from $\mathcal{K}_{d_2} = \{Q(x)\}$ to $\mathcal{K}'$ maps x to a. Now we need to check whether there exists an extension $\hat{h}'$ of h from $\{(Q(a), 0.6), (R(a, y), 0.5)\}$ to $\mathcal{K}'$. Indeed, such an extension (mapping y to a and observing $0.5 \le 0.6$) exists. We therefore conclude that $\mathcal{K}' = \{(P(a), 0.5), (Q(a), 0.6), (R(a, a), 0.6)\}$ satisfies all dependencies in Σ and $\mathcal{J} = \mathcal{K}' \setminus \mathcal{I} = \{(R(a, a), 0.6)\}$ is a solution of $\mathcal{I}$.   ◇

In the above example, the chase sequence starting with tgd d_1 leads to a smaller solution than the chase sequence starting with d_2. In other words, the chase sequence starting with d_2 introduces a redundancy. In this simple example, the redundancy could be avoided by the "right" choice of the chase sequence. However, Fagin et al. [10] showed that, in general, the introduction of redundancies by the chase cannot be avoided – no matter in which order the tgds and egds are applied. Therefore, Fagin et al. proposed the redundancy elimination as a post-processing procedure by reducing the chase result $\mathcal{J}$ to the so-called *core*. Formally, we thus search for a proper subinstance $\mathcal{J}' \subset \mathcal{J}$ with $\mathcal{J} \rightarrow \mathcal{J}'$ (i.e., $\mathcal{J}'$ is a proper subinstance of $\mathcal{J}$ and there exists a homomorphism from $\mathcal{J}$ to $\mathcal{J}'$). The core $\mathcal{J}^*$ of an instance $\mathcal{J}$ is a subinstance that cannot be further "shrunk" by a proper endomorphism, i.e., there does not exist a proper subinstance $\hat{\mathcal{J}} \subset \mathcal{J}^*$ such that $\mathcal{J} \rightarrow \hat{\mathcal{J}}$. The core is unique up to isomorphism. In [11] it was shown that the core can be computed in polynomial time (data complexity) for schema mappings consisting of s-t tgds, a weakly acyclic set of target tgds, and egds.

Extending this sophisticated core computation algorithm to our fuzzy setting is left for future work.

Deciding the Data Exchange Problem. In the classical setting, it was shown that the chase can be used to decide the data exchange problem [9], i.e., if the chase fails, we may conclude that no solution at all exists; if the chase succeeds, it ends with a universal solution. Below we carry this result over to the fuzzy setting. As in the classical case, we start with the following crucial lemma:

Lemma 1. *Given a fuzzy chase step $\mathcal{K}_1 \xrightarrow{d,h} \mathcal{K}_2$, where $\mathcal{K}_2 \neq \bot$ and d is either a tgd or an egd. If $\mathcal{K}$ is an f-instance such that:*

- *$\mathcal{K}$ satisfies d and*
- *there exists an f-homomorphism $h_1 \colon \mathcal{K}_1 \to \mathcal{K}$,*

then there exists an f-homomorphism $h_2 \colon \mathcal{K}_2 \to \mathcal{K}$.

With the previous lemma, the main result of this section is obtained by an easy induction argument.

Theorem 1. *Given a fuzzy data exchange setting $(\mathbf{S}, \mathbf{T}, \Sigma)$, where Σ consists of s-t tgds, a weakly acyclic set of target tgds and target egds. Let $\mathcal{I}$ be a source instance and let $\mathcal{K}$ be the result of a finite chase of $\langle \mathcal{I}, \varnothing \rangle$ with Σ. Then we have:*

1. *If $\mathcal{K} \neq \bot$, then there exists an f-instance $\mathcal{J}$, such that $\mathcal{K} = \langle \mathcal{I}, \mathcal{J} \rangle$, and $\mathcal{J}$ is a fuzzy universal solution.*
2. *If $\mathcal{K} = \bot$, then there exists no fuzzy solution for $\mathcal{I}$.*

Proof: The proof is analogous to the one given in [9]. For instance, consider the case $\mathcal{K} \neq \bot$. The result of a successful finite chase of course is a fuzzy solution. It remains to show that it is a fuzzy *universal* solution. Let $\mathcal{J}'$ be another solution. We show that after every chase step we have an instance $\mathcal{K}_i = \langle \mathcal{I}, \mathcal{J}_i \rangle$ such that there exists an f-homomorphism $\mathcal{K}_i \to \mathcal{K}'$ with $\mathcal{K}' = \langle \mathcal{I}, \mathcal{J}' \rangle$. Since source and target schema are disjoint, it follows that there is an f-homomorphism from $\mathcal{J}_i$ to $\mathcal{J}'$. Initially, we have $\mathcal{K}_0 = \langle \mathcal{I}, \mathcal{J}_0 \rangle$ where $\mathcal{J}_0 = \varnothing$. Here, the identity is an f-homomorphism $\mathcal{J}_0$ to $\mathcal{J}'$. We prove this property for every i by induction on the length of the chase sequence, applying Lemma 1 for the induction step. □

5 Query Answering

In data exchange, the semantics of query answering is given by the notion of certain answers, i.e., for a given source instance $\mathcal{I}$ and query q, we are interested in those tuples that are obtained by evaluating q over target instance $\mathcal{J}$ for *every solution* $\mathcal{J}$ of $\mathcal{I}$. In fuzzy data exchange, we consider weighted k-ary queries with $k \geq 0$, i.e., pairs $(q(\mathbf{x}), \lambda)$, where $\mathbf{x}$ is the k-ary vector of free variables in q and $\lambda \in L$. Below, we restrict ourselves to the most important class of queries, namely conjunctive queries (CQs), i.e., $q(\mathbf{x})$ is of the form $\exists \mathbf{y} R_1(\mathbf{x}, \mathbf{y}) \wedge \cdots \wedge R_m(\mathbf{x}, \mathbf{y})$.

For an f-instance $\mathcal{J}$ and a weighted k-ary CQ $(q(\mathbf{x}), \lambda)$, one important set is $q(\mathcal{J}, \lambda)$. For $k > 0$ this set is formed by the k-tuples $\mathbf{t}$, such that there exists an f-homomorphism h from $\mathcal{K}_d = \{R_1(\mathbf{x}, \mathbf{y}) \wedge \cdots \wedge R_m(\mathbf{x}, \mathbf{y})\}$ to $\mathcal{J}$ with the properties that (1) the confidence value associated with $\mathcal{K}_d^h$ is $\vartheta_{\mathcal{K}_d^h}$; (2) we have $\lambda \preceq \vartheta_{\mathcal{K}_d^h}$; and (3) $h(\mathbf{x}) = \mathbf{t}$. For $k = 0$, the only possible answer tuple is the empty tuple $\langle \rangle$ and, therefore, $q(\mathcal{J}, \lambda)$ is either $\{\langle \rangle\}$ or $\varnothing$. In the former case, we say $q(\mathcal{J}, \lambda)$ is true and in the latter case, $q(\mathcal{J}, \lambda)$ is false. Once the meaning of a weighted query and $q(\mathcal{J}, \lambda)$ have been introduced, the definition of certain answers in fuzzy data exchange can be given.

Definition 6. *Given a fuzzy data exchange setting* $(\mathbf{S}, \mathbf{T}, \Sigma)$, *a weighted k-ary query* $q(\mathbf{x}, \lambda)$ *over the target schema* $\mathbf{T}$, *a source f-instance* $\mathcal{I}$ *and a confidence value* $\lambda \in L$. *The set of* certain answers *of* $q(\mathbf{x})$ *with respect to* $\mathcal{I}$ *at level* λ, *denoted by* $\mathrm{certain}(q(\mathbf{x}), \mathcal{I}, \lambda)$, *is defined as follows*

- *If $k > 0$, $\mathrm{certain}(q(\mathbf{x}), \mathcal{I}, \lambda)$ is the set of all k-tuples $\mathbf{t}$ of constants from $\mathcal{I}$ such that for every solution $\mathcal{J}$ of $\mathcal{I}$, we have that $\mathbf{t} \in q(\mathcal{J}, \lambda)$.*
- *If $k = 0$, then $\mathrm{certain}(q(\mathbf{x}), \mathcal{I}, \lambda) = true$ if $q(\mathcal{J}, \lambda) = true$ for all solutions $\mathcal{J}$ of $\mathcal{I}$, and otherwise $\mathrm{certain}(q(\mathbf{x}), \mathcal{I}, \lambda) = false$.* ◇

For $k > 0$, an interesting subset of $q(\mathcal{J}, \lambda)$ is formed by those k-tuples $\mathbf{t}$ which consist of constants only (i.e., no labelled nulls). This subset of $q(\mathcal{J}, \lambda)$ will be denoted as $q(\mathcal{J}, \lambda)_\downarrow$. For Boolean queries q, we simply have $q(\mathcal{J}, \lambda)_\downarrow = q(\mathcal{J}, \lambda)$ (hence, true or false). In classical data exchange, universal solutions can be used to compute the certain answers of conjunctive queries. This result naturally extends to fuzzy data exchange. We thus get the following theorem.

Theorem 2. *Given a fuzzy data exchange setting* $(\mathbf{S}, \mathbf{T}, \Sigma)$, *a source instance* $\mathcal{I}$, *and a weighted CQ* $(q(\mathbf{x}), \lambda)$ *over the target schema* $\mathbf{T}$. *Let* $\mathcal{J}$ *be an arbitrary universal f-solution of* $\mathcal{I}$. *Then* $\mathrm{certain}(q(\mathbf{x}), \mathcal{I}, \lambda) = q(\mathcal{J}, \lambda)_\downarrow$.

Moreover, if $\mathcal{J}$ *is a fuzzy solution of* $\mathcal{I}$ *such that, for every weighted CQ* $(q(\mathbf{x}), \lambda)$ *over* $\mathbf{T}$ *the equality* $\mathrm{certain}(q(\mathbf{x}), \mathcal{I}, \lambda) = q(\mathcal{J}, \lambda)_\downarrow$ *holds, then* $\mathcal{J}$ *is an f-universal solution of* $\mathcal{I}$.

Proof: The proof is analogous to the proof of the corresponding theorem in classical data exchange [9]. □

6 Conclusion and Future Work

In this article, we have extended all of the basic concepts of data exchange to a fuzzy setting. We have thus shown that data exchange can be nicely extended to fuzzy data exchange. A lot of work remains to be done in this area.

First, further properties of fuzzy data exchange should be investigated. This line of research includes, for instance, the study of the relationship with probabilistic data exchange and a detailed analysis of the algorithmic aspects of fuzzy data exchange: above all, the complexity of query answering. Second, our extension of classical data exchange to a fuzzy setting should include further concepts and methods. We have already mentioned the computation of the so-called *core* [10,11] to get a minimal solution as one interesting task for future work. In [8], after extending classical data exchange to probabilistic data exchange by considering probability spaces of source and target instances, also the mappings themselves are considered as probabilistic. Analogously, we should investigate the extension of fuzzy exchange as presented here to fuzzy dependencies, i.e., also the tgds and egds have to be augmented with confidence values [7,12].

References

1. Afrati, F.N., Kolaitis, P.G.: Answering aggregate queries in data exchange. In: PODS, pp. 129–138. ACM (2008)
2. Arenas, M., Barceló, P., Libkin, L., Murlak, F.: Relational and XML Data Exchange. Morgan & Claypool Publishers (2010)
3. Arenas, M., Barceló, P., Reutter, J.L.: Query languages for data exchange: beyond unions of conjunctive queries. In: ICDT, pp. 73–83. ACM (2009)
4. Arenas, M., Libkin, L.: Xml data exchange: Consistency and query answering. J. ACM 55(2) (2008)
5. Barceló, P., Pérez, J., Reutter, J.L.: Schema mappings and data exchange for graph databases. In: ICDT, pp. 189–200. ACM (2013)
6. Beeri, C., Vardi, M.Y.: A proof procedure for data dependencies. J. ACM 31(4), 718–741 (1984)
7. Viegas Damásio, C., Moniz Pereira, L.: Monotonic and residuated logic programs. In: Benferhat, S., Besnard, P. (eds.) ECSQARU 2001. LNCS (LNAI), vol. 2143, pp. 748–759. Springer, Heidelberg (2001)
8. Fagin, R., Kimelfeld, B., Kolaitis, P.G.: Probabilistic data exchange. J. ACM 58(4), 15 (2011)
9. Fagin, R., Kolaitis, P.G., Miller, R.J., Popa, L.: Data exchange: semantics and query answering. Theor. Comput. Sci. 336(1), 89–124 (2005)
10. Fagin, R., Kolaitis, P.G., Popa, L.: Data exchange: getting to the core. ACM Trans. Database Syst. 30(1), 174–210 (2005)
11. Gottlob, G., Nash, A.: Efficient core computation in data exchange. J. ACM 55(2) (2008)
12. Medina, J., Ojeda-Aciego, M., Vojtáš, P.: Similarity-based unification: a multi-adjoint approach. Fuzzy Sets and Systems 146, 43–62 (2004)

Fuzzy Relational Compositions Based on Generalized Quantifiers

Martin Štěpnička and Michal Holčapek

Institute for Research and Applications of Fuzzy Modeling
University of Ostrava
Centre of Excellence IT4Innovations
30. dubna 22, 701 03 Ostrava 1, Czech Republic
{Martin.Stepnicka,Michal.Holcapek}@osu.cz
http://irafm.osu.cz/

Abstract. Fuzzy relational compositions have been extensively studied by many authors. Especially, we would like to highlight initial studies of the fuzzy relational compositions motivated by their applications to medical diagnosis by Willis Bandler and Ladislav Kohout. We revisit these types of compositions and introduce new definitions that directly employ generalized quantifiers. The motivation for this step is twofold: first, the application needs for filling a huge gap between the classical existential and universal quantifiers and second, the already existing successful implementation of generalized quantifiers in so called divisions of fuzzy relations, that constitute a database application counterpart of the theory of fuzzy relational compositions. Recall that the latter topic is studied within fuzzy relational databases and flexible querying systems for more than twenty years. This paper is an introductory study that should demonstrate a unifying theoretical framework and introduce that the properties typically valid for fuzzy relational compositions are valid also for the generalized ones, yet sometimes in a weaken form.

1 Introduction

Fuzzy relational compositions are widely used in many areas of fuzzy mathematics, including the formal constructions of fuzzy inference systems [1, 2] , medical diagnosis [3] or architectures for information processing and protection of IT systems [4]. Since late 70's and early 80's when Willis Bandler and Ladislav Kohout studied classical relational compositions and extended the concept in order to deal with fuzzy relational compositions, these area became deeply elaborated by numerous researchers. Let us recall mainly Radim Bělohlávek's book [5], an article by Bernard De Baets and Etienne Kerre [6] and finally an exhaustive investigation in the so called Fuzzy Class Theory [7] by Libor Běhounek and Martina Daňková [8].

As one may find in the very early articles [3, 9], the fuzzy relational compositions came to live as a very natural generalization of well motivated compositions of classical relations. For example, the basic sup-T compositions is nothing else

A. Laurent et al. (Eds.): IPMU 2014, Part II, CCIS 443, pp. 224–233, 2014.

but a generalization of the basic composition of two classical binary relations. It is sufficient to consider binary fuzzy relations and to deal with the operations that serve as interpretations of fuzzy connectives, which involves the t-norm T in the formula that gave rise to the notion "sup-T composition". However, what remained untouched and up to the best knowledge of the authors, never generalized so far, is the nature of the quantifiers that are used in the definitions of the compositions. Particularly, sup-T compositions that use the operation of supremum, implicitly employ the existential quantifier while inf-R compositions that use the operation of infimum, implicitly employ the universal quantifier. The fact that there is nothing in between the option of the existential quantifier where just one element is enough to result the truth and the other option of the universal quantifier where all elements have to fulfill a given formula in order to result the truth, may be very limiting in distinct applications. Therefore, the introduction of fuzzy relational compositions based on generalized quantifiers, such as 'Most' or 'Many', is a well motivated natural step. This has been noticed by many authors who deal with so called *flexible query answering systems* or simply *fuzzy relational databases* where the use of so called *soft quantifiers* in evaluation of quantified sentences or mainly in fuzzy relational divisions attracted a great interest of many scholars, see e.g. [10–13].

2 Relational Compositions and Fuzzy Relational Compositions

2.1 Relational Compositions

Let us consider three non-empty finite universes X, Y, Z of elements. Following the work of Willis Bandler and Ladislav Kohout [3], for the sake of illustrative nature, we can assume that X is a finite set of patients, Y is a finite set of symptoms and Z is a finite set of diseases.

Let us be given two binary relations $R \subseteq X \times Y$ and $S \subseteq Y \times Z$, i.e., if a pair $(x, y) \in X \times Y$ belongs to relation R then it means that patient x has symptom y and similarly, if a pair $(y, z) \in Y \times Z$ belongs to relation S then it means that symptom y belongs to disease z. Both relations are usually at disposal since R can be easily obtained by asking patients or by measuring symptoms (body temperature, cholesterol, blood pressure etc.) and S constitutes an expert medical knowledge that is at disposal e.g. in literature. The usual diagnosis task of a physician is from the mathematical point of view nothing else but a composition of these two relation in order to obtain a relation between patients and diseases. In order words, to state what are the potential diseases of a given patient. Obviously, formally, a similar job may be done by a relational composition @ which gives a binary relation $R \circ S$ on $X \times Z$:

$$\frac{\begin{array}{ll} R & \subseteq X \times Y \\ S & \subseteq Y \times Z \end{array}}{R @ S \subseteq X \times Z.}$$

We will consider the four main compositions @ and denote them as [4] by $\circ, \lhd, \rhd$ and $\square$. The composed relations are given as follows

$$R \circ S = \{(x, z) \in X \times Z \mid \exists\, y \in Y : (x, y) \in R \,\&\, (y, z) \in S\}, \tag{1}$$

$$R \lhd S = \{(x, z) \in X \times Z \mid \forall\, y \in Y : (x, y) \in R \Rightarrow (y, z) \in S\}, \tag{2}$$

$$R \rhd S = \{(x, z) \in X \times Z \mid \forall\, y \in Y : (x, y) \in R \Leftarrow (y, z) \in S\}, \tag{3}$$

$$R \square S = \{(x, z) \in X \times Z \mid \forall\, y \in Y : (x, y) \in R \Leftrightarrow (y, z) \in S\} \tag{4}$$

and they are called *basic (circlet)* composition, *Bandler-Kohout subproduct*, *Bandler-Kohout superproduct* and *Bandler-Kohout square product*, respectively.

The relation $R \circ S$ then expresses a sort of suspicion of a disease for a particular patient. It is a basic relation – for each patient it is sufficient to have only a single symptom related to a particular disease in order to detect the suspicion. Thus, a patient having a very general symptom related to many diseases is immediately suspicious of having all these diseases.

The "triangle" and square compositions (2)-(4) provide a sort of more accurate specification or a strengthening of the initial suspicion [3]. The Bandler-Kohout (abb. BK) subproduct is defined as a relation of patients and diseases such that, for all symptoms that a given patient has, it holds, that they belong to the given disease. The BK superproduct is defined as a relation of patients and diseases such that, for all symptoms that belong to a given disease, it holds, that the given patient has them. Finally, the BK square product models an ideal example when a given patient has all the symptoms of a given disease and all the symptoms of the patient belong to the given disease.

Using the fact that the existential and universal quantifiers may be interpreted by the operations of supremum and infimum, respectively, formulas (1)-(4) may be rewritten into the following functional form:

$$\chi_{(R \circ S)}(x, z) = \bigvee_{y \in Y} (\chi_R(x, y) \wedge \chi_S(y, z)), \tag{5}$$

$$\chi_{(R \lhd S)}(x, z) = \bigwedge_{y \in Y} (\chi_R(x, y) \Rightarrow \chi_S(y, z)), \tag{6}$$

$$\chi_{(R \rhd S)}(x, z) = \bigwedge_{y \in Y} (\chi_R(x, y) \Leftarrow \chi_S(y, z)), \tag{7}$$

$$\chi_{(R \square S)}(x, z) = \bigwedge_{y \in Y} (\chi_R(x, y) \Leftrightarrow \chi_S(y, z)), \tag{8}$$

where χ_R, χ_S and $\chi_{(R @ S)}$ denote characteristic functions of relations R, S and $R @ S$, respectively, symbol $\wedge$ denotes the minimum, symbol $\Rightarrow$ expresses the binary operation of the classical implication and finally, symbol $\Leftrightarrow$ denotes the operation of the classical equivalence.

2.2 Compositions of Fuzzy Relations

Since usual symptoms such as *high temperature*, *increased cholesterol* or *very high blood pressure* are basically vaguely specified and imprecisely measured (all these values oscillate during a day) and very often some symptoms do not clearly or necessarily belong to a given disease however, they might belong to it under some assumptions or conditions, the extension of the compositions for fuzzy relations $R \subseteq X \times Y$ and $S \subseteq Y \times Z$ was highly desirable. Obviously, since such an extension causes that we deal with fuzzy relations which contain pairs of elements up to some degrees from the interval $[0,1]$, we have to take into account appropriate operations. Basically, it is appropriate to deal with a residuated lattice as the underlying algebraic structure and the used operations will be left-continuous t-norms and their residual (bi)implications [14].

In this article, we only briefly recall the basic definitions of the fuzzy relational compositions as introduced by Willis Bandler and Ladislav Kohout.

Definition 1. *Let X, Y, Z be non-empty universes, let $R \subseteq X \times Y$, $S \subseteq Y \times Z$ and let $\rightarrow$ be a residual implication derived from a left-continuous t-norm $*$. Then the compositions $\circ, \triangleleft, \triangleright, \square$ of fuzzy relations R and S are fuzzy relations on $X \times Z$ defined as follows:*

$$(R \circ S)(x, z) = \bigvee_{y \in Y} (R(x,y) * S(y,z)),$$

$$(R \triangleleft S)(x, z) = \bigwedge_{y \in Y} (R(x,y) \rightarrow S(y,z)),$$

$$(R \triangleright S)(x, z) = \bigwedge_{y \in Y} (R(x,y) \leftarrow S(y,z)),$$

$$(R \square S)(x, z) = \bigwedge_{y \in Y} (R(x,y) \leftrightarrow S(y,z)),$$

for all $x \in X$ and $z \in Z$.

Since $*$ is a t-norm, often denoted by the capital T, the sup-$*$ composition is also called the sup-T composition. Similarly, the Bandler-Kohout products, since being constructed with help of the infimum and the residual operation, are called inf-R compositions or more specifically, inf-$\rightarrow$, inf-$\leftarrow$ or inf-$\leftrightarrow$.

Remark 1. Note, that for $x \in X$ such that $R(x,y) = 0$ for all $y \in Y$, the composed relation $(R \triangleleft S)(x, z) = 1$ for any $z \in Z$. Similarly, for $z \in Z$ such that $S(y, z) = 0$ for all $y \in Y$, the composed relation $(R \triangleright S)(x, z) = 1$ for any $x \in X$. Bernard De Baets and Etienne Kerre in [6] approached this problem by a redefinition of the original inf-R compositions where an existence of joining element (symptom) $y \in Y$ is assumed. In this preliminary investigation we stay stuck to the original definitions and we leave the investigation of the later modification for further studies.

As we may see from Definition 1, the generalizations focus on the use of fuzzy relations and the internal operations only. However, the definitions still deal with the supremum and the infimum and thus, inherently use the existential and universal quantifiers, respectively. It is obvious that this may be a potential drawback for applications as there is big gap between these quantifiers. Particularly, we can meet situations when all patients are suspicious of having all diseases in a high degree when using ∘, but if we want to strengthen the suspicion with help of □, no patients are suspicious of having any disease in a high degree anymore because not for all but only for majority of symptoms, the inside implication present in the definitions of inf-R compositions are valid.

3   Generalized Quantifiers

3.1   Generalized Quantifiers Based on Fuzzy Measures

In the above sections, we have recalled relational compositions and fuzzy relational compositions. Motivated by the theoretical drawback that has been observed by many researchers and "solved" by the use of generalized quantifiers, see e.g. [12, 13], we directly employ the so called *monadic quantifiers of the type* $\langle 1 \rangle$ *determined by fuzzy measures* [15]. First of all, let us recall some basic definitions.

Definition 2. *Let* $U = \{u_1, \ldots, u_n\}$ *be a finite universe,* $\mathcal{P}(U)$ *denote the power set of* U *and* $\mu : \mathcal{P}(U) \to [0,1]$ *be a normalized fuzzy measure, i.e., a monotone mapping with* $\mu(\emptyset) = 0$ *and* $\mu(U) = 1$. *We say that the fuzzy measure* μ *is invariant with respect to cardinality, if the following condition holds:*

$$\forall A, B \in \mathcal{P}(U) : |A| = |B| \Rightarrow \mu(A) = \mu(B)$$

where $| \cdot |$ *denotes the cardinality of a set.*

Example 1. The measure called *relative cardinality*, given by

$$\mu_{rc}(A) = \frac{|A|}{|U|} , \tag{9}$$

is invariant w.r.t. cardinality. If $f : [0,1] \to [0,1]$ is a non-decreasing mapping with $f(0) = 0$ and $f(1) = 1$ then μ defined as $\mu(A) = f(\mu_{rc}(A))$ is again a fuzzy measure that is invariant w.r.t. cardinality.

Note, that all the models of linguistic evaluative expressions [16] of the type Big and modified by arbitrary linguistic hedge (e.g. More or less, Very, Roughly, Extremely etc.) are fuzzy sets on $[0,1]$ that fulfill the boundary conditions and thus, may be used in order to modify the original relative cardinality.

In the sequel, we will deal only with such fuzzy measures that are created by a modification of the relative cardinality by an appropriate fuzzy set (cf. Definition 3.7 in [15]).

Definition 3. *Let* U *be non-empty finite universe and* μ *be a fuzzy measure on* U *that is invariant w.r.t. cardinality. A mapping* $Q : \mathcal{F}(U) \to [0,1]$ *defined by*

$$Q(C) = \bigvee_{D \in \mathcal{P}(U) \setminus \{\emptyset\}} \left(\left(\bigwedge_{u \in D} C(u) \right) * \mu(D) \right), \quad C \in \mathcal{F}(U) \tag{10}$$

where $*$ *is a left-continuous t-norm, is called* fuzzy quantifier determined by fuzzy measure μ.

Example 2. Let us assume that the fuzzy measures μ defined as follows

$$\mu^{\forall}(D) = \begin{cases} 1 & D \equiv U \\ 0 & \text{otherwise,} \end{cases} \qquad \mu^{\exists}(D) = \begin{cases} 0 & D \equiv \emptyset \\ 1 & \text{otherwise.} \end{cases} \qquad (11)$$

Then the derived quantifiers are the classical universal and existential quantifiers.

One can immediately see, that formula (10) is not very appropriate from the computational point of view as it requires calculation over all sets from $\mathcal{P}(U) \setminus \{\emptyset\}$. However, we may use the property of fuzzy measure being invariant w.r.t. cardinality and show that the fuzzy quantifier may be efficiently calculated.

Theorem 1. *Let Q be a fuzzy quantifier on U determined by a fuzzy measure μ that is invariant w.r.t. cardinality. Then*

$$Q(C) = \bigvee_{i=1}^{n} C(u_{\pi(i)}) * \mu(\{u_1, \ldots, u_i\}), \quad C \in \mathcal{F}(U) \qquad (12)$$

where π is a permutation on U such that $C(u_{\pi(1)}) \geq C(u_{\pi(2)}) \geq \cdots \geq C(u_{\pi(n)})$.

Proof. Let C be an arbitrary fuzzy set. It is easy to see that, for any $i = 1, \ldots, n$ and $D \in \mathcal{P}(U)$ with $|D| = i$, it holds

$$C(u_{\pi(i)}) = \bigwedge_{u \in \{u_{\pi(1)}, \ldots, u_{\pi(i)}\}} C(u) \geq \bigwedge_{u \in D} C(u).$$

The statement immediately follows from the invariance of μ w.r.t. cardinality. $\square$

Theorem 1 shows that the fuzzy quantifier defined by Definition 3 can be equivalently expressed by means of the Sugeno fuzzy integral, which is neither surprising nor undesirable fact, cf. [17].

In other words, if we again apply the fuzzy measure μ^f that is constructed from the relative cardinality by some modifying fuzzy set f, i.e. $\mu^f = f(\mu_{rc})$, then formula (12) turns into the following equality:

$$Q(C) = \bigvee_{i=1}^{n} C(u_{\pi(i)}) * f(i/n) \qquad (13)$$

which is very easy to be calculated.

3.2   Fuzzy Relational Compositions Based on Generalized Quantifiers

In this part of the text, we directly apply the above introduced theory of generalized quantifiers to our problem of fuzzy relational compositions.

Let us recall, e.g., the definition of the Bandler-Kohout subproduct of two classical relations R and S given by formula (2). The universal quantifier $\forall$ may be replaced by a generalized quantifier, say, e.g., the quantifier 'Most'. Then the modified composition gives a set of pairs of patients and diseases such that for most of the symptoms, that a given patient has, it holds, that they belong to the given disease.

Definition 4. *Let X, Y, Z be non-empty finite universes, let $R \subseteq X \times Y$, $S \subseteq Y \times Z$, let $*$ be a left-continuous t-norm and $\rightarrow$ be its residual implication. Let μ be a fuzzy measure on Y that is invariant w.r.t. cardinality and let Q be a fuzzy quantifier on Y determined by the fuzzy measure μ. Then the $\circ^Q, \triangleleft^Q, \triangleright^Q, \square^Q$ compositions of fuzzy relations R and S are fuzzy relations on $X \times Z$ defined as follows:*

$$(R \circ^Q S)(x, z) = \bigvee_{D \in \mathcal{P}(Y) \setminus \{\emptyset\}} \left(\left(\bigwedge_{y \in D} R(x, y) * S(y, z) \right) * \mu(D) \right),$$

$$(R \triangleleft^Q S)(x, z) = \bigvee_{D \in \mathcal{P}(Y) \setminus \{\emptyset\}} \left(\left(\bigwedge_{y \in D} R(x, y) \rightarrow S(y, z) \right) * \mu(D) \right),$$

$$(R \triangleright^Q S)(x, z) = \bigvee_{D \in \mathcal{P}(Y) \setminus \{\emptyset\}} \left(\left(\bigwedge_{y \in D} R(x, y) \leftarrow S(y, z) \right) * \mu(D) \right),$$

$$(R \square^Q S)(x, z) = \bigvee_{D \in \mathcal{P}(Y) \setminus \{\emptyset\}} \left(\left(\bigwedge_{y \in D} R(x, y) \leftrightarrow S(y, z) \right) * \mu(D) \right),$$

for all $x \in X$ and $z \in Z$.

Though the definition is general and enables to use any quantifier Q, obviously, for application purposes, when dealing with the inf-R compositions $\triangleleft^Q, \triangleright^Q, \square^Q$ quantifiers weakening the universal quantifier such as 'Most' or 'Many' are expected to be applied. These quantifiers may be applied for example using the fuzzy sets modeling the meaning of evaluative linguistic expression Very Big or Roughly Big [16]. In the case of sup-T composition $\circ^Q$, we should apply quantifiers that slightly strengthen the expectations from the existential quantifiers, such as 'A Few' that may be modeled by a fuzzy sets representing the meaning of the expression Not Very Small.

Corollary 1. *Let μ be a fuzzy measure that is constructed from the relative cardinality by the modification using function f. Then for all $x \in X$ and $z \in Z$:*

$$(R \circ^Q S)(x, z) = \bigvee_{i=1}^{n} \left((R(x, y_{\pi(i)}) * S(y_{\pi(i)}, z)) * f(i/n) \right),$$

$$(R \triangleleft^Q S)(x, z) = \bigvee_{i=1}^{n} \left((R(x, y_{\pi(i)}) \rightarrow S(y_{\pi(i)}, z)) * f(i/n) \right),$$

$$(R \triangleright^Q S)(x, z) = \bigvee_{i=1}^{n} \left((R(x, y_{\pi(i)}) \leftarrow S(y_{\pi(i)}, z)) * f(i/n) \right),$$

$$(R \square^Q S)(x, z) = \bigvee_{i=1}^{n} \left((R(x, y_{\pi(i)}) \leftrightarrow S(y_{\pi(i)}, z)) * f(i/n) \right),$$

where π is a permutation such that (putting $\circledast \in \{*, \rightarrow, \leftarrow, \leftrightarrow\}$)

$$R(x, y_{\pi(i)}) \circledast S(y_{\pi(i)}, z) \geq R(x, y_{\pi(i+1)}) \circledast S(y_{\pi(i+1)}, z), \quad i = 1, \ldots, n-1.$$

Indeed, the original compositions are special cases of the newly defined ones. Using the fuzzy measure $\mu^{\vee}$ given by (11), one may easily check that $R \triangleleft S \equiv R \triangleleft^{\vee} S$ and similarly that $R \triangleright S \equiv R \triangleright^{\vee} S$, $R \square S \equiv R \square^{\vee} S$. Indeed, since $f(i/n) = 0$ for all $i < n$ and $f(1) = 1$ then

$$(R \triangleleft^{\vee} S)(x, z) = \left(R(x, y_{\pi(n)}) \rightarrow S(y_{\pi(n)}, z) \right) * f(n/n)$$

which due to the fact that

$$R(x, y_{\pi(n)}) \rightarrow S(y_{\pi(n)}, z) = \bigwedge_{i=1}^{n} (R(x, y_i) \rightarrow S(y_i, z))$$

proves $R \triangleleft S \equiv R \triangleleft^{\vee} S$. The other equalities may be proved analogously.

4   Properties

Many appropriate properties were proved for the original classical as well as fuzzy relational compositions, see e.g. [8]. In this section, we face the question whether the same or similar properties may be valid also for the compositions based on generalized quantifiers. As we will show, the answer is rather positive.

Theorem 2. Let X, Y, Z, U are finite universes and let $R_1, R_2 \subseteq X \times Y$, $S_1, S_2 \subseteq Y \times Z$ and $T \subseteq Z \times U$. Furthermore, let $\cup, \cap$ denote the Gödel union and intersection, respectively. Then

1. $R \circ^Q (S \circ^Q T) = (R \circ^Q S) \circ^Q T$
2. $R \square^Q S \leq (R \triangleleft^Q S) \cap (R \triangleright^Q S)$
3. $R_1 \leq R_2 \Rightarrow (R_1 \circ^Q S) \subseteq (R_2 \circ^Q S)$ and $S_1 \leq S_2 \Rightarrow (R \circ^Q S_1) \subseteq (R \circ^Q S_2)$
4. $R_1 \leq R_2 \Rightarrow (R_1 \triangleleft^Q S) \supseteq (R_2 \triangleleft^Q S)$ and $(R_1 \triangleright^Q S) \subseteq (R_2 \triangleright^Q S)$
5. $(R_1 \cup R_2) \circ^Q S = (R_1 \circ^Q S) \cup (R_2 \circ^Q S)$
6. $(R_1 \cap R_2) \triangleleft^Q S = (R_1 \triangleleft^Q S) \cup (R_2 \triangleleft^Q S)$
7. $(R_1 \cup R_2) \triangleright^Q S = (R_1 \triangleright^Q S) \cup (R_2 \triangleright^Q S)$
8. $(R_1 \cap R_2) \circ^Q S \leq (R_1 \circ^Q S) \cap (R_2 \circ^Q S)$
9. $(R_1 \cup R_2) \triangleleft^Q S \leq (R_1 \triangleleft^Q S) \cap (R_2 \triangleleft^Q S)$
10. $(R_1 \cap R_2) \triangleright^Q S \leq (R_1 \triangleright^Q S) \cap (R_2 \triangleright^Q S)$

Sketch of the proof: All the properties are proved based on the properties of left-continuous t-norms and their residual implications on a linearly order set $[0,1]$, i.e., using

$$(a \wedge b) * c = (a * c) \wedge (b * c), \qquad\qquad (a \vee b) * c = (a * c) \vee (b * c),$$
$$(a \wedge b) \to c = (a \to c) \vee (b \to c), \qquad\qquad (a \vee b) \to c = (a \to c) \wedge (b \to c),$$
$$a \to (b \wedge c) = (a \to b) \wedge (a \to c), \qquad\qquad a \to (b \vee c) = (a \to b) \vee (a \to c),$$
$$(a \leftrightarrow b) = (a \to b) \wedge (a \leftarrow b), \quad \bigvee_i ((a_i * b) \wedge (a_i * c)) \leq \bigvee_i (a_i * b) \wedge \bigvee_i (a_i * c)$$

and the antitonicity and the isotonicity of $\to$ in its first and second argument, respectively. Furthermore, the monotonicity properties *3.-4.* are extensively used in proving the latter properties.

$\square$

Remark 2. Obviously, items *5.-10.* may be also read as follows:

12. $R \circ^Q (S_1 \cup S_2) = (R \circ^Q S_1) \cup (R \circ^Q S_2)$
13. $R \triangleright^Q (S_1 \cap S_2) = (R \triangleright^Q S_1) \cup (R \triangleright^Q S_2)$
14. $R \triangleleft^Q (S_1 \cup S_2) = (R \triangleleft^Q S_1) \cup (R \triangleleft^Q S_2)$
15. $R \circ^Q (S_1 \cap S_2) \leq (R \circ^Q S_1) \cap (R \circ^Q S_2)$
16. $R \triangleright^Q (S_1 \cup S_2) \leq (R \triangleright^Q S_1) \cap (R \triangleright^Q S_2)$
17. $R \triangleleft^Q (S_1 \cap S_2) \leq (R \triangleleft^Q S_1) \cap (R \triangleleft^Q S_2)$

5   Conclusions

We have recalled classical and fuzzy relational composition. The big gap between the classical existential and universal quantifiers brings practical weakness into the theory of fuzzy relational compositions. This can be elegantly solved by the use of generalized quantifiers. Moreover, this fact has been observed by the community of researchers dealing with flexible querying systems, where the compositions, investigated under the name fuzzy relational divisions, have been "equipped" with generalized quantifiers a certain time ago. We stem from this motivation and involve generalized quantifiers into the fuzzy relational compositions as well. Particularly, they allow us to define fuzzy relational compositions with help of linguistically very natural quantifiers such as 'A Few', 'Many', 'Majority' or 'Most'. The whole topic is however, studied in the view of fuzzy relational compositions and not in the view of fuzzy relational databases. The reason lies in the older origin of compositions and in the goal to bring the nice ideas from fuzzy relational databases back to the mathematical theory. Thus, we follow the previous works on fuzzy relational compositions and investigate their basic properties. It is shown that many of the appreciated properties of the standard fuzzy relational compositions are preserved at least in a weaken form. This gives a promising potential to employ the fuzzy relational compositions based on generalized quantifiers in many other areas of application such as inference systems where the compositions stand for the main theoretical pilots.

Acknowledgments. This investigation was mainly supported by the European Regional Development Fund in the IT4Innovations Centre of Excellence project (CZ.1.05/1.1.00/02.0070).

References

1. Pedrycz, W.: Applications of fuzzy relational equations for methods of reasoning in presence of fuzzy data. Fuzzy Sets and Systems 16, 163–175 (1985)
2. Štěpnička, M., De Baets, B., Nosková, L.: Arithmetic fuzzy models. IEEE Transactions on Fuzzy Systems 18, 1058–1069 (2010)
3. Bandler, W., Kohout, L.J.: Fuzzy relational products and fuzzy implication operators. In: Proc. Int. Workshop on Fuzzy Reasoning Theory and Applications. Queen Mary College, London (1978)
4. Bandler, W., Kohout, L.J.: Relational-product architectures for information processing. Information Sciences 37, 25–37 (1985)
5. Bělohlávek, R.: Fuzzy relational systems: Foundations and principles. Kluwer Academic, Plenum Press, Dordrecht, New York (2002)
6. De Baets, B., Kerre, E.: Fuzzy relational compositions. Fuzzy Sets and Systems 60, 109–120 (1993)
7. Běhounek, L., Cintula, P.: Fuzzy class theory 154(1), 34–55 (2005)
8. Běhounek, L., Daňková, M.: Relational compositions in fuzzy class theory. Fuzzy Sets and Systems 160(8), 1005–1036 (2009)
9. Bandler, W., Kohout, L.J.: Fuzzy relational products as a tool for analysis and synthesis of the behaviour of complex natural and artificial systems. In: Wang, S.K., Chang, P.P. (eds.) Fuzzy Sets: Theory and Application to Policy Analysis and Information Systems, pp. 341–367. Plenum Press, New York (1980)
10. Pivert, O., Bosc, P.: Fuzzy preference queries to relational databases. Imperial College Press, London (2012)
11. Dubois, D., Prade, H.: Semantics of quotient operators in fuzzy relational databases. Fuzzy Sets and Systems 78, 89–93 (1996)
12. Delgado, M., Sanchez, D., Vila, M.A.: Fuzzy cardinality based evaluation of quantified sentences. International Journal of Approximate Reasoning 23, 23–66 (2000)
13. Bosc, P., Liétard, L., Pivert, O.: Flexible database querying and the division of fuzzy relations. Scientia Iranica 2, 329–340 (1996)
14. Baczyński, M., Jayaram, B.: Fuzzy Implications. STUDFUZZ, vol. 231. Springer, Heidelberg (2008)
15. Dvořák, A., Holčapek, M.: L-fuzzy quantifiers of type $\langle 1 \rangle$ determined by fuzzy measures. Fuzzy Sets and Systems 160(23), 3425–3452 (2009)
16. Novák, V.: A comprehensive theory of trichotomous evaluative linguistic expressions. Fuzzy Sets and Systems 159(22), 2939–2969 (2008)
17. Bosc, P., Liétard, L., Pivert, O.: Sugeno fuzzy integral as a basis for the interpretation of flexible queries involving monotonic aggregates. Information Processing & Management 39(2), 287–306 (2003)

A Functional Approach to Cardinality
of Finite Fuzzy Sets

Michal Holčapek

University of Ostrava
Institute for Research and Applications of Fuzzy Modeling
Centre of Excellence IT4Innovations
30. dubna 22, 701 03 Ostrava, Czech Republic
michal.holcapek@osu.cz
http://irafm.osu.cz/

Abstract. In this contribution, we present a functional approach to the cardinality of finite fuzzy sets, it means an approach based on one-to-one correspondences between fuzzy sets. In contrast to one fixed universe of discourse used for all fuzzy sets, our theory is developed within a class of fuzzy sets which universes of discourse are countable sets, and finite fuzzy sets are introduced as fuzzy sets with finite supports. We propose some basic operations with fuzzy sets as well as two constructions - fuzzy power set and fuzzy exponentiation. To express the fact that two finite fuzzy sets have approximately the same cardinality we propose the concept of graded equipollence. Using this concept we provide graded versions of several well-known statements, including the Cantor-Bernstein theorem and the Cantor theorem.

Keywords: Fuzzy sets, fuzzy classes, graded equipollence, cardinal theory of finite fuzzy sets.

1 Introduction

In the classical set theory, we can recognize two approaches to the cardinality of sets. One of them is a functional approach that uses one-to-one correspondences between sets to compare their sizes. More precisely, we say that two sets a and b are equipollent (equipotent or have the same cardinality) and write $a \sim b$ if there exists a one-to-one mapping of a onto b. The relation "being equipollent" is an equivalence on the class of all sets and is called equipollence. Note that special objects expressing the power of sets are introduced in the second approach to the cardinality of sets. These objects are called cardinal numbers and are defined as equivalence classes with respect to the relation of equipollence, or by initial ordinal numbers of these classes, if it is possible (see, e.g., [7, 10]).

The equipollence of (finite) fuzzy sets has been investigated primarily by S. Gottwald [2, 3] and M. Wygralak [13–16] (see also [8]). S. Gottwald proposed a graded approach to the equipollence of fuzzy sets defined using the uniqueness of fuzzy mappings in his set theory for fuzzy sets of higher level. Additionally,

A. Laurent et al. (Eds.): IPMU 2014, Part II, CCIS 443, pp. 234–243, 2014.
© Springer International Publishing Switzerland 2014

a graded generalization of equipollence suggesting that fuzzy sets have approximately the same number of elements has been noted by M. Wygralak in [14], but a substantial development of cardinal theory based on this type of equipollence has not been realized yet.

In [6] (see also [5]) we proposed a new approach to the equipollence of fuzzy sets over a universe of sets (e.g., a universe of all finite or countable sets, all sets, or a Grothendieck universe). Analogously to Gottwald's approach, a graded equipollence is considered, where the degrees of being equipollent are obtained more simply than in Gottwald's approach. More precisely, this approach is based on a graded one-to-one correspondence between fuzzy sets, where the crisp mappings between universes are considered.

The aim of this short contribution is to present a functional approach to the cardinality of finite fuzzy sets based on this type of equipollence. The definition of graded equipollence for finite fuzzy sets generalizes the definition suggested in [6] in such a way that we use the multiplication of algebra of truth values as an alternative operation to the infimum. We will show that the graded equipollence with respect to a more general operation is a fuzzy similarity relation on the class of all finite fuzzy sets, and well-known statements of the cardinal theory of sets (including the Cantor-Bernstein theorem and the Cantor theorem stating the different cardinalities for sets and their power sets) can mostly be proved in a graded design, where if-then formulas are replaced by the inequalities between the degrees in which the antecedent and consequent are satisfied.

This contribution is structured as follows. In the next section, we recall the definition of residuated lattice as an algebraical structure of membership degrees of fuzzy sets and fuzzy classes. In Section 3, we introduce basic concepts concerning of the theory of (finite) fuzzy sets. The graded equipollence of finite fuzzy sets is proposed, and some of their properties are analyzed in Section 4. In Section 5, a functional approach to cardinal theory of finite fuzzy sets based on the graded equipollence is elaborated. The last section concludes.

2   Algebras of Truth Values

In this contribution, we assume that the truth values are interpreted in a *residuated lattice*, i.e., in an algebra $\mathbf{L} = \langle L, \wedge, \vee, \otimes, \rightarrow, \bot, \top \rangle$ with four binary operations and two constants satisfying the following conditions:

(i) $\langle L, \wedge, \vee, \bot, \top \rangle$ is a bounded lattice, where $\bot$ is the least element and $\top$ is the greatest element of L,

(ii) $\langle L, \otimes, \bot \rangle$ is a commutative monoid,

(iii) the pair $\langle \otimes, \rightarrow \rangle$ forms an adjoint pair, i.e.,

$$\alpha \leq \beta \rightarrow \gamma \quad \text{if and only if} \quad \alpha \otimes \beta \leq \gamma$$

hold for each $\alpha, \beta, \gamma \in L$ ($\leq$ denotes the corresponding lattice ordering).

The operations $\otimes$ and $\rightarrow$ are called the *multiplication* and *residuum*, respectively. We will say that a residuated lattice is *complete* (*linearly ordered*), if $\langle L, \wedge, \vee, \bot, \top \rangle$ is a complete (linearly ordered) lattice.

Example 1. Let T be a left continuous t-norm; we define $\to_T$ by

$$\alpha \to_T \beta = \bigvee \{\gamma \in [0,1] \mid T(\alpha, \gamma) \leq \beta\}.$$

Then, the algebra $\mathbf{L} = \langle [0,1], \min, \max, T, \to_T, 0, 1 \rangle$ is a complete residuated lattice. If T is the Łukasiewicz conjunction, i.e., $T(\alpha, \beta) = \max(\alpha + \beta - 1, 0)$ $(\alpha \to_T \beta = \min(1 - \alpha + \beta, 1))$, then we will use $\mathbf{L_L}$ to denote the Łukasiewicz algebra.

Let us define the following additional operations for any $\alpha, \beta \in L$ and set $\{\alpha_i \mid i \in I\}$ of elements from L over a countable (possibly empty) index set I:

$$\alpha \leftrightarrow \beta = (\alpha \to \beta) \wedge (\beta \to \alpha), \qquad \qquad (biresiduum)$$

$$\neg \alpha = \alpha \to \bot, \qquad \qquad (negation)$$

$$\bigotimes_{i \in I} \alpha_i = \begin{cases} \top, & I = \emptyset, \\ \bigwedge_{K \in \mathrm{Fin}(I)} \bigotimes_{i \in K} \alpha_i, & \text{otherwise}, \end{cases} \qquad (countable\ multiplication)$$

where $\mathrm{Fin}(I)$ denotes the set of all finite subsets of I.[1] In order to integrate some alternative constructions based on the operations of $\wedge$ ($\bigwedge$) and $\otimes$ ($\bigotimes$), in the sequel, we will use the common symbol $\odot$ ($\bigodot$).

Example 2. The residuum and negation in the Łukasiewicz algebra $\mathbf{L_L}$ is defined as $\alpha \leftrightarrow \beta = 1 - |\alpha - \beta|$ and $\neg \alpha = 1 - \alpha$, respectively.

3   Fuzzy Sets in $\mathfrak{Count}$

Intuitively, finite fuzzy sets are fuzzy sets apparently defined in finite universes of discourse. Nevertheless, for an expression of their cardinalities using fuzzy cardinals it is useful to consider fuzzy sets with the set of natural numbers as their universe of discourse (see [4, 5, 15–17]). This motivates us to use the proper class $\mathfrak{Count}$ of all countable sets as a common framework for the functional approach based on the relation of equipollence and the approach based on fuzzy cardinals.

Let us suppose that a complete residuated lattice $\mathbf{L}$ is given. A fuzzy set in $\mathfrak{Count}$ is defined as follows.

Definition 1. *A mapping $A : x \to L$ is called a fuzzy set in $\mathfrak{Count}$ if $x \in \mathfrak{Count}$. The class of all countable fuzzy sets will be denoted by $\mathfrak{FCount}$.*

Let us denote by $\mathrm{Dom}(A)$ the domain or also *universe of discourse* of A, $\mathrm{Ran}(A)$ the range of A and $\mathrm{Supp}(A) = \{z \in \mathrm{Dom}(A) \mid A(z) > \bot\}$ the *support* of fuzzy set A.

In the literature on fuzzy sets (see, e.g., [1, 9, 12]), a fuzzy set assigning $\bot$ to each element of its universe of discourse is usually referred to the empty fuzzy set, and a fuzzy set assigning $\alpha > \bot$ to only one element of its universe to a singleton (fuzzy set). In our theory, we will use a different interpretation of the empty fuzzy set and singleton in $\mathfrak{Count}$ as follows.

---

[1] Note that the countable multiplication in that form has been consider in [11].

Definition 2. *The empty mapping* $\emptyset : \emptyset \to L$ *is called the* empty fuzzy set. *A fuzzy set* A *is called a* singleton *if* $\mathrm{Dom}(A)$ *contains only one element.*

We say that two fuzzy sets are the *same* and write $A = B$ if they coincide on their domains, i.e., $A(x) = B(x)$ for any $x \in \mathrm{Dom}(A) = \mathrm{Dom}(B)$. An essential predicate in our theory is a binary relation that extends the concept of being the same fuzzy sets and states that two fuzzy sets coincide on their supports.

Definition 3. *We say that fuzzy sets* A *and* B *are* equivalent *(symbolically,* $A \equiv B$*) if* $\mathrm{Supp}(A) = \mathrm{Supp}(B)$ *and* $A(x) = B(x)$ *for any* $x \in \mathrm{Supp}(A)$. *The class of all equivalent fuzzy sets with* A *is denoted by* $\mathrm{cls}(A)$.

Since we are dealing with fuzzy sets which have countable universes of discourse in general, we provide the following definition of finite fuzzy set.

Definition 4. *We say that a fuzzy set* A *from* $\mathfrak{FCount}$ *is* finite *if there exists* $A' \in \mathrm{cls}(A)$ *such that* $\mathrm{Dom}(A')$ *is a finite set. The class of all finite fuzzy sets in* $\mathfrak{Count}$ *is denoted by* $\mathfrak{Ffin}$.

If a fuzzy set A has a finite universe, then we will use the following simple notation

$$A = \{\alpha_1/x_1, \ldots, \alpha_n/x_n\},$$

where $\mathrm{Dom}(A) = \{x_1, \ldots, x_n\}$ and $\alpha_i \in L$ for any $i = 1, \ldots, n$. The basic operations with fuzzy sets are introduced as follows.

Definition 5. *Let* $A, B \in \mathfrak{Fcount}$, $x = \mathrm{Dom}(A) \cup \mathrm{Dom}(B)$ *and* $A' \equiv A$, $B' \equiv B$ *such that* $\mathrm{Dom}(A') = \mathrm{Dom}(B') = x$. *Then,*

- *the* union *of* A *and* B *is the mapping* $A \cup B : x \to L$ *defined by*

$$(A \cup B)(a) = A'(a) \vee B'(a)$$

for any $a \in x$,
- *the* intersection *of* A *and* B *is the mapping* $A \cap B : x \to L$ *defined by*

$$(A \cap B)(a) = A'(a) \wedge B'(a)$$

for any $a \in x$,
- *the* difference *of* A *and* B *is the mapping* $A \setminus B : x \to L$ *defined by*

$$(A \setminus B)(a) = A'(a) \otimes (B'(a) \to \bot) = A'(a) \otimes \neg B'(a)$$

for any $a \in x$.

Definition 6. *Let* $A, B \in \mathfrak{Fcount}$, $x = \mathrm{Dom}(A) \times \mathrm{Dom}(B)$ *and* $y = \mathrm{Dom}(A) \sqcup \mathrm{Dom}(B)$ *(the disjoint union). Then,*

- *the* product *of* A, B *is the mapping* $A \times B : x \to L$ *defined by*

$$(A \times B)(a, b) = A(a) \wedge B(b)$$

for any $(a, b) \in x$,

- the strong product *of A, B is the mapping $A \otimes B : x \to L$ defined by*

$$(A \otimes B)(a, b) = A(a) \otimes B(b)$$

for any $(a, b) \in x$,
- the disjoint union *of A, B is the mapping $A \sqcup B : y \to L$ defined by*

$$(A \sqcup B)(a, i) = \begin{cases} A(a, i), & \text{if } i = 1, \\ B(a, i), & \text{if } i = 2, \end{cases}$$

for any $(a, i) \in y$.

Definition 7. *The fuzzy set $\overline{A} = \mathrm{Dom}(A) \setminus A$ is called the* complement *of A.*

It is easy to see that the relation "being equivalent" is a congruence for all mentioned operations except the complement. Moreover, if the proposed operations are applied on finite fuzzy sets, the resulting fuzzy set is again finite.

Example 3. Let $\mathbf{L_L}$ be the Łukasiewicz algebra and $A = \{1/a, 0.4/b\}$ and $B = \{0.6/a, 0.2/c\}$. Then, we have

$$A \cup B = \{1/a, 0.4/b, 0.2/c\},$$
$$A \cap B = \{0.6/a, 0/b, 0/c\},$$
$$A \setminus B = \{0.4/a, 0.4/b, 0/c\},$$
$$A \times B = \{0.6/(a, a), 0.2/(a, c), 0.4/(b, a), 0.2/(b, c)\},$$
$$A \otimes B = \{0.6/(a, a), 0.2/(a, c), 0/(b, a), 0/(b, c)\},$$
$$A \sqcup B = \{1/(a, 1), 0.4/(b, 1), 0.6/(a, 2), 0.2/(c, 2)\},$$
$$\overline{A} = \{0/a, 0.6/b\}.$$

It is not easy to say what the power set means for fuzzy sets. We propose the following simple definition that straightforwardly generalizes the classical approach to the concept of power set. We will use χ_x to denote the characteristic function of a set x, i.e., $\chi_x(y) = \top$, if $y \in x$, and $\chi_x(y) = \bot$, otherwise.

Definition 8. *Let $A \in \mathfrak{FCount}$ and $x = \{y \mid y \subseteq \mathrm{Dom}(A)\}$. The fuzzy set $\mathbf{P}(A) : x \to L$ defined by*

$$\mathbf{P}(A)(y) = \bigwedge_{z \in \mathrm{Dom}(A)} (\chi_y(z) \to A(z))$$

is called the fuzzy power set *of A.*

Example 4. Let $\mathbf{L_L}$ be the Łukasiewicz algebra and $A = \{1/a, 0.4/b\}$. Then,

$$\mathbf{P}(A) = \{1/\emptyset, 1/\{a\}, 0.4/\{b\}, 0.4/\{a, b\}\}.$$

Further, we introduce the concept of exponentiation for fuzzy sets. Recall that if x, y are sets, then the exponentiation x^y is the set of all mappings of y to x. We propose the following definition, which generalizes the classical one.

Definition 9. *Let $A, B \in \mathfrak{F}\mathfrak{Count}$ and put $x = \mathrm{Dom}(A)$ and $y = \mathrm{Dom}(B)$. The fuzzy set $B^A : y^x \to L$ defined by*

$$B^A(f) = \bigwedge_{z \in x} (A(z) \to B(f(z)))$$

is called the exponentiation of A to B.

Example 5. Let $\mathbf{L_L}$ be the Łukasiewicz algebra and $A = \{1/a, 0.4/b\}$, $B = \{0.6/a, 0.2/c\}$. Obviously, the domain of B^A is the set of all mappings of $\{a, b\}$ to $\{a, c\}$. For example, if $f \in \mathrm{Dom}(B^A)$ is defined as $f(a) = f(b) = a$, then

$$B^A(f) = (1 \to 0.6) \wedge (0.4 \to 0.6) = 0.6 \wedge \top = 0.6.$$

Finally, we introduce the concept of fuzzy class. Although our attention is focused on fuzzy sets the concept of fuzzy class will help us to denote special objects that are related to our theory but that are not fuzzy sets (because they are too large).

Definition 10. *A mapping $A : x \to L$ is called a fuzzy class in $\mathfrak{Count}$ if $x \subseteq \mathfrak{Count}$.*

4   Graded Equipollence of Finite Fuzzy Sets

In [6], we introduced the concept of graded equipollence using the degrees in which a one-to-one mapping between sets is a one-to-one mapping between fuzzy sets. This idea practically generalizes one of Cantor's approaches to the cardinality of sets, namely, that two sets have the same cardinality (are equipollent) if there exists a one-to-one mapping of one set onto the second one. Thus, to check the same cardinality of two sets, one needs to construct a one-to-one correspondence between them. However, in the fuzzy case, the situation is more complicated by the membership degrees, and intuitively, not all one-to-one correspondences between universes of fuzzy sets are appropriate to assert that fuzzy sets have the same cardinality. Moreover, if two fuzzy sets are very similar (consider $[A \approx B] = 0.999$) but there is no one-to-one correspondence between them (consider $f^{\to}(A) = B$), then it seems to be advantageous to say that these sets have approximately the same cardinality. Thus, the graded equipollence gives a degree in which two fuzzy sets have approximately the same cardinality, and this degree is derived from degrees in which one may construct one-to-one correspondences between fuzzy sets.

Let us start with the concept of one-to-one mapping between fuzzy sets in a degree.

Definition 11. *Let $A, B \in \mathfrak{F}\mathrm{fin}$, $x, y \in \mathrm{Count}$, and let $f : x \to y$ be a one-to-one mapping of x onto y in Count. We will say that f is a one-to-one mapping of A onto B in the degree α with respect to $\odot$ if $\mathrm{Supp}(A) \subseteq x \subseteq \mathrm{Dom}(A)$ and $\mathrm{Supp}(B) \subseteq y \subseteq \mathrm{Dom}(B)$ and*

$$\alpha = \bigodot_{z \in x} (A(z) \leftrightarrow B(f(z))).$$

We write $[A \sim_f^\odot B] = \alpha$ if f is a one-to-one mapping of A onto B in the degree α with respect to $\odot$.

As could be seen above not all one-to-one mappings are considered to specify the degree in which a mapping is a one-to-one mappings between fuzzy sets. The following establishes the set of all important one-to-one mappings between fuzzy sets.

Definition 12. Let $A, B \in \mathfrak{F}\text{fin}$. A mapping $f : x \to y$ belongs to the set $\mathrm{Bij}(A, B)$ if f is a one-to-one mapping of x onto y, $\mathrm{Supp}(A) \subseteq x \subseteq \mathrm{Dom}(A)$, and $\mathrm{Supp}(B) \subseteq y \subseteq \mathrm{Dom}(B)$.

Now we can proceed to the definition of graded equipollence.

Definition 13. Let $A, B \in \mathfrak{F}\text{fin}$. We will say that A is equipollent with B (or A has the same cardinality as B) in the degree α with respect to $\odot$ if there exist fuzzy sets $C \in \mathrm{cls}(A)$ and $D \in \mathrm{cls}(B)$ such that

$$\alpha = \bigvee_{f \in \mathrm{Bij}(C,D)} [C \sim_f^\odot D] \tag{1}$$

and, for each $A' \in \mathrm{cls}(A)$, $B' \in \mathrm{cls}(B)$ and $f \in \mathrm{Bij}(A', B')$, there is $[A' \sim_f^\odot B'] \leq \alpha$.

Let $A, B \in \mathfrak{F}\text{fin}$ such that $|\mathrm{Dom}(A)| = |\mathrm{Dom}(B)|$. We will use $\mathrm{Perm}(A, B)$ to denote the set of all $f \in \mathrm{Bij}(A, B)$ such that $\mathrm{Dom}(f) = \mathrm{Dom}(A)$ and $\mathrm{Ran}(f) = \mathrm{Dom}(B)$.[2] The following theorem shows how to find the degree of equipollence α between A and B.

Theorem 1. Let $A, B \in \mathfrak{F}\text{fin}$. Then,

$$[A \sim^\odot B] = \bigvee_{f \in \mathrm{Perm}(C,D)} [C \sim_f^\odot D]. \tag{2}$$

for any $C \in \mathrm{cls}(A)$ and $D \in \mathrm{cls}(B)$ such that $|\mathrm{Dom}(C)| = |\mathrm{Dom}(D)| = m$.

Similarly to the equipollence of sets (or fuzzy sets), the graded equipollence of fuzzy set is a $\otimes$-similarity relation (i.e., reflexive, symmetric and $\otimes$-transitive) on the class of all finite fuzzy sets as the following theorem shows.

Theorem 2. The fuzzy class relation $\sim^\odot$: $\mathfrak{F}\text{fin} \times \mathfrak{F}\text{fin} \to L$ is a $\otimes$-similarity relation on $\mathfrak{F}\text{fin}$, i.e.,

$$[A \sim^\odot A] = \top,$$
$$[A \sim^\odot B] = [B \sim^\odot A],$$
$$[A \sim^\odot B] \otimes [B \sim^\odot C] \leq [A \sim^\odot C],$$

which holds for arbitrary fuzzy sets $A, B, C \in \mathfrak{F}\text{fin}$.

---

[2] Although f is not a permutation on a universe in general, we use the denotation Perm, because to each pair of universes of fuzzy sets A and B we can define a common universe and fuzzy sets A' and B' equivalent to A and B, respectively, such that each mapping from $\mathrm{Perm}(A', B')$ is a permutation on this common universe.

5   Graded Versions of Fundamental Results of Set Theory

The most familiar theorem in set theory is the Cantor-Bernstein theorem (CBT). One of its forms states that if a, b, c, d are sets such that $b \subseteq a$ and $d \subseteq c$ and $a \sim d$ and $b \sim c$, then $a \sim c$. Unfortunately, we cannot prove its graded form in a full generality and have to restrict ourselves to the case of $\odot = \wedge$ and assume the linearity of residuated lattice.

Theorem 3 (Cantor-Bernstein theorem). *Let* $\mathbf{L}$ *be a linearly ordered residuated lattice and* $A, B, C, D \in \mathfrak{F}\mathrm{fin}$ *such that* $B \subseteq A$ *and* $D \subseteq C$. *Then*

$$[A \sim^\wedge D] \wedge [C \sim^\wedge B] \leq [A \sim^\wedge C].$$

A consequence of Theorem 3 is the following form of the graded Cantor-Bernstein theorem (cf. Corollary 4.8 in [16]), which generalizes a more commonly used version of the Cantor-Bernstein theorem.

Corollary 1. *Let* $\mathbf{L}$ *be a linearly ordered residuated lattice and* $A, B, C \in \mathfrak{F}\mathrm{fin}$ *such that* $A \subseteq B \subseteq C$. *Then,*

$$[A \sim^\wedge C] \leq [A \sim^\wedge B] \wedge [B \sim^\wedge C].$$

Let a, b, c, d be sets such that $a \sim c$ and $b \sim d$. Then, it is well-known that $a \cup b \sim c \cup d$, whenever $a \cap b = \emptyset$ and $c \cap d = \emptyset$, $a \times b \sim c \times d$, $a \sqcup b \sim c \sqcup d$. Graded versions of these and two further statements are presented in the following theorem.

Theorem 4. *Let* $A, B, C, D \in \mathfrak{F}\mathrm{fin}$. *Then,*

$$[A \sim^\circ B] \leq [\overline{A} \sim^\circ \overline{B}],$$
$$[A \sim^\circ B] \otimes [C \sim^\circ D] \leq [A \otimes C \sim^\circ B \otimes D],$$
$$[A \sim^\circ B] \otimes [C \sim^\circ D] \leq [A \times C \sim^\circ B \times D],$$
$$[A \sim^\circ B] \otimes [C \sim^\circ D] \leq [A \sqcup C \sim^\circ B \sqcup D],$$

if $\mathrm{Supp}(A) \cap \mathrm{Supp}(B) = \mathrm{Supp}(C) \cap \mathrm{Supp}(D) = \emptyset$, *then*

$$[A \sim^\circ C] \otimes [B \sim^\circ D] \leq [A \cup B \sim^\circ C \cup D].$$

If a, b are sets and $a \sim b$, then $\mathbf{P}(a) \sim \mathbf{P}(b)$, where $\mathbf{P}(a)$ and $\mathbf{P}(b)$ denote the power sets of a and b, respectively. The following theorem provides a graded version of this classical statement, where we have to restrict ourselves to the operation $\wedge$ in the computation of the degree of graded equipollence between fuzzy power sets.

Theorem 5. *Let* $A, B \in \mathfrak{F}\mathrm{fin}$. *Then,* $[A \sim^\circ B] \leq [\mathbf{P}(A) \sim^\wedge \mathbf{P}(B)]$.

One of the significant Cantor's theorems states that a is not equipollent with its power, i.e., $a \not\sim \mathbf{P}(a)$. The following is a generalization of this statement for finite fuzzy sets, stating that A and $\mathbf{P}(A)$ cannot have the same number of elements.

Theorem 6 (Cantor's theorem). *Let* $A \in \mathfrak{F}$fin. *Then,* $[A \sim^\circ \mathbf{P}(A)] < \top$.

Example 6. If $A = \{1/a, 0.4/b\}$ and $\mathbf{P}(A) = \{1/\emptyset, 1/\{a\}, 0.4/\{b\}, 0.4/\{a, b\}\}$ are from Example 4, then with $C = \{1/a, 0.4/b, 0/c, 0/d\}$, one may simply check that

$$[A \sim^\wedge \mathbf{P}(A)] = [C \sim^\wedge \mathbf{P}(A)] = (1 \leftrightarrow 1) \wedge (1 \leftrightarrow 0.4) \wedge (0.4 \leftrightarrow 0) \wedge (0.4 \leftrightarrow 0) =$$
$$1 \wedge 0.4 \wedge 0.6 \wedge 0.6 = 0.4.$$

Hence, we have $0 < [A \sim^\wedge \mathbf{P}(A)] < 1$. Moreover, we obtain $[A \sim^\otimes \mathbf{P}(A)] = 0$.

If a, b, c, d are sets such that $a \sim c$ and $b \sim d$, then $b^a \sim d^c$. The following theorem is a graded version of this statement, where again we have to restrict ourselves to the operation $\wedge$ in the computation of the degree of graded equipollence between fuzzy exponentials.

Theorem 7. *Let* $A, B, C, D \in \mathfrak{F}$fin *such that* $|\mathrm{Dom}(A)| = |\mathrm{Dom}(C)| = m$ *and* $|\mathrm{Dom}(B)| = |\mathrm{Dom}(D)| = n$. *Then,*

$$[A \sim^\circ C] \otimes [B \sim^\circ D] \leq [B^A \sim^\wedge C^D].$$

If a, b, c are sets, then it holds $c^{a \times b} \sim (c^b)^a$. The following theorem generalizes this relation for finite fuzzy sets.

Theorem 8. *Let* $A, B, C \in \mathfrak{F}$fin *such that their universes are finite. Then,*

$$[C^{A \otimes B} \sim^\circ (C^B)^A] = \top.$$

Remark 1. Note that an analogous relation to $\mathbf{P}(a) \sim 2^a$ cannot be proved for fuzzy sets.

6   Conclusion

In this contribution, we presented a cardinal theory of finite fuzzy sets based on the concept of graded equipollence. Fuzzy sets are propose to be inside the class of all countable sets. In contrast to the standard approach to fuzzy sets, we do not suppose a fixed universe for all fuzzy sets. A basic theory of fuzzy sets in the universe of all countable sets was introduced, including constructions such as fuzzy power sets and exponentiation. Graded equipollence was defined as a fuzzy class relation assigning a degree to each pair of fuzzy sets, expressing the fact that these fuzzy sets have approximately the same cardinality (are approximately equipollent). The graded equipollence is derived from degrees in which one-to-one mappings between sets may be considered to be one-to-one correspondences between fuzzy sets. With this concept, a functional approach to cardinal theory of finite fuzzy sets was developed, and several well-known statements, including the Cantor-Bernstein theorem and the Cantor theorem stating different cardinalities for sets and their power sets, were generalized in a graded design.

Acknowledgments. This work was supported by the European Regional Development Fund in the IT4Innovations Centre of Excellence project (CZ.1.05/1.1.00/02.0070).

References

1. Dubois, D., Prade, H. (eds.): Fundamentals of fuzzy sets. Foreword by Lotfi A. Zadeh. The Handbooks of Fuzzy Sets Series 7, vol. xxi, p. 647. Kluwer Academic Publishers, Dordrecht (2000)
2. Gottwald, S.: Fuzzy uniqueness of fuzzy mappings. Fuzzy Sets and Systems 3, 49–74 (1980)
3. Gottwald, S.: A note on fuzzy cardinals. Kybernetika 16, 156–158 (1980)
4. Holčapek, M.: An axiomatic approach to fuzzy measures like set cardinality for finite fuzzy sets. In: Hüllermeier, E., Kruse, R., Hoffmann, F. (eds.) IPMU 2010, Part I. CCIS, vol. 80, pp. 505–514. Springer, Heidelberg (2010)
5. Holčapek, M.: Graded equipollence and fuzzy c-measures of finite fuzzy sets. In: Proc. of 2011 IEEE International Conference on Fuzzy Systems, pp. 2375–2382. DnE, Taiwan (2011)
6. Holčapek, M., Turčan, M.: Graded equipollence of fuzzy sets. In: Carvalho, J.P., Dubois, D., Kaymak, D.U., Sousa, J.M.C. (eds.) Proceedings of IFSA/EUSFLAT 2009, pp. 1565–1570. Universidade Técnica de Lisboa, Lisbon (2009)
7. Jech, T.J.: Set Theory. Springer, Berlin (1997)
8. Klaua, D.: Zum kardinalzahlbegriff in der mehrwertigen mengenlehre. In: Asser, G., Flashmayers, J., Rinow, W. (eds.) Theory of Sets and Topology, pp. 313–325. Deutsher Verlag der Wissenshaften, Berlin (1972)
9. Klir, G.J., Yuan, B.: Fuzzy Sets and Fuzzy Logic: Theory and Applications. Prentice Hall, New Jersey (1995)
10. Levy, A.: Basic set theory. Dover Books on Mathematics. Dover Publications (2002)
11. Mesiar, R., Thiele, H.: On T-quantifiers and S-quantifiers. In: Discovering the World with Fuzzy Logic, pp. 310–326. Physica-Verlag, Heidelberg (2000)
12. Novák, V.: Fuzzy Sets and Their Application. Adam-Hilger, Bristol (1989)
13. Wygralak, M.: Generalized cardinal numbers and operations on them. Fuzzy Sets and Systems 53(1), 49–85 (1993)
14. Wygralak, M.: Vaguely defined objects. Representations, fuzzy sets and nonclassical cardinality theory. Theory and Decision Library. Series B: Mathematical and Statistical Methods, vol. 33. Kluwer Academic Publisher, Dordrecht (1996)
15. Wygralak, M.: Fuzzy sets with triangular norms and their cardinality theory. Fuzzy Sets and Systems 124(1), 1–24 (2001)
16. Wygralak, M.: Cardinalities of Fuzzy Sets. Kluwer Academic Publisher, Berlin (2003)
17. Zadeh, L.A.: A computational approach to fuzzy quantifiers in natural languages. Comp. Math. with Applications 9, 149–184 (1983)

Piecewise Linear Approximation of Fuzzy Numbers Preserving the Support and Core

Lucian Coroianu[1], Marek Gagolewski[2,3], Przemysław Grzegorzewski[2,3],
M. Adabitabar Firozja[4], and Tahereh Houlari[5]

[1] Department of Mathematics and Informatics, University of Oradea,
1 Universitatii Street, 410087 Oradea, Romania
[2] Systems Research Institute, Polish Academy of Sciences,
Newelska 6, 01-447 Warsaw, Poland
[3] Faculty of Mathematics and Information Science, Warsaw University of Technology,
Koszykowa 75, 00-662 Warsaw, Poland
[4] Department of Mathematics, Qaemshahr Branch, Islamic Azad University, Iran
[5] School of Mathematics and Computer Sciences, Damghan University, Iran
lcoroianu@uoradea.ro, {gagolews,pgrzeg}@ibspan.waw.pl,
mohamadsadega@yahoo.com

Abstract. A reasonable approximation of a fuzzy number should have a simple membership function, be close to the input fuzzy number, and should preserve some of its important characteristics. In this paper we suggest to approximate a fuzzy number by a piecewise linear 1-knot fuzzy number which is the closest one to the input fuzzy number among all piecewise linear 1-knot fuzzy numbers having the same core and the same support as the input. We discuss the existence of the approximation operator, show algorithms ready for the practical use and illustrate the considered concepts by examples. It turns out that such an approximation task may be problematic.

Keywords: Approximation of fuzzy numbers, core, fuzzy number, piecewise linear approximation, support.

1 Introduction

Complicated membership functions generate many problems in processing imprecise information modeled by fuzzy numbers including problems with calculations, computer implementation, etc. Moreover, handling complex membership functions entails difficulties in interpretation of the results too. This is the reason why a suitable approximation of fuzzy numbers is so important. So we usually try to substitute the original "input" membership functions by the "output" which is simpler or more regular and hence more convenient for further tasks. We expect that a desired approximation will reveal the following priorities:

(P.1) simplicity of a membership function,
(P.2) closeness to the input fuzzy number,
(P.3) preservation of some important characteristics.

A. Laurent et al. (Eds.): IPMU 2014, Part II, CCIS 443, pp. 244–253, 2014.

The simplest possible shape of a membership function is acquired by linear sides. Hence, (P.1) is fulfilled by the trapezoidal approximation. One may do it in many ways but it seems that the desired approximation output should be as close as possible to the input. Thus, (P.1) and (P.2) may lead to the approximation of a fuzzy number by the closest trapezoidal one. However, such an approximation does not guarantee automatically any other interesting properties. Therefore, we often look for the approximation that has some additional properties like the invariance of the expected interval (see, e.g. [1,6,5,8]). It seems that the core and the support belong to the most important characteristics of fuzzy numbers. It is quite obvious since these very sets are the only ones which are connected with our "sure" knowledge. Actually, the core contains all the points which surely belong to the fuzzy set under study. On the other hand, the complement of the support consists of the points that surely do not belong to given fuzzy set. The belongingness of all other points to the fuzzy set under discussion is just a matter of degree described quantitatively by the membership function. Hence, one may easily agree that both the support and core play a key role in fuzzy set analysis. However, if we try to approximate a fuzzy number by a trapezoidal one that preserves both the support and core of the input, the approximation problem simplifies too much since we obtain the unique solution just by joining the borders of the support and core by the straight lines. Unfortunately, the output of such approximation may be significantly distant from the input. The way out from this dilemma is to consider the approximation by a trapezoidal fuzzy number which is as close as possible to the input and preserves either the core or the support. This way we obtain a procedure which fulfills all the desired conditions (P.1)-(P.3). However, one may easily indicate examples where the output of the approximation with fixed core has the support significantly different than the support of the input. And conversely, the output of the approximation with fixed support may have the core significantly different than the core of the input.

This discussion shows that usually we cannot obtain a satisfying trapezoidal approximation of an arbitrary fuzzy number that fulfills the nearness criterion and preserves both the support and core. In this paper we propose to consider the 1-knot piecewise linear fuzzy numbers (see [2]) as a reasonable solution of the approximation problem satisfying requirements (P.1)-(P.3). More precisely, we suggest to approximate a fuzzy number by the closest piecewise linear 1-knot fuzzy number having the same core and the same support as the input.

2   Piecewise Linear 1-Knot Fuzzy Numbers

Fuzzy numbers are particular cases of fuzzy sets of the real line. The membership function of a fuzzy number A is given by:

$$A(x) = \begin{cases} 0 & \text{if} \quad x < a_1, \\ l_A(x) & \text{if} \quad a_1 \leq x < a_2, \\ 1 & \text{if} \quad a_2 \leq x \leq a_3, \\ r_A(x) & \text{if} \quad a_3 < x \leq a_4, \\ 0 & \text{if} \quad x > a_4, \end{cases} \tag{1}$$

where $a_1, a_2, a_3, a_4 \in \mathbb{R}$, $l_A : [a_1, a_2] \longrightarrow [0,1]$ is a nondecreasing upper semi-continuous function, $l_A(a_1) = 0$, $l_A(a_2) = 1$, called the *left side* of the fuzzy number, and $r_A : [a_3, a_4] \longrightarrow [0,1]$ is a nonincreasing upper semicontinuous function, $r_A(a_3) = 1$, $r_A(a_4) = 0$, called the *right side* of the fuzzy number A. The α-*cut* of A, $\alpha \in (0,1]$, is a crisp set defined as: $A_\alpha = \{x \in \mathbb{R} : A(x) \geq \alpha\}$. The *support* or 0-cut, A_0, of a fuzzy number is defined as

$$\text{supp}(A) = A_0 = \overline{\{x \in \mathbb{R} : A(x) > 0\}}.$$

It is easily seen that for each $\alpha \in (0,1]$ every α-cut of a fuzzy number is a closed interval $A_\alpha = [A_L(\alpha), A_U(\alpha)]$, where $A_L(\alpha) = \inf\{x \in \mathbb{R} : A(x) \geq \alpha\}$ and $A_U(\alpha) = \sup\{x \in \mathbb{R} : A(x) \geq \alpha\}$.

Moreover, if the sides of the fuzzy number A are strictly monotone, then A_L and A_U are inverse functions of l_A and r_A, respectively. The 1-cut of A will be called the core of A and we use the notation

$$A_1 = \text{core}(A) = \{x \in \mathbb{R} : A(x) = 1\}.$$

From now on, we denote by $\mathbb{F}(\mathbb{R})$ the set of all fuzzy numbers. However, in practice, e.g. when calculations of arithmetic operations is performed, fuzzy numbers with simple membership functions are often preferred. The most commonly used subclass of $\mathbb{F}(\mathbb{R})$ is formed by so-called *trapezoidal fuzzy numbers*, i.e. fuzzy numbers with linear sides. The set of all trapezoidal fuzzy numbers is denoted by $\mathbb{F}^T(\mathbb{R})$. Trapezoidal fuzzy numbers are often used directly for modeling vague concepts or for approximating more complicated fuzzy numbers due to their simplicity. Unfortunately, in some situations such simple description may appear too limited. In some cases we are interested in specifying the membership function in one (or more) additional α-cuts other than 0 or 1. Thus in [2] a generalization of the trapezoidal fuzzy numbers was proposed by considering fuzzy numbers with piecewise linear side functions each consisting of two segments.

Definition 1. *For any fixed $\alpha_0 \in (0,1)$ an α_0-piecewise linear 1-knot fuzzy number S is a fuzzy number with the following membership function*

$$S(x) = \begin{cases} 0 & \text{if} \quad x < s_1, \\ \alpha_0 \frac{x - s_1}{s_2 - s_1} & \text{if} \quad s_1 \leq x < s_2, \\ \alpha_0 + (1 - \alpha_0) \frac{x - s_2}{s_3 - s_2} & \text{if} \quad s_2 \leq x < s_3, \\ 1 & \text{if} \quad s_3 \leq x \leq s_4, \\ \alpha_0 + (1 - \alpha_0) \frac{s_5 - x}{s_5 - s_4} & \text{if} \quad s_4 < x \leq s_5, \\ \alpha_0 \frac{s_6 - x}{s_6 - s_5} & \text{if} \quad s_5 < x \leq s_6, \\ 0 & \text{if} \quad x > s_6, \end{cases}$$

where $\mathbf{s} = (s_1, \ldots, s_6)$ *such that* $s_1 \leq \cdots \leq s_6$.

Since any α_0-piecewise linear 1-knot fuzzy number is completely defined by its knot α_0 and six real numbers $s_1 \leq \cdots \leq s_6$, hence it will be denoted as $S = S(\alpha_0, \mathbf{s})$. An example of an α_0-piecewise linear 1-knot fuzzy number is given in Fig. 1.

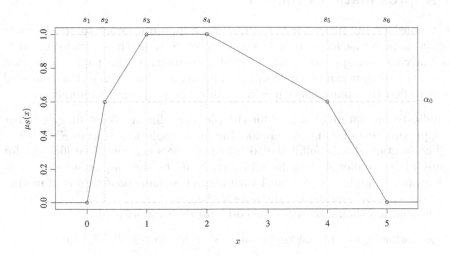

Fig. 1. The membership function of $S = S(0.6, (0, 0.3, 1, 2, 4, 5))$

Alternatively, an α_0-piecewise linear 1-knot fuzzy number may be defined using its α-cut representation, i.e.

$$
S_L(\alpha) = \begin{cases} s_1 + (s_2 - s_1)\frac{\alpha}{\alpha_0} & \text{for} \quad \alpha \in [0, \alpha_0), \\ s_2 + (s_3 - s_2)\frac{\alpha - \alpha_0}{1 - \alpha_0} & \text{for} \quad \alpha \in [\alpha_0, 1] \end{cases} \tag{2}
$$

and

$$
S_U(\alpha) = \begin{cases} s_5 + (s_6 - s_5)\frac{\alpha_0 - \alpha}{\alpha_0} & \text{for} \quad \alpha \in [0, \alpha_0), \\ s_4 + (s_5 - s_4)\frac{1 - \alpha}{1 - \alpha_0} & \text{for} \quad \alpha \in [\alpha_0, 1]. \end{cases} \tag{3}
$$

Let us denote the set of all such fuzzy numbers by $\mathbb{F}^{\pi(\alpha_0)}(\mathbb{R})$. By setting $\mathbb{F}^{\pi(0)}(\mathbb{R}) = \mathbb{F}^{\pi(1)}(\mathbb{R}) := \mathbb{F}^T(\mathbb{R})$ we also include the cases $\alpha_0 \in \{0, 1\}$. Please note that the inclusion $\mathbb{F}^T(\mathbb{R}) \subseteq \mathbb{F}^{\pi(\alpha_0)}(\mathbb{R})$ holds for any $\alpha_0 \in [0, 1]$.

Moreover, to simplify notation, let $\mathbb{F}^{\pi[a,b]}(\mathbb{R})$ denote the set of all α-piecewise linear 1-knot fuzzy numbers, where $\alpha \in [a, b]$ for some $0 \le a \le b \le 1$, i.e.

$$
\mathbb{F}^{\pi[a,b]}(\mathbb{R}) := \bigcup_{\alpha \in [a,b]} \mathbb{F}^{\pi(\alpha)}(\mathbb{R}).
$$

In many problems an adequate metric over the space of fuzzy numbers should be considered. The flexibility of the space of fuzzy numbers allows for the construction of many types of metric structures over this space. In the area of fuzzy number approximation the most suitable metric is an extension of the *Euclidean* (L^2) *distance* d defined by (see, e.g., [4])

$$
d^2(A, B) = \int_0^1 (A_L(\alpha) - B_L(\alpha))^2 d\alpha + \int_0^1 (A_U(\alpha) - B_U(\alpha))^2 d\alpha. \tag{4}
$$

3   Approximation Problem

Let us consider any fuzzy number $A \in \mathbb{F}(\mathbb{R})$. Suppose we want to approximate A by an α_0-piecewise linear 1-knot fuzzy number S. In [2] the piecewise linear 1-knot fuzzy number approximation is broadly discussed. In this paper we suggest another type of approximation. Keeping in mind postulates (P.1)–(P.3), our goal now is to find the approximation which fulfills the following requirements:

1. Indicate the optimal knot α_0 for the piecewise linear 1-knot fuzzy number approximation of A, i.e. we are looking for the solution $S(A)$ in $\mathbb{F}^{\pi[0,1]}(\mathbb{R})$.
2. The solution should fulfill the so-called nearness criterion (see [6]), i.e. for any fuzzy number A the solution $S(A)$ should be the α_0-piecewise linear 1-knot fuzzy number nearest to A with respect to some predetermined metric. In our case we consider the distance d given by (4).
3. The solution should preserve the core and the support of A.

More formally, we are looking for such $S^* = S^*(A) \in \mathbb{F}^{\pi[0,1]}(\mathbb{R})$ that

$$d(A, S^*) = \min_{S \in \mathbb{F}^{\pi[0,1]}(\mathbb{R})} d(A, S), \tag{5}$$

which satisfies the following constraints:

$$\mathrm{core}(S^*) = \mathrm{core}(A), \tag{6}$$
$$\mathrm{supp}(S^*) = \mathrm{supp}(A). \tag{7}$$

At first, let us investigate whether the above problem has at least one solution for every $A \in \mathbb{F}(\mathbb{R})$. For that we will use the property that the space $(F(\mathbb{R}), d, +, \cdot)$ can be embedded in the Hilbert space $\left(L^2[0,1] \times L^2[0,1], \tilde{d}, \oplus, \odot \right)$ (see e.g. [2]). Therefore, we have $d(A, B) = \tilde{d}(A, B)$, $A + B = A \oplus B$ and $\lambda \cdot A = \lambda \odot A$, for all $A, B \in \mathbb{F}(\mathbb{R})$ and $\lambda \in [0, \infty)$. By Proposition 4 in [2] it is known that $\mathbb{F}^{\pi[0,1]}(\mathbb{R})$ is a closed subset of $L^2[0,1] \times L^2[0,1]$ in the topology generated by $\tilde{d}$. Unfortunately, it may happen that the set

$$CS(A) = \left\{ S \in \mathbb{F}^{\pi[0,1]}(\mathbb{R}) : \mathrm{core}(S) = \mathrm{core}(A), \mathrm{supp}(S) = \mathrm{supp}(A) \right\} \tag{8}$$

would not be closed in $\mathbb{F}^{\pi[0,1]}(\mathbb{R})$. Indeed, suppose that $A_\beta = [\beta^3, 1]$, $\beta \in [0, 1]$. Then let us consider a sequence $(\mathrm{S}(\alpha_n, \mathbf{s}_n))_{n \geq 1}$, $\mathbf{s}_n = (s_{n,1}, ..., s_{n,6})$ in $CS(A)$, where for each $n \geq 1$ we have $\alpha_n = (n-1)/n$, $s_{n,1} = s_{n,2} = 0$ and $s_{n,3} = ... = s_{n,6} = 1$. It is immediate that $(\tilde{d}) \lim_{n \to \infty} \mathrm{S}(\alpha_n, \mathbf{s}_n) = (d) \lim_{n \to \infty} \mathrm{S}(\alpha_n, \mathbf{s}_n) = [0, 1]$ and since $\mathrm{core}([0,1]) \neq \mathrm{core}(A)$ it results that the set $CS(A)$ is not closed in $L^2[0,1] \times L^2[0,1]$, nor in $F(\mathbb{R})$. Therefore, it is an open question whether problem (5)-(7) has a solution for any $A \in \mathbb{F}(\mathbb{R})$.

Interestingly, the solution always exists if we consider a local approximation problem. Suppose that $0 < a < b < 1$ and let us consider the set $\mathbb{F}^{\pi[a,b]}(\mathbb{R}) = \{\mathrm{S}(\alpha, \mathbf{s}) \in \mathbb{F}^{\pi[0,1]}(\mathbb{R}) : a \leq \alpha \leq b\}$. Now let us consider the following set

$$CS_{a,b}(A) = \left\{ S \in \mathbb{F}^{\pi[a,b]}(\mathbb{R}) : \mathrm{core}(S) = \mathrm{core}(A), \mathrm{supp}(S) = \mathrm{supp}(A) \right\}. \tag{9}$$

We are looking for such $S^* = S^*(A) \in CS_{a,b}(A)$ that

$$d(A, S^*) = \min_{S \in CS_{a,b}(A)} d(A, S). \tag{10}$$

Obviously, there is a sequence $(S(\alpha_n, \mathbf{s}_n))_{n \geq 1}$, in $CS_{a,b}(A)$, such that

$$\lim_{n \to \infty} d(A, S(\alpha_n, \mathbf{s}_n)) = \inf_{S \in CS_{a,b}(A)} d(A, S) := m. \tag{11}$$

Let $n_0 \in \mathbb{N}$ be such that $d(A, S(\alpha_n, \mathbf{s}_n)) \leq m + 1$ for all $n \geq n_0$. This implies that $d(0, S(\alpha_n, \mathbf{s}_n)) \leq d(0, A) + d(A, S(\alpha_n, \mathbf{s}_n)) \leq d(0, A) + m + 1$ for all $n \geq n_0$. Therefore, the sequence $(S(\alpha_n, \mathbf{s}_n))_{n \geq 1}$ is bounded with respect to metric d and hence with respect to $\tilde{d}$. By Lemma 2 (iii) in [2] it results that each sequence $(c_{n,i})_{n \geq 1}$, $i = 1, \ldots, 8$, is bounded, where

$$s_{n,1} = c_{n,1}, \qquad s_{n,2} = c_{n,2} \cdot \alpha_n + c_{n,1}, \qquad s_{n,3} = c_{n,3} + c_{n,4},$$
$$s_{n,4} = c_{n,7} + c_{n,8}, \qquad s_{n,5} = c_{n,5} + c_{n,6} \cdot \alpha_n, \qquad s_{n,6} = c_{n,5}.$$

Without loss of generality let us suppose that $\lim_{n \to \infty} \alpha_n = \alpha_0$ (obviously we have $\alpha_0 \in [a, b]$) and $\lim_{n \to \infty} c_{n,i} = c_i$, $i = 1, \ldots, 8$. Letting $n \to \infty$ in the above equations and denoting $\mathbf{s} = (s_1, \ldots, s_6)$, where $s_i = \lim_{n \to \infty} s_{n,i}$, $i = 1, \ldots, 6$, it easily results that $S(\alpha_0, \mathbf{s}) \in \mathbb{F}^{\pi[a,b]}(\mathbb{R})$. Then, since $S(\alpha_n, \mathbf{s}_n) \in CS_{a,b}(A)$ for all $n \geq 1$, it follows that $s_{n,1} = A_L(0)$, $s_{n,3} = A_L(1)$, $s_{n,4} = A_U(1)$ and $s_{n,6} = A_U(0)$ and therefore we easily obtain that $S(\alpha_0, \mathbf{s})$ preserves the core and support of A and hence $S(\alpha_0, \mathbf{s}) \in CS_{a,b}(A)$. On the other hand, by Lemma 3 in [2] (making some suitable substitutions) we also obtain that $(\tilde{d}) \lim_{n \to \infty} S(\alpha_n, \mathbf{s}_n) = S(\alpha_0, \mathbf{s})$. This property together with relation (11) and the continuity of d, implies that $d(A, S(\alpha_0, \mathbf{s})) = m$. Hence we have just proved that problem (10) has at least one solution. Note that one can easily prove that $CS_{a,b}(A)$ is not convex in $L^2[0, 1] \times L^2[0, 1]$ which means that the solution of problem (10) may not be unique. All these results are summarized in the following theorem.

Theorem 1. *If $A \in \mathbb{F}(\mathbb{R})$ and $0 < a < b < 1$, then there exists at least one element $S^* = S^*(A) \in \mathbb{F}^{\pi[a,b]}(\mathbb{R})$ such that $d(A, S^*) = \min_{S \in CS_{a,b}(A)} d(A, S)$.*

4   Algorithm

Let us show how to find a solution to problem (10). We have to minimize the function

$$f(\alpha, x, y) = \int_0^\alpha \left(A_L(\beta) - \left(A_L(0) + (x - A_L(0)) \cdot \frac{\beta}{\alpha} \right) \right)^2 d\beta$$
$$+ \int_\alpha^1 \left(A_L(\beta) - \left(x + (A_L(1) - x) \cdot \frac{\beta - \alpha}{1 - \alpha} \right) \right)^2 d\beta$$
$$+ \int_0^\alpha \left(A_U(\beta) - \left(y + (A_U(0) - y) \cdot \frac{\alpha - \beta}{\alpha} \right) \right)^2 d\beta$$
$$+ \int_\alpha^1 \left(A_U(\beta) - \left(A_U(1) + (y - A_U(1)) \cdot \frac{1 - \beta}{1 - \alpha} \right) \right)^2 d\beta$$

subject to $A_L(0) \leq x \leq A_L(1)$ and $A_U(1) \leq y \leq A_U(0)$.

This problem may have more than one solution and, in addition, it seems to be difficult to be solved analytically in this form since the equation $f'_\alpha(\alpha, x, y) = 0$ cannot be solved in general as we are forced to work with functions where we cannot separate α from the integral. Therefore, we will start by considering the knot $\alpha = \alpha_0$ being fixed. For some $\alpha_0 \in (0, 1)$ we want to minimize the function $g_{\alpha_0}(x, y) = f(\alpha_0, x, y)$ with the same restrictions as above. Obviously we can split this problem into two independent sub-problems. Firstly, we have to minimize the function

$$x \mapsto \int_0^{\alpha_0} \left(A_L(\beta) - \left(A_L(0) + (x - A_L(0)) \cdot \frac{\beta}{\alpha_0} \right) \right)^2 d\beta$$
$$+ \int_{\alpha_0}^1 \left(A_L(\beta) - \left(x + (A_L(1) - x) \cdot \frac{\beta - \alpha_0}{1 - \alpha_0} \right) \right)^2 d\beta$$

on the interval $[A_L(0), A_L(1)]$ and then we have to minimize the function

$$y \mapsto \int_0^{\alpha_0} \left(A_U(\beta) - \left(y + (A_U(0) - y) \cdot \frac{\alpha_0 - \beta}{\alpha_0} \right) \right)^2 d\beta$$
$$+ \int_{\alpha_0}^1 \left(A_U(\beta) - \left(A_U(1) + (y - A_U(1)) \cdot \frac{1 - \beta}{1 - \alpha_0} \right) \right)^2 d\beta$$

on the interval $[A_U(1), A_U(0)]$. Obviously, the above functions are quadratic functions of one variable and after some simple calculations we obtain their unique minimum points on $\mathbb{R}$ as

$$x_m = 3 \int_0^{\alpha_0} \left(A_L(\beta) - A_L(0) \cdot \frac{\alpha_0 - \beta}{\alpha_0} \right) \cdot \frac{\beta}{\alpha_0} d\beta$$
$$+ 3 \int_{\alpha_0}^1 \left(A_L(\beta) - A_L(1) \cdot \frac{\beta - \alpha_0}{1 - \alpha_0} \right) \cdot \frac{1 - \beta}{1 - \alpha_0} d\beta$$

and

$$y_m = 3 \int_0^{\alpha_0} \left(A_U(\beta) - A_U(0) \cdot \frac{\alpha_0 - \beta}{\alpha_0} \right) \cdot \frac{\beta}{\alpha_0} d\beta$$
$$+ 3 \int_{\alpha_0}^1 \left(A_U(\beta) - A_U(1) \cdot \frac{\beta - \alpha_0}{1 - \alpha_0} \right) \cdot \frac{1 - \beta}{1 - \alpha_0} d\beta.$$

From here we easily obtain the solutions of our two sub-problems as

$$x_0 = \begin{cases} A_L(0) & \text{if} \quad x_m < A_L(0), \\ A_L(1) & \text{if} \quad x_m > A_L(1), \\ x_m & \text{if} \quad A_L(0) \leq x_m \leq A_L(1) \end{cases} \tag{12}$$

and

$$y_0 = \begin{cases} A_U(1) & \text{if} \quad y_m < A_U(1), \\ A_U(0) & \text{if} \quad y_m > A_U(0), \\ y_m & \text{if} \quad A_U(1) \leq y_m \leq A_U(0). \end{cases} \tag{13}$$

When a computer implementation is needed, in most of the cases, x_m and y_m may be easily calculated via numeric integration, cf. [3].

In conclusion, we have just proved for fixed α the existence and uniqueness of the piecewise linear 1-knot approximation which preserves the core and the support. More exactly we have the following approximation result.

Theorem 2. *Suppose that $\alpha_0 \in (0,1)$ and for some fuzzy number A let us define the set*

$$CS_{\alpha_0}(A) = \{S \in \mathbb{F}^{\pi(\alpha_0)}(\mathbb{R}) : \mathrm{core}(S) = \mathrm{core}(A) \text{ and } \mathrm{supp}(S) = \mathrm{supp}(A)\}.$$

Then there exists a unique best approximation (with respect to metric d) of A relatively to the set $CS_{\alpha_0}(A)$. This approximation is $S_{\alpha_0}(A) = \mathbf{S}(\alpha_0, \mathbf{s}(A))$, $\mathbf{s}(A) = (s_1(A), ..., s_6(A))$, where

$$s_1(A) = A_L(0), \quad s_2(A) = x_0, \quad s_3(A) = A_L(1),$$
$$s_4(A) = A_U(1), \quad s_5(A) = y_0, \quad s_6(A) = A_U(0),$$

and x_0, y_0 are given by (12) and (13) respectively.

We will use the previous theorem to approach a solution $S^*(A) \in CS_{a,b}(A)$ of problem (10). We will construct a sequence $(S_{\alpha_n}(A))_{n \geq 1}$ in $CS_{a,b}(A)$ such that (d) $\lim_{n \to \infty} S_{\alpha_n}(A) = S^*(A)$. Here, $S_{\alpha_n}(A)$ is the unique best approximation of A relatively to the set $CS_{\alpha_n}(A)$.

5 Some Numerical Examples

Example 1. Consider a fuzzy number A with supp $= [0,5]$, core $= [3,4]$ and $A_L(\alpha) = 3\,\mathrm{qbeta}_A(\alpha; 2, 1)$, $A_U(\alpha) = 5 - \alpha^3$, where $\mathrm{qbeta}(x; a, b)$ denotes the quantile function of the Beta distribution $B(a, b)$. Let $\alpha_0 = 0.2$. The best piecewise linear approximation $A'_{0.2}$ of A preserving the support and core is defined by $\mathbf{s}' = (0, 1.53, 3, 4, 5, 5)$. We have $d(A, A'_{0.2}) \simeq 0.212$. On the other hand, for the best piecewise linear approximation $A''_{0.2}$ obtained using algorithm from [2] and given by $\mathbf{s}'' = (0.34, 1.46, 3.09, 4.26, 5.08, 5.08)$, we get $d(A, A''_{0.2}) \simeq 0.105$. The discussed fuzzy numbers are depicted in Fig. 2a.

Given a method to obtain $S^*_{\alpha_0}(A)$, i.e. best α_0-piecewise linear approximation of A preserving its support and core, we may find $S^*(A)$ using some general one-dimensional optimization technique, like the Brent algorithm implemented in R' `optimize()` function. This may be done for a fixed fuzzy number A by finding the argument minimizing the function $D_A(\alpha) = d(A, S^*_\alpha(A))$, $\alpha \in [0,1]$.

Example 2. Let us go back to the fuzzy number A from Example 1. Fig. 2b depicts the distance function D_A. By applying the Brent algorithm we find that for $\alpha^* \simeq 0.3$ we get the solution to our problem, with $d(A, S^*_{\alpha^*}) \simeq 1.999$.

Example 3. Consider a fuzzy number B with supp $= [-1, 1]$ and core $= \{0\}$, $B_L(\alpha) = \mathrm{qbeta}(\alpha; 2, 2)$, $B_U(\alpha) = 1 - \mathrm{qbeta}(\alpha; 2, 2)$. Fig. 3a shows the distance function D_B having two minima near 0 and 1 that are hardly to find numerically.

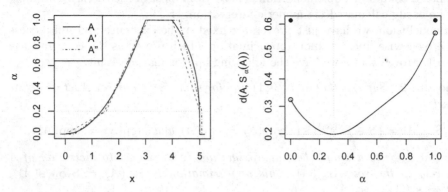

Fig. 2. (a) Fuzzy number A from Example 1 and its best approximation A' preserving its support and core as well as its best approximation A'' with no such constraints; (b) distance function $D_A(\alpha)$

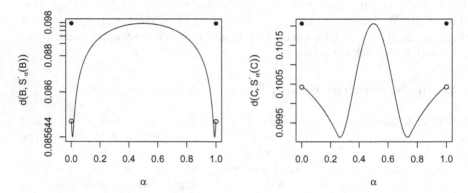

Fig. 3. (a) Distance function $D_B(\alpha)$ for a fuzzy number B from Example 3 (b) Distance function $D_C(\alpha)$ for a fuzzy number C from Example 4

Example 4. Consider a fuzzy number C with supp $= [0,1]$ and core $= \{1\}$ with

$$C_L(\alpha) = \begin{cases} \alpha & \text{for} & \alpha \in [0, 0.25), \\ 0.5 & \text{for} & \alpha \in [0.25, 0.75], \\ \alpha & \text{for} & \alpha \in (0.75, 1]. \end{cases}$$

The distance function D_C is depicted in Fig. 3b. It has two minima at $\alpha = 0.25$ and $\alpha = 0.75$.

Moreover, an example may easily be constructed for which D_X has a local minimum that is not its global minimum. This information is important when using numerical optimization techniques, as an algorithm may fall into a suboptimal solution.

6 Conclusions

In this paper we have considered a fuzzy number approximation by a piecewise linear 1-knot fuzzy number which is the closest one to the input fuzzy number among all piecewise linear 1-knot fuzzy numbers having the same core and the same support as the input. We have indicated and discussed problems related to the existence and uniqueness of the solution.

One may easily notice that the approximation method suggested in this paper works nicely if the input fuzzy number is more or less symmetrical. However, if the left and the right side of a fuzzy number differ a lot, e.g. one is convex while the other is concave, our approximation method is not so convincing. It seems that in such a case a much more natural approach is to treat both sides of a fuzzy number separately.

Moreover, for particular fuzzy numbers (e.g. when the derivative in the neighborhood of the borders of the support or core is close to 0) it might be reasonable to consider weakened restrictions on the support and core than (6)–(7), just like suggested in [7], i.e. $\mathrm{supp}(S(A)) \subseteq \mathrm{supp}(A)$ and $\mathrm{core}(A) \subseteq \mathrm{core}(S(A))$.

Acknowledgments. The contribution of Lucian Coroianu was possible with the support of a grant of the Romanian National Authority for Scientific Research, CNCS-UEFISCDI, project number PN-II-ID-PCE-2011-3-0861. Marek Gagolewski's research was supported by the FNP START 2013 Scholarship.

References

1. Ban, A.I.: Approximation of fuzzy numbers by trapezoidal fuzzy numbers preserving the expected interval. Fuzzy Sets and Systems 159, 1327–1344 (2008)
2. Coroianu, L., Gagolewski, M., Grzegorzewski, P.: Nearest piecewise linear approximation of fuzzy numbers. Fuzzy Sets and Systems 233, 26–51 (2013)
3. Gagolewski, M.: FuzzyNumbers Package: Tools to deal with fuzzy numbers in R (2014), http://FuzzyNumbers.rexamine.com/
4. Grzegorzewski, P.: Metrics and orders in space of fuzzy numbers. Fuzzy Sets and Systems 97, 83–94 (1998)
5. Grzegorzewski, P.: Trapezoidal approximations of fuzzy numbers preserving the expected interval - algorithms and properties. Fuzzy Sets and Systems 159, 1354–1364 (2008)
6. Grzegorzewski, P., Mrówka, E.: Trapezoidal approximations of fuzzy numbers. Fuzzy Sets and Systems 153, 115–135 (2005)
7. Grzegorzewski, P., Pasternak-Winiarska, K.: Natural trapezoidal approximations of fuzzy numbers. Fuzzy Sets and Systems (2014), http://dx.doi.org/10.1016/j.fss.2014.03.003
8. Yeh, C.T.: Trapezoidal and triangular approximations preserving the expected interval. Fuzzy Sets and Systems 159, 1345–1353 (2008)

Characterization of the Ranking Indices
of Triangular Fuzzy Numbers

Adrian I. Ban and Lucian Coroianu

Department of Mathematics and Informatics, University of Oradea,
Universităţii 1, 410087 Oradea, Romania
{aiban,lcoroianu}@uoradea.ro

Abstract. We find necessary and sufficient conditions for a ranking index defined on the set of triangular fuzzy numbers as a linear combination of its components to rank effectively. Then, based on this result, we characterize the class of ranking indices which generates orderings on triangular fuzzy numbers satisfying the basic requirements by Wang and Kerre in a slightly modified form.

Keywords: Fuzzy number, Triangular fuzzy number, Ranking.

1 Introduction

The main approach in the ordering of fuzzy numbers is based on so called ranking indices. They are functions from fuzzy numbers to real values, the ordering between fuzzy numbers being generated by a procedure based on the standard ordering of real numbers (see, e.g., [1], [2], [4]-[7], [10], [14], [15], [17]). A good choice of the ranking is very important in many applications related with decision theory, optimization, artificial intelligence, approximate reasoning, socioeconomic systems, etc.

Starting from the reasonable properties proposed in [17], we characterize rankings over triangular fuzzy numbers obtained from ranking indices. The main result of the paper is based on the finding of necessary and sufficient conditions such that the requirements in [17] to be satisfied by a particular class which include the most important ranking indices as expected value, value, ambiguity and linear combinations of them. The present study continues our contribution [3] dedicated to trapezoidal case.

2 Fuzzy Numbers and Ranking

We recall some definitions related to fuzzy numbers and basic requirements for ordering fuzzy numbers.

Definition 1. *(see [11]) A fuzzy number A is a fuzzy subset of the real line, $A : \mathbb{R} \to [0,1]$, where $A(x)$ denotes the value of the membership function of A in x, satisfying the following properties:*

A. Laurent et al. (Eds.): IPMU 2014, Part II, CCIS 443, pp. 254–263, 2014.

(i) A is normal (i. e. there exists $x_0 \in \mathbb{R}$ such that $A(x_0) = 1$);

(ii) A is fuzzy convex (i. e. $A(\lambda x_1 + (1 - \lambda) x_2) \geq \min(A(x_1), A(x_2))$, for every $x_1, x_2 \in \mathbb{R}$ and $\lambda \in [0,1]$);

(iii) A is upper semicontinuous in every $x_0 \in \mathbb{R}$ (i. e. $\forall \varepsilon > 0, \exists \delta > 0$ such that $A(x) - A(x_0) < \varepsilon$, whenever $|x - x_0| < \delta$);

(iv) $supp(A)$ is bounded, where $supp(A) = cl\{x \in \mathbb{R} : A(x) > 0\}$ and $cl(M)$ denotes the closure of the set M.

The α−cut, $\alpha \in (0,1]$, of a fuzzy number A is a crisp set defined as $A_\alpha = \{x \in \mathbb{R} : A(x) \geq \alpha\}$. Every α−cut, $\alpha \in [0,1]$, of a fuzzy number A is a closed interval

$$A_\alpha = [A_L(\alpha), A_U(\alpha)],$$

where

$$A_L(\alpha) = \inf\{x \in \mathbb{R} : A(x) \geq \alpha\}, \tag{1}$$
$$A_U(\alpha) = \sup\{x \in \mathbb{R} : A(x) \geq \alpha\} \tag{2}$$

for any $\alpha \in (0,1]$, with the convention $A_0 = [A_L(0), A_U(0)] := supp\, A$.

Fuzzy numbers with simple membership functions are preferred in practice. Often used are triangular fuzzy numbers, that is fuzzy numbers with α−cuts given by

$$\Delta_\alpha = [x_0 - \sigma + \sigma\alpha, x_0 + \beta - \beta\alpha] \tag{3}$$

where $x_0, \sigma, \beta \in \mathbb{R}$, $\sigma \geq 0, \beta \geq 0$. We denote by $\Delta = [x_0, \sigma, \beta]$ a such fuzzy number. We observe that

$$\Delta_0 = [x_0 - \sigma, x_0 + \beta] := supp\Delta. \tag{4}$$

Another α−cut representation of a triangular fuzzy number is given by

$$[t_1 + (t_2 - t_1)\alpha, t_3 - (t_3 - t_2)\alpha]. \tag{5}$$

In this case we denote $\Delta = (t_1, t_2, t_3)$. Comparing these representations we have

$$t_1 = x_0 - \sigma \tag{6}$$
$$t_2 = x_0 \tag{7}$$
$$t_3 = x_0 + \beta. \tag{8}$$

Throughout this paper we denote by $F(\mathbb{R})$ the set of all fuzzy numbers and by $F^\Delta(\mathbb{R})$ the set of all triangular fuzzy numbers.

Let $A, B \in F(\mathbb{R})$ and $\lambda \in \mathbb{R}$. The addition $A + B$ and the scalar multiplication $\lambda \cdot A$ are defined by

$$(A + B)_\alpha = A_\alpha + B_\alpha = [A_L(\alpha) + B_L(\alpha), A_U(\alpha) + B_U(\alpha)],$$

$$(\lambda \cdot A)_\alpha = \lambda \cdot A_\alpha = \begin{cases} [\lambda A_L(\alpha), \lambda A_U(\alpha)], \text{ for } \lambda \geq 0, \\ [\lambda A_U(\alpha), \lambda A_L(\alpha)], \text{ for } \lambda < 0. \end{cases}$$

The below basic requirements for the binary relation $\succeq$ on the set $\mathcal{S} \subseteq F(\mathbb{R})$ were introduced in [17], under the assumption that one of $A \succ B$ or $B \succ A$ or $A \sim B$ is true for every $(A, B) \in \mathcal{S}^2$ and with the notation $A \succeq B$ when $A \succ B$ or $A \sim B$.

$\mathbb{A}_1$) $A \succeq A$ for any $A \in \mathcal{S}$.

$\mathbb{A}_2$) For any $(A, B) \in \mathcal{S}^2$, from $A \succeq B$ and $B \succeq A$ results $A \sim B$.

$\mathbb{A}_3$) For any $(A, B, C) \in \mathcal{S}^3$, from $A \succeq B$ and $B \succeq C$ results $A \succeq C$.

$\mathbb{A}_4$) For any $(A, B) \in \mathcal{S}^2$, from $\inf supp(A) \geq sup\ supp(B)$ results $A \succeq B$.

$\mathbb{A}'_4$) For any $(A, B) \in \mathcal{S}^2$, from $\inf supp(A) > sup\ supp(B)$ results $A \succ B$.

$\mathbb{A}'_4$) For any $(A, B) \in \mathcal{S}^2$, from $\inf supp(A) > sup\ supp(B)$ results $A \succ B$.

$\mathbb{A}_5$) Let $\mathcal{S}' \subseteq F(\mathbb{R})$ and $A, B \in \mathcal{S} \cap \mathcal{S}'$. $A \succ B$ on $\mathcal{S}$ if and only if $A \succ B$ on $\mathcal{S}'$.

$\mathbb{A}_6$) Let $A, B, A+C$ and $B+C$ be elements of $\mathcal{S}$. If $A \succeq B$, then $A+C \succeq B+C$.

$\mathbb{A}'_6$) Let $A, B, A+C$ and $B+C$ be elements of $\mathcal{S}$. If $A \succ B$, then $A+C \succ B+C$.

$\mathbb{A}_7$) For any $(A, B) \in \mathcal{S}^2$ and $\lambda \in \mathbb{R}$ such that $\lambda \cdot A, \lambda \cdot B \in \mathcal{S}$, from $A \succeq B$ results $\lambda \cdot A \succeq \lambda \cdot B$ if $\lambda \geq 0$ and $\lambda \cdot A \preceq \lambda \cdot B$ if $\lambda \leq 0$.

Remark 1. In [3] a justification of the replacing of the corresponding property from [17] introduced as

"$\widehat{\mathbb{A}_7}$) Let $A, B, A \cdot C$ and $B \cdot C$ be elements of $\mathcal{S}$ and $C \geq 0$. If $A \succeq B$ then $A \cdot C \succeq B \cdot C$"

with $\mathbb{A}_7$) is given.

In most of the cases, ranking fuzzy numbers assumes to associate for each fuzzy number a real number and so fuzzy numbers are ranked through these real values. In this way, if $\mathcal{S}$ is a subset of $F(\mathbb{R})$ then a ranking index $P : \mathcal{S} \to \mathbb{R}$ generates on $\mathcal{S}$ a ranking in the following way:

  (*i*) $A \succ_P B$ if and only if $P(A) > P(B)$,

  (*ii*) $A \prec_P B$ if and only if $P(A) < P(B)$,

  (*iii*) $A \sim_P B$ if and only if $P(A) = P(B)$,

  (*iv*) $A \succ_P B$ or $A \sim_P B$ if and only if $P(A) \geq P(B)$,

  (*v*) $A \prec_P B$ or $A \sim_P B$ if and only if $P(A) \leq P(B)$.

When $A \succ_P B$ or $A \sim_P B$ we denote $A \succeq_P B$ and, when $A \prec_P B$ or $A \sim_P B$, we denote $A \preceq_P B$.

Remark 2. In some recent papers (see e.g. [1], [2], [15]) the following requirement is considered as an important reasonable property for a ranking index $P : \mathcal{S} \to \mathbb{R}$.

  $\mathbb{A}''_4$) For any $A \in \mathcal{S}$, $P(A) \in supp(A)$.

In [3], Theorem 9, under the assumption $\mathbb{R} \subset \mathcal{S}$, we prove that if $P : \mathcal{S} \to \mathbb{R}$ is a ranking index on $\mathcal{S}$ such that $\succeq_P$ satisfies $\mathbb{A}'_4$ (and simultaneously $\mathbb{A}_4$) and P is continuous on $\mathbb{R}$, then there exists a ranking index $R : \mathcal{S} \to \mathbb{R}$ which satisfies $\mathbb{A}''_4$ and $\succeq_R$ is equivalent of $\succeq_P$, that is, for any $A, B \in \mathcal{S}$, from $A \succeq_P B$ it results $A \succeq_R B$ and from $A \succ_P B$ it results $A \succ_R B$. In the present paper, $\mathbb{R} \subset \mathcal{S}$ and the continuity on $\mathbb{R}$ are assumed for every ranking index $P : \mathcal{S} \to \mathbb{R}$.

3   Characterization of Valuable Ranking Indices on Triangular Fuzzy Numbers

We recall, the expected value $EV(A)$, the ambiguity $Amb(A)$ and the value $Val(A)$ of a fuzzy number A are given by (see [9], [12], [13])

$$EV(A) = \frac{1}{2} \int_0^1 (A_L(\alpha) + A_U(\alpha))\, d\alpha,$$

$$Amb(A) = \int_0^1 \alpha(A_U(\alpha) - A_L(\alpha))d\alpha,$$

$$Val(A) = \int_0^1 \alpha(A_U(\alpha) + A_L(\alpha))d\alpha.$$

After some simple calculations we get

$$EV(\Delta) = x_0 - \frac{1}{4}\sigma + \frac{1}{4}\beta, \tag{9}$$

$$Amb(\Delta) = \frac{1}{6}\sigma + \frac{1}{6}\beta, \tag{10}$$

$$Val(\Delta) = x_0 - \frac{1}{6}\sigma + \frac{1}{6}\beta, \tag{11}$$

where $\Delta = [x_0, \sigma, \beta]$ is a triangular fuzzy number. The expected value, value and linear combinations of ambiguity and value are considered as ranking indices (see [9], [16]), therefore it is justified to study the following set

$$\Omega = \{R : F^\Delta(\mathbb{R}) \to \mathbb{R} \,|\, R\,([x_0, \sigma, \beta]) = ax_0 + b\sigma + c\beta\}.$$

On the other hand, we introduce the sets

$$Ran^\Delta = \{R : F^\Delta(\mathbb{R}) \to \mathbb{R} \,|\, R \text{ satisfies } \mathbb{A}_4''$$
$$\text{and } \succeq_R \text{ satisfies } \mathbb{A}_1, \mathbb{A}_2, \mathbb{A}_3, \mathbb{A}_5, \mathbb{A}_6, \mathbb{A}_6', \mathbb{A}_7\}$$

and

$$\widehat{Ran^\Delta} = \{R : F^\Delta(\mathbb{R}) \to \mathbb{R} \,|\, \succeq_R \text{ satisfies } \mathbb{A}_1, \mathbb{A}_2, \mathbb{A}_3, \mathbb{A}_4, \mathbb{A}_4', \mathbb{A}_5, \mathbb{A}_6, \mathbb{A}_6', \mathbb{A}_7 \}.$$

We prove that not exist ranking indices in Ran^Δ and in $\widehat{Ran^\Delta}$ (making abstraction of equivalent orders over $F^\Delta(\mathbb{R})$, see Remark 2), which do not belong to Ω. We mention here that the trapezoidal case was already treated in [3].

Therefore, it is very important to study in detail the set Ω and we do that by finding necessary and sufficient conditions for a, b, c such that $R \in \Omega$ can be used to rank effectively triangular fuzzy numbers. Because the order $\succeq_R, R \in \Omega$ is generated by a ranking index then $\mathbb{A}_1, \mathbb{A}_2, \mathbb{A}_3$ and $\mathbb{A}_5$ hold. Since one can easily prove that $R(\Delta + \Delta') = R(\Delta) + R(\Delta')$ for all $R \in \Omega$ and $\Delta, \Delta' \in F^\Delta(\mathbb{R})$, it results that $\mathbb{A}_6$ and $\mathbb{A}_6'$ hold too. Therefore, it remains to find necessary and sufficient conditions such that properties $\mathbb{A}_4, \mathbb{A}_4', \mathbb{A}_4''$ and $\mathbb{A}_7$ would hold.

Theorem 1. *Let* $R \in \Omega, R([x_0, \sigma, \beta]) = ax_0 + b\sigma + c\beta$. *The order* $\succeq_R$ *satisfies* A_4 *on* $F^{\triangle}(\mathbb{R})$ *if and only if*

$$a \geq c \geq 0 \tag{12}$$

and

$$a \geq -b \geq 0. \tag{13}$$

Proof. ($\Rightarrow$) We consider particular cases for $\triangle = [x_0, \sigma, \beta]$ and $\triangle' = [x_0', \sigma', \beta']$ such that *inf supp*$(\triangle) \geq$*sup supp*$(\triangle')$ is satisfied, until we obtain that (12) and (13) are satisfied. Note that since we have supposed that A_4 holds it results that $\triangle \succeq_R \triangle' \Leftrightarrow R(\triangle) \geq R(\triangle')$. Also note that for all the choices of $\triangle$ and $\triangle'$ we obtain proper triangular fuzzy numbers.

Firstly, let us consider the particular case when $x_0 > x_0' > 0$ and $\sigma = \beta = \sigma' = \beta' = 0$. Since $R(\triangle) \geq R(\triangle')$ implies $ax_0 \geq ax_0'$ and since $x_0 > x_0' > 0$, it is immediate that $a \geq 0$.

Now, let us consider the particular case when $x_0 = \sigma = \beta = x_0' = \beta' = 0$. Since $R(\triangle) \geq R(\triangle')$ implies $b\sigma' \leq 0$, it is immediate that $b \leq 0$.

Consider now the case when $x_0 = \sigma = x_0' = \sigma' = \beta' = 0$ and $\beta > 0$. Since $R(\triangle) \geq R(\triangle')$ implies $c\beta \geq 0$, we obtain $c \geq 0$.

Now, we consider the case when $x_0 = \beta' = 1$ and $\sigma = x_0' = \sigma' = \beta = 0$. It is immediate that from $R(\triangle) \geq R(\triangle')$ we get $a - c \geq 0$.

Finally, we consider the case when $x_0 = \sigma = 1, x_0' = \sigma' = \beta' = \beta = 0$. It is immediate that from $R(\triangle) \geq R(\triangle')$ we get $a + b \geq 0$.

Collecting the inequalities obtained in the particular cases from above, we obtain that (12) and (13) hold.

($\Leftarrow$) Let a, b, c be real numbers satisfying (12) and (13) and let $\triangle = [x_0, \sigma, \beta]$ and $\triangle' = [x_0', \sigma', \beta']$ denote two arbitrary triangular fuzzy numbers such that *inf supp*$(\triangle) \geq$*sup supp*$(\triangle')$. This immediately implies that $x_0 - x_0' \geq \sigma + \beta' \geq 0$. Noting that the hypothesis imply $-b\sigma' \geq 0$ and $c\beta \geq 0$, by direct calculations we get

$$\begin{aligned} R(\triangle) &- R(\triangle') \\ &= a(x_0 - x_0') + b(\sigma - \sigma') + c(\beta - \beta') \\ &\geq a(\sigma + \beta') + b\sigma - c\beta' \\ &= \sigma(a + b) + \beta'(a - c) \geq 0. \end{aligned}$$

This implies $\triangle \succeq_R \triangle'$ and the theorem is proved.

Theorem 2. *Let* $R \in \Omega, R([x_0, \sigma, \beta]) = ax_0 + b\sigma + c\beta$. *The order* $\succeq_R$ *satisfies* A_4' *if and only if*

$$a \geq c \geq 0 \tag{14}$$
$$a \geq -b \geq 0 \tag{15}$$

and

$$a > 0. \tag{16}$$

Proof. ($\Rightarrow$) In [3] we obtain that $\mathbb{A}'_4$ is satisfied by $\succeq_R$ implies that $\mathbb{A}_4$ is satisfied by $\succeq_R$. Since $\mathbb{A}_4$ holds, by the previous theorem it results that for the direct implication it suffices to prove that $a > 0$. For this purpose let us consider $\Delta = [x_0, \sigma, \beta]$ and $\Delta' = [x'_0, \sigma', \beta']$ in the particular case when $x_0 = 1$ and $\sigma = \beta = x'_0 = \sigma' = \beta' = 0$. Clearly we have *inf supp*$(\Delta) >$*sup supp*(Δ'), therefore $R(\Delta) > R(\Delta')$, that is $a > 0$.

($\Leftarrow$) Let a, b, c be real numbers such that (14)-(16) are satisfied and let $\Delta = [x_0, \sigma, \beta]$, $\Delta' = [x'_0, \sigma', \beta']$ denote two arbitrary triangular fuzzy numbers such that *inf supp*$(\Delta) >$*sup supp*(Δ'). Then it is easy to check that $x_0 - x'_0 > \sigma + \beta' \geq 0$. Then, from $a > 0$ we obtain

$$a(x_0 - x'_0) > a(\sigma + \beta')$$

and this implies

$$
\begin{aligned}
&R(\Delta) - R(\Delta') \\
&= a(x_0 - x'_0) + b(\sigma - \sigma') + c(\beta - \beta') \\
&> a(\sigma + \beta') + b(\sigma - \sigma') + c(\beta - \beta') \\
&\geq a(\sigma + \beta') + b\sigma - c\beta' \\
&= \sigma(a + b) + \beta'(a - c) \geq 0.
\end{aligned}
$$

We obtain $R(\Delta) > R(\Delta')$, therefore $\Delta \succ_R \Delta'$ and the proof is complete.

Theorem 3. *The ranking index $R \in \Omega, R([x_0, \sigma, \beta]) = ax_0 + b\sigma + c\beta$ satisfies $\mathbb{A}''_4$ if and only if*

$$a = 1 \tag{17}$$
$$b \in [-1, 0] \tag{18}$$

and

$$c \in [0, 1]. \tag{19}$$

Proof. ($\Rightarrow$) Let us notice that if $\mathbb{A}''_4$ holds then it is immediate that $\mathbb{A}'_4$ holds too. Therefore, comparing conditions (14)-(16) and (17)-(19), it follows that for the direct implication of the present theorem it suffices to prove $a = 1$. Since we have supposed that $\mathbb{A}''_4$ holds, it results that for any triangular fuzzy number $\Delta = [x_0, \sigma, \beta]$, we have

$$x_0 - \sigma \leq ax_0 + b\sigma + c\beta \leq x_0 + \beta. \tag{20}$$

Take $x_0 = 1$ and $\sigma = \beta = 0$. Replacing in (20) we get $1 \leq a \leq 1$ and thus we obtain (17).

($\Leftarrow$) Let a, b, c be real numbers such that (17)-(19) are satisfied and let $\Delta = [x_0, \sigma, \beta]$ denote a triangular fuzzy number. Conditions (17)-(19) imply

$$
\begin{aligned}
R(\Delta) &= x_0 + b\sigma + c\beta \\
&\leq x_0 + c\beta \leq x_0 + \beta
\end{aligned}
$$

and

$$R(\Delta) = x_0 + b\sigma + c\beta$$
$$\geq x_0 + b\sigma \geq x_0 - \sigma.$$

From the two inequalities from above results that $R(\Delta) \in supp\ \Delta$ and the theorem is proved.

Theorem 4. Let $R \in \Omega, R([x_0, \sigma, \beta]) = ax_0 + b\sigma + c\beta$. The order $\succeq_R$ satisfies $\mathbb{A}_7$ if and only if

$$b + c = 0. \tag{21}$$

Proof. ($\Rightarrow$) Let us choose $\Delta = [0, 1, 1]$ and $O = [0, 0, 0]$. If $R(\Delta) \geq R(O)$ then the hypothesis imply $R(-\Delta) \leq R(O)$. Since $R(O) = 0$ and $-\Delta = \Delta$, it immediately follows that $R(\Delta) = 0$ that is $b + c = 0$. If $R(\Delta) \leq R(O)$ then the reasoning is similar therefore we omit the details.

($\Leftarrow$) Let us consider the reals a, b, c such that $b + c = 0$. If $\Delta = [x_0, \sigma, \beta]$ then $R(\Delta) = ax_0 - c\sigma + c\beta$. To prove that $\mathbb{A}_7$ holds it suffices to prove that the operator R is scale invariant. Now, if $\lambda \geq 0$ then one can easily prove that $R(\lambda \cdot \Delta) = \lambda R(\Delta)$. Because

$$R(-\Delta) = -ax_0 - c\beta + c\sigma = -(ax_0 - c\sigma + c\beta) = -R(\Delta),$$

we immediately obtain $R(\lambda \cdot \Delta) = \lambda R(\Delta)$ for every $\Delta \in F^\Delta(\mathbb{R})$ and $\lambda \in \mathbb{R}$.

Example 1. Delgado, Vila and Voxman [9] proposed the ranking index given by

$$ri(\lambda, \delta)(A) = \lambda Val(A) + \delta Amb(A),$$

where $\lambda \in [0, 1]$ and $\delta \in [-1, 1]$ are given such that $|\delta| \ll \lambda$, δ representing the decision-maker's attitude against the uncertainty. If $\Delta = [x_0, \sigma, \beta]$ is a triangular fuzzy number then we get

$$ri(\lambda, \delta)(\Delta) = \lambda x_0 + \frac{\delta - \lambda}{6}\sigma + \frac{\lambda + \delta}{6}\beta$$

and it is immediate that $ri(\lambda, \delta) \in \Omega$ for all choices of λ and δ. From Theorems 1 and 2 we obtain $\succeq_{ri(\lambda, \delta)}$ satisfies $\mathbb{A}_4$ and $\mathbb{A}'_4$ for every λ and δ. From Theorems 3 and 4 we obtain $\succeq_{ri(\lambda, \delta)}$ satisfies $\mathbb{A}''_4$ if and only if $\lambda = 1$ and $\mathbb{A}_7$ if and only if $\delta = 0$.

Example 2. A method of ranking fuzzy numbers with integral value as index was proposed [14]. It becomes

$$I^k_\Delta([x_0, \sigma, \beta]) = x_0 + \frac{k - 1}{2}\sigma + \frac{k}{2}\beta,$$

where $k \in [0, 1]$ represents the degree of optimism. The order generated by I^k_Δ satisfies $\mathbb{A}_7$ if and only if $k = \frac{1}{2}$. If $k \neq \frac{1}{2}$ then the effectiveness of the proposed ranking method is at least debatable.

The following important theorems, immediate consequences of the above theoretical results, characterize the element of Ω which are in Ran^Δ or $\widehat{Ran^\Delta}$.

Corollary 1. *Let* $R \in \Omega, R([x_0, \sigma, \beta]) = ax_0 + b\sigma + c\beta$. *Then* $R \in Ran^\Delta$ *if and only if*

$$a = 1 \tag{22}$$
$$b + c = 0 \tag{23}$$

and

$$c \in [0, 1]. \tag{24}$$

Proof. It is immediate taking into account that (17)-(19) and (21) have to be satisfied simultaneously.

Corollary 2. *Let* $R \in \Omega, R([x_0, \sigma, \beta]) = ax_0 + b\sigma + c\beta$. *Then* $R \in \widehat{Ran^\Delta}$ *if and only if* $a > 0, b = -c$ *and* $a \geq c \geq 0$.

Proof. It is immediate taking into account that (14)-(16) and (21) have to be satisfied simultaneously.

Example 3. $ri(\lambda, \delta) \in \widehat{Ran^\Delta}$ and hence $\succeq_{ri(\lambda,\delta)}$ satisfies $\mathbb{A}_1, \mathbb{A}_2, \mathbb{A}_3, \mathbb{A}_4, \mathbb{A}_4', \mathbb{A}_5,$ $\mathbb{A}_6, \mathbb{A}_6', \mathbb{A}_7$ if and only if $\delta = 0$. Moreover, if $\lambda < 1$ then $ri(\lambda, \delta) \in \widehat{Ran^\Delta} \setminus Ran^\Delta$ and if $\lambda = 1$ then $ri(1, 0) = Val \in Ran^\Delta$.

According to Corollary 1 it is easy to deduce that some already introduced ranking indices are elements of Ran^Δ and implicitly they satisfy $\mathbb{A}_1, \mathbb{A}_2, \mathbb{A}_3,$ $\mathbb{A}_4, \mathbb{A}_4', \mathbb{A}_5, \mathbb{A}_6, \mathbb{A}_6', \mathbb{A}_7$.

Example 4. Let us consider the function $EV : F^\Delta(\mathbb{R}) \to \mathbb{R}$ which for any triangular fuzzy number $\Delta = [x_0, \sigma, \beta]$ associates its expected value that is $EV(T) = x_0 - \frac{1}{4}\sigma + \frac{1}{4}\beta$. It is immediate that $EV \in Ran^\Delta$.

Example 5. In [10] a ranking procedure is proposed via so called valuation functions. The authors consider a strictly monotonous function (valuation) $f : [0, 1] \to [0, \infty)$ and a ranking index which on triangular fuzzy numbers becomes $R : F^\Delta(\mathbb{R}) \to \mathbb{R}$,

$$R([x_0, \sigma, \beta]) = (2 - \omega) x_0 - \frac{1 - \omega}{2}\sigma + \frac{1 - \omega}{2}\beta,$$

where

$$\omega = \frac{\int_0^1 \alpha f(\alpha) d\alpha}{\int_0^1 f(\alpha) d\alpha}.$$

Since $0 < \omega < 1$ for any valuation f, by Theorems 1, 2 and 4 respectively, we easily observe that R generates an order which satisfies $\mathbb{A}_4, \mathbb{A}_4'$ and $\mathbb{A}_7$ and hence $R \in \widehat{Ran^\Delta}$ for any valuation f. On the other hand by Theorem 3 we observe that $R \notin Ran^\Delta$. It is easily seen that $R_* \in Ran^\Delta$ where $R_* = \frac{1}{2-\omega} R$ and $\succeq_{R_*}$ is equivalent with $\succeq_R$. We also observe that by taking $f(\alpha) = \alpha, \alpha \in [0, 1]$ we obtain $R_* = ri(1, 0) = Val$.

Having in mind the results from above, it would be important to know whether there exists any other ranking index $R \in Ran^{\triangle}$ which does not belong to Ω. The answer to this question is negative and hence we can give now the main result of this paper. The idea of the proof is the same as in the trapezoidal case [3].

Theorem 5. *Let us consider a ranking index* $R : F^{\triangle}(\mathbb{R}) \to \mathbb{R}$. *Then* $R \in Ran^{\triangle}$ *if and only if there exists* $c \in [0,1]$ *such that for some* $\triangle \in F^{\triangle}(\mathbb{R}), \triangle = [x_0, \sigma, \beta]$, *we have*

$$R(\triangle) = x_0 - c\sigma + c\beta. \tag{25}$$

Proof. Taking into account Corollary 1, it is easily seen that we can obtain the desired conclusion by proving that $R \in Ran^{\triangle}$ implies $R \in \Omega$. Firstly, let us observe that from Theorem 22 in [3] it results that R is linear on $F^{\triangle}(\mathbb{R})$. Let $\triangle \in F^{\triangle}(\mathbb{R})$, $\triangle = (t_1, t_2, t_3)$ where we used the form (5) because it is more suitable for this proof. Now, let us consider the triangular fuzzy numbers

$$v_1 = (0, 0, 1),$$
$$v_2 = (0, 1, 1),$$
$$v_3 = (1, 1, 1).$$

Having in mind the addition and the scalar multiplication of fuzzy numbers we get that $\triangle = t_1 v_3 + (t_2 - t_1)v_2 + (t_3 - t_2)v_1$. The linearity of R implies $R(\triangle) = t_1 R(v_3) + (t_2 - t_1)R(v_2) + (t_3 - t_2)R(v_1)$. Returning now to the other parametric representation of $\triangle$, that is $\triangle = [x_0, \sigma, \beta]$ and taking into account (6)-(8), we obtain $R(\triangle) = x_0 R(v_3) + \sigma(R(v_2) - R(v_3)) + \beta R(v_1)$. Clearly, this last relation implies that $R \in \Omega$ and the proof is complete.

4   Conclusion

In applied sciences most often triangular or trapezoidal fuzzy numbers are used. This is an important reason why we should investigate the theoretical properties of these classes especially from application oriented purposes. In this paper, we characterized the class of ranking indices over the set of triangular fuzzy numbers which satisfy the reasonable criteria of Wang and Kerre [17]. This study continues our interest on ranking of fuzzy numbers which started with paper [3], where among others, the trapezoidal case was investigated. Recently, fuzzy numbers of higher dimension have been proposed. In the paper [8] a special class of piecewise linear fuzzy numbers was considered. These fuzzy numbers depend on 6 parameters and recently a more general approach was proposed considering piecewise linear fuzzy numbers depending on n parameters. Therefore it would be interesting to extend the present study for such classes of fuzzy numbers.

Acknowledgments. The work of authors was supported by a grant of the Romanian National Authority for Scientific Research, CNCS–UEFISCDI, project number PN-II-ID-PCE-2011-3-0861.

References

1. Abbasbandy, S., Hajjari, T.: A new approach for ranking of trapezoidal fuzzy numbers. Computers and Mathematics with Applications 57, 413–419 (2009)
2. Asady, B., Zendehman, A.: Ranking fuzzy numbers by distance minimization. Applied Mathematical Modelling 31, 2589–2598 (2007)
3. Ban, A.I., Coroianu, L.: Simplifying the search for effective ranking of fuzzy numbers. IEEE Transactions on Fuzzy Systems (in press), doi:10.1109/TFUZZ.2014.2312204
4. Chen, S.: Ranking fuzzy numbers with maximizing set and minimizing set. Fuzzy Sets and Systems 17, 113–129 (1985)
5. Chen, L.H., Lu, H.W.: An approximate approach for ranking fuzzy numbers based on left and right dominance. Computers and Mathematics with Applications 41, 1589–1602 (2001)
6. Choobineh, F., Li, H.: An index for ordering fuzzy numbers. Fuzzy Sets and Systems 54, 287–294 (1993)
7. Chu, T.-C., Tsao, C.-T.: Ranking fuzzy numbers with an area between the centroid point and original point. Computers and Mathematics with Applications 43, 111–117 (2002)
8. Coroianu, L., Gagolewski, M., Grzegorzewski, P.: Nearest piecewise linear approximation of fuzzy numbers. Fuzzy Sets and Systems 233, 26–51 (2013)
9. Delgado, M., Vila, M.A., Voxman, W.: On a canonical representation of a fuzzy number. Fuzzy Sets and Systems 93, 125–135 (1998)
10. Detynicki, M., Yagger, R.R.: Ranking fuzzy numbers using $\alpha-$weighted valuations. International Journal of Uncertainty, Fuzziness and Knowledge-Based Systems 8, 573–592 (2001)
11. Dubois, D., Prade, H.: Operations on fuzzy numbers. International Journal of Systems Science 9, 613–626 (1978)
12. Dubois, D., Prade, H.: The mean value of a fuzzy number. Fuzzy Sets and Systems 24, 279–300 (1987)
13. Heilpern, S.: The expected value of a fuzzy number. Fuzzy Sets and Systems 47, 81–86 (1992)
14. Liou, T.-S., Wang, M.-J.: Ranking fuzzy numbers with integral value. Fuzzy Sets and Systems 50, 247–255 (1992)
15. Saeidifar, A.: Application of weighting functions to the ranking of fuzzy numbers. Computers and Mathematics with Applications 62, 2246–2258 (2011)
16. Yager, R.R.: A procedure for ordering fuzzy subsets of the unit interval. Information Sciences 24, 143–161 (1981)
17. Wang, X., Kerre, E.E.: Reasonable properties for the ordering of fuzzy quantities (I). Fuzzy Sets and Systems 118, 375–385 (2001)

New Pareto Approach for Ranking Triangular Fuzzy Numbers

Oumayma Bahri[1,2], Nahla Ben Amor[1], and Talbi El-Ghazali[2]

[1] LARODEC Laboratory, ISG Tunis, Le Bardo, Tunisie
oumayma.b@gmail.com, nahla.benamor@gmx.fr
[2] INRIA Laboratory, LIFL/CNRS, Villeneuve d'Ascq, Lille, France
el-ghazali.talbi@lifl.fr

Abstract. Ranking fuzzy numbers is an important aspect in dealing with fuzzy optimization problems in many areas. Although so far, many fuzzy ranking methods have been discussed. This paper proposes a new Pareto approach over triangular fuzzy numbers. The approach is composed of two dominance stages. In the first stage, mono-objective dominance relations are introduced and tested with some examples. In the second stage, a Pareto dominance is defined for multi-objective optimization and then applied to solve a vehicle routing problem (VRP).

Keywords: Fuzzy ranking, Pareto approach, Triangular fuzzy numbers, Mono-objective dominance, Multi-objective optimization, VRP.

1   Introduction

Real-world optimization problems are often subject to some uncertainty, which must be taken into account. This uncertainty can takes different forms of representation, such as fuzzy numbers which are widely used in many applications, since they offer a suitable and natural way for expressing uncertain values. Nevertheless, in practical applications, the comparison of fuzzy numbers becomes an important and necessary procedure for decision makers. Therefore, several methods for ranking fuzzy numbers have been suggested in literature. A review and comparison of existing methods can be found in [8].

In addition, real optimization problems usually involve the simultaneous satisfaction of more than one objective function and these objectives are generally conflicting. In such problems, commonly known as multi-objective, a set of best solutions, called Pareto optimal, has to be determined. Over years, a significant number of resolution methods and techniques have been developed for multi-objective optimization [16], while only few studies have been focused on extending multi-objective concepts to the uncertain optimization context [11]. This paper addresses multi-objective optimization problems under uncertainty, in particularly, those containing fuzzy data represented by triangular fuzzy numbers. To this end, a new Pareto approach for ranking the generated triangular fuzzy solutions of such problems is proposed. The proposed approach is based firstly

A. Laurent et al. (Eds.): IPMU 2014, Part II, CCIS 443, pp. 264–273, 2014.

on the definition of new mono-objective dominance for ranking the triangular solutions of each objective function and secondly on the use of this mono-objective
dominance for identifying the Pareto optimality in multi-objective case.

The remainder of this paper is organized as follows. Section 2 recalls the main
concepts on which the proposed approach is based. Section 3 presents the new
Pareto approach between triangular fuzzy numbers in both mono-objective and
multi-objective cases. The proposed approach is first examined with numerical examples for mono-objective case and then illustrated on a multi-objective
vehicle routing problem with fuzzy demands, in Section 4.

2   Background

This section first defines the fundamental notions of triangular fuzzy numbers
and then provides a brief background on multi-objective optimization.

2.1   Triangular Fuzzy Numbers

Among the different shapes of fuzzy number, triangular fuzzy number (TFN) is
the most popular one, frequently used in various real world applications [21]. A
triangular fuzzy number A can be defined as a normal fuzzy set represented by
a triplet $[\underline{a}, \widehat{a}, \overline{a}]$, where $[\underline{a}, \overline{a}]$ is the interval of possible values called its bounded
support and $\widehat{a}$ denotes its kernel value (the most plausible value). This represen-
tation is characterized by a membership function $\mu_A(x)$ which assigns a value
within $[0, 1]$ to each element in A. Its mathematical definition is given by:

$$\mu_A(x) = \begin{cases} \frac{x - \underline{a}}{\widehat{a} - \underline{a}}, & \underline{a} \le x \le \widehat{a} \\ 1, & x = \widehat{a} \\ \frac{\overline{a} - x}{\overline{a} - \widehat{a}}, & \widehat{a} \le x \le \overline{a} \\ 0, & \text{otherwise.} \end{cases} \tag{1}$$

However, in practical use of triangular fuzzy numbers, a ranking procedure
needs to be applied for decision-making. In other words, one triangular fuzzy
number needs to be evaluated and compared with the others in order to make
a choice among them. Indeed, all possible topological relations between two
triangular fuzzy numbers $A = [\underline{a}, \widehat{a}, \overline{a}]$ and $B = [\underline{b}, \widehat{b}, \overline{b}]$ may be covered by
only four different situations, which are: Fuzzy disjoint, Fuzzy weak overlap-
ping, Fuzzy overlapping and Fuzzy inclusion [2]. These situations, illustrated
in Fig. 1 should be taken into account for ranking triangular fuzzy numbers.
In the literature, several different methods of fuzzy rankings have been proposed [1,3,4,5,7,9,10,19,20]. However, almost each ranking method may contain
some shortcomings, such as inconsistency with human intuition, requirement of
complicated calculations, difficulty of interpretation or indiscrimination in some
situations.

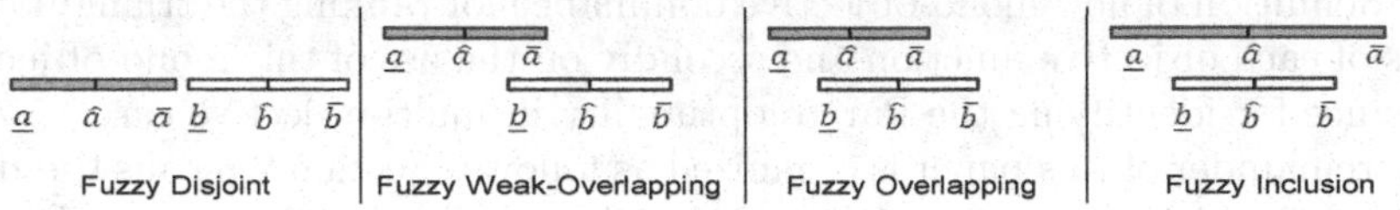

Fig. 1. Possible topological situations for two TFNs

2.2 Multi-objective Optimization

Multi-objective optimization aims to optimize several conflicting objective functions, simultaneously. Contrary to mono-objective optimization, multi-objective optimization does not restrict to find a unique single solution, but a set of best solutions known as *Pareto optimal set* or *Pareto front* [16]. Therefore, obtaining the *Pareto front* is the the main goal of solving a given multi-objective problem.

A classical multi-objective optimization problem (MOP), defined in the sense of minimization of all the objectives, is often formalized as follows:

$$Min\ F(x) = (f_1(x), \ldots, f_n(x))\ subject\ to\ x \in S. \tag{2}$$

where x is a feasible solution from the decision space S and $F(x)$ is the vector of n independent objectives to be minimized. This vector can be defined as a cost function in the objective space by assigning an objective vector $\overrightarrow{y} = (y_1, \ldots, y_n)$ that represents the quality of each solution x.

An objective vector $\overrightarrow{u} = (u_1, \ldots, u_n)$ is said to *Pareto dominate* another objective vector $\overrightarrow{u}' = (u'_1, \ldots, u'_n)$ (denoted by $\overrightarrow{u} \prec_p \overrightarrow{u}'$) if and only if no component of $\overrightarrow{u}'$ is smaller than the corresponding component of $\overrightarrow{u}$ and at least one component of $\overrightarrow{u}$ is strictly smaller:

$$\forall i \in 1, \ldots, n : u_i \le u'_i \wedge \exists i \in 1, \ldots, n : u_i < u'_i. \tag{3}$$

A solution $x^* \in S$ is said *Pareto optimal* (also known as non-dominated or non-inferior) if for every $x \in S$, $F(x)$ does not dominate $F(x^*)$, that is, $F(x) \nprec_p F(x^*)$. Thus, the *Pareto optimal set* P^* is defined as:

$$P^* = \{x \in X / \exists x' \in X, F(x') \nprec_p F(x)\}. \tag{4}$$

The image of this *Pareto optimal set* P^* in the objective space is the *Pareto front* PF^* defined as:

$$PF^* = \{F(x), x \in P^*\}. \tag{5}$$

The above concepts are meant especially for deterministic multi-objective optimization where the solutions are exact values. Therefore, in the case of uncertain multi-objective optimization, where the solutions become often uncertain, these concepts will be addressed. Yet, most of proposed approaches for handling multi-objective problems under uncertainty have been limited to treat such problems in mono-objective context by considering the set of objectives as if there is only one [13] or even in multi-objective context but with ignoration of uncertainty propagation to the considered objectives using statistical properties like expectation values [17]. Only few works have been focused on treating the problem as-is without erasing any of its characteristics [11].

The issue we are interested in this paper is how handling multi-objective problems with fuzzy data, consequently with triangular-valued objectives. To this end, we introduce a new Pareto approach, refined from our previous work [14], in order to rank the generated triangular-valued solutions of such problems.

3   New Pareto Approach for Ranking Triangular Fuzzy Numbers (TFNs)

In this section, we first present new mono-objective dominance relations between two TFNs. Then, based on these mono-objective dominance, we define a new Pareto dominance between vectors of TFNs, for multi-objective case. Note that, the minimization sense is considered in all our definitions.

3.1   Mono-objective Dominance Relations

In mono-objective case, three dominance relations over triangular fuzzy numbers are defined: Total dominance ($\prec_t$), Partial strong-dominance ($\prec_s$) and Partial weak-dominance ($\prec_w$).

Definition 1 *Total dominance*
   Let $y = [\underline{y}, \widehat{y}, \overline{y}] \subseteq \mathbb{R}$ and $y' = [\underline{y}', \widehat{y}', \overline{y}'] \subseteq \mathbb{R}$ be two triangular fuzzy numbers. y dominates y' totally or certainly (denoted by $y \prec_t y'$) if: $\overline{y} < \underline{y}'$.
This dominance relation represents the fuzzy disjoint situation between two triangular fuzzy numbers and it imposes that the upper bound of y is strictly inferior than the lower bound of y' as shown by case (1) in Fig. 2.

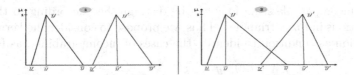

Fig. 2. Total dominance and Partial strong-dominance

Definition 2 *Partial strong-dominance*
   Let $y = [\underline{y}, \widehat{y}, \overline{y}] \subseteq \mathbb{R}$ and $y' = [\underline{y}', \widehat{y}', \overline{y}'] \subseteq \mathbb{R}$ be two triangular fuzzy numbers. y strong dominates y' partially or uncertainly (denoted by $y \prec_s y'$) if:

$$(\overline{y} \geq \underline{y}') \wedge (\widehat{y} \leq \underline{y}') \wedge (\overline{y} \leq \widehat{y}').$$

This dominance relation appears when there is a fuzzy weak-overlapping between both triangles and it imposes that firstly there is at most one intersection between them and secondly this intersection should not exceed the interval of their kernel values $[\widehat{y}, \widehat{y}']$, as shown by case (2) in Fig.2.

Definition 3 *Partial weak-dominance*

Let $y = [\underline{y}, \widehat{y}, \overline{y}] \subseteq \mathbb{R}$ and $y' = [\underline{y}', \widehat{y}', \overline{y}'] \subseteq \mathbb{R}$ be two triangular fuzzy numbers. y weak dominates y' partially or uncertainly (denoted by $y \prec_w y'$) if:

1. *Fuzzy overlapping*

$$[(\underline{y} < \underline{y}') \wedge (\overline{y} < \overline{y}')]\wedge$$
$$[((\widehat{y} \leq \underline{y}') \wedge (\overline{y} > \widehat{y}')) \vee ((\widehat{y} > \underline{y}') \wedge (\overline{y} \leq \widehat{y}')) \vee ((\widehat{y} > \underline{y}') \wedge (\overline{y} > \widehat{y}'))].$$

2. *Fuzzy Inclusion*

$$(\underline{y} < \underline{y}') \wedge (\overline{y} \geq \overline{y}').$$

In this dominance relation, the two situations of fuzzy overlapping and inclusion may occur. Fig. 3 presents four examples of possible cases, where in (3) and (5) y and y' are overlapped, while, in (4) and (6) y' is included in y.

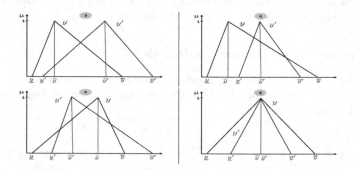

Fig. 3. Partial weak-dominance

Yet, the partial weak-dominance relation cannot discriminate all possible cases and leads often to some incomparable situations as for cases (5) and (6) in Fig. 3. These incomparable situations can be distinguished according to the kernel value positions in fuzzy triangles. Thus, we propose to consider the kernel values configuration as condition to identify the cases of incomparability, as follows:

$$\widehat{y} - \widehat{y}' = \begin{cases} < 0, & y \prec_w y' \\ \geq 0, & y \text{ and } y' \text{ can be incomparable.} \end{cases}$$

Subsequently, to handle the identified incomparable situations (with kernel condition $\widehat{y} - \widehat{y}' \geq 0$), we introduce another comparison criterion, which consists in comparing the discard between both fuzzy triangles as follows:

$$y \prec_w y' \Leftrightarrow (\underline{y}' - \underline{y}) \leq (\overline{y}' - \overline{y})$$

Similarly, it is obvious that: $y' \prec_w y \Leftrightarrow (\underline{y}' - \underline{y}) > (\overline{y}' - \overline{y})$.

It is easy to check that in the mono-objective case, we obtain a total pre-order between two triangular fuzzy numbers, contrarily to the multi-objective case, where the situation is more complex and it is common to have some cases of indifference.

3.2   Pareto Dominance Relations

In multi-objective case, we propose to use the mono-objective dominance relations, defined previously, in order to rank separately the triangular fuzzy solutions of each objective function. Then, depending to the types of mono-objective dominance founded for all the objectives, we define the Pareto dominance between the vectors of triangular fuzzy solutions. In this context, two Pareto dominance relations: *Strong Pareto dominance* ($\prec_{SP}$) and *Weak Pareto dominance* ($\prec_{WP}$) are introduced.

Definition 4 *Strong Pareto dominance*
Let $\vec{y}$ and $\vec{y}'$ be two vectors of triangular fuzzy numbers. $\vec{y}$ strong Pareto dominates $\vec{y}'$ (denoted by $\vec{y} \prec_{SP} \vec{y}'$) if:

(a) $\forall i \in 1,\dots,n : y_i \prec_t y_i' \vee y_i \prec_s y_i'$.

(b) $\exists i \in 1,\dots,n : y_i \prec_t y_i' \wedge \forall j \neq i : y_j \prec_s y_j'$

(c) $\exists i \in 1,\dots,n : (y_i \prec_t y_i' \vee y_j \prec_s y_j') \wedge \forall j \neq i : y_j \prec_w y_j'$.

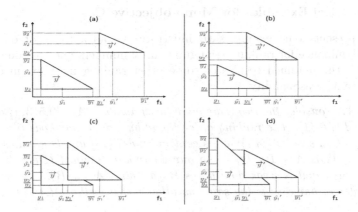

Fig. 4. Strong Pareto dominance

The strong Pareto dominance holds if either y_i total dominates or partial strong dominates y_i' in all the objectives (Fig.4-(a): $y_1 \prec_t y_1'$ and $y_2 \prec_t y_2'$), either y_i total dominates y_i' in one objective and partial strong dominates it in another (Fig.4-(b): $y_1 \prec_s y_1'$ and $y_2 \prec_t y_2'$), or at least y_i total or partial strong dominates y_i' in one objective and weak dominates it in another (Fig.4-(c),(d): $y_1 \prec_s y_1'$ and $y_2 \prec_w y_2'$).

Definition 5 *Weak Pareto dominance*
Let $\vec{y}$ and $\vec{y}'$ be two vectors of triangular fuzzy numbers. $\vec{y}$ weak Pareto dominates $\vec{y}'$ (denoted by $\vec{y} \prec_{WP} \vec{y}'$) if: $\forall i \in 1,\dots,n : y_i \prec_w y_i'$.

The weak Pareto dominance holds if y_i weak dominates y_i' in all the objectives (Fig.5-(e)). Yet, a case of indifference (defined below) can occur if there is a weak dominance with inclusion type in all the objectives (Fig.5-(f)).

Definition 6 *Case of indifference*

Two vectors of triangular fuzzy numbers are indifferent or incomparable (denoted by $\overrightarrow{y} \| \overrightarrow{y}'$) if: $\forall i \in 1, \ldots, n : y_i \subseteq y_i'$.

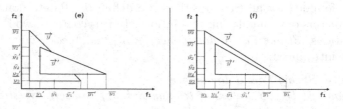

Fig. 5. Weak Pareto dominance and Case of indifference

4 Results Analysis

4.1 Numerical Examples for Mono-objective Case

Here, we present some examples to illustrate the advantages of our mono-objective dominance for ranking triangular fuzzy numbers and we compare our results with the obtained results of some other ranking methods in order to demonstrate its reasonability.

Example 1. *Consider the two triangular fuzzy numbers $A = [0.5, 3, 7]$ and $B = [1, 6, 10]$ in Fig.6-(1). The ranking order found by most of methods like Cheng's distance [4], Chu's index [5], Wang's centroid index [19] and kaufman's left and right scores [9], is $A \prec B$. By using our dominance method (Definition 3), it is easy to check that A weak dominates B partially ($A \prec_w B$). Therefore, the ranking order in our case is the same as other tested methods ($A \prec B$).*

Example 2. *Consider the two triangular fuzzy numbers $A = [0.1, 0.6, 0.8]$ and $B = [0.2, 0.5, 0.9]$ from [20] (see Fig.6-(2)). By using some ranking methods such as Yao's signed distance [20], Chu's index [5] and Abbas's sign distance [1], the ranking order is $A \approx B$. This is the shortcoming of previous methods that rank two different fuzzy numbers equally. However, by applying our dominance method, we observe at the first step, that the discrimination between A and B is not possible using Definition 3, since the kernel condition gives $0.6 - 0.5 \geq 0$. At this level, we use the discard criterion ($0.2 - 0.1 = 0.9 - 0.8$) which leads to conclude that A partial weak dominates B, and consequently $A \prec B$. The same result is obtained by [10].*

Example 3. *Consider the two triangular fuzzy numbers $A = [3, 6, 9]$ and $B = [5, 6, 7]$ from [1] (see Fig. 6-(3)). Almost the majority of ranking methods such as [1,5,19,20] failed to discriminate two fuzzy numbers having the same symmetrical spread, as for this example $A \approx B$, whereas Ezzati et al. [7] prefer the ranking order $B \prec A$ and consider this choice as reasonable result, since it agrees with*

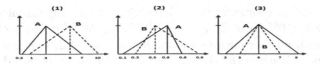

Fig. 6. Three examples of triangular fuzzy numbers A and B

human intuition. By using our dominance method, we conclude that A partial weak dominates B, since the discard criterion gives: $5 - 3 > 7 - 9$. Thus, we obtain the same rational result $B \prec A$.

From these examples, we may conclude that our simple dominance method can effectively rank triangular fuzzy numbers and produces reasonable and intuitive results to the well-defined problems of indiscrimination, that have failed to be ranked by some previous ranking methods.

4.2 Application to Multi-objective VRP with Fuzzy Demands

The Vehicle Routing Problem (VRP) is a difficult and very well-known combinatorial optimization problem which has a large number of real-life applications [18]. The classical VRP consists of finding the optimal routes used by a set of identical vehicles, stationed at a central depot, to serve a given set of customers geographically distributed, with known demands. In this paper, we are interested in a particular variant of VRP, the so-called Multi-objective VRP with Time Windows and Uncertain Demands (MO-VRPTW-UD), in which a time window is imposed on the visit of each customer and the customer's demands are supposed to be uncertain, expressed in our case by triangular fuzzy numbers $dm = [\underline{dm}, \widehat{dm}, \overline{dm}]$. The objective functions to be minimized are respectively, the total traveled distance and total tardiness time. These two objectives will be disrupted in our case by the used fuzzy form of uncertainty and so obtained as triangular results. Therefore, in order to handle the generated triangular results, we need to use our Pareto approach over triangular fuzzy numbers.

To solve the MO-VRPTW-UD, two well-known algorithms: the Strength Pareto Evolutionary Algorithm SPEA2 [22] and the Non-dominated Sorting Genetic Algorithm NSGAII [6], are adapted to our uncertain context by integrating the proposed Pareto approach. These algorithms are implemented with the version 1.3-beta of ParadisEO framework under Linux [12]. Subsequently, to validate our approach, we choose to use the Solomon's benchmark, considered as a basic reference for the evaluation of several VRP methods [15]. More precisely, six different Solomon's instances are used in our experimentation, namely, C101, C201, R101, R201, RC101 and RC201. Yet, in these instances, all the input values are exact and so the uncertainty of customer demands is not taken into account. At this level, we propose to generate for each instance the triangular fuzzy version of crisp demands in the following manner. Firstly, the kernel value $(\widehat{dm})$ for each triangular fuzzy demand dm is kept the same as the current crisp demand dm_i of the instance. Then, the lower $(\underline{dm})$ and upper

($\overline{dm}$) bounds of this triangular fuzzy demand are uniformly sampled at random in the intervals [50%dm, 95%dm] and [105%dm, 150%dm], respectively. This fuzzy generation manner ensures the quality and reliability of generated fuzzy numbers. Finally, each of the six sampled fuzzy instances is tested on the both algorithms, executed 30 times. Therefore, since 30 runs have been performed on each algorithm, we obtained for each instance, 30 sets of optimal solutions that represent the Pareto fronts of our problem. Each solution shows the minimal traveled distance and tardiness time, which are represented by triangular fuzzy numbers. Examples of Pareto fronts obtained for one execution of the instance C101 using each algorithm, are shown in Fig. 7, where the illustrated Pareto fronts are composed by a set of triangles, such that each triangle represents one Pareto optimal solution. For instance, the bold triangular represents an optimal solution with minimal distance (the green side) equal to [2413, 2515, 2623] and tardiness time (the red side) equal to [284312, 295280, 315322].

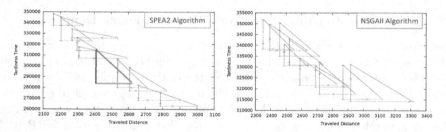

Fig. 7. Examples of Pareto fronts for C101 instance

5 Conclusion

In this paper, a new Pareto approach for handling multi-objective problems with triangular fuzzy data is introduced. The approach is based on the definition of new mono-objective dominance relations, that have been validated with some examples. As practical application, a bi-objective vehicle routing problem with uncertain demands is solved. For a future work, we intend to extend the proposed approach for ranking different fuzzy numbers (i.e. trapezoidal).

References

1. Abbasbandy, S., Asady, B.: Ranking of fuzzy numbers by sign distance. Information Sciences 176(16), 2405–2416 (2006)
2. Boukezzoula, R., Galichet, S., Foulloy, L.: MIN and MAX operators for fuzzy intervals and their potential use in aggregation operators. IEEE Transactions on Fuzzy Systems 15(6), 1135–1144 (2007)
3. Chen, S.: Ranking fuzzy numbers with maximizing set and minimizing set. Fuzzy Sets and Systems 17(3), 113–129 (1985)

4. Cheng, C.: A new approach for ranking fuzzy numbers by distance method. Fuzzy Sets and Systems 95(3), 307–317 (1998)
5. Chu, T., Tsao, C.: Ranking fuzzy numbers with an area between the centroid point and original point. Computers and Mathematics with Applications 43(1), 111–117 (2002)
6. Deb, K., et al.: A Fast Elitist Non-Dominated Sorting Genetic Algorithm for Multi-Objective Optimization: NSGAII. In: Deb, K., Rudolph, G., Lutton, E., Merelo, J.J., Schoenauer, M., Schwefel, H.-P., Yao, X. (eds.) PPSN 2000. LNCS, vol. 1917, pp. 849–858. Springer, Heidelberg (2000)
7. Ezzati, R., Allahviranloo, T., Khezerloo, M.: An approach for ranking of fuzzy numbers. Expert Systems with Applications 43(1), 690–695 (2012)
8. Bortolan, G., Degam, R.: A review of some methods for ranking fuzzy subsets. Fuzzy Sets Systems 15, 1–19 (1985)
9. Kaufmann, A., Gupta, M.: Fuzzy Mathematical Models in Engineering and Management Science. Elsevier Science, New York (1988)
10. Boulmakoul, A., Laarabi, M.H., Sacile, R., Garbolino, E.: Ranking Triangular Fuzzy Numbers Using Fuzzy Set Inclusion Index. In: Masulli, F. (ed.) WILF 2013. LNCS (LNAI), vol. 8256, pp. 100–108. Springer, Heidelberg (2013)
11. Limbourg, P., Daniel, E.S.: An Optimization Algorithm for Imprecise Multi-Objective Problem Functions. Evolutionary Computation 1, 459–466 (2005)
12. Liefooghe, A., Basseur, M., Jourdan, L., Talbi, E.-G.: ParadisEO-MOEO: A Framework for Evolutionary Multi-objective Optimization. In: Obayashi, S., Deb, K., Poloni, C., Hiroyasu, T., Murata, T., et al. (eds.) EMO 2007. LNCS, vol. 4403, pp. 386–400. Springer, Heidelberg (2007)
13. Paquet, L.F.: Stochastic Local Search algorithms for Multi-objective Combinatorial Optimization: A review. In: Handbook of Approximation Algorithms and Metaheuristics, vol. 13 (2007)
14. Oumayma, B., Nahla, B.A., Talbi, E.G.: A Possibilistic Framework for Solving Multi-objective Problems under Uncertainty. In: IPDPSW, pp. 405–414. IEEE (2013)
15. Solomon, M.M.: Algorithms for the vehicle Routing and Scheduling Problem with Time Window Constraints. Operations Research 35(2), 254–265 (1987)
16. Talbi, E.-G.: Metaheuristics: from design to implementation, vol. 74, pp. 309–373. John Wiley & Sons (2009)
17. Teich, J.: Pareto Front Exploration with Uncertain Objectives. In: Zitzler, E., Deb, K., Thiele, L., Coello Coello, C.A., Corne, D.W. (eds.) EMO 2001. LNCS, vol. 1993, pp. 314–328. Springer, Heidelberg (2001)
18. Toth, P., Vigo, D.: The vehicle routing problem, vol. 9. Siam (2002)
19. Wang, Y.: Centroid defuzzification and the maximizing set and minimizing set ranking based on alpha level sets. Computers and Industrial Engineering 57(1), 228–236 (2009)
20. Yao, J., Wu, K.: Ranking fuzzy numbers based on decomposition principle and signed distance. Fuzzy Sets and Systems 116(2), 275–288 (2000)
21. Zadeh, L.A.: Fuzzy Sets. Information and Control 8(3), 338–353 (1965)
22. Zitzler, E., Laumans, M., Thiele, L.: SPEA2: Improving the strength Pareto evolutionary algorithm (2001)

MI-groups: New Approach[*]

Martin Bacovský

University of Ostrava
Faculty of Science
Department of Mathematics
30. dubna 22, 701 03 Ostrava, Czech Republic
martin.bacovsky@osu.cz
http://www.osu.cz

Abstract. The notion of MI-group introduced in [1], [2] and later on elaborated in [3] is redefined and its structure analysed. In our approach, the role of the "Many Identities" set is replaced by an involutive anti-automorphism. Every finite MI-group coincides with some classical group, whilst infinite MI-groups comprise two parts: a group part and a semigroup part.

Keywords: many identities group, algebraic structures, fuzzy numbers, involutive anti-automorphism.

1 Introduction

In the work [1] of Holčapek and Štěpnička, the MI-algebras (MI stands for "Many Identities") were introduced. Their axioms are generalization of arithmetics with extensional numbers (for more details, again see [1]). The term MI refers to expressions $a+(-a)$, which are not generally equal to zero element. These elements are then grouped together to create the set E of pseudoidentities. Many results similar to those known from the classical group theory were proven, so that one might think about the real difference between MI-groups and groups. Our paper aims to answer this question.

In doing so, it is convenient to redefine the notion of MI-groups as introduced in [3] to simplify the reasoning about its properties. After redefinition we will show that an MI-group can be viewed as a conglomeration of some group and some semigroup. The number of MI-groups constructed over groups depends solely on the number of involutive anti-automorphisms that can be constructed over them. For the case of finite cyclic groups, these were described in [4] as an answer to problems arising in the analysis of processor networks (see e.g. [5]). Surprisingly, many of the techniques utilized in [4] are also very fruitful in our approach. Next, we find out that the only pseudoinversion applicable to the most prominent monoid present in almost every fuzzy numbers model, namely $\mathbb{R}_0^+$, is the identity. In the last part, one example illustrates usefulness of such approach.

---

[*] This work was supported by grant SGS13/PřF/2014 of the University of Ostrava.

A. Laurent et al. (Eds.): IPMU 2014, Part II, CCIS 443, pp. 274–283, 2014.

By $\mathbb{N}$, $\mathbb{Z}$ and $\mathbb{R}$ we denote the set of positive integers, integers and real numbers, respectively. Subscript 0 means the neutral element is added, e.g. $\mathbb{N}_0$ is the set of non-negative integers, and the superscript $+$ denotes the restriction to positive numbers, e.g. $\mathbb{R}^+$ denotes the set of positive reals. When integer m divides integer n we write $m|n$, by $a \equiv b$ mod m is denoted that a is congruent to b mod m, when k is the result of l modulo m we write $k = l$ mod m. The subtraction of sets A and B is denoted by $A - B$. For a non-empty set A, all finite sequences over A are denoted by A^f.

2 Redefinition of MI-group

We start this section with the redefinition of MI-group notion as introduced in [3].

Definition 1. *A quadruplet* $\mathbf{G} = (G, \circ, \nu, e)$ *is said to be an* MI-group *if for all* $x, y, z \in G$ *the following hold:*

(M1) $x \circ (y \circ z) = (x \circ y) \circ z$,
(M2) $x \circ e = e \circ x = x$,
 (I1) $\nu(x \circ y) = \nu(y) \circ \nu(x)$,
 (I2) $\nu(\nu(x)) = x$,
(CL) *if* $x \circ z = y \circ z$ *or* $z \circ x = z \circ y$ *then* $x = y$,

i.e., if triplet $(G, \circ, e)$ *is a monoid (axioms (M1), (M2)),* ν *is an involutive anti-automorphism (axioms (I1), (I2)) and the cancellation law is satisfied for* $\circ$ *(axiom (CL)).*

The axioms in the definition try to mimic the properties of the classical group (see [6]) with inverse replaced by pseudoinverse, i.e., the $\nu(a) \circ a$ and $a \circ \nu(a)$ may differ from the identity e.

Proposed definition simplifies the old one in the way that it allows more objects to be considered as MI-groups (see ex. 4). It is also more convenient, since when we try to construct an MI-group, we usually start first with some given carrier set G and operation $\circ$ and we are looking for a fitting ν. Then the set E of pseudoidentities (see [3]) can be defined appropriately with respect to ν. In fact, E can be viewed just as a constraint restricting the set of all applicable ν. The definition 1 is used throughout this paper as a definition of MI-group. We will omit the symbol $\circ$ in $x \circ y$ and write the multiplication simply as xy even for $\circ$ restricted to some $A \subset G$. ν will be often called a *pseudoinversion*.

The structure of MI-groups can be very close to a group one, thus we extract the greatest group-like structure and study its properties with respect to the whole MI-group. We will denote

$$\Gamma_{\mathbf{G}} = \{x \in G : (\exists y \in G)(x \circ y = e)\}$$

and

$$\Sigma_{\mathbf{G}} = G - \Gamma_{\mathbf{G}} = \{x \in G : (\forall y \in G)(x \circ y \neq e)\}.$$

Then we can enrich $\Gamma_\mathbf{G}$ by the classical group inversion, while $\Sigma_\mathbf{G}$ represents the semigroup part of $\mathbf{G}$.

Notice, that by associativity (M1) the result of $\circ$ does not depend on the operations order, e.g. $a(b(cd)) = (ab)(cd)$, so we simply write $abcd$ without explicitly stating the order. Next, $e \notin \Sigma_\mathbf{G}$, $\nu(e) = e$ $(e\nu(x) = \nu(x) = \nu(xe) = \nu(e)\nu(x)$ and apply the cancellation law) and when $xy = e$, then $yx = e$, i.e., both $x, y \in \Gamma_\mathbf{G}$ $(xy = e$, $(yx)^2 = y(xy)x = yx$ and the proposition again follows from the cancellation law). When the MI-group $\mathbf{G}$ is not important, we will write only Γ and Σ.

Lemma 1. Γ *is closed under $\circ$ and ν.*

Proof. Let $x, y \in \Gamma$ and $\tilde{x}, \tilde{y} \in \Gamma$ be their group inversions, respectively. Then

$$(xy)(\tilde{y}\tilde{x}) = xy\tilde{y}\tilde{x} = xe\tilde{x} = x\tilde{x} = e,$$

hence $xy \in \Gamma$. For ν it follows from $\nu(x)\nu(\tilde{x}) = \nu(\tilde{x}x) = \nu(e) = e$, thus $\nu(x) \in \Gamma$. $\square$

Lemma 2. Σ *is closed under $\circ$, ν and multiplication by elements from Γ. In symbols: $\Sigma\Sigma \subseteq \Sigma$, $\nu(\Sigma) = \Sigma$ and $\Sigma\Gamma \cup \Gamma\Sigma \subseteq \Sigma$.*

Proof. Let $x, y \in \Sigma$. To show that $xy \in \Sigma$, let us consider the set $Y = \{yz : z \in G\}$. Note that $e \notin Y$. From $x \in \Sigma$ it follows that $xw \neq e$ for $w = yz \in Y$ and by associativity $xw = x(yz) = (xy)z \neq e$ for all $z \in G$.

For $x \in \Gamma, y \in \Sigma$ and all $z \in G$ we have: $(yx)z = y(xz) \neq e$ and $(xy)z = x(yz) \neq e$, since by lemma 1 the group inversion of x lies in Γ, but yz from the preceding step lies in Σ.

For $x \in \Sigma$, $y, z \in G$ such that $y = \nu(z)$, we compute

$$\nu(x)y = \nu(x)\nu(z) = \nu(zx) \neq e.$$

Hence Σ is closed under ν. $\square$

By the preceding lemmas, group and semigroup part of an MI-group are self-contained algebraic structures, while their interactions $\Sigma\Gamma$ always fall into semigroup part. This property substantially separates MI-groups from groups and semigroups and thus justifies studying MI-groups.

Lemma 3. *For $x \in \Sigma$ it holds that $x^n \neq x$ and $x^n \neq \nu(x)$ for $n \in \mathbb{N}, n > 1$.*

Proof. Let $x \in \Sigma$ and $n \in \mathbb{N}$ such that $x^n = x$. Then by the cancellation law $x^{n-1} = e$ and from $x \in \Sigma$ it follows that $n = 1$.

For the second part, suppose that $\nu(x) = x^n$. Then

$$x = \nu(\nu(x)) = \nu(x^n) = \nu(x)^n = x^{n^2},$$

and by the first part we get $n = 1$. $\square$

Corollary 1. *When Σ is nonempty, it has infinitely many elements.*

To study genuine MI-groups (those that can not be viewed as a group), one shall restrict her studies to structures with group and semigroup parts non-empty, i.e., by the corollary, necessarily to infinite MI-groups. However, we will in this paper also consider finite MI-groups, which can always be viewed as groups, to find out the differences between the two.

Following definition of homomorphisms differs from that in [3] in the fact there are no constraints regarding a set of pseudoidentities E.

Definition 2. *A mapping $f : G \to H$ is said to be an* MI-group homomorphism *of MI-groups* $\mathbf{G} = (G, \circ_{\mathbf{G}}, \mu, e_{\mathbf{G}})$ *and* $\mathbf{H} = (H, \circ_{\mathbf{H}}, \nu, e_{\mathbf{H}})$ *(shortly, a homomorphism) if it satisfies following conditions for all $x, y \in G$:*

(H1) $f(x \circ_{\mathbf{G}} y) = f(x) \circ_{\mathbf{H}} f(y)$,
(H2) $f(\mu(x)) = \nu(f(x))$.

When f is an injective mapping, we will call it a monomorphism.

Lemma 4. *A homomorphism $f : G \to H$ restricted to $\Gamma_{\mathbf{G}}$ maps into $\Gamma_{\mathbf{H}}$ and, moreover, denoting $x^{-1_{\mathbf{A}}}$ a group inversion in $\Gamma_{\mathbf{A}}$ for some MI-group $\mathbf{A}$,*

$$f(x^{-1_{\mathbf{G}}}) = f(x)^{-1_{\mathbf{H}}}.$$

Proof. Let $x \in \Gamma_{\mathbf{G}}$. Then

$$f(e_{\mathbf{G}}) = f(xx^{-1_{\mathbf{G}}}) = f(x)f(x^{-1_{\mathbf{G}}}) = e_{\mathbf{H}},$$

thus $f(x) \in \Gamma_{\mathbf{H}}$ and the last statement follows directly.

Lemma 5. *$f : \Gamma_{\mathbf{G}} \to \Gamma_{\mathbf{H}}$ is a monomorphism if and only if $f(x) = e_{\mathbf{H}}$ implicates $x = e_{\mathbf{G}}$.*

Proof. The *only if* direction is clear, we prove the *if* direction. If $f(x) = f(y)$, then also

$$e_{\mathbf{H}} = f(x)f(y)^{-1_{\mathbf{H}}} = f(x)f(y^{-1_{\mathbf{G}}}) = f(xy^{-1_{\mathbf{G}}}),$$

thus $xy^{-1_{\mathbf{G}}} = e_{\mathbf{G}}$ and $x = y$.

In [3] a definition of a kernel homomorphism was proposed. Without going into much detail, in usual cases it can be considered as preimages of elements of the set $\{x\mu(x) : x \in H\}$. Nevertheless, considering $\mathbf{G} = (\mathbb{N}_0, +, id, 0)$ and $\mathbf{G}_2 = (2\mathbb{N}_0, +, id, 0)$ together with mapping $f : \mathbb{N}_0 \to 2\mathbb{N}_0$ given by prescription $f(x) = 4x$, f maps onto $\{4x : x \in \mathbb{N}_0\}$ and is clearly a monomorphism, but the kernel is the whole $\mathbb{N}_0$ and thus, we can not use kernels defined in [3] to determine whether a homomorphism is injective. The problem of introducing similar notion to that of kernel is left open.

We end this section by the very basic examples of MI-groups.

Example 1. $\mathbf{G}_k = (k\mathbb{N}_0, +, id, 0)$, where $k \in \mathbb{N}_0$ and $k\mathbb{N}_0 = \{kn \mid n \in \mathbb{N}_0\}$. For $k > 0$ the $\mathbf{G}_k$ is an MI-group, but can not be considered as a group. There always exists a non-unique monomorphism between $\mathbf{G}_k$ and $\mathbf{G}_l$ for $k, l \neq 0$, namely $f : k\mathbb{N}_0 \to l\mathbb{N}_0$, $f(k) = nl$ for a fixed $n \in \mathbb{N}$.

Every group is an MI-group, and every abelian group has at least two possible pseudoinversions (when they do not coincide, e.g. $\mathbb{Z}_2$) the classical inversion and the identity one. Other examples will be elicited in next sections.

3   MI-groups over Finite Cyclic Groups

In this section, we consider MI-groups constructed over finite cyclic groups $\mathbb{Z}_n$, i.e., with different pseudoinversions. In this case, every homomorphism and consequently every pseudoinversion is completely described when the image of one of the group generators is known, w.l.o.g. we will take 1. Define $\xi : \mathbb{N} \to \{-1, 0, 1\}$ by

$$\xi(n) = \begin{cases} 1 & \text{if } 8|n, \\ -1 & \text{if } 2|n \text{ and } 4 \nmid n, \\ 0 & \text{otherwise.} \end{cases}$$

In [4] the following theorem was proved.

Theorem 1. *Let $n = p_1^{m_1} p_2^{m_2} \ldots p_r^{m_r}$ be a prime decomposition of n with $p_i < p_{i+1}$ for $1 \leq i \leq r - 1$, and $m_i > 0$ for $1 \leq i \leq r$. Then the number of involutive automorphisms of $\mathbb{Z}_n$ equals $2^{r+\xi(n)}$.*

For $\mathbb{Z}_8$ there are four pseudoinversions: $\nu_1(1) = 1$, $\nu_2(1) = 3$, $\nu_3(1) = 5$, $\nu_4(1) = 7$. ν_1 is the identity, ν_4 the group inversion, and ν_2 with ν_3 are new pseudoinversions. We will denote $\mathbb{Z}_{n,l}$ an MI-group constructed over the group $\mathbb{Z}_n$ with pseudoinversion $\nu(1) = l$. Note, that by the corollary 1 finite MI-groups always allow such a pseudoinversion that is in fact an inversion, as is the case of ν_4 for $\mathbb{Z}_8$.

It is well known (see e.g. [6]), that two finite cyclic groups $\mathbb{Z}_m$ and $\mathbb{Z}_n$ for $m < n$ have a monomorphism $f : \mathbb{Z}_m \to \mathbb{Z}_n$ iff $m|n$, namely $f(1) = \frac{n}{m}$. In the case of MI-groups, this condition is necessary but not sufficient.

Lemma 6. *A monomorphism $f : \mathbb{Z}_{m,k} \to \mathbb{Z}_{n,l}$ exists if and only if $m|n$ and $k \equiv l \pmod{m}$.*

Proof. The monomorphism must map $1 \mapsto \frac{n}{m}$ for the same reasons as in the case of classical groups. Denote by ν_k and ν_l pseudoinversions of $\mathbb{Z}_{m,k}$ and $\mathbb{Z}_{n,l}$, respectively. We have to verify, that (H2) from the definition of homomorphism is satisfied, i.e., that

$$f(\nu_k(1)) = f(k) = k\frac{n}{m}, \quad \nu_l(f(1)) = \nu_l\left(\frac{n}{m}\right) = l\frac{n}{m}$$

are equal modulo n. The equivalence

$$n|(k-l)\frac{n}{m} \Leftrightarrow k - l \equiv 0 \pmod{m}$$

completes the proof.

Lemma 7. *For an MI-group $\mathbb{Z}_{n,l}$ all injectively insertable MI-groups into $\mathbb{Z}_{n,l}$ constructed over finite cyclic group form a lattice $(L, \vee, \wedge, \mathbf{0}, \mathbf{1})$, where*

– L is the set of all MI-groups $\mathbb{Z}_{m,k}$ such that $m|n$ and $k \equiv l \mod m$,
– $\mathbb{Z}_{m_1,k_1} \vee \mathbb{Z}_{m_2,k_2} = \mathbb{Z}_{m,\, l \bmod m}$, where $m = \mathrm{lcm}(m_1, m_2)$,
– $\mathbb{Z}_{m_1,k_1} \wedge \mathbb{Z}_{m_2,k_2} = \mathbb{Z}_{m,\, l \bmod m}$, where $m = \gcd(m_1, m_2)$,
– $\mathbf{0} = \mathbb{Z}_{1,0}$, the trivial group,
– $\mathbf{1} = \mathbb{Z}_{n,l}$.

Proof. From the lemma 6 and the Chinese Remainder Theorem (see [6]), given the image MI-group $\mathbb{Z}_{n,l}$ and the domain group $\mathbb{Z}_m$, there is at most one possible pseudoinversion $\nu(1) = k$ $(k < m)$ such that a monomorphism $f : \mathbb{Z}_{m,k} \to \mathbb{Z}_{n,l}$ exists, namely $k = l \bmod n$. Hence for every divisor m of n there is exactly one k such that $\mathbb{Z}_{m,k}$ is injectively insertable into $\mathbb{Z}_{n,l}$.

The statement then follows from the well known fact that the divisors of m compose a lattice with respect to operations gcd and lcm.

Example 2. For the group $\mathbb{Z}_{3 \cdot 5 \cdot 7}$ there are eight possible pseudoinversions:

$$\nu(1) = 1, 29, 34, 41, 64, 71, 76, 104.$$

In the Fig. 1, the lattice from the previous lemma is drawn for the choice $\nu(1) = 41$. It is readily seen, that there is a one-to-one correspondence between elements of this lattice and the lattice of all subgroups of the group $\mathbb{Z}_{3 \cdot 5 \cdot 7}$. For other pseudoinversions the lattices are similar. From this point of view, the structure of a finite MI-group is not essentially different from its group structure.

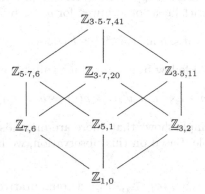

Fig. 1. Lattice of all injectively insertable MI-groups into $\mathbb{Z}_{3 \cdot 5 \cdot 7, 41}$ with groups underlined

4   MI-groups with Non-empty Semigroup Part

In this section, properties of monoids important from the MI-groups point of view are presented.

Lemma 8. *For given MI-group* $\mathbf{G}$, *if* $\Sigma_{\mathbf{G}}$ *is nonempty, then* ν *has at least one fix point in* $\Sigma_{\mathbf{G}}$.

Proof. Let $x \in \Sigma_{\mathbf{G}}$. Then

$$\nu(x\nu(x)) = \nu(\nu(x))\nu(x) = x\nu(x).$$

Since $\Sigma_{\mathbf{G}}$ is closed under ν and multiplication, $x\nu(x) \in \Sigma_{\mathbf{G}}$.

Following lemma has left and right variant for non-commutative monoids. We prove only the left variant, the right can be proved similarly.

Lemma 9. *Let* $\mathbf{G} = (G, \circ, e)$ *be a monoid and* ν *a pseudoinversion such that whenever* $\nu(a) = a \circ c$ *for some* $a, c \in G$, *then* $a \neq d \circ \nu(a)$ *for any* $d \in G, d \neq e$. *Then* $c = e$.

Proof. When $\nu(a) = a \circ c$, then also $\nu(\nu(a)) = a = \nu(c) \circ \nu(a)$ and by the assumption must be $\nu(c) = e$, which concludes that c must be in fact e.

Corollary 2. *The only pseudoinversion on* $\mathbb{R}_0^+$ *is the identity.*

Proof. Take $a \in \mathbb{R}_0^+$, then either $\nu(a) \geq a$ or $a \geq \nu(a)$, where $\geq$ is a natural ordering on $\mathbb{R}_0^+$. In the first case, there exists exactly one $b \in \mathbb{R}_0^+$ so that $\nu(a) = a + b$ holds. Then lemma 9 implies $b = 0$. For the second case put $a' = \nu(a)$ and apply the first part for a'.

Lemma 9 may seem to be inappropriate for the preceding corollary. On $\mathbb{R}_0^+$ one could also define strict linear ordering $<$ for $a \neq b$ by

$$a < b \text{ iff } \exists c \in \mathbb{R}^+ : b = a + c$$

and the statement would follow from (c is some suitable element from $\mathbb{R}^+$)

$$a < \nu(a) \Leftrightarrow \nu(a) = a + c \Leftrightarrow a = \nu(a) + \nu(c) \Leftrightarrow a > \nu(a), \text{ a contradiction.}$$

However, following example shows that there are monoids on which such linear ordering is not realizable. Based on this observation, we have chosen presented form of lemma 9.

Example 3. Let $\mathbf{G}_\alpha = (\mathbb{R}_0^+ \cup \{e\}, \oplus_\alpha, e)$ be a commutative monoid with

$$a \oplus_\alpha b = a + b + \alpha, \ e \oplus_\alpha a = a \oplus_\alpha e = a,$$

where $a, b \in \mathbb{R}_0^+$ and $\alpha \in \mathbb{R}^+$. Then take $a, b \in \mathbb{R}_0^+$ such that $|a - b| < \alpha$. Neither $a = b \oplus_\alpha c = b + c + \alpha$ nor $b = a \oplus_\alpha d = a + d + \alpha$ may hold for any $c, d \in \mathbb{R}_0^+$. On the other hand, it is easily seen, that when $\nu(a) = a \oplus_\alpha b$ for some $a, b \in \mathbb{R}_0^+ \cup \{e\}$ it can not hold $a = \nu(a) \oplus_\alpha b$ for $b \neq e$ and, subsequently, $b = e$ and ν is the identity mapping on $\mathbb{R}_0^+$. This together with satisfied cancellation law (CL) shows that $\mathbf{G}_\alpha$ can be extended to an MI-group with only one possible pseudoinversion — the identity.

Although $\mathbb{R}_0^+$ makes possible only identity pseudoinversion, on the product $\mathbb{R}_0^+ \times \mathbb{R}_0^+$ with component-wise addition and neutral element $(0,0)$ one can besides identity pseudoinversion define a swapping pseudoinversion $\nu((a,b)) = (b,a)$, since e.g. $(1,2)$ and $(2,1)$ are not convertible to each other by addition on $\mathbb{R}_0^+ \times \mathbb{R}_0^+$.

Next example shows that there exist non-commutative MI-groups for which need not be $\forall a \in G : a \circ \nu(a) = \nu(a) \circ a$ (axiom (G2) in [3] that is absent in our definition).

Example 4. Consider finite sequences over a set $\{0,1\}$ with an operation $\circ$ as concatenation. Then $\circ$ is associative, non-commutative and allows a cancellation. The identity element can be added as an empty word ϵ. To complete its structure as MI-group, we have to define involutive anti-automorphism. The identity $id(s) = s$ for $s \in \{0,1\}^f$ is not suitable, since

$$id(01) = 01 \neq 10 = id(1)\,id(0).$$

On the other hand, it is easily seen that it suffices to define ν just on generators $0, 1$ and extend it anti-homomorphically to other sequences. The only options are "negation" $\nu_n(0) = 1$, $\nu_n(1) = 0$ and "identity" $\nu(x) = x$ for $x \in \{0,1\}$. Thus, we constructed MI-group $\mathbf{W} = (\{0,1\}^f \cup \{\epsilon\}, \circ, \nu_n, \epsilon)$ with $a \circ \nu_n(a) \neq \nu_n(a) \circ a$, e.g.

$$0 \circ \nu_n(0) = 01 \neq \nu_n(0) \circ 0 = 10.$$

By lemma 8, there is at least one fix point in $\{0,1\}^f$ for ν_n. Indeed, sequences 01 and 10 are such fix points.

One may wonder, if the statement of lemma 9 would be true in the case not $a \neq d \circ \nu(a)$, but $a \neq \nu(a) \circ d$ for all $d \neq e$. Following example disproves this idea.

Example 5. Consider finite sequences with elements from $\{a, b, c, d\}$, the operation concatenation and neutral element as in the previous example with two relations

$$dc = abcd, \quad ab = badc.$$

We mean that one can instead of a sequence dc write $abcd$, i.e., a sequence twice longer in the number of elements. Define ν as $\nu(a) = c$ and $\nu(b) = d$ and extend it anti-automorphically to all finite sequences. Then $\nu(ab) = dc = abcd$ and $ab \neq \nu(ab) \circ \alpha$ for any finite sequence α, since $\nu(ab) \circ \alpha = abcd \circ \alpha$ and there is no way how to get rid of cd on the right-hand side. However, $ab = \beta \circ \nu(ab)$ has a solution, since $ab = badc$, thus put $\beta = ba$ and the equality follows.

5   Trapezoidal Fuzzy Numbers as MI-groups

As a main example we choose trapezoidal fuzzy numbers, since they generalize interval fuzzy numbers together with triangular fuzzy numbers, the two most common fuzzy number models. Trapezoidal fuzzy numbers compose in fact an MI-group with non-empty semigroup part $(\Sigma \neq \emptyset)$ and as it will be shown it is this part, that plays the role of vagueness.

For $a, b, c, d \in \mathbb{R}$ and $a \leq b \leq c \leq d$ such a fuzzy number $A = (a, b, c, d)$ is defined as

$$A(x) = \begin{cases} 0 & \text{if } x < a \text{ or } d \leq x, \\ (x - a)/(b - a) & \text{if } a \leq x \leq b, \\ 1 & \text{if } b \leq x \leq c, \\ (d - x)/(d - c) & \text{if } c \leq x < d. \end{cases}$$

Addition is defined component-wise:

$$A +_\tau E = (a, b, c, d) +_\tau (e, f, g, h) = (a + e, b + f, c + g, d + h),$$

$\mathbf{0}_\tau = (0, 0, 0, 0)$ is the neutral element and pseudoinversion $-_\tau(a, b, c, d) = (-d, -c, -b, -a)$. Using the bijection $(a, b, c, d) \mapsto (\frac{b+c}{2}, c - b, b - a, d - c)$ which maps $\{(a, b, c, d) \mid a \leq b \leq c \leq d, \ a, b, c, d \in \mathbb{R}\} \subset \mathbb{R}^4$ onto $T = \mathbb{R} \times (\mathbb{R}_0^+)^3$ we can view trapezoidal fuzzy numbers as MI-group $\mathbf{T} = (T, +_T, -_T, \mathbf{0}_T)$ with

$$A +_T E = (a, b, c, d) +_T (e, f, g, h) = (a + e, b + f, c + g, d + h),$$
$$-_T A = -_T(a, b, c, d) = (-a, d, c, b).$$

Notice that $-_T$ is now not constructable as a product of corresponding pseudoinversions acting only on $\mathbb{R}$ and $\mathbb{R}_0^+$, since it also swaps components.

The $\mathbb{R}$ part is indeed a group, while the rest $(\mathbb{R}_0^+)^3$ is a monoidal part with no chance to introduce a group inversion. The set

$$S = \{(0, a, b, a) \mid a, b \in \mathbb{R}_0^+\}$$

consists of symmetrical elements, i.e., for any $x \in S$ the $-_T x = x$ holds. In the literature sometimes the equivalence $\sim_T$ for $a, b \in T$

$$a \sim_T b \Leftrightarrow \exists s_a, s_b \in S : \ a +_T s_a = b +_T s_b$$

is proposed (see [7]). In this case of $\mathbf{T}$ and S the equivalence classes with respect to $\sim_T$ are isomorphic to group $\mathbb{R} \times \mathbb{R}$ via the isomorphism $(a, b, c, d) \mapsto (a, b - d)$ (remember that we can take arbitrary representant of the equivalence class). Thence, vagueness corresponding to b and d is inserted into the additive group $\mathbb{R}$ by $b - d$ and vagueness represented by c is completely forgotten.

6   Conclusion

We redefined the notion of an MI-group to obtain more insight into its structure. Theoretically, it may be described partly as a group and partly as a semigroup. Practically, semigroup structure is often present in fuzzy numbers models and as such it is a source of vagueness. Since MI-groups are generalization of groups, some analogue theorems from groups are provable only under very restrictive conditions. Instead of trying to make these MI-groups as similar as possible to groups, we tried to point out the main differences. For the finite MI-groups constructed over cyclic groups these were not significant. The situation is more interesting for MI-groups with infinite carrier set. Since there is just one pseudoinversion for the most prominent monoid utilized in fuzzy numbers, namely $\mathbb{R}_0^+$, our next research will concern other models of fuzzy numbers and pseudoinversions not constructable as direct products of elementary pseudoinversions over the underlying carrier sets, as shown in the penultimate section.

References

1. Holčapek, M., Štěpnička, M.: Arithmetics of extensional fuzzy numbers – part I: Introduction. In: Proc. IEEE Int. Conf. on Fuzzy Systems, Brisbane, pp. 1517–1524 (2012)
2. Holčapek, M., Štěpnička, M.: Arithmetics of extensional fuzzy numbers – part II: Algebraic framework. In: Proc. IEEE Int. Conf. on Fuzzy Systems, Brisbane, pp. 1525–1532 (2012)
3. Holčapek, M., Štěpnička, M.: MI-algebras: A new framework for arithmetics of (extensional) fuzzy numbers. Fuzzy Sets and Systems (2014), http://dx.doi.org/10.1016/j.fss.2014.02.016
4. Hage, J., Harju, T.: On Involutive Anti-Automorphisms of Finite Abelian Groups. Technical UU WINFI Informatica en Informatiekunde (2007)
5. Hage, J., Harju, T.: The size of switching classes with skew gains. Discrete Math. 215, 81–92 (2000)
6. Dummit, D.S., Foote, R.M.: Abstract Algebra. Wiley (2003)
7. Mareš, M.: Computation over Fuzzy Quantities. CRC Press, Boca Raton (1994)

On Combining Regression Analysis and Constraint Programming*

Carmen Gervet and Sylvie Galichet

LISTIC, Laboratoire d'Informatique, Systems, Traitement de l'Information et de la
Connaissance - Universit de Savoie, BP 80439
74944 Annecy-Le-Vieux Cedex, France
{gervetec,sylvie.galichet}@univ-savoie.fr

Abstract. Uncertain data due to imprecise measurements is commonly specified as bounded interval parameters in a constraint problem. For tractability reasons, existing approaches assume independence of the parameters. This assumption is safe, but can lead to large solution spaces, and a loss of the problem structure. In this paper we propose to combine the strengths of two frameworks to tackle parameter dependency effectively, namely constraint programming and regression analysis. Our methodology is an iterative process. The core intuitive idea is to account for data dependency by solving a set of constraint models such that each model uses data parameter instances that satisfy the dependency constraints. Then we apply a regression between the parameter instances and the corresponding solutions found to yield a possible relationship function. Our findings show that this methodology exploits the strengths of both paradigms effectively, and provides valuable insights to the decision maker by accounting for parameter dependencies.

Keywords: Data uncertainty, Constraint Programming, regression analysis.

1 Introduction

Data uncertainty due to imprecise measurements or incomplete knowledge is ubiquitous in many real world applications, network design, renewable energy investment planning, and inventory management, to name a few. Regression analysis is one of the most widely used statistical techniques to model and represent the relationship among variables to describe or predict phenomena for a given context. Recently, models derived from fuzzy regression have been defined to represent incomplete and imprecise measurements in a contextual manner, using intervals [2]. Such models apply to problems in finance or complex systems analysis in engineering whereby a relationship between crisp or fuzzy measurements is sought.

Constraint Programming (CP) on the other hand, is a powerful paradigm used to solve decision and optimization problems in areas as diverse as planning, scheduling, routing. The CP paradigm models a decision problem using

---

* This research was partly support by the Marie Curie CIG grant, FP7-332683.

A. Laurent et al. (Eds.): IPMU 2014, Part II, CCIS 443, pp. 284–293, 2014.

constraints to express the relations between variables, and propagates any information gained from a constraint onto other constraints. When data imprecision is present, forms of uncertainty modeling have been embedded into constraint models using bounded intervals to represent such parameters. For instance if we model traffic in a network, the flow over links between two routers cannot be measured with precision as it depends on the collected traffic volume data at each router which cannot be synchronized. As a result, the data information obtained is erroneous. We use the sigcomm4 network given in Fig. 1 as a running example taken from [8]. On link $A \to C$, for example, the traffic volume might measure as 565 at A and as 637 at C, whereas the true value, equal at both nodes, is presumably somewhere in between. Thus, the flow between A and C will be specified by 565.0..637.0. The mean values are shown in the figure. Such models are used to find the flow matrix between any pair of routers. Commonly flow distribution is considered between 30 ad 70 percent. Thus on the link $A \to C$ the flow constraint is specified by: $0.3..0.7 * F_{AC} + 0.3..0.7 * F_{BC} + 0.3..0.7 * F_{AB} + 0.3..0.7 * F_{AD} = 565.0..637.0$. A variable F_{AD} denotes the traffic (unknonwn) from router A to router D.

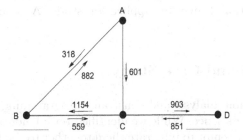

Fig. 1. Sigcomm4 network topology and mean values for link traffic

Traditional models either omit any routing uncertainty for tractability reasons, and consider solely the shortest path routing or embed the uncertain parameters but with no dependency relationships. Values for the flow variables are derived by computing bounded intervals, which are safe enclosing of all possible solutions. Such intervals enclose the solution set without relating to the various instances of the parameters. For instance, the traffic between A and C can also pass through the link $A \to B$. Thus the flow constraint on this link also contains $0.3..0.7 * F_{AC}$. However, the parameter constraint stating that the sum of the coefficients of the traffic F_{AC} in both constraints should be equal to 1 should also be present. Assuming independence of the parameters for tractability reasons, leads to safe computations, but at the potential cost of a very large solution set, even if no solution actually holds. The problem structure is lost. Also, there is not insight as to how the potential solutions evolve given instances of the data.

The question remains as to how can this information be embedded in a constraint model that would remain tractable. To our knowledge this issue has not been addressed in the general case. It is the purpose of this work.

In this paper we propose a new methodology and efficient process to account for data dependency constraints in decision problems. We aim to more closely model the actual problem structure, refine the solutions produced and add accuracy to the decision making process. We use regression analysis to show the relationship among various instances of the uncertain data parameters and the solutions produced. The basic process is to extract the parameter constraints from the model, solve them to obtain tuple solutions over the parameters. Then we run simulations on the constraint models using the parameter tuples as data instances that embed the dependencies. In the example above this would imply for the two constraints given, that if one parameter takes the value 0.3, the other one would take the value 0.7. A set of constraint models can thus be solved efficiently, matching a tuple of consistent parameters to a solution, to determine whether there are solutions once dependencies are taken into account or not. Finally, we run a regression analysis between the parameter values and solutions produced to determine the regression function, i.e. see how potential solutions relate to parameter variations.

The paper is structured as follows. Section 2 summarizes the related work and gives a small illustrative case study. Section 3 describes the approach and algorithms, and Section 4 gives an application study. A conclusion is given in Section 5.

2   Background and Case Study

The fields of regression analysis and constraint programming are both well established in computer science. While we identified both fields as complementary, there has been little attempt to integrate them together to the best of our knowledge. The reason is, we believe, that technology experts tackle the challenges in each research area separately. However, each field has today reached a level of maturity shown by the dissemination in academic and industrial works, and their integration would bring new research insights and a novel angle in tackling real-world optimization problems with measurement uncertainty. There has been some research in Constraint Programming (CP) to account for data uncertainty, and similarly there has been some research in regression modeling to use optimization techniques. We give an account of the state of the art against our main objective to integrate both paradigms.

CP is a paradigm within Artificial Intelligence that proved effective and successful to model and solve difficult combinatorial search and optimization problems from planning and resource management domains [9]. Basically it models a given problem as a Constraint Satisfaction Problem (CSP), which means: a set of variables, the unknowns for which we seek a value, the range of values allowed for each variable , and a set of constraints which define restrictions over the variables. Constraint solving techniques have been primarily drawn from Artificial Intelligence (constraint propagation and search), and more recently Operations Research (graph algorithms, Linear Programming). A solution to a constraint model is a complete consistent assignment of a value to each decision variable.

In the past 15 years, the growing success of constraint programming technology to tackle real-world combinatorial search problems, has also raised the question of its limitations to reason with and about uncertain data, due to incomplete or imprecise measurements, (e.g. energy trading, oil platform supply, scheduling). In the past 10 years, the generic CSP formalism has been extended to account for forms of uncertainty: e.g. numerical, mixed, quantified, fuzzy, uncertain CSP and CDF-interval CSPs [3]. The fuzzy and mixed CSP [7] coined the concept of parameters, as uncontrollable variables, meaning they can take a set of values, but their domain is not meant to be reduced to one value during problem solving (unlike decision variables). Constraints over parameters, or uncontrollable variables, can be expressed and thus some form of data dependency modeled. However, there is a strong focus on discrete data, and the consistency techniques used are not always effective to tackle large scale or optimization problems.

Frameworks such as *numerical, uncertain, or CDF-interval* CSPs, extend the classical CSP to approximate and reason with continuous uncertain data represented by intervals; see the real constant type in Numerica [11] or the bounded real type in ECLiPSe [4]. Our previous work introduced the *uncertain and CDF-interval* CSP [12,10]. The goal was then to derive efficient techniques to compute reliable solution sets that ensure that each possible solution corresponds to at least one realization of the data. In this sense they compute an enclosure of the set of solutions. Even though we identified the issue of having a large solution set, the means to relate different solutions to instances of the uncertain data parameters and their dependencies were not thought of. On the other hand, in the field of regression analysis, the main challenges have been in the definition of optimization functions to build a relevant regression model, and the techniques to do so efficiently. Regression analysis evaluates the functional relationship, often of a linear form, between input and output parameters in a given environment. Here we are interested in using regression to seek a possible relation between uncertain constrained parameters in a constraint problem, e.g. distribution of traffic among two routers on several routes and the solutions computed according to the parameter instances. We note also that methods such as sensitivity analysis in Operations Research allow to analyze how solutions evolve relative parameter changes. However, such models assume independence of the parameters. In the following case study we show how our approach can establish relationships with the solutions and uncertain parameters while accounting for dependencies.

Case study. We present a small case study to give the intuition of our approach. The core element is to go around the solving of a constraint optimization problem with uncertain parameter constraints by first solving the parameter constraints alone. This way we handle uncertainty in an efficient and tractable manner. We then substitute solution tuples of these parameters to solve a set of constraint optimization problems (now without parameter constraints). Finally to provide insight to the solutions to the uncertain constraint problem, we run a regression between the solutions produced and the corresponding parameter tuple instances.

Consider the following fictitious constraint between two unknown positive variables, X and Y ranging in $0.0..1000.0$, with uncertain data parameters A, B taking their values in the real interval $[0.1..0.7]$:

$$A * X + B * Y = 150$$

The objective is to compute values for X and Y in the presence of uncertain parameters (A, B). Without any parameter dependency a constraint solver based on interval propagation techniques with bounded coefficients, derives the ranges $[0.0..1000.0]$ for both variables X and Y [4]. Let us add to the model a parameter constraint over the uncertain parameters A and B: $A = 2 * B$. Without adding this parameter constraint to the model, since it is not handled by the solver, we can manually refine the bounds of the uncertain parameters in the constraint model such that the parameter constraint holds over the bounds, thus accounting partially for the dependency. We obtain the constraint system:

$$[0.2..0.7] * X + [0.1..0.35] * Y = 150$$

The solution returned to the user is a solution space: $X \in [0.0..750.0], Y \in [0.0..1000.0]$. The actual polyhedron describing the solution space is depicted in Fig. 2.

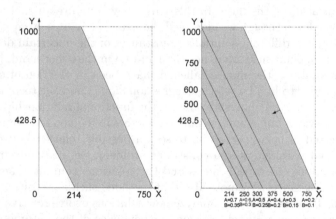

Fig. 2. Left: Solution space. Tight and certain bounds for the decision variables: $[0, 750]$ $[0, 1000]$. Right: Solutions vectors of problem instances with consistent parameter solutions.

We now give the intuition of our approach. The idea is to first solve the parameter dependency constraints alone to obtain solution tuples, not intervals. To do so we use a traditional branch and bound algorithm. We obtain a set of tuples for A and B such that for each tuple the constraint $A = 2 * B$ holds. The idea is to have a well distributed sample of solutions for this parameter constraint.

We obtain a set of tuples that satisfy the parameter constraint, in this case for instance $(0.2, 0.1), (0.3, 0.15), (0.4, 0.2), (0.5, 0.25), (0.6, 0.3), (0.7, 0.35)$. We then substitute each tuple in the uncertain constraint model rendering it a standard constraint problem, and solve each instance. We record the solution matching each parameter instance. The issue now is that we have a set of solutions for each tuple of paramers. There is no indication how the solutions are related and evolve. The idea is to apply a regression analysis between both. The regression function obtained includes the solution bounds obtained by the standard approach, but mainly shows the trends between the data parameters and the solutions. In this small example we can visualize how the solution evolves with the data, see Fig. 2 on the right. In the case of much larger data sets, a tool like Matlab can be used to compute the regression function and display the outcome. The algorithm and complexity analysis are given in the next section.

3   Process and Algorithms

Our methodology is a three-steps iterative process: 1) Extract the uncertain parameter constraints from the uncertain optimization problem and run branch and bound to produce a set of tuple solutions, 2) solve a sequence of standard constraint optimization problems where the tuples are being substituted to the uncertain parameters. This is a simulation process that produces, if it exists, one solution per tuple instance. And finally, 3) run a regression analysis on the parameter instances and their respective solution, to identify the relationship function showing how the solution evolves with consistent parameter constraints. The overall algorithmic process is given in Fig. 3, where the outcomes at each step are highlighted in italic bold. A constraint satisfaction and optimisation problem, or CSOP, is a constraint satisfaction problem (CSP) that seeks complete and consistent instantiations optimizing a cost function. We use the notion of uncertain CSOP, or UCSOP first introduced in [12]. It extends a classical CSOP with uncertain parameters.

3.1   Uncertain CSOP and Uncertain Parameter Constraints

Recall that a CSOP is commonly specified as a tuple $(\mathcal{X}, \mathcal{D}, \mathcal{C}, f)$, where $\mathcal{X}$ is a finite set of variables, $\mathcal{D}$ is the set of corresponding domains, $\mathcal{C} = \{c_1, \ldots, c_m\}$ is a finite set of constraints, and f is the objective function (min or max of a given expression over a subset of the variables).

Definition 1 (UCSOP). *An uncertain constraint satisfaction and optimisation problem is a classical CSOP in which some of the constraints may be uncertain, and is specified by the tuple $(\mathcal{X}, \mathcal{D}, \mathcal{C}_{\mathcal{X}}, \Lambda, \mathcal{U}, f)$. The finite set of parameters is denoted by Λ, and the set of ranges for the parameters by $\mathcal{U}$. A solution to a UCSOP is a solution space enclosing safely the set of possible solutions.*

Example 1. Let $X_1 \in D_1$ and $X_2 \in D_2$ both have domains $D_1 = D_2 = [1.0..7.0]$. Let λ_1 and λ_2 be parameters with uncertainty sets $U_1 = [2.0..4.0]$ and $U_2 = [1.0..6.0]$ respectively. Consider three constraints:

$$C_1 : X_1 > \lambda_1, \ C_2 : X_1 = X_2 + \lambda_2, \ C_3 : X_2 > 2, \ C_4 : \lambda_2 = \lambda_1 + 3$$

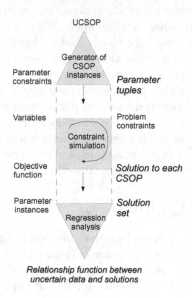

Fig. 3. Process

and the objective function to maximize $f(X_1, X_2) = X_1 + X_2$. Writing $\mathcal{X} = \{X_1, X_2\}$, $\mathcal{D} = \{D_1, D_2\}$, $\Lambda = \{\lambda_1, \lambda_2\}$, $\mathcal{U} = \{U_1, U_2\}$, and $\mathcal{C}_\mathcal{X} = \{C_1, C_2, C_3\}$, then $(\mathcal{X}, \mathcal{D}, \mathcal{C}_\mathcal{X}, \Lambda, \mathcal{U}, f)$ is a UCSOP. Note that C_3 is a certain constraint; C_1 and C_2 are both uncertain constraints because they contain uncertain parameters. $\mathcal{C}_\Lambda = \{C_4\}$ is the set of parameter constraints.

3.2   Constraint Simulation

We now account for the parameters constraints by transforming the UCSOP into a set of tractable CSOPs instances. More formally, consider a UCSOP $(\mathcal{X}, \mathcal{D}, \mathcal{C}_\mathcal{X} \cup \mathcal{C}_\Lambda, \Lambda, \mathcal{U}, f)$.

Definition 2 (instance of UCSOP). *We denote n the number of variables, m the number of uncertain parameters, p the number of parameter constraints, and $inst(\mathcal{U}_i)$ a value within the range of an uncertainty set. An instance of an UCSOP is a certain CSOP $(\mathcal{X}, \mathcal{D}, \mathcal{C}_\mathcal{X})$ such that for each uncertain constraint $C_i(X_1..X_m, \lambda_1, ..\lambda_m)$, we have $\lambda_j = inst(\mathcal{U}_j)$, such that $\forall k \in \{1, .., p\}$, the parameter constraint $C_k(\lambda_1, ..\lambda_m)$ is satisfied.*

Example 2. Continuing example 1, the UCSOP has two possible instances such that the parameter constraint $\lambda_2 = \lambda_1 + 3$ holds, given that $\lambda_1 \in \mathcal{U}_1, \lambda_2 \in \mathcal{U}_2$. The valid tuples (λ_1, λ_2) are $(2, 5)$, and $(3, 6)$. The CSOP instances are:

$$C_1 : X_1 > 2, C_2 : X_1 = X_2 + 5, C_3 : X_2 > 2$$

and
$$C_1 : X_1 > 3, C_2 : X_1 = X_2 + 6, C_3 : X_2 > 2$$
with the same objective function to maximize $f = X_1 + X_2$.

The generator of CSOP instances extracts the parameter constraints, polynomial in the number of constraints in the worst case, then produces a set of parameter tuples that satisfy the parameter constraints. We can use a branch and bound search on the parameter constraints of the UCSOP. The constraint simulation then substitutes the tuple solutions onto the original UCSOP to search for a solution to each optimization problem, that is each CSOP. This is polynomial in the complexity of the UCSOP. The process is depicted in Algorithm 1.

Algorithm 1. Generate and solve CSOPs from one UCSOP

   Input: A UCSOP $(\mathcal{X}, \mathcal{D}, \mathcal{C}_{\mathcal{X}} \cup \mathcal{C}_\Lambda, \Lambda, \mathcal{U}, f)$
   Output: Solutions to the CSOPs
1 $SolsTuples \leftarrow \emptyset$
2 extract($\mathcal{C}_\Lambda$)
3 $Tuples \leftarrow solveBB(\Lambda, \mathcal{U}, \mathcal{C}_\Lambda)$
4 **for** $T_i \in Tuples$ **do**
5 | substitute Λ with T_i in $(\mathcal{X}, \mathcal{D}, \mathcal{C}_{\mathcal{X}}, \Lambda, f)$
6 | $S_i \leftarrow solveOpt(\mathcal{X}, \mathcal{D}, \mathcal{C}_{\mathcal{X}}, T_i, f)$
7 | $SolsTuples \leftarrow SolsTuples \cup \{(S_i, T_i)\}$
8 **return** $SampleSols$

---

4   Application

We illustrate the benefits of our approach by solving an uncertain constraint optimization problem, the traffic matrix estimation for the sigcomm4 problem, given in Fig. 1. The topology and data values can be found in [8,12]. Given traffic measurements over each network link, and the traffic entering and leaving the network at the routers, we search the actual flow routed between every pair of routers. To find out how much traffic is exchanged between every pair of routers, we model the problem as an uncertain optimization problem that seeks the min and max flow between routers such that the traffic link and traffic conservation constraints hold. The traffic link constraints state that the sum of traffic using the link is equal to the measured flow. The traffic conservation constraints, two per router, state that the traffic entering the network must equal the traffic originating at the router, and the traffic leaving the router must equal the traffic whose destination is the router.

We compare three models. The first one does not consider any uncertain parameters and simplifies the model to only the variables in bold with coefficient 1. The traffic between routers takes a single fixed path, as implemented in [8]. The second model extends the first one with uncertain parameters but without the parameter dependency constraints. The third one is our approach with the parameter dependency constraints added. A parameter constraint, over the flow F_{AB}, for instance, states that the coefficients representing one given route of traffic from A to B take the same value; and the sum of coefficients corresponding to different routes equals to 1. Note that the uncertain parameter equality constraints are already taken into account in the link traffic constraints. The uncertain parameters relative to flow distributions are commonly assumed between 30 and 70 % [12].

Link traffic constraints:

$$[\lambda_{1_{AB}}, \lambda_{1_{AC}}, \lambda_{1_{AD}}, \lambda_{2_{AB}}, \lambda_{2_{AC}}, \lambda_{2_{AD}}] \in 0.3..0.7$$

$$
\begin{array}{lll}
A \to B & \lambda_{1_{AB}} * \mathbf{F_{AB}} + \lambda_{1_{AC}} * F_{AC} + \lambda_{1_{AD}} * F_{AD} & = 309.0..328.0 \\
B \to A & \mathbf{F_{BA}} + \mathbf{F_{CA}} + \mathbf{F_{DA}} + \lambda_{1_{BC}} * F_{BC} & = 876.39..894.35 \\
A \to C & \lambda_{2_{AC}} * \mathbf{F_{AC}} + \lambda_{2_{AD}} * \mathbf{F_{AD}} + \lambda_{2_{AB}} * F_{AB} + \lambda_{1_{BC}} * F_{BC} & = 591.93..612.34 \\
B \to C & \lambda_{2_{BC}} * \mathbf{F_{BC}} + \mathbf{F_{BD}} + \lambda_{1_{AC}} * F_{AC} + \lambda_{1_{AD}} * F_{AD} & = 543.30..562.61 \\
C \to B & \lambda_{2_{AB}} * F_{AB} + \mathbf{F_{CB}} + \mathbf{F_{CA}} + \mathbf{F_{DA}} + \mathbf{F_{DB}} & = 1143.27..1161.06 \\
C \to D & \mathbf{F_{CD}} + \mathbf{F_{BD}} + \mathbf{F_{AD}} & = 896.11..913.98 \\
D \to C & \mathbf{F_{DC}} + \mathbf{F_{DB}} + \mathbf{F_{DA}} & = 842.09..861.35
\end{array}
$$

Parameter constraints

$$\lambda_{1_{AB}} + \lambda_{2_{AB}} = 1, \lambda_{1_{AC}} + \lambda_{2_{AC}} = 1, \lambda_{1_{AD}} + \lambda_{2_{AD}} = 1, \lambda_{1_{BC}} + \lambda_{2_{BC}} = 1$$

Traffic conservation constraints

$$
\begin{array}{lll}
A \ origin & F_{AD} + F_{AC} + F_{AB} & = 912.72..929.02 \\
A \ destination & F_{DA} + F_{CA} + F_{BA} & = 874.70..891.00 \\
B \ origin & F_{BD} + F_{BC} + F_{BA} & = 845.56..861.86 \\
B \ destination & F_{DB} + F_{CB} + F_{AB} & = 884.49..900.79 \\
C \ origin & F_{CD} + F_{CB} + F_{CA} & = 908.28..924.58 \\
C \ destination & F_{DC} + F_{BC} + F_{AC} & = 862.53..878.83 \\
D \ origin & F_{DC} + F_{DB} + F_{DA} & = 842.0..859.0 \\
D \ destination & F_{CD} + F_{BD} + F_{AD} & = 891.0..908.0
\end{array}
$$

We first run the initial model and reproduced the results of [12] in constant time. By adding the uncertain prameters the solution bounds got much larger as the space of potential solutions expanded. However when we run simulations using our approach and the linear EPLEX solver, and we were not able to find any solution to the model with dependency constraints. This shows the importance of taking into account such dependencies, indicating that the data provided match a single path routing algorithm for the sigcomm4 topology. After enlarging the interval bounds of the input data we were able to find a solution with a 50 % split of traffic, but none with $40 - 60$ or other combinations. Our approach showed the effectiveness and strong impact of taking into account dependency constraints with simulations.

5 Conclusion

In this paper we introduced an approach to account for dependency constraints among data parameters in an uncertain constraint problem. The approach follows an iterative process that first satisfies the dependency constraints using a branch and bound search. The solutions are then embedded to generate a set of CSPs to be solved. However this does not indicate the relationship between the dependent consistent parameters and possible solutions. We propose to use regression analysis to do so. The current case study showed that by embedding constraint dependencies only one instance had a solution. This was valuable information on its own, but limited the use of regression analysis. Further experimental studies are underway with applications in inventory management, problems clearly permeated with data uncertainty. Even though our approach has been applied to traditional constraint problems in mind, its benefits could be stronger on data mining applications with constraints [6].

References

1. Benhamou, F., Goualard, F.: Universally quantified interval constraints. In: Dechter, R. (ed.) CP 2000. LNCS, vol. 1894, pp. 67–82. Springer, Heidelberg (2000)
2. Boukezzoula, R., Galichet, S., Bisserier, A.: A Midpoint Radius approach to regression with interval data. International Journal of Approximate Reasoning 52(9) (2011)
3. Brown, K., Miguel, I.: Uncertainty and Change. In: Handbook of Constraint Programming, ch. 21. Elsevier (2006)
4. Cheadle, A.M., Harvey, W., Sadler, A.J., Schimpf, J., Shen, K., Wallace, M.G.: ECLiPSe: An Introduction. Tech. Rep. IC-Parc-03-1, Imperial College London, London, UK
5. Chinneck, J.W., Ramadan, K.: Linear programming with interval coefficients. J. Operational Research Society 51(2), 209–220 (2000)
6. De Raedt, L., Mannila, H.: O'Sullivan and Van Hentenryck P. organizers. Constraint Programming meets Machine Learning and Data Mining. Dagstuhl seminar (2011)
7. Fargier, H., Lang, J., Schiex, T.: Mixed constraint satisfaction: A framework for decision problems under incomplete knowledge. In: Proc. of AAAI (1996)
8. Medina, A., Taft, N., Salamatian, K., Bhattacharyya, S., Diot, C.: Traffic Matrix Estimation: Existing Techniques and New Directions. In: Proceedings of ACM SIGCOMM 2002 (2002)
9. Rossi, F., van Beek, P., Walsh, T.: Handbook of Constraint Programming. Elsevier (2006)
10. Saad, A., Gervet, C., Abdennadher, S.: Constraint Reasoning with Uncertain Data using CDF-Intervals. In: Lodi, A., Milano, M., Toth, P. (eds.) CPAIOR 2010. LNCS, vol. 6140, pp. 292–306. Springer, Heidelberg (2010)
11. Van Hentenryck, P., Michel, L., Deville, Y.: Numerica: a Modeling Language for Global Optimization. The MIT Press, Cambridge (1997)
12. Yorke-Smith, N., Gervet, C.: Certainty Closure: Reliable Constraint Reasoning with Uncertain Data. ACM Transactions on Computational Logic 10(1) (2009)

Graph-Based Transfer Learning for Managing Brain Signals Variability in NIRS-Based BCIs

Sami Dalhoumi[1], Gérard Derosiere[2], Gérard Dray[1],
Jacky Montmain[1], and Stéphane Perrey[2]

[1] Laboratoire d'Informatique et d'Ingénierie de Production (LGI2P), Ecole des Mines d'Alès
Parc Scientifique G. Besse, 30035 Nîmes, France
{name.surname}@mines-ales.fr
[2] Movement to Health (M2H), Montpellier 1-University, Euromov,
700 Avenue du Pic Saint-Loup, 34090 Montpellier, France
{name.surname}@univ-montp1.fr

Abstract. One of the major limitations to the use of brain-computer interfaces (BCIs) based on near-infrared spectroscopy (NIRS) in realistic interaction settings is the long calibration time needed before every use in order to train a subject-specific classifier. One way to reduce this calibration time is to use data collected from other users or from previous recording sessions of the same user as a training set. However, brain signals are highly variable and using heterogeneous data to train a single classifier may dramatically deteriorate classification performance. This paper proposes a transfer learning framework in which we model brain signals variability in the feature space using a bipartite graph. The partitioning of this graph into sub-graphs allows creating homogeneous groups of NIRS data sharing similar spatial distributions of explanatory variables which will be used to train multiple prediction models that accurately transfer knowledge between data sets.

Keywords: Brain-computer interface (BCI), near-infrared spectroscopy (NIRS), brain signals variability, transfer learning, bipartite graph partitioning.

1 Introduction

A brain-computer interface (BCI) is a communication system that allows people suffering from severe neuromuscular disorders to interact with their environment without using peripheral nervous and muscular system, by directly monitoring electrical or hemodynamic activity of the brain [1]. Recently, near-infrared spectroscopy (NIRS) has been investigated for use in BCI applications [2]. NIRS-based BCIs employ near-infrared light to characterize alterations in cerebral metabolism during neural activation. During neural activation in a specific region of the brain, hemodynamic concentration changes in oxyhemoglobin (oxy-Hb) increase while those in deoxyhemoglobin (deoxy-Hb) decrease slightly [3] (see figure 1).

A. Laurent et al. (Eds.): IPMU 2014, Part II, CCIS 443, pp. 294–303, 2014.

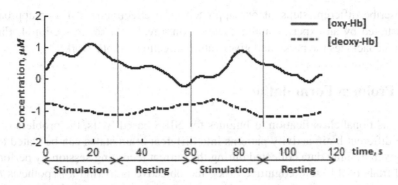

Fig. 1. Prototypical brain activity pattern using NIRS technology. Measured values of concentrations are relative and not absolute.

A BCI is considered as a pattern recognition system that classifies different brain activity patterns into different brain states according to their spatio-temporal characteristics [4]. The relevant signals that decode brain states may be hidden in highly noisy data or overlapped by signals from other brain states. Extracting such information is a very challenging issue. To do so, a long calibration time is needed before every use of the BCI in order to extract enough data used for feature selection and classifier training [5]. Because calibration is time-consuming and boring even for healthy users, several works in BCIs based on electroencephalography [1] addressed this problem by developing new data processing and pattern classification methods. [6] proposed a semi-supervised support vector machine (SVM) classifier designed to accurately classify brain signals with small training set. [7] used data recorded from the same subject during past recording sessions in order to determine prototypical spatial filters which have better generalization properties than session-specific filters. Other authors [5, 8-11] developed different subject-transfer frameworks to reduce calibration time before each use of a BCI. It consists of using data recorded from several users that performed the same experiment as a training set for a classifier that will be used to predict brain activity patterns of a new user. In NIRS-based BCIs, the problem of long calibration time has not been well addressed. Recent studies highlighted the need of changing data processing and classifiers design strategies in order to conceive more robust and practical BCIs [12-14]. To our best knowledge, [15] is the only work that addressed this issue by designing an adaptive classifier based on multiple-kernel SVM.

In this paper, we introduce a novel approach inspired from graph theory to address this problem. The novelty of our contribution compared to the works cited above lies in the fact of using a bipartite graph to have prior knowledge about variability in the feature space between brain signals recorded from different users. This prior knowledge allows designing a prediction model that adaptively chooses the best users set and features representation to accurately classify data of a new user and consequently reduce calibration time. The reminder of this paper is organized as follows. In section 2, we formalize the problem in the context of transfer learning between heterogeneous data. In section 3, we briefly review the background of bipartite graph partitioning and describe different steps of our approach. The effectiveness of our approach is demonstrated by an experimental evaluation on a real data set in section 4. Finally, section 5 concludes the paper and gives future directions of our work.

2   Problem Formulation

Using traditional classification techniques for NIRS-based BCIs, the problem of classifying different brain activity patterns into different brain states can be stated as follows: given training data collected during the current recording session by performing several trials of different cognitive tasks, the objective is to find a hypothesis h that allows good prediction of class labels corresponding to each cognitive task for the reminder of trials performed during the same session. Because classification performance of h increases with the size of training data, a long calibration time is necessary before each use of the BCI. In the context of transfer learning [16], the problem can be reformulated as follows. Given

1. A small amount of labeled NIRS signals recorded during the current session.
2. A large amount of labeled NIRS signals recorded during previous sessions of the same subject or from other subjects that performed the same experiment.

We want to exploit relatedness between these data sets in order to find a hypothesis h that allows good prediction of the class labels for the reminder of trials for the current session. Item 1 is called training set and item 2 is called support set. A single hypothesis h can achieve good classification performance when the training and support sets are assumed to be drawn from the same feature space and the same distribution. Such assumption may be too strong for our application because of high variability of NIRS signals collected from different subjects during different recording sessions which affects mostly the spatial distribution of explanatory variables.

In this paper, we design a prediction model which learns multiple hypotheses instead of one in order to overcome the problem of NIRS signals variability and accurately transfer knowledge between data of different individuals and different recording sessions. To do so, we divide our support set into multiple subsets $\{S_1, S_2,..., S_K\}$ and we learn a hypothesis h_i for each subset S_i. The choice of the most appropriate hypothesis to predict class label y for a new trial x in the current session depends on the number of explanatory features shared between the training set T and each subset S_i. The general architecture of our approach can be expressed in a probabilistic manner as follows:

$$P(y/x) = \sum_{i=1}^{K} P(h_i /T)P(y/x, h_i) \tag{1}$$

In the next section, we illustrate how bipartite graph partitioning allows creation of multiple hypotheses which minimizes the influence of NIRS signals variability in the feature space on classification performance.

3   Transfer Learning Framework Based on Bipartite Graph Partitioning

In order to design a transfer learning framework for heterogeneous NIRS data, we model the spatial variability of explanatory features between different data sets. To do so, we propose to borrow the bipartite graph partitioning from graph theory [17-18]. This technique allows performing simultaneous grouping of instances and features and consequently mapping of data into richer space. This is important for reducing the effect spatial variability of brain signals on classification performance because each hypothesis h_i is drawn from explanatory variables in subset S_i.

3.1   Bipartite Graph Partitioning

Before describing different steps of our approach, we start with relevant terminology related to bipartite graph partitioning.

A bipartite graph $G = (D, F, E)$ is defined by two sets of vertices $D = \{d_1, d_2, \ldots, d_N\}$ and $F = \{f_1, f_2, \ldots, f_M\}$, and a set of edges $E = \{(d_i, f_j) \mid d_i \in D \text{ and } f_j \in F\}$. In this paper, we assume that an edge $E_{ij} = (d_i, f_j)$ exists if vertices d_i and f_j are related (i.e., $E_{ij} \in \{0, 1\}$).

Assume that the set of vertices F is grouped into disjoint clusters $\{F_1, F_2, \ldots, F_K\}$. The set D can be clustered as follows: a vertex d_i ($i = 1 \ldots N$) belongs to the cluster D_p ($p = 1 \ldots K$) if its association with the cluster F_p is greater than its association with any other cluster in the vertex set F. This can be written as:

$$D_p = \{di \, / \sum_{j \in F_p} E_{ij} \geq \sum_{j \in F_l} E_{ij}, \forall \, l = 1, \ldots, K\} \tag{2}$$

Given disjoint clusters $D_1, \ldots, D_K$, the set of vertices F can be clustered similarly. As illustrated in [18], the optimal clustering of the two sets of vertices can be achieved when:

$$cut(D_1 \cup F_1, \ldots, D_K \cup F_K) = min_{V_1, \ldots, V_K} cut(V_1, \ldots, V_K) \tag{3}$$

where $V_1, \ldots, V_K$ is a K-partition of the bipartite graph (V_k, $k = 1, \ldots, K$ are sub-graphs of the graph G) and

$$cut(V_1, \ldots, V_K) = \sum_{i<j} \sum_{p \in D(V_i), q \in F(V_j)} E_{pq} \tag{4}$$

It is well known that graph partitioning is a NP-complete problem. There are many heuristics that were introduced to give better global solutions and reduce complexity of bipartite graph partitioning. In this work, we use spectral clustering which is an effective heuristic that uses properties of graph Laplacian matrix to solve this problem. Because of space limitations, we will not show details of this heuristic. For more information, see [18].

3.2   Overview of Our Graph-Based Transfer Learning Framework

In our support set, the first set of vertices $D = \{d_1, d_2, ..., d_N\}$ corresponds to different data sets of NIRS signals recorded from different subjects during different sessions and the second set of vertices $F = \{f_1, f_2, ..., f_M\}$ corresponds to M measurement channels placed on the same locations of participants' heads. A recording session consists of performing several trials of two cognitive tasks T_1 and T_2. Our approach is accomplished in three steps (figure 2).

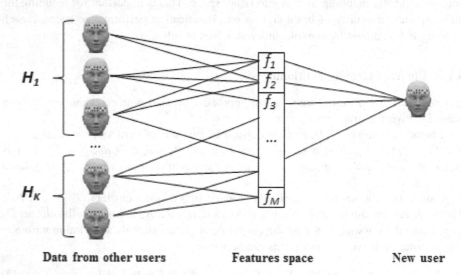

Data from other users   Features space   New user

Fig. 2. Bipartite graph model for characterizing brain signals variability in the features space between different users

Heterogeneous NIRS Signals Partitioning

For each data set d_i, we perform features selection in order to find channels that allow detection of signals amplitude changes between different cognitive tasks. Then, we assign the number 1 to explanatory channels and 0 to the rest (*i.e.*, $E_{ij} \in \{0,1\}$). This channel selection procedure allows us to create an N by M co-occurrences matrix of data sets and channels and consequently create a representation of spatial variability of brain activity patterns in heterogeneous NIRS data. The creation of different groups of data sets $\{D_1, ..., D_K\}$ sharing similar spatial distributions of brain activity patterns $\{F_1, ..., F_K\}$ is performed by applying bipartite graph partitioning to the co-occurrences matrix.

Classifiers Training

After creation of several groups of NIRS signals having local features representations, a single classifier is trained on each group. The training performance of each classifier is evaluated using leave-one-out cross validation. The global performance of our prediction model is the average of all classifiers performance. If it is below the required performance, the bipartite graph partitioning step is repeated with different number of partitions.

New NIRS Signals Classification Using the Multiple-Hypotheses Prediction Model

Once prediction model training is finished, NIRS signals recorded during a new session will be classified as follows: first, we find the group of data sets sharing the most similar spatial distribution of brain activity patterns and then use the hypothesis trained on that data to predict class labels of each trial in the new session. In real time conditions, assuming that spatial distribution of brain activity patterns do not vary significantly during the same session, only the first few trials (*i.e.*, training set T) are used to find the closest co-cluster in our support set.

In the probabilistic interpretation of our transfer learning framework given in (1), $P(h_i/T)$ is calculated using "the winner takes all" rule and consequently $P(y/x)$ will be determined using only one hypothesis H_{i^*}:

$$P(h_i/T) = \begin{cases} 1, & if \ \sum_{j \in F_i} E_{Tj} \geq \sum_{j \in F_l} E_{Tj}, \forall \ l = 1, ..., K \\ 0, & otherwise \end{cases} \tag{5}$$

Then, $\exists! \, i^* \, / \, P(h_{i^*}/T) = 1$ and

$$P(y/x) = \sum_{i=1}^{K} P(h_i/T)P(y/x, h_i) = P(y/x, h_{i^*}) \tag{6}$$

4   Experimental Evaluation

In this section, our approach is evaluated on a real NIRS-based BCI data set using a linear discriminant analysis (LDA) classifier, which is the most widely used classifier in BCI applications [4], as a base learner. Then, its performance is compared to a single LDA classifier.

4.1   Data Set Description

To evaluate our approach, we used the publicly available data set described in [15]. It is composed of NIRS signals recorded from seven healthy subjects using 16 measurement channels. The study consisted of two experiments, each one lasted four sessions. The aim of the first experiment was to discern brain activation patterns related to imagery movement of right forearm from the activation patterns related to relaxed state, denoted respectively t_1 and b. While the aim of the second one was to discern brain activation patterns related to imagery movement of left forearm t_2 from the activation patterns related to relaxed state b. During each session, participants performed three trials of b and three trials of t_1 for experiment 1 and three trials of b and three trials of t_2 for experiment 2. Thus, in each experiment, the first set of vertices D corresponds to 7 subjects × 4 sessions (*i.e.*, $D = \{d_1, d_2, ... d_{28}\}$), while the second set of vertices corresponds to *16* channels *(i.e., F = {f$_1$, f$_2$, ..., f$_{16}$})*. Because the majority of NIRS-based BCIs studies reported that deoxy-Hb does not necessarily show significant changes in activated areas of the brain [3], we focused only on oxy-Hb changes.

4.2   Data Preprocessing

NIRS signals going through human brain may be overlapped by many physiological *(e.g.,* respiration, heart beat) and experimental *(e.g.,* motion artifacts) sources of noise [3]. In order to minimize the effect of these sources of noise on classifiers performance, we applied a 5th-oder Butterworth low-pass filter with cut-of frequency of *0.5* Hz [15]. Another problem that we may encounter when we classify heterogeneous NIRS signals is the difference in amplitudes of hemodynamic brain activity between subjects, sessions or even trials performed during the same session [14]. To overcome this problem, we performed zero-mean and unit-variance normalization on time series data of each trial.

4.3   Results

In this study, we used the Wilcoxon signed-rank test, which is a non-parametric statistical hypothesis test, to compare mean amplitude oxy-Hb time series averaged over time windows of imagery movement and relaxed state. Among *28* data sets in each experiment, *15* data sets in experiment *1* illustrated a significant difference in oxy-Hb amplitudes between time windows of imagery movement and relaxed state and *22* data sets in experiment 2. Data sets in which no channel discerned activity patterns related to each brain state were removed resulting in *15* by *16* co-occurrences matrix for the first experiment and *22* by *16* co-occurrences matrix for the second one. Inter-subjects and inter-sessions variability in our data set are illustrated in figure 3.

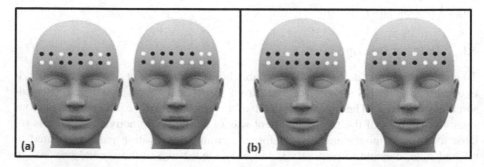

Fig. 3. Brain signals variability in NIRS-based BCIs. (a) Inter-sessions variability of explanatory channels for subject 5 in experiment 1. (b) Inter-subjects variability of explanatory channels for subjects 3 and 4 in experiment 2. White dots represent explanatory channels and black dots represent non-explanatory channels.

Because the number of trials in each session is not enough to evaluate our approach in online fashion, we tested it in a batch mode and compared classification results to a single LDA. As performance measurement, we used sensitivity and specificity which are respectively the probability that the activity pattern is classified as movement imagination given that the participant effectively performed movement imagination and the probability that the activity pattern is classified as relaxed state given that the participant was resting. The comparison results are illustrated in table 1.

Table 1. Comparison of classification performance of our approach and a single LDA classifier. The number of partitions is 2 in experiment 1 and 3 in experiment 2.

	Experiment 1		Experiment 2	
	Sensitivity	Specificity	Sensitivity	Specificity
LDA	0.60	0.71	0.65	0.68
Our approach	0.95	0.94	0.92	0.87

As expected, a single LDA classifier trained on heterogeneous data sets cannot find a hyper plane that separates well instances from different classes because each dataset's decision boundary lies in a different subspace. In contrast, projecting each group of data sets on the subspace spanned by their shared explanatory features and training different classifiers separately allows building a prediction model that significantly outperforms the former prediction model in classifying data recorded from a new user.

5   Conclusions and Directions for Future Work

This paper proposed a graph-based transfer learning framework for accurately classifying heterogeneous NIRS signals recorded from different subjects during different sessions. It consists of modeling brain signals variability in the feature space using a bipartite graph and adaptively choosing the best users set and features representation used to train a classifier that will predict class labels of brain signals recorded during new session. The experimental evaluation of our approach compared to a single LDA classifier showed that our approach accurately transfers knowledge between different data sets despite the high variability of spatial distributions of explanatory channels between different users and different sessions of the same user.

Although first results are promising, many issues should be considered in future work. Boolean representation of the edges in our bipartite graph model may be not suitable for distinguishing robust features from non-robust ones. In fact, in NIRS-based BCIs there is a phenomenon of saturation which means that after performing several trials of different cognitive tasks the changes in oxy-hemoglobin and deoxy-hemoglobin concentrations between different brain states become non-significant in some regions of the brain [14]. Thus, features weighting may be important for designing robust classifiers which maintain good classification performance for long periods of time. Another important issue related to our transfer learning framework is the classifiers aggregation method given in (5). Choosing only one hypothesis may be non-suitable when the data set of current user shares many significant features with more than one partition in the support set. Other classifiers aggregation methods like weighted sum [19] and fuzzy integrals [20] will be investigated in future work.

Acknowledgement. We would like to thank Berdakh Abibullaev for providing the NIRS data set available on the website: http://kernelx.net/fnirs-bci-data/.

References

1. Nicolas-Alonso, L.F., Gomez-Gil, J.: Brain Computer Interfaces, a Review. Sensors 12, 1211–1279 (2012)
2. Coyle, S., Ward, T., Markham, C., McDarby, G.: On the suitability of near-infrared (NIR) systems for next-generation brain–computer interfaces. Physiological Measurement 25, 815–822 (2004)
3. Sitaram, R., Caria, A., Birbaumer, N.: Hemodynamic brain-computer interfaces for communication and rehabilitation. Neural Networks 22, 1320–1328 (2009)
4. Lotte, F., Congedo, M., Lécuyer, A., Lamarche, F., Arnaldi, B.: A Review of Classification Algorithms for EEG-based Brain-Computer Interfaces. Journal of Neural Engineering 4, R1–R13 (2007)
5. Tu, W., Sun, S.: A subject transfer framework for EEG classification. Neurocomputing 82, 109–116 (2011)
6. Li, Y., Guan, C., Li, H., Chin, Z.: A self-training semi-supervised SVM algorithm and its application in an EEG-based brain computer interface speller system. Pattern Recognition Letters 29, 1285–1294 (2008)
7. Krauledat, M., Tangermann, M., Blankertz, B., Muller, K.R.: Towards Zero Training for Brain-Computer Interfacing. Plos One 3(8), e2967 (2008)
8. Lotte, F., Guan, C.: Learning from other subjects helps reducing brain-computer interface calibration time. In: International Conference on Audio Speech and Signal Processing (ICASSP), pp. 614–617 (2010)
9. Falzi, S., Grozea, C., Danoczy, M., Popescu, F., Blankertz, B., Muller, K.R.: Subject independent EEG-based BCI decoding. In: Neural Information Processing Systems Conference (NIPS), pp. 513–521 (2009)
10. Samek, W., Meinecke, F.C., Muller, K.R.: Transferring Subspaces Between Subjects in Brain-Computer Interfacing. IEEE Transactions on Biomedical Engineering 60(8), 2289–2298 (2013)
11. Lu, S., Guan, C., Zhang, H.: Unsupervised Brain Computer Interface Based on Intersubject Information and Online Adaptation. IEEE Transactions on Neural Systems and Reabilitation Engineering 17(2), 135–145 (2009)

12. Sato, H., Fushino, Y., Kiguchi, M., Katura, T., Maki, A., Yoro, T., Koizumi, H.: Intersubject variability of near-infrared spectroscopy signals during sensorimotor cortex activation. Journal of Biomedical Optics 10(4), 44001 (2005)

13. Power, S.D., Kushki, A., Chau, T.: Intersession Consistency of Single-Trial Classification of the Prefrontal Response to Mental Arithmetic and the No-Control State by NIRS. Plos One 7(7), e37791 (2012)

14. Holper, L., Kobashi, N., Kiper, D., Scholkmann, F., Wolf, M., Eng, K.: Trial-to-trial variability differentiates motor imagery during observation between low versus high responders : A functional near-infrared spectroscopy study. Bihavioural Brain Research 229, 29–40 (2012)

15. Abibullaev, B., An, J., Jin, S.H., Lee, S.H., Moon, J.I.: Minimizing Inter-Subject Variability in fNIRS-based Brain-Computer Interfaces via Multiple-Kernel Support Vector Learning. Medical Engineering and Physics, S1350-4533(13)00183-5 (2013)

16. Pan, S.J., Yang, Q.: A Survey on Transfer Learning. IEEE Transactions on Knowledge and Data Engineering 22(10), 1345–1359 (2010)

17. Zha, H., He, X., Ding, C., Simon, H., Gu, M.: Bipartite Graph Partitioning and Data Clustering. In: CIKM 2001, Atlanta, Georgia, USA (2001)

18. Dhillon, I.S.: Co-clustering documents and words using Bipartite Spectral Graph Partitioning. In: KDD, San Francisco, California, USA (2001)

19. Kittler, J., Hatef, M., Duin, R.P.W., Matas, J.: On Combining Classifiers. IEEE Transactions on Pattern Analysis and Machine Intelligence 20(3), 226–239 (1998)

20. Pizzi, N.J., Pedrycz, W.: Aggregating multiple classification results using fuzzy integration and stochastic feature selection. International Journal of Approximate Reasoning 51(8), 883–894 (2010)

Design of a Fuzzy Affective Agent Based on Typicality Degrees of Physiological Signals

Joseph Onderi Orero[1] and Maria Rifqi[2]

[1] Faculty of Information Technology, Strathmore University, Kenya
jorero@strathmore.edu
[2] LEMMA, University Panthéon-Assas, France
maria.rifqi@lip6.fr

Abstract. Physiology-based emotionally intelligent paradigms provide an opportunity to enhance human computer interactions by continuously evoking and adapting to the user experiences in real-time. However, there are unresolved questions on how to model real-time emotionally intelligent applications through mapping of physiological patterns to users' affective states.

In this study, we consider an approach for design of fuzzy affective agent based on the concept of typicality. We propose the use of typicality degrees of physiological patterns to construct the fuzzy rules representing the continuous transitions of user's affective states. The approach was tested on experimental data in which physiological measures were recorded on players involved in an action game to characterize various gaming experiences. We show that, in addition to exploitation of the results to characterize users' affective states through typicality degrees, this approach is a systematic way to automatically define fuzzy rules from experimental data for an affective agent to be used in real-time continuous assessment of user's affective states.

Keywords: Machine learning, fuzzy logic, prototypes, typicality degrees, affective computing, physiological signals.

1 Introduction

Emotions play a significant role in normal human relations. Other than speech and text channels, humans communicate very easily through implicit emotion expressions. In many cases, humans attach importance on *how* something is done or said and not only what was said or done. Communication is continually modulated and enhanced through and by emotions. On the contrary, in human computer interaction, computers rely mainly on the input and output, whose effectiveness is solely measured by the ability to execute and not *how* it was executed. There is a need to develop methodologies for assessing user's emotional experience while interacting with computers. As a result, affective computing [16] has become a major research interest in Human Computer Interaction (HCI). The main concern in this domain is how to enhance the quality of interaction

A. Laurent et al. (Eds.): IPMU 2014, Part II, CCIS 443, pp. 304–313, 2014.
© Springer International Publishing Switzerland 2014

between the user and the computer, making it more enjoyable by automatically recognizing and adapting to the user's emotional states.

In this context, among a vast range of possible ways to continuously assess a user's emotional responses such as facial gesture or voice recognition through video and audio recording, physiological measures have a key advantage. In addition to their ability to be measured continuously in real-time, physiological data are a record of involuntary autonomic nervous system processes and therefore not culture specific like other modes.

In the recent past, scientific works have demonstrated the enormous prospects in developing systems equipped with the ability to assess user emotional states through the analysis of physiological data [8,2,14]. However, there are unresolved questions on how to model real-time emotionally intelligent applications through mapping of physiological patterns to users' affective states. In this study, we consider a model that employs typicality degrees to construct fuzzy rules that summarize cardinal physiological properties of users' affective states. We show that, in addition to systematic exploitation of the data to characterize affective states through typicality degrees, the results can be extended to construct fuzzy rules for an affective agent to be used in real-time to assess user's affective states.

The rest of the paper is organized as follows: in Section 2, we give the state of the art on physiological and affective computing. In Section 3, we give details of our approach. Then, in Section 4, we outline the details of the experimental data for the design of affective controller and give analysis of the test results. Finally, we give conclusions and future perspectives in Section 5.

2   State of the Art

2.1   Modeling User's Affective States

In the field of psychophysiology, many studies have been conducted to discover relationships between certain psychological states and physiological activity [3,8,2,14, gives a comparative study]. The experimental conceptualization has heavily relied either on the emotional dimensional theory or the basic emotions theory. On the one hand, dimensions theory considers emotions as dynamical, boundaries-free states mainly represented in terms of arousal and valence [22,9]. On the other hand, in the basic emotions theory, emotions are conceptualized as static, biologically-rooted states or classes such as anger, disgust, fear, joy, sadness and surprise [7]. Other than variation in the methods of evoking emotions and physiological measures used, there is a large volume of work to discriminate emotions through physiological measures represented as the distinct classes [17,18,3, among others]. For example, in [18] an attempt was made to prove that the basic emotions of fear, anger, sadness and happiness are associated with distinct patterns of cardiorespiratory activity.

In relation to gameplay design, the user affective states are modeled mainly based on theory of *flow* [5]. In relation to *flow*, experiments are designed so as to discriminate between three states based on game challenge [12,4]: easy game (boredom), medium (engagement/flow) and very difficult (anxiety/frustration).

The agenda is to devise paradigms that enable real-time game adaptation according to the player's affective states. For example, in the study by [4], participants played Tetris and the player's affective states were hypothesized according to three difficulty levels to give rise to: *boredom, engagement/comfort* and *anxiety/frustration* states. In this work, we model the affective states of an affective agent based on these three states of game challenge (low challenge, medium challenge and very high challenge).

2.2   Assessment User's Affective States through Physiology

Various methods have been used in physiological modeling of basic emotions such as k-nearest neighbors algorithm, discriminant analysis, support vector machines, bayesian networks and decision trees [17,22,19,9,3]. A compressive comparative study on these methods has been given in [14]. Specifically, in modeling gameplay through physiology, similar methods have been proposed such artificial neural networks [23], support vector machines [4], decision trees [12,15,11]. Although some of these models may have proved to successful in discriminating affective states, there has been less focus on how the extracted results should be easily integrated in design of a real-life affective systems.

In this context, [12]'s work is closer to our current study. In their study, a decision trees based framework was used to map physiological measures to three levels of anxiety based on difficulty: easy, moderately difficult and very difficult. The extracted rules from the decision tree was then used to construct a controller for an adaptive game. In our study, we consider a derivation of fuzzy rules from the physiological data for the affective controller.

2.3   Fuzzy Model in Physiological Computing

To begin with, modeling of affective states through physiology involves aggregation and fusion of various physiological measures and features. But the physiological data from sensors is itself imperfect, such that it is difficult to express the results in crisp terms [1]. As such, it is necessary to express in fuzzy terms the relationship between physiological data and affective markers. Also, emotions are conceptual quantities with indeterminate fuzzy boundaries [2]. As such changes from one psychological state to the next can be gradual rather than abrupt, owing to overlapping of class boundaries.

Models based on fuzzy logic are naturally most appropriate to represent the continuous transitions, uncertainties and imperfections associated with physiological data. In a fuzzy controller changes from one rule to another is gradual with fuzzy values $[0, 1]$ instead of crisp values $\{0, 1\}$ in classical controller. Therefore, in addition to robust decision making, fuzzy logic based models enable a continuous assessment of these states to evoke the appropriate response.

In this domain, a fuzzy expert system has been proposed by [13]. In their study, an expert knowledge derived from psychophysiological studies literature was used to directly define fuzzy rules for each physiological signal. However, automatic construction of fuzzy rules through induction of the experimental data has not been explored before. Automatically defining fuzzy rules based on induction from experimental data has key advantages as it avoids prior definition of rules by an expert, which are often subjective and difficult to validate. Moreover, it enables us to validate and compare the results with existing psychophysiological studies.

In this study, we propose typicality-based approach to automatically construct fuzzy rules from the experimental data. The aim is to discover typical psychophysiological patterns through induction of the collected physiological data. We extract the most typical physiological patterns that best describe a given user affective state and use them as prototypes to construct fuzzy rules relating to the affective states.

3 Our Approach

3.1 Typicality and Prototypes from Cognitive Science Perspective

The concept of typicality and prototypes has been studied in the field cognitive science and psychology, initially by Rosch and Mervis [21]. According to their study, *typicality* relies on the notion that some members of the same category are more characteristic of the category they belong to than others. This is contrary to the traditional thoughts that have treated category membership of items as possessing a full and equal degree of membership. Some members are more characteristic (*typical*) of the category they belong i.e have features that can be said to be most descriptive of that category. Thus, subjects can belong to the same category but differing in their level of typicality. Typicality degree of an item depends on two factors [21]; *internal resemblance*: an object's resemblance to the other members of the same category, and *external dissimilarity*; its dissimilarity to the members of the other categories.

The concept of typicality can be used to define *prototypes* for a given group or category as an object that summarizes the characteristics of the group. In this case, a prototype of a given category is the object with the highest typicality in that category.

3.2 Typicality Degrees

The aim is to discover pertinent psychophysiological characteristics based on the concept of typicality. Since the typicality degree of an object indicates the extent to which it resembles the members of the same group and differs from the members not in the same group, we can measure its power to characterize i.e its ability to summarize the cardinal properties of a group. Specifically, we consider Rifqi's formalism [20] that computes the typicality degrees of objects to automatically construct fuzzy prototypes.

Formally, let X be a data set composed of m instances in n dimensional space and labeled to belong to a particular state or class, $k \in C$ and $C = \{1, \ldots k, \ldots c\}$ where c is the number of all possible states or classes.

It computes for each example, $x \in X$, belonging to a given class, k, its internal resemblance, $R(x)$, the aggregate of similarity to the members in the same class and its external dissimilarity, $D(x)$, the aggregate of dissimilarity to members not in the same class. The typicality degree, $T(x)$, of x is then computed as the aggregate of these two quantities given as:

$$R(x) = \frac{1}{|k|} \sum_{y \in k} r(x, y) \tag{1}$$

$$D(x) = \frac{1}{|X \backslash k|} \sum_{z \notin k} d(x, z) \tag{2}$$

$$T(x) = t(R(x), D(x)) \tag{3}$$

Where r is a similarity measure for computing internal resemblance, d is a dissimilarity measure for computing external dissimilarity, and t is an aggregation operator for aggregating resemblance and dissimilarity. y is used to designate examples belonging to the same class while z designates examples not belonging to the same class as the given example x.

The choice of similarity measures, dissimilarity measures and aggregation operators depends on the nature of the desired properties and have been studied in detail [6,10]. In this study, we choose to use the normalized euclidian distance as dissimilarity measure in Equation 2 and its complement as a similarity measure in Equation 1. This ensures that both the internal resemblance and external dissimilarity on a comparative scale. Then, to compute typicality degrees, as an aggregation of internal resemblance and external dissimilarity in Equation 3, we chose to use the symmetric sum. The symmetric sum has a reinforcement property [6]. In such a case, if both the similarity and the dissimilarity are high, the aggregated value becomes higher than any of the two and if both are low, the aggregate becomes lower than any of the two values. This ensures that the aggregation is high only if both the similarity and the dissimilarity are high and vise versa.

We consider a prototype formulated by computing typicality degrees attribute by attribute. In our context, we exploit this concept of typicality to determine the characterization power of a given physiological feature from the typicality degree of its prototypes. If an attribute typicality degree is high, then it follows that the attribute is relevant in characterizing the given state. On the contrary, if the typicality is low, then the attribute alone, can not be used as reference for characterizing the given states.

3.3   Defining Fuzzy Rules Based on Typicality Degrees

In this work, we wish to extend the concept of typicality degrees to construct fuzzy rules for each state or class. We define fuzzy values based on typicality of a particular attribute. We make the most typical example to have a full membership of 1 and decreases until zero where the membership of the other state is 1.

Formally, given $v_1, v_2, \ldots, v_k \ldots v_s$ as the most typical values for states $c_1, c_2, \ldots, c_k, \ldots, c_s$ respectively and $v_1 < v_2 < \cdots < v_k < \cdots < v_s$. We use triangular functions (for state c_k) and trapezoidal functions (R-function for state c_1 and L-function for state c_s). Therefore we have a membership function, μ_k, for a given state, c_k, in respect to values, ϑ of a given attribute given by:

$$\mu_1(\vartheta) = \begin{cases} 0 & \text{if } \vartheta > v_2 \\ \frac{v_2 - \vartheta}{v_2 - v_1} & \text{if } v_1 \leq \vartheta \leq v_2 \\ 1 & \text{if } \vartheta < v_1 \end{cases} \tag{4}$$

$$\mu_k(\vartheta) = \begin{cases} 0 & \text{if } \vartheta < v_{k-1} \\ \frac{\vartheta - v_{k-1}}{v_k - v_{k-1}} & \text{if } v_{k-1} \leq \vartheta < v_k \\ \frac{v_{k+1} - \vartheta}{v_{k+1} - v_k} & \text{if } v_k \leq \vartheta \leq v_{k+1} \\ 0 & \text{if } \vartheta > v_{k+1} \end{cases} \tag{5}$$

$$\mu_c(\vartheta) = \begin{cases} 0 & \text{if } \vartheta < v_{s-1} \\ \frac{\vartheta - v_{s-1}}{v_s - v_{s-1}} & \text{if } v_{s-1} \leq \vartheta \leq v_s \\ 1 & \text{if } \vartheta > v_s \end{cases} \tag{6}$$

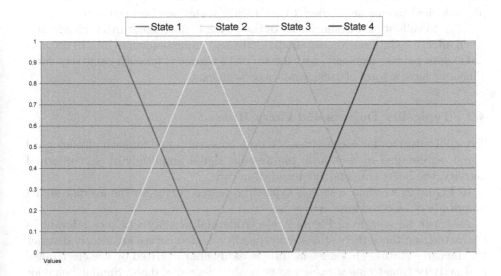

Fig. 1. Representation of fuzzy memberships against signal values

In Figure 1, we show a representation of fuzzy memberships against signal values. After constructing the fuzzy rules for each physiological feature, we use the weighted mean based on the typicality degree of each attribute (feature) as aggregation operator.

4   Experimental Data

4.1   Training Data Construction

To compute the typicality degrees for physiological features, we use data from an experimental study in which physiological measures were recorded on players involved in an action game as detailed in [11]. During this experiment, participants played successively four game sequences. Three of the four sequences that were distinguished in terms of level of challenge and user's affective states we classify the players' experiences in relation to appraisal of challenge in three distinct categories: boredom (due to an insufficient challenge), flow/comfort (due to comfortable level of challenge) and frustration (due to very high challenge).

We use three physiological measures recorded on thirty (30) participants during the experiment: Electrodermal activity measure (EDA), Heart Rate (HR) and Respiration Rate (RESP). To minimize the effect of the transition periods, we used only the physiological recordings of the last two minutes of each game sequence.

For each participant, normalization was done for each signal using the minimum and the maximum value of the signal for that participant. This is because physiological signals are subject to significant variations between individuals.

Also, to validate the homogeneity of the physiological signatures throughout a given affective state session, we subdivided these game sequences into 10 seconds (2000 data points) segments, with a total of 12 segments for each game sequence. Thus, we have a total of 1080 samples or segments.

4.2   Typicality Degrees and Fuzzy Rules

Typicality degrees for each segment/sample was computed for each physiological signal, as detailed in Section 3: the average signal amplitude of electrodermal activity, the heart rate and the respiration rate.

In Figure 2, we show a sample typicality degree curves for the average signal amplitude of electrodermal activity.

As it can be seen from Figure 2, electrodermal activity was very relevant in characterizing these three states: its typicality degree curves are clearly distinct for the three states. The *boredom* state is easily characterized by low electrodermal activity (*most typical value* has typicality degree of 0.85). Similar behavior is seen in the case of *frustration* state. The heart rate also had three distinct patterns for the three states but typicality degrees were lower than those of electrodermal activity. However, respiration rate's characterization power is low for all the three states i.e: the prototype typicality degree is less than 0.6 for all the three states and their curves are almost horizontal.

As detailed in Section 3, the fuzzy affective agent consists of a set of fuzzy rules of these features that are used to determine the user's affective state from experimental data. In this study, we focus on the affective states in relation to the *Flow* in the game. The affective states are as presented before: *Boredom*, *Comfort* (Flow) and *Frustration*.

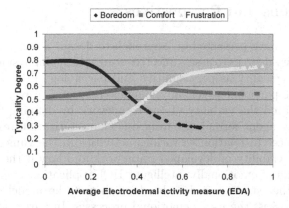

Fig. 2. Flow states typical values for average electrodermal activity

In this case, we have $v_1, v_k = v_2, v_s = v_3$ as the most typical values for states c_1, c_2 and c_3 respectively and $v_1 < v_2 < v_3$. We use triangular functions (for state c_2) and trapezoidal functions (R-function for state c_1 and L-function for state c_3) as given in Equation 5, Equation 4 and Equation 6. The values for Figure 2 above were: $v_1 = 0.088, v2 = 0.420, v3 = 0.954$. Based on Equation 4, Equation 5 and Equation 6, respectively we have the fuzzy rules shown on Figure 3. For example, *if the EDA* < 0.088 *then Boredom* = 1, *Comfort* = 0 *and Frustration* = 0, ... *if the EDA* > 0.95 *then Boredom* = 0, *Comfort* = 0 *and Frustration* = 1 .

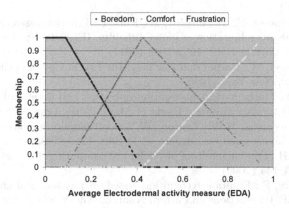

Fig. 3. Fuzzy rules in flow states for average electrodermal activity

As it can be seen, the constructed fuzzy rules are clearly realigned with the typicality degrees of the signal in various affective states. The constructed fuzzy rules represent a more objective representation of psychophysiological relations as they are induced from real experimental data.

5   Conclusions and Perspectives

In this study, we have presented an approach to design a fuzzy affective agent based on typicality degrees to assess affective states through physiological signals. We considered typicality as per cognitive and psychology principles of categorization to define pertinent psychophysiological relations.

We demonstrated the viability of our framework by defining fuzzy rules for design of a fuzzy agent formulated from our characterization results to assess player's experiences in relation to various activities in the game. This demonstrates both the viability of developed framework and how it the model can be used in the design of emotionally intelligent HCI applications.

However, further studies need to be done to test the model on other real-life scenarios to assess the user's emotional processes. In particular, to consider multi-modal fusion of measures such as audio-visual and various physiological measures.

References

1. Bouchon-Meunier, B.: Aggregation and Fusion of Imperfect Information. Physica-Verlag, Spring-Verlag Company (1998)
2. Calvo, R.A., D'Mello, S.: Affect detection: an interdisciplinary review of models, methods, and their applications. IEEE Transactions on Affective Computing 1, 18–37 (2010)
3. Chanel, G., Kierkels, J.J.M., Soleymani, M., Pun, T.: Short-term emotion assessment in a recall paradigm. International Journal of Human-Computer Studies 67, 607–627 (2009)
4. Chanel, G., Rebetez, C., Btrancourt, M., Pun, T.: Emotion assessment from physiological signals for adaptation of game difficulty. IEEE Transactions on Systems, Man, and Cybernetics 41(6), 1052–1063 (2011)
5. Csikszentmihalyi, M.: Harper and row flow: the psychology of optimal experience. Harper & Row, New York (1990)
6. Detyniecki, M.: Mathematical aggregation operators and their application to video querying. PhD thesis, Université Pierre et Marie Curie, France (2001)
7. Ekman, P., Levenson, R.W., Friesen, W.V.: Autonomic nervous system activity distinguishes among emotions. Science 221, 1208–1210 (1983)
8. Fairclough, S.H.: Fundamentals of physiological computing. Interacting With Computers 21, 133–145 (2009)
9. Kim, J., André, E.: Emotion recognition based on physiological changes in music listening. IEEE Transactions on Pattern Analysis And Machine Intelligence 30(12), 2067–2083 (2008)
10. Lesot, M.-J., Rifqi, M., Bouchon-Meunier, B.: Fuzzy prototypes: from a cognitive view to a machine learning principle. In: Bustince, H., Herrera, F., Montero, J. (eds.) Fuzzy sets and Their Extensions: Representation, Aggregation and Models. STUDFUZZ, vol. 220, pp. 431–452. Springer, Heidelberg (2008)
11. Levillain, F., Orero, J.O., Rifqi, M., Bouchon-Meunier, B.: Characterizing player's experience from physiological signals using fuzzy decision trees. In: IEEE Conference on Computational Intelligence and Games (2010)

12. Liu, C., Agrawal, P., Sarkar, N., Chen, S.: Dynamic difficulty adjustment in computer games through real-time anxiety-based affective feedback. International Journal of Human-Computer Interaction 25(6), 506–529 (2009)
13. Mandryk, R.L., Atkins, M.S.: A fuzzy physiological approach for continuously modeling emotion during interaction with play technologies. International Journal of Human-Computer Studies 65(4), 329–347 (2007)
14. Novak, D., Mihelj, M., Munih, M.: A survey of methods for data fusion and system adaptation using autonomic nervous system responses in physiological computing. Interacting with Computers 24, 154–172 (2012)
15. Orero, J.O., Levillain, F., Damez-Fontaine, M., Rifqi, M., Bouchon-Meunier, B.: Assessing gameplay emotions from physiological signals: a fuzzy decision trees based model. In: International Conference on Kansei Engineering and Emotion Research (2010)
16. Picard, R.W.: Affective computing. The MIT Press, Cambridge (1997)
17. Picard, R.W., Vyzas, E., Healey, J.: Toward machine emotional intelligence: Analysis of affective physiological state. IEEE Transactions Pattern Analysis and Machine Intelligence 23, 1175–1191 (2001)
18. Rainville, P., Bechara, A., Naqvi, N., Damasio, A.R.: Basic emotions are associated with distinct patterns of cardiorespiratory activity. International Journal of Psychophysiology 61, 5–18 (2006)
19. Rani, P., Sarkar, N., Adams, J.: Anxiety-based affective communication for implicit human machine interaction. Advanced Engineering Informatics 21, 323–334 (2007)
20. Rifqi, M.: Constructing prototypes from large databases. In: International Confrence on Information Processing and Management of Uncertainity in Knowledge-Based System, IPMU (1996)
21. Rosch, E., Mervis, C.: Family resemblance: studies of the internal structure of categories. Cognitive Psychology 7, 573–605 (1975)
22. Wagner, J., Kim, J., André, E.: From physiological signals to emotions: implementing and comparing selected methods for feature extraction and classification. In: IEEE International Conference in Multimedia and Expo, pp. 940–943 (2005)
23. Yannakakis, G.N., Hallam, J.: Entertainment modeling through physiology in physical play. International Journal of Human-Computer Studies 66, 741–755 (2008)

Multilevel Aggregation of Arguments in a Model Driven Approach to Assess an Argumentation Relevance

Olivier Poitou and Claire Saurel

ONERA
Toulouse, France
{first name.last name}@onera.fr

Abstract. Figuring out which hypothesis best explain an observed on-going situation can be a critical issue. This paper introduces a generic model based approach to support users during this task. It then focuses on an hypothesis relevance scoring function that helps users to efficently build a convincing argumentation towards hypothesis. This function uses a multi-level extension of Yager's aggregation algorithm, exploiting both the strength of the components of an argumentation, and the confidence the user puts in them. The presented work was illustrated on a maritime surveillance application.

Keywords: decision support, argumentation, model-based approach, multilevel multicriteria aggregation.

1 Introduction

As far as sensor, vision, information, communication or intelligence technologies have been widely developed and improved, a worthy challenge is to exploit these capacities to efficiently support surveillance activities in operational situations. In lots of critical domains surveillance is still performed by human operators only equipped with paper and pencils. In those conditions, dealing with a huge amount of uncertain and partial data, more or less reliable information sources, and evolving and imperfectly defined potential threats within a sometime very short time period is almost unfeasible or at least error prone. Maritime domain provided such an initial case study [1] for the theoretical work presented here[1].

One aim is so to provide users with support in identifying ongoing abnormal behaviours in order to ease their decision making to face up to them. Automatic model-based approaches often fail because it is difficult to capture all characteristics of all abnormal behaviours in models : either new occurrences of a generic behaviour have non modelled specificities, or not yet modelled behaviours appear. So human intervention seems necessary in a model-based situation recognition process. However, we claim that a model-based *assistance* can help users

---

[1] Our work has been funded by several projects since 2009: TaMaris (French Research Agency , ANR), SisMaris (FUI): http://www.sismaris.org/ and now I2C (EU): http://www.i2c.eu/, and also by ONERA fundings.

A. Laurent et al. (Eds.): IPMU 2014, Part II, CCIS 443, pp. 314–323, 2014.

focusing their information gathering and analysis process, hence improving the surveillance process efficiency. Moreover, as an assistance process keeps the user in the loop it should ease users' acceptance and promotion.

The generic approach we propose aims at : guiding users' investigation, providing them with argumented hypothesis about ongoing situation identification, giving them a mean to compare argumentations. Users can then compare hypothesis relevance to explain the observed situation, before choosing which hypothesis appear to them as the most likely ones, then taking suited decision.

Our approach is based on models of behaviours which can explain most of abnormal or suspicious situations that are to be interpreted and recognized. Starting from alarms automatically raised about an observed situation[2] it consists first in providing users with an initial set of abnormal behaviour hypothesis which might explain the alarms, then in assisting them into further hypothesis assessment, to get the hypothesis which best explain the ongoing situation. An hypothesis assessment is defined as the score of its argumentation – this argumentation being continuously built from information collected by users on the observed situation according to the contents of the associated model. A multilevel aggregation algorithm is then used to assess hypothesis relevance to explain the ongoing situation, by merging some characteristics of pro or cons arguments.

This paper focuses on information modeling and processing techniques underlying the behaviour hypothesis generation and assessment support.

Section 2 introduces the principles of the behaviour model-based approach. Behaviour models and their formalization are described in section 3. In section 4 we explain how behaviour models generate behaviour hypothesis, and how an argumentation is implicitly created for each hypothesis, based upon user's beliefs collected about the situation. Then in section 5 we make a focus on a multilevel aggregation algorithm we have developed, in order to assess the behaviour hypothesis matching to the ongoing situation. This algorithm here deals with two ingredients of the quality of its argumentation : argument strength, and user's confidence in its arguments. Section 6 concludes by positioning this work.

2   An Abnormal Behaviour Model-Driven Approach

Our approach relies on an abnormal behaviour models base which collects structured business knowledge, gathering some features which are often tied to occurrences of the described behaviour or, conversely, which generally disclaim the described behaviour. When an alarm is thrown, it comes in with a set of information about the ongoing situation. The interpretation support component then goes through all the abnormal behaviour models, and if some alarm information matches, even partially, a model contents, then this model is retained as an hypothesis of ongoing situation identification. From this initial set of hypothesis, the aim is then to support users in further assessment : by using the model contents to guide them in their collection of additional relevant information on

---

[2] Meaning for instance that some physical parameters, as reported by sensors, have abnormal values according to business rules.

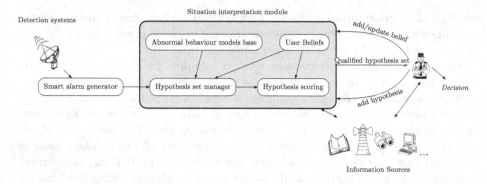

Fig. 1. A model-driven approach for inquiry assistance

the ongoing situation, and by evaluating how each hypothesis is likely to explain the observed situation at any state of the users' knowledge acquisition (see Figure 1).

The process ends when the user decides to stop the hypothesis investigation, generally once he or she has got some hypothesis with a fairly high relevance evaluation score.

3 Abnormal Behaviour Models: Concepts and Structure

Abnormal behaviour models describe specific features of situations in a given application domain, according to users. These features are structured as sets of conditions which roughly correspond to important clues of the ongoing occurence or non occurence of the behaviour. Conditions are modelled as a user understandable label and its pro or cons impact on the identification of the behaviour. These conditions may be associated to a list of sources with a default confidence value.

The contents of a behaviour model is represented within the following formal frame.

Let $\mathbb{P}$ be the set of all domain available propositions, supposed to be logically independent.[3]

Let $\mathbb{A}$ be the set of argumentary forces $\{-\mu, -3, -2, -1, 0, 1, 2, 3, \mu\}$, where μ and $-\mu$ are pure symbolic values.

Let $\mathbb{M}$ be the abnormal behaviour model set, with $\forall m \in \mathbb{M}$, $m =< id_m, \mathcal{C}_m >$ stating that :

- $\mathcal{P}_m \subset \mathbb{P}$, the set of propositions considered by users to be relevant for the model m definition

---

[3] Proposition labels will all be expressed with affirmative interrogative sentences, since negative sentences are not user-friendly. For instance, in a context of maritime surveillance : *is the ship older than 15 years old ?*

- $C_m = \{(p, f, cf) \mid p \in \mathcal{P}_m, f = f_{prop_m}(p), cf = cf_{prop_m}(p)\}$, being the clue set associated to the model m
- $f_{prop_m} : \mathcal{P}_m \mapsto \mathbb{A}$, $cf_{prop_m} : \mathcal{P}_m \mapsto \mathbb{A}$, two functions respectively giving the argumentary force (used if proposition is assessed to be true) and counter force (used if proposition is assessed to be false) of a proposition towards the hypothesis that would derive from the model.

$\mathbb{A}$ represents the set of argumentary forces values going from $-\mu$ (the proposition can be considered as a proof that the hypothesis is WRONG) to μ (the proposition can be considered as a proof of the hypothesis being RIGHT).

Note that the numerical values of the set $\{-3,... 3\}$ are not significative from a numerical point of view : the aim is only to provide a preference order between argumentary forces. What is important is to get a scale which seems convenient for users (depending on applicative context) .

In the context of an hypothesis h based upon an abnormal behaviour model m being assessed by an user, a proposition associated to its truth value (and corresponding argumentary force ou counter force) will become an argument (or counter argument) towards h.

4   From Abnormal Behaviour Models to Hypothesis

Abnormal behaviour models are used in order to be compared with the user's belief set, formally defined hereafter.

let $\mathbb{B}$ denote $\{True, False\}$, a set of truth values.

let $interp_e : \mathbb{P} \mapsto \mathbb{B}$ the partial function that indicates which truth value is associated to a proposition in an e user's mind,

let $\mathcal{P}_e = \{p \in \mathbb{P}, interp_e(p) \in \mathbb{B}\}$. $\mathcal{P}_e$ denotes the subset of propositions on which an user e has beliefs.

let $confid_e : \mathbb{P} \mapsto]0, 1]$ the real function that, given a proposition, returns the confidence a user e puts in the belief that the proposition has the truth value returned by $interp_e$ (a value of 1 being reported as an absolute confidence).

Given a function $interp_e$, the belief set B_e of the user e is then defined as

$$B_e = \{(p, i, c) \mid p \in \mathcal{P}_e, i = interp_e(p), c = confid_e(p)\}$$

Given an ongoing observed situation, there are at least two ways of defining the contents of B_e : alarms produced from sensors[4], and beliefs acquisition about the situation.

Alarms are supposed to be reliable. They are expressed as a set B_A of beliefs, each of them being an n-uplets $< p, v, c >$ where $p \in \mathbb{P}$, v is the truth value of p, and $c = 1$ represents the maximal confidence for (p, v). We state : $B_e \supset B_A$ (informally user fully accepts information coming from alarms).

Abnormal behaviour hypothesis are generated from B_A and $\mathbb{M}$ (the behaviour hypothesis models set) : given an behaviour hypothesis model m, if a belief from

---

[4] This process is outside the scope of this paper: see [Brax, N. and al, 2012], or [Ray, C. and al, 2013].

B_A would become a positive argument for the hypothesis that would derive from m, then m is actually derived into an hypothesis h potentially explaining the situation. Formally, m is derived into h if and only if

$$\exists < p, v, 1 >\in B_A,$$
$$p \in \mathcal{P}_m \wedge [(v = True \wedge f_{prop_m}(p) > 0) \vee (v = False \wedge cf_{prop_m}(p) > 0)]$$

Note that at any time the user may also declare a behaviour hypothesis model as generating a candidate hypothesis, as far this model belongs to the model base.

Once a set H of hypothesis has been defined, the mission of the user is to furthermore assess hypothesis relevance to provide good explanations on the observed situation. The contents of each hypothesis may be used as a guideline to support user in his beliefs acquisition, by suggesting propositions on which it is worth to make an opinion.

Thus we define the argument set of an hypothesis h as the set of its propositions that can find a truth value in the beliefs of the user e.

Given a function $interp_e$ defined on $\mathbb{P}$,

$$\mathcal{A}_{h,e} = \{(p, c, f) \mid p \in \mathcal{P}_h \cap \mathcal{P}_e, c = confid_e(p), f = f_{arg_{h,e}}(p)\}$$

with $f_{arg_{h,e}}(p) : \mathbb{P} \mapsto \mathbb{A}$ being $\begin{cases} f_{prop_h}(p) & \text{when } interp_e(p) = True \\ cf_{prop_h}(p) & \text{when } interp_e(p) = False \end{cases}$

For a given hypothesis h and an user e lets state that,

– $f_{arg_{h,e}}(p)$ is the *argument force* for p
– the sign of an argument $f_{arg_{h,e}}(p)$ gives the *direction* of the argument, p is said to be *positive* (resp. *negative*) if $f_{arg_{h,e}}(p) > 0$ (resp. < 0).
– an argument p such that $f_{arg_{h,e}}(p) \in \{\mu, -\mu\}$ is said to be a definitive argument (or proof).

Given an argument set of an hypothesis and a user, the relevance of this hypothesis to explain the ongoing situation is evaluated through the quality and relevance score of the argumentation based on these arguments. This score is computed with a multicriteria aggregation algorithm which is described in the next section.

5   Hypothesis Evaluation

The aim is to evaluate the *relevance* of an abnormal behaviour hypothesis, that means its ability to explain the observed situation (represented by user's beliefs). We assess an hypothesis relevance by assessing the argumentation produced towards the hypothesis as described above, based upon its arguments' features. Arguments of hypothesis are characterized by two dimensions: argument force (pro or cons the described behaviour) and confidence, each of these dimensions having several intensity levels. Thus we considered that defining a global relevance score of an hypothesis was a multicriteria aggregation process. Hence the scoring function had to be written as a multicriteria aggregation function.

5.1   Quick Review of Algorithms

The problematic characteristics suggest that the chosen algorithm should have the following properties.

1. **Monotony against argument:** if an argument a is added then the global score evolves in the same direction as a
2. **Monotony against argument metainformation:** if argument metadata improves (resp. downgrades) then the global score increases (resp. decreases)
3. **Absorbance:** if the belief set contains a definitive argument with a full confidence then the corresponding hypothesis must be evaluated as "sure" (if the argument is positive) or "sure that not" (if the argument is negative).
4. **Inconsistency detection:** if the argument set of an hypothesis contains both definitive positive and definitive negative arguments, both associated to a full confidence, then the state is inconsistent and no score should be computed (formally the codomain of hsf is extended with the symbolic value $\perp$ that is returned in this situation)[5].
5. **Limited criteria compensability:** The compensability is the ability to obtain a high global evaluation even when having a low evaluation on a criteria (intuitively: several good marks compensate a bad one).
6. **No arbitrary weigth numerical values**

Min and Max aggregation algorithm family does not meet the strict monotony properties (1 and 2). Most usual and simple aggregation methods, such as averages and sums, are fully compensatory and then not suitable (property 5). Some more complex weighted techniques try to overcome this difficulty: but, in this family of functions, static numeric weights that are associated to each criteria (or criteria combination) are difficult to choose and control (property 6). Limits to these weighted approaches are reached for example with Choquet integrals [2] and its 2^n weights for n criteria. Techniques to "guess" weight values then seem to be necessary and they do not match the approach adopted here (no historical data available and lower user acceptance).

In addition, none of above mentionned algorithms meets properties 3 and 4.

Prioritized aggregation functions [3] try to address these difficulties by relying on a preference relation between criteria instead of on some arbitrary numeric weights[6]. Criteria weights values are then no longer static and depend on each alternative to evaluate (see figure 2). During evaluation of each alternative, the weight that is to be associated to each criteria will be computed, based onto the rank of the criteria class in the preference relation and the scores obtained in user preferred criteria classes.

Practically a 1 value is given to the weight w_1 associated with the most preferred criteria class, and all criteria scores v are such that $0 \leq v \leq 1$. Then the next weight value is computed by the formula $w_{i+1} = w_i \times m_i$.

---

[5] Such a situation should never occur, when it does, either the user has done something wrong or the model itself is wrong.

[6] Given two criteria c1 and c2, $c1 > c2$ means that the user considers c1 more important than c2, $c1 \sim c2$ means that c1 and c2 are in the same class.

$$Score(a) = \sum_{i=1}^{N} score_i(a) \times w_i(a)$$

where :

$$score_i(a) = \frac{\sum\limits_{c \in cc_i} score_c(a)}{\|cc_i\|}$$

$w_1(a) = 1$ and $w_{i+1}(a) = w_i(a) \times m_i(a)$

$m_i(a) = \min_{c \in cc_i} score_c(a)$

cc_i being the ith criteria class

Fig. 2. An implementation of Yager's algorithm

Thus, an alternative obtaining a low score for an important criteria, will result in small weights for less important criteria, hence reducing the impact of less preferred criteria scores on the overall evaluation score of the alternative.

This approach may be particularly efficient from the compensability limitation point of view [4] and may even offer absorbing elements opportunity. Indeed if the score corresponding to a given importance level criteria is null, then all lower importance criteria will inherit a null weight : that means that they will not be taken into account at all (this is what is previously referred to as absorbance property). This is particularly powerful when applied to high importance level : applying this at the highest level corresponds to consider the concerned criteria as a necessary condition which must be filled for the alternative to be worthwhile to be considered.

The ability of Yager's algorithm to tend to overcome unwanted compensation effects is not the only advantage it provides. Other advantages are that its processing doesn't require neither lots of experimental input data, nor somewhat arbitrary numerical weights; it relies mostly on symbolic notions, as user's preferences between criteria. These advantages are particularly relevant for applications where a sufficient reference data set is unavailable, for instance because of confidentiality constraints.

5.2   Algorithm Description

Our aggregation method is based upon the principles of prioritized aggregation functions. However, we do not have a unique total preference relation on the whole boolean criteria set. Instead we have distinct sets of criteria, so called dimensions, on which such a relation exists. In our initial case study, dimensions are confidence in belief and argument force, in each of them a criteria corresponds to an intensity degree. In order to keep compensation control property on every dimensions, our approach was to keep dimensions separated, recursively considering each of them as a level of importance inside of which the next one also represents a level of importance (i.e. defining a preference relation between the dimensions). We are so defining a multi-level version of prioritized aggregation functions, where each level is processed using a similar prioritized aggregation algorithm : thus the criteria non compensability property is favoured at every level (see figure 3).

As a first introduction to the algorithm, let's describe the two dimensions version from the case study, and arbitrarily suppose that confidence have been preferred to argument force.

Following Yager, at the highest first level (confidence, in this case), the highest confidence degree is associated to a weight of $w_{1_1} = 1$. Suppose we have n_1 degrees in this first dimension, the global evaluation of hypothesis h will be:

$$Score_1(\mathcal{A}_{h,e}) = \sum_{k_1=1}^{n_1} w_{1_{k_1}} \times Score_2(\mathcal{A}_{h,e_{<k_1>}})$$

where $Score_2(\mathcal{A}_{h,e_{<k_1>}})$ is a computed score for the set of arguments having the k_1^{th} degree according to the first dimension.

At the lowest level, the highest force intensity degree is associated with a maximal weight $w_{2_1} = 1$. Suppose we have n_2 degrees in this second dimension.

Value $Score_2$ of a given confidence level k_1 will then be computed with the formula :

$$Score_2(\mathcal{A}_{h,e_{<k_1>}}) = \sum_{k_2=1}^{n_2} w_2(k_2) \times localScore(\mathcal{A}_{h,e_{<k_1,k_2>}})$$

where $localScore(\mathcal{A}_{h,e_{<k_1,k_2>}})$ is the score of the subset of arguments having a k_1 confidence degree and a k_2 force degree for the hypothesis h here defined as proportional to its cardinality. Each argument contributes to the score positively or negatively according to its direction.

Note that numerical values of argument strength and argument confidence are only used through the preference order they induce. That means for instance, that arguments having 3 as argument strength value is considered by experts as more important than arguments having 2. So numerical values of argument strength and argument confidence have no direct impact in the final score, as far as their value scale is on accordance with the intended expert's preference order.

5.3   Algorithm Additional Tricks

As definitive arguments better correspond to a fully symbolic reasoning approach, their process is made outside of the previously described computation. Thus maximal positive and negative scores as well as inconsistency are discovered and dealt with before trying to estimate a numerical score.

Normalization of a positive score is realized by computing the maximal score (based on a virtual belief set making all propositions produce positive arguments for the hypothesis[7]), then dividing the score computed with the real belief set

---

[7] Or null if the maximum between force and counter force is zero.

$$(\forall d : 1..N) \quad \mathbf{score_d}(A_{<k_1,..,k_{d-1}>}) = \mathbf{norm_d}\Big(\sum_{k_d=1}^{n_d} \mathbf{score_{d+1}}(A_{<k_1,..,k_d>}) \times w_{d_{k_d}}\Big)$$

with :

$$A_{<k1,..,kq>} = \begin{cases} A \text{ for q=0} \\ \{(p, k_1, .., k_q, v_{q+1}, ..., v_N) \in A\} \text{ elsewhere} \end{cases}$$
$$w_{d_1} = 1 \text{ and } w_{d_{j+1}} = \mathbf{impact}(w_{d_j}, \mathbf{score}(A_{<k_1,..,k_{d-1}>})) - naturalGradient_{dim_i}$$
$$\mathbf{score_N}(A_{<k_1,..,k_N>}) = \mathbf{norm_{local}}(\mathbf{localScore}(A_{<k_1,..,k_N>}))$$

and :

$A = \{(p, v_1, ...v_N), p \in \mathbb{P}\}$ each v_d being the set of evaluations of p against the criteria from the criteria classes of the dimension d

$k_d = c$ meaning that k represents the criteria class c of the dimension d

Fig. 3. Proposed N-dimensional generic algorithm

by this maximum. Conversely negative score normalization is based on a virtual belief set producing only negative arguments for the hypothesis.

We also introduce a minimum weight decrease between levels so that in no case a lower level has the same weight as a higher one (*naturalGradient*).

6   Results and Perspectives

We have proposed a generic model-based approach to support users in any investigation process where potential decisions are modeled in a clue list way. We then enable them to get argumented hypothesis as well as a mean to compare hypothesis relevance based on their argumentation.

The multi-level version of the Yager prioritized aggregation algorithm we have developed to score hypothesis argumentation has several advantages :

- it only relies on preferences between criteria and between criteria dimensions,
- it does not suffer of somewhat arbitrary numerical choices,
- it does not require a lot of historical data (no learning process),
- and its results satisfy usually wishable properties.

This algorithm is generic and would apply to any context where aggregating several criteria dimensions is required. It is particularly relevant when sufficient reference data set is unavailable to compute advanced weights (as with Choquet or Sugeno integral algorithms[5]).

Some frameworks have been proposed in the literature to formalize argumentation [6], and several types of interactions between component arguments have been identified (for instance : support, aggregation, conflict). More recently some works have exhibited the impact of some argument meta-information, such as strength or validity in time, on the value of an argumentation [7] [8] [9]. Our concepts of argument strength and user's argument confidence are such useful meta-information (close to graded truth uncertainty sources from [10]). Our work can be thus compared to this family of works [9]. However our argumentations are only structured with aggregation and conflict interactions between arguments, since we suppose all component arguments are logically independent. The multi-level aggregation algorithm we have developed can be considered as a generic operator which combines two (or more) kinds of argument meta-information (in the illustration: strength and confidence). As such it can be easily integrated in many processes dealing, for instance, with imperfect information reasoning.

Our approach and algorithm are implemented [1] in a full maritime surveillance and investigation chain (I2C project, http://www.i2c.eu/). Other projects of the same domain have similar issues [11]. But as far as we know, they do not include an assistance module as what we propose.

References

1. Poitou, O., Saurel, C.: Supporting situation assessment by threat modeling and belief analysis. In: OCOSS Ocean and Coastal Observation: Sensors and observing Systems, Numerical Models and Information. Nice (2013)
2. Choquet, G.: Theory of capacities. In: Annales de l'Institut Fourier – tome 5, pp. 131–295 (1953)
3. Yager, R.: Prioritized aggregation operators. International Journal of Approximate Reasoning 48, 263–274 (2008)
4. da Costa Pereira, C., Dragoni, M., Pasi, G.: A prioritized "and" aggregation operator for multidimensional relevance assessment. In: Serra, R., Cucchiara, R. (eds.) AI*IA 2009. LNCS (LNAI), vol. 5883, pp. 72–81. Springer, Heidelberg (2009)
5. Grabisch, M.: Fuzzy Measures and Integrals: Theory and Applications. Springer-Verlag New York, Inc., Secaucus (2000)
6. Dung, P.: On the acceptability of arguments and its fundamental role in nonmonotonic reasoning, logic programming and n-person games. Artificial Intelligence 77, 321–357 (1995)
7. Bench-Capon, T.: Value-based argumentation frameworks, pp. 7–22. Taylor & Francis (2002)
8. Pollock, J.: Defeasible reasoning and degrees of justification. In: Argument and Computation, vol. 1(1), pp. 7–22. Taylor & Francis (2010)
9. Maximiliano, C.D., Budan, M.J., Gomez Lucero, G.R.S.: Modeling reliability varying over time through a labeled argumentative framework. In: IJCAI 2013 workshop on weighted logics for Artificial Intelligence, WL4AI 2013 (2013)
10. Demolombe, R.: Graded trust. AAMAS Trust, pp. 1–12 (2009)
11. Ray, C., Granger, A., Thibaud, R., Etienne, L.: Temporal rule-based analysis of maritime traffic. In: OCOSS Ocean and Coastal Observation: Sensors and Observing Systems, Numerical Models and Information, pp. 171–178. Nice (2013)

Analogical Proportions and Square
of Oppositions

Laurent Miclet[1] and Henri Prade[2]

[1] University of Rennes 1 – Irisa, Lannion, France
[2] CNRS/IRIT – University of Toulouse, Toulouse, France
laurent.miclet@univ-rennes1.fr, prade@irit.fr

Abstract. The paper discusses analogical proportions in relation with
the square of oppositions, a classical structure in Ancient logic which is
related to the different forms of statements that may be involved in deductive syllogisms. The paper starts with a short reminder on the logical
modeling of analogical proportions, viewed here as Boolean expressions
expressing similarities and possibly differences between four items, as in
the statement "a is to b as c is to d". The square of oppositions and its
hexagon-based extension is then restated in a knowledge representation
perspective. It is observed that the four vertices of a square of oppositions
form a constrained type of analogical proportion that emphasizes differences. In fact, the different patterns making an analogical proportion true
can be covered by a square of oppositions or by a "square of agreement",
leading to disjunctive expressions of the analogical proportion. Besides,
an "analogical octagon" is shown to capture the general construction of
an analogical proportion from two sets of properties. Since the square of
oppositions offers a common setting relevant for syllogisms and analogical proportions, it also provides a basis for the discussion of the possible
interplay between deductive arguments and analogical arguments.

1 Introduction

Deductive reasoning and analogical reasoning are two ways of drawing inferences.
Both were already clearly identified in the Antiquity. Syllogistic reasoning was
then the basis for deduction. Related to syllogisms is the square of oppositions
which dates back to the same time. In its original form, the square of oppositions
was associated with universal and existential statements and their negations,
which corresponds to the different statements encountered in syllogisms.

While deductive reasoning may involve both universal and existential statements, analogical reasoning only considers instantiated statements. A basic figure
of analogical reasoning is analogical proportion, i.e. a comparative statement of
the form "a is to b as c is to d". A logical view of analogical proportion has been
recently proposed [6]. Then the analogical proportion reads "a differs from b as
c differs from d and b differs from a as d differs from c", which involves negation
in the logical writing. This fact, together with the quaternary nature of such
statements, leads us to establish a connection between the analogical proportion
and the square of oppositions, whose meaning is investigated in the following.

A. Laurent et al. (Eds.): IPMU 2014, Part II, CCIS 443, pp. 324–334, 2014.
© Springer International Publishing Switzerland 2014

The paper first briefly recalls the logical view of the analogical proportion in Section 2, and its relevance to analogical reasoning. Then the square of oppositions and its hexagonal extension to a triple of similar squares is restated in Section 3, pointing out its interest in the analysis of the relations between categorical statements. Section 4 shows the relevance of the square of oppositions for discussing analogical proportions as well, while Section 5 discusses the interplay of arguments based on structures of opposition, before concluding in Section 6.

2   The Propositional View of the Analogical Proportion

An analogical proportion "a is to b as c is to d", denoted $a : b :: c : d$, is supposed to satisfy two characteristic properties: i) $a : b :: c : d$ is equivalent to $c : d :: a : b$ (symmetry), and ii) $a : b :: c : d$ is equivalent to $a : c :: b : d$ (central permutation). There are in fact 8 equivalent forms obtained by applying symmetry and permutation. Moreover, $a : b :: a : b$ always holds (reflexivity). Viewing a, b, c and d as Boolean variables, and $a : b :: c : d$ as a quaternary connective forces the analogical proportion to be true for the following 6 patterns $0 : 1 :: 0 : 1, 1 : 0 :: 1 : 0, 1 : 1 :: 0 : 0, 0 : 0 :: 1 : 1, 1 : 1 :: 1 : 1, 0 : 0 :: 0 : 0$. A logical expression, which is true *only* for these 6 patterns (among $2^4 = 16$ possible entries) has been proposed in [6]:

$$a : b :: c : d = ((a \land \neg b) \equiv (c \land \neg d)) \land ((\neg a \land b) \equiv (\neg c \land d)).$$

This remarkable expression makes clear that a differs from b as c differs from d and, conversely, b differs from a as d differs from c. This clearly covers the patterns $0 : 1 :: 0 : 1, 1 : 0 :: 1 : 0$, as well as the 4 remaining patterns where a and b are identical on the one hand and c and d are also identical on the other hand. It can be easily checked that under this logical view the analogical proportion indeed satisfies symmetry, central permutation, and reflexivity. It is also transitive. Moreover $a : b :: \neg b : \neg a$ also holds. Such a logical view of an analogical proportion can be advocated as being the genuine symbolic counterpart of the notion of numerical proportion [10,11].

This view easily extends to Boolean vectors a, b, c, d, whose components can be thought as binary attribute values, each vector describing a particular situation. Let x_i be a component of a vector x. Then an analogical proportion $a : b :: c : d$ between such vectors can be defined *componentwise*, i.e., for each component i, the analogical proportion $a_i : b_i :: c_i : d_i$ holds. Then the differences between a and b pertain to the same attribute(s) as the difference between c and d and are oriented in the same way (e.g., if from a_i to b_i, one goes from 1 to 0, it is the same from c_i to d_i). When there is no difference between a_i and b_i, then c_i and d_i are as well identical. This modeling of analogical proportions has been successfully applied to learning [5] and to the solving of IQ tests [8].

This vector-based view can be easily related to a set-based view originally proposed in [4], later proved to be equivalent (see [6] for details). In the set-based view each situation is associated with the set of attributes that are true in it. Let us denote A, B, C, and D the sets thus associated with a, b, c, and d, respectively. Then we shall write $A : B :: C : D$ if $a : b :: c : d$ holds.

Note that in particular, for any pair of sets A and B, $A : B :: \overline{B} : \overline{A}$ holds (where $\overline{A}$ denotes the set complement of A).

The case of attributes on discrete domains with more than 2 values can be handled as easily as the binary case. Indeed, consider a finite attribute domain $\{v_1, \cdots, v_m\}$. This attribute can be binarized by means of the m properties "having value v_i, or not". Consider the partial description of objects a, b, c, and d with respect to this attribute. Assume, for instance, that objects a and c have value v_1, while objects b and d have value v_2. Then it can be checked that an analogical proportion *holds true between the four objects for* each *of the m binary property*, and in the example, can be more compactly encoded as an analogical proportion between the attribute values themselves, namely here: $v_1 : v_2 :: v_1 : v_2$. More generally, x and y denoting possible values of a considered attribute, the analogical proportion between objects a, b, c, and d holds for this attribute iff the 4-tuple of their values wrt this attribute is equal to a 4-tuple having one of the three forms (s, s, s, s), (s, t, s, t), or (s, s, t, t).

3   The Square of Oppositions

Let us start with a refresher on the classical square of opposition [7]. This square involves four logically related statements exhibiting universal or existential quantifications: it has been noticed that a statement (**A**) of the form "every x is p" is negated by the statement (**O**) "some x is not p", while a statement like (**E**) "no x is p" is clearly in even stronger opposition to the first statement (**A**). These three statements, together with the negation of the last one, namely (**I**) "some x is p", give birth to the Aristotelian square of opposition in terms of quantifiers **A** : $\forall x\ p(x)$, **E** : $\forall x\ \neg p(x)$, **I** : $\exists x\ p(x)$, **O** : $\exists x\ \neg p(x)$, pictured in Figure 1 (where it is assumed that there are some x in order that the square makes sense).

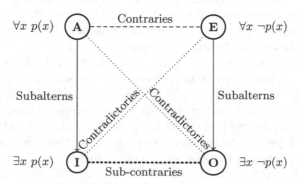

Fig. 1. Square of opposition

Such a square is usually denoted by the letters **A**, **I** (affirmative half) and **E**, **O** (negative half). The names of the vertices come from a traditional Latin reading: Aff**I**rmo, n**E**g**O**). As can be seen, different relations hold between the vertices:
- (a) **A** and **O** are the negation of each other, as well as **E** and **I**;
- (b) **A** entails **I**, and **E** entails **O** ;

- (c) **A** and **E** cannot be true together, but may be false together;
- (d) **I** and **O** cannot be false together, but may be true together.

Viewing the square in a Boolean way, where **A, I, E,** and **O** are now associated with Boolean variables, i.e. **A, I, E,** and **O** are the truth values of statements of the form $\forall x S(x) \to P(x)$, $\forall x S(x) \to \neg P(x)$, $\exists x S(x) \wedge P(x)$, and $\exists x S(x) \wedge \neg P(x)$ respectively. Then, the following can be easily checked.

The link between **A** and **E** represents the symmetrical relation of *contrariety*, whose truth table is given in Figure 2(a).
We recognize the mutual exclusion, i.e. $\neg\mathbf{A} \vee \neg\mathbf{E}$ holds.

A E	A _ _ _E
0 0	1
0 1	1
1 0	1
1 1	0

(a) Contrariety

I O	I.............O
0 0	0
0 1	1
1 0	1
1 1	1

(b) Subcontrariety

A I	A ⟶ I
0 0	1
0 1	1
1 0	0
1 1	1

(c) Implication

A O	A.........O
0 0	0
0 1	1
1 0	1
1 1	0

(d) Contradiction

Fig. 2. The four relations involved in the square of oppositions

The link between **I** and **O** represents the symmetrical relation of *subcontrariety*, whose truth table is is given in Figure 2(b).
We recognize the disjunction, i.e. $\mathbf{I} \vee \mathbf{O}$ holds.
The vertical arrows represent *implication* relations $\mathbf{A} \to \mathbf{I}$ and $\mathbf{E} \to \mathbf{O}$, whose truth table is given in Figure 2(c).
The diagonal links represent the symmetrical relation of *contradiction*, whose truth table is given in Figure 2(d).
We recognize the exclusive or, i.e. $\neg(\mathbf{A} \equiv \mathbf{O}) = (\mathbf{A} \wedge \neg\mathbf{O}) \vee (\neg\mathbf{A} \wedge \mathbf{O}) = (\mathbf{A} \vee \mathbf{O}) \wedge (\neg\mathbf{A} \vee \neg\mathbf{O})$ holds, or if we prefer we have $\mathbf{A} \equiv \neg\mathbf{O}$.
Thus, note that $\neg\mathbf{A} \vee \neg\mathbf{E}$ and $\mathbf{I} \vee \mathbf{O}$ are consequences of $\mathbf{A} \equiv \neg\mathbf{O}$, $\mathbf{A} \to \mathbf{I}$ and $\mathbf{E} \to \mathbf{O}$ in the square, but not $\mathbf{E} \equiv \neg\mathbf{I}$. Moreover $\mathbf{I} \vee \mathbf{O}$ and $\mathbf{E} \to \mathbf{O}$ are not enough for entailing $\mathbf{E} \equiv \neg\mathbf{I}$.

4 The Analogical Proportion and the Square of Opposition

Let us first continue to consider the Boolean square of oppositions where **A, E, I** and **O** are binary variables. What are the joint assignments of values for these variables that are compatible with the logical square of oppositions? It is easy to prove that there are only 3 valid squares:

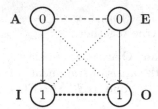 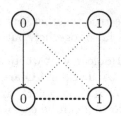 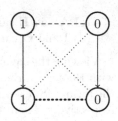

One can immediately conclude that if **A, E, I** and **O** are the (Boolean-valued) vertices of a square of oppositions, then they form an analogical proportion when taken in this order, i.e. **A : E :: I : O**, since $0 : 0 :: 1 : 1$, $0 : 1 :: 0 : 1$ and $1 : 0 :: 1 : 0$ are 3 of the 6 patterns that make an analogical proportion true.

This is fully consistent with an empirical reading of the traditional square of oppositions. Indeed one can say that "$\forall x P(x)$ is to $\forall x \neg P(x)$ as $\exists x P(x)$ is to $\exists x \neg P(x)$". A less academic example of square of oppositions associated with the analogical proportion "mice are to shrews as mammals except shrews are to mammals except mice" is pictured below.

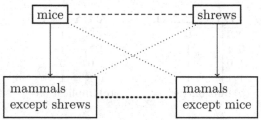

Remark 1. Consistency of analogical proportions with an empirical reading of the square: Traditionally, the square is based on the empirical notions of *quality* (affirmative or negative) and *quantity* (universal or existential) which can be used for describing the vertices of the square:

- **A** corresponds to a *U*niversal *A*ffirmative statement,
- **E** corresponds to a *U*niversal *N*egative statement,
- **I** corresponds to a *E*xistential *A*ffirmative statement and
- **O** corresponds to a *E*xistential *N*egative statement.

Thus in the square of oppositions each vertex can be described by means of a 2-component vector, the first component being the quantity (value 0 if quantity is existential, 1 if it is universal) the second component being the quality (value 1 if the quality is affirmative, 0 if it is negative). The square thus obtained is the superposition of the first and the third previous squares, as can be seen.

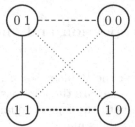

4.1   The Analogical Proportion and Its Two Squares

As recalled in Section 2, the analogical proportion a $\quad:$ $\quad b$ $\quad::$ $\quad\neg b$ $\quad:$ $\quad\neg a$ always holds. It clearly gives birth to 4 of the 6 patterns that makes an analogical proportion true, namely $1 : 1 :: 0 : 0$, $0 : 0 :: 1 : 1$, $0 : 1 :: 0 : 1$ and $1 : 0 :: 1 : 0$. However, the first one ($1 : 1 :: 0 : 0$) is forbidden when an analogical proportion is stated in terms of a square of oppositions, since in the following square a and b form a square of oppositions together with their complements if and only if their conjunction is false (indeed a and b cannot be true in the same time). Thus, the square of oppositions appears to be *a strict restriction* of the analogical proportion.

What about the 3 other patterns making true an analogical proportion that are left aside, namely $1 : 1 :: 0 : 0$, $1 : 1 :: 1 : 1$ and $0 : 0 :: 0 : 0$? Interestingly enough, they can also be organized in another square that might be called *square of agreement*. It is pictured below (the double arrow represents equivalence). In this square, the following holds $a \equiv b$, $c \equiv d$, $c \to a$, $c \to b$, $d \to a$, and $d \to b$. It corresponds to the pattern $a : a :: b : b$ under the constraint $b \to a$. Under these constraints, the square has only 3 possible instantiations, namely $1 : 1 :: 0 : 0$, $1 : 1 :: 1 : 1$ and $0 : 0 :: 0 : 0$.

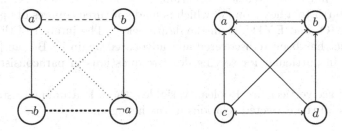

Fig. 3. Square of opposition and square of agreement

Moreover it can be noticed that the three patterns involved in the square of opposition ($a : b :: c : d = 0 : 0 :: 1 : 1$, $0 : 1 :: 0 : 1$ or $1 : 0 :: 1 : 0$) satisfy constraints $a \to c$ and $b \to d$, while the three patterns involved in the square of agreement ($a : b :: c : d = 1 : 1 :: 0 : 0$, $1 : 1 :: 1 : 1$ or $0 : 0 :: 0 : 0$) satisfy the constraints $c \to a$ and $d \to b$.

Since analogical proportions are both a matter of dissimilarity and similarity [11], it should come as no surprise that one "half" of it satisfies the square of opposition, while the other "half" satisfies a square of agreement. Moreover, we can see that that $(a \not\equiv d) \wedge (b \not\equiv c)$ is true only for the 4 patterns $1 : 1 :: 0 : 0$, $0 : 0 :: 1 : 1$, $0 : 1 :: 0 : 1$ and $1 : 0 :: 1 : 0$, and thus $((a \not\equiv d) \wedge (b \not\equiv c) \wedge (\neg a \vee \neg b))$ is true only for the last 3 patterns, and thus corresponds to the square of opposition. Similarly, $((a \equiv b) \wedge (c \equiv d)$ is true only for the 4 patterns $0 : 0 :: 1 : 1$, $1 : 1 :: 0 : 0$, $1 : 1 :: 1 : 1$ and $0 : 0 :: 0 : 0$, while $((a \equiv b) \wedge (c \equiv d) \wedge (c \to b) \wedge (d \to a))$ is true only for the last 3 and corresponds to the square of agreement ($\wedge (d \to a)$ can be deleted and is put here only for emphasizing symmetry). This leads to three noticeable, equivalent, disjunctive expressions of the analogical proportion:

$$a : b :: c : d = ((a \not\equiv d) \wedge (b \not\equiv c)) \vee ((a \equiv b) \wedge (c \equiv d))$$

$$a : b :: c : d = ((a \equiv b) \wedge (c \equiv d)) \vee ((a \equiv c) \wedge (b \equiv d))$$

$$a : b :: c : d = (((a \equiv c \wedge b \equiv d)) \vee ((a \not\equiv d) \wedge (b \not\equiv c))$$

(since $(a \equiv c) \wedge (b \equiv d)$ is true only for the 4 patterns $1 : 0 :: 1 : 0, 0 : 1 :: 0 : 1,$ $1 : 1 :: 1 : 1$ and $0 : 0 :: 0 : 0$). A counterpart of the second expression can be found in [13] in their factorization-based view of analogical proportion.

4.2   Analogical Proportion and the Hexagon of Oppositions

As proposed and advocated by Blanché [2], it is always possible to complete a classical square of opposition into a hexagon by adding the vertices $\mathbf{Y} =_{def} \mathbf{I} \wedge \mathbf{O}$, and $\mathbf{U} =_{def} \mathbf{A} \vee \mathbf{E}$. It fully exhibits the logical relations inside a structure of oppositions generated by the three mutually exclusive situations $\mathbf{A}$, $\mathbf{E}$, and $\mathbf{Y}$, where two vertices linked by a diagonal are contradictories, $\mathbf{A}$ and $\mathbf{E}$ entail $\mathbf{U}$, while $\mathbf{Y}$ entails both $\mathbf{I}$ and $\mathbf{O}$. Moreover $\mathbf{I} = \mathbf{A} \vee \mathbf{Y}$ and $\mathbf{O} = \mathbf{E} \vee \mathbf{Y}$. Conversely, three mutually exclusive situations playing the roles of $\mathbf{A}$, $\mathbf{E}$, and $\mathbf{Y}$ always give birth to a hexagon [3], which is made of three squares of opposition: $\mathbf{AEOI}$, $\mathbf{YAUO}$, and $\mathbf{EYIU}$, as in the figure below. The interest of this hexagonal construct has been rediscovered and advocated again by Béziau [1] in the recent years in particular for solving delicate questions in paraconsistent logic modeling.

When the six vertices are Boolean variables, there is a unique instantiation that satisfies all the expected relations in the hexagon:

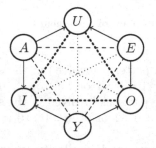
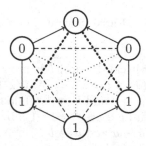

Thus the hexagon of oppositions is made of three squares of oppositions. Each of these squares exactly corresponds to one of the three possible analogical patterns. Indeed, the square $\mathbf{AEOI}$ corresponds to the analogical proportion $\mathbf{A} : \mathbf{E} :: \mathbf{I} : \mathbf{O}$, i.e., $0 : 0 :: 1 : 1$, the square $\mathbf{YAUO}$ to $\mathbf{Y} : \mathbf{A} :: \mathbf{O} : \mathbf{U}$, i.e., $1 : 0 :: 1 : 0$, and the square $\mathbf{EYIU}$ to $\mathbf{E} : \mathbf{Y} :: \mathbf{U} : \mathbf{I}$, i.e., $0 : 1 :: 0 : 1$.

Since a structure of oppositions is generated by any three mutually exclusive situations $\mathbf{A}$, $\mathbf{E}$, and $\mathbf{Y}$, taking any pair of subsets R and S such that $R \subset S$ (in order to insure that one cannot be in R and in $\neg S$ in the same time), one gets the hexagon below where the following analogical proportions hold $R : \neg S :: S : \neg R$, $\neg R \cap S : R :: \neg R : R \cup \neg S$, and $\neg S : \neg R \cap S :: R \cup \neg S : S$.

It is possible as well to build a more general hexagon from a pair of unconstrained subsets R and S, with proportions $R : \overline{R} \cap \overline{S} :: R \cup S : \overline{R}$, $\overline{R} \cap S : R ::$ $\overline{R} : R \cup \overline{S}$, and $\overline{R} \cap \overline{S} : \overline{R} \cap S :: R \cup \overline{S} : R \cup S$. An illustration in terms of Boolean variables may be obtained by taking R as "speak English" and S as "speak Spanish", where the hexagon structures the logical relations between 6 possible epistemic states regarding the competence of an agent wrt these languages.

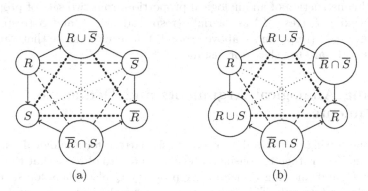

(a)   (b)

Fig. 4. Two hexagons constructed from a square of oppositions

4.3 From an Hexagon to an Octagon

A way to display the construction of a general analogical proportion from unconstrained subsets R and S is to start from the hexagon of Figure 4(a) and to

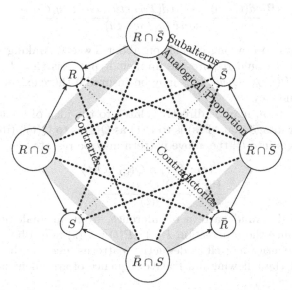

Fig. 5. The analogical octagon constructed from R and S

add one node $R \cap S$ between R and S, as well as one node $\bar{R} \cap \bar{S}$ between nodes $\bar{R}$ and $\bar{S}$, and finally to turn the node $R \cup \bar{S}$ into the node $R \cap \bar{S}$ (see Figure 5). The following square in this octagon is a complete analogical proportion:

$$R \cap S \ : \ R \cap \bar{S} \ :: \ \bar{R} \cap S \ : \ \bar{R} \cap \bar{S}$$

Note that the nodes $R \cap S$ and $\bar{R} \cap \bar{S}$, from one side, and $\bar{R} \cap S$ and $R \cap \bar{S}$, from the other side, are not contradictories. This new figure is not a full octagon of oppositions, but could rather be called an *analogical octagon*, since it captures the general construction of an analogical proportion from two sets of properties R and S. Taking R (resp. $\bar{R}$) as "aerial" (resp. "aquatic"), and S (resp. $\bar{S}$) as "move on ground" (resp. "move above ground"), it leads to state that "ants are to birds as crabs are to fishes" for instance.

5   Mixing Analogical Arguments and Deductive Arguments

Stated in the setting of first order logic, a basic pattern for analogical reasoning (see, e.g., e.g. [12]) is then to consider 2 terms s and t, to observe that they share a property P, and knowing that another property Q also holds for s, to infer that it holds for t as well. This is known as the "analogical jump" and can be described with the following inference pattern:

$$\frac{P(s) \ P(t) \ Q(s)}{Q(t)}$$

A typical instance of this kind of inference would be:

$$\frac{isBird(Coco) \ isBird(Tweety) \ canFly(Coco)}{canFly(Tweety)}$$

leading (possibly) to a wrong conclusion about Tweety. Making such an inference pattern valid would require the implicit hypothesis that P determines Q inasmuch as $\nexists u \, P(u) \wedge \neg Q(u)$. This may be ensured if there exists an underlying functional dependency.

The above pattern, may be directly related to the idea of analogical proportion. Taking advantage that "P(s) is to P(t) as Q(s) is to Q(t)" (indeed they are similar changing s into t), the above pattern may be restated as

$$\frac{P(s) : P(t) :: Q(s) : Q(t)}{\frac{P(s), P(t), Q(s)}{Q(t)}}$$

which is a logically valid pattern of inference, from an analogical proportion point of view (since the proposition $P(s) : P(t) :: Q(s) : Q(t)$ holds) [9,11].

Analogical patterns, as well as deductive patterns, may be the basis of arguments. Consider the following illustrative sequence of arguments:

"P's are Q's"

"s is a P"

then "s is a Q"

This is a deductive (syllogistic) argument in favor of $Q(s)$. Assume the input of the following information

"t is also a Q"

then "t should be a P"

by virtue of the analogical pattern, as t is a Q, s is a P and a Q. This an analogical argument in favor of $P(t)$. Now assume the following claim is made

"s is to t as u is to v"

based on the following facts: s, t are P's and Q's, while u, v are P's and $\neg Q$'s (then, for the 4-tuple (s, t, u, v), we have the following patterns $1 : 1 :: 1 : 1$ for P and $1 : 1 :: 0 : 0$ for Q).

Then, an opposition takes place, and

"it is wrong that P's are Q's".

Indeed we have a new argument that questions the first premise ("P's are Q's'). This little sketch intends to point out the fact analogy as deduction may be the basis of arguments as well as counterarguments.

An analogical proportion-based statement may also *support* a deductive argument: "P's are Q's", indeed "s is a P and a Q as t is a P and a Q", then knowing that "r is a P", we conclude "r is a Q", which is a form of syllogism called *epicherem* (i.e. the basic syllogism is enriched by a supportive argument).

6   Concluding Remarks

This discussion paper has shown that squares of opposition may play a role in the analysis of analogical proportions as much as they are encountered in various other forms of reasoning. Squares of opposition can indeed be viewed as a constrained form of analogical proportions (since analogical proportions both encompass the ideas of disagreement and agreement). The paper has also pointed out the relevance of other structures of oppositions (hexagon, octagon) in the analysis of analogical proportions. This may also help classifying various forms of arguments.

References

1. Béziau, J.-Y.: New light on the square of oppositions and its nameless corner. Logical Investigations 10, 218–233 (2003)
2. Blanché, R.: Structures Intellectuelles. Essai sur l'Organisation Systématique des Concepts. Vrin, Paris (1966)
3. Dubois, D., Prade, H.: From Blanché's hexagonal organization of concepts to formal concept analysis and possibility theory. Logica Univers. 6, 149–169 (2012)
4. Lepage, Y.: De l'analogie rendant compte de la commutation en linguistique. Habilit. à Diriger des Recher. Univ. J. Fourier, Grenoble (2003)
5. Miclet, L., Bayoudh, S., Delhay, A.: Analogical dissimilarity: definition, algorithms and two experiments in machine learning. JAIR 32, 793–824 (2008)
6. Miclet, L., Prade, H.: Handling analogical proportions in classical logic and fuzzy logics settings. In: Sossai, C., Chemello, G. (eds.) ECSQARU 2009. LNCS (LNAI), vol. 5590, pp. 638–650. Springer, Heidelberg (2009)

7. Parsons, T.: The traditional square of opposition. In: Zalta, E.N. (ed.) The Stanford Encyclopedia of Philosophy (Fall 2008 Edition) (2008)

8. Prade, H., Richard, G.: Analogy-making for solving IQ tests: A logical view. In: Ram, A., Wiratunga, N. (eds.) ICCBR 2011. LNCS (LNAI), vol. 6880, pp. 241–257. Springer, Heidelberg (2011)

9. Prade, H., Richard, G.: Cataloguing/analogizing: A non monotonic view. Int. J. Intell. Syst. 26(12), 1176–1195 (2011)

10. Prade, H., Richard, G.: Homogeneous logical proportions: Their uniqueness and their role in similarity-based prediction. In: Brewka, G., Eiter, T., McIlraith, S.A. (eds.) Proc. 13th Int. Conf. on Principles of Knowledge Representation and Reasoning (KR 2012), Roma, June 10-14, pp. 402–412. AAAI Press (2012)

11. Prade, H., Richard, G.: From analogical proportion to logical proportions. Logica Universalis 7(4), 441–505 (2013)

12. Russell, S.J.: The use of Knowledge in Analogy and Induction. Pitman, UK (1989)

13. Stroppa, N., Yvon, F.: Analogical learning and formal proportions: Definitions and methodological issues. Technical report (June 2005)

Towards a Transparent Deliberation Protocol Inspired from Supply Chain Collaborative Planning

Florence Bannay and Romain Guillaume

IRIT, Toulouse University, France

Abstract. In this paper we propose a new deliberation process based on argumentation and bipolar decision making in a context of agreed common knowledge and priorities together with private preferences. This work is inspired from the supply chain management domain and more precisely by the "Collaborative Planning, Forecasting and Replenishment" model which aims at selecting a procurement plan in collaborative supply chains.

Keywords: Decision process, Argumentation, Supply Chain Management.

1 Introduction

In the supply chain management (SCM) domain, collaborative planning is the process of finding a production plan which is suitable for every agent involved in the supply chain (SC). In this domain, a useful requirement is that the agents have agreed about some general conventions concerning the eligible criteria wrt production plans (avoiding more discussions about criteria). From Artificial Intelligence point of view, the SCM collaborative plan problem is a particular case of a collaborative decision given a background consensual knowledge and the particular preferences of each agent. This decision is the selection of a candidate according to a global convention and according to the precise arguments uttered by them about each candidate. In the classification of dialogs given by Walton and Krabbe [17], we are facing a deliberation process, since the initial situation is a dilemma (the agents need to find a good plan among several options), each agent aims at coordinating its goals and actions with the others in order to decide the best course of action that will be beneficial for all the participants.

In existing process, supply chain participants are compelled to provide numerical data (which may require some expensive computation, or may be based on some debatable numerical evaluations). Our main goal is to design a formal framework in which (qualitative) arguments in favor or against a plan maybe expressed and may interact with each-other. This framework will enable to conceive systems that could facilitate and guide the agents to converge rationally towards a set of consensual plans. Moreover, the production plan that will be chosen should have a guaranteed quality: if the selection leads to a set of plans that are under an admissibility threshold then the process is aborted and an exception is raised. Otherwise, if several plans are admissible then the customer (or the most important actor playing a role similar to a customer wrt a supplier ([13]) has usually the last choice. Thus, our aim is twofold, enlarge the expressive power and readability of the decision protocols concerning collaborative planning in

A. Laurent et al. (Eds.): IPMU 2014, Part II, CCIS 443, pp. 335–344, 2014.

SCM domains, import these ideas to AI field in order to propose a new rational model of collaborative decision making (or deliberation) under clear admissibility criteria.

More precisely we propose a new deliberation process based on bipolar argumentation where the arguments, their incompatibilities (called attacks) and their importance levels are commonly agreed before the deliberation. We directly apply this process for collaborative planning in supply chain using CPFR® (see Section 2.3). In AI literature some models have already been proposed based on a bipolar view of alternatives. Indeed, it is often the case that human people evaluate the possible alternatives considering positive and negative aspects separately [16]. Moreover, argumentation has already been proposed to govern decision making in a negotiation context (see for instance [2] and [14] for a survey). But, as far as we know, the use of bipolar argumentation in order to govern a deliberation process under instantiated arguments had never been studied. Moreover, in context of supply chain the agents only require to obtain admissible production plan and do not necessarily need to class them (in the worst case the best plan can be inadmissible, in other cases every plan may be admissible). Hence, we propose and study the rationality and the calculability of the admissible set. Furthermore, the notion of efficiency and simplicity that are central in SCM domain can bring a new perspective for modeling deliberation process in AI.

We first recall the basis of qualitative bipolar decision, classical argumentation and supply chain planning. In a second step, we define a common decision making structure gathering the concepts of bipolarity and argumentation, then we propose admissibility thresholds. We apply this deliberation process to an example of collaborative planning.

2 Background

2.1 Qualitative Bipolar Decision

In this paper we focus on the problem of qualitative bipolar decision, this problem can be formalized as follows. Let C be a finite set of potential choices and $\mathcal{A}$ be a set of arguments (or criteria) viewed as decision attributes ranging on a bipolar scale $\{-,+\}$. More precisely if $a \in \mathcal{A}$ and $c \in C$ then $a(c)$ is either an argument in favor of the choice c (when $a(c) = +$) or against it ($a(c) = -$). Let us consider a totally ordered scale *level* expressing the relative importance of arguments: $\forall a_1, a_2 \in \mathcal{A}$, if $level(a_1) > level(a_2)$ then a_2 is more important than a_1 (the best level is 1). In this paper, a set $A \subseteq \mathcal{A}$ of arguments is called a **bipolar leveled set of arguments** abbreviated *bla*, and $A^+ = \{a \mid a(c) = +\}$ and $A^- = \{a \mid a(c) = -\}$ denote respectively the set of arguments in favor and against the choice c. The problem of rank-ordering the possible choices has been well studied in the literature. For this purpose, decision rules which build preference relations between decisions were proposed [4].

Most approaches in the literature focus on preferences but do not discus the problem of admissibility threshold. In this paper we propose such kind of thresholds.

Note that classical models do not take into account the conflicts that may occur between arguments, indeed the presence of some arguments may decrease the validity of other arguments. In this paper we propose a model that takes into account both the importance of arguments and the relations between them.

2.2 Classical Argumentation

In the abstract argumentation theory, Dung [8] has defined several ways, called "semantics", to select admissible arguments. Given a graph (X,R), called argumentation system where X is a set of arguments and R is a binary relation on X called attack, the selected sets of arguments $S \in X$, called extensions, should be conflict free (i.e. should not contain internal attacks). Most of the extensions proposed by Dung are based on a defense notion[1] that is not very intuitive in our context. Including defended arguments would amount to accept *every* positive argument as soon as one positive argument is not attacked (and similarly for negative arguments). Moreover, as already noticed by Amgoud and Vesic [2] "Dung's framework cannot be used for decision making since it simply selects groups of arguments containing at last, one of the strongest (positive or negative) arguments". Hence, Dung's framework is not useful in our context: it will give the same results with a more complex computation, as the results obtained by simply selecting the unattacked arguments.

2.3 Supply Chain Planning

A supply chain is "a network of connected and interdependent organizations mutually and cooperatively working together to control, manage and improve the flow of materials and information from suppliers to end users" [5]. Handling a supply chain is called supply chain management and can be defined as "the management of upstream and downstream relationships with suppliers and customers to deliver superior customer value at less cost to the supply chain as a whole" [5]. Two types of supply chains can be distinguished: *decentralized* supply chains (SC where actors are independents) and *centralized* supply chains (SC where decision authority and information is hold by one single party) [1]. In this paper we focus on *decentralized* supply chains. The key of successful SCM for *decentralized* supply chains is the collaborative process.

The collaborative processes are usually characterized by a set of point-to-point customer / supplier relationships with partial information sharing [7]: one or several procurement plans are built and propagated through the supply chain using negotiation processes. One of the most popular collaborative processes has been standardized under the name of "Collaborative Planning, Forecasting and Replenishment" (CPFR®) [10]. It is also a registered trademark of Voluntary Inter-industry Commerce Standards (VICS). The CPFR® is a hierarchical process (strategic, tactical and operational) and is decomposed into four main tasks: *Strategy and Planning*, *Demand and Supply Management*, *Execution* and *Analysis*. In this paper, we are interested by the collaborative task of *Demand and Supply Management*, and more precisely by *Order planning/forecasting* and *replenishment planning* which are tasks that aim to determine future product ordering and some delivery requirements based upon the sales forecast, inventory positions, transit lead times, and other factors. In the literature of *production planning*, the collaborative order planning and replenishment planning are based on linear programming and/or mixed integer programming where the actors negotiate about a cost function

---

[1] A set of arguments Y defends an argument x iff for any argument y s.t. $(y,x) \in R$ there exists an argument $z \in Y$ such that $(z,y) \in R$.

which aggregates the inventory, the capacity, the production consequences ... (see [7] and [1] for a survey). Since some coefficients of the cost function are difficult to estimate for these actors, the analyses of resulting plans become difficult.

Another method, is to propose a set of plans to the supplier. Then the supplier evaluates the plans in terms of risk of back-ordering [9] and chooses the less risky plan. However, the risk of back-ordering is still poor information regarding the multi-dimension nature of evaluation of a production planning process [12].

Another part of SCM literature focuses on the actor planning process and on how to take into account the multi-objective nature of the problem and the qualitative evaluation of a production plan. For these purposes, the fuzzy goal programming approach [11] and the fuzzy multi-objective approach [3] have been proposed. The fuzzy goal programming approach consists in defining goals in terms of fuzzy intervals, the more the result belongs to the goal the more pleased is the decision maker. The fuzzy multi-objective approach takes into account the fuzziness of the importance of the criteria. For the problem of production distribution planning, a collaborative fuzzy programming approach has been developed in [15] which proposes a hierarchy of priority levels for the objective functions of the model, taking into account the dominance of each firm.

Those models do not take into account the bipolarity of decision. Our proposal aims to generalize the approach of [9] by taking into account the multi-objectives and qualitative nature of the problem using a bipolar argumentative framework.

3   A Formal Framework for Bipolar Argumentative Deliberation

A collaborative deliberation process involves several agents that want to select a candidate (or an option) in a "rational" manner. In order to do this selection, the agents are going to present arguments in favor or against some candidates. Then the system will aggregate their arguments and return the winner(s). The idea, borrowed from collaborative planning, is that the agents should first agree on general criteria in terms of arguments that may be used in favor or against a decision, and in terms of dominance/defeat relations between those arguments. After this agreement stage which is done once and for all, they can proceed to the selection of the candidates by giving arguments in favor or against some of them according to their private knowledge and preferences.

3.1   Bipolar Leveled Features Set

Let $C = \{c_1, \ldots c_n\}$ be a set of candidates (choices) and $V = \{v_1, \ldots v_n\}$ a set of voters, we are going to consider the arguments in favor or against the different candidates from the point of view of each voter. Arguments are entities representing a particular feature that may characterize some candidates, and influence their selection.

Example 1. In supply chain planning the problem is to find a production plan which satisfies both customers and suppliers. Thus, the set of candidates is the set of possible production plans and the customers and the suppliers are the voters. In this example, we consider three plans c_1, c_2 and c_3 and two voters v_1 (the supplier) and v_2 (the customer).

Definition 1 (instantiated argument). *An* instantiated argument $a_v(c)$ *is a predicate based on an argument a, given a voter v and a candidate c. Let I be the set of all instantiated arguments[2], namely, $I = \{a_v(c) \mid v \in V, c \in C, a \in \mathcal{A}\}$. Moreover for any $c \in C$, $I(c) = \{a_v(c) \mid a \in \mathcal{A}, v \in V\}$ denotes the set of instantiated arguments about c.*

Given a set of arguments the universe is the structure that contains those arguments together with their interactions.

Definition 2 ((instantiated) universe). *A* universe *(respectively* instantiated universe*) (A_U, R_U) is a graph whose vertices are arguments $A_U \subseteq \mathcal{A}$ (respectively instantiated arguments $A_U \subseteq I$) and arcs are conflict relations between arguments such that $R_U \subseteq A_U \times A_U$.*

If R is an attack relation then $(a, b) \in R$ means that the argument a defeats the argument b. In our context a is a possible feature, thus $(a, b) \in R$ means that if a characterizes a candidate then the fact that b also holds for this candidate has no interest nor influence on the final result (b is invalidated by a).

Note that a universe is a Dung argumentation system [8]. Now, let us consider that the protagonists have agreed about a consensual universe. This consensual agreement concerns the features that will be used in arguments, it concerns also the way they are divided into two sets (features "in favor of" and features "against" the candidates). Finally the agreement between the participants concerns also the importance accorded to each feature, it is translated by a level, the higher level corresponds to the features that play the smaller role in the decision.

Note that the different features in A^+ are arguments in favor of a candidate hence they are, by nature, opposed to the arguments against this candidate i.e. features in A^- (and conversely). This is why a *bla* will be associated with a graph in which each argument in A^+ is attacking any less important argument in A^- (i.e. belonging to a greater level).

For arguments situated at the same level λ, the relation is not so systematic. Let us denote by A_λ the set of arguments at level λ, we consider that a relation R_λ should be defined in order to express the attacks between A_λ^+ and A_λ^-. These attacks are not necessarily symmetric. We can be confronted to three relationships between a positive and a negative feature of the same level:

- they may have a completely independent impact on the accuracy to select a candidate having these features: no attack (e.g. in the supply chain management domain, the carbon impact and the stability of the production plan)
- they may have an opposite impact: hence when they appear together the attack is symmetric (e.g., a plan that may be considered both satisfactory and not satisfactory for personal reason)
- they may be related in a way that the presence of the first one outperforms the other (although they have the same level of importance) for instance a plan that may imply a high inventory level for an actor of the supply chain is bad even if this plan makes a quick flow for an other actor.

This is why the relation R_λ should be given for each level λ.

---

[2] In the literature, the instantiated arguments could be called *practical arguments* while generic arguments are more related to *epistemic arguments*.

Definition 3 (universe associated to a bla). *Let A be a bla with l levels. The uni-*
verse associated to A given a family of l binary relations on A, $(R_\lambda)_{\lambda \le l}$ such that
$\forall \lambda \in [\![1,l]\!], R_\lambda \subseteq (A_\lambda^+ \times A_\lambda^-) \cup (A_\lambda^- \times A_\lambda^+)$, *is the graph (A,R) with*
$R = \{(x,y) \in (A^+ \times A^-) \cup (A^- \times A^+) \text{ and } level(x) < level(y)\} \cup \bigcup_{\lambda \in [\![1,l]\!]} R_\lambda$
    The instantiated universe associated to a bla A given R and given a set of instantiated
arguments I included in $I_{A,V,C}$ is a graph (I, R_I) where R_I is s.t. $(a_{1v_1}(c_1), a_{2v_2}(c_2)) \in R_I$
iff $c_1 = c_2$ and $a_1, a_2 \in A$ and $(a_1, a_2) \in R$.

In other words, an instantiated universe is a graph of instantiated arguments. Each
argument is a feature relative to a given candidate according to the opinion of a voter.
Attack relations are induced between two arguments that concern the same candidate
when the features described in those arguments were consensually said incompatible.

Example 2. This example is inspired from the literature on multi-criteria production
planning. The supply chain characteristics and objectives give us concrete features that
are used in industrial domain. Here, each argument corresponds to a given industrial
objective, then, an argument is enabled for a given production plan if the plan achieves
this objective. The definition of arguments is done at the same time as the definition of
the bla. We consider that the customer and the supplier deliberate on the basis of the
following table A (which is a bla) containing the features that may characterize a plan[3]:

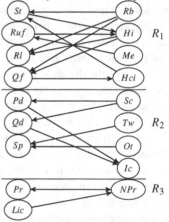

A^+	A^-
St: stable load	Rb: risk of back-ordering
Ruf: robust under failure	Hi: high inventory level
Rl: regrouping lot	Me: maintenance expensive
Qf: quick prod. flow	Hci: high capacity not used
Pd: periodic delivery date	Sc: requires subcontract
Qd: constant quantity	Tw: necessity of temporary
delivered	workers
Sp: stable plan	Ot: necessity of overtime
	Ic: important impact carbon
Pr: satisf. for pers. reason	NPr: non satisfactory
Lic: low impact carbon	for personal reason

In this bla A, the set of arguments is A = {St, Rb, Ruf, Hi, Rl, Me, Qf, Hci, Pd, Sc,
Qd, Ot, Sp, Tw, Ic, Pr, NPr, Lic}. This bla has three levels of importance. The family
of attack relations $(R_\lambda)_{\lambda \le 3}$ between arguments of the same level is s.t. $R_1 = \{(St, Hi),$
(Rb, St), (Rb, Rl), (Rb, Qf), (Ruf, Hi), (Hi, Rl), (Hi, Qf), (Me, Ruf), (Qf, Hci), (Hci,
St)}, $R_2 = \{(Pd, Ic), (Sc, Pd), (Sc, Qd), (Qd, Ic), (Tw, Sp), (Ot, Sp)\}$ and $R_3 = \{(Pr,$
NPr), (NPr, Pr), (Lic, NPr)}. For sake of clarity, the other attack relations from any argument towards each opposite argument of a greater level are not shown on the picture.
    The three leveled *bla* proposed in this example is consistent with the standard ideas
used in the supply chain management domain. This is why we consider at the most
important level that the supply chain has to satisfy the final customer (hence avoid the

---

[3] In real life, the establishment of this table is more complex and depends on the particular
objectives and priorities of the suppliers and the customers.

risk of back-ordering Rb). The supply chain should also have a quick flow of production (Qf) (hence the inventory level should not be high: Hi). A good production plan should ensure a proper management of the storage (i.e a stable load St and no waste in the storage capacities Hci). Moreover, the supply chain is supposed to have a low risk of failure (Ruf) and to be cost effective (hence avoid expensive maintenance (Me) and allow for regrouping lots Rl). At the second level, it is interesting (but not crucial) to have stability on production and delivery (a stable plan Sp, a constant quantity delivered Qd and periodic delivery dates Pd) in order to satisfy the workers unless this could increase the cost or induce some errors. Besides, here, the actors are considering that subcontracting (Sc) increases the risk of back-ordering, hence they prefer not to use overtime (Ot) because it is expensive. Temporary workers (Tw) decrease the productivity because they need some learning time. Since the environmental impact has to be integrated in the decision process (according to the idea of "Green SCM"), the actors should consider carbon impact (Ic). The third level takes into account the good impact on environment (Lic) and the actors personal preferences (a plan may be satisfactory - or not- for personal reason Pr - NPr) witch are not mandatory.

Let us justify the attack relation R_1 of the highest level of this bla, Rb attacks St, Rl and Qf since the primary objective of the supply chain is to fulfill the demand. Rb does not attack Ruf because this feature limits the risk to increase the level of back-ordering. Hi attacks Rl because the actors want to enforce regrouping the lots unless the inventory level becomes to high. The attack between Hi and Qf is different because for a voter the two arguments may coexist (or may be exclusive). So this attack can be described as follow "if the plan makes quick flow for one of the voter then it is acceptable only if this plan does not deal with a high inventory level for an other voter". Hi is attacked by St and Ruf because St or Ruf may justify Hi. Me attacks Ruf since a robust plan is a benefit only if this robustness is not too expensive to guarantee. Hci is attacked by Qf since it is better to have a quick flow of production than to under-exploit the storage capacity. An example of set of instantiated arguments I_B for c_1, c_2 and c_3 is given below together with a picture of the part of the instantiated bla obtained for candidate c_3:

	$I^+(c_1)$	$I^-(c_1)$	$I^+(c_2)$	$I^-(c_2)$	$I^+(c_3)$	$I^-(c_3)$
1	Qf_{v_2}		$Ruf_{v_1}, Qf_{v_1},$ St_{v_1}	Hci_{v_2}		
2	Sp_{v_1}		Pd_{v_1}, Qd_{v_1}	Ic_{v_2}	Qd_{v_2}, Sp_{v_2}	Tw_{v_1}, Ic_{v_1}
3	$Pr_{v_1}, Pr_{v_2},$ Lic_{v_2}		Pr_{v_1}, Lic_{v_1}	NPr_{v_2}	Pr_{v_2}	NPr_{v_1}

Part of the bla relative to $c3$:

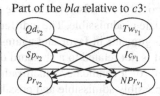

Proposition 1 (compactness of inputs). *When the bla contains at least two levels, the framework used to express arguments concerning a set of candidates from the point of view of a set of voters with all the corresponding attacks is more compact[4] in the bla framework than in a classical argumentation system.*

3.2   Decision Protocol

Given a bla A, the set of possible candidates C is given to all voters then each voter v:
- divides privately C into two sets C_v^+ and C_v^-,

---

[4] In terms of number of symbols used to represent the same information in the worst case.

- instantiates the features concerning each candidate[5] c, collected in $I_v(c)$,
- and finally chooses for each candidate the arguments that he wants to present.

For instance, in our example, v_1's private preferences are $C_{v_1}^+ = \{c_1, c_2\}$, $C_{v_1}^- = \{c_3\}$ and $v2$'s ones are $C_{v_2}^+ = \{c_1, c_3\}$, $C_{v_2}^- = \{c_2\}$. In order to select admissible candidates, we can focus on the existence of non attacked arguments that are in favor and against them. Given a candidate c, the set of non-attacked arguments concerning c, are denoted $S(c)$ (for "sources"), they are the vertices in $I(c)$ that have no predecessors. We may consider six main cases:

Definition 4 (admissibility status). *Given a bla A with an associated instantiated universe* (I, R_I)*, and a candidate c, let* $S(c) = \{x \in I(c) \mid \forall y, (y, x) \notin R\}$*, the status of c is*
- *necessary admissible* (N_{ad}) *if* $\varnothing \subset S(c) \subseteq I^+(c)$
- *possibly admissible* (Π_{ad}) *if* $S(c) \cap I^+(c) \neq \varnothing$
- *indifferent* (Id_{ad}) *if* $S(c) = \varnothing$
- *controversial* (Ct_{ad}) *if* $S(c) \cap I^+(c) \neq \varnothing$ *and* $S(c) \cap I^-(c) \neq \varnothing$
- *possibly inadmissible* $(\Pi_{\neg ad})$ *if* $S(c) \cap I^-(c) \neq \varnothing$
- *necessary inadmissible* $(N_{\neg ad})$ *if* $\varnothing \subset S(c) \subseteq I^-(c)$

In other words, a necessary admissible candidate has an argument in its favor that is unattacked and no unattacked argument against it; a possibly admissible candidate has at least one unattacked argument in its favor. An indifferent candidate has no unattacked arguments in its favor or against it, while a controversial candidate is both supported and criticized by unattacked arguments. A candidate is possibly inadmissible if there is an unattacked argument against it and necessary inadmissible if there is an unattacked argument against it and no unattacked argument in favor of it.

Example 3. The arguments given by v_1 and v_2 about c_1 belong only to I^+, thus $c_1 \in N_{ad}$, they seem to agree to select candidate c_1. c_2 is also in N_{ad}, c_3 is only possibly admissible ($c_3 \in \Pi_{ad}$) because Qd_{v_2} and Tw_{v_1} are not attacked.

The above definition is related to possibility theory [18,6], where necessary (resp. possibly) admissible could be understood as it is certain (resp. possible) that the candidate is admissible. The indifference case is linked to an impossibility to have unattacked arguments in favor and against a candidate, thus an impossibility to decide. However it is not related to a standard definition of possibilistic ignorance about the admissibility of a candidate, which rather corresponds to a controversial candidate that is both possibly admissible and possibly inadmissible. With these definitions we get:

Proposition 2. *Inclusion and Duality:*
- $N_{ad} \subseteq \Pi_{ad}$ *and* $N_{\neg ad} \subseteq \Pi_{\neg ad}$
- $Ct_{ad} = \Pi_{ad} \cap \Pi_{\neg ad}$ *and* $Id_{ad} = (C \setminus \Pi_{ad}) \cap (C \setminus \Pi_{\neg ad})$
- $N_{ad} = C \setminus (\Pi_{\neg ad} \cup Id_{ad})$ *and* $N_{\neg ad} = C \setminus (\Pi_{ad} \cup Id_{ad})$.

The following property reveals that if a candidate is acceptable for all the voters then it is necessary admissible, similarly, if a candidate is unacceptable for all the voters then it is necessary inadmissible.

---

[5] In this paper we suppose that a voter v uses the following *basic strategy*: if $c \in C_v^+$ then he only gives arguments in $I^+(c)$, if $c \in C_v^-$ then he only gives arguments in $I^-(c)$.

Proposition 3 (unanimity). *Considering a set of voters V using the basic strategy, for any candidate $c \in C$*

- *If $\forall v \in V, c \in C^+(v)$ and $I^+(c) \neq \varnothing$ then $c \in N_{ad}$.*
- *If $\forall v \in V, c \in C^-(v)$ and $I^-(c) \neq \varnothing$ then $c \in N_{\neg ad}$.*

With these admissibility status, we can propose 3 thresholds of admissibility (from the strongest to the weakest): the first threshold contains the candidates such that all the unattacked arguments about them are positive, the second threshold can be divided into two sets: 2a is the set of candidates under the first threshold together with the candidates for which no unattacked argument concerning them is available (neither positive nor negative), the set 2b tolerates candidates that are concerned by negative unattacked argument provided that they are also concerned at least by one positive unattacked argument. The third threshold is the union of the sets 1, 2a and 2b as shown in the figure below:

- threshold 1: $c \in N_{ad}$
- threshold 2a: $c \in N_{ad} \cup Id_{ad}$ (or in $C \setminus \Pi_{\neg ad}$)
- threshold 2b: $c \in \Pi_{ad}$ (or in $N_{ad} \cup Ct_{ad}$)
- threshold 3: $c \in \Pi_{ad} \cup Id_{ad}$ (or in $C \setminus N_{\neg ad}$)

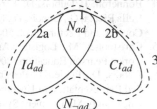

Note that qualitative explanations of the selection can be drawn on a graph, they are simple and clear compared to quantitative explanations (that may result from complex and obscure numeric computation and that are not necessarily based on a consensual agreement).

How to compute the admissible sets? Given a candidate c, we first consider the set of unattacked arguments concerning c (i.e. $S(c) = \{x \in I(c) \mid \forall y, (y, x) \notin R\}$) and then according to this set, we may assign one of the six status (necessary/possibly admissible/inadmissible, indifferent/controversial) to c by using Def. 4. Building the sets N_{ad}, Π_{ad}, Id_{ad}, $N_{\neg ad}$ and $\Pi_{\neg ad}$ can be done in polynomial time wrt $|I|$.

4 Conclusion

This paper introduces a new approach for SCM collaborative planning which is the basis of a new framework for collaborative group decision making. The proposed process is a kind of bipolar argumentative vote with more expressive power than standard vote procedures but also a more compact representation of an argumentation system which focus on admissibility of candidate and not on a ranking.

There are numerous benefits of this new decision process. First, our framework follows the standard principles defined by CPFR® that are commonly used by industrialists. Moreover our proposal handles multi-criteria expression which is often not the case for classical approaches or done roughly by using arbitrary aggregation functions. Furthermore, it guarantees a succinct expression. And finally, the qualitative aspect of this approach enables to justify the decision in an understandable and clear way to the actors. However there are some drawbacks, for instance the need to define a protocol for reaching a bla and a common definition of conflicts and priorities which maybe a difficult task (but it has only to be done once). This difficulty may be increased in domains with more open criteria than in the SCM field where standard criteria are already well established. In order to complete the definition of this bipolar deliberation protocol, a study is still needed to establish that the admissible sets are rational regarding classical decision rules such as *Pareto*, *BiPoss* and *BiLexi* [4].

References

1. Albrecht, M.: Supply Chain Coordination Mechanisms: New Approaches for Collaborative Planning. Lecture Notes in Eco. and Math. Systems, vol. 628. Springer, Heidelberg (2010)
2. Amgoud, L., Vesic, S.: A formal analysis of the role of argumentation in negotiation dialogues. Journal of Logic and Computation 22, 957–978 (2012)
3. Baykasoglu, A., Gocken, T.: Multi-objective aggregate production planning with fuzzy parameters. Advances in Engineering Software 41(9), 1124–1131 (2010)
4. Bonnefon, J., Dubois, D., Fargier, H.: An overview of bipolar qualitative decision rules. In: Della Riccia, G., Dubois, D., Kruse, R., Lenz, H.-J. (eds.) Preferences and Similarities. CISM Courses and Lectures, vol. 504, pp. 47–73. Springer (2008)
5. Christopher, M.: Logistics And Supply Chain Management: Creating Value-Adding Networks. Pearson Education (2005)
6. Dubois, D., Prade, H.: Possibility theory: qualitative and quantitative aspects. In: Quantified Representation of Uncertainty and Imprecision. Handbook of Defeasible Reasoning and Uncertainty Management Systems, vol. 1, pp. 169–226. Kluwer Academic (1998)
7. Dudek, G.: Collaborative Planning in Supply Chains: A Negotiation-based Approach. Springer, Heidelberg (2009)
8. Dung, P.M.: On the acceptability of arguments and its fundamental role in nonmonotonic reasoning, logic programming and n-person games. Artificial Intelligence 77, 321–357 (1995)
9. Guillaume, R., Grabot, B., Thierry, C.: Management of the risk of backorders in a MTO-ATO/MTS context under imperfect requirements. In: Applied Mathematical Modelling (2013)
10. Ireland, R.K., Crum, C.: Supply Chain Collaboration: How To Implement CPFR And Other Best Collaborative Practices. Integrated Business Management Series. J. Ross Publishing, Incorporated (2005)
11. Jamalnia, A., Soukhakian, M.: A hybrid fuzzy goal programming approach with different goal priorities to aggregate production planning. Comp. Ind. Eng. 56(4), 1474–1486 (2009)
12. Lu, T.-P., Trappey, A.J.C., Chen, Y.-K., Chang, Y.-D.: Collaborative design and analysis of supply chain network management key processes model. Journal of Network and Computer Applications (2013)
13. Marcotte, F., Grabot, B., Affonso, R.: Cooperation models for supply chain management. International Journal of Logistics Systems and Management 5(1), 123–153 (2009)
14. Rahwan, I., Ramchurn, S.D., Jennings, N.R., McBurney, P., Parsons, S., Sonenberg, L.: Argumentation-based negotiation. Knowledge Engineering Review 18(4), 343–375 (2003)
15. Selim, H., Araz, C., Ozkarahan, I.: Collaborative production–distribution planning in supply chain: A fuzzy goal programming approach. Transportation Research Part E: Logistics and Transportation Review 44(3), 396–419 (2008)
16. Slovic, P., Finucane, M., Peters, E., MacGregor, D.: Rational actors or rational fools? Implications of the affect heuristic for behavioral economics. The Journal of Socio-Economics 31, 329–342 (2002)
17. Walton, D.N., Krabbe, E.C.W.: Commitment in Dialogue: Basic Concepts of Interpersonal Reasoning. State University of New York Press, Albany (1995)
18. Zadeh, L.A.: Fuzzy Sets as a Basis for a Theory of Possibility. Memorandum: Electronics Research Laboratory. College of Eng., University of California (1977)

Encoding Argument Graphs in Logic

Philippe Besnard, Sylvie Doutre, and Andreas Herzig

IRIT-CNRS, University of Toulouse, France
{besnard,doutre,herzig}@irit.fr

Abstract. Argument graphs are a common way to model argumentative reasoning. For reasoning or computational purposes, such graphs
may have to be encoded in a given logic. This paper aims at providing
a systematic approach for this encoding. This approach relies upon a
general, principle-based characterization of argumentation semantics.

1 Introduction

In order to provide a method to reason about argument graphs [1], Besnard and
Doutre first proposed encodings of such graphs and semantics in propositional
logic [2]. Further work by different authors following the same idea was published
later, e.g. [3–7]. However, all these approaches were devoted to specific cases
in the sense that for each semantics, a dedicated encoding was proposed from
scratch. We aim here at a generalization, by defining a *systematic* approach to
encoding argument graphs (which are digraphs) and their semantics in a logic
⊢. Said differently, our objective is to capture the extensions under a given
semantics of an argument graph in a given logic (be it propositional logic or
any other logic). We hence generalize the approach originally introduced in [2]
by parametrizing the encoding in various ways, including principles defining a
given semantics.

  We consider abstract arguments first, and then provide guidelines to extend
the approach to structured arguments (made up of a support that infers a
conclusion).

2 Argument Graph and Semantics

2.1 Reminder

The notion of an argument graph has been introduced by Dung in [1][1].

Definition 1. *An* argument graph *is a couple* $G = (\mathcal{A}, \mathfrak{R})$ *such that* $\mathcal{A}$ *is a
finite set and* $\mathfrak{R} \subseteq \mathcal{A} \times \mathcal{A}$ *is a binary relation over* $\mathcal{A}$.

The elements of the set of vertices $\mathcal{A}$ are viewed as a set of abstract arguments,
the origin and the structure of which are unspecified. The edges $\mathfrak{R}$ represent
attacks: $(a, b) \in \mathfrak{R}$, also written $a\mathfrak{R}b$, means that a attacks b. A set of arguments
S attacks an argument a if a is attacked by some element of S.

---

[1] Dung uses the term argumentation framework instead of argument graph.

A. Laurent et al. (Eds.): IPMU 2014, Part II, CCIS 443, pp. 345–354, 2014.

Dung introduced several semantics to define which sets of arguments can be considered as collectively acceptable: the admissible, stable, grounded, preferred and complete semantics. The application of a semantics to a given argument graph results in a set of acceptable sets, called extensions. As an example of a semantics, one may consider the stable semantics [1].

Definition 2. *Given an argument graph $G = (\mathcal{A}, \mathfrak{R})$, a stable extension $S \subseteq \mathcal{A}$ is a set that satisfies the following two conditions:*

1. *it does not exist two arguments a and b in S such that $a\mathfrak{R}b$;*
2. *for each argument $b \notin S$, there exists $a \in S$ such that $a\mathfrak{R}b$ (any argument outside the extension is attacked by the extension).*

More generally, a *semantics* gives a formal definition of a method ruling the argument evaluation process. Extensions under a given semantics σ are called σ-*extensions*. $\mathcal{E}_\sigma(G)$ denotes the set of the σ-extensions of an argument graph G. Following Dung, a huge range of semantics have been defined (see [8] for a comprehensive overview). For these semantics, the following notions are essential (where an argument graph $G = (\mathcal{A}, \mathfrak{R})$ is assumed).

A set $S \subseteq \mathcal{A}$ is *conflict-free* iff $\nexists a, b \in S$ such that $a\mathfrak{R}b$.

An argument $a \in \mathcal{A}$ is *defended* by $S \subseteq \mathcal{A}$ iff $\forall b$ such that $b\mathfrak{R}a, \exists c \in S$ such that $c\mathfrak{R}b$.

A set of extensions $\mathcal{E} \subseteq \mathcal{E}_\sigma(G)$ is *inclusive-maximal* iff $\forall E_1, E_2 \in \mathcal{E}$, if $E_1 \subseteq E_2$ then $E_1 = E_2$.

An *admissible set* is a conflict-free set that defends all its elements.

A stable extension is an admissible set, but not all admissible sets are stable extensions.

The set of *preferred extensions* of an argument graph is the inclusive-maximal set of its admissible sets.

A number of complexity results have been established for decision problems in abstract argument graphs [9]. Two such problems are:

Verification VER$_\sigma$. Given a semantics σ, an argument graph $G = (\mathcal{A}, \mathfrak{R})$ and a set $S \subseteq A$, is S a σ-extension of G?

Existence EX$_\sigma$. Given a semantics σ and an argument graph $G = (\mathcal{A}, \mathfrak{R})$, does G have at least one σ-extension?

For instance, as regards the verification problem [9]: VER$_{stable}$ is in P, but VER$_{preferred}$ is coNP-complete. As regards the existence problem, the question of the existence of a stable extension, EX$_{stable}$, is NP-complete.

2.2   Encoding

Given any semantics σ, our objective is to capture the σ-extensions of an argument graph $(\mathcal{A}, \mathfrak{R})$ in a logic $\vdash$. The only requirements for this logic are that it should contain all the Boolean connectives (in order to capture "not", "and", and "or").

There are two ways to achieve our objective:

(α) *By providing a formula θ_σ whose models characterize the set $\mathcal{E}_\sigma(G)$ of σ-extensions of $G = (\mathcal{A}, \mathfrak{R})$. So the set $\mathrm{Mod}(\theta_\sigma)$ of the models of θ_σ is isomorphic to the set of σ-extensions of $(\mathcal{A}, \mathfrak{R})$: every model of θ_σ determines a σ-extension of $(\mathcal{A}, \mathfrak{R})$ and vice-versa.* [2]

(β) *By providing a formula $\theta_{\sigma,S}$, depending on a subset S of $\mathcal{A}$, that is satisfiable if and only if S is a σ-extension of $(\mathcal{A}, \mathfrak{R})$.*

Adopting terminology from [2], we call (α) "the model checking approach" and (β) "the satisfiability approach" (answering the verification problem VER$_\sigma$).

In (α), we must provide a means to identify extensions in the encoding. (There might for instance be non-effective ways for a model to coincide with an extension.) In the rest of the paper, we will focus on the (β) approach.

An additional issue (γ) may be to find a formula (in the logic $\vdash$) that is satisfiable iff there exists a σ-extension for the argument graph (existence problem EX$_\sigma$). This issue is of interest for the stable semantics for instance, but not for other admissibility-based semantics (preferred, complete, grounded...), the empty set being always an admissible set.

3 Encoding Methodology

Now, we provide a methodology for encoding the σ-extensions of an argument graph in a given logic, following the satisfiability approach previously introduced. We are going to illustrate it by a case study in Section 5.

3.1 Encoding Extensions

At the abstract level, given a set of abstract arguments $\mathcal{A} = \{a_1, a_2, \ldots\}$ and an argument graph $G = (\mathcal{A}, \mathfrak{R})$, in order to construct $\theta_{\sigma,S}$, the following questions should be answered:

1. How to represent a subset S of the set of arguments $\mathcal{A}$? For instance, it could be:

$$\chi_S = \bigwedge_{a_i \in S} a_i \wedge \bigwedge_{a_j \notin S} \neg a_j$$

2. How to define that S is a σ-extension of G? For instance, if σ is the stable semantics then we might have:

$$S \text{ is a stable extension of } G \text{ iff } \chi_S \models \bigwedge_{a \in \mathcal{A}} \left(a \leftrightarrow \bigwedge_{b \in \mathcal{A}: b \mathfrak{R} a} \neg b \right)$$

In [2], it was shown that $\bigwedge_{a \in \mathcal{A}}(a \leftrightarrow \bigwedge_{b \in \mathcal{A}: b \mathfrak{R} a} \neg b) = \theta_{stable}$. More generally, we are aiming at constructing $\theta_{\sigma,S}$ enjoying the following equivalence: S is a σ-extension of G iff $\chi_S \models \theta_\sigma$, i.e.,

$$\theta_{\sigma,S} \text{ is satisfiable iff } \chi_S \models \theta_\sigma$$

---

[2] In the case that $\mathrm{Mod}(\theta_\sigma)$ is isomorphic to $\mathcal{E}_\sigma(G)$ then the following consequence holds: $\theta_\sigma \vdash \varphi$ iff φ encodes a $\vdash$-definable property of G.

3. When a semantics involves a notion of maximality or minimality, how to capture the corresponding sets?

3.2 Encoding Set-Theoretic Relations

Our methodology for systematic encodings $\theta_{\sigma,S}$ where S *is the subset to be tested for being an extension* relies on several building bricks and a rule as follows.

– **Rule**

An encoding is of the form

$$\theta_{\sigma,S} = \varphi_S \wedge \overline{\varphi}_S \wedge \Psi_S$$

where φ_S encodes the necessary conditions for membership in S, $\overline{\varphi}_S$ encodes the sufficient conditions, and Ψ_S is a Boolean combination over basic building bricks (intuitively, Ψ_S expresses that S enjoys σ).

– **Basic Building Bricks**[3]

- *Membership in a subset of the arguments:* a is an argument in $X \subseteq \mathcal{A}$ is encoded as

$$\varphi_{(a \in X)} = \begin{cases} \varphi_{(a \in S)} & \text{if } X = S \\ \bigvee_{a = x \in X} \top & \text{if } X \neq S \end{cases}$$

where for each $a \in \mathcal{A}$, we assume a generic formula[4] $\varphi_{(a \in S)}$ expressing that "**a is in the set of arguments S**".

- *Subset X of the arguments:*

$$\varphi_X = \bigwedge_{a \in X} \varphi_{(a \in X)}$$

- *Complement of X in the set of arguments:*

$$\overline{\varphi}_X = \bigwedge_{a \notin X} \neg \varphi_{(a \in X)}$$

Intuitively, φ_S expresses that S contains all the elements of S whereas $\overline{\varphi}_S$ expresses that S contains only elements of S.

---

[3] Remember that an empty conjunction, i.e., a conjunction $\bigwedge_{\mathfrak{C}(x)} \gamma[x]$ whose condition $\mathfrak{C}(x)$ holds for no x, amounts to $\top$. An empty disjunction, i.e., a disjunction $\bigvee_{\mathfrak{C}(x)} \gamma[x]$ whose condition $\mathfrak{C}(x)$ holds for no x, amounts to $\bot$.

[4] By generic formula, we mean a formula that is constructed in a systematic way, by contrast to ad-hoc formulas with no common form. For example, $\varphi_{(a \in S)}$ can be a (provided that, for all arguments a, there is an atom a in the language). This is generic because all such formulas have the same form: an atom naming an argument.

- **Intermediate Building Bricks**
  1. $X \subseteq Y$
     This is captured as
     $$\bigwedge_{a \in \mathcal{A}} \left(\varphi_{(a \in X)} \rightarrow \varphi_{(a \in Y)} \right)$$

  2. X *is maximal such that* Ψ_X *holds*
     This, which amounts to Ψ_X & $\forall Y \supseteq X \ (\Psi_Y \rightarrow Y \subseteq X)$, is captured as

     $$\Psi_X \wedge \bigwedge_{X \subseteq Y \in 2^{\mathcal{A}}} \left(\Psi_Y \rightarrow \bigwedge_{a \in \mathcal{A}} (\varphi_{(a \in Y)} \rightarrow \varphi_{(a \in X)}) \right)$$

  3. X *is minimal such that* Ψ_X *holds*
     This, which amounts to Ψ_X & $\forall Y \subseteq X \ (\Psi_Y \rightarrow X \subseteq Y)$, is captured as

     $$\Psi_X \wedge \bigwedge_{X \supseteq Y \in 2^{\mathcal{A}}} \left(\Psi_Y \rightarrow \bigwedge_{a \in \mathcal{A}} (\varphi_{(a \in X)} \rightarrow \varphi_{(a \in Y)}) \right)$$

Please bear in mind that, as a consequence of the first building brick, the following holds: In all of the above clauses, whenever $X \neq S$ (and similarly for Y and Z), $\varphi_{(a \in X)}$ *must* be encoded as $\bigvee_{a = x \in X} \top$ else it is encoded as $\varphi_{(a \in S)}$.

As an illustration, here are the details for the case of maximality:

max: Let us assume that E satisfies Ψ' (hence Ψ'_E holds). Then, for all $S \subset E$,

$$\theta_{\sigma,S} \overset{\text{def}}{=} \varphi_S \wedge \overline{\varphi}_S \wedge \Psi'_S \wedge \bigwedge_{S \subseteq Y \in 2^{\mathcal{A}}} \left(\Psi'_Y \rightarrow \bigwedge_{a \in \mathcal{A}} (\varphi_{(a \in Y)} \rightarrow \varphi_{(a \in S)}) \right)$$

is not satisfiable *because* the conjunct $\overline{\varphi}_S$ is contradicted by means of $\Psi'_Y \rightarrow \varphi_Y$ (for the case $Y = E$); for $a \in E \setminus S$, it happens that $\overline{\varphi}_S$ entails $\neg \varphi_{(a \in S)}$ whereas $\Psi'_E \rightarrow \varphi_E$ yields $\varphi_{(a \in S)}$ (remember, Ψ'_E holds).

4 Encoding Semantic Principles

Baroni and Giacomin have shown in [10] that the existing argumentation semantics satisfy a number of principles. They have provided a comprehensive list of such principles. From some subsets of these principles, it is possible to characterize existing semantics. Based on such a general characterization of a semantics, the objective in this section is to encode into formulas the principles $P_1 \ldots P_n$ that define the semantics.

This requires two things. One is that we must prepare, from the building bricks listed in Section 3.2, encodings of statements (and their denials) such as:

- "a is in E"
- "a attacks b" (in symbols, $a \Re b$) and set versions thereof

- "S is maximal such that..."
- ...

The other thing is to provide a concrete list of such principles P. In Section 2 we have already mentioned conflict-freeness, inclusion-maximality, and admissibility. We recall here some of the list in [10]:

Conflict-Free Principle. A semantics σ satisfies the C-F principle iff $\forall G, \forall E \in \mathcal{E}_\sigma(G)$, E is conflict-free.

From the building bricks in Section 3.2, conflict-freeness can be encoded as[5]

$$\bigwedge_{b \Re a} \neg (\varphi_{(a \in S)} \wedge \varphi_{(b \in S)})$$

Inclusion-Maximality Criterion. A semantics σ satisfies the I-M criterion iff $\forall G$, $\mathcal{E}_\sigma(G)$ is inclusive-maximal.

Here, an encoding has been already explicitly given in Section 3.2.

Encoding defence (for all b such that $b \Re a$, there exists $c \in S$ such that $c \Re b$) is achieved by

$$\bigwedge_{b \Re a} \bigvee_{c \Re b} \varphi_{(c \in S)}$$

which can be used in the encoding of the next three principles that are based on defence, as follows.

Admissibility Criterion. A semantics σ satisfies the admissibility criterion iff $\forall G$, $\forall E \in \mathcal{E}_\sigma(G)$, if $a \in E$ then a is defended by E.

The admissibility criterion can be captured through

$$\varphi_{(a \in S)} \to \left(\bigwedge_{b \Re a} \bigvee_{c \Re b} \varphi_{(c \in S)} \right)$$

Reinstatement Criterion. A semantics σ satisfies the reinstatement criterion iff $\forall G$, $\forall E \in \mathcal{E}_\sigma(G)$, if a is defended by E then $a \in E$.

The reinstatement criterion can be captured by means of

$$\left(\bigwedge_{b \Re a} \bigvee_{c \Re b} \varphi_{(c \in S)} \right) \to \varphi_{(a \in S)}$$

Conflict-Free Reinstatement Criterion. A semantics σ satisfies the CFR criterion iff $\forall G$, $\forall E \in \mathcal{E}_\sigma(G)$, if a is defended by E and $E \cup \{a\}$ is conflict-free then $a \in E$.

Now, the CFR criterion can be captured just as the reinstatement criterion, only adding (in the antecedent) a conjunct for conflict-freeness (see above the conflict-free principle).

---

[5] Please observe that S instead of E occurs in the specification of the formula because we construct a formula (parameterized by S) which is satisfiable iff S is an extension.

In addition to the criteria listed in [10], we propose the following one:

Complement Attack Criterion. A semantics σ satisfies this criterion iff $\forall G$, $\forall E \in \mathcal{E}_\sigma(G)$ it holds that $\forall b \in \mathcal{A}$ if $b \notin E$ then $\exists a \in E$ such that $a\Re b$. The complement attack criterion can be captured by means of

$$\bigwedge_{a \in \mathcal{A}} \left(\neg\varphi_{(a \in S)} \rightarrow \left(\bigvee_{b\Re a} \varphi_{(b \in S)} \right) \right)$$

Another encoding is to be found in the example detailed in Section 5.

The stable semantics is characterized by the conflict-free principle and the complement attack criterion. The stable semantics satisfies the admissibility criterion as well.

5   A Case Study

There are two approaches, (a) and (f). One approach (a) introduces dedicated atoms in the object language to represent the attack relation. This means that there is a list of fresh Att_{xy} atoms, one for each ordered pair of nodes (x, y) in the graph. The other approach (f) dispenses from such atoms, and the properties to be encoded must be expressed in such a way that reference to attacks can be captured as a range over conjunction or disjunction.

Example

Let us find $\theta_{\sigma,S}$ for $\sigma = stable$. The definition for a set of arguments S being conflict-free is

$$\forall a \in S \; \nexists b \in S \quad b\Re a$$

According to the (f)-approach, i.e., no dedicated atom is used, we must reformulate the condition as follows: *for all a in S, it is not the case that there exists some b attacking a such that b is in S*. The (f)-encoding for being conflict-free is

$$\bigwedge_{a \in S} \neg \bigvee_{b\Re a} \varphi_{(b \in S)}$$

or, equivalently,

$$\bigwedge_{a \in S} \bigwedge_{b\Re a} \neg\varphi_{(b \in S)}$$

In the case of stable extensions, we need to additionally encode the property of S attacking its complement:

$$\forall a \notin S \exists b \in S \quad b\Re a$$

According to the (f)-approach, we must reformulate the condition in the form: *for all a not in S, there exists some b attacking a such that b is in S*. Hence the (f)-encoding for S attacking its complement is

$$\bigwedge_{a \notin S} \bigvee_{b\Re a} \varphi_{(b \in S)}$$

Combining both conditions, we obtain the building brick Ψ_S (introduced above: the general case) for stable extensions:

$$\Psi_S \;=\; \Big(\bigwedge_{a\in S}\bigwedge_{b\Re a}\neg\varphi_{(b\in S)}\Big) \;\wedge\; \Big(\bigwedge_{a\notin S}\bigvee_{b\Re a}\varphi_{(b\in S)}\Big)$$

Conjoining with φ_S and $\overline{\varphi}_S$, we obtain:

$$\varphi_S \wedge \overline{\varphi}_S \wedge \Psi_S \;=\; \bigwedge_{a\in S}\Big(\varphi_{(a\in S)}\wedge\bigwedge_{b\Re a}\neg\varphi_{(b\in S)}\Big) \;\wedge\; \bigwedge_{a\notin S}\Big(\neg\varphi_{(a\in S)}\wedge\bigvee_{b\Re a}\varphi_{(b\in S)}\Big)$$

which is exactly the formula encoding stable extensions in Proposition 10 of [2] where $\varphi_{(x\in S)}$ is the atom x.

6   Towards Encoding Graphs of Structured Arguments

The aim of this section is to indicate how the approach, designed for abstract argument graphs, may be extended to graphs with structured arguments. Our exposition follows [11].

The very first issue is to choose how to represent nodes of an argumentation graph in this case. Then, arises the question of the representation of edges, and next, of how to define the extensions. Moreover, the properties of the logic underlying the arguments play a role now.

We assume that nodes in the graph (i.e., arguments) enjoy a minimal amount of structure as pairs (A_i, c_i) where A_i is a set of formulas, the premises of the argument, and c_i is a formula, the claim of the argument. Importantly,

$$A_i \Vdash c_i$$

In any case, our approach involves the following steps.

1. Representing nodes, with two options:
   - Either every two arguments of the form (A_i, c_i) and (A_i, c_i') are treated as two distinct entities,
   - or they are treated as equals.

   The former might be justified on the grounds that we have some (granted, rudimentary) structure within an argument (it is less abstract) and there must be the possibility to detail the content of the attack relation. We choose to leave these two options open.
2. Representing edges, with the question:
   - Is it necessary that the attack relation be captured at the object level, i.e., by a formula of the logic?

   Whether representing or not (i.e., simply capturing constraints and conditions to be satisfied by the attack relation), we would need the language to include something capturing consistency (and presumably, inference, hence the need for the deduction theorem). E.g., there could be a modal possibility operator $\Diamond$ and a naming device $\lceil\cdot\rceil$. All these would express consistency of support of two arguments in an extension, as for example in

$$\Diamond(A_i \wedge A_j) \to \neg\lceil i\Re j\rceil.$$

Another option would be to use QBF, as done in [12] to characterize the complexity, or to use Dynamic Logic of Propositional Assignments, as done in [13] to provide a dynamic account of the construction of extensions.

3. Representing extensions.

As a start, we must decide whether $\neg A_i$ is to mean that (A_i, c_i) fails to be in the extensions being tested. The answer determines whether or not an argument with the same support as an argument in the extension at hand has to be in the extension.

Another point is that extensions are defined as conditions using the attack relation. This may mean building bricks, other than those we have examined, from which various notions of extensions can be defined.

4. Lastly, attention must be paid to properties of the logic that can play a role. For example, contraposition and transitivity make the inference constraint to turn the conflict between arguments i and j into $A_i \Vdash \neg A_j$ as follows:

$$A_i \Vdash \neg c_j \qquad \text{(assumption)}$$
$$A_j \Vdash c_j \qquad \text{(assumption)}$$
$$\neg c_j \Vdash \neg A_j \text{ (contraposition)}$$
$$A_i \Vdash \neg A_j \qquad \text{(transitivity)}$$

7   Conclusion

We have proposed in this paper a methodology to encode an argument graph and a semantics in a given logic. Few constraints are imposed on the logic. We have considered abstract arguments, and we have given guidelines to extend the approach to structured, non-abstract arguments.

In this methodology, we have made the assumption that a semantics is defined by a set of principles. Even though evaluation principles have been put forward by [8, 10], the characterization of a semantics by a set of principles is an issue that has not been addressed yet. The example of the complement attack criterion (mandatory to characterize the stable semantics), shows that in current approaches, criteria are missing to obtain such a characterization.

In line with such a general characterization, it can be noticed that [14] contains a general recursive schema that captures all of Dung's semantics and that is able to capture other admissibility-based, or non-admissibility-based, semantics. However, the schema embeds a "base function", that basically characterizes the extensions of an argument graph made of only one strongly connected component; that is, the semantics is not described by the principles it is based on.

As regards computational issues, [15] surveys implementations of abstract argument graphs. Many abstract semantics have been implemented, with various techniques, but none of these implementations is built from a principle-based characterization of the semantics.

We have encoded in this paper a number of semantic principles, but others remain to be defined (and encoded) in order to characterize the existing semantics, and possibly, new semantics as well.

Moreover, the guidelines given to extend the approach to structured, non-abstract arguments, will be further explored.

Acknowledgements. This work benefited from the support of the AMANDE project (ANR-13-BS02-0004) of the French National Research Agency (ANR).

References

1. Dung, P.M.: On the acceptability of arguments and its fundamental role in non-monotonic reasoning, logic programming and n-person games. Artificial Intelligence 77(2), 321–357 (1995)
2. Besnard, P., Doutre, S.: Checking the acceptability of a set of arguments. In: Delgrande, J.P., Schaub, T. (eds.) Proc. NMR 2004, pp. 59–64 (2004)
3. Egly, U., Gaggl, S.A., Woltran, S.: Answer-set programming encodings for argumentation frameworks. Argument and Computation 1(2), 147–177 (2010)
4. Amgoud, L., Devred, C.: Argumentation frameworks as constraint satisfaction problems. In: Benferhat, S., Grant, J. (eds.) SUM 2011. LNCS, vol. 6929, pp. 110–122. Springer, Heidelberg (2011)
5. Walicki, M., Dyrkolbotn, S.: Finding kernels or solving SAT. Discrete Algorithms 10, 146–164 (2012)
6. Dyrkolbotn, S.K.: The same, similar, or just completely different? Equivalence for argumentation in light of logic. In: Libkin, L., Kohlenbach, U., de Queiroz, R. (eds.) WoLLIC 2013. LNCS, vol. 8071, pp. 96–110. Springer, Heidelberg (2013)
7. Nofal, S., Atkinson, K., Dunne, P.E.: Algorithms for decision problems in argument systems under preferred semantics. Artificial Intelligence 207, 23–51 (2014)
8. Baroni, P., Giacomin, M.: Semantics of abstract argument systems. In: Simari, G., Rahwan, I. (eds.) Argumentation in Artificial Intelligence, pp. 25–44. Springer (2009)
9. Dunne, P.E., Wooldridge, M.: Complexity of abstract argumentation. In: Simari, G., Rahwan, I. (eds.) Argumentation in Artificial Intelligence, pp. 85–104. Springer (2009)
10. Baroni, P., Giacomin, M.: On principle-based evaluation of extension-based argumentation semantics. Artificial Intelligence 171(10), 675–700 (2007)
11. Besnard, P., Garcia, A., Hunter, A., Modgil, S., Prakken, H., Simari, G., Toni, F.: Special issue: Tutorials on structured argumentation. Argument and Computation 5(1) (2014)
12. Arieli, O., Caminada, M.W.: A QBF-based formalization of abstract argumentation semantics. Journal of Applied Logic 11(2), 229–252 (2013)
13. Doutre, S., Herzig, A., Perrussel, L.: A dynamic logic framework for abstract argumentation. In: Baral, C., De Giacomo, G. (eds.) Proc. KR 2014. AAAI Press (2014)
14. Baroni, P., Giacomin, M., Guida, G.: SCC-recursiveness: A general schema for argumentation semantics. Artificial Intelligence 168(1), 162–210 (2005)
15. Charwat, G., Dvořák, W., Gaggl, S.A., Wallner, J.P., Woltran, S.: Implementing abstract argumentation — A survey. Technical Report DBAI-TR-2013-82, Technische Universität Wien, Fakultät für Informatik, Vienna, Austria (2013)

On General Properties of Intermediate Quantifiers

Vilém Novák and Petra Murinová

University of Ostrava, Institute for Research and Applications of Fuzzy Modeling
Centre of Excellence IT4Innovations
30 dubna 22, 701 03 Ostrava 1, Czech Republic
Vilem.Novak@osu.cz

Abstract. In this paper, we will first discuss fuzzy generalized quantifiers and their formalization in the higher-order fuzzy logic (the fuzzy type theory). Then we will briefly introduce a special model of intermediate quantifiers, classify them as generalized fuzzy ones and prove that they have the general properties of isomorphism invariance, extensionality and conservativity. These properties are characteristic for the quantifiers of natural language.

1 Introduction

Intermediate quantifiers form a class of linguistic quantifiers that characterize size of a class of elements fulfilling a given property and whose meaning lays between the classical general ($\forall$) and existential ($\exists$) ones. Typical examples of intermediate quantifiers are "most, many, almost all, few", and many other ones. These quantifiers were studied from the point of view of their semantics and general logical properties in the book [15]. A reasonable formalization of them within a special formal theory of higher-order fuzzy logic (using the syntax of fuzzy type theory (FTT)) was first given in [11]. The intermediate quantifiers are there defined as classical ones $\forall$ or $\exists$ whose universe of quantification is modified by means of special evaluative linguistic expressions.

It is clear that intermediate quantifiers form a special subclass of generalized quantifiers in the sense of their theory presented, e.g., in [5,14]. Therefore, it is interesting to know whether our formalization fulfills some of the basic properties, namely conservativity, extensionality, isomorphism invariance because as stated, e.g., in [16], linguistic quantifiers are supposed to have just these properties. Recall that the fuzzy quantifiers were included in the theory of generalized quantifiers in [3,8] and more precisely especially in [2,4].

In this paper, we will work in the Łukasiewicz fuzzy type theory. On many places, however, our results hold in an arbitrary one. Therefore, we will use the short FTT to denote arbitrary fuzzy type theory and Ł-FTT to denote the Łukasiewicz one. Note that the Łukasiewicz logic preserves most of the important properties of the classical logic but still is highly non-trivial many-valued one. Thus, the Ł-FTT proved to be an effective tool for modeling of fuzzy quantifiers. We proved that our model of the intermediate quantifiers, besides other

A. Laurent et al. (Eds.): IPMU 2014, Part II, CCIS 443, pp. 355–364, 2014.
© Springer International Publishing Switzerland 2014

properties, makes it possible to construct the generalized square of opposition (cf. [7]) based on them.

2   Preliminaries

The formal theory of intermediate quantifiers is developed within special higher-order fuzzy logic — the Łukasiewicz fuzzy type theory Ł-FTT because this theory is strong enough to fit our requirements. The algebra of truth values is a linearly ordered MV_Δ-algebra

$$\mathcal{E}_\Delta = \langle E, \vee, \wedge, \otimes, \rightarrow, \mathbf{0}, \mathbf{1}, \Delta \rangle$$

Because of the lack of space, we refer the reader to [9] for the details. Let us only mention, that basic syntactical objects of Ł-FTT are classical (cf. [1]), namely the concepts of *type* and *formula*. The atomic types are ϵ (elements) and o (truth values). General types are denoted by Greek letters $\alpha, \beta, \ldots$. The set of all types is denoted by *Types*. The *language* of Ł-FTT denoted by J, consists of variables $x_\alpha, \ldots$, special constants $c_\alpha, \ldots$ ($\alpha \in$ *Types*), the symbol λ, and brackets.

Definition 1. *Let J be a language of FTT and $(M_\alpha)_{\alpha \in Types}$ be a system of sets called* basic frame *such that M_o, M_ϵ are sets and for each $\alpha, \beta \in$ Types, $M_{\beta\alpha} \subseteq M_\beta^{M_\alpha}$, i.e. it is a set of weakly extensional functions[1] from M_α to M_β. The* frame *is a tuple*

$$\mathcal{M} = \langle (M_\alpha, \overset{\circ}{=}_\alpha)_{\alpha \in Types}, \mathcal{E}_\Delta \rangle \tag{1}$$

such that the following holds:

(i) *The $\mathcal{E}_\Delta$ is a structure of truth values (a linearly ordered MV_Δ-algebra). We put $M_o = E$ and assume that each set $M_{oo} \cup M_{(oo)o}$ contains all the operations from $\mathcal{E}$. The fuzzy equality on truth values is the biresiduation $\overset{\circ}{=}_o := \leftrightarrow$.*

(ii) *The set M_ϵ is the set of* individuals. *The fuzzy equality $\overset{\circ}{=}_\epsilon$ on individuals is a separated[2] fuzzy equality on M_ϵ.*

(iii) *If $\alpha \neq o, \epsilon$ then $\overset{\circ}{=}_\alpha$ is a fuzzy equality on M_α. We assume that $\overset{\circ}{=}_\alpha \in M_{(o\alpha)\alpha}$ for every $\alpha \in$ Types.*

We often put $M_\epsilon = \mathbb{R}$ (set of real numbers).

Recall that interpretation of a formula A_α of type α in $\mathcal{M}$ is determined with respect to an assignment $p \in \mathrm{Asg}(\mathcal{M})$ of elements from the sets M_α to variables where $\mathrm{Asg}(\mathcal{M})$ is a set of all assignments. The assignment p assures that all free variables occurring in the formula A_α are assigned specific elements so that the whole formula A_α can be evaluated in $\mathcal{M}$. The value of a formula A_α in a model $\mathcal{M}$ is denoted by $\mathcal{M}(A_\alpha)$. Clearly, $\mathcal{M}(A_\alpha) \in M_\alpha$ for all $\alpha \in$ *Types*.

By a fuzzy set in M_α we understand a function $A : M_\alpha \longrightarrow E$, i.e., it is obtained as interpretation of a formula $A_{o\alpha}$ in the model $\mathcal{M}$. We will use the

---

[1] By currying, we may confine only to unary functions.

[2] For all $m, m' \in M_\epsilon$, $\overset{\circ}{=}_\epsilon (m, m') = \mathbf{1}$ implies $m = m'$.

symbol $A \subseteq M_\alpha$ to stress that A is a fuzzy set in M_α. If x_α is a variable occurring free in A_α and $p(x_\alpha) = m \in M_\alpha$ then $\mathcal{M}_p(A_{o\alpha}x_\alpha) \in M_o$ is a membership degree of m in the fuzzy set $\mathcal{M}_p(A_{o\alpha})$ (recall that $M_o = E$).

The subsethood between fuzzy sets is on the level of syntax defined as a formula

$$A_{o\alpha} \subseteq B_{o\alpha} := (\forall u_\alpha)(A_{o\alpha}u_\alpha \Rightarrow B_{o\alpha}u_\alpha) \tag{2}$$

where $A_{o\alpha}, B_{o\alpha}$ are formulas whose interpretation are fuzzy sets (the symbol ":=" means "is defined as").

We say that a model $\mathcal{M}^1 = \langle (M_\alpha^1, \doteq_\alpha^1)_{\alpha \in Types}, \mathcal{L}_\Delta \rangle$ is a *submodel* of $\mathcal{M}^2 = \langle (M_\alpha^2, \doteq_\alpha^2)_{\alpha \in Types}, \mathcal{L}_\Delta \rangle$, in symbols $\mathcal{M}^1 \subset \mathcal{M}^2$, if $M_\alpha^1 \subseteq M_\alpha^2$, $\doteq_\alpha^1$ is a restriction of $\doteq_\alpha^2$ to the set M_α^1 for all $\alpha \in Types$ and

$$m_{o\alpha}^1(m_\alpha^1) = m_{o\alpha}^2(m_\alpha^1)$$

holds true for all $m_{o\alpha}^1 \in M_{o\alpha}^1, m_{o\alpha}^2 \in M_{o\alpha}^2$ and $m_\alpha^1 \in M_\alpha^1$, $\alpha \in Types$. Note that $M^1 \subseteq M^2$ can be expressed using an identity function $f : M^1 \longrightarrow M^2$. If $\mathcal{M}^1 \subset \mathcal{M}^2$, A_α is a quantifier-free formula and $p \in Asg(\mathcal{M}^1)$ an assignment to variables then

$$\mathcal{M}_p^1(A_\alpha) = \mathcal{M}_{p \circ f}^2(A_\alpha) \tag{3}$$

where $p \circ f$ is an assignment defined by $p \circ f(x_\alpha) = f(p(x_\alpha))$.

We say that two models $\mathcal{M}^1, \mathcal{M}^2$ are *isomorphic*, in symbols $\mathcal{M}^1 \cong \mathcal{M}^2$, if there is a set of bijections

$$\mathfrak{f} = \{f^\alpha : M_\alpha^1 \longrightarrow M_\alpha^2 \mid \alpha \in Types\} \tag{4}$$

such that for each $\alpha, \beta \in Types$ the following diagram commutes:

$$
\begin{array}{ccc}
M_\alpha^1 & \xrightarrow{\ f^\alpha\ } & M_\alpha^2 \\
\scriptstyle m_{\beta\alpha}^1 \downarrow & & \downarrow \scriptstyle m_{\beta\alpha}^2 = f^{\beta\alpha}(m_{\beta\alpha}^1) \\
M_\beta^1 & \xrightarrow{\ f^\beta\ } & M_\beta^2
\end{array}
$$

where $m_{\beta\alpha}^i \in M_{\beta\alpha}^i$, $i = 1, 2$ are functions and for each constant $c_\alpha \in J$, $\alpha \in Types$, it holds that $f^\alpha(\mathcal{M}^1(c_\alpha)) = \mathcal{M}^2(c_\alpha)$. The following theorem shows that interpretation of a formula in two isomorphic models is the same.

Theorem 1. *Let $\mathcal{M}^1 \cong \mathcal{M}^2$ be an isomorphism and A_α a formula of type α. Then*

$$f^\alpha(\mathcal{M}_p^1(A_\alpha)) = \mathcal{M}_{p \circ f}^2(A_\alpha)$$

for any assignment $p \in Asg(\mathcal{M}^1)$ and the corresponding assignment $p \circ \mathfrak{f} \in Asg(\mathcal{M}^2)$.

More details on model theory in FTT can be found in [12].

Note that the tautology

$$\vdash (A \Rightarrow B) \equiv ((A \wedge B) \equiv A). \tag{5}$$

is provable in (arbitrary) FTT.

The theory of intermediate quantifiers (denoted by T^{IQ}) is an extension of a special formal theory T^{Ev} of L-FTT that is a theory of the meaning of evaluative linguistic expressions. These are expressions of natural language such as *small, medium, big, about fourteen, very short, more or less deep, quite roughly strong,* etc. To develop the theory of their meaning, we must first introduce the concept of *context*[3].

The concept of the context for evaluative expressions can be reduced to a triple of numbers $\langle v_L, v_S, v_R \rangle$ where v_L is the leftmost and v_R the rightmost value that has sense to consider. For example, for heights of people, $v_L = 40$ cm and $v_R = 250$ cm. The v_S a typical medium value; in our example $v_S = 170$ cm.

The meaning of the evaluative expression is construed as a special formula representing *intension* whose interpretation in a model is a function from the set of contexts into a set of fuzzy sets. Given a context, the intension determines a corresponding extension that is a fuzzy set in a universe determined by the context. In this paper, we will deal with the context $\langle 0, 0.5, 1 \rangle$ only. Extensions of the

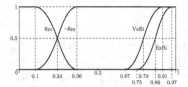

Fig. 1. Model of extensions of selected evaluative expressions

considered evaluative expressions are fuzzy sets of the shape depicted in Fig. 1. All the details, justification of the formal theory T^{Ev} including its special axioms and motivation can be found in [10].

In this paper, we do not need to introduce this theory in detail; it is sufficient to consider special formulas $Ev \in Form_{oo}$ that represent intensions of selected evaluative linguistic expressions.

3   Generalized Fuzzy Quantifiers

3.1   Generalized Fuzzy Quantifiers in FTT

In the classical theory of generalized quantifiers, the type $\langle k_1, \ldots, k_n \rangle$ denotes a quantifier which acts on n formulas, each having k_i free variables, $i = 1, \ldots, n$ (cf. [5,14]). This corresponds to semantic interpretation of a generalized quantifier as a relation among subsets of certain sets $M_1^{k_1}, \ldots, M_n^{k_n}$. Generalization of this concept to fuzzy generalized quantifiers was introduced in [8]. In fact, it replaces sets and relations by the fuzzy ones. Both classical as well as generalized definitions relate to first-order (fuzzy) logics. This means that all variables occurring in formulas have the same type — namely, they represent objects in a given universe.

In fuzzy type theory, however, this situation is more complicated because formulas may contain variables of various types. At the same time, types are included in the syntax of FTT and so, we can use them when characterizing a generalized quantifier.

---

[3] A more general concept used in logic is *possible world*.

To simplify slightly the notation, let us use boldface Greek letters to denote complex types, i.e. we write $\boldsymbol{\alpha}$ instead of $\alpha_1 \ldots \alpha_k$ where k depends on the formula in consideration. Similarly, the symbol $\mathbf{x}_{\boldsymbol{\alpha}}$ means a sequence of variables $x_{\alpha_1} \cdots x_{\alpha_k}$. Then we can introduce the following definition.

Definition 2. *Let $\mathcal{M}$ be a model of FTT. A generalized fuzzy quantifier is a special formula $\mathbf{Q}$ of type*

$$o(o\boldsymbol{\alpha}_1) \ldots (o\boldsymbol{\alpha}_n), \tag{6}$$

together with a symbol Q, defined as follows:

$$\mathbf{Q}_{o(o\boldsymbol{\alpha}_1)\ldots(o\boldsymbol{\alpha}_n)} := \lambda u_{o\boldsymbol{\alpha}_1} \ldots \lambda u_{o\boldsymbol{\alpha}_n} (Q\mathbf{x}_{\boldsymbol{\alpha}_1} \cdots \mathbf{x}_{\boldsymbol{\alpha}_n})(u_{o\boldsymbol{\alpha}_1}\mathbf{x}_{\boldsymbol{\alpha}_1}, \ldots, u_{o\boldsymbol{\alpha}_n}\mathbf{x}_{\boldsymbol{\alpha}_n}) \tag{7}$$

where $(Q\mathbf{x}_{\boldsymbol{\alpha}_1} \cdots \mathbf{x}_{\boldsymbol{\alpha}_n})(u_{o\boldsymbol{\alpha}_1}\mathbf{x}_{\boldsymbol{\alpha}_1}, \ldots, u_{o\boldsymbol{\alpha}_n}\mathbf{x}_{\boldsymbol{\alpha}_n})$ is a formula of type o. Interpretation of the quantifier (7) in the model $\mathcal{M}$ is

$$\mathcal{M}(\mathbf{Q}) \in M_o^{M_{o\boldsymbol{\alpha}_1} \times \cdots \times M_{o\boldsymbol{\alpha}_n}}. \tag{8}$$

Hence, a type of quantifier is just FTT-type of a special formula (7). We can introduce a simplification related to the classical theory by saying that $\mathbf{Q}$ is a *quantifier of type* $\langle \boldsymbol{\alpha}_1, \ldots, \boldsymbol{\alpha}_n \rangle$ understanding that it is a formula of type (6). Then a quantifier $\mathbf{Q}$ of type $\langle \alpha_1, \ldots, \alpha_n \rangle$ is called *monadic*. This means that in formula (7), single variables occur instead of sequences of them. If all formulas in the quantifier contain the same (known) variable x_α, then we can also write the type classically as $\langle 1, \ldots, 1 \rangle$.

The interpretation $\mathcal{M}(\mathbf{Q})$ in (8) is obtained using a special function $\mathbf{Q}_\mathcal{M}$ assigning to each n-tuple of fuzzy relations from the sets $M_{o\boldsymbol{\alpha}_1}, \ldots, M_{o\boldsymbol{\alpha}_n}$ a truth value. If the quantifier is monadic then $\mathbf{Q}_\mathcal{M}$ assigns a truth value to an n-tuple of fuzzy sets. A *global fuzzy generalized quantifier* is a functional assigning to each model $\mathcal{M}$ a function $\mathbf{Q}_\mathcal{M}$.

If $A_{o\boldsymbol{\alpha}_1}, \ldots, A_{o\boldsymbol{\alpha}_n}$ are formulas of the given types then the result of λ-conversion of $\mathbf{Q}_{o(o\boldsymbol{\alpha}_1)\ldots(o\boldsymbol{\alpha}_n)}$ will be written classically as

$$(Q\mathbf{x}_{\boldsymbol{\alpha}_1} \cdots \mathbf{x}_{\boldsymbol{\alpha}_n})(A_{o\boldsymbol{\alpha}_1}\mathbf{x}_{\boldsymbol{\alpha}_1}, \ldots, A_{o\boldsymbol{\alpha}_n}\mathbf{x}_{\boldsymbol{\alpha}_n}). \tag{9}$$

3.2 Intermediate Quantifiers

A special case of the generalized fuzzy quantifiers in the sense of Definition 2 are intermediate quantifiers. They are defined within a special theory denoted by $T^{IQ}[\mathcal{S}]$ where $\mathcal{S}$ is a set of distinguished types that must be considered to avoid possible difficulties with interpretation of the formula μ representing a measure (see below). The set $\mathcal{S}$ is supposed not to include too complex types α that would correspond to sets of very large, possibly non-measurable cardinalities. The theory $T^{IQ}[\mathcal{S}]$ is a special theory of L-FTT obtained as a certain extension of the theory of evaluative linguistic expressions T^{Ev} (for the precise definitions — see [6,11]).

A special role in T^{IQ} is played by a formula $\mu \in Form_{o(o\alpha)(o\alpha)}$ of the type $o(o\alpha)(o\alpha)$[4] representing measure of fuzzy sets that has the following properties:

---

[4] Eeach type requires one specific formula μ.

(M1) $T^{IQ} \vdash \mathbf{\Delta}(x_{o\alpha} \equiv z_{o\alpha}) \equiv ((\mu z_{o\alpha}) x_{o\alpha} \equiv \top)$,

(M2) $T^{IQ} \vdash \mathbf{\Delta}(x_{o\alpha} \subseteq z_{o\alpha}) \, \& \, \mathbf{\Delta}(y_{o\alpha} \subseteq z_{o\alpha}) \, \& \, \mathbf{\Delta}(x_{o\alpha} \subseteq y_{o\alpha}) \Rightarrow ((\mu z_{o\alpha}) x_{o\alpha} \Rightarrow (\mu z_{o\alpha}) y_{o\alpha})$,

(M3) $T^{IQ} \vdash \mathbf{\Delta}(z_{o\alpha} \neq \emptyset_{o\alpha}) \, \& \, \mathbf{\Delta}(x_{o\alpha} \subseteq z_{o\alpha}) \Rightarrow ((\mu z_{o\alpha})(z_{o\alpha} - x_{o\alpha}) \equiv \neg(\mu z_{o\alpha}) x_{o\alpha})$,

(M4) $T^{IQ} \vdash \mathbf{\Delta}(x_{o\alpha} \subseteq y_{o\alpha}) \, \& \, \mathbf{\Delta}(x_{o\alpha} \subseteq z_{o\alpha}) \, \& \, \mathbf{\Delta}(y_{o\alpha} \subseteq z_{o\alpha}) \Rightarrow ((\mu z_{o\alpha}) x_{o\alpha} \Rightarrow (\mu y_{o\alpha}) x_{o\alpha})$.

Interpretation of μ in a model $\mathcal{M}$ is a function $\mathcal{M}_p(\mu) : M_{o\alpha} \times M_{o\alpha} \longrightarrow E$. In our case, we consider measures defined for fuzzy subsets of a fixed fuzzy set B. Therefore, we will denote the measure $\mathcal{M}_p(\mu)$ by μ_B. The properties (M1)–(M4) are then the following: $\mu_B(B) = 1$, $\mu_B(Z_1) \leq \mu_B(Z_2)$ for all $Z_1 \subseteq Z_2 \subseteq B$, $\mu_B(B - Z) = \neg\mu_B(Z)$ and $\mu_{B_1}(Z) \geq \mu_{B_2}(Z)$ for all $Z \subseteq B_1 \subseteq B_2$.

Definition 3. *Let $T^{IQ}[\mathcal{S}]$ be a theory of intermediate quantifiers and $Ev \in Form_{oo}$ be an intension of some evaluative expression. Furthermore, let $z \in Form_{o\alpha}$, $x \in Form_\alpha$ be variables and $A, B \in Form_{o\alpha}$ be formulas representing measurable fuzzy sets. An intermediate quantifier interpreting the sentence*

"$\langle Quantifier \rangle$ B are A"

is one of the following formulas:

$$(Q^\forall_{Ev} x)(Bx, Ax) := (\exists z)((\mathbf{\Delta}(z \subseteq B) \, \& \, (\forall x)(zx \Rightarrow Ax)) \wedge Ev((\mu B)z)), \quad (10)$$

$$(Q^\exists_{Ev} x)(Bx, Ax) := (\exists z)((\mathbf{\Delta}(z \subseteq B) \, \& \, (\exists x)(zx \wedge Ax)) \wedge Ev((\mu B)z)). \quad (11)$$

Both quantifiers are, with respect to the definition above, of type $\langle \alpha, \alpha \rangle$. Indeed, the function $\mathbf{Q}$ obtained after interpretation formulas (10) and (11) is the following:

$$Q^\forall_{Ev,M}(B, A) = \bigvee \left\{ \bigwedge_{m \in M} (Z(m) \to A(m)) \wedge Ev(\mu_B(Z)) \, \middle| \, Z \underset{\sim}{\subseteq} M, Z \subseteq B \right\} \quad (12)$$

$$Q^\forall_{Ev,M}(B, A) = \bigvee \left\{ \bigvee_{m \in M} (Z(m) \wedge A(m)) \wedge Ev(\mu_B(Z)) \, \middle| \, Z \underset{\sim}{\subseteq} M, Z \subseteq B \right\} \quad (13)$$

where $A, B \underset{\sim}{\subseteq} M$ are given fuzzy sets, $\mu_B(Z)$ is a measure of the fuzzy set Z w.r.t. the fuzzy set B and Ev is interpretation of the corresponding evaluative expression in the context $\langle 0, 0.5, 1 \rangle$ (cf. [6,10,11] and Fig. 1). Let us remark that when referring to intermediate quantifiers we will in the sequel mean their model in the sense of Definition 3.

The following is an example of two concrete intermediate quantifiers:

T: Most B are $A := Q^\forall_{Bi\,Ve}(B, A) \equiv$
$$(\exists z)((\mathbf{\Delta}(z \subseteq B) \, \& \, (\forall x)(zx \Rightarrow Ax)) \wedge (Bi\,Ve)((\mu B)z)) \quad (14)$$

P: Almost all B are $A := Q^\forall_{Bi\,Ex}(B, A) \equiv$
$$(\exists z)((\mathbf{\Delta}(z \subseteq B) \, \& \, (\forall x)(zx \Rightarrow Ax)) \wedge (Bi\,Ex)((\mu B)z)), \quad (15)$$

The formulas *Bi Ve* and *Bi Ex* represent the evaluative expressions "very big" and "extremely big", respectively. Interpretation of (14) and (15) in a model is as follows: "all *B*s have *A*s" holds in a "very big" or an "extremely big" part (in the sense of the measure μ) of the universe, respectively.

4   General Properties of Intermediate Quantifiers

In this section we will discuss three general properties of the fuzzy general quantifiers, namely extensionality, isomorphism invariance, and conservativity and show that the intermediate quantifiers have all of them.

4.1   Extensionality

Extensionality of the classical generalized quantifiers of type $\langle 1, 1 \rangle$ is defined as follows: If $M \subseteq M'$ then

$$Q_M(B, A) \quad \text{iff} \quad Q_{M'}(B, A) \tag{16}$$

for all $A, B \subseteq M$.

Extensionality of the intermediate quantifiers in L-FTT must be defined w.r.t. models.

Theorem 2. *Let $\mathcal{M}^1 \subset \mathcal{M}^2$ be models of T^{IQ}. Let $A_{o\alpha}, B_{o\alpha}$ be quantifier-free formulas. Then the following holds:*

$$\mathcal{M}_p^1((Q_{Ev}^\forall x)(Bx, Ax)) = \mathcal{M}_{pof}^2((Q_{Ev}^\forall x)(Bx, Ax)), \tag{17}$$

$$\mathcal{M}_p^1((Q_{Ev}^\exists x)(Bx, Ax)) = \mathcal{M}_{pof}^2((Q_{Ev}^\exists x)(Bx, Ax)) \tag{18}$$

which means that intermediate quantifiers (in the sense of Definition 3) are extensional.

Proof. To simplify the notation, we denote $A = \mathcal{M}_p^1(A_{o\alpha})$, $B = \mathcal{M}_p^1(B_{o\alpha})$, $Z = \mathcal{M}_p^1(z_{o\alpha})$ and $Ev(\mu_B(Z)) = \mathcal{M}_p^1(Ev((\mu B)z))$.

First, note that from $A \subseteq M_\alpha^1 \subseteq M_\alpha^2$ we have that $A(m) = 0$ for all $m \in M_\alpha^2 - M_\alpha^1$ and similarly for B and Z. Furthermore, $\mathcal{M}_p^1(Ev((\mu B)z)) = \mathcal{M}_{pof}^2(Ev((\mu B)z))$ because $Z \subseteq B \subseteq M_\alpha^1 \subseteq M_\alpha^2$ and so, the measure of the fuzzy set Z is taken w.r.t. the fuzzy set B that is the same both in M_α^1 as well as in M_α^2. Hence,

$$\mathcal{M}_{pof}^2((\forall x)(zx \Rightarrow Ax)) = \bigwedge_{m \in M_\alpha^1} (Z(m) \rightarrow A(m)) \wedge$$

$$\bigwedge_{m \in M_\alpha^2 - M_\alpha^1} (Z(m) \rightarrow A(m)) = \mathcal{M}_p^1((\forall x)(zx \Rightarrow Ax))$$

because $Z(m) \rightarrow A(m) = 1$ for all $M_\alpha^2 - M_\alpha^1$. From it follows using (12) that

$$\mathcal{M}^2_{po\,f}((Q^\forall_{Ev}\,x)(Bx,Ax)) =$$

$$\bigvee \left\{ \bigwedge_{m\in M^2_\alpha} (Z(m) \to A(m)) \wedge Ev(\mu_B(Z)) \,\Big|\, Z \underset{\sim}{\subseteq} M^2_\alpha \text{ and } Z \subseteq B \underset{\sim}{\subseteq} M^1_\alpha \subseteq M^2_\alpha \right\} =$$

$$= \mathcal{M}^1_p((Q^\forall_{Ev}\,x)(Bx,Ax)).$$

Analogously we proceed for $(Q^\exists_{Ev}\,x)(Bx,Ax)$.

4.2   Invariance w.r.t. Isomorphism

Isomorphism of the classical generalized quantifiers of type $\langle 1,1 \rangle$ is defined as follows: If $f : M \longrightarrow M'$ is a bijection then

$$Q_M(B,A) \quad \text{iff} \quad Q_{M'}(B,A) \tag{19}$$

for all $A, B \subseteq M$. In FTT, however, the condition that the functions (4) are bijections is not sufficient. The theorem on isomorhpism invariance is thus stronger.

Theorem 3. *Let* $\mathcal{M}^1 \cong \mathcal{M}^2$ *be models of* T^{IQ} *and* $A, B \in Form_{o\alpha}$ *be formulas. Then the following holds:*

$$\mathcal{M}^1_p((Q^\forall_{Ev}\,x)(Bx,Ax)) = \mathcal{M}^2_{po\,f}((Q^\forall_{Ev}\,x)(Bx,Ax)), \tag{20}$$

$$\mathcal{M}^1_p((Q^\exists_{Ev}\,x)(Bx,Ax)) = \mathcal{M}^2_{po\,f}((Q^\exists_{Ev}\,x)(Bx,Ax)) \tag{21}$$

which means that intermediate quantifiers are isomorphism invariant.

Proof. This is an immediate consequence of Theorem 1.

4.3   Conservativity

Conservativity of the classical generalized quantifiers of type $\langle 1,1 \rangle$ is defined as follows:

$$Q_M(B,A) \quad \text{iff} \quad Q_M(B,A \cap B) \tag{22}$$

for all $A, B \subseteq M$. In case of fuzzy generalized quantifiers, we require equality of truth values in (22) instead of the equivalence.

Theorem 4. *The following is provable in* T^{IQ}:

(a) $T^{IQ} \vdash (Q^\forall_{Ev}\,x)(Bx,Ax) \equiv (Q^\forall_{Ev}\,x)(Bx,Ax \wedge Bx).$
(b) $T^{IQ} \vdash (Q^\exists_{Ev}\,x)(Bx,Ax) \equiv (Q^\forall_{Ev}\,x)(Bx,Ax \wedge Bx).$

This means that intermediate quantifiers are conservative.

Proof. The reflexivity

$$T^{IQ} \vdash (Q^\forall_{Ev}\,x)(Bx,Ax) \equiv (Q^\forall_{Ev}\,x)(Bx,Ax) \tag{23}$$

is provable. Hence, this formula is true in an arbitrary model $\mathcal{M} \models T^{\mathrm{IQ}}$. Let us check formula $\Delta(z \subseteq B)$ being a subformula of (10). Since the algebra of truth values is linear, either $\mathcal{M}(\Delta(z \subseteq B)) = 0$ or $\mathcal{M}(\Delta(z \subseteq B)) = 1$. In the former case, interpretation of both sides of (23) in the model $\mathcal{M}$ is equal to 0 so that interpretation of (23) in $\mathcal{M}$ is equal to 1. Otherwise, by (2) and (5) we have

$$\mathcal{M}_p(zx \equiv (zx \wedge Bx)) = 1 \tag{24}$$

for arbitrary assignment p which gives $\mathcal{M}_p(zx) = \mathcal{M}_p(zx \wedge Bx)$. From this, taking into account again definition (10) and tautology (5), we conclude that

$$\mathcal{M}_p((\forall x)(zx \Rightarrow Ax)) = \mathcal{M}_p((\forall x)(z\,x \Rightarrow Bx \wedge Ax))$$

which implies $T^{\mathrm{IQ}} \vdash (\forall x)(zx \Rightarrow Ax) \equiv (\forall x)(z\,x \Rightarrow Bx \wedge Ax)$. From this, using rule (R) we derive (a) from (23).

Analogously we prove also (b).

5 Conclusion

In this paper, we focused on three basic general properties of the intermediate quantifiers: conservativity, extensionality and isomorphism invariance. According to [14,16], quantifiers of natural language should fulfill all these three properties. Using formal properties of L-FTT and model theory of higher-order fuzzy logics we proved that our model of intermediate quantifiers fulfills them which supports the proclaim that (10) and (11) provide a reasonable mathematical model of the meaning of the intermediate quantifiers used in natural language. Other properties of them supporting further this claim can be found in the cited literature (cf. [6,7,13]).

Acknowledgments. The research was supported by the European Regional Development Fund in the IT4Innovations Centre of Excellence project (CZ.1.05/ 1.1.00/02.0070).

References

1. Andrews, P.: An Introduction to Mathematical Logic and Type Theory: To Truth Through Proof. Kluwer, Dordrecht (2002)
2. Dvořák, A., Holčapek, M.: L-fuzzy quantifiers of the type $\langle 1 \rangle$ determined by measures. Fuzzy Sets and Systems 160, 3425–3452 (2009)
3. Glöckner, I.: Fuzzy Quantifiers: A Computational Theory. STUDFUZZ, vol. 193. Springer, Berlin (2006)
4. Holčapek, M.: Monadic L-fuzzy quantifiers of the type $\langle 1^n, 1 \rangle$. Fuzzy Sets and Systems 159, 1811–1835 (2008)
5. Lindström, P.: First order predicate logic with generalized quantifiers. Theoria 32, 186–195 (1966)
6. Murinová, P., Novák, V.: A formal theory of generalized intermediate syllogisms. Fuzzy Sets and Systems 186, 47–80 (2012)

7. Murinová, P., Novák, V.: Analysis of generalized square of opposition with intermediate quantifiers. Fuzzy Sets and Systems 242, 89–113 (2014)
8. Novák, V.: Antonyms and linguistic quantifiers in fuzzy logic. Fuzzy Sets and Systems 124, 335–351 (2001)
9. Novák, V.: On fuzzy type theory. Fuzzy Sets and Systems 149, 235–273 (2005)
10. Novák, V.: A comprehensive theory of trichotomous evaluative linguistic expressions. Fuzzy Sets and Systems 159(22), 2939–2969 (2008)
11. Novák, V.: A formal theory of intermediate quantifiers. Fuzzy Sets and Systems 159(10), 1229–1246 (2008)
12. Novák, V.: Elements of model theory in higher order fuzzy logic. Fuzzy Sets and Systems 205, 101–115 (2012)
13. Novák, V., Murinová, P.: Intermediate quantifiers, natural language and human reasoning. In: Wang, G., Zhao, B., Li, Y. (eds.) Quantitative Logic and Soft Computing, pp. 684–692. World Scientific, New Jersey (2012)
14. Peters, S., Westerståhl, D.: Quantifiers in Language and Logic. Claredon Press, Oxford (2006)
15. Peterson, P.: Intermediate Quantifiers. Logic, linguistics, and Aristotelian semantics. Ashgate, Aldershot (2000)
16. Westerståhl, D.: Quantifiers in formal and natural languages. In: Gabbay, D., Guenthner, F. (eds.) Handbook of Philosophical Logic, vol. IV, pp. 1–131. D. Reidel, Dordrecht (1989)

A Note on Drastic Product Logic

Stefano Aguzzoli[1], Matteo Bianchi[1], and Diego Valota[2]

[1] Department of Computer Science, Università degli Studi di Milano, via Comelico 39/41, 20135, Milano, Italy
aguzzoli@di.unimi.it, matteo.bianchi@unimi.it
[2] Dipartimento di Scienze Teoriche e Applicate, Università degli Studi dell'Insubria, via Mazzini 5, 21100, Varese, Italy
diego.valota@gmail.com

Abstract. The drastic product $*_D$ is known to be the smallest t-norm, since $x *_D y = 0$ whenever $x, y < 1$. This t-norm is not left-continuous, and hence it does not admit a residuum. So, there are no drastic product t-norm based many-valued logics, in the sense of [7]. However, if we renounce standard completeness, we can study the logic whose semantics is provided by those MTL chains whose monoidal operation is the drastic product. This logic is called S_3MTL in [17]. In this note we justify the study of this logic, which we rechristen DP (for drastic product), by means of some interesting properties relating DP and its algebraic semantics to a weakened law of excluded middle, to the Δ projection operator and to discriminator varieties. We shall show that the category of finite DP-algebras is dually equivalent to a category whose objects are multisets of finite chains. This duality allows us to classify all axiomatic extensions of DP, and to compute the free finitely generated DP-algebras.

1 Introduction and Motivations

The *drastic product* t-norm $*_D \colon [0,1]^2 \to [0,1]$ is defined as follows: $x *_D y = 0$ if $x, y < 1$, $x *_D y = \min\{x, y\}$ otherwise (see Fig. 1). It is clear from the definition that $*_D$ is the smallest t-norm, in the sense that for any t-norm $*$ and for each $x, y \in [0,1]$ it holds that $x *_D y \leq x * y$. For this reason it is considered one of the fundamental t-norms (see, *e.g.* [13]). This notwithstanding, there is no drastic product t-norm-based logic, in the sense of [7], since $*_D$ is not left-continuous, and hence it has no associated residuum.

In [18] Schweizer and Sklar introduce a class of t-norms which arise as modifications of the drastic product t-norm in such a way to render them border continuous. In this paper the authors explicitly state *"The result is a t-norm which coincides with [the drastic product] over most of the unit square"*. In [11], Jenei introduced left-continuous versions of the above mentioned t-norms, which he called *revised drastic product* t-norms, as an example of an ordinal sum of triangular subnorms, namely the ordinal sum of the subnorm which is constantly 0 with the t-norm $\min\{x, y\}$. The logic RDP based on these t-norms has been studied by Wang in [19], where, by way of motivation, the author recalls the argument of [18] about RDP t-norms as good approximators of the drastic product.

A. Laurent et al. (Eds.): IPMU 2014, Part II, CCIS 443, pp. 365–374, 2014.

As RDP is a prominent extension of the logic of weak nilpotent minimum WNM, in [4] Bova and Valota introduce a categorical duality for finite RDP-algebras, as a step towards a duality for the case of WNM-algebras.

As it has already been pointed out for [18] and [19], one justification held for the study of RDP is that revised drastic product t-norms make good approximations of $*_D$, in the sense that the graph of such a t-norm can be chosen to coincide with $*_D$ up to a subset of $[0,1]^2$ of euclidean measure as small as desired.

A simple observation will show that RDP t-norms are as good approximators of $*_D$ as t-norms isomorphic (as ordered commutative semigroups) with Łukasiewicz t-norm. Consider, for instance, the parameterised family of t-norms introduced by the same Schweizer and Sklar in [18], defined as follows (here we consider only positive real values for the parameter λ): $x *_{SS}^{\lambda} y := \max\{0, x^\lambda + y^\lambda - 1\}^{1/\lambda}$. These t-norms, being continuous and nilpotent, are all isomorphic to Łukasiewicz t-norm, which is obtained by choosing $\lambda = 1$. It is easy to verify that, for each $c \in (0,1)$, the unique RDP t-norm $*_c$ having c as negation fixpoint (that is, $\sim c = c$), has its zeroset $\{(x,y) \in [0,1]^2 \mid x *_c y = 0\}$ properly included in the zeroset of $*_{SS}^{\lambda}$ for each $\lambda \geq \log_{1/c} 2$. See Fig. 1 for an example.

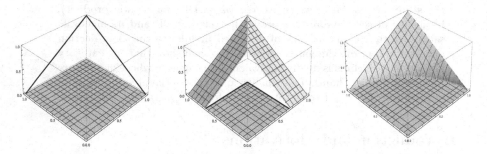

Fig. 1. The DP t-norm, the RDP t-norm $*_{2/3}$ and the Schweizer-Sklar t-norm $*_{SS}^{\log_{3/2} 2}$

Moreover, t-norms isomorphic with the Łukasiewicz one are continuous functions over $[0,1]^2$, while RDP t-norms are only left-continuous.

On the other hand, if we do not require a MTL logic to be standard complete, that is, complete with respect to a set of *standard* algebras (algebras $([0,1], *, \Rightarrow, \min, \max, 0, 1)$, where $*$ is a t-norm, and $\Rightarrow$ its associated residuum), we can naturally study the logic of residuated drastic product chains, whose class, clearly, does not contain any standard algebra. Needless to say, the drastic product chains defined on subsets of $[0,1]$ coincide with the drastic product t-norm over their whole universes. Further, the logic of these chains is nicely axiomatised by a slightly weakened version of the law of the excluded middle.

It turns out that the logic of all residuated drastic product chains is the logic called S3MTL in [17, 10], where some of its properties are stated and proved.

In this note we shall justify the study of S3MTL, that we rename DP for Drastic Product logic, in the light of several interesting logico-algebraic properties. Further, we introduce a category dually equivalent to finite DP-algebras and utilise it to classify all schematic extensions of DP, and to characterise the finitely generated free DP-algebras.

2   Preliminaries

We assume that the reader is acquainted with many-valued logics in Hájek's sense, and with their algebraic semantics. We refer to [9, 6] for any unexplained notion. We recall that MTL is the logic, on the language $\{\&, \wedge, \rightarrow, \bot\}$, of all left-continuous t-norms and their residua, and that its associated algebraic semantics in the sense of Blok-Pigozzi [3] is the variety MTL of MTL-algebras $(A, *, \Rightarrow, \sqcap, \sqcup, 0, 1)$, that is, prelinear, commutative, bounded, integral, residuated lattices [6]. Derived connectives are negation $\neg\varphi := \varphi \rightarrow \bot$, top element $\top := \neg\bot$, lattice disjunction $\varphi \vee \psi := ((\varphi \rightarrow \psi) \rightarrow \psi) \wedge ((\psi \rightarrow \varphi) \rightarrow \varphi)$. The connectives $\&, \wedge, \vee$ are modeled by the monoidal and lattices operations $*, \sqcap, \sqcup$, while $\rightarrow$ by the residuum $\Rightarrow$ and $\bot, \top$ by the elements 0,1. On the algebraic side: $\sim x := x \Rightarrow 0$.

Every axiomatic extension L of MTL has its associate algebraic semantics: a subvariety $\mathbb{L}$ of MTL such that a formula φ is a theorem of L iff the equation $\varphi = \top$ holds in any algebra of $\mathbb{L}$. BL is axiomatized as MTL plus $\varphi \wedge \psi = \varphi\&(\varphi \rightarrow \psi)$, MV as BL plus $\neg\neg\varphi = \varphi$, and MV_3 as MV plus $\varphi\&\varphi = (\varphi\&\varphi)\&\varphi$. *Drastic product chains* are MTL-chains $(A, *, \Rightarrow, \sqcap, \sqcup, 0, 1)$ s.t., for all $x, y \in A$,

$$x * y := \begin{cases} 0 & \text{if } x, y < 1, \\ \min\{x, y\} & \text{otherwise.} \end{cases} \tag{1}$$

We denote by $\mathbb{DP}$ the subvariety of MTL generated by all drastic product chains. The members of $\mathbb{DP}$ are called *drastic product algebras*.

Noguera points out in [17, Page 108] that $\mathbb{DP}$ coincides with the variety named S_3MTL. Each logic in the hierarchy S_kMTL (for $2 \leq k \in \mathbb{Z}$) is axiomatised by a generalised form of the *excluded middle* law: $\varphi \vee \neg(\varphi^{k-1})$ (where $\varphi^0 := \top$ and $\varphi^n := \varphi^{n-1}\&\varphi$), whence the logic of drastic product DP is axiomatised by

$$\varphi \vee \neg(\varphi^2). \tag{DP}$$

Clearly, the logic S_2MTL, axiomatised as MTL plus $\varphi \vee \neg\varphi$ is just classical Boolean logic, as the latter axiom is *the* excluded middle law.

The basic example of a drastic product chain is, for any real c, with $0 < c < 1$, the algebra $[0,1]^c := ([0, c] \cup \{1\}, *, \Rightarrow, \sqcap, \sqcup, 0, 1)$ where $*$ is defined as in (1), while

$$x \Rightarrow y := \begin{cases} 1 & \text{if } x \leq y, \\ c & \text{if } 1 > x > y, \\ y & \text{if } x = 1. \end{cases} \tag{2}$$

Equations (1) and (2) express the operations of any DP-chain. Indeed:

Lemma 1. *A non-trivial MTL-chain $\mathcal{A} = (A, *, \Rightarrow, \sqcap, \sqcup, 0, 1)$ is a DP-chain iff it has a coatom c, and $x * x = 0$ for all $1 > x \in A$. If $\mathcal{A}$ is a DP-chain then $*$ and $\Rightarrow$ are defined as in (1) and (2). Moreover, if $c > 0$ then $c =\sim c$ is its only negation fixpoint.*

Proof. Assume $\mathcal{A}$ is a DP-chain: if $\mathcal{A} \cong \{0,1\}$ the claim trivially holds. Assume then $|A| > 2$. Clearly, $x * x = 0$ for all $1 > x \in A$. Since $*$ is non-decreasing, $x * y = 0$ for all $x, y < 1$. Take $y < x < 1 \in A$, and let $c = x \Rightarrow y$. By the properties of residuum, $c < 1$. Take now any $z < 1$ in A. Since $x * z = 0$ we have $z \leq c$. Hence c is the coatom of A. It is now easy to check that (1) and (2) define $*$ and $\Rightarrow$. For the other direction notice that if $\mathcal{A}$ is an MTL-chain satisfying the two assumptions, then $x * x \in \{0,1\}$ for all $x \in A$, and hence $\mathcal{A}$ satisfies (DP). The lemma follows noting that if $c > 0$ then $c * 1 = c$ and $c * c = 0$, hence $c = \sim c$ (an MTL-chain may have at most one negation fixpoint). $\blacksquare$

Remark 1. Lemma 1 shows that $\mathbb{DP}$ does not contain any *standard algebra*, that is a chain whose lattice reduct is $([0,1], \leq_{\mathbb{R}})$ (where $\leq_{\mathbb{R}}$ denotes the restriction of the standard order of real numbers). It must be stressed, however, that for any $0 < c < 1$, the operation $*$ of the algebra $[0,1]^c$ coincides with the drastic product t-norm $*_D$ wherever defined.

3   DP, RDP and WNM

We recall that WNM and RDP are the subvarieties of MTL respectively satisfying the identities (wnm), and both (wnm) and (rdp), given below.

$$\neg(\varphi \& \psi) \vee ((\varphi \wedge \psi) \to (\varphi \& \psi)) = \top. \tag{wnm}$$

$$(\varphi \to \neg\varphi) \vee \neg\neg\varphi = \top. \tag{rdp}$$

Proposition 1. $\mathbb{DP} \subset \mathbb{RDP} \subset \mathbb{WNM}$.

Proof. Lemma 1 shows immediately that DP-chains satisfy both identities.

We recall that a variety $\mathbb{V}$ is *locally finite* whenever every finitely generated subalgebra of an algebra in $\mathbb{V}$ is finite. Equivalently, free finitely generated algebras are finite. In a locally finite variety the three classes of finitely generated, finitely presented, and finite algebras coincide. Now, since WNM is locally finite (see [17, Proposition 9.15]), from Proposition 1 we obtain:

Corollary 1. $\mathbb{DP}$ *is locally finite and is generated by the class of all finite DP-chains.*

Proposition 2. $\mathbb{DP} \cap \mathbb{BL} = \mathbb{NM} \cap \mathbb{BL} = \mathbb{MV}_3$. $\mathbb{RDP} \cap \mathbb{BL} = \mathbb{WNM} \cap \mathbb{BL} = \mathbb{MV}_3 \oplus \mathbb{G}$, *where* $\mathbb{NM}$ *(nilpotent minimum) is* WNM *plus* $\neg\neg\varphi = \varphi$, *and* $\mathbb{MV}_3 \oplus \mathbb{G}$ *is the variety generated by the ordinal sum of the 3-element MV-chain with the standard Gödel algebra.*

Proof. The first two equalities are shown in [17, 10]. For the other two equalities, a direct inspection shows that a BL-chain satisfies (wnm) iff it is isomorphic to an ordinal sum whose first component is an MV-chain with no more than three elements, and the others (if present) are isomorphic to $\{0,1\}$. Finally, note that (rdp) holds in a BL-chain iff it is isomorphic to an ordinal sum whose first component is an MV-chain with no more than three elements. So, if a BL-chain models (wnm), then it satisfies also (rdp).

4   Canonical Completeness

We have seen that DP is axiomatised from MTL by a weakened form of excluded middle law, and that algebras $[0,1]^c$ are, defensibly, good approximators of the drastic product t-norm, in the sense of [18, 19]. In this section we show that each algebra $[0,1]^c$ is a canonical model of the logic DP.

Recall that an extension L of MTL is *standard complete* if $\mathbb{L}$ is generated by a set of standard algebras (see Remark 1).

Notice that DP is not standard complete since Lemma 1 shows that each DP-chain must have a coatom. However, it must be noticed that the form of completeness DP enjoys is precisely the same that is enjoyed by classical propositional logic, which technically is not a standard complete logic either. To stress this fact we propose here the following notion of completeness, which strengthens the notion of single-chain completeness (an extension L of MTL is *single-chain complete* whenever its associated variety $\mathbb{L}$ is generated by a chain, [16]).

Definition 1. A schematic extension L of MTL is *canonically complete* if it is complete with respect to a single algebra $\mathcal{A}$, called *canonical model* of L, such that:

- The lattice reduct of $\mathcal{A}$ is a sublattice of $\langle [0,1], \leq_{\mathbb{R}} \rangle$.
- For every L-chain $\mathcal{B}$ whose lattice reduct is a sublattice of $\langle [0,1], \leq_{\mathbb{R}} \rangle$, there is $\mathcal{A}' \cong \mathcal{A}$ such that $\langle B, \leq_B \rangle$ is a sublattice of $\langle A', \leq_{A'} \rangle$.

In other terms, $\mathcal{A}$ generates $\mathbb{L}$ and is (up to isomorphism of MTL-algebras) lattice-inclusion-maximal among the algebras in $\mathbb{L}$ whose lattice reduct is a sublattice of $\langle [0,1], \leq_{\mathbb{R}} \rangle$. With this definition in place, we list some examples:

- Classical propositional logic is canonically complete even though not standard complete;
- Gödel, product and Łukasiewicz logic are both canonically and standard complete, as it is BL (w.r.t. the ordinal sum of ω copies of the standard MV-algebra, for instance) [6, 9, 15].
- MTL is standard complete, but it is not known whether it is canonically complete, nor single chain complete ([16]). The same applies to IMTL.
- The logic WCBL ([17, Ch. 7.2,7.3]) obtained extending BL with the weak cancellativity axiom $\neg(\varphi \& \psi) \vee ((\varphi \to (\varphi \& \psi)) \to \psi)$ is standard complete, being complete w.r.t. the set formed by the standard MV-algebra and the standard product algebra, but it is not canonically complete. Indeed, an MTL-chain belongs to this variety iff it is a product or an MV-chain: hence WCBL is not single chain complete.
- Each logic BL^n (with $n \geq 2$), axiomatised as BL plus the n-contraction axiom $\varphi^n \to \varphi^{n+1}$ (see [2]), is neither standard nor canonically complete. In the associated variety the only standard algebra is the standard Gödel algebra, but there are no generic chains at all.

Lemma 2. *All the algebras of the form $[0,1]^c$ are isomorphic.*

Proof. By Lemma 1.

Theorem 1. $\mathbb{DP}$ *is generated by any infinite DP-chain.*

Proof. By Corollary 1, we have that if an equation fails in some DP-algebra, then it fails in some finite DP-chain. Take now an infinite DP-chain $\mathcal{A}$. Denote by c its coatom. By Lemma 1, we have that every finite DP-chain $\mathcal{B}$ embeds into $\mathcal{A}$: Trivially, $\{0, 1\} \hookrightarrow \mathcal{A}$. If $|\mathcal{B}| > 2$, call b its coatom. Then every injective order preserving mapping ϕ from $\mathcal{B}$ to $\mathcal{A}$ such that $\phi(0) = 0$, $\phi(1) = 1$, and $\phi(b) = c$ (note that such a map always exists, since $\mathcal{B}$ is finite) is such that $\phi \colon \mathcal{B} \hookrightarrow \mathcal{A}$. Hence every equation that fails in some finite DP-chain also fails in $\mathcal{A}$.

Theorem 1 together with Corollary 1 proves the following result:

Theorem 2. $\mathbb{DP}$ *is generated by any set of DP-chains of unbounded cardinality.*

Theorem 3. *The logic DP enjoys the strong completeness w.r.t.* $[0, 1]^c$*, with* $c \in (0, 1)$*. Moreover, DP is not standard complete but it is canonically complete.*

Proof. By [5, Theorem 3.5] it is enough to show that every countable DP-chain embeds into $[0, 1]^c$. Let $\mathcal{B}$ be a countable DP-chain. Reasoning as in the proof of Theorem 1 we find the desired embedding $\phi \colon \mathcal{B} \hookrightarrow [0, 1]^c$ (preserving $0, 1$ and the coatom). Such ϕ exists because $[0, c]$ is an uncountable dense linear order. The latter statement follows from Remark 1 and Lemma 2.

5 S_nMTL and the Definability of the Δ Operator

DP coincides with S_3MTL. In this section we shall recall some interesting properties of S_nMTL-algebras from [17, 10, 14], and relate them to the definability of the Δ projection operator [1].

We recall that a variety is *semisimple* if all its subdirectly irreducible algebras are simple, and it is a *discriminator* variety if the ternary discriminator t ($t(x, x, z) = z$, while $t(x, y, z) = x$ if $x \neq y$), is definable on every subdirectly irreducible algebra.

Theorem 4 ([17, 10, 14]). *Let* $\mathbb{L}$ *be a variety of MTL-algebras. Then the following are equivalent:*

- $\mathbb{L}$ *is semisimple.*
- $\mathbb{L}$ *is a discriminator variety.*
- $\mathbb{L}$ *is a subvariety of* S_nMTL *for some* $n \geq 2$.
- *Every chain in* $\mathbb{L}$ *is simple and n-contractive (i.e. it satisfies* $x^n = x^{n-1}$*), for some* $n \geq 2$.

Given a schematic extension L of MTL we write L_Δ for the extension/expansion of L with the Δ unary projection connective, axiomatised as follows:

$$\Delta\varphi \vee \neg\Delta\varphi, \qquad \Delta\varphi \to \varphi, \qquad \Delta\varphi \to \Delta\Delta\varphi,$$
$$\Delta(\varphi \to \psi) \to (\Delta\varphi \to \Delta\psi), \quad \Delta(\varphi \vee \psi) \to (\Delta\varphi \vee \Delta\psi).$$

Recall that on every MTL-chain A and every $x \in A$, the identities associated with these axioms model the operation $\Delta x = 1$ if $x = 1$, while $\Delta x = 0$ if $x < 1$.

Proposition 3. *Let* $\mathbb{L}$ *be a variety of MTL-algebras. Then* $\mathbb{L}_\Delta$ *is semisimple.*

Proof. We prove that each chain in $\mathbb{L}_\Delta$ is simple.

Take a chain $\mathcal{A} \in \mathbb{L}_\Delta$. For every non-trivial congruence θ on $\mathcal{A}$, it holds that $\langle a, b \rangle \in \theta$, for some $a \neq b$. Then exactly one of $a \Rightarrow b$ and $b \Rightarrow a$ is 1, say $a \Rightarrow b$. Hence $\langle b \Rightarrow a, 1 \rangle \in \theta$, and $\langle \Delta(b \Rightarrow a), \Delta(1) \rangle \in \theta$. That is, $\langle 0, 1 \rangle \in \theta$. Since θ is a congruence of the lattice reduct of $\mathcal{A}$, all elements between 0 and 1 are in the same class, which means that $\mathcal{A}$ is simple.

Theorem 5. *Let* $\mathbb{L}$ *be a variety of MTL-algebras. Then* $\mathbb{L}_\Delta = \mathbb{L}$ *iff* $\mathbb{L}$ *is a subvariety of* $\mathbb{S}_k\text{MTL}$ *for some integer* $k > 1$.

Proof. It is immediate to check that the Δ operator is definable in each variety $\mathbb{S}_k\text{MTL}$ as $\Delta x = x^{k-1}$.

For the other direction, assume Δ is definable in a variety $\mathbb{L}$ of MTL-algebras. Then $\mathbb{L}$ is semisimple by Proposition 3. By Theorem 4, $\mathbb{L}$ is a subvariety of $\mathbb{S}_k\text{MTL}$ for some integer $k > 1$.

Remark 2. Theorem 5 shows that if Δ is definable in some subvariety $\mathbb{L}$ of MTL-algebras, then it is always definable as $\Delta x = x^k$ for some $k \geq 1$. Further, $\mathbb{L}$ is a discriminator variety iff it defines Δ and, in this case, $(\Delta(x \leftrightarrow y) \wedge z) \vee (\neg \Delta(x \leftrightarrow y) \wedge x)$ is the discriminator term.

Notice that the assumptions of Theorem 4 cannot be generalised to extensions/expansions of MTL. For instance, consider the logic $G_\sim$ introduced in [8], which is Gödel logic extended/expanded with an independent involutive negation $\sim$. It is an exercise to check that Δ is definable in $G_\sim$ as $\Delta x = \neg \sim x$, but not as x^k for any integer k. Hence $G_\sim$ is a discriminator variety, but $G_\sim$ is not an extension/expansion of any $\mathbb{S}_k\text{MTL}$.

Corollary 2. *A variety* $\mathbb{L}$ *of BL-algebras coincides with* $\mathbb{L}_\Delta$ *iff it is a variety of MV-algebras generated by a finite set of finite MV-chains.*

Proof. By [17, Corollary 8.16], $\mathbb{S}_n\text{MTL} \cap \mathbb{BL}$ is generated by the set of all MV-chains with at most n elements, for every integer $n > 1$. The result follows from Theorem 5.

The following result is needed in the next section.

Lemma 3. *The classes of simple WNM-chains and of simple RDP-chains both coincide with the class of DP-chains.*

Proof. The result follows from Proposition 1 and from Theorem 4, since every WNM-chain satisfies $x^3 = x^2$.

6   A Dual Equivalence

In [4] a dual equivalence between the category of finite RDP-algebras and homomorphisms and the category HF of *finite hall forests* is proven. We recall here that a finite hall forest is a finite multiset whose elements are pairs (T, J), where T is a finite *tree* (that is, a poset with minimum such that the downset of each element is a chain) and J is a (possibily empty) finite chain, while a morphism $h\colon \{(T_i, J_i)_{i\in I}\} \to \{(T_k, J_k)_{k\in K}\}$ is a family of pairs $\{(f_i, g_i)\}_{i\in I}$ such that for each $i \in I$ there is $k \in K$ such that $f_i\colon T_i \to T_k$, and $g_i\colon J_i \to J_k$ are order-preserving downset preserving maps, with the additional constraint that $g_i(\max J_i) = \max J_k$. In case J_k is empty it is stipulated that g_i is the partial, nowhere defined, map.

For each integer $k > 0$ let $\mathbf{k}$ denote the k-element chain.

Definition 2. Let MC be the category whose objects are finite multisets of (nonempty) finite chains, and whose morphisms $h\colon C \to D$, are defined as follows. Display C as $\{C_1, \dots, C_m\}$ and D as $\{D_1, \dots, D_n\}$. Then $h = \{h_i\}_{i=1}^m$, where each h_i is an order preserving surjection $h_i\colon C_i \twoheadrightarrow D_j$ for some $j = 1, 2, \dots, n$. Let $\mathsf{MC}^\top$ be the nonfull subcategory of MC whose morphisms $h\colon C \to D$ satisfy the following additional constraint: for each $i = 1, 2, \dots, m$, if the target D_j of h_i is not isomorphic with $\mathbf{1}$, then $h_i^{-1}(\max D_j) = \{\max C_i\}$.

Theorem 6. *The category* $\mathsf{MC}^\top$ *is equivalent to the full subcategory of* HF *whose objects have the form* $\{(\mathbf{1}, J_i)\}_{i\in I}$. *Whence,* $\mathsf{MC}^\top$ *is dually equivalent to the category* $\mathbb{DP}_{fin}$ *of finite DP-algebras and their homomorphisms.*

Proof. Note that the only map from $\mathbf{1}$ to itself is the identity id_1. Then direct inspection shows that the functor $Tr\colon \mathsf{MC}^\top \to$ HF, defined on objects as $Tr(\{C_1, \dots, C_m\}) = \{(\mathbf{1}, C_1 \setminus \{\max C_1\}), \dots, (\mathbf{1}, C_m \setminus \{\max C_m\})\}$, and on morphisms as $Tr(\{h_i\}_{i=1}^m) = \{(id_1, h_i \restriction C_i \setminus \{\max C_i\})\}_{i=1}^m$ (we agree that $h_i \restriction \emptyset$ is the nowhere defined map), implements the equivalence stated in the first statement. Observe that the dual of a simple RDP-algebra in the category HF is a hall forest of the form $\{(\mathbf{1}, J)\}$. The last statement then follows from Lemma 3. $\square$

Clearly, the multiset $\{\mathbf{1}\}$ is the terminal object of $\mathsf{MC}^\top$. Applying the first equivalence of Theorem 6 one can verify the following constructions, as they are carried over to $\mathsf{MC}^\top$ from HF. Given two objects $C, D \in \mathsf{MC}^\top$, the coproduct object $C \uplus D$ of C and D is just the disjoint union of the multisets C and D; the product object $C \times D$ is computed using the following $\mathsf{MC}^\top$ isomorphisms. First, products distribute over coproducts: $C \times (D \uplus E) \cong (C \times D) \uplus (C \times E)$. Given $C \in \mathsf{MC}^\top$, let $C^\top$ denote the object obtained adding to each chain in C a fresh maximum. Then $\{\mathbf{i}\} \times \{\mathbf{1}\} \cong \{\mathbf{i}\}$ and $\{\mathbf{i+1}\} \times \{\mathbf{2}\} \cong \{\mathbf{i+1}\}$. Moreover,

$$\{\mathbf{i+2}\} \times \{\mathbf{j+2}\} \cong ((\{\mathbf{i+2}\} \times \{\mathbf{j+1}\}) \uplus (\{\mathbf{i+1}\} \times \{\mathbf{j+1}\}) \uplus (\{\mathbf{i+1}\} \times \{\mathbf{j+2}\}))^\top.$$

Denote by $MC\colon \mathbb{DP}_{fin} \to \mathsf{MC}^\top$ the functor implementing the dual equivalence. The following lemma is straightforward.

Lemma 4. *For any integer* $i > 0$, $MC^{-1}\{\mathbf{i}\}$ *is the DP-chain with* $i+1$ *elements.*

Let $\mathbf{F}_k$ denote the free DP-algebra over a set of k many free generators. In the following we write nC for the nth copower of $C \in \mathsf{MC}^\top$.

Theorem 7. $MC\,\mathbf{F}_1$ *is the multiset of chains* $\{\mathbf{1},\mathbf{3},\mathbf{2},\mathbf{1}\}$. *Hence,* $\mathbf{F}_1$ *has exactly* $2^2 \cdot 3 \cdot 4 = 48$ *elements. More generally,*

$$MC\,\mathbf{F}_k \cong 2^k\{\mathbf{1}\} \uplus (3^k - 2^k)\{\mathbf{2}\} \uplus \biguplus_{h=3}^{k+2} \left(\sum_{i=0}^{h-2}(-1)^i \binom{h-2}{i}(h+1-i)^k \right)\{\mathbf{h}\}. \quad (3)$$

Whence, the cardinality of $\mathbf{F}_k$ *is given by*

$$2^{2^k} \cdot 3^{3^k - 2^k} \cdot \prod_{h=3}^{k+2}(h+1)^{\sum_{i=0}^{h-2}(-1)^i\binom{h-2}{i}(h+1-i)^k}. \quad (4)$$

Proof. For what regards $\mathbf{F}_1$ the proof follows at once by Theorem 6, Lemma 4 and [4, Proposition 6]. For the general case, recall that $\mathbf{F}_k$ is the kth copower of $\mathbf{F}_1$, hence $MC\,\mathbf{F}_k$ is given by the kth power of $\{\mathbf{1},\mathbf{3},\mathbf{2},\mathbf{1}\}$. Proceeding by induction on k, we denote by $a_i^{(k)}$ the coefficient multiplier of $\{\mathbf{i}\}$ in Eq. (3). Using the fact that $\{\mathbf{k}\} \times \{\mathbf{3}\} \cong (k-1)\{\mathbf{k+1}\} \uplus (k-2)\{\mathbf{k}\}$, the computation of $MC\,\mathbf{F}_k \times \{\mathbf{1},\mathbf{3},\mathbf{2},\mathbf{1}\}$ gives the recurrences $a_1^{(k+1)} = 2a_1^{(k)}$, $a_2^{(k+1)} = a_1^{(k)} + 3a_2^{(k)}$, $a_3^{(k+1)} = a_1^{(k)} + a_2^{(k)} + 4a_3^{(k)}$, and $a_h^{(k+1)} = (h-2)a_{h-1}^{(k)} + (h+1)a_h^{(k)}$ for $h > 3$, whose solutions finally yield Eq. (3). By Lemma 4, Eq. (4) yields the cardinality of $\mathbf{F}_k$.

Notice that by replacing in Eq. (4) each base number with the DP-chain with the same cardinality one gets the decomposition of $\mathbf{F}_k$ as direct product of chains.

By Theorem 2, any proper subvariety of $\mathbb{DP}$ is generated by a necessarily finite family of finite chains. We can then classify all subvarieties of $\mathbb{DP}$.

Theorem 8. *Each proper subvariety of* $\mathbb{DP}$ *is generated by a finite chain. Moreover, two finite chains of different cardinality generate distinct subvarieties.*

Proof. First note that each non-trivial finite DP-chain embeds into any DP-chain of greater cardinality. Whence, each proper subvariety $\mathbb{V}$ of $\mathbb{DP}$ is generated by a finite chain, which is the chain with maximum cardinality among any given set of chains generating $\mathbb{V}$. For each object $C \in \mathsf{MC}^\top$ let its *height* $H(C)$ be the maximum cardinality of its chains. It is easy to see that $H(C \uplus D) = \max\{H(C), H(D)\}$ and that given maps $C_1 \hookrightarrow D_1$ and $C_2 \twoheadrightarrow D_2$, it holds that $H(C_1) \leq H(D_1)$ and $H(D_2) \leq H(C_2)$. It follows by Thm. 6, Lemma 4 and the HSP theorem that all finite algebras in the variety generated by the k-element DP-chain must have dual of height $\leq k - 1$. Let $\mathbb{V}_h$ be the subvariety generated by the h-element chain. Then, if $h < k$, $(\mathbb{V}_h)_{fin}$ is properly included in $(\mathbb{V}_k)_{fin}$. By [12, Cor. VI.2.2] and Cor. 1, $\mathbb{V}_h$ is properly included in $\mathbb{V}_k$.

One may wonder whether there are classes of residuated lattices dually equivalent to MC. The answer is in the positive, for it is easy to prove that the class of finite algebras in $\mathbb{G}_\Delta$, the variety of Gödel algebras plus Δ, is the category sought for. It turns out that $\mathbb{DP}$ is equivalent to a non-full subcategory of $\mathbb{G}_\Delta$, with the same objects, but with fewer morphisms.

Even if we do not have analyzed the first-order case, there is a result that may have some interest. The logic DP∀ enjoys strong completeness w.r.t. $[0,1]^c$, with $c \in (0,1)$. This can be proved via a modification of a construction described in [5], that allows to embed every countable DP-chain into a chain isomorphic to $[0,1]^c$ (for some $c \in (0,1)$), by preserving all inf's and sup's. By [5, Theorems 5.9, 5.10] this suffices to prove DP∀ is strongly complete w.r.t. $[0,1]^c$.

References

[1] Baaz, M.: Infinite-valued Gödel logics with 0-1-projections and relativizations. In: Gödel 1996: Logical Foundations of Mathematics, Computer Science and Physics – Kurt Gödel's Legacy, pp. 23–33. Springer, Berlin (1996)

[2] Bianchi, M., Montagna, F.: n-contractive BL-logics. Arch. Math. Log. 50(3-4), 257–285 (2011)

[3] Blok, W., Pigozzi, D.: Algebraizable logics. Memoirs of The American Mathematical Society, vol. 77(396). American Mathematical Society (1989)

[4] Bova, S., Valota, D.: Finite RDP-algebras: duality, coproducts and logic. J. Log. Comp. 22(3), 417–450 (2012)

[5] Cintula, P., Esteva, F., Gispert, J., Godo, L., Montagna, F., Noguera, C.: Distinguished algebraic semantics for t-norm based fuzzy logics: methods and algebraic equivalencies. Ann. Pure Appl. Log. 160(1), 53–81 (2009)

[6] Cintula, P., Hájek, P., Noguera, C.: Handbook of Mathematical Fuzzy Logic, vol. 1 and 2. College Publications (2011)

[7] Esteva, F., Godo, L.: Monoidal t-norm based logic: Towards a logic for left-continuous t-norms. Fuzzy Sets Syst. 124(3), 271–288 (2001)

[8] Esteva, F., Godo, L., Hájek, P., Navara, M.: Residuated fuzzy logics with an involutive negation. Arch. Math. Log. 39(2), 103–124 (2000)

[9] Hájek, P.: Metamathematics of fuzzy logic. Trends in Logic, vol. 4. Kluwer Academic Publishers (1998)

[10] Horčík, R., Noguera, C., Petrík, M.: On n-contractive fuzzy logics. Math. Log. Q. 53(3), 268–288 (2007)

[11] Jenei, S.: A note on the ordinal sum theorem and its consequence for the construction of triangular norms. Fuzzy Sets Syst. 126(2), 199–205 (2002)

[12] Johnstone, P.T.: Stone spaces. Cambridge Studies in Advanced Mathematics. Cambridge University Press (1982)

[13] Klement, E.P., Mesiar, R., Pap, E.: Triangular norms. Trends in Logic, vol. 8. Kluwer Academic Publishers (2000)

[14] Kowalski, T.: Semisimplicity, EDPC and Discriminator Varieties of Residuated Lattices. Stud. Log. 77(2), 255–265 (2004)

[15] Montagna, F.: Generating the variety of BL-algebras. Soft Comput. 9(12), 869–874 (2005)

[16] Montagna, F.: Completeness with respect to a chain and universal models in fuzzy logic. Arch. Math. Log. 50(1-2), 161–183 (2011)

[17] Noguera, C.: Algebraic study of axiomatic extensions of triangular norm based fuzzy logics. Ph.D. thesis, IIIA-CSIC (2006), http://ow.ly/rV2sL

[18] Schweizer, B., Sklar, A.: Associative functions and abstract semigroups. Publ. Math. Debrecen 10, 69–81 (1963)

[19] Wang, S.: A fuzzy logic for the revised drastic product t-norm. Soft Comput. 11(6), 585–590 (2007)

Fuzzy State Machine-Based Refurbishment Protocol for Urban Residential Buildings

Gergely I. Molnárka[1] and László T. Kóczy[2]

[1] Department of Building Constructions and Architecture, Széchenyi István
University, Egyetem tér 1, Győr, Hungary
[2] Department of Automation, Széchenyi István University,
Egyetem tér 1, Győr, Hungary

Abstract. The urban-type residential houses built before World War
Two represent a large part of the built environment in Hungary. Due
to their physical condition and low energy-efficiency the retrofit of these
buildings is very much advisable nowadays. In this paper we propose
an approach based on fuzzy signatures and state machines, that helps
decision support for determining the renovation scenario concerning necessity, cost efficiency and quality. Using the knowledge obtained from
diagnostic surveys done during the previous decades by architect experts, and technical guides and the available database of contractors
billing, a protocol for the preparation for optimized refurbishment is proposed, based on the concept of an extended fuzzy state machine model.
In this combined model the theoretical concepts of finite-state machine
and fuzzy state machine, and also the principles of fuzzy signatures are
applied.

Keywords: urban-type residential houses, building refurbishment protocol, fuzzy signature state machine.

1 Introduction

1.1 The EPBD Recast Requirements

Due to the new version of the European Energy Performance of Buildings Directive (EPBD), the EPBD Recast (Directive 2010/31/EU) newly erected buildings
in Europe may consume "nearly zero" energy from 2021. This regulation also
influences the future of the existing building stock. In Hungary, the adaptability of recast requirements in connection with several types of existing buildings
was examined recently. Among other statements, the final report [1] verifies that
even though due to settlement and other architectural attributes urban-type
residential houses (former tenement houses) in Budapest are not able to meet
these requirements; however, the improvement of their energy efficiency is still
strongly advisable.

In addition to this statement, a large-scale retrofit process of existing residential houses may be an effective response to climate change and for reducing CO_2
emission [2].

A. Laurent et al. (Eds.): IPMU 2014, Part II, CCIS 443, pp. 375–384, 2014.

1.2   The Actual Physical State of Urban-Type Residential Houses

The above mentioned former tenement houses represent a significant part of existing Hungarian apartment buildings: 11% of all apartments are located in these urban-type residential houses (in Budapest this ratio is 27%) built before WW2 [3]. Due to several causes (age of buildings, highly fragmented ownership structure (the former tenants became the owners), lack of capital, missing regular maintenance processes during the past, incompetence in the maintenance at present, etc.) the average physical condition of this kind of residential buildings is below standard. In fact, they are unsuitable for direct energy efficiency development. The symptoms of overall physical obsolescence and deterioration are clearly observable even in representative urban areas. The types of building failures that resulted in the current state are determined by the types of the subject residential houses.

1.3   The Subgroups of Budapest Urban-Type Residential Houses

From building construction and architectural aspects in Budapest two main types of urban-type residential houses exist. The major part of these buildings was built between 1880 and 1920 in Academic Style; the remaining part between 1928 and 1944. The actual research focuses on the Academic Style residential houses.

The criteria of this building type are the existence of courtyard and air-shafts; the presence of apartments differentiated by size and orientation in the same building; the traditional masonry of the load bearing structures, the varying and innovative slab systems and the ventilated wooden structure pitched roofs. A typical classic urban-type residential house is presented in Fig.1.

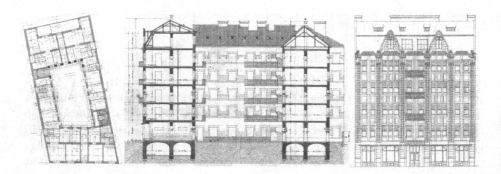

Fig. 1. Floor plan, section and elevation of an urban-type residential house with inner court in Academic Style (copy of blueprints, property of Budapest Archives)

As a general overview of the deterioration of Academic Style residential houses the following are ascertainable:

- Due to the lack of wall and plinth damp proof course (DPC), the soil moisture oozes up into the brick walls, resulting in mold (damp air) in ground level apartments and peeling the plaster layer off;
- Slab systems have an uncertain load-bearing performance (that also relates to the outside corridors);
- The common mechanical (drainage systems, pipelines, elevators, etc.) and electrical systems (including the lighting and the feeble current systems) are out-dated, and may cause accidental bursts;
- The building envelope, floor covering and other additional elements of the staircases and other common areas (entrance hall, court, outside corridors, etc.) are broken and crumbling;
- The roof tiling and flashing are deteriorated (cracked, punctured), and as a result leaking occurs, especially in roof valleys and around the external supplementary element fastenings;
- Low energetic performance of air-shaft partitioned wall;
- The finishing of the façade is detached or missing in greater areas, especially on the surface of courtyard façades and firewalls.

Although in specific situations several other failures may be observed, the above mentioned defects determine the actual state of the examined type of building stock as it is shown in the series of [4], [5] and other building renovation guides. Because of this these areas of deterioration are generally considered as the main targets of the maintenance procedures (the refurbishment protocol). It should be noted that the proposed method focuses on the building elements owned jointly by the flat owners and on constructions, which are determined by the Householders' Act (Hungarian Act of Residential Houses. 2003/CXXXIII).

1.4 Uncertainties in Decision-Making in the Refurbishment Schedule

At present, the present owners (former tenants), as stakeholders, are responsible for the stability of their property only. However, this role can be interpreted widely: in case of developing the physical condition of the building constructions a thorough refurbishment schedule has to be planned and accepted by the owners.

In practice, several unprofessional factors influence the joint decision-making. Among others, the community of the owners cannot interpret and handle the information in the standard building diagnostic surveys that report on the failures and give general recommendations for a refurbishment. The more so as these surveys date from different period between the 1950s and the 2000s, and their depth, precision, detail, etc. are very heterogeneous. Since the statements of surveys do not give an explicit refurbishment protocol, the owners may have the feeling that decisions about the steps of maintenance may be determined arbitrarily. However, although in some cases the accidentally determined individual repairs may help develop the overall physical condition of the whole structure, in general, only a consequent and well organized maintenance procedure might help determine the necessary decisions about the refurbishment steps, offering answers to questions on both financing and scheduling.

1.5   Factors Influencing the Refurbishment Schedule

For designing such a decision support system, first the steps of maintenance have to be examined, taking into consideration their respective importance, interrelations and other non-constructional aspects. In our approach, the main goal of the system is to examine each maintenance step that is necessary for renovating the given building from its initial (present) state to the acceptable (final) state.

There are several factors influencing the schedule of the refurbishment procedure. Without any details, the most important such factors are: the *grade of danger* caused by the given failure, the *interrelations* among deteriorated constructions and decays, and the presence of a protocol for the repair, the *financial* schedulability, some complex *logistics* aspects, etc.

2   A Model Proposed for the Refurbishment Protocol

The proposed approach to be introduced here for tackling the problem described above is a model and an attempt for solution based on the principles of fuzzy signatures (or simply vector valued fuzzy sets) and fuzzy state machines combined into what we will call *Fuzzy Signature State Machines.*

The most important reason for using fuzzy signatures here as the starting point is the fact that the structure of the surveys follows the architectural and civil engineering common sense, where the sub-structures and components of each building are arranged in hierarchical tree-like structures, where the whole building might be represented by the root of the tree and each major sub-component is a first level branch, with further sub-branches describing sub-sub-components, etc. as it will be shown in the next sections.

At this point we revisit the definition of fuzzy signature. Fuzzy sets of a universe of discourse X are defined by

$A = \{X, \mu_A\}$, where $\mu_A : X \to [0, 1]$.
Vector Valued Fuzzy Sets (VVFS) [6] are a simple extension that may be considered as a special case of L-fuzzy sets [7]:

$A_n = \{X, \mu_{A_n}\}$, where $\mu_A n : X \to [0, 1]^n$. Thus a membership degree is a multi-component value here, e.g. $[\mu_1, \mu_2, \cdots, \mu_n]^T$.

Fuzzy Signatures represent a further extension of VVFS as here any component might be a further nested vector, and so on [10], [8]:
$A_{fs} = \{X, \mu_{A_{fs}}\}$, where $\mu_{A_{fs}} : X \to M_1 \times M_2 \times \cdots \times M_n$, where $M_i = [0, 1]$ or $[M_{i_1} \times M_{i_2} \times \cdots \times M_{i_n}]^T$.

The next is a very simple example:

$$\mu_{A_{fs}} = [\mu_1, \mu_2, [\mu_{3_1}, \mu_{3_2}, [\mu_{3_{3_1}}, \mu_{3_{3_2}}, \mu_{3_{3_3}}]], \mu_4, [\mu_{5_1}, \mu_{5_2}], \mu_6]^T$$

The first advantage of using fuzzy signatures rather than VVFS is that here any closer grouping and sub-grouping of fuzzy features may be given. Fuzzy signatures are associated with an aggregation system. Each sub-component set may be aggregated by its respective aggregation operation, thus reducing the sub-component to one level higher. The above example has the following associated aggregation structure:

$\{a_0\{a_3\{a_{3_3}\}\{a_5\}\}\}$, where each a_o denotes an aggregation, particularly the one associated with the child node x_o associated with μ_o, thus the following example signature might be reduced upwards as follows:

$$\mu_{A_{fs}} \Rightarrow [\mu_1, \mu_2, [\mu_{3_1}, \mu_{3_2}, \mu_{3_3} = a_{3_3}(\mu_{3_{3_1}}, \mu_{3_{3_2}}, \mu_{3_{3_3}}), \mu_4, \mu_5 = a_5(\mu_{5_1}, \mu_{5_2}), \mu_6]^T$$
$$\Rightarrow [\mu_1, \mu_2, \mu_3 = a_3(\mu_{3_1}, \mu_{3_2}, \mu_{3_3}), \mu_4, \mu_5, \mu_6]^T \Rightarrow \mu_0 = a_0(\mu_1, \mu_2, \mu_3, \mu_4, \mu_5, \mu_6)$$

This kind of membership degree reductions is necessary when the data are partially of different structure, e.g. some of the sub-components are missing. Then operations among fuzzy signatures with partially different structure may be carried out, by finding the *largest common sub-structure* and reducing all signatures up to that substructure. This might be necessary if the surveys referred to in this paper are considered as often their depth and detail are different. As an example, maybe in survey one the roof is considered as a single component of the house and is evaluated by a single linguistic quality label, while in survey two this is done in detail and tiles, beams, tinwork, chimneys are described separately.

In our previous work we applied vector-valued fuzzy sets and fuzzy signatures [10] for describing sets of objects with uncertain features, especially when an internal theoretical structure of these features could be established. In [11] we presented an approach where the fuzzy signatures could be deployed for describing existing (typically old) residential houses in order to support decisions concerning when and how these buildings should be renovated (or, if necessary demolished). In that research a series of theoretically arrangeable features were taken into consideration and eventually a single aggregated fuzzy membership value could be calculated on the basis of available detailed expert evaluation sheets. In that model, however, the available information does not support any decision strategy concerning actual sequence of the measures leading to complete renovation; and it is also insufficient to optimize the sequence from the aspect of local or global cost efficiency.

In the following section, the mathematical model of the proposed refurbishment protocol will be introduced.

2.1   Application of Fuzzy Finite State Machines in the Modelling

Finite State Machines are determined by the sets of input states X, internal states Q, and the transition function f. The latter determines the transition that will occur when a certain input state change triggers state transition. There are several alternative (but mathematically equivalent) models known from the literature. For simplicity the following is assumed as the starting point of our new model:

$$A = \langle X, Q, f \rangle \tag{1}$$

$$f : X \times Q \to Q \tag{2}$$

where $X = \{x_i\}$ and $Q = \{q_i\}$.

Thus, a new internal state is determined by the transition function as follows:

$$q_{i+1} = f(x_i, q_i) \tag{3}$$

In matrix form:

$$F = \begin{bmatrix} f(x_1, q_1) & f(x_2, q_1) & \cdots & f(x_n, q_1) \\ f(x_1, q_2) & & & f(x_n, q_2) \\ \vdots & & & \vdots \\ f(x_1, q_m) & f(x_2, q_m) & \cdots & f(x_n, q_m) \end{bmatrix} \tag{4}$$

The transition function/matrix maybe interpreted with help of a relation R on $X \times Q \times Q$, where

$$R(x_i, q_j, q_k) = 1 \tag{5}$$

if

$$f(x_i, q_j) = q_k \tag{6}$$

and

$$R(x_i, q_j, q_k) = 0 \tag{7}$$

if

$$f(x_i, q_j) \neq q_k \tag{8}$$

The states of the finite state machine are elements of Q. In the present application an extension to fuzzy states is considered in the following sense. Every aspect of the phenomenon to model is represented by a state universe of sub-states Q_i. The states themselves are (fuzzy) subsets of the universe of discourse state sets, so that within Q_i a frame of cognition is determined (its fineness depending on the application context and on the requirements toward the optimisation algorithm), so that typical states like "Totally intact", "Slightly damaged", "Medium condition", etc., up to "Dangerous for life" are considered. Any transition from one state to the other (improvement of the condition, refurbishment or renovation) involves a certain cost c. In the case of a transition from q_i to q_j it is expressed by a membership value $\mu_{ij} = c(q_i, q_j)$. In our model the added cost $\Sigma \mu_{i_j}$ along a path $q_{i_1} \to q_{i_2} \to \cdots \to q_{i_n}$ is not usually equivalent with the cost of the transition μ_{i_n} along the edge $q_{i_l} \to q_{i_n}$. This is in accordance with the non-additivity property of the fuzzy (possibility) measure and is very convenient in our application, as it is also not additive in the case of serial renovations.

As a simple example the Fig. 2 depicts the possible transitions among the states of a specific sub-state: Q_0^1 represents the initial (present) state, while Q_9^1 represents the acceptable (final) state.

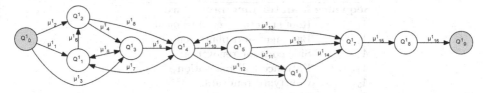

Fig. 2. Diagram representation of a sub-state space and the possible transitions among the states

In the case of fuzzy signature machines each of the leaves contains a sub-automaton with the above property. The parent leave of a certain sub-graph is constructed from the child leaves, so that the sub-automaton

$$A_i = A_{i_1} \times A_{i_2} \times \cdots \times A_{i_m}, \text{ and thus the states of } A_i \text{ are}$$

$$Q_i = Q_{i_1} \times Q_{i_2} \times \cdots \times Q_{i_n}, \text{ so that the transition } Q_{j_1} \to Q_{j_2} \text{ in this case}$$
means the parallel (or subsequent) transitions

$$q_{j_{1_1}} \to q_{j_{1_2}} \times q_{j_{2_1}} \to q_{j_{2_2}} \times \cdots \times q_{j_{n_1}} \to q_{j_{n_2}}$$

A special aggregation is associated with each leaf; similarly as it is in the fuzzy singatures, however, in this case the aggregation calculates the resulting cost $\mu_{j_{1_2}}$ of the transition $q_{j_1} \to q_{j_2}$, so that

$$\mu_{j_{1_2}} = c(q_{j_1}, q_{j_2}) = a_j(c(q_{j_{1_1}}, q_{j_{1_2}}), c(q_{j_{2_1}}, q_{j_{2_2}}), \cdots, c(q_{j_{n_1}}, q_{j_{n_2}}))$$

where a stands for the respective aggregation. (Note that these aggregations sometimes do not satisfy the symmetricity property of the general axiom structure of aggregations, thus it may be referred as a "non-symmetric aggregation".)

2.2 Modelling the Repair Procedure

As an initial state an overall visual diagnostic survey of the given residential building is supposed. This evaluation sheet gives detailed determination and state description of each building component, disclosing the relation between causes and consequences. The obtained information helps determining the maintenance steps; observing their influencing attributes additional data can be given to each step. With the knowledge of professional rules as they are clearly described in [12], [13], the general maintenance procedure can be decomposed into distinct sequences that correspond to the previously mentioned supposition. In Table 1, the sequences as sub-automaton of this procedure are represented by A_1 to A_8.

Table 1. Distinct sequences of general refurbishment of Academic Style residential houses

Sequence	Maintenance procedure
A_1	Roof structure refurbishment
A_2	Basement and plinth renovation
A_3	Side corridor renovation
A_4	Entrance hall renovation
A_5	Courtyard renovation
A_6	Staircase renovation
A_7	Mechanical system refurbishment
A_8	Electrical system refurbishment

In further analyses, these sub-automata represent parent nodes of building component groups that constitute independent or quasi-independent maintenance processes. As an example, the Table 2 represents the sub-sub automata of A_1 process.

Table 2. Maintenance subsequences in A_1 process (Roof structure refurbishment)

Subsequences	Maintenance sequence
A_{1_1}	Chimney shafts repair
A_{1_2}	Removing unnecessary supplementary elements (e.g. aerials)
A_{1_3}	Replacement of dangerous catwalks
A_{1_4}	Replacement of damaged flashings
A_{1_5}	Timber roof structure repair
A_{1_6}	Firewall renderings
A_{1_7}	Mounting subsidiary sheeting and its components
A_{1_8}	Roof tiling and necessary tinsmith work

The complexity of the renovation procedure can be demonstrated with the example of chimney shaft repair maintenance step (subsequence A_{1_1}). In case of Academic Style urban-type residential buildings there are numerous chimneys (or groups of chimneys) that have to be handled separately: depending on its physical condition and necessity each chimney shaft can be repaired, rebuilt or demolished. In addition to this, the stakeholders' financial abilities may also determine that this maintenance process is executed in one step or in individual steps considering the importance of intervention. In the proposed model these facts can be considered with the implementation of signature structure: in this case the Q_1^1 vector as parent node can be determined as nested vectors of child nodes $(Q_{1_1}^1, Q_{1_1}^1, \ldots, Q_{1_n}^1$ vectors). In the evaluation process, the importance of several factors (e.g. necessity, financial circumstances, etc.) can be taken into consideration in the aggregation procedure with the application of weighted relevance aggregation operator (WRAO) [14]. It is clearly visible that in major cases the decomposition of vectors can help modelling the maintenance procedure; in some cases this operation is omissible. Following the instructions of major technical literatures of building renovation the state space of every defined component machine can be easily determined. With the support of official billing contractors' database and with the basis of collected information about stakeholders' intention in renovation scenario, the input states that influence the operation of component machines can be determined in each state. In major cases, several alternative solutions for repair steps can be defined: these solutions can be different in price (with the linguistic categories of cheap, reasonable, expensive) quality (good, acceptable and poor) and life span (short, average and long). Among these the optimal solution can be chosen with the support of fuzzy evaluation. The transition degrees that determine the next internal states in the state space depend on the mentioned financial factors and technical adaptability.

The state space of A_{1_1} sub-automata is illustrated in Fig. 2, where Q_0^1 to Q_9^1 represent the initial, the internal and final states; the μ_1^1 to μ_{16}^1 represent the transition degrees. As a result of the evaluation of the transitions among the states of A_{1_1} the optimum solution for the most effective renovation process of chimney shaft can be obtained, concerning the renovation costs.

3 Conclusion and Future Work

The presented determination of the physical condition and the refurbishment sequences of Academic Style residential houses give a general state description and provide the necessary maintenance steps for the mentioned building type only. Another seriously affected building type that has similar difficulties in refurbishment procedure is the residential house that was built in the interwar era. The experienced defects are quite different, therefore the renovation sequences of Modern Style residential buildings have to be determined separately.

In general case the extended structure of a vectorial fuzzy state machine (with theoretically) unbounded number of components results in difficulties in optimization. In practice, the optimization of the refurbishment procedure of any sort of residential buildings always has a limited number of sequences; however this number might be rather high. Thus, our intention for the future is to find a proper heuristics at the basic approach to this optimization task, which is able to provide a quasi-optimal solution for every concrete problem, or a very lightly optimal solution for every problem in a manageable time. These methods seem to be various population-based evolution algorithms, which may be combined with local search cycles (evolutionary memetics), e.g. bacterial or particle swarm algorithms.

Acknowledgments. The authors wish to record their gratitude to Szilveszter Kovács for his advices and professional assistance during each phase of the project.

The research was supported by TÁMOP-4.2.2.A-11/1/KONV-2012-0012 and Hungarian Scientific Research Fund (OTKA) K105529, K108405.

References

1. Csoknyai, T., Kalmar, F., Szalai, Z., Talamon, A., Zöld, A.: Basic standards of nearly zero energy consuming buildings with renewable energy sources. Technical report, University of Debrecen, Hungary (2012) (in Hungarian)
2. Hrabovszky-Horváth, S., Pálvölgyi, T., Csoknyai, T., Talamon, A.: Generalized residential building typology for urban climate change mitigation and adaptation strategies: The case of Hungary. Energy and Buildings 62, 475–485 (2013)
3. Csizmady, A., Hegedüs, J., Kravalik, Z., Teller, N.: Long-term housing conception and mid-term housing program of Budapest, Hungary. Technical report, Local Government of Budapest, Hungary (2005) (in Hungarian)
4. Schild, Oswald, Roger, Schweikert: Weak Points 1-4. Bauverlag GmbH (1980) (in German)
5. Arató, A.: Directives and techincal guide to evaluate load-bearing constuctions of old buildings 1-5. Technical report, TTI, Budapest (1987) (in Hungarian)
6. Kóczy, L.T.: Vector Valued Fuzzy Sets. J. BUSEFAL 4, 41–57 (1980)
7. Goguen, J.: L-fuzzy sets. Journal of Mathematical Analysis and Applications 18, 145–174 (1967)
8. Pozna, C., Minculete, N., Precup, R.-E., Kóczy, L.T., Ballagi, Á.: Signatures: Definitions, operators and applications to fuzzy modelling. Fuzzy Sets and Systems 201, 86–104 (2012)
9. Ruspini, E.H.: A new approach to clustering. Information and Control 15, 22–32 (1969)
10. Kóczy, L.T., Vámos, T., Biró, G.: Fuzzy Signatures. In: EUROFUSE-SIC 1999, pp. 210–217 (1999)
11. Molnárka, G.I., Kóczy, L.T.: Decision Support System for Evaluating Existing Apartment Buildings Based on Fuzzy Signatures. IJCCC 6, 442–457 (2011)
12. de Freitas, V.P. (ed.): A State-of-the-Art Report on Building Pathology (CIB W086). Technical report, Porto University-Faculty of Engineering, Porto, Portugal (2013)
13. Harris, S.Y.: Building pathology: deterioration, diagnostics, and intervention. Wiley, New York (2001)
14. Mendis, B.S.U., Gedeon, T.D., Botzheim, J., Kóczy, L.T.: Generalised Weighted Relevance Aggregation Operators for Hierarchical Fuzzy Signatures. In: International Conference on Computational Intelligence for Modelling, Control and Automation (CIMCA 2006) (2006)

Exploring Infinitesimal Events through MV-algebras and non-Archimedean States

Denisa Diaconescu[1], Anna Rita Ferraioli[2],
Tommaso Flaminio[2], and Brunella Gerla[2]

[1] Department of Computer Science, Faculty of Mathematics and Computer Science,
University of Bucharest, Romania
ddiaconescu@fmi.unibuc.ro

[2] DiSTA - Department of Theoretical and Applied Science,
University of Insubria, Italy
{annarita.ferraioli,tommaso.flaminio,brunella.gerla}@uninsubria.it

Abstract. In this paper we use tools from the theory of MV-algebras and MV-algebraic states to study infinitesimal perturbations of classical (i.e. Boolean) events and their non-Archimedean probability. In particular we deal with a class of MV-algebras which can be roughly defined by attaching a cloud of infinitesimals to every element of a finite Boolean algebra and for them we introduce the class of Chang-states. These are non-Archimedean mappings which we prove to be representable in terms of a usual (i.e. Archimedean) probability measure and a positive group homomorphism capable to handle the infinitesimal side of the MV-algebras we are dealing with. We also study in which relation Chang-states are with MV-homomorphisms taking value in a suitable perfect MV-algebra.

Keywords: MV-algebras, non-Archimedean states, probability measures.

1 Motivation

Consider a Boolean algebra B. In classical probability theory every element $b \in B$ determines an uncertain, but precisely defined, statement about the world for which, through a probability measure p on B, we want to estimate how likely it is to happen, i.e. b is an *event*. Imagine now a situation in which the statements for which we want to measure their uncertain value are imprecise, vague, and hence such that they are loosely modeled in the realm of Boolean algebras. In this paper we are interested in providing a model to deal and probabilistically estimate those peculiar imprecise events which arise from classical, precisely determined, ones by perturbing each of them by infinitesimal values. In particular we shall deal with algebraic structures which are built starting from any Boolean algebra B and by attaching, to each element of B, a *cloud* of infinitesimals. In this generalized realm we shall introduce a notion of measure which we prove to be a variant of classical probability measures which can further cope with the infinitesimal information brought by such models.

In order to introduce the algebraic models we want to deal with, we are going to use the following ingredients: a Boolean algebra and an abelian ℓ-group G,

A. Laurent et al. (Eds.): IPMU 2014, Part II, CCIS 443, pp. 385–394, 2014.

the former being the domain of precise events and the latter being a suitable environment for infinitesimals. Although there might be several ways to combine B and G to obtain a model for imprecise events, we shall define a construction which allows to show that the resulting structure actually is a peculiar kind of MV-algebra [8]. Those are algebraic structures which are intimately related to the infinite valued Łukasiewicz calculus and for which, in the last years, a generalization of probability theory (i.e. state theory [7,8]) have been developed. For these reasons MV-algebras are an appropriate setting for our investigation. It is also worth to remark the following:

(1) If, instead of an arbitrary Boolean algebra B, we restrict to the Boolean chain **2**, the construction we are going to introduce, and which consists in combining B with an abelian ℓ-group G, produces *perfect MV-algebras* [1]. Those form a well-known class of MV-algebras which contains, in particular, Chang MV-algebra C. This structure, although it is usually defined in terms of the group-theoretical construction of lexicographic product [6], can also be described in our framework. C can intuitively be regarded as a perturbation of the Boolean chain **2** by infinitesimals from the abelian ℓ-group $\mathbb{Z}$ of integers.

(2) MV-algebraic states, being $[0, 1]$-valued functions, do not preserve positive infinitesimals. For this reason, and since the algebras we are going to introduce do have positive infinitesimal values, we shall need to slightly change the notion of state by considering, rather than $[0, 1]$-valued functions, mappings ranging on a suited defined MV-subalgebra of a non-trivial ultrapower $^*[0, 1]$ of $[0, 1]$.

The paper is structured as follows: in the next section we introduce all the necessary background on MV-algebras with a particular focus on the class $\mathfrak{BG}$ of structures arising by adding a cloud of infinitesimals around Boolean elements. Also the notion of state of an MV-algebra is given together with its relation with states of ℓ-groups. In the third section the main definitions and results are stated: the notion of Chang-state over algebras in $\mathfrak{BG}$ is given and a characterization of Chang-states in terms of Boolean probabilities and ℓ-states is proved.

2   Preliminaries

2.1   MV-algebras and $\mathfrak{BG}$-algebras

Definition 1. *An MV-algebra is a structure $(A, \oplus, {}^*, 0)$, where $\oplus$ is a binary operation, * is a unary operation and 0 is a constant such that the following conditions are satisfied for any a, $b \in A$:*

(MV1) $(A, \oplus, 0)$ is an abelian monoid, (MV2) $(a^*)^* = a$,
(MV3) $0^* \oplus a = 0^*$ (MV4) $(a^* \oplus b)^* \oplus b = (b^* \oplus a)^* \oplus a$.

We can define a new constant $1 = 0^*$ and an auxiliary operation $\odot$ as $x \odot y = (x^* \oplus y^*)^*$.

The class of MV-algebras forms a variety that we shall denote by $\mathbb{MV}$.

In any MV-algebra a lattice order is defined by setting $x \vee y = (x^* \oplus y)^* \oplus y$. The most important example of MV-algebra is the unit interval $[0, 1]$ equipped with the operations $x^* = 1 - x$ and $x \oplus y = \min(x + y, 1)$. This algebra, denoted $[0, 1]_{MV}$, is called the *standard MV-algebra* and it generates MV as a variety and as quasi-variety [2].

The largest Boolean subalgebra of an MV-algebra A, called the *Boolean skeleton of A*, is based on $\{x \in A \mid x \oplus x = x\}$ and it is denoted $B(A)$.

A non-empty subset I of an MV-algebra A is said to be an *ideal* if $x \in I$ and $y \leq x$, then $y \in I$, and if $x, y \in I$, then $x \oplus y \in I$. An ideal I is said *maximal* if it is proper and it is not contained in any proper ideal of A. For a given MV-algebra A, $Rad(A)$ (the *radical* of A) denotes the intersection of all maximal ideals of A. The radical of an MV-algebra A contains all the infinitesimal elements of A. In fact, for every $x \in A$, $x \in Rad(A)$ iff, for every $n \in \mathbb{N}$, $nx \leq x^*$, where nx is an abbreviation for $x \oplus \ldots \oplus x$ (n-times). Dually, the co-radical $Rad(A)^* = \{x^* \mid x \in Rad(A)\}$ of an MV-algebra A is the set of all co-infinitesimal elements of A and is denoted by $Rad(A)^*$.

An abelian *ℓ-group* is a structure $\mathcal{G} = (G, +, -, 0, \vee, \wedge)$ such that $(G, \vee, \wedge)$ is a lattice, $(G, +, -, 0)$ is an abelian group and $+$ is order preserving. Let G be an ℓ-group and let $u \in G$, $u > 0$. We call u a *strong unit* if for each $x \in G$, there exists $n \in \omega$ such that $x \leq nu$. Throughout this paper ℓ-groups will always be abelian. Mundici proved the existence of an equivalence functor Γ between the category of MV-algebras and the category of ℓ-groups with strong unit [2]. For every ℓ-group with strong unit $(G, +, 0, u)$, we obtain the MV-algebra $\Gamma(G, u) = [0, u] = \{x \in G \mid 0 \leq x \leq u\}$ by equipping the unit interval $[0, u]$ with the following operations: $x \oplus y = u \wedge (x + y)$, $x^* = u - x$.

A relevant subclass of MV-algebras is that of *perfect MV-algebras*, i.e. MV-algebras that are generated by their radical. Indeed, every perfect MV-algebra P can be displayed as $Rad(P) \cup Rad(P)^*$, all its elements being either infinitesimals or co-infinitesimals. Although a complete treatment of these structures is beyond the scope of this paper, it is worth mentioning that in [6] Di Nola and Lettieri established a categorical equivalence between perfect MV-algebras and ℓ-groups which can be roughly summarized as follows:

1. Given an ℓ-group G, the Di Nola-Lettieri functor Δ assigns to G, the perfect MV-algebra $P \cong \Gamma(\mathbb{Z} \times_{lex} G, (1, 0))$ where $\times_{lex}$ is the lexicographic product between ℓ-groups.
2. For every perfect MV-algebra P, $Rad(P)$ uniquely generates an ℓ-group G (denoted $\Delta^{-1}(\langle Rad(P) \rangle)$, where $\langle Rad(P) \rangle$ denotes the perfect MV-algebra $Rad(P) \cup Rad(P)^*$ generated by $Rad(P)$) and such that $Rad(P)$ coincided with the positive cone G^+ of G.

The well known Chang MV-algebra C is hence the perfect MV-algebra $\Delta(\mathbb{Z}) = \Gamma(\mathbb{Z} \times_{lex} \mathbb{Z}, (1, 0))$.

Let $B = (B, \vee, \wedge, {}^*, 0, 1)$ be a Boolean algebra. We recall that an *ideal* of B is a subset $I \subseteq B$ such that for every $x \in I$ and $y \leq x$ also $y \in I$ and if $x, y \in I$ then $x \vee y \in I$. I is *maximal* if it is proper and it is not contained in any proper ideal of B. Note that M is a maximal ideal of B if and only if it is an ideal and for every $x \in B$ either $x \in M$ or $x^* \in M$. We denote by $Max(B)$ the set of maximal ideals of B.

Let M be a maximal ideal of B, let $G = (G, +, 0,)$ be an ℓ-group and let A be a subset of the cartesian product $B \times G$ defined as follows:

$$A = \{M \times G^+\} \cup \{\overline{M} \times G^-\}$$

where $G^+ = \{g \in G \mid g \geq 0\}$ is the positive cone of G, $G^- = \{g \in G \mid g \leq 0\}$ the negative cone and $\overline{M}$ is the set-complement of M. We stress that $0 \in G^+ \cap G^-$.

Let us define over A the following two operations, $\oplus$ and $\neg$:

$$(b_1, g_1) \oplus (b_2, g_2) = \begin{cases} (b_1 \vee b_2, g_1 + g_2) & \text{if } b_1, b_2 \in M \\ (b_1 \vee b_2, 0) & \text{if } b_1, b_2 \in \overline{M} \\ (b_1 \vee b_2, (g_1 + g_2) \wedge 0) & \text{otherwise} \end{cases}$$

$$\neg(b, g) = (b^*, -g).$$

We hence have

$$(b_1, g_1) \odot (b_2, g_2) = \begin{cases} (b_1 \wedge b_2, g_1 + g_2) & \text{if } b_1, b_2 \in \overline{M} \\ (b_1 \wedge b_2, 0) & \text{if } b_1, b_2 \in M \\ (b_1 \wedge b_2, (g_1 + g_2) \vee 0) & \text{otherwise} \end{cases} \qquad (1)$$

We denote by $B \times_M G$ the structure $(A, \oplus, \neg, (0,0))$. Then we have:

Proposition 1 ([4]). *For each Boolean algebra B, maximal ideal M of B and ℓ-group G, the structure $A = B \times_M G$ is an MV-algebra such that $B(A) = B \times \{0\}$ and $Rad(A) = \{0\} \times G^+$.*

Let us denote by $\mathfrak{BG}$ the class of all MV-algebras isomorphic to $B \times_M G$, for some finite[1] Boolean algebra B, $M \in Max(B)$ and ℓ-group G.

Example 1.
1. Each finite Boolean algebra B is in $\mathfrak{BG}$. Indeed, take $G = \{0\}$ and M be any maximal ideal of B, then $B \cong B \times_M \{0\}$.
2. Each perfect MV-algebra P is in $\mathfrak{BG}$. Indeed, let $B = \mathbf{2}$, $M = \{0\}$ and G be the ℓ-group associated with P in the Di Nola-Lettieri functor, i.e. $G = \Delta^{-1}(\langle Rad(A) \rangle)$ then $P \cong \mathbf{2} \times_{\{0\}} G$. In particular, the perfect MV-algebra $\Delta(\mathbb{R}) = \Gamma(\mathbb{Z} \times_{lex} \mathbb{R}, (1,0))$ is the MV-algebra in $\mathfrak{BG}$ given by $B = \mathbf{2}$, $M = \{0\}$ and $G = \mathbb{R}$.
3. Consider the Boolean algebra $B_4 = \mathbf{2}^2$ and take $M = \{(0,0), (1,0)\}$ as a maximal ideal of B_4. Then $B_4 \times_M \mathbb{Z} = (M \times \mathbb{Z}^+) \cup (\overline{M} \times \mathbb{Z}^-)$ that is isomorphic to $B_2 \times C$.

Proposition 2 ([4]). *For every $A \in \mathfrak{BG}$ and $(b, g) \in A$, we have*

$$(b, g) = \begin{cases} (b, 0) \oplus (0, g) & \text{if } b \in M \\ (b, 0) \odot (1, g) & \text{if } b \notin M. \end{cases}$$

---

[1] This class of algebras can be defined in general, getting rid of the restriction on B to be finite. For the purpose of this paper, we shall just focus on algebras in $\mathfrak{BG}$ whose Boolean skeleton is finite.

2.2 States of MV-algebras and States of ℓ-groups

The notion of state of an MV-algebra was introduced by Mundici [7]:

Definition 2. *If A is an MV-algebra, a state of A is a function $s : A \to [0,1]$ satisfying:*

- *$s(1) = 1$,*
- *for all $a, b \in A$ such that $a \odot b = 0$, $s(a \oplus b) = s(a) + s(b)$.*

A state s is faithful *if $s(a) = 0$ implies $a = 0$, for every $a \in A$.*

Every MV-homomorphism of an arbitrary MV-algebra A into $[0,1]_{MV}$ is a state and, moreover, every state is a limit, in the product topology of $[0,1]^A$, of convex combinations of homomorphisms [7].

It is well know (cf. [7]) that states do not preserve infinitesimals of the MV-algebra they are defined on. This behavior is particularly drastic in the case of perfect MV-algebra: each perfect MV-algebra admits only one trivial state which sends the infinitesimal elements to 0 and the co-infinitesimal elements to 1. As an example, it is worth noticing that Chang MV-algebra has only one *trivial* state. i.e. a map $s : C \to [0,1]$ such that $s(x) = 0$, for all $x \in Rad(C)$, and $s(x) = 1$, for all $x \in Rad(A)^*$.

We recall that a state of an ℓ-group G is a function $m : G \to \mathbb{R}$ which is additive and positive. If (G, u) is an ℓu-group, then a state m of G is *normalized* if $m(u) = 1$. In the sequel, we will call a state of an ℓ-group simply by an ℓ-*state* and we will call an ℓu-*state* any normalized ℓ-state.

The following result establishes the relation between ℓu-states and states of MV-algebras:

Theorem 1 ([7]). *Let (G, u) be an ℓu-group and let $A = \Gamma(G, u)$. Then there is a one-one correspondence between ℓu-states of G and states of A.*

3 Chang-States

In this section we are going to introduce a notion of state for the algebras in $\mathfrak{B}\mathfrak{G}$ that we call Chang-state.[2] As every MV-algebra $A \in \mathfrak{B}\mathfrak{G}$ has a non-trivial radical, states of A make all the infinitesimal elements collapse so that basically they coincide with Boolean probabilities over the Boolean skeleton. Therefore, we now present a particular MV-algebra, denoted $\mathcal{L}(\mathbb{R}_\varepsilon)$, which provides, in our opinion, a natural codomain for Chang-states.

Let $^*[0,1]$ be a non-trivial ultrapower of the standard MV-algebra $[0,1]_{MV}$, and let $\varepsilon \in {}^*[0,1]$ be any positive infinitesimal. Then we denote by $\mathcal{L}(\mathbb{R}_\varepsilon)$ that

---

[2] The name *Chang-state* comes from the following observation: the MV-algebras in $\mathfrak{B}\mathfrak{G}$ are particular cases of structures which can be framed in the algebraic variety $\mathbb{V}(C)$ generated by Chang MV-algebra C. Therefore, although we are not working in the whole $\mathbb{V}(C)$, we are confident that the intuition behind our definition of Chang-states also applies in generalizing this notion to the whole $\mathbb{V}(C)$.

MV-algebra which is generated, in $^*[0,1]$, by the reals in $[0,1]$ plus all elements εr for $r \in \mathbb{R}$. In symbols

$$\mathcal{L}(\mathbb{R}_\varepsilon) = \langle [0,1] \cup \{r\varepsilon \mid r \in \mathbb{R}\}\rangle_{*[0,1]}.$$

Remark 1. $\mathcal{L}(\mathbb{R}_\varepsilon)$ can be equivalently presented as the MV-algebra that Mundici's functor Γ associates to the ℓ-group with strong unit $(\mathbb{R} \times_{lex} \mathbb{R}, (1,0))$. This kind of representation somehow justifies the symbol $\mathcal{L}(\mathbb{R}_\varepsilon)$, and the name "lexicographic $\mathbb{R}$". Let us also notice that the radical $\mathcal{R}$ of $\mathcal{L}(\mathbb{R}_\varepsilon)$ coincides with $\{r\varepsilon \mid r \in \mathbb{R}^+\}$.

Definition 3. *Let* $A \in \mathfrak{BG}$. *A map* $s : A \to \mathcal{L}(\mathbb{R}_\varepsilon)$ *is a* Chang-state *of* A *if*

1. $s(1) = 1$,
2. *if* $a \odot b = 0$, *then* $s(a \oplus b) = s(a) + s(b)$,
3. $s \restriction_{B(A)}$ *is a faithful probability measure.*

Lemma 1. *For every* $A \in \mathfrak{BG}$ *and for every Chang-state* s *of* A, *the following properties hold:*

1. $s(a^*) = 1 - s(a) = s(a)^*$,
2. $s(0) = 0$,
3. *if* $a \le b$ *then* $s(a) \le s(b)$,
4. *if* $y \in Rad(A)$ *then* $s(y) \in \mathcal{R}$.

Proof. *(1)-(3)* The proof can be easily obtained by adapting the analogous result [8, Proposition 10.2] proved for states of MV-algebras.

(4) Since $y \in Rad(A)$ we get $ny \le y^*$, for any $n \in \mathbb{N}$, so, from (3) $s(ny) \le s(y^*)$, and hence, from (1), $s(ny) \le s(y)^*$ for any $n \in \mathbb{N}$. From $(ny) \odot y = 0$, for any $n \in \mathbb{N}$, it follows that $s(ny) = ns(y)$ for any $n \ge 1$. Hence $ns(y) \le s(y)^*$, for any $n \in \mathbb{N}$, which implies that $s(y) \in \mathcal{R}$.   □

Theorem 2. *Let* $A \in \mathfrak{BG}$ *and a map* $s : A \to \mathcal{L}(\mathbb{R})$. *The following are equivalent:*

1. s *is a Chang-state of* A,
2. *there exists a faithful probability measure* p *on* $B(A)$ *and an* ℓ-state γ *of* $\Delta^{-1}(\langle Rad(A)\rangle)$ *such that*

$$s(b, g) = p(b) + \varepsilon\gamma(g).$$

Proof. We can uniquely display A as $A = (M \times G^+) \cup (\overline{M} \times G^-)$, where M is a maximal ideal of a Boolean algebra B and G is an ℓ-group. Note that $B(A) = B \times \{0\}$ and $Rad(A) = \{0\} \times G^+$.

$(1) \Rightarrow (2)$. Let s be a Chang-state of A and denote by p the faithful probability measure obtained as restriction of s to $B(A)$. Further, for every $(0, g) \in Rad(A)$ (where $g \ge 0$), set

$$\gamma_1(0, g) = \frac{s(0, g)}{\varepsilon} \in \mathbb{R}^+.$$

Then γ_1 can be extended to a local state of the perfect MV-algebra $P(A) = \langle Rad(A) \rangle$ and, by Lemma 13 of [5], the map $\gamma : G \to R$ such that for every $g \geq 0$

$$\gamma(g) = \gamma_1(0, g) = \frac{s(0, g)}{\varepsilon}$$

is a state of the ℓ-group G.

If $(b, g) \in A$ with $b \in M$ and $g \geq 0$, by Proposition 2 we have

$$s(b, g) = s((b, 0) \oplus (0, g)) = s(b, 0) + s(0, g) = p(b) + \varepsilon\gamma(g)$$

since $(b, 0) \odot (0, g) = (0, 0)$.

On the other hand, if (b, g) is such that $b \notin M$ and $g \leq 0$, then

$$(b, g) = (b, 0) \odot (1, g) = \neg(\neg(b, 0) \oplus (1, g)) = \neg((b^*, 0) \oplus (0, -g))$$

where $(b^*, 0) \odot (0, -g) = (0, 0)$ and

$$s(b, g) = 1 - (s(b^*, 0) + s(0, -g)) = 1 - (1 - p(b) + \varepsilon\gamma(-g)) = p(b) + \varepsilon\gamma(g).$$

$(2) \Rightarrow (1)$. Let p be a faithful probability measure on $B(A)$ and let γ be an ℓ-state on $G = \Delta^{-1}(\langle Rad(A) \rangle)$. Let hence, for each $(b, g) \in A$, $s(b, g)$ be defined as

$$s(b, g) = p(b) + \varepsilon\gamma(g).$$

Since p is faithful, for every $b \in B(A) \setminus \mathbf{2}$, we have $0 < p(b) < 1$ and hence, being ε a positive infinitesimal, $0 < s(b, g) < 1$ too. Moreover, it is easy to see that $s(1, 0) = 1$ and $s(0, 0) = 0$, whence $s(b, g) \in \mathcal{L}(\mathbb{R}_\varepsilon)$ for each $(b, g) \in A$.

Let $(b, g), (b', g') \in A$ such that $(b, g) \odot (b', g') = (0, 0) = 0$. This means, from (1), that, either $b \wedge b' = 0$ and $g + g' = 0$ if $b, b' \in \overline{M}$, or $b \wedge b' = 0$ if $b, b' \in M$, or $b \wedge b' = 0$ and $g + g' \leq 0$ otherwise. Let hence enter a case distinction:

1. If $b, b' \in \overline{M}$, then $s((b, g) \oplus (b', g')) = s(b \vee b', 0) = p(b \vee b') + \varepsilon\gamma(0) = p(b) + p(b') + \varepsilon\gamma(g + g') = p(b) + p(b') + \varepsilon\gamma(g) + \varepsilon\gamma(g') = s(b, g) + s(b', g')$.
2. If $b, b' \in M$, then $s((b, g) \oplus (b', g')) = s(b \vee b', g + g') = p(b \vee b') + \varepsilon\gamma(g + g') = s(b, g) + s(b, g')$.
3. If finally $b \in M$ and $b' \in \overline{M}$, then $g \in G^+$ and $g' \in G^-$, hence either $g + g' \geq 0$, or $g + g' \leq 0$. On the other hand, since $(b, g) \odot (b', g') = (b \wedge b', (g + g') \vee 0) = (0, 0)$, the former is never the case and hence $g + g' \leq 0$. Then $s((b, g) \oplus (b', g')) = s(b \vee b', g + g') = p(b \vee b') + \varepsilon\gamma(g + g') = p(b) + p(b') + \varepsilon\gamma(g) + \varepsilon\gamma(g') = s(b, g) + s(b, g')$.

Hence s is a Chang-state and the proof is complete. $\qquad\square$

3.1   On the Space of Chang-States

It is known [7] that, given any MV-algebra A, each homomorphism of A into the standard MV-algebra $[0, 1]_{MV}$ is a state.

As we set $\mathcal{L}(\mathbb{R}_\varepsilon)$ as a natural codomain for Chang-states, we are interested now in studying the relation between Chang-states and MV-homomorphisms of any $\mathfrak{BG}$-algebra A into $\mathcal{L}(\mathbb{R}_\varepsilon)$. Notice that, since every MV-homomorphism $h : A \to \mathcal{L}(\mathbb{R}_\varepsilon)$ actually maps $B(A)$ into $B(\mathcal{L}(\mathbb{R}_\varepsilon)) = \mathbf{2}$, this is the same as considering homomorphisms of A into the $\mathfrak{BG}$-algebra $\Delta(\mathbb{R})$ we briefly introduced in Example 1 (2).

It is easy to see that, if $A \in \mathfrak{BG}$ and h is any MV-homomorphism of A into $\Delta(\mathbb{R})$, then h is not a Chang-state. In fact, if h was a Chang-state of A, its restriction $h \restriction_{B(A)}$ on the Boolean skeleton $B(A)$ of A, would be faithful (as Boolean homomorphism). But $h \restriction_{B(A)}$ is a homomorphism of $B(A)$ into $\mathbf{2}$ and hence, unless $B(A) = \mathbf{2}$, there are elements $b, b^* \in B(A) \setminus \mathbf{2}$ for which either $h \restriction_{B(A)} (b) = 1$ or $h \restriction_{B(A)} (b^*) = 1$.

Faithful probability measures can be characterized as follows:

Proposition 3. *Let B be a finite Boolean algebra. Then $p : B \to [0,1]$ is a faithful probability measure iff*

$$p \in \mathrm{int}(\mathrm{conv}(\mathcal{H}(B, \mathbf{2}))).$$

That is, p is a faithful probability measure iff p belongs to the interior of the convex polytope generated by the (finitely many) homomorphisms of B into $\mathbf{2}$.

Proof. Let B be a finite Boolean algebra and let $\alpha_1, \ldots, \alpha_t$ be its atoms. Then, every probability measure on B, when restricted to the α_i's, defines a probability distribution, and every probability measure on B arises in this way. Moreover it is well known (see [9]) that the class of all probability measures on B coincides with the convex polytope $\mathrm{conv}(\mathcal{H}(B, \mathbf{2}))$. In fact, since atoms of B and homomorphisms of B into $\mathbf{2}$ are in bijection, every p is uniquely expressible as $\sum_{j=1}^t p(\alpha_j)h_j$ (where h_j is the unique Boolean homomorphism corresponding to α_j). Moreover, the interior of the convex polytope generated by $h_1, \ldots, h_t$ is characterized by those convex combinations $\sum_j \lambda_j h_j$ where the parameters satisfy $\lambda_j > 0$ for every j. Then it is left to show that p is faithful iff, for every $j = 1, \ldots, t$, $p(\alpha_j) > 0$.

($\Rightarrow$) Assume, by way of contradiction that there exists an α_j for which $p(\alpha_j) = 0$, then, being $\alpha_j \neq 0$, p cannot be faithful.

($\Leftarrow$) Conversely, assume $p(\alpha_j) > 0$ for every α_j. Then, if $p(b) = \sum_h p(\alpha_j)h(b) = 0$, necessarily $h(b) = 0$ for every $h \in \mathcal{H}(B, \mathbf{2})$, that is, $b = 0$ and p is faithful. $\square$

Now, let $A \in \mathfrak{BG}$, let h be a Boolean homomorphism of $B(A)$ into $\mathbf{2}$, and let γ be an ℓ-state on $G = \Delta^{-1}(\langle Rad(A) \rangle)$. Then, if we define k on A as

$$k(b, g) = h(b) + \varepsilon \gamma(g) \quad \text{for every } (b, g) \in A,$$

then k is not an MV-homomorphism since, for some $(b, g) \in A$ it can be the case that $k(b, g) > 1$. Indeed, k maps A into $\mathbb{Z} \times_{lex} \mathbb{R}$. On the other hand, these mappings provides a peculiar characterization for Chang-states in terms of convex combinations.

Proposition 4. *Let $A \in \mathfrak{BG}$ and let $s : A \to \mathcal{L}(\mathbb{R}_\varepsilon)$. The following are equivalent:*

1. *s is a Chang-state,*
2. *for every Boolean homomorphism $h : B(A) \to \mathbf{2}$ there exists an ℓ-state $\gamma_h : \Delta^{-1}(\langle Rad(A)\rangle) \to \mathbb{R}$ and a real number λ_h such that $\sum_h \lambda_h = 1$, $\lambda_h > 0$ and for every $(b, g) \in A$*

$$s(b, g) = \sum_{h \in \mathcal{H}(B(A), \mathbf{2})} \lambda_h (h(b) + \varepsilon \gamma_h(g)).$$

Proof. ($\Rightarrow$). Let $s : A \to \mathcal{L}(\mathbb{R}_\varepsilon)$ be a Chang-state. Then from Theorem 2, there exists a faithful probability measure $p : B(A) \to [0,1]$ and an ℓ-state $\gamma : \Delta^{-1}(\langle Rad(A)\rangle) \to \mathbb{R}$ such that, for all $(b, g) \in A$ one has $s(b, g) = p(b) + \varepsilon\gamma(g)$. Now, Proposition 3 ensures that, being p faithful and $B(A)$ being finite, the existence, for every $h \in \mathcal{H}(B(A), \mathbf{2})$, of a real number $\lambda_h > 0$ such that $\sum_h \lambda_h = 1$ and, for all $(b, g) \in A$,

$$\begin{aligned}
s(b, g) &= p(b) + \varepsilon\gamma(g) \\
&= \left(\textstyle\sum_{h \in \mathcal{H}(B(A), \mathbf{2})} \lambda_h h(b)\right) + \varepsilon\gamma(g) \\
&= \left(\textstyle\sum_{h \in \mathcal{H}(B(A), \mathbf{2})} \lambda_h h(b) + \varepsilon\frac{\gamma(g)}{\lambda_h}\right).
\end{aligned}$$

Then the claim follows setting $\gamma_h = \gamma/\lambda_h$ and observing that, for each positive λ_h, γ/λ_h is an ℓ-state of $\Delta^{-1}(\langle Rad(A)\rangle)$.

($\Leftarrow$). Conversely, for each $h \in \mathcal{H}(B(A), \mathbf{2})$, let $\lambda_h > 0$ such that $\sum_h \lambda_h = 1$ and γ_h a ℓ-state of $\Delta^{-1}(\langle Rad(A)\rangle)$. Then, setting for every $(b, g) \in A$, $s(b, g) = \sum_{h \in \mathcal{H}(B(A), \mathbf{2})} \lambda_h(h(b) + \varepsilon\gamma_h(g))$ we get

$$\begin{aligned}
s(b, g) &= \left(\textstyle\sum_{h \in \mathcal{H}(B(A), \mathbf{2})} \lambda_h h(b)\right) + \varepsilon\left(\textstyle\sum_{h \in \mathcal{H}(B(A), \mathbf{2})} \lambda_h \gamma_h(g)\right) \\
&= p(b) + \varepsilon\gamma(g)
\end{aligned}$$

where p is the faithful probability measure whose existence is endured by Proposition 3, and γ is defined through $\gamma = \sum_{h \in \mathcal{H}(B(A), \mathbf{2})} \lambda_h \gamma_h$. Obviously γ is a ℓ-state of $\Delta^{-1}(\langle Rad(A)\rangle)$ since ℓ-states are closed under convex combinations. □

In order to understand the general relationship among Chang-states and MV-homomorphisms between two $\mathfrak{BG}$-algebras, we show the following proposition:

Proposition 5. *Let $A, A_1 \in \mathfrak{BG}$ with $A = B \times_M G$ and $A_1 = B_1 \times_{M_1} G_1$ and let $f : A \to A_1$ be an MV-algebra homomorphism. Then either f maps M to M_1 or all the elements of A are mapped into Boolean elements of A_1.*

Proof. We know that f maps the Boolean skeleton of A into the Boolean skeleton of A_1. Suppose there is $\bar{b} \in M$ with $f(\bar{b}, 0) = (b_1, 0)$ and $b_1 \notin M_1$, and let for every $g \in G^+$, $f(\bar{b}, g) = (b_1, h(g))$ (it must be $h(g) \leq 0$). Then:

$$f(\bar{b}, g) \odot f(\bar{b}^*, 0) = (\bar{b}_1, h(g)) \odot (f(\bar{b}, 0)^*) = (\bar{b}_1, h(g)) \odot (\bar{b}_1^*, 0) = (0, 0)$$

while on the other side:

$$f(\overline{b},g) \odot f(\overline{b}^*,0) = f((\overline{b},g) \odot (\overline{b}^*,0)) = f(0,g)$$

hence, for every $g \in G^+$, $f(0,g) = (0,0)$. Since $Rad(A) = \{(0,g) \mid g \in G^+\}$ we have that f maps $Rad(A)$ into the element $(0,0)$, and $Rad(A)^*$ into $(1,0)$. Now let $(b,g) \in A$ with $b \in M$. By Proposition 2, $(b,g) = (b,0) \oplus (0,g)$ hence $f(b,g) = f(b,0) \oplus f(0,g) = f(b,0) \in B_1$. If $(b,g) \in A$ with $b \notin M$ then $(b,g) = (b,0) \odot (1,g)$ and $f(b,g) = f(b,0) \odot (1,0) = f(b,0) \in B_1$. $\square$

Let $A \in \mathfrak{BG}$ with $A = (M \times G^+) \cup (\overline{M} \times G^-)$ and consider an MV-homomorphisms f of A into $\Delta(\mathbb{R})$. Then, apart from the case in which $f(A) = \mathbf{2}$, it must be $f(M) = 0$ and $f(\overline{M}) = 1$ and hence the MV-homomorphism is completely determined on its Boolean elements. This means that, in Proposition 3, the Boolean homomorphisms h are not restrictions of Chang-states.

4   Conclusion and Future Research

We generalized the notion of state given in [7] in order to deal with infinitesimal events, continuing the work done in [5] and in [3]. We started our investigation with a class $\mathfrak{BG}$ of MV-algebras that can be roughly described as obtained by adding infinitesimals around Boolean elements [4], and we defined states taking values in an ultrapower of $[0,1]$. We gave a characterization in terms of homomorphisms that is not a translation of the classical case, mainly due to the behaviour of homomorphisms of MV-algebras in $\mathfrak{BG}$. This is a first step towards the formulation of a notion of state for all non-semisimple MV-algebras, keeping the nature of the infinitesimal elements.

References

1. Belluce, P., Di Nola, A., Lettieri, A.: Local MV-algebras. Rend. Circ. Mat. Palermo 42, 347–361 (1993)
2. Cignoli, R., D'Ottaviano, I.M.L., Mundici, D.: Algebraic Foundations of Many-valued Reasoning. Kluwer, Dordrecht (2000)
3. Diaconescu, D., Flaminio, T., Leuştean, I.: Lexicographic MV-algebras and lexicographic states (submitted)
4. Di Nola, A., Ferraioli, A.R., Gerla, B.: Combining Boolean algebras and ℓ-groups in the variety generated by Chang's MV-algebra. Mathematica Slovaca (accepted)
5. Di Nola, A., Georgescu, G., Leustean, I.: States on Perfect MV-algebras. In: Novak, V., Perfilieva, I. (eds.) Discovering the World With Fuzzy Logic. STUDFUZZ, vol. 57, pp. 105–125. Physica, Heidelberg (2000)
6. Di Nola, A., Lettieri, A.: Perfect MV-algebras are categorically equivalent to abelian ℓ-groups. Studia Logica 53, 417–432 (1994)
7. Mundici, D.: Averaging the truth-value in Łukasiewicz logic. Studia Logica 55(1), 113–127 (1995)
8. Mundici, D.: Advanced Łukasiewicz calculus and MV-algebras. Trends in Logic, vol. 35. Springer (2011)
9. Paris, J.B.: The uncertain reasoner's companion: A mathematical perspective. Cambridge University Press (1994)

Accelerating Effect of Attribute Variations: Accelerated Gradual Itemsets Extraction

Amal Oudni[1,2], Marie-Jeanne Lesot[1,2], and Maria Rifqi[3]

[1] Sorbonne Universités, UPMC Univ Paris 06, UMR 7606
LIP6, F-75005, Paris, France
[2] CNRS, UMR 7606, LIP6, F-75005, Paris, France
{amal.oudni,marie-jeanne.lesot}@lip6.fr
[3] Université Panthéon-Assas - Paris 02, LEMMA, F-75005, Paris, France
maria.rifqi@u-paris2.fr

Abstract. Gradual itemsets of the form *"the more/less A, the more/less B"* summarize data through the description of their internal tendencies, identified as correlation between attribute values. This paper proposes to enrich such gradual itemsets by taking into account an acceleration effect, leading to a new type of gradual itemset of the form *"the more/less A increases, the more quickly B increases"*. It proposes an interpretation as convexity constraint imposed on the relation between A and B and a formalization of these accelerated gradual itemsets, as well as evaluation criteria. It illustrates the relevance of the proposed approach on real data.

Keywords: Gradual Itemset, Acceleration, Enrichment, Convexity.

1 Introduction

Information extraction can take many forms, leading to various types of knowledge which are then made available to experts. This paper focuses on gradual itemsets which can be illustrated by the example *"the closer the wall, the harder the brakes are applied"*. Initially introduced in the fuzzy implication formalism [1–3], gradual itemsets have then been interpreted as expressing constraints on the attribute covariations. Several interpretations of the constraints have been proposed, as regression [4], correlation of induced order [5, 6] or identification of compatible object subsets [7, 8]. Each interpretation is associated with the definition of a support to quantify the validity of gradual itemsets and to methods for the identification of the itemsets that are frequent according to these support definitions.

Furthermore, several types of enrichments have been proposed: in the case of categorical or fuzzy data clauses, clauses linguistically introduced by the expression "all the more" lead to so-called strengthened gradual itemsets [9]. They can be illustrated by an example such as *"the closer the wall, the harder the brakes are applied, all the more the higher the speed"*. For numerical data, an enrichment by characterization clauses [10] adds a clause linguistically introduced by the expression "especially if": characterized gradual itemsets can be illustrated

A. Laurent et al. (Eds.): IPMU 2014, Part II, CCIS 443, pp. 395–404, 2014.

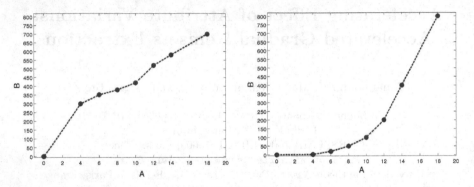

Fig. 1. Two data sets, leading to *"the more A, the more B"* where an acceleration effect is observed for the right data set and not for the left one

by a sentence as *"the closer the wall, the harder the brakes are applied, especially if the distance to the wall $\in [0, 50]\,m$"*.

In this paper, we consider a new type of enrichment in the case of numerical data, to capture a new type of information: the aim is to express how fast the values of some attributes vary as compared to others, as illustrated by the two data sets represented in Figure 1. In both cases, a covariation constraint is satisfied, which justifies the extraction of the gradual itemset *"the more A, the more B"*. However, on the right-hand example, the speed of B augmentation appears to increase, making it possible to enrich the gradual itemset to *"the more A increases, the more quickly B increases"*.

This paper addresses the task of extracting such accelerated gradual itemsets. The principle of acceleration is naturally understood as speed variation increase, which can be translated as a convexity constraint on the underlying function associating the considered attributes. This constraint can be modelled as an additional covariation constraint, leading to the definition of a criterion called *accelerated support* to assess the validity of such accelerated gradual itemsets.

The paper is organized as follows: Section 2 recalls the formalism of gradual itemsets and details the existing types of enrichment. Section 3 discusses the proposed interpretation of accelerated gradual itemsets and its formalization. Section 4 defines the criteria proposed for the evaluation of this new type of itemsets. Section 5 illustrates and analyses the experimental results obtained on real data.

2   Typology of Gradual Itemset Enrichments

This section first recalls the notations and definitions of gradual items and itemsets [9, 8] as well as the support definition based on compatible data subsets [8]. It then describes the existing enrichments of gradual itemsets.

2.1   Gradual Itemset Definitions

Let $\mathcal{D}$ denote the data set. A *gradual item* A^* is made of an attribute A and a variation $* \in \{\geq, \leq\}$, which represents a comparison operator. A *gradual itemset* is then defined as a set of gradual items $M = \{(A_j, *_j), j = 1..k\}$, interpreted as their conjunction. It induces a pre-order, $\preceq_M$, defined as $o \preceq_M o'$ iff $\forall\, j \in [1, k]\, A_j(o) *_j A_j(o')$ where $A_j(o)$ represents the value of attribute A_j for object o.

As briefly recalled in the introduction, there exists several interpretations of gradual itemsets [1–8]. In this paper, we consider the interpretation of co-variation constraint by identification of compatible subsets [7, 8]: it consists in identifying subsets D of $\mathcal{D}$, called *paths*, that can be ordered so that all data pairs of D satisfy the pre-order induced by the considered itemset. More formally, for an itemset $M = \{(A_j, *_j), j = 1..k\}$, $D = \{o_1, ..., o_m\} \subseteq \mathcal{D}$ is a path if and only if there exists a permutation π such that $\forall l \in [1, m-1], o_{\pi_l} \preceq_M o_{\pi_{l+1}}$. Gradual itemsets thus depend on the order induced by the attribute values, not on the values themselves.

Such a path is called *complete* if no object can be added to it without violating the order constraint imposed by M. $\mathcal{L}(M)$ denotes the set of complete paths associated to M. The set of maximal complete paths, i.e. complete paths of maximal length, is denoted $\mathcal{L}^*(M) = \{D \in \mathcal{L}(M)/\forall D' \in \mathcal{L}(M)\, |D| \geq |D'|\}$.

The gradual support of M, $GS_{\mathcal{D}}(M)$, is then defined as the length of its maximal complete paths divided by the total number of objects [7]:

$$GS_{\mathcal{D}}(M) = \frac{1}{|\mathcal{D}|} \max_{D \in \mathcal{L}(M)} |D| \tag{1}$$

2.2   Existing Enrichments

Two enrichment types for gradual itemsets have been proposed, namely through characterization [10] and strengthening [9]. Both are based on a principle of increased validity when the data are restricted to a subset: the gradual support of the considered itemset must increase when it is computed on the data subset only.

More precisely, in the case of characterization [10], the restriction is defined as a set of intervals: characterized gradual itemsets are linguistically of the form *"the more/less A, the more/less B, especially if $J \in R$"*, where J is a set of attributes belonging to $A \cup B$ and R is a set of intervals defined for each attribute in J. R defines the data subset, it applies only in the case of numerical data.

In the strengthening case [9], the restriction is defined by a presence, possibly in a fuzzy weighted way, of values required by the strengthening clause: the (fuzzy) data subset only contains objects possessing the required values. Strengthened gradual itemsets are linguistically of the form *"the more/less A, the more/less B, all the more C"*, where C is the strengthening clause that consists of values of categorical attributes or fuzzy modalities of fuzzy attributes.

As opposed to the existing enrichments, the peculiarities of the proposed enrichment are mentioned in the following section.

2.3   Characteristics of the Proposed Acceleration Enrichment

The main difference between accelerated gradual itemsets and the previous gradual itemset enrichments comes from the nature of the additional clause: both for

characterization and strengthening the enriching clause has a presence semantics, insofar as the additional constraint leads to a data restriction defined by
the presence of specific values (in the interval R or in the clause C) on which
the itemset validity must increase. On the contrary, as detailed in the next sections, the semantics of the acceleration clause is gradual, depending not on the
attribute values but on the order they induce. It thus has the same nature as
the considered itemset.

It must be underlined that the accelerated gradual itemsets apply to numerical
data, excluding the categorical case.

3   Formalization of Accelerated Gradual Itemsets

This section presents the interpretation and the principle of gradual itemset
acceleration, as well as the proposed formalization.

3.1   Principle of Accelerated Gradual Itemsets

As already mentioned in the introduction, Figure 1 represents two data sets with
the same cardinality described by two attributes, A (x-axis) and B (y-axis). In
both cases, the data sets lead to the same gradual itemset $M = A^{\geq}B^{\geq}$ supported
by all data points: the gradual support is 100% in both cases. Now it can be
noticed that the covariation between A and B is different: an acceleration effect
of B values with respect A values can be observed for the right data set, whereas
it does not hold for the left-hand data set.

Accelerated gradual itemsets aim at capturing this difference. It must be underlined that it breaks the symmetry property, distinguishing the cases "the
more A, the more quickly B" and "the more B, the more quickly A", whereas
the gradual itemset is "the more A, the more B" in both cases.

Mathematically, the acceleration effect corresponds to a convexity property
of the function that associates B values to A values, imposing that its graph
is "turned up" as illustrated on the right part of Figure 1, meaning that the
line segment between any two points on the graph of the function lies above the
graph. Convex growth means "increasing at an increasing rate (but not necessarily proportionally to current value)" which is equivalent to desired acceleration
effect. Differentiable functions are convex if and only if their derivative is monotonically non-decreasing.

Now, data sets from which accelerated gradual itemsets must be extracted
do not give access to the mathematical function relating A and B values, hence
its derivative cannot be computed. Therefore we propose to consider a rough
discretization, defined as the quotient of the successive differences $\frac{\Delta B}{\Delta A}$ when
data are ordered with respect to their A values.

We thus propose to interpret the acceleration effect as an increase of $\left(\frac{\Delta B}{\Delta A}\right)$. It
must be noticed that this interpretation does not take into account the shape of
the convex function: for instance no difference is made whether the underlying
function is quadratic or exponential.

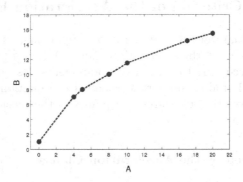

Fig. 2. $A^{\geq} B^{\geq} \left(\frac{\Delta B}{\Delta A}\right)^{\leq}$ with deceleration effect

3.2 Formalization

To address the principle presented in the previous section, we propose to formalize an accelerated gradual itemset as a triplet: $A^{*_1} B^{*_2} \left(\frac{\Delta B}{\Delta A}\right)^{*_3}$, where $A^{*_1} B^{*_2}$ represents a gradual itemset, and $\left(\frac{\Delta B}{\Delta A}\right)^{*_3}$ represents the acceleration clause that compares the variations of B with that of A. $*_1$ determines whether "the more A increases" ($*_1 =\geq$) or "the more A decreases" ($*_1 =\leq$). $*_2$ plays the same role for B. $*_3$ determines whether acceleration or deceleration is considered: $*_3 =\geq$ leads to "the more quickly" and $*_3 =\leq$ leads to "the less quickly" or equivalently "the more slowly" .

This paper focuses on the acceleration effect, i.e. attributes for which values increase "quickly", i.e. the case $*_3 =\geq$. It corresponds to the convex curve case. The case where $*_3 = \leq$ corresponds to a deceleration effect, as illustrated on Figure 2, which can be described as "the more A, the more slowly B increases". It can be noticed that this is equivalent to "the more B increases, the more quickly A increases", i.e. $A^{\geq} B^{\geq} \left(\frac{\Delta A}{\Delta B}\right)^{\geq}$. Thus considering only $*_3 =\geq$ is not a limitation.

3.3 Generalization

The previous definition focuses on the case of itemsets containing two attributes. In the general case, the itemset to enrich may be composed of several attributes, as well as the acceleration clause.

Now the notion of convex function is also mathematically defined for functions depending on several variables, based on properties of their Hessian matrices. Similarly, a discretization based on the available data may be computed for a given data set, leading to accelerated gradual itemsets made on several attributes, which may be written schematically $M_1 M_2 \frac{\Delta M_2}{\Delta M_1}$.

4   Evaluation Criterion of the Acceleration Effect

An accelerated gradual itemset contains two components, the classical gradual itemset $M = A^{*_1}B^{*_2}$ and the acceleration clause $M_a = \left(\frac{\Delta B}{\Delta A}\right)^{*_3}$. It must therefore be evaluated according to these two components. Its quality is measured both by the classical gradual support as recalled in Equation (1) and an accelerated gradual support that measures the quality of the acceleration, as defined below.

4.1   Order Induced by the Acceleration Clause

The itemset M induces a pre-order on objects as defined in Section 2; the acceleration clause $\left(\frac{\Delta B}{\Delta A}\right)^{*_3}$ induces a pre-order on pairs of objects denoted $\preceq_{M_a}$: for any o_1, o_2, o_3 and o_4

$$(o_1, o_2) \preceq_{M_a} (o_3, o_4) \Leftrightarrow \frac{B(o_2) - B(o_1)}{A(o_2) - A(o_1)} *_3 \frac{B(o_4) - B(o_3)}{A(o_4) - A(o_3)}. \tag{2}$$

where $A(o)$ and $B(o)$ respectively represent the value of attributes A and B for object o.

4.2   Definition of the Accelerated Support

The quality of the candidate accelerated gradual itemset MM_a is high if there exists a subset of data that simultaneously satisfies the order induced by M and that induced by M_a. Therefore the acceleration quality first requires to identify a data subset that satisfies $\preceq_M$. To that aim, the GRITE algorithm [7] can be used to identify candidate gradual itemsets as well as their set of maximal complete support paths $\mathcal{L}^*(M)$.

For any $D \in \mathcal{L}^*(M)$, the computation of the accelerated support then consists first in identifying subsets of D so that the constraint $\left(\frac{\Delta B}{\Delta A}\right)^{*_3}$ is verified simultaneously.

We denote φ the function that identifies a maximal subset of objects from D such that

$$\forall o_1, o_2, o_3 \in \varphi(D), (o_1 \preceq_M o_2 \preceq_M o_3 \Rightarrow (o_1, o_2) \preceq_{M_a} (o_2, o_3))$$

The accelerated gradual support of MM_a is then computed as:

$$GS_a = \frac{1}{|D| - 1} \max_{D \in \mathcal{L}^*(M)} |\varphi(D)| \tag{3}$$

where $|D|$ denotes the size of any maximal complete path in $\mathcal{L}^*(M)$, as, by definition of $\mathcal{L}^*(M)$, they all have the same size. $|D| - 1$ is then the maximal possible value of $\varphi(D)$ and thus the normalizing factor. Indeed, $\left(\frac{\Delta B}{\Delta A}\right)^{*_3}$ does not have the same definition set as classical gradual itemsets: it applies to pairs of successive objects.

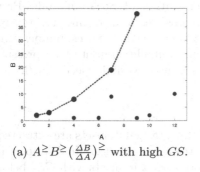

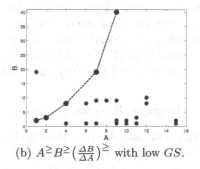

(a) $A^{\geq}B^{\geq}(\frac{\Delta B}{\Delta A})^{\geq}$ with high GS.   (b) $A^{\geq}B^{\geq}(\frac{\Delta B}{\Delta A})^{\geq}$ with low GS.

Fig. 3. Two data sets for which $A^{\geq}B^{\geq}(\frac{\Delta B}{\Delta A})^{\geq}$ holds with different GS and the same $GS_a = 100\%$

Combination of the Quality Criterion. The classical validity definition is then extended to integrate the condition on GS_a: an accelerated gradual itemset MM_a is valid if $GS \geq s$ and $GS_a(MM_a) \geq s_a$ where s_a is a threshold for the accelerated gradual support and s the threshold of classical gradual support. It is worth noticing that both GS and GS_a are necessary to assess the quality of an accelerated gradual itemset. Figure 3 illustrates the case of two datasets leading to the same $GS_a = 100\%$ but with different GS: GS equals 45% for the data set on the left and 22% on the right. Indeed GS_a is computed relatively to the path size whereas GS takes into account the total number of points. When combining the two components, a priority is given to GS: for a given GS level, accelerated gradual itemsets are compared in terms of GS_a.

Computational Cost. The computational time of the extraction of accelerated gradual itemsets depends on the number of objects and attributes of the data set, as well as on the gradual support threshold. The experiments described in the next section show that the most expensive step corresponds to the extraction of the basic gradual itemsets and that the step of acceleration clause identification only adds a much smaller computational cost. More precisely, 85% of the total time necessary for the extraction is used in the step of the basic gradual itemset extraction and only 15% is used in the step of the acceleration clause extraction.

5   Experimental Study

This section describes the experiments carried out using the proposed method of accelerated gradual itemset extraction on a real data set. The analysis of the results is based on the number of extracted gradual itemsets and their quality.

5.1   Considered Data

We use a real data set called *weather* downloaded from the site http://www.meteo-paris.com/ile-de-france/station-meteo-paris/pro:

these data come from the Parisian weather station of St-Germain-des-Prés. They contain 2164 meteorological observations realized during eight days (December 20th to 27th 2013), described by 8 numerical attributes: temperature (°C), wind chill (°C), wind run (km), rain (mm), outside humidity (%), pressure (hPa), wind speed (km/hr) and wind gusts measured as high speed (km/hr).

5.2   Results: Extracted Itemsets

Setting a gradual support threshold $s = 20\%$, 153 gradual itemsets are extracted, two of them with 100%. Figure 4 represents the accelerated gradual support of all identified gradual itemsets. It can be observed that itemsets with GS_a below 20% are not numerous and almost 30% have GS_a above 50%. When setting the accelerated support threshold $s_a = 20\%$, represented by the horizontal line on Figure 4, 130 itemsets are considered as enriched by an acceleration clause, which corresponds to more than 85%.

According to the criteria combination with priority discussed in the previous section, the most interesting accelerated gradual itemset is then the one corresponding to point A on the graph. It represents the itemset "the more the wind speed increases, the more quickly the wind run increases: $GS = 100\%$ and $GS_a = 90\%$". This corresponds to an expected result from the proposed definition of accelerated itemsets: the underlying linear relation between these two attributes corresponds to the limit case of acceleration and thus gets a high accelerated support.

The next most interesting itemsets are then the two points in region B in the graph, that respectively correspond to the itemsets

- the more the temperature decreases, the more quickly the rain accumulation increases: $GS = 100\%$ and $GS_a = 32\%$.
- the more the humidity decreases, the more quickly the temperature increases $GS = 94.73\%$, $GS_a = 34\%$.

The middle points in region C in the graph show a trade-off between GS and GS_a. The two ones with highest GS_a correspond to

- the more the wind gusts increase, the more quickly the wind run increases: $GS = 54.9\%$ and $GS_a = 51\%$.
- the more the wind gusts increase, the more quickly the wind speed increases: $GS = 57.3\%$ and $GS_a = 48\%$.

Finally, it can be observed that the majority of extracted gradual itemsets have a gradual support slightly above the threshold 20%, many of them reaching a high accelerated support. Examples with highest GS_a in region D in the graph, include

- the more the humidity increases, the more quickly the wind run increases: $GS = 22.69\%$ and $GS_a = 81\%$.
- the more the pressure decreases, the more quickly the humidity increases: $GS = 20.93\%$ and $GS_a = 88\%$.

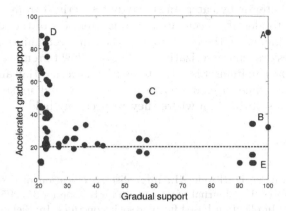

Fig. 4. Gradual support and accelerated gradual support, for each of the 153 extracted gradual itemsets

– the more the wind chill increases, the more quickly the temperature increases: $GS = 22.88\%$ and $GS_a = 86\%$.

It is also interesting to look at an example without accelerating effect: the gradual itemset represented by point E corresponds to

– the more the rain accumulation decreases, the more quickly the wind chill increases: $GS = 94.72\%$ and $GS_a = 10\%$.

Accelerated gradual itemsets thus make it possible to extract rich meteorological knowledge from the individual weather station observations.

6 . Conclusion and Future Work

In this paper we proposed an approach to enrich gradual itemsets, using an acceleration clause linguistically expressed by the expression "quickly", so as to extract more information summarizing data sets. The extraction of these accelerated gradual itemsets relies on the identification of attributes occurring in the considered gradual itemset for which the speed increase augments compared with other attributes values. The constraint is interpreted in terms of convexity and leads to the definition of a quality criterion to evaluate the acceleration effect.

Ongoing works include complementary experimentations taking into account both computation efficiency (time and memory) and use of other real data where expert advice can be given on the understanding and interest of extracted accelerated gradual itemsets.

Future works also include the combination of the acceleration effect with other enrichment principles, applied to the acceleration clauses: it would be interesting to identify restriction of the data sets on which the acceleration effect particularly holds. In particular, in the case of meteorological data, restriction induced by a temporal attribute, or by categorical attributes derived from the date, could make it possible to identify accelerated gradual itemsets of the form "the more the temperature increases, the more quickly the rain accumulation decreases, in the summer". Besides, characterization could also allow to remove the ambiguity that may exist when an itemset is extracted with an acceleration and deceleration effect at the same time: a characterization clause would make it possible to identify the subsets of the data where they respectively hold.

References

1. Galichet, S., Dubois, D., Prade, H.: Imprecise specification of illknown functions using gradual rules. Int. Journal of Approximate Reasoning 35, 205–222 (2004)
2. Hüllermeier, E.: Implication-based fuzzy association rules. In: Siebes, A., De Raedt, L. (eds.) PKDD 2001. LNCS (LNAI), vol. 2168, pp. 241–252. Springer, Heidelberg (2001)
3. Dubois, D., Prade, H.: Gradual inference rules in approximate reasoning. In: Proc. of the Int. Conf. on Fuzzy Systems, vol. 61, pp. 103–122 (1992)
4. Hüllermeier, E.: Association rules for expressing gradual dependencies. In: Elomaa, T., Mannila, H., Toivonen, H. (eds.) PKDD 2002. LNCS (LNAI), vol. 2431, pp. 200–211. Springer, Heidelberg (2002)
5. Berzal, F., Cubero, J.C., Sanchez, D., Vila, M.A., Serrano, J.M.: An alternative approach to discover gradual dependencies. Int. Journal of Uncertainty, Fuzziness and Knowledge-Based Systems 15, 559–570 (2007)
6. Laurent, A., Lesot, M.-J., Rifqi, M.: GRAANK: Exploiting rank correlations for extracting gradual itemsets. In: Andreasen, T., Yager, R.R., Bulskov, H., Christiansen, H., Larsen, H.L. (eds.) FQAS 2009. LNCS, vol. 5822, pp. 382–393. Springer, Heidelberg (2009)
7. Di Jorio, L., Laurent, A., Teisseire, M.: Fast extraction of gradual association rules: a heuristic based method. In: Proc. of the 5th Int. Conf. on Soft Computing as Transdisciplinary Science and Technology, pp. 205–210 (2008)
8. Di-Jorio, L., Laurent, A., Teisseire, M.: Mining frequent gradual itemsets from large databases. In: Adams, N.M., Robardet, C., Siebes, A., Boulicaut, J.-F. (eds.) IDA 2009. LNCS, vol. 5772, pp. 297–308. Springer, Heidelberg (2009)
9. Bouchon-Meunier, B., Laurent, A., Lesot, M.-J., Rifqi, M.: Strengthening fuzzy gradual rules through "all the more" clauses. In: Proc. of the Int. Conf. on Fuzzy Systems, pp. 1–7 (2010)
10. Oudni, A., Lesot, M.-J., Rifqi, M.: Characterisation of gradual itemsets through "especially if" clauses based on mathematical morphology tools. In: EUSFLAT, pp. 826–833 (2013)

Gradual Linguistic Summaries

Anna Wilbik and Uzay Kaymak

Information Systems
School of Industrial Engineering
Eindhoven University of Technology
Eindhoven, The Netherlands
{A.M.Wilbik,U.Kaymak}@tue.nl

Abstract. In this paper we propose a new type of protoform-based linguistic summary – the gradual summary. This new type of summaries aims in capturing the change over some time span. Such summaries can be useful in many domains, for instance in economics, e.g., "prices of X are getting smaller", in eldercare, e.g., "resident Y is getting less active", in managing production, e.g. "production is dropping" or "delays in deliveries are getting smaller".

Keywords: linguistic summaries, fuzzy logic, computing with words, protoforms.

1 Introduction

Recent increasing attention in the field of big data, means that more and more data are created and stored and need to be analyzed. However very often the amount of available data is beyond human cognitive capabilities and comprehension skills. Acknowledging this problem, creating summaries of data has been goal of the artificial intelligence and computational intelligence community for many years. This was especially visible in the context of written texts. However also other data like images and sensor data were summarized.

In this paper we are dealing with linguistic summaries of numerical data. This topic has been widely investigated, cf. [4,16,17,24]. For instance, Dubois and Prade [4] proposed representation and reasoning for gradual inference rules for linguistic summarization in the form "the more X is F, the more/ the less Y is G" that could summarize various relationships. Such rules expressed a progressive change of the degree to which the entity Y satisfies the gradual property G when the degree to which the entity X satisfies the gradual property F is modified. Rasmussen and Yager [17] discussed the benefit of using fuzzy sets in data summaries based on generalized association rules. Gradual functional dependences were investigated, and a query language called SummarySQL was proposed. Bosc et al. [1] discussed the use of fuzzy cardinalities for linguistic summarization. The SAINTETIQ model [16] provides the user synthetic views of groups of tuples over the database. In [18], the authors proposed a summarization procedure to describe long-term trends of change in human behavior, e.g., "the

A. Laurent et al. (Eds.): IPMU 2014, Part II, CCIS 443, pp. 405–413, 2014.

quality of the 'wake up' behavior has been decreasing in the last month" or "the quality of the 'morning routine' is constant but has been highly unstable in the last month."

In this paper we follow the approach of Yager [24], in the form "Q objects in Y are S,". This approach was considerably advanced and implemented by Kacprzyk [5], Kacprzyk and Yager [12], Kacprzyk et al. [13,15,14]. Those summaries have been applied in different areas, e.g. in financial data [10,11,2,3], eldercare [23,22,21]. However we notice that they describe more static situation, and even if dealing with time series, they do not focus on change in time.

In the paper, inspired by gradual inference rules by Dubois and Prade [4], we propose a new type of protoform-based linguistic summary, that we call gradual linguistic summary. This new type of summaries aims in capturing the change in time. Such summaries can be useful in many domains, for instance in economics, e.g., "prices of X are getting smaller", in eldercare, e.g., "resident Y is getting less active", in managing production, e.g. "production is dropping" or "problems are solved faster".

2   Gradual Linguistic Summaries

Linguistic summaries are protoform [26] (template-) based quasi-natural language sentences that capture the information hidden in data. Yager [24] proposed two protoforms, simple and extended one, which can be written, respectively as

$$Qy\text{'s are } P \qquad\qquad QRy\text{'s are } P. \qquad (1)$$

Q is a linguistic quantifier, P is the summarizer, R is the qualifier and y's are the objects that are to be summarized. Every sentence may be evaluated using different criteria, however the most important and basic one is the degree of truth ($\mathcal{T}$) which is the degree to which the summary is valid. The truth value, indication of how compatible it is with the database, is calculated using Zadeh's calculus of quantified propositions [25] and hence with the following formulas:

$$\mathcal{T}(Qy\text{'s are } P) = \mu_Q\left(\frac{1}{n}\sum_{i=1}^{n}\mu_P(y_i)\right) \qquad (2)$$

$$\mathcal{T}(QRy\text{'s are } P) = \mu_Q\left(\frac{\sum_{i=1}^{n}\mu_P(y_i)\wedge\mu_R(y_i)}{\sum_{i=1}^{n}\mu_R(y_i)}\right) \qquad (3)$$

This approach turned out to be useful e.g. in small retail [9,8] or eldercare [23,22,21]. Moreover the protoforms were modified to fit other type of data, e.g. time series [10,11,2], texts [19].

The drawback of the above mentioned solution is that they describe somewhat static situation, and do not capture the change very well. To overcome this drawback, we introduce a new type of linguistic summaries, namely gradual summaries.

Gradual linguistic summaries are defined with the following protoform

$$y\text{'s are getting } P_G \tag{4}$$

where y's are the objects that are summarized and P_G is gradual summarizer, e.g. faster. Examples of such summaries are "resident Y is getting less active", "problems are solved faster", "the price is getting smaller".

Naturally it is possible to extend the above summary by adding a qualifier, e.g. "problems of high impact are solved faster", "the prices of luxiorious goods are getting higher". However we won't consider this case in this paper.

With the gradual summary we wish to capture the decreasing (or increasing) trend within the data points, assuming that we consider a single variable at a time. One option could be to use some kind of linear approximation, however we would like to have a high truth value also in case of nonlinear trends. Therefore we assume for our purpose that the decreasing trend is present when in most cases points are smaller than the previous ones.

We evaluate the degree to which most points fulfill the property that in most cases data point has a smaller value than any previous point and bigger value than any following point from some specified neighborhood of the point. The neighborhood we define with two parameters k_1 and k_2, $k_1 > k_2$. For a point y_i generally are the points that have at most k_1 and no less than k_2 points between with point y_i, i.e. $\{y_{i-k_1}, y_{i-k_1-1}, \ldots, y_{i-k_2}, y_{i+k_2}, y_{i+k_2+1}, \ldots, y_{i+k_1}\}$. To avoid a situation when point has no known neighborhood we add a condition, that all points except the first and last point have at least one neighbor preceding and one following, and if there are no points in the neighborhood we consider then first or last data point as the neighborhood. Then, the truth value is calculated as

$$T = \mu_{most}\left\{ \frac{1}{n}\left[\sum_{i=1}^{n} \mu_{most}(d_i) \right] \right\} \tag{5}$$

$$d_i = \frac{\sum_{j=\max(i-k_1,0)}^{\max(i-k_2,0)} \mu_d\left(y_j - y_i\right) + \sum_{j=\min(i+k_2,n)}^{\min(i+k_1,n)} \mu_d\left(y_i - y_j\right)}{\min(i - k_2, k_1 - k_2) + \min(n - k_2, k_1 - k_2)} \tag{6}$$

where μ_{most} is the membership function of the quantifier *most*. In this formula two different quantifiers may be used, however we decided to keep it simple and use the same quantifier. n is the number of points in the data set. y_i is the value of i-th data point, and we assume that the points are ordered with respect to the time. $\mu_d\left(y_j - y_i\right)$ is the degree to which value of y_j is bigger than value of y_i.

For every point we calculate first the number of preceding points from the neighborhood that are bigger and following that point that are smaller. For this purpose we use the Σ-count. Next we divide this number by the number of considered neighbors ($\min(i - k_2, k_1 - k_2) + \min(n - k_2, k_1 - k_2)$) to normalize it, and this value is denoted as d_i. Then we compute the degree to which this proportion is "most". This value can be understood as the degree to which the point is "in trend" within a given neighborhood. We compute those degrees of being "in trend" for each point, and then we aggregate them using quantifier based aggregation, since we wish to have most points "in trend". Note some similarity with the soft degree of consensus [6,7] that is a degree to which most important experts agree on most important issues.

3 Examples

The method was tested on several artificial data sets as well as on real data. In this section we will present the results shortly.

3.1 Artificial Examples

We generated 6 sets of data, each containing 100 points. We introduced also the noise by adding a randomly chosen value from normal distribution $N(0,1)$.

The first data set is a set of uniformly decreasing points from 100 to 1, They can be described by a formula $y = 100 - x$. Second data set was generated with the equation $y = 100 * e^{\left(\frac{-x}{20}\right)}$ for x=1:100. The third data set contains constant trend. In the fourth case, half of the points are decreasing uniformly, and the next half is increasing to create a "V" pattern. The fifth data set contains points with a decreasing pattern sawtooth, if x is odd $y(x) = y(x-1) - 2$ else $y(x) = y(x-1) + 0.5$. The last, sixth data set can be described as short quickly decreasing trend, that is followed by long slowly increasing trend, and finished again by short quickly increasing trend. All the data sets with added noise are shown in Figure 1.

For those data set we compute the truth value of the sentence "the values are getting smaller" with the two methods described above. Those methods require defining the linguistic quantifier $most$, which is in our case always defined with trapezoidal membership function as $\mu_{most} = Trap[0.3, 0.8, 1, 1]$. We define notions "value a is $bigger$ than b" and "value b is $smaller$ than a" as a fuzzy set of value $a - b$ with the following membership value $\mu_d = Trap[-0.5, 0.5, \infty, \infty]$.

We varied also the size of the neighborhood and used following values of parameters:

- $k_1 = 100$ and k_2=1, 30 or 99
- $k_1 = 30$ and k_2=1, 10 or 29
- $k_1 = 10$ and k_2=1, 5 or 9
- $k_1 = 2$ and $k_2 = 1$

For the data set 1 without and with noise and in all cases the truth value was equal to 1. Same results were also obtained for the data set 2 without noise. When noise was added the truth value was equal to 1, except for the case when $k_1 = 2$ and $k_2 = 1$, in which the truth value was equal to 0.86, so also a high value. In case of data set 3 without noise all values of the truth were equal to 0.2. In case with the noise, all truth values were from range $0.17 - 0.35$, except for one extreme case for $k_1 = 100$ and $k_2 = 99$, in which only first and last point

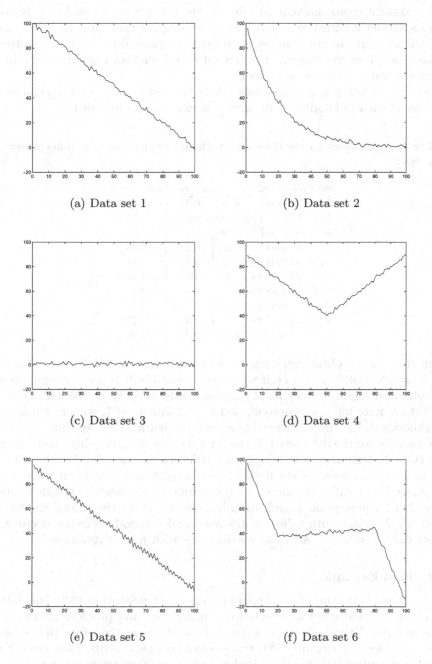

(a) Data set 1

(b) Data set 2

(c) Data set 3

(d) Data set 4

(e) Data set 5

(f) Data set 6

Fig. 1. Artificial data sets used in the example

were taken into consideration. In this case the truth value is equal to 1, because in this dataset it turned out that first value is higher than the last one. In case of data set 4, all the truth values were low from range $0.17 - 0.44$, no matter if noise or no noise was present. For the data set 5 without and with noise in all cases the truth value was equal to 1.

Set 6 is an interesting, tricky case where the results vary a lot depending on the size of the neighborhood. We show the exact values in Tab 1.

Table 1. Truth values for the data set 6 with and without noise for different sizes of neighborhood

neighborhood	no noise	with noise
$k_1 = 100$ and $k_2 = 1$	0.33	0.48
$k_1 = 100$ and $k_2 = 30$	0.61	0.66
$k_1 = 100$ and $k_2 = 99$	1	1
$k_1 = 30$ and $k_2 = 1$	0.29	0.43
$k_1 = 30$ and $k_2 = 10$	0.28	0.41
$k_1 = 30$ and $k_2 = 29$	0.4	0.48
$k_1 = 10$ and $k_2 = 1$	0.3	0.48
$k_1 = 10$ and $k_2 = 5$	0.31	0.56
$k_1 = 10$ and $k_2 = 9$	0.32	0.49
$k_1 = 2$ and $k_2 = 1$	0.52	0.60

In this case we obtain very big value of truth value equal 1 just in one case for $k_1 = 100$ and $k_2 = 99$, i.e if we consider just the first and last data point. We obtain moderate values for 2 neighborhoods when $k_1 = 100$ and $k_2 = 30$, so distant quite big neighborhood, and $k_1 = 2$ and $k_2 = 1$, so very small close neighborhood. For the other neighborhoods the truth values are small.

Generally, for the data sets 1, 2 and 5 we have received very high truth values, therefore the summary "the values are getting smaller" is valid. For the data set 3 and 4 we have received small values of the truth value, and hence the above summary is not valid. In case of set 6 situation is not clear, although in many cases human perception would towards assuming that "the values are getting smaller". Also the truth value for different neighborhoods were not decisive, in many case close to 0.5. All those results agree with human intuition.

3.2   Real Example

The real example comes from Volvo IT Belgium, published in [20]. The data is an event log, that contains events from an incident and problem management system called VINST. In this example we analyze only the sets that contain closed problems. There are 6660 events in the log that describe 1487 cases. Each event is described by several attributes such as problem number, its status and sub-status, time stamp and many others.

In our simple case we are only interested how long the problems were solved. Especially we would like to compute the truth of the sentence "the problems are solved faster".

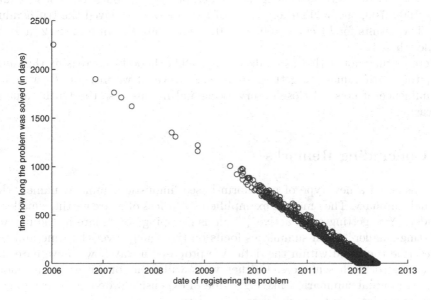

Fig. 2. Volvo data set – time of solving the problem (in days) vs the first entry in the event log of that problem

In the Figure 2 we plot the time of solving the problem vs the first entry in the event log of that problem. From this figure it is clearly that the analyzed summary is true.

Here we defined "value a is solved *faster* than b" as a fuzzy set of value $a - b$ with the following membership value $\mu_d = Trap\,[-5, 5, \infty, \infty]$. We assume that if the difference in time needed to solve a problem is bigger than a week (5 days) then the problem is solved definitelly faster (with membership degree equal 1). This definition was chosen arbitrary. *Most* was defined in the same way as in previous case.

Table 2. Truth values for the Volvo IT data set for different sizes of neighborhood

neighborhood	truth value
$k_1 = 100$ and $k_2 = 1$	0.77
$k_1 = 100$ and $k_2 = 50$	0.86
$k_1 = 100$ and $k_2 = 70$	0.89
$k_1 = 50$ and $k_2 = 1$	0.62
$k_1 = 50$ and $k_2 = 10$	0.65
$k_1 = 50$ and $k_2 = 30$	0.71
$k_1 = 10$ and $k_2 = 1$	0.37
$k_1 = 10$ and $k_2 = 5$	0.41
$k_1 = 5$ and $k_2 = 1$	0.32
$k_1 = 2$ and $k_2 = 1$	0.30

In this case we were also changing the size of the neighborhood. For k_1 equal 2000, 1000, 700, 500, 300 for any value of k_2 we always received the truth value of 1. The results for $k1$ equal 100, 50, 10, 5 and 2 are shown in Tab. 2 and are smaller than 1.

Here we may notice that as we decrease neighborhood by decreasing the value of k_1 the truth values are getting smaller. However if we increase k_2 we limit the influence of noise of close observations and in this case the truth value is increasing.

4 Concluding Remarks

We considered a new type of protoform-based linguistic summary, namely the gradual summary. They may be exemplified by 'prices of X are getting smaller", "resident Y is getting more active", "sale is dropping" or "share in the market is getting smaller". Such summaries focus on the change over the time and try to describe the trend within the data. We proposed a way how to evaluate the truth value of such sentences. Further work will focus on investigating more complex gradual summaries, like e.g. "sale of expensive products is dropping", where *expensive* may be modeled as a fuzzy set.

References

1. Bosc, P., Dubois, D., Pivert, O., Prade, H., Calmes, M.D.: Fuzzy summarization of data using fuzzy cardinalities. In: Proceedings of the IPMU 2002 Conference, pp. 1553–1559 (2002)
2. Castillo-Ortega, R., Marín, N., Sánchez, D.: Time series comparison using linguistic fuzzy techniques. In: Hüllermeier, E., Kruse, R., Hoffmann, F. (eds.) IPMU 2010. LNCS, vol. 6178, pp. 330–339. Springer, Heidelberg (2010)
3. Castillo-Ortega, R., Marín, N., Sánchez, D.: Linguistic local change comparison of time series. In: 2011 IEEE International Conference on Fuzzy Systems (FUZZ), pp. 2909–2915 (June 2011)
4. Dubois, D., Prade, H.: Gradual rules in approximate reasoning. Information Sciences 61, 103–122 (1992)
5. Kacprzyk, J.: Intelligent data analysis via linguistic data summaries: a fuzzy logic approach. In: Decker, R., Gaul, W. (eds.) Classification and Information Processing at the Turn of Millennium, pp. 153–161. Springer, Heidelberg (2000)
6. Kacprzyk, J., Fedrizzi, M.: "Soft" consensus measures for monitoring real consensus reaching processes under fuzzy preferences. Control and Cybernetics 15, 309–323 (1986)
7. Kacprzyk, J., Fedrizzi, M.: A 'human-consistent' degree of consensus based on fuzzy logic with linguistic quantifiers. Mathematical Social Sciences 18, 275–290 (1989)
8. Kacprzyk, J., Strykowski, P.: Linguistic data summaries for intelligent decision support. In: Felix, R. (ed.) Proceedings of EFDAN 1999-4th European Workshop on Fuzzy Decision Analysis and Recognition Technology for Management, pp. 3–12 (1999)

9. Kacprzyk, J., Strykowski, P.: Linguistic summaries of sales data at a computer retailer: a case study. In: Proceedings of IFSA 1999, vol. 1, pp. 29–33 (1999)

10. Kacprzyk, J., Wilbik, A., Zadrożny, S.: Linguistic summarization of time series using a fuzzy quantifier driven aggregation. Fuzzy Sets and Systems 159(12), 1485–1499 (2008)

11. Kacprzyk, J., Wilbik, A., Zadrożny, S.: An approach to the linguistic summarization of time series using a fuzzy quantifier driven aggregation. International Journal of Intelligent Systems 25(5), 411–439 (2010)

12. Kacprzyk, J., Yager, R.R.: Linguistic summaries of data using fuzzy logic. International Journal of General Systems 30, 33–154 (2001)

13. Kacprzyk, J., Yager, R.R., Zadrożny, S.: A fuzzy logic based approach to linguistic summaries of databases. International Journal of Applied Mathematics and Computer Science 10, 813–834 (2000)

14. Kacprzyk, J., Zadrożny, S.: Fuzzy linguistic data summaries as a human consistent, user adaptable solution to data mining. In: Gabrys, B., Leiviska, K., Strackeljan, J. (eds.) Do Smart Adaptive Systems Exist? STUDFUZZ, vol. 173, pp. 321–340. Springer, New York (2005)

15. Kacprzyk, J., Zadrożny, S.: Linguistic database summaries and their protoforms: toward natural language based knowledge discovery tools. Information Sciences 173, 281–304 (2005)

16. Raschia, G., Mouaddib, N.: SAINTETIQ: a fuzzy set-based approach to database summarization. Fuzzy Sets and Systems 129, 137–162 (2002)

17. Rasmussen, D., Yager, R.R.: Finding fuzzy and gradual functional dependencies with SummarySQL. Fuzzy Sets and Systems 106, 131–142 (1999)

18. Ros, M., Pegalajar, M., Delgado, M., Vila, A., Anderson, D.T., Keller, J.M., Popescu, M.: Linguistic summarization of long-term trends for understanding change in human behavior. In: Proceedings of the IEEE International Conference on Fuzzy Systems, FUZZ-IEEE 2011, pp. 2080–2087 (2011)

19. Szczepaniak, P., Ochelska, J.: Linguistic summaries of standardized documents. In: Last, M., Szczepaniak, P.S., Volkovich, Z., Kandel, A. (eds.) Advances in Web Intelligence and Data Mining. SCI, vol. 23, pp. 221–232. Springer, Heidelberg (2006)

20. doi:10.4121/c2c3b154-ab26-4b31-a0e8-8f2350ddac11

21. Wilbik, A., Keller, J.M., Bezdek, J.C.: Linguistic prototypes for data from eldercare residents. IEEE Transactions on Fuzzy Systems (2013) (in press)

22. Wilbik, A., Keller, J.M.: A distance metric for a space of linguistic summaries. Fuzzy Sets and Systems 208, 79–94 (2012)

23. Wilbik, A., Keller, J.M., Alexander, G.L.: Linguistic summarization of sensor data for eldercare. In: Proceedings of the IEEE International Conference on Systems, Man, and Cybernetics (SMC 2011), pp. 2595–2599 (2011)

24. Yager, R.R.: A new approach to the summarization of data. Information Sciences 28, 69–86 (1982)

25. Zadeh, L.A.: Toward a theory of fuzzy information granulation and its centrality in human reasoning and fuzzy logic. Fuzzy Sets and Systems 9(2), 111–127 (1983)

26. Zadeh, L.A.: A prototype-centered approach to adding deduction capabilities to search engines – the concept of a protoform. In: Proceedings of the Annual Meeting of the North American Fuzzy Information Processing Society (NAFIPS 2002), pp. 523–525 (2002)

Mining Epidemiological Dengue Fever Data from Brazil: A Gradual Pattern Based Geographical Information System

Yogi Satrya Aryadinata[1], Yuan Lin[2], C. Barcellos[3],
Anne Laurent[1], and Therese Libourel[2]

[1] LIRMM, Montpellier, France
{aryadinata, laurent}@lirmm.fr
[2] UMR ESPACE-DEV (IRD-UM2), Montpellier, France
therese.libourel@univ-montp2.fr, yuan.lin@ird.fr
[3] Fundaćõ Oswaldo Cruz, Rio de Janeiro, Brazil
xris@fiocruz.br

Abstract. Dengue fever is the world's fastest growing vector-borne disease. Studying such data aims at better understanding the behaviour of this disease to prevent the dengue propagation. For instance, it may be the case that the number of cases of dengue fever in cities depends on many factors, such as climate conditions, density, sanitary conditions. Experts are interested in using geographical information systems in order to visualize knowledge on maps. For this purpose, we propose to build maps based on gradual patterns. Such maps provide a solution for visualizing for instance the cities that follow or not gradual patterns.

Keywords: Epidemiological Data, Data Mining, Geographic Information Systems, Gradual Patterns.

1 Introduction

There are approximately 50 millions new dengue cases each year, and approximately 2.5 billion people live in endemic countries [1] located in the tropical zone between the latitudes of 35° N and 35° [2]. The vector for dengue infection is Aedes aegypti, which has a strictly synanthropic lifestyle [3]. The proliferation of these mosquitoes is supported by both weather patterns and the contemporary style of human life in large cities, where large amounts of water are deposited in the environment and become potential breeding grounds for mosquito reproduction [4]. These factors have been shown to influence the occurrence and spread of dengue infection over the last 50 years.

The incidence of dengue infection has a characteristic seasonal movement in almost all regions of the world, where periods of high transmissibility are experienced during certain months of the year. This phenomenon has been explained by the close relationship between the density variation in winged forms of the vector and climatic conditions [5, 6], such as rainfall, temperature and relative humidity.

Since 1986, Brazil has been affected by dengue epidemics that have reached dramatic proportions. This country of continental dimensions has a wide territorial range of tropical and subtropical climates (hot and humid), with an average annual temperature above

A. Laurent et al. (Eds.): IPMU 2014, Part II, CCIS 443, pp. 414–423, 2014.

20°C and rainfall exceeding 1,000 mm per year. These characteristics provide suitable abiotic conditions for the survival of Aedes aegypti [7], the main dengue vector present in the Americas.

From 1990 to 2010, 5.98 million cases of dengue were reported, and autochthonous cases have been recorded in 80% of the 5,565 Brazilian municipalities. The period of greatest risk for dengue occurrence has been shown to be during or immediately following the rainy season [5,6], and there is a reduced incidence during the remaining months of the year. However, epidemiological studies on the relationship between dengue infection and climate variables in Brazil are scarce.

Studies in wide (national) geographical scales and considering the interactions between spatial and time are still scarce in dengue literature. [3] revealed rapid travelling waves of DHF crossing Thailand emanating from Bangkok every 3 years. Inversely, in Cambodia seasonal propagation waves are originated in poor rural areas being their propagation conditioned by road traffic [8]. In Peru, dengue spatial and temporal dynamics was influenced by the different sociodemographic and environmental among eco-regions [9]. The recent spread of dengue in Brazil is equally related to human mobility across cities network and leaving remote country regions relatively protected [10]. However, unlike contagious diseases, dengue transmission is constrained the environmental substrate on which vector must reproduce and infect people. Thus, the presence and abundance of vector are necessary but not sufficient condition to dengue transmission.

Climate and environmental changes may exacerbate the present distribution of vector borne diseases as well as extend transmission to new niches and populations [11]. Both trends underline the role of health surveillance systems in order to detect and conduct preventive actions in unusual transmission contexts. Climate changes affect populations in different ways and intensities according to the vulnerability of social groups, which is associated to their insertion in place and society. Spatial analysis offers important tools to describe measure and monitor health impact on vulnerable populations under possible scenarios. Brazilian territory presents a wide variety of temperature ranges and rainfall regimes. In addition to climatic variations, unequal urban infrastructure among cities and differential territory occupation patterns increases the complexity of dengue nationwide dynamics.

The important questions arising from experts debate and studies are :

1. What factors are determinants to explain dengue distribution? Which are the most important? Climate? Sanitary conditions? Human mobility?
2. Which years are typical and regular (the same patterns appear along all years)? And which are abnormal (patterns are different for one atypical year)? For example, extreme climate events, el Niño ?
3. Mapping the patterns. Are patterns concentrated (spatial clusters)? Where are located the different (and contradictory) patterns?
4. Is this spatial pattern related to other geographical features? (relief, ecosystems, roads, rivers, urban regions, etc.).

In this paper, we propose a method using a gradual pattern mining to analyze and to discover the patterns of behavior in dengue fever cases in Brazil. This method also allows us to produce a gradual map that can directly be used to see the behavior of the cases of dengue fever from the geographic approach.

In Section 2, we introduce the method to find the gradual patterns, which can be used to create binary and gradual map that described in Section 3. Section 4 presents the data and indicators in terms of analysis and how to produce the binary map (Section 4.2) and the gradual map (Section 4.3). Section 5 concludes the paper and gives the perspectives of our research.

2 Gradual Patterns

Gradual pattern mining has been recently introduced as the topic addressing the automatic discovery of gradual patterns from large databases. Such databases are structured over several attributes which domains are totally ordered, considering a relation $\leq$. For instance Table 1 reports an example of such a database.

We consider a toy database containing the information about a disease taken on five cities. Each tuple from the table 1 corresponds to a city, and reports the number of cases for this disease (last column) studied by respect with the number of inhabitants from the city (in thousands), the average humidity (in percentage), and the average income (in K euros).

Table 1. Database D describing a toy example for a disease in 5 cities

Id	Nb Inhabitants ($\times 1000$)	Humidity	Income	Nb Cases ($\times 1000$)
C_1	110	53	30	10
C_2	202	71	61	28
C_3	192	64	62	43
C_4	233	83	81	41
C_5	225	75	73	39

An example of gradual patterns is *The higher the number of inhabitants, the higher the degree of Humidity, the higher the number of cases of the disease.*

We remind below some concepts of the literature on gradual patterns.

Definition 1. *Gradual item. A gradual item is a pair (i, v) where i is an item and v is variation $v \in \{\uparrow, \downarrow\}$. $\uparrow$ stands for an increasing variation while $\downarrow$ stands for a decreasing variation.*

Example 1. $(NbInhabitants, \uparrow)$ is a graduel item.

Definition 2. *Gradual Pattern (also known as Gradual Itemset). A gradual pattern is a set of gradual items, denoted by $GP = \{(i_1, v_1), \ldots, (i_n, v_n)\}$. The set of all gradual patterns that can be defined is denoted by GP.*

Example 2. $\{(NbInhabitants, \uparrow), (Humidity, \uparrow), (NbCases, \uparrow)\}$ is a gradual itemset.

Gradual pattern mining aims at extracting the frequent patterns, as in the classical data mining framework for itemsets and association rules.

Definition 3. *Given a threshold of a minimum support σ, a gradual pattern GP is said to be frequent if $supp(GP) \geq \sigma$.*

For describing what *frequent* means in the context of gradual patterns, several supports have been proposed in the literature. All these materials are based on the idea that it takes the number / proportion of transactions in the database (e.g., the number / proportion of cities in our example) that respect the pattern. For being counted, a transaction must behave adequately with respect to other cities. For example, in Table 2, we see that the number of cases and the number of inhabitants increase together for cities 1 and 2, as $110 < 202$ and $10 < 28$. If the variation is decreasing ($\downarrow$) then the numbers must follow it. For instance for cities 3 and 4, the number of inhabitants increases $((Inhabitants, \uparrow))$ and the number of cases $((cases, \downarrow))$ decreases as $192 < 233$ and $43 > 41$.

One of the support proposed in the literature [12] is based on the length of the longest path of cities that can be built using this idea.

Definition 4. *The support of gradual pattern GP is given by the following formula : $support(GP) = \frac{max_{L \in l}(|L|)}{(|R|)}$, where L is the set of rows the support of a gradual pattern GP and R is the set of rows of the database D.*

For determining the longest path, we build the precedence graph for the pattern being considered (Fig. 1). It can be the case that several paths can be built for a pattern, as shown below when trying to order the cities with respect with the pattern P $\{(NbInhabitants, \uparrow), (Humidity, \uparrow), (NbCases, \uparrow)\}$. Two orderings are possible: $L1$ and $L2$.

Table 2. Two list obtained of $\{(NbInhabitants, \uparrow), (Humidity, \uparrow), (NbCases, \uparrow)\}$: $L1$ and $L2$ (from left to right)

Id	Nb Inhabitants	Humidity	Nb Cases	Id	Nb Inhabitants	Humidity	Nb Cases
C_1	110	53	10	C_1	110	53	10
C_2	202	71	28	C_2	202	71	28
C_4	233	83	41	C_5	225	75	39

Precedence graphs can also be represented in the form of binary matrices as shown in Table 3 for the pattern $\{(NbInhabitants, \uparrow), (Humidity, \uparrow), (NbCases, \uparrow)\}$. C_2 precedes C_4 and C_5 (value 1 of the matrix), but not C_1 and C_3 (value 0 of the matrix). We have here: support $(\{(NbInhabitants, \uparrow), (Humidity, \uparrow), (NbCases, \uparrow)\}) = \frac{3}{5}$ as pattern P is taken by the maximum list of cities $< C_1, C_2, C_4 >$ and $< C_1, C_2, C_5 >$.

For representing a pattern on the whole database, we consider precedence graphs, as shown by Fig. for the pattern $\{(NbInhabitants, \uparrow), (Humidity, \uparrow), (NbCases, \uparrow)\}$. Such a graph can be represented in a binary matrix, which allows to optimize the

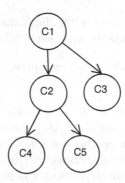

Fig. 1. Precedence graph associated to the pattern $\{(NbInhabitants, \uparrow), (Humidity, \uparrow), (NbCases, \uparrow)\}$

Table 3. Binary matrix associated to the pattern $\{(NbInhabitants, \uparrow), (Humidity, \uparrow), (NbCases, \uparrow)\}$

City	C_1	C_2	C_3	C_4	C_5
C_1	1	1	1	1	1
C_2	0	1	0	1	1
C_3	0	0	1	0	0
C_4	0	0	0	1	0
C_5	0	0	0	0	1

computations. For instance, there are two longest paths in this example: $< C_1, C_2, C_4 >$ and $< C_1, C_2, C_5 >$. The support is thus equal to $\frac{3}{5}$.

In our work, we aim at displaying such gradual patterns computing from geographical data on maps. For this purpose, we propose to display cities in a form (e.g., shade intensity or size) that depends on whether it contributes or not to some gradual pattern. For instance, if the city 1 does not behave as all the other cities for pattern $(Humidity, \uparrow), (NbCases, \uparrow)$ then it will be displayed as squares, while the other cities will be represented as a triangle.

3   Building Binary and Gradual Maps

In this section, we present novel methods for visualizing gradual patterns on maps. We propose two methods of visualization. Both these methods rely on the calculation of the support with the longest path in the precedence graph (Fig. 4).

Our idea is to produce maps starting from the extraction of gradual patterns. We will explain this part with a simple example of a region that contains five cities (C_1, C_2, C_3, C_4, C_5). Based on data from these cities, we believe we have an interesting pattern that we can apply in a map. We then want to know, for each city, how it contributes or not to identify spatial pattern. Then we can represent this information to the user on the maps which he is accustomed. The principle is that each pattern corresponds to a layer

Table 4. The list of longest paths associated to Fig. 1

List	Length
C_1, C_3	2
C_1, C_2, C_4	3
C_1, C_2, C_5	3

in the GIS. Cities are then represented differently depending on their contribution to the cause, i.e., a dichotomous variable.

We consider two approaches. The first approach, called "binary map" is to represent by a form (e.g., triangle) the cities that contribute to maximum path length on the pattern, and with other form (e.g. square) the other cities. The second approach is to represent cities by the intensity of the shade that are in the longest path and less intense shade for cities that are in the shortest path. This is called "gradual map", i.e., a continuous variable.

3.1 Binary Map

In this binary map, we identify items through its participation in the gradual pattern (between 0 and 1) we use the following steps:

1. Calculating each item of the support in line with the binary matrix
2. Extract lines that have the maximum support
3. Identification (triangle (1)) cities that are included in the set of itemset length of the maximum path.
4. Identification (square (0)) of the cities that are not in the set of itemset length of the maximum path.

By looking at Figure 1, we know the cities which respect the gradual interesting pattern. For example, Figure 2 shows that the city C_3 as a square, do not fully respect the gradual pattern. In contrast, other cities are triangles because they are in the longest path.

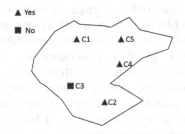 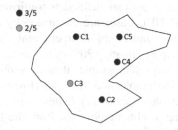

Fig. 2. Binary Map   **Fig. 3.** Gradual Map

3.2 Gradual Map

In order to realize a more detailed maps of the binary map. To do this, we propose
a *gradual* map to visualize the value of the support of each item which corresponds
to a proportion of belonging to a support item. To realize this map, we consider the
following steps:

1. Calculate the support of each item per line in the binary matrix
2. Take the maximal support size
3. Calculate the intensity of each item in order to identify the importance of which
   item depends on its intensity value

$$Intensity(v) = \frac{The\ length\ of\ support\ of\ v}{The\ maximum\ path\ length}$$

4. From the intensity value make the classification using the shade intensity as indicator on the map.

With this map, we can identify the cities that are more important than others concerning the classification defined in the figure. Thus, the city be more or less *illuminated* depending on the length of the maximum path in which it appears. For example,
in Figure 3, the cities C_1, C_4 and C_5 are the most illuminated because they belong to
the maximum path length.

4 Experimentation

Therefore, in our experimentation we use our methods with the dengue cases in Brazil.
Firstly, we analyse the data sources and indicators that described in Section 4.1 and
Section 4.2 and 4.3 to build the binary and gradual maps.

4.1 Data Sources and Indicators

Dengue fever (DF) notifications from 2001 to 2012 were summarized by year of
symptoms upset and municipality of residence. Data were obtained from the Notifi-
able Diseases Information System (SINAN acronym in Portuguese), organized by the
Brazilian Ministry of Health and freely available in Health Information Department
(Datasus). Cases are defined as confirmed Dengue Fever (DF) or Dengue Hemorrhagic
Fever (DHF). Cases are confirmed by clinical and laboratory according to standard
procedures and submitted to epidemiological investigation by local health surveillance
teams. Approximately 30% of the cases of dengue are also laboratory-confirmed.

The socio-demographic data were obtained from the website of the Brazilian Institute of Geography and Statistics (Instituto Brasileiro de Geografia e Estatstica IBGE;
http://mapas.ibge.gov.br). Cities were categorized according to the climate classification map of climate obtained from the Brazilian Institute of Geography and Statistics
(IBGE). There are four types of variables on this data: Temperature regime was classified in three categories, Rainfall regime was categorized into four classes, Humidity was
categorized into four classes, Sanitary conditions were summarized by the combination
of three variables, Mobility was evaluated by means of two variables.

Individual variables were ranked and the result of summing the ranks was used to categorize into four classes: Very high, High, Medium and Low. All indicators were geocoded and mapped using the coordinates of the city as the center of gravity on the common 5506 existing in 2010. This position was used to create maps and assign information from other layers, such as climate, in a geographic information system (GIS).

During the extraction process of gradual patterns, we cleaned the data so that they do not contain any missing values, false values, etc. In Section 2, we introduced methods for research on dengue epidemic. Then, we will apply these methods in order to obtain better results.

For simplicity, we choose 25 cities in the state of Rio de Janeiro, Brazil. The State of Rio de Janeiro (RJ) is located east of the southeast region. The capital is Rio de Janeiro. This state has an area of 43 909 km^2 with about 14,367,000 mainly concentrated along the coast. We choose the state of Rio de Janeiro because of the high frequency of dengue outbreaks in the region. In addition, this region presents a wide morphological diversity (mountains, beaches, dunes, lagoons, etc..). In general, it is divided into three major geographical subregions: the metropolitan lowlands (often called Baixada Fluminense), coastal elevations and northern lowlands. The climate is tropical and the average annual temperature is 23 °C. With this data, we extract the gradual patterns with the method of extraction of conventional gradual patterns on the Section 2.

Table 5 displays some of the interesting gradual patterns in the case of epidemic dengue. After looking at the Table 5 and patterns found, we can infer that the climate (the drought, temperature, humidity) plays the most important role in the case of the dengue epidemic, followed by the level of mobility and sanitation state level.

Table 5. Example of extracted gradual patterns (Support = 0.25)

Motifs	Support	Longest Path
$[Temp \uparrow NbCases \uparrow]$	19	(330450 330550 330320 330270 330555 330510 330250 330370 330430 330455 330190 330414 330490 330227 330600 330200 330220 330330)
$[Drought \uparrow Humid \downarrow NbCases \downarrow]$	15	(330330 330200 330227 330030 330490 330190 330430 330250 330510 330180 330610 330280 330320 330550 330450)
$[Drought \uparrow Temp \downarrow Humid \downarrow NbCases \downarrow]$	15	(330330 330200 330227 330414 330190 330430 330250)
$[Mobility \uparrow NbCases \downarrow]$	9	(330360 330220 330600 330490 330455 330250 330510 330555 330270)
$[Temp \uparrow Mobility \downarrow NbCases \downarrow]$	7	(330330 330190 330600 330227 330370 330320 330550)

Finally, we can make a binary map and a gradual map that take into account the gradual patterns on the dengue epidemic in Brazil in Fig. 4 and Fig. 5 for the pattern $[Temp \uparrow Mobility \uparrow NbCases \downarrow]$.

4.2 Binary Map

Fig. 4 show that the most of the cities have high dengue fever cases, followed to the pattern $(Temperature, \uparrow)$, $(Mobility, \downarrow)$, $(NbCases, \downarrow)$, displayed as triangles.

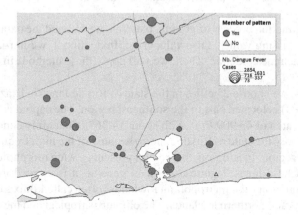

Fig. 4. Binary Map $(Temperature, \uparrow)$, $(Mobility, \downarrow)$, $(NbCases, \downarrow)$

4.3 Gradual Map

In Fig. 5, we present the support level $(0-1)$ of the cities using the intensity color. This map shows more detail information than binary map,Which cities are more related to the pattern $(Temperature, \uparrow)$, $(Mobility, \downarrow)$, $(NbCases, \downarrow)$. We can see that important emerging epidemic dengue fever mostly appeared in cities located around the coast (shown with the green intensity).

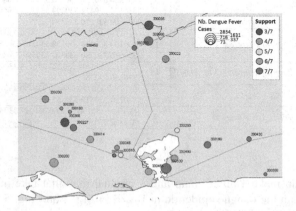

Fig. 5. Gradual Map $(Temperature, \uparrow)$, $(Mobility, \downarrow)$, $(NbCases, \downarrow)$

The difficulty of this step is the production of maps from found gradual patterns, but it is important to retrieve all the items in each pattern corresponding to the length of maximum path.

5 Conclusion

In this article, we study how gradual patterns can help to produce maps in geographical information systems. We apply our method in the context of dengue epidemics in Brazil.

Our main perspectives are to merge more criteria for building such maps and to study how a large volume of gradual maps can be displayed to end-users in a user-friendly way by using the so-called layers in geographical information system tools.

References

1. Organization, W.H.: The World Health Report 2006: Working together for health (2006)
2. Gubler, D., Ooi, E., Vasudevan, S., Farrar, J.: Dengue and Dengue Hemorrhagic Fever. CABI (2013)
3. Gubler, D.J., et al.: Dengue, urbanization and globalization: The unholy trinity of the 21st century. Tropical Medicine and Health 39(4 suppl.), 3–11 (2011)
4. Kovats, R., Campbell-Lendrum, D., McMichael, A., Woodward, A., Cox, J.: Early effects of climate change: do they include changes in vector-borne disease? Philos. Trans. R. Soc. Lond. B. Biol. Sci. 356(1411), 1057–1068 (2001)
5. Souza-Santos, R.: The factors associated with the occurrence of immature forms of aedes aegypti in Ilha do Governador, Rio de Janeiro, Brazil. Rev. Soc. Bras. Med. Trop. 32(4), 373–382 (1999)
6. Souza-Santos, R., Carvalho, M.: Spatial analysis of aedes aegypti larval distribution in the Ilha do Governador neighborhood of Rio de Janeiro, Brazil. Cad. Saude Publica 16(1) (2000)
7. Yang, H.M., Macoris, M.L., Galvani, K.C., Andrighetti, M.T., Wanderley, D.M.: Dinâmica da transmissao da dengue com dados entomológicos temperatura-dependentes. Tema–Tend. Mat. Apl. Comput. 8(1), 159 (2007)
8. Teurlai, M., Huy, R., Cazelles, B., Duboz, R., Baehr, C., Vong, S.: Can human movements explain heterogeneous propagation of dengue fever in Cambodia? PLoS Negl. Trop. Dis. 6(12), e1957 (2012)
9. Chowell, G., Cazelles, B., Broutin, H., Munayco, C.V.: The influence of geographic and climate factors on the timing of dengue epidemics in Peru, 1994-2008. BMC Infectious Diseases 11, 164 (2011)
10. Catão, R.D.C., Guimarães, R.B.: Mapeamento da reemergência do dengue no Brasil– 1981/82-2008. Hygeia 7(13) (2011)
11. McMichael, A., Lindgren, E.: Climate change: present and future risks to health, and necessary responses. J. Intern. Med. 270(5), 401–413 (2011)
12. Di-Jorio, L., Laurent, A., Teisseire, M.: Mining frequent gradual itemsets from large databases. In: Adams, N.M., Robardet, C., Siebes, A., Boulicaut, J.-F. (eds.) IDA 2009. LNCS, vol. 5772, pp. 297–308. Springer, Heidelberg (2009)

A New Model of Efficiency-Oriented Group Decision and Consensus Reaching Support in a Fuzzy Environment

Dominika Gołuńska[1,2,*], Janusz Kacprzyk[2], and Sławomir Zadrożny[2]

[1] Department of Automatic Control and Information Technology,
Cracow University of Technology, ul. Warszawska 24, 31-155 Cracow, Poland
[2] Systems Research Institute, Polish Academy of Sciences
ul. Newelska 6, 01-447 Warsaw, Poland
dominika.golunska@pk.edu.pl
kacprzyk@ibspan.waw.pl

Abstract. We present a novel comprehensive model of a consensus reaching support system in the fuzzy context. We assume the individual fuzzy preferences, a fuzzy majority in group decision making, as proposed by Kacprzyk [9], some fuzzy majority based solution concepts in group decision making, notably fuzzy cores (cf. Kacprzyk [9]) and their choice function based representations by Kacprzyk and Zadrożny [15],[16], a soft degree of consensus by Kacprzyk and Fedrizzi [10],[11]. Using as a point of departure Kacprzyk and Zadrożny's [18] approach of the use of linguistic data summaries to support the running of a consensus reaching process, we develop and implement a novel approach that synergistically combines the tools and techniques mentioned above. We assume that moderated consensus reaching process which is run in the group of agents by a special agent called a moderator, is the most effective and efficient solution. We attempt to facilitate the work of a moderator, by some useful guidelines and additional indicators. We extend this idea and finally, we present a new implementation followed by a numerical evaluation of the new model proposed.

Keywords: consensus reaching, group decision support systems, fuzzy preference relations, soft degree of consensus, linguistic quantifier, fuzzy cores.

1   Introduction

Decision making processes usually lead to better decisions if are run in a goal-directed way [6], and in a collective decision making setting to help reach a decision better reflecting the opinion of the entire group, which is normally easier and more effective and efficient if the agents involved arrive first at a consensus.

The problem considered in this article concerns a consensus reaching process, and its related group decision making process, in a (small) group of autonomous agents

* Dominika Gołunska's work was supported by the Foundation for Polish Science under Inter national PhD Projects in Intelligent Computing financed from The European Union within the Innovative Economy Operational Programme 2007-2013 and European Regional Development Fund.

A. Laurent et al. (Eds.): IPMU 2014, Part II, CCIS 443, pp. 424–433, 2014.
© Springer International Publishing Switzerland 2014

(individuals, decision makers, experts, …) who present their testimonies as fuzzy (graded) preference relation defined on a set of options that usually significantly differ in the beginning. The consensus reaching process boils down to step-by-step changes of the testimonies of the particular agents until they become close enough, i.e., until the group arrives at a sufficient agreement to expressed through a degree of consensus assumed here due to Kacprzyk and Fedrizzi [10,11]. The willingness to change the testimonies by agents is clearly a prerequisite for consensus reaching. Recent developments in IT/ICT tools and techniques, notably in the area of human computer interfaces (HCI), have made possible to use for this the decision support systems (DSSs) [25].

We are concerned with consensus reaching in a group of agents and our purpose is to develop and implement a novel model using modern tools for the *manipulation of testimonies*. We assume a moderated consensus reaching process run by a "superagent" called a *moderator*. This process may be substantially enhanced and accelerated by employing modern concepts and techniques (Kacprzyk and Zadrożny [20],[21]]) based on the use of additional information and insight into which agents and their pairs, options, etc. are critical by having a considerable influence on a degree of consensus, and should be dealt with by the moderator, e.g., by trying to persuade to change the testimonies of specified agents with respect to pairs of specified options, etc.

This general approach boils down to a clearly efficiency-oriented scenario in which usually only a few of the most promising agents and options are taken into account as this may imply the best increase of the degree of consensus.

2   A General Framework for a Moderated Consensus Reaching

We briefly present a general framework for the consensus reaching support proposed by Fedrizzi, Kacprzyk and Zadrożny [5] in which the process is moderator-run/assisted, with a special "superagent", a *moderator*, being explicitly involved, as shown in Fig. 1.

Basically, this is an interactive and iterative process, cf. [2], with a special role of a moderator who constantly measures distances (dissimilarities) between agents and checks whether (a proper degree of) consensus is reached or not, and supports the *discussion* by stimulating the exchange of information, suggesting arguments (pros and cons), convincing appropriate changes of preferences, focusing the discussion on issues which may resolve conflicts of opinions within the group, etc. This is repeated until the group arrives at a state of sufficient agreement or until we reach some time limit [11].

3   A Model of Consensus Reaching in a Fuzzy Environment

The first basic element of the proposed approach is the use of some fuzzy logic-based representations of preferences. We assume a finite set of $m \geq 2$ agents (individuals, experts), $E = \{e_1, e_2, ..., e_m\}$, and a set of $n \geq 2$ options (alternatives, issues), $S = \{s_1, s_2, ..., s_n\}$. Each agent $e_k \in E$ expresses his or her testimonies as to the particular pairs of options from S which are assumed to be individual fuzzy preference relation, R_k, defined over the set of options S (that is, in $S \times S$) [9].

An individual   fuzzy preference relation of agent e_k, R_k, is given by its membership function $\mu_{R_k} : S \times S \rightarrow [0,1]$. Namely, $\mu_{R_k}(s_i, s_j) > 0.5$ indicates the preference degree of option s_i over option s_j. $\mu_{R_k}(s_i, s_j) < 0.5$ denotes the preference degree of s_j over s_i. Finally, $\mu_{R_k}(s_i, s_j) = 0.5$ determines the indifference between s_i and s_j.

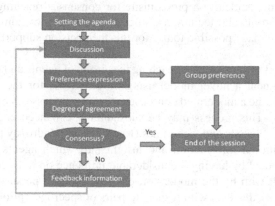

Fig. 1. A general scheme of supporting consensus reaching run by the moderator [14]

The relevant element of our consensus reaching model is the use of linguistic quantifiers (using Zadeh's calculus [26]) which has been proposed by Kacprzyk [9] as a model of a fuzzy majority in either a soft measure of consensus or a group decision making solution (cf. Kacprzyk [9], Kacprzyk and Fedrizzi [10]. Kacprzyk and Zadrożny [15-16]). Linguistic quantifiers are represented by terms like most, much more than a half, almost all, etc. and represented by fuzzy sets in [0, 1]. Basically, we will use such relative fuzzy quantifiers in order to represent the concept of a fuzzy majority.

A linguistically quantified statement, exemplified by "most individuals are satisfied", can be written as

$$Q y's \, are \, F \tag{1}$$

where Q is a linguistic quantifier (e.g., most), $Y = \{y\}$ is a set of objects (e.g., agents) and F is a property (e.g., satisfied).

The degree of truth of the linguistically quantified statement (1) is determined using Zadeh's calculus, with a fuzzy linguistic quantifier Q assumed to be a fuzzy set in [0,1]. For instance, $Q = "most"$ may be defined as

$$\mu_{"most"}(x) = \begin{cases} 1 & for \quad x > 0.8 \\ 2x - 0.6 & for \, 0.3 \leq x \leq 0.8 \\ 0 & for \quad x < 0.3. \end{cases} \tag{2}$$

Property F is defined as a fuzzy set in the set of objects Y, and if $Y = \{y_1,..., y_p\}$, then the truth value of $y_i \, is \, F$ is identified with $\mu_F(y_i)$ [10]. Now, we have:

$$r = \frac{1}{p} \sum_{i=1}^{p} \mu_F(y_i) \tag{3}$$

$$truth(Qy's \ are \ F) = \mu_Q(r). \tag{4}$$

We adopt the concept of a soft measure (degree) of consensus (Kacprzyk and Fedrizzi [10],[11]), that is, the degree to which: "most of agents agree with their preferences to most of the options". A classic concept of consensus is when "all the agents agree with their preferences to all of the options". However, this "full and unanimous agreement" is unrealistic in practice, because agents usually reveal significant differences in their viewpoints, flexibility, aversion to change opinions, etc. [5].

The soft degree of consensus is now derived in three steps [6]:

1) for each pair of agents we calculate a degree of agreement as to their preferences between all the pairs of options,

2) we aggregate these degrees to derive a degree of agreement of each pair of individuals as to their preferences between Q_1 (e.g., "most") pairs of options,

3) we combine these degrees to obtain a degree of agreement of Q_2 (a linguistic quantifier similar to Q_1) pairs of individuals as to their preferences between Q_1 pairs of options and this is meant to be the degree of consensus.

We start with the degree of a strict agreement between agents e_{k1} and e_{k2} as to their preferences between options s_i and s_j :

$$v_{ij}(k_1, k_2) = \begin{cases} 1 & if \quad r_{ij}^{k_1} = r_{ij}^{k_2} \\ 0 & otherwise \end{cases} \tag{5}$$

where, $k_1 = 1,...,m-1$, $k_2 = k_1 + 1,...,m$, $i = 1,...,n-1$, $j = i+1,...,n$.

The relevance of options is given by a fuzzy set over the set of options, B , such that $\mu_B(s_i) \in [0,1]$ is a degree of relevance of option s_i, from 0 for fully irrelevant to 1 for fully relevant, through all intermediate values. The relevance b_{ij} of each pair of options $(s_i, s_j) \in S \times S$, may be defined, as

$$b_{ij}^B = \frac{1}{2}[\mu_B(s_i) + \mu_B(s_j)] \tag{6}$$

for each i, j , where $i \neq j$. Evidently $b_{ij}^B = b_{ji}^B$, for each i, j ; for simplicity, the relevance of options is not accounted for, ie.e. $\mu_B(s_i) = 1$ in (6), for each $s_i \in S$.

Then, the degree of agreement between agents k_1 and k_2 as to their preferences between all the pairs of options is (* stands for a t-norm):

$$v(k_1,k_2,S) = \sum_{i=1}^{n-1} \sum_{j=i+1}^{n} v_{ij}(k_1,k_2) * b_{ij}^{B} \left/ \sum_{i=1}^{n-1} \sum_{j=i=1}^{n} b_{ij}^{B} \right. \qquad (7)$$

Next, the degree of agreement between agents k_1 and k_2 as to their preferences between Q_1 pairs of options is:

$$v_{Q_1}(k_1,k_2,S) = \mu_{Q_1}(v(k_1,k_2,S)). \qquad (8)$$

The degree of agreement of all pairs of agents as to their preferences between Q_1 pairs of options is:

$$v_{Q_1}(E,S) = \frac{2}{m(m-1)} \sum_{k_1=1}^{m-1} \sum_{k_2=k_1+1}^{m} v_{Q_1}(k_1,k_2). \qquad (9)$$

Finally, the degree of agreement of Q_2 pairs of agents as to their preferences between Q_1 pairs of options, called the *degree of consensus*, is:

$$con_{Q_1,Q_2}(E,S) = \mu_{Q_2}(v_{Q_1}(E,S)). \qquad (10)$$

3.1   Discussion Guidance Using Additional Indicators

An important component of the consensus reaching support system is a set of indicators measuring the current state of agreement between agents and with respect to specific pairs of options. We are concerned how to determine those agents and pairs of options that are the main obstacles to reaching consensus, which preference matrix may be a candidate for a consensual one, etc. [22]. Those indicators may be both given as traditional numeric values or linguistic values (summaries), for instance expressed by using natural language generation techniques (cf. Kacprzyk and Zadrożny [18][21]).

In addition to many indicators proposed by, e.g. Kacprzyk and Zadrożny [14,18], here we propose the following new effective and efficient indicators:

1) *Response to omission* of agent $e_k \in E$, $RTO(k) \in [-1,1]$, is defined as a difference between the consensus degree for the whole group (10) and the consensus degree for the group without taking into account agent e_k:

$$RTO(k) = con_{Q_1,Q_2}(E,S) - con_{Q_1,Q_2}(E - \{e_k\},S). \qquad (11)$$

This measure makes it possible to estimate the influence of a given agent on the agreement in the whole group. The range of values is from -1, for an absolutely negative impact, through 0 for a lack of effect, to +1 for a definitely positive influence. Hence, this indicator conveys how important is the participation of agent k in the group for consensus reaching. Its positive value indicates a "consensus-constructive" position of a given agent while its negative value indicates a "consensus-destructive" position.

2)*Personal consensus degree* of agent $e_k \in E$, $PCD(k) \in [0,1]$, is the degree of truth of the following proposition: "Preferences of an agent e_k as to the most pairs of options are in agreement with the preference of most agents" what is:

$$PCD(k_1) = \mu_{Q_2}[\frac{1}{(m-1)} \sum_{k_2 \neq k_1}^{m} v_{Q_1}(k_1, k_2, S)] . \tag{12}$$

PCD takes values from 0 for an agent who is the most isolated in his or her opinions as compared to the rest of the group; to 1 for an agent whose preferences are shared by most agents; through intermediate values for agents with opinions of a varying degree of agreement with the opinions of other agents. This indicator points out how "typical" the opinion of an agent e_k is and helps identify agents who may be called "outsiders".

3)*Option consensus degree* for option s_i, $OCD(s_i) \in [0,1]$, is the degree of truth value of the statement: "Most pair of agents agree with their preferences in respect to option s_i" which can be expressed as:

$$s_{i,Q}(k_1, k_2) = \mu_{Q_1}[\frac{1}{n-1} \sum_{j \neq i}^{n} v_{ij}(k_1, k_2)] \tag{13}$$

$$OCD(s_i) = \mu_{Q_2}[\frac{2}{m(m-1)} \sum_{k_1=1}^{m-1} \sum_{k_2=k_1+1}^{m} s_{i,Q}(k_1, k_2)] . \tag{14}$$

The OCD may take values from 0, that the preferences of most agents differ substantially for a given option; to 1, that there is a substantial agreement among the agents as to this option. This indicator may help us omit some options from a further discussion and thus better focus the consensus reaching.

4)*Response to exclusion of option* $s_i \in S$, $RTE(s_i) \in [-1,1]$, is the difference between the consensus degree for the whole set of options (10) and without option s_i.

$$RTE(s_i) = con_{Q_1,Q_2}(E,S) - con_{Q_1,Q_2}(E_1, S - \{s_i\}) . \tag{15}$$

This measure makes it possible to estimate the influence of a given option on the consensus degree, and may be interpreted similarly as RTO.

3.2   Group Decision Making Solution

Essentially, two lines of calculation may be followed while considering solution concepts of group decision making (cf. Nurmi and Kacprzyk [24]. Kacprzyk, Zadrożny, Fedrizzi, and Nurmi [22]):

1) A direct approach: $\{R_1,...,R_m\} \rightarrow solution$, that is, a solution is obtained directly just from the set of individual fuzzy preference relations.

2) An indirect approach: $\{R_1,...,R_m\} \rightarrow R \rightarrow solution$, that is, from the set of individual fuzzy preference relations we calculate first a group (social) fuzzy preference relation, R, which is then employed to find a final solution.

In our model we use the direct approach, i.e. by using the individual fuzzy preference relations and the concept of a fuzzy core C defined as a set of non-ndominated options, i.e. those not defeated in pairwise comparisons by a required majority $\theta \leq m$, i.e.

$$C = \{s_j \in S : \neg \exists s_i \in S \ r_{ij}^k > 0.5 \text{ for at least } \theta \text{ agents } e_k\}. \tag{16}$$

To employ a fuzzy majority to extend (fuzzify) the core, we start by determining [22]

$$h_{ij}^k = \begin{cases} 1 & if \ r_{ij}^k < 0.5 \\ 0 & otherwise \end{cases} \tag{17}$$

where $i, j = 1, \ldots, n$ and $k = 1, \ldots, m$. So, h_{ij}^k reveals if s_j dominates (in pairwise comparison) s_i ($h_{ij}^k = 1$) or not ($h_{ij}^k = 0$).

Next, we compute

$$h_j^k = \frac{1}{n-1} \sum_{i=1, i \neq j}^{n} h_{ij}^k \tag{18}$$

which is the degree, from 0 to 1, to which agent e_k is not against option s_j, where 0 denotes definitely against and 1 definitely not against, through all intermediate values.

Then we calculate

$$h_j = \frac{1}{m} \sum_{k=1}^{m} h_j^k \tag{19}$$

which defines to which degree all the agents are not against option s_j.

Finally, we compute

$$v_Q^j = \mu_Q(h_j) \tag{20}$$

which is the extent to which Q (e.g., most) agents are not against option s_j.

The *fuzzy Q - core* is now defined as a fuzzy set in S with a membership function, $\mu_{C_Q}(s_i) = v_Q^i$, i.e., as a fuzzy set of options that are not defeated by Q (most) agents.

4   A Numerical Example

Consider Example 3.1 from Carlson, Fedrizzi, and Fuller [1], with the number of options, $n = 4$, and agents, $m = 4$, and the fuzzy preference relations:

$$R_1 = \begin{pmatrix} 0.4 & 0.7 & 0.1 \\ & 0.8 & 0.2 \\ & & 0.7 \end{pmatrix} \quad R_2 = \begin{pmatrix} 0.4 & 0.5 & 0.0 \\ & 0.8 & 0.2 \\ & & 0.7 \end{pmatrix}$$

$$R_3 = \begin{pmatrix} 0.4 & 0.4 & 0.3 \\ & 0.8 & 0.2 \\ & & 0.7 \end{pmatrix} \quad R_4 = \begin{pmatrix} 0.4 & 0.7 & 0.1 \\ & 0.7 & 0.1 \\ & & 0.7 \end{pmatrix}$$

Initially, the degree of consensus in the group of four agents is $con_{most,most}(E,S) = 0.42$, and suppose that this is not satisfactory so that we use indicators defined in the paper:

1) The value of the indicator $PCD(k)$ points out that agent 4 is the most isolated with his opinion with the rest of the group.
2) Moreover, value of the indicator $RTO(k)$ for agent 4 confirms that he or she has a negative influence on the agreement in the group.
3) On the contrary, the values of indicators $PCD(k)$ and $RTO(k)$ point out that the preferences of agent 1 are shared by most of agents (his or her fuzzy preference relation matrix may be a candidate for a consensual one).
4) The value of the indicator $OCD(s_i)$ points out that option 2 is the most promising direction for a further discussion.

Table 1. Values of PCD and RTO, for each agent, and OCD, for each option

PCD(k)	Value	RTO(k)	Value	OCD(s$_i$)	Value
PCD(1)	0.86	RTO(1)	+0.42	OCD(2)	0.47
PCD(2)	0.42	RTO(2)	-0.09	OCD(3)	0.42
PCD(3)	0.42	RTO(3)	-0.09	OCD(4)	0.42
PCD(4)	0	RTO(4)	-0.44	OCD(1)	0

In conclusion, a change of the preferences as to option 2 may be suggested to agent 4 towards the preferences expressed by agent 1, and if accepted, his or her new fuzzy preference relation matrix may be as follows:

$$R_4 = \begin{pmatrix} 0.4 & 0.7 & 0.1 \\ & 0.8 & 0.2 \\ & & 0.7 \end{pmatrix}.$$

Hence, the changes in the preference matrix are: $r_{23} = 0.7 \rightarrow 0.8$, $r_{24} = 0.1 \rightarrow 0.2$.

These small changes in two values cause the new value of group consensus degree to be now: $con_{most,most}(E,S) = 0.96$. This degree of consensus is satisfactory and the session is finished.

So, the *fuzzy* Q - *core* (21) is: $\mu_{C_Q}(s_1)=0; \mu_{C_Q}(s_2)=0.73; \mu_{C_Q}(s_3)=0.23; \mu_{C_Q}(s_4)=0.73$. Notice that two options are now pointed out as the best group choice: options 2 and 4.

5   Concluding Remarks

We proposed a new solution to the support of a group consensus reaching process under fuzzy preferences and a fuzzy majority. We assumed the moderated consensus reaching process which is clearly an efficiency oriented strategy. The main task here was to obtain the best increase either in the degree of consensus or in the speed of its reaching.

We emphasized the role of a moderator who facilitates obtaining a (satisfactory degree of) consensus by stimulating the exchange of information, suggesting arguments, convincing appropriate decision makers to change their preferences, focusing the discussion on the most promising directions, and the like. We developed some useful additional indicators to facilitate the process.

Our further research is to extend the model using some fairness oriented indicators, and by explicitly employing linguistic summaries, in particular by using tools of natural language generation (cf. Kacprzyk and Zadrożny [21]).

References

1. Carlsson, C., Fedrizzi, M., Fuller, R.: Group Decision Support Systems. In: Carlsson, C., Fedrizzi, M., Fuller, R. (eds.) Fuzzy Logic in Management, vol. 66, ch. 3, Springer (2004)
2. Consensus Decision Making,
   http://www.uhc.org.uk/projects/toolbox/
   meetings_and_organisation/consensus_short.htm
3. Fedrizzi, M., Kacprzyk, J., Nurmi, H.: Consensus degrees under fuzzy majorities and fuzzy preferences using OWA (ordered weighted average) operators. Control and Cybernetics 22, 71–80 (1993)
4. Fedrizzi, M., Kacprzyk, J., Owsiński, J.W., Zadrożny, S.: Consensus reaching via a GDSS with fuzzy majority and clustering of preference profiles. Annals of Operations Research 51, 127–139 (1994)
5. Fedrizzi, M., Kacprzyk, J., Zadrożny, S.: An interactive multi-user decision support system for consensus reaching process using fuzzy logic with linguistic quantifiers. Decision Support Systems 4(3), 313–327 (1988)
6. Gołuńska, D., Kacprzyk, J.: The Conceptual Framework of Fairness in Consensus Reaching Process Under Fuzziness. In: Proceedings of the 2013 Joint IFSA World Congress NAFIPS Annual Meeting, Edmonton, Canada, June 24-28, pp. 1285–1290 (2013)

7. Herrera-Viedma, E., García-Lapresta, J.L., Kacprzyk, J., Fedrizzi, M., Nurmi, H., Zadrożny, S. (eds.): Consensual Processes. STUDFUZZ, vol. 267. Springer, Heidelberg (2011)

8. Herrera-Viedma, E., Cabrerizo, F.J., Kacprzyk, J., Pedrycz, W.: A review of soft consensus models in a fuzzy environment. Information Fusion 17, 4–13 (2014)

9. Kacprzyk, J.: Group decision making with a fuzzy linguistic majority. Fuzzy Sets and Systems 18, 105–118 (1986)

10. Kacprzyk, J., Fedrizzi, M.: A 'soft' measure of consensus in the setting of partial (fuzzy) preferences. European Journal of Operational Research 34, 315–325 (1988)

11. Kacprzyk, J., Fedrizzi, M.: A 'human-consistent' degree of consensus based on fuzzy logic with linguistic quantifiers. Mathematical Social Sciences 18, 275–290 (1989)

12. Kacprzyk, J., Fedrizzi, M., Nurmi, H.: Group decision making and consensus under fuzzy preferences and fuzzy majority. Fuzzy Sets and Systems 49, 21–31 (1992)

13. Kacprzyk, J., Nurmi, H., Fedrizzi, I.M. (eds.): Consensus under Fuzziness, pp. 55–83. Kluwer Academic Publishers, Boston (1996)

14. Kacprzyk, J., Zadrożny, S.: On the use of fuzzy majority for supporting consensus reaching under fuzziness. In: Proceedings of FUZZ-IEEE 1997 - Sixth IEEE International Conference on Fuzzy Systems, Barcelona, Spain, vol. 3, pp. 1683–1988 (1997)

15. Kacprzyk, J.J., Zadrożny, S.: Computing with words in decision making through individual and collective linguistic choice rules. International Journal of Uncertainty, Fuzziness and Knowledge – Based Systems 9, 89–102 (2001)

16. Kacprzyk, J., Zadrożny, S.: Collective choice rules in group decision making under fuzzy preferences and fuzzy majority: a unified OWA operator based approach. Control and Cybernetics 31(4), 937–948 (2002)

17. Kacprzyk, J., Zadrożny, S.: An Internet-based group decision support system. Management VII(28), 4–10 (2003)

18. Kacprzyk, J., Zadrożny, S.: Supporting consensus reaching in a group via fuzzy linguistic data summaries. In: IFSA 2005 World Congress, Beijing, pp. 1746–1751. Tsinghua University Press/Springer (2005)

19. Kacprzyk, J., Zadrożny, S.: Towards a general and unified characterization of individual and collective choice functions under fuzzy and nonfuzzy preferences and majority via the Ordered Weighted Average Operators. International Journal of Intelligent Systems 24(1), 4–26 (2009)

20. Kacprzyk, J., Zadrożny, S.: Soft computing and Web intelligence for supporting consensus reaching. Soft Computing 14(8), 833–846 (2010)

21. Kacprzyk, J., Zadrożny, S.: Computing with words is an implementable paradigm: fuzzy queries, linguistic data summaries and natural language generation. IEEE Transactions on Fuzzy Systems 18(3), 461–472 (2010)

22. Kacprzyk, J., Zadrożny, S., Fedrizzi, M., Nurmi, H.: On Group Decision Making, Consensus Reaching, Voting and Voting Paradoxes under Fuzzy Preferences and a Fuzzy Majority: A Survey and some Perspectives. In: Bustince, H., et al. (eds.) Fuzzy Sets and Their Extensions: Representations, Aggregation and Models. STUDFUZZ, vol. 220, pp. 263–295. Springer, Heidelberg (2008)

23. Kacprzyk, J., Zadrożny, S., Raś, Z.W.: How to support consensus reaching using action rules: a novel approach. International Journal of Uncertainty, Fuzziness and Knowledge-Based Systems 18(4), 451–470 (2010)

24. Nurmi, H., Kacprzyk, J.: On fuzzy tournaments and their solution concepts in group decision making. European Journal of Operational Research 51, 223–232 (1991)

25. Turban, E., Aronson, J.E., Liang, T.P.: Decision Support Systems and Intelligent Systems, 6th edn., pp. 11–19, 94-101.C. Prentice Hall (2005)

26. Zadeh, L.A.: A computational approach to fuzzy quantifiers in natural languages. Computers and Mathematics with Applications 9, 149–184 (1983)

Aggregation of Uncertain Qualitative Preferences for a Group of Agents

Paolo Viappiani[1,2]

[1] Sorbonne Universités, UPMC Univ Paris 06, UMR 7606, LIP6
[2] CNRS, UMR 7606, LIP6, F-75005, Paris, France
paolo.viappiani@lip6.fr

Abstract. We consider aggregation of partially known qualitative preferences for a group of agents, considering necessary and potentially optimal choices with respect to different notions of optimality (consensus, extreme choices, Pareto optimality) and provide a theoretical characterization. We report statistics (obtained with simulations with synthetic data) about the cardinality of the sets of possible and necessarily optimal choices for the different cases. Finally we introduce preliminary ideas on a qualitative notion of *fairness* and on interactive elicitation.

1 Introduction

In this paper we consider aggregation of partially known qualitative preferences of different agents. By *qualitative* we mean that preferences are explicitly and directly encoded by binary relations [6,1]. We are in a setting of *strict* uncertainty, meaning that no probabilistic assumption is made. Preference information is incomplete, meaning that we only know a fraction of the binary preferences. We assume that any consistent extension of the currently known preference relations is considered possible. In particular, we consider possible and necessary Pareto optimality and study the relations with other notions of optimality.

This work is an effort towards the direction of effective, practical methods for preference assessment (preference elicitation) with purely qualitative statements. Our work is similar to [4]; however our setting differs since we assume qualitative preferences expressed as orders (while they assume numeric utility functions).

2 Model

General Assumptions. We assume a set $\mathcal{C}$ of m items or elements (choices) and a set $\mathcal{G} = \{1, ..., n\}$ of n agents; preferences are explicitly modeled by binary relations. For each agent the *preference relation* $\succeq_i$ (for each agent i) models the fact that, if $o \succeq_i o'$ holds, option o is at least as preferred as option o' for agent i. As usual [5] a preference relation $\succeq_i$ induces a preference structure $(\succ_i, \approx_i, \sim_i)$ for each agent, where $\succ_i$ is the *strict preference relation*, $\approx_i$ is the *indifference relation* and $\sim_i$ is the *incomparability relation*: $o \succ_i o'$ if $o \succeq_i o'$ and $o' \not\succeq_i o$, $o \approx_i o'$ if $o \succeq_i o'$ and $o' \succeq_i o$ and $o \sim_i o'$ if $o \not\succeq_i o'$ and $o' \not\succeq_i o$.

A. Laurent et al. (Eds.): IPMU 2014, Part II, CCIS 443, pp. 434–443, 2014.

For a given relation $\succeq$, we use the operator top to denote the maximum element (if it exists):

$$\text{top}(\succeq) = \{o \in \mathcal{C} | \forall o' \in \mathcal{C}, o' \neq o : o \succ o'\}. \tag{2.1}$$

Notice that the cardinality of $\text{top}(\succeq)$ is at most 1. The operator max denotes maximal elements:

$$\max(\succeq) = \{o \in \mathcal{C} | \nexists o' \in \mathcal{C} : o' \succ o\}. \tag{2.2}$$

Obviously it holds $\text{top}(\succeq) \subseteq \max(\succeq)$. If the preferences of an agent over the available items constitute a chain (i.e. $\succeq_i$ is a total order), then his optimal option is $\text{top}(\succeq_i) = \max(\succeq_i)$.

As usual in multi-agent systems and multi-objective decision analysis, an option o dominates another option o' if it holds $o \succeq_i o'$ for all i, and for at least one j it holds $o \succ_j o'$ (strict preference); the *Pareto optimal* choices for the group of agents are the those choices that are not dominated by any other choice. We can conveniently express Pareto Optimality using the notion of aggregate *group dominance* relation $\succeq_\mathcal{G}$, defined as $o \succeq_\mathcal{G} o'$ iff $\forall i \in \mathcal{G}, o \succeq_i o'$. A consensus choice, if it exists, it is an option that is the best preferred item for all agents.

$$\text{Cons}(\succeq_1, ..., \succeq_n) = \text{top}(\succeq_\mathcal{G}) = \text{top}(\bigcap_{i \in \mathcal{G}} \succeq_i). \tag{2.3}$$

By representing relations in their extensive form as subset of the Cartesian product $\mathcal{C} \times \mathcal{C}$, the group dominance relation can be conveniently written as $\succeq_\mathcal{G} = \bigcap_i \succeq_i$. Then, Pareto Optimal choices with respect to $\succeq_\mathcal{G}$ are the maximal elements of the aggregated group preference order:

$$\text{ParetoOpt}(\succeq_1, ..., \succeq_n) = \max(\succeq_\mathcal{G}) = \max(\bigcap_{i \in \mathcal{G}} \succeq_i). \tag{2.4}$$

A particular kind of solutions (that are also Pareto Optimal) are those that are *extreme* in the sense that they are the best choice for one (or more) agents but might not be good choices for the other agents. The set of these *extreme choices* is

$$\text{Ext}_\mathcal{G}(\succeq_1, ..., \succeq_n) = \bigcup_{i \in \mathcal{G}} \text{top}(\succeq_{i \in \mathcal{G}}). \tag{2.5}$$

Partial Knowledge of Preference Orders. We suppose that only partial information about the agents' preferences (the order relation) is known. We consider that only a subset of each preference order is known, meaning that we are aware of some pairwise preferences but not others. Since we assume a context where preferences are transitive, we assume that also the known preference relation is transitive, i.e. it is a partial order[1]. Let $\succeq_i^*$ be the true preference order for

---

[1] If preferences are given in terms of pairwise comparison statements, we consider the transitive closure of the binary relation containing all such pairwise comparisons.

agent i (unknown to the system); the knowledge about the agent's preferences is encoded by a partial order $\succeq_i \subseteq \succeq_i^*$ for each agent i. While the true maximal elements of each agent i are $\max(\succeq_i^*)$, we can only deduce a set $\max(\succeq_i)$ of "current" maximal items. Of course, current maximal items might not be maximal in the "true" complete preference model. It holds $\max(\succeq_i^*) \subseteq \max(\succeq_i)$, since a maximal element according to the $\succeq_i^*$ must be also maximal for the "sparser" relation $\succeq_i$.

From these partial orders, we consider the "known" aggregate relation of dominance for group $\mathcal{G}$, $\succeq_\mathcal{G} = \bigcap_{i \in \mathcal{G}} \succeq_i$. Notice that $\succeq_\mathcal{G}$ is also a subset of $\succeq_\mathcal{G}^*$, as it is the intersection of partial orders that are included in the underlying true orders: $\succeq_\mathcal{G} \subseteq \succeq_i^*$ as $\succeq_\mathcal{G} = \bigcap_{i \in \mathcal{G}} \succeq_i \subseteq \bigcap_{i \in \mathcal{G}} \succeq_i^* = \succeq_\mathcal{G}^*$.

The true Pareto optimal choices are $\text{ParetoOpt}(\succeq_1^*, ..., \succeq_n^*)$ and the extrema $\text{Ext}_\mathcal{G}(\succeq_1^*, ..., \succeq_n^*)$ (one simply substitute $\succeq_i$ with $\succeq_i^*$ in Equation 2.4 and in Equation 2.5); however we only know the "current" Pareto Optimal choices are $\text{ParetoOpt}(\succeq_1, ..., \succeq_n)$ and the current extrema choices are $\text{Ext}_\mathcal{G}(\succeq_1, ..., \succeq_n)$,.

When considering aggregation, the ideal case (but usually rare in practice) is that all agents agree on the best option (consensus); the agents share a common maximum element. If we have incomplete knowledge about the individual preferences, we might wonder if, given the current information, a consensus might exist. Hence, we define the notion of possible and necessary consensus for a group of agents, and similarly for Pareto Optimal and Extreme choice.

3   Consensus, Extreme and Pareto Optimal Items: The Possible and the Necessary

Definition 1. *Given* $(\succeq_1, ..., \succeq_n)$, *the* possible *consensus choices* $\text{PossCons}_\mathcal{G}$ *for group* $\mathcal{G}$ *are those for which there exists a set of total orders* $(\succeq_1', ..., \succeq_n')$, *with each* $\succeq_i'$ *extending* $\succeq_i$, *such that they are a maximum element of the derived aggregate dominance relation* $\succeq_\mathcal{G}' = \bigcap_i \succeq_i'$.

$$\text{PossCons}_\mathcal{G}(\succeq_1, ..., \succeq_n) = \quad (3.1)$$

$$\{o \in \mathcal{C} \mid \exists(\succeq_1', ..., \succeq_n') : (\succeq_1' \supseteq \succeq_1), ..., (\succeq_n' \supseteq \succeq_n) \land o \in \text{Cons}(\succeq_1', ..., \succeq_n')\} = \quad (3.2)$$

$$= \bigcup_{\succeq_i' \supseteq \succeq_i \forall i \in \mathcal{G}} \text{top}(\bigcap \succeq_i') = \bigcup_{\succeq_i' \supseteq \succeq_i \forall i \in \mathcal{G}} \text{top}(\succeq_\mathcal{G}'). \quad (3.3)$$

The necessary *consensus choices* $\text{NecCons}_\mathcal{G}$ *for group* $\mathcal{G}$ *are those that are a maximum element for all complete orders extending* $\succeq_\mathcal{G}$.

$$\text{NecCons}_\mathcal{G}(\succeq_1, ..., \succeq_n) = \quad (3.4)$$

$$= \{o \in \mathcal{C} \mid \forall(\succeq_1', ..., \succeq_n') : (\succeq_1' \supseteq \succeq_1), ..., (\succeq_n' \supseteq \succeq_n) \land o \in \text{top}(\succeq_\mathcal{G}')\} = \quad (3.5)$$

$$= \bigcap_{\succeq_i' \supseteq \succeq_i \forall i \in \mathcal{G}} \text{top}(\succeq_\mathcal{G}'). \quad (3.6)$$

Proposition 1. *Necessary* and *possible* *consensus for* $\mathcal{G}$ *can be formulated as:*

1. $\text{NecCons}_{\mathcal{G}} = \bigcap_{i \in \mathcal{G}} \text{top}(\succeq_i)$
2. $\text{PossCons}_{\mathcal{G}} = \bigcap_{i \in \mathcal{G}} \max(\succeq_i)$

Proof. 1) $\text{NecCons}_{\mathcal{G}}$: If a choice belongs to $\bigcap_i \text{top}(\succeq_i)$ then (by definition) it means that it dominates all other choices for each agent i; it will still be the maximum in any extension of the preference relations. Moreover, if an option is dominated in a preference relation $\succeq_i$, it will still dominated in any extension. Therefore $\text{NecCons}_{\mathcal{G}}$ is exactly the intersection of all $\text{top}(\succeq_i)$.
2) $\text{PossCons}_{\mathcal{G}}$: If an option o_1 is dominated wrt agent i (e.g. there is a choice o_2 such that $o_2 \succ_i o_1$), it will also be dominated in any extension of $\succ_i$, and cannot be a possible consensus. Therefore only options that are maximal elements for all agents can potentially be a consensus. Consider an option $o \in \bigcap_{i \in \mathcal{G}} \max(\succeq_i)$; we construct an extension of the preference order by breaking all incomparabilities in favor of o. Since o is a consensus in the extension, the argument follows.

While theoretically interesting, the concept of possible/necessary consensus can be of limited interest, as frequently agents will have conflicting preferences. We therefore consider weaker notions of optimality.

Definition 2. *The* possible *extreme choices* $\text{PossExt}_{\mathcal{G}}$ *for group* $\mathcal{G}$ *are those for which there is a set of total orders* $(\succeq'_1, ..., \succeq'_n)$, *with each* $\succeq'_i$ *extending* $\succeq_i$, *for which they are an extreme choice.*

$$\text{PossExt}_{\mathcal{G}} = \{o \mid \exists (\succeq'_1, ..., \succeq'_n) : (\succeq'_1 \supseteq \succeq_1), ..., (\succeq'_n \supseteq \succeq_n) \wedge o \in \text{Ext}(\succeq'_1, ..., \succeq'_n)\} \quad (3.7)$$

The necessary *extreme choices* $\text{NecExt}_{\mathcal{G}}$ *for group* $\mathcal{G}$ *are those that are extreme for all total orders* $(\succeq'_1, ..., \succeq'_n)$, *with each* $\succeq'_i$ *extending* $\succeq_i$.

$$\text{NecExt}_{\mathcal{G}} = \{o \mid \forall (\succeq'_1, ..., \succeq'_n) : (\succeq'_1 \supseteq \succeq_1), ..., (\succeq'_n \supseteq \succeq_n), o \in \text{Ext}(\succeq'_1, ..., \succeq'_n)\} \quad (3.8)$$

Proposition 2. *Necessary* extreme *and* possible *extreme choices for a group* $\mathcal{G}$ *can be rewritten as follows:*

- $\text{NecExt}_{\mathcal{G}} = \bigcup_{i \in \mathcal{G}} \text{top}(\succeq_i)$
- $\text{PossExt}_{\mathcal{G}} = \bigcup_{i \in \mathcal{G}} \max(\succeq_i)$

We now consider *potential* and *necessary* Pareto optimal choices.

Definition 3. *A choice is a* possible Pareto optimal *for group* $\mathcal{G}$ *if there exists a set of total orders* $(\succeq'_1, ..., \succeq'_n)$, *with each* $\succeq'_i$ *extending* $\succeq_i$, *for which they are a Pareto optimal choice.*

$$\text{PossParetoOpt}_{\mathcal{G}} = \{o \mid \exists (\succeq'_1, ..., \succeq'_n) : (\succeq'_1 \supseteq \succeq_1), ..., (\succeq'_n \supseteq \succeq_n) \wedge o \in \max(\bigcap_{i \in \mathcal{G}} \succeq'_i)\}$$

A choice is a necessary Pareto optimal *for group* $\mathcal{G}$ *if it is a maximal element of the aggregate preference relation with respect to all extensions of the current preference orders.*

$$\text{NecParetoOpt}_{\mathcal{G}} = \{o \mid \forall (\succeq'_1, ..., \succeq'_n) : (\succeq'_1 \supseteq \succeq_1), ..., (\succeq'_n \supseteq \succeq_n), o \in \max(\bigcap_{i \in \mathcal{G}} \succeq'_i)\}$$

An equivalent statement is the following: a choice is necessary Pareto Optimal if there is no option that *possibly* dominates it. An option o_1 is a possible dominator for o_2 if there is a consistent extension of the known preference orders that make o_1 dominate o_2. *Possible dominance* can be formalized accordingly:

Definition 4. *The relation $\succeq_{\mathcal{G}}^{PD}$ of possible dominance for a group $\mathcal{G}$ is such that*

$$o_1 \succeq_{\mathcal{G}}^{PD} o_2 \text{ iff } \exists(\succeq_1', ..., \succeq_n') : o_1 \succ_{\mathcal{G}}' o_2 \tag{3.9}$$

with each $\succeq_i'$ extending $\succeq_i$ and $\succeq_{\mathcal{G}}' = \bigcap_i \succeq_i'$ being the associated aggregate relation.

Given this definition, an item o is a necessary Pareto Optimal if o is a maximal element of $\succ^{PD}$ (the induced strict relation). The relation $\succeq^{PD}$ can be characterized in terms of the currently known preferences in the following way: an option o_1 is a potential dominator of o_2 if, for every agent i, o_2 is not strictly preferred to o_1, and o_1 and o_2 are not equally preferred for all agents.

Proposition 3. *The relation of possible dominance ($\succeq_{\mathcal{G}}^{PD}$) can be written as:*

$$\succeq_{\mathcal{G}}^{PD} = \left[\bigcap_{i \in \mathcal{G}} \not\prec_i\right] - \bigcap_{i \in \mathcal{G}} \approx_i = \left[\bigcap_{i \in \mathcal{G}} \succeq_i \cup \sim_i\right] - \bigcap_{i \in \mathcal{G}} \approx_i \tag{3.10}$$

In the case of underlying linear orders, two options are never equally preferred, and the expression simplifies to

$$\succeq_{\mathcal{G}}^{PD} = \left[\bigcap_{i \in \mathcal{G}} \succeq_i \cup \sim_i\right] \tag{3.11}$$

One could make a similar reasoning and define a relation of *necessary dominance* $\succ_{\mathcal{G}}^{ND}$. Notice that, if $o_1 \succeq_{\mathcal{G}} o_2$ then $o_1 \succeq_{\mathcal{G}}^{ND} o_2$, therefore $\succeq_{\mathcal{G}} \subseteq \succ_{\mathcal{G}}^{ND}$. Moreover it can be shown that it holds exactly that $\succeq_{\mathcal{G}}^{ND} = \succeq_{\mathcal{G}}$.

We can now characterize the sets PossParetoOpt and NecParetoOpt of possible and necessary Pareto Optimal choices in the following way.

Proposition 4. *The set of Possible Pareto Optimal choices coincides with the set of the current undominated (Pareto Optimal) options given the known preference orders $\succeq_i$.*

$$\text{PossParetoOpt}(\succeq_1, ..., \succeq_n) = \max(\succeq_{\mathcal{G}}) = \max\left(\bigcap_{i \in \mathcal{G}} \succeq_i\right) \tag{3.12}$$

The set of Necessary Pareto Optimal choices coincides with the maximal choices with respect to the strict relation of possible dominance.

$$\text{NecParetoOpt}(\succeq_1, ..., \succeq_n) = \max(\succ_{\mathcal{G}}^{PD}) \tag{3.13}$$

Proof. 1) PossParetoOpt$_{\mathcal{G}}$: We show the argument by constructions. Let choice o_1 belong to the Pareto set of the currently known preference orders; $o_1 \in \max(\succeq_{\mathcal{G}})$. Then for all preference orders $\succeq_i$ construct a complete (linear) order $\succeq_i'$ extending $\succeq_i$ such that, for any item o_2, if the preference between o_1

and o_2 is not known (they are incomparable in $\succeq_i$) the pair $o_1 \succeq_i o_2$ is added into $\succeq'_i$. For pairs not involving o_1, pick any assignment consistent with transitivity. o_1 is a Pareto Optimal choice in $\succeq'_i$ by construction, hence it belongs to PossParetoOpt$_\mathcal{G}$. This shows that $\max(\succeq_\mathcal{G}) \subseteq$ PossParetoOpt. To show that only Pareto optimal choices wrt $\succeq_i$ belong to PossParetoOpt$_\mathcal{G}$, it is enough to realize that adding new pairwise preferences cannot break domination ; hence dominated options will remain such in any extensions, and cannot be possible Pareto optimal choices. Therefore if $o_1 \notin \max(\succeq_\mathcal{G}) \rightarrow o_1 \notin$ PossParetoOpt, meaning that PossParetoOpt $\subseteq \max(\succeq_\mathcal{G})$. Hence, PossParetoOpt $= \max(\succeq_\mathcal{G})$.

2) NecParetoOpt$_\mathcal{G}$: straightforward from the previous considerations on $\succeq_\mathcal{G}^{PD}$ and Proposition 3.

We are able now able to characterize the relationship between NecCons, PossCons, NecExt, PossExt, NecParetoOpt and PossParetoOpt.

Proposition 5. *We derive the following taxonomy:*
1. NecCons$_\mathcal{G} \subseteq$ NecExt$_\mathcal{G} \subseteq$ NecParetoOpt$_\mathcal{G}$
2. PossCons$_\mathcal{G} \subseteq$ PossExt$_\mathcal{G} \subseteq$ PossParetoOpt$_\mathcal{G}$
3. NecCons$_\mathcal{G} \subseteq$ PossCons$_\mathcal{G}$, NecExt$_\mathcal{G} \subseteq$ PossExt$_\mathcal{G}$ *and*
   NecParetoOpt$_\mathcal{G} \subseteq$ PossParetoOpt$_\mathcal{G}$.

Proof. 1. NecCons$_\mathcal{G} = \bigcap_{i \in \mathcal{G}} \text{top}(\succeq_i) \subseteq \bigcup_{i \in \mathcal{G}} \text{top}(\succeq_i) = $ NecExt$_\mathcal{G} \subseteq \max(\succ_\mathcal{G}^{PD}) =$
NecParetoOpt$_\mathcal{G}$; the last inclusion inequality holding as an element o of NecExt$_\mathcal{G}$ belongs to top$(\succeq_j)$ for a given $j \in \mathcal{G}$ (by definition); o cannot be possibly dominated (wrt $\succ^{PD} = \bigcap_{i \in \mathcal{G}} \succeq_i \cup \sim_i$) since it is the maximum in $\succeq_j$, therefore it has to be a maximal element of $\succ_\mathcal{G}^{PD}$.
2. PossCons$_\mathcal{G} = \bigcap_{i \in \mathcal{G}} \max(\succeq_i) \subseteq \bigcup_{i \in \mathcal{G}} \max(\succeq_i) =$ PossExt$_\mathcal{G} \subseteq \max(\succeq_\mathcal{G}) =$
$=$ PossParetoOpt$_\mathcal{G}$
3. Straightforward from definition of possible and necessary: if an element is, respectively, a consensus, extreme, or Pareto optimal for all extensions of the preference orders, then in particular there exists an extension for which it is, respectively, a consensus, extreme or Pareto optimal.

Example 1. Consider the following case with $\mathcal{C} = \{o_1, o_2, o_3\}$. It is known that agent 1 prefers option o_1 to o_2, and option o_2 to o_3 (that is a linear order, or chain, $o_1 \succ_1 o_2 \succ_1 o_3$). We also know that agent 2 prefers o_1 to o_3 and also o_2 to o_3 ($o_1 \succ_2 o_3$ and $o_3 \succ_2 o_3$), but nothing is known about his preference between o_1 and o_2. For agent 3 we only know $o_1 \succ o_3$, meaning that he prefers o_1 to o_3.

There is only one maximal element for agent 1 (o_1, that is also the maximum element for this agent), while o_1 and o_2 are maximal for agents 2 and 3. The intersection is $\{o_1\}$ and therefore o_1 is the only possible consensus choice. There is no necessary optimal choice as only the preference order of agent 1 has a maximum. There is one necessary extreme, o_1, and the possible extreme options are o_1, o_2 (the union of maximal elements). Then the group relation $\succ_\mathcal{G}$ is such that $o_1 \succ_\mathcal{G} o_3$ (the pair o_1 and o_2 is incomparable with respect to $\succ_\mathcal{G}$, as well as the pair o_2 and o_3). Then o_1 and o_2 are the maximal elements for $\succ_\mathcal{G}$ and the possible Pareto optimal choices for the agents. The relation of potential dominance is $\succeq^{PD}$ that in this case is a linear order $o_1 \succ^{PD} o_2 \succ^{PD} o_3$; option o_1 is the only necessary Pareto optimal choice.

The taxonomy that we obtained (Figure 1) is perhaps not very surprising: the stricter the notion of optimality (consensus, extreme, Pareto) the smaller the sets. Sets of possible optimal items are included in the sets of necessary optimal items. However, our theoretical characterisation is useful as it allows to compute the possible and necessary optimal items reasoning only with respect to the current available preference information (the $\succeq_i$) without the need to individually consider all possible extensions of the currently available preference information. In particular, *the set of possible Pareto optimal choices* PossParetoOpt *coincides with the current Pareto optimal set*, the set of maximal items with respect to the group dominance relation computed with the currently available preferences. The set of necessary Pareto optimal choices are those items that are non dominated with respect to the strict relation of possible dominance ($\succ^{PD}$), that can be expressed in a convenient way thanks to Equation 3.13. Furthermore the intersection of the maximal elements of the $\succeq_i$ coincides PossCons and the union of the maximal elements of the $\succeq_i$ coincides with PossExt.

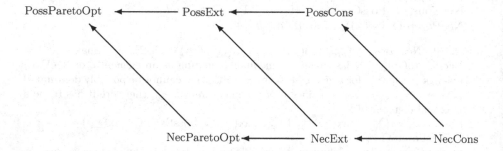

Fig. 1. Inclusion membership between the different classes

4   Cardinality

The previous section provided a theoretical characterization of different kinds of "optimality" (consensus, extreme choices and Pareto Optimal choices) when dealing with partial binary preference information, providing a mathematical formulation of possible and necessary optimal choices under the different semantics. Here we perform a number of simulations in order to assess the cardinality of the sets NecCons, PossCons, NecExt, PossExt, NecParetoOpt and PossParetoOpt in practical circumstances.

- We randomly generate (uniformly) a permutation of the n elements for each agent of the m agents, this is assumed to be their true preference ranking.
- For each agent, from the incidence matrix representing the preference relation, we randomly cancel a fraction f of pairs in relation. We compute the transitive closure of the relation.
- From the obtained partial orders, we compute the sets NecCons, PossCons, NecExt, PossExt, NecParetoOpt and PossParetoOpt.

In the table below we report the cardinalty of these sets averaged over 100 runs, for some values of n, m and f. Further studies might consider cardinality under different ranking probability models (for instance considering the models in [3]).

n	m	f	|NecCons|	|PossCons|	|NecExt|	|PossExt|	|NecParetoOpt|	|PossParetoOpt|
2	3	0.20	0.40	0.40	1.60	1.60	1.78	1.78
2	3	0.40	0.19	0.63	1.16	2.02	1.34	2.13
2	3	0.60	0.15	0.56	1.21	2.08	1.36	2.25
2	3	0.80	0.00	1.33	0.00	2.67	0.20	2.86
2	3	0.90	0.00	1.40	0.00	2.60	0.16	2.78
2	3	0.95	0.00	1.42	0.00	2.58	0.15	2.83
2	4	0.20	0.24	0.43	1.37	1.96	1.75	2.31
2	4	0.40	0.12	0.52	1.12	2.24	1.47	2.60
2	4	0.60	0.02	0.80	0.51	2.71	0.70	3.20
2	4	0.80	0.00	1.26	0.00	3.29	0.10	3.64
2	4	0.90	0.00	2.18	0.00	3.82	0.00	3.94
2	4	0.95	0.00	2.29	0.00	3.71	0.00	3.89
2	5	0.20	0.10	0.25	1.53	2.12	1.97	2.62
2	5	0.40	0.02	0.47	0.92	2.64	1.34	3.29
2	5	0.60	0.00	0.81	0.08	3.55	0.32	4.09
2	5	0.80	0.00	2.08	0.00	4.42	0.02	4.75
2	5	0.90	0.00	3.25	0.00	4.75	0.00	4.93
2	5	0.95	0.00	3.19	0.00	4.81	0.00	4.94

n	m	f	|NecCons|	|PossCons|	|NecExt|	|PossExt|	|NecParetoOpt|	|PossParetoOpt|
3	3	0.20	0.09	0.09	2.12	2.12	2.40	2.40
3	3	0.40	0.02	0.25	1.68	2.48	1.89	2.69
3	3	0.60	0.01	0.20	1.58	2.55	1.97	2.78
3	3	0.80	0.00	0.83	0.00	2.91	0.43	2.97
3	3	0.90	0.00	0.87	0.00	2.88	0.38	2.98
3	3	0.95	0.00	0.94	0.00	2.87	0.34	2.94
3	4	0.20	0.06	0.19	1.86	2.46	2.39	2.94
3	4	0.40	0.04	0.26	1.39	2.83	2.06	3.25
3	4	0.60	0.00	0.38	0.61	3.25	1.32	3.65
3	4	0.80	0.00	0.76	0.00	3.60	0.44	3.90
3	4	0.90	0.00	1.76	0.00	3.90	0.01	3.95
3	4	0.95	0.00	1.61	0.00	3.91	0.01	3.99
3	5	0.20	0.03	0.07	2.04	2.72	2.89	3.56
3	5	0.40	0.00	0.14	1.10	3.42	2.02	4.07
3	5	0.60	0.00	0.39	0.24	4.11	0.90	4.72
3	5	0.80	0.00	1.37	0.00	4.66	0.03	4.95
3	5	0.90	0.00	2.57	0.00	4.95	0.00	5.00
3	5	0.95	0.00	2.64	0.00	4.92	0.00	4.99

n	m	f	|NecCons|	|PossCons|	|NecExt|	|PossExt|	|NecParetoOpt|	|PossParetoOpt|
4	3	0.20	0.07	0.07	2.28	2.28	2.57	2.57
4	3	0.40	0.00	0.17	1.82	2.63	2.14	2.81
4	3	0.60	0.01	0.17	1.82	2.62	2.15	2.82
4	3	0.80	0.00	0.63	0.00	2.98	0.69	3.00
4	3	0.90	0.00	0.64	0.00	2.93	0.66	2.99
4	3	0.95	0.00	0.69	0.00	2.95	0.72	3.00
4	4	0.20	0.00	0.04	2.37	2.95	3.14	3.57
4	4	0.40	0.00	0.06	1.79	3.29	2.63	3.67
4	4	0.60	0.00	0.17	0.72	3.61	1.86	3.88
4	4	0.80	0.00	0.31	0.00	3.85	0.83	3.98
4	4	0.90	0.00	1.31	0.00	3.96	0.08	4.00
4	4	0.95	0.00	1.29	0.00	3.98	0.04	4.00
4	5	0.20	0.00	0.03	2.56	3.18	3.73	4.27
4	5	0.40	0.01	0.04	1.60	3.85	2.93	4.61
4	5	0.60	0.00	0.14	0.30	4.48	1.35	4.88
4	5	0.80	0.00	0.89	0.00	4.92	0.07	5.00
4	5	0.90	0.00	2.02	0.00	4.99	0.00	5.00
4	5	0.95	0.00	2.06	0.00	4.98	0.00	5.00

5   Current Works

This section discusses current work dealing with elicitation of qualitative preferences and with the identification of "fair" choices.

Elicitation. We are interested in interactive settings, where the agents provide new information to the system (statements of the type $o_1 \succeq o_2$) at each step; this results in an update of the preference order $\succeq_i := \succeq_i \cup (o_1 \succeq o_2)$ for the agent i who entered new information. The challenge is to define effective strategies for elicitation, that pick the items to compare in some smart way. Of course, we want to ask agents to compare two items that are currently incomparable given the available information. One heuristic strategy would be to consider queries about maximal items in the currently known preference relation. When choosing to which agent ask queries, one heuristic could consist in targeting the agent whose current preference relation is sparser (the lowest number of pairwise comparison is known).

We believe that the development of efficient query strategies, as well as practical evaluation and comparison of different techniques, is an important next step involving a substantial research effort. One challenge would be the definition of suitable measures of the value of information [7] of a candidate query in this intrinsically qualititative setting.

Fairness. We aim to provide a characterisation of fairness in a qualitative way (in a way similar to [2]). Intuitively, a choice is *more fair* than another if it is less "extreme" in the satisfaction of the agents. Fairness is a well developed concept in numerical approaches, where one can consider the least satisfied agent, or more refined aggregators such as OWA. However, typical fairness measures are not meaningful in this context, as we deal with qualitative preferences. One could of course map choices according to their position in each agent's ranking, and then consider fairness using numerical methods, but then the advantages of a qualitative approach would disappear.

We want to work on partial preference orders without the need of a distance-from-equality measure. In order to define what equity means in this qualitative setting, we propose to use the notion of *reference point*. The reference point e expresses an *intermediate level of preference satisfaction*, encoding a somewhat intermediate level of satisfaction. Note that the option at which the level e is assessed, can be different for each agent (for instance, the e might corresponds to the second best choice for agent 1, while to the third choice for agent 2). This point is such that, if each agent could achieve e, this would correspond to a maximally fair situation, as all agents will be equality satisfied.

Definition 5. *A choice o_1 is more fair than choice o_2 with respect to agent i and j, written as $o_1 \succeq_{(i,j)}^{F} o_2$, if $o_1 \succeq_i o_2 \succeq_i e$ and if $e \succeq_j o_2 \succeq_j o_2$*

From $\succeq_{(i,j)}^{F}$ a single fairness relation $\succeq^{F}$ for all agents can be obtained by intersection: $\succeq^{F} = \bigcap_{(i,j) \in \mathcal{G}} \succeq_{(i,j)}^{F}$. In order to obtain a practical way to rank items accounting for both preferences and fairness, we now combine the two preference relations $\succeq_{\mathcal{G}}$ and $\succeq^{F}$ into a single preference order. A choice o_1 is preferred to another choice o_2 if it is either that o_1 dominates o_2 or if o_1 is more fair than o_2.

Definition 6. *The* combined dominance-fairness *preference relation $\succeq^{C}$ is defined as $\succeq^{C} = \succeq_{\mathcal{G}} \cup \succeq^{F}$.*

We propose to consider the Pareto optimal choices wrt the obtained order $\succeq^{C}$: these solutions might be considered candidate choices for the groupwise decision making. These solutions can be considered the qualitative counterpart of the concept of well balanced solutions in multi-objective optimization (Lorentz dominance). Future works consist in an investigation of the mathematical properties of the relation $\succeq^C$ and in practical evaluation (including simulations).

6 Conclusions

In this paper we considered the case of partially known preferences of different agents, considering purely qualitative preferences (partial preference relations). We derived a mathematical characterisation and a taxonomy of the possible and necessary optimal choices, according to three different notions of optimality: shared optimality, extrema and Pareto optimality. In simulations we reported the cardinality measures of these sets in a number of different settings.

Notice that we have considered a setting with multiple agents, but this work can be very easily adapted to a multi-criteria setting: in this other setting each i would refer to a different criteria and $\mathcal{G}$ to the group of criteria. This work is related to works in computational social choice theory, interested in establishing how hard is to compute necessary and possible winners given common election rules [8]. Researchers in Operations Research [4] have also considered necessary and possibly optimal items with several feasible utility functions.

Acknowledgments. The author is supported by the projects BR4CP (contract n.ANR-11-BS02-008) and LARDONS (contract n. ANR-10-BLAN-0215) financed by the Agence Nationale de la Recherche (ANR).

References

1. Aleskerov, F., Bouyssou, D., Monjardet, B.: Utility Maximization, Choice and Preference, 2nd edn. Studies in economic theory. Springer (April 2007) (Anglais)
2. Cowell, F., Flachaire, E.: Inequality with ordinal data (2012),
   http://www.hec.ca/iea/seminaires/130226_emmanuel_flachaire.pdf
   (manuscript available online)
3. Critchlow, D.E., Fligner, M.A., Verducci, J.S.: Probability models on rankings. Journal of Mathematical Psychology 35(3), 294–318 (1991)
4. Greco, S., Mousseau, V., Slowinski, R.: Ordinal regression revisited: Multiple criteria ranking using a set of additive value functions. European Journal of Operational Research 191(2), 416–436 (2008)
5. Kaci, S.: Working with preferences: Less is more. Cognitive Technologies. Springer (2011)
6. Roubens, M., Vincke, P.: Preference modelling. Lecture Notes in Economics and Mathematical Systems (250) (1985)
7. Viappiani, P., Boutilier, C.: Optimal Bayesian recommendation sets and myopically optimal choice query sets. In: Advances in Neural Information Processing Systems 23 (NIPS), pp. 2352–2360. MIT Press (2010)
8. Xia, L., Conitzer, V.: Determining possible and necessary winners under common voting rules given partial orders. In: Proceedings of the Twenty-Third AAAI Conference on Artificial Intelligence (AAAI), pp. 196–201 (2008)

Choquet Expected Utility Representation of Preferences on Generalized Lotteries

Giulianella Coletti[1], Davide Petturiti[2], and Barbara Vantaggi[2]

[1] Dip. Matematica e Informatica, Università di Perugia, Italy
coletti@dmi.unipg.it
[2] Dip. S.B.A.I., Università di Roma "La Sapienza", Italy
{barbara.vantaggi,davide.petturiti}@sbai.uniroma1.it

Abstract. The classical von Neumann–Morgenstern's notion of lottery is generalized by replacing a probability distribution on a finite support with a belief function on the power set of the support. Given a partial preference relation on a finite set of generalized lotteries, a necessary and sufficient condition (weak rationality) is provided for its representation as a Choquet expected utility of a strictly increasing utility function.

Keywords: Generalized lottery, preference relation, belief function, probability envelope, Choquet expected utility.

1 Introduction

In decision problems as well as in automated reasoning, especially in situations of incomplete and revisable information, both in the classical expected utility (EU) [20,9,15] and in the Choquet expected utility (CEU) [11,17,18,6] (see also [1,14,22,21]) frameworks it can be difficult to construct the utility function u and even to test if the preferences agree with an EU (or a CEU). In fact, to find the utility u the classical methods ask for comparisons between "lotteries" and "certainty equivalent" or, in any case, comparisons among particular large classes of lotteries (for a discussion in the EU framework see [12]). For that, the decision maker is often forced to make comparisons which have little or nothing to do with the given problem, having to choose between risky prospects and certainty.

In [2], referring to the EU model, a different approach (based on a "rationality principle") is proposed: it does not need all these non-natural comparisons but, instead, it can work by considering only the (few) lotteries and comparisons of interest. Moreover, when new information is introduced, the same principle assures that the preference relation can be extended maintaining rationality, and, even more, the principle suggests how to extend it. The "rationality principle" can be summarized as follows: *it is not possible to obtain the same lottery by combining in the same way two groups of lotteries, if the first ones are strictly preferred to the second ones.*

The aim of this paper is to propose a similar approach for the CEU model by generalizing the usual definition of lottery. In detail, a *generalized lottery L*

A. Laurent et al. (Eds.): IPMU 2014, Part II, CCIS 443, pp. 444–453, 2014.

(or *g-lottery* for short) is a random quantity with a finite support X_L endowed with a Dempster-Shafer *belief function* Bel_L [5,16,19] (or, equivalently, a *basic assignment* m_L) defined on the power set $\wp(X_L)$. We recall that, in particular probabilistic inferential problems, the belief function can be obtained either as a lower envelope of a family of probabilities (see next Example 1, generalizing the famous Ellsberg "paradox" [7]) or as a coherent lower extension of a probability related to a set of lotteries with different support (see for instance [5,3,8,13,4]).

The "weak rationality principle" proposed here is not the direct generalization of the "rationality principle" given for probability, simply obtained by changing "lotteries" with "generalized lotteries": actually, the condition derived in this way is only necessary for the representability of the preference by a CEU. Such a principle is based on the following property: if the elements of the set $X = \{x_1, \ldots, x_n\}$ resulting by the union of the supports of the considered g-lotteries is totally ordered as $x_1 < \ldots < x_n$, then for every g-lottery L the Choquet integral of any strictly increasing utility function $u : X \to \mathbb{R}$, not only is a weighted average (as observed in [10]), but the weights have a clear meaning. In fact, for the least preferred prize x_1 the weight is the sum of the values of m_L on the events implied by the event $\{L = x_1\}$, for x_2 it is the sum of the values of m_L on the events implied by the event $\{L = x_2\}$ but not by $\{L = x_1\}$, and so on. This allows to map every g-lottery L to a "standard" lottery whose probability distribution is constructed (following a pessimistic approach) through the *aggregated basic assignment* M_L.

The "weak rationality principle" turns out to be a necessary and sufficient condition for the existence of a strictly increasing $u : X \to \mathbb{R}$ whose CEU represents our preferences on a finite set $\mathcal{L}$ of g-lotteries, under a natural assumption of agreement of the preference relation with the order of X.

2   A Motivating Example

To motivate the topic dealt with in this paper we introduced the following example, which is inspired to the well-known Ellsberg's paradox [7].

Example 1. Consider the following hypothetical experiment. Let us take two urns, say U_1 and U_2, from which we are asked to draw a ball each. U_1 contains $\frac{1}{3}$ of white (w) balls and the remaining balls are black (b) and red (r), but in a ratio entirely unknown to us, analogously, U_2 contains $\frac{1}{4}$ of green (g) balls and the remaining balls are yellow (y) and orange (o), but in a ratio entirely unknown to us.

In light of the given information, the composition of U_1 singles out a class of probability measures $\mathbf{P}^1 = \{P^\theta\}$ on the power set $\wp(S_1)$ of $S_1 = \{w, b, r\}$ s.t. $P^\theta(\{w\}) = \frac{1}{3}$, $P^\theta(\{b\}) = \theta$, $P^\theta(\{r\}) = \frac{2}{3} - \theta$, with $\theta \in \left[0, \frac{2}{3}\right]$. Analogously, for the composition of U_2 we have the class $\mathbf{P}^2 = \{P^\lambda\}$ on $\wp(S_2)$ with $S_2 = \{g, y, o\}$ s.t. $P^\lambda(\{g\}) = \frac{1}{4}$, $P^\lambda(\{y\}) = \lambda$, $P^\lambda(\{o\}) = \frac{3}{4} - \lambda$, with $\lambda \in \left[0, \frac{3}{4}\right]$.

Concerning the ball drawn from U_1 and the one drawn from U_2, the following gambles are considered:

	w	b	r
L_1	100€	0€	0€
L_2	0€	0€	100€
L_3	0€	100€	100€
L_4	100€	100€	0€

	g	y	o
G_1	100€	10€	10€
G_2	10€	10€	100€
G_3	10€	100€	100€
G_4	100€	100€	10€

If we express the strict preferences $L_2 \prec L_1$, $L_4 \prec L_3$, then for no value of θ there exists a function $u : \{0, 100\} \to \mathbb{R}$ s.t. its expected value on the L_i's w.r.t. P^θ represents our preferences on the L_i's. Indeed, putting $w_1 = u(0)$ and $w_2 = u(100)$, both the following inequalities must hold $\frac{1}{3}w_1 + \theta w_1 + \left(\frac{2}{3} - \theta\right) w_2 < \frac{1}{3}w_2 + \theta w_1 + \left(\frac{2}{3} - \theta\right) w_1$ and $\frac{1}{3}w_2 + \theta w_2 + \left(\frac{2}{3} - \theta\right) w_1 < \frac{1}{3}w_1 + \theta w_1 + \left(\frac{2}{3} - \theta\right) w_2$, from which, summing memberwise, we get $w_1 + w_2 < w_1 + w_2$, i.e., a contradiction. The same can be proven if we express the strict preferences $G_2 \prec G_1$, $G_4 \prec G_3$.

Now take $\underline{P}^1 = \min \mathbf{P}^1$ and $\underline{P}^2 = \min \mathbf{P}^2$, where the minimum is intended pointwise on the elements of $\wp(S_1)$ and $\wp(S_2)$, obtaining:

$\wp(S^1)$	$\emptyset$	$\{w\}$	$\{b\}$	$\{r\}$	$\{w,b\}$	$\{w,r\}$	$\{b,r\}$	S_1
$\underline{P}^1$	0	$\frac{1}{3}$	0	0	$\frac{1}{3}$	$\frac{1}{3}$	$\frac{2}{3}$	1

$\wp(S_2)$	$\emptyset$	$\{g\}$	$\{y\}$	$\{o\}$	$\{g,y\}$	$\{g,o\}$	$\{y,o\}$	S_2
$\underline{P}^2$	0	$\frac{1}{4}$	0	0	$\frac{1}{4}$	$\frac{1}{4}$	$\frac{3}{4}$	1

It is easily verified that both $\underline{P}^1$ and $\underline{P}^2$ are belief functions, i.e., n-monotone Choquet capacities, for every $n \geq 2$ (see next equation 1).

In this case, for any strictly increasing function $u : \{0, 100\} \to \mathbb{R}$ we have that the Choquet integral of u on the L_i's w.r.t. $\underline{P}^1$ represents our preferences $L_2 \prec L_1$, $L_4 \prec L_3$ and coincides with the minimum of the expected utilities of u on the L_i's w.r.t. the class $\mathbf{P}^1$. Indeed, denoting $w_1 = u(0)$ and $w_2 = u(100)$ and $\mathrm{CEU}(L_i) = \int_{Ch} u(L_i) \mathrm{d}\underline{P}^1$ we get $\mathrm{CEU}(L_1) = \mathrm{CEU}(L_4) = \frac{2}{3}w_1 + \frac{1}{3}w_2$, $\mathrm{CEU}(L_2) = w_1$, $\mathrm{CEU}(L_3) = \frac{1}{3}w_1 + \frac{2}{3}w_2$, thus $\mathrm{CEU}(L_2) < \mathrm{CEU}(L_1)$ and $\mathrm{CEU}(L_4) < \mathrm{CEU}(L_3)$ hold whenever $w_1 < w_2$. The same can be proven for the preferences $G_2 \prec G_1$, $G_4 \prec G_3$.

3   Numerical Model of Reference

Let X be a finite set of states of nature and denote by $\wp(X)$ the power set of X. We recall that a *belief function* Bel [5,16,19] on an algebra of events $\mathcal{A} \subseteq \wp(X)$ is a function such that $Bel(\emptyset) = 0$, $Bel(X) = 1$ and satisfying the n-monotonicity property for every $n \geq 2$, i.e., for every $A_1, \ldots, A_n \in \mathcal{A}$,

$$Bel\left(\bigcup_{i=1}^n A_i\right) \geq \sum_{\emptyset \neq I \subseteq \{1,\ldots,n\}} (-1)^{|I|+1} Bel\left(\bigcap_{i \in I} A_i\right). \tag{1}$$

A belief function Bel on $\mathcal{A}$ is completely singled out by its Möbius inverse, defined for every $A \in \mathcal{A}$ as

$$m(A) = \sum_{B \subseteq A} (-1)^{|A \setminus B|} Bel(B).$$

Such a function, usually called *basic (probability) assignment*, is a function m : $\mathcal{A} \to [0,1]$ satisfying $m(\emptyset) = 0$ and $\sum_{A \in \mathcal{A}} m(A) = 1$, and is such that for every $A \in \mathcal{A}$

$$Bel(A) = \sum_{B \subseteq A} m(B). \tag{2}$$

A set A in $\mathcal{A}$ is a *focal element* for m (and so also for the corresponding Bel) whenever $m(A) > 0$.

In the classical von Neumann–Morgenstern theory [20] a *lottery* L consists of a *probability distribution* on a finite *support* X_L, which is an arbitrary finite set of *prizes* or *consequences*.

In this paper we adopt a generalized notion of lottery L, by assuming that a *belief function* Bel_L is assigned on the power set $\wp(X_L)$ of X_L.

Definition 1. *A generalized lottery, or **g-lottery** for short, on a finite set X_L is a pair $L = (\wp(X_L), Bel_L)$ where Bel_L is a belief function on $\wp(X_L)$.*

Let us notice that, a g-lottery $L = (\wp(X_L), Bel_L)$ could be equivalently defined as $L = (\wp(X_L), m_L)$, where m_L is the basic assignment associated to Bel_L. We stress that this definition of g-lottery generalizes the classical one in which $m_L(A) = 0$ for any $A \in \wp(X_L)$ with card $A > 1$.

For example, a g-lottery L on $X_L = \{x_1, x_2, x_3\}$ can be expressed as

$$L = \begin{pmatrix} \{x_1\} & \{x_2\} & \{x_3\} & \{x_1, x_2\} & \{x_1, x_3\} & \{x_2, x_3\} & \{x_1, x_2, x_3\} \\ b_1 & b_2 & b_3 & b_{12} & b_{13} & b_{23} & b_{123} \end{pmatrix}$$

where the belief function Bel_L on $\wp(X_L)$ is such that $b_I = Bel_L(\{x_i : i \in I\})$ for every $I \subseteq \{1, 2, 3\}$. Notice that as one always has $Bel_L(\emptyset) = m_L(\emptyset) = 0$, the empty set is not reported in the tabular expression of L. An equivalent representation of previous g-lottery is obtained through the basic assignment m_L associated to Bel_L (where $m_I = m_L(\{x_i : i \in I\})$ for every $I \subseteq \{1, 2, 3\}$)

$$L = \begin{pmatrix} \{x_1\} & \{x_2\} & \{x_3\} & \{x_1, x_2\} & \{x_1, x_3\} & \{x_2, x_3\} & \{x_1, x_2, x_3\} \\ m_1 & m_2 & m_3 & m_{12} & m_{13} & m_{23} & m_{123} \end{pmatrix}.$$

Given a finite set $\mathcal{L}$ of g-lotteries, let $X = \bigcup \{X_L : L \in \mathcal{L}\}$. Then, any g-lottery L on X_L with belief function Bel_L can be rewritten as a g-lottery on X by defining a suitable extension Bel'_L of Bel_L.

Proposition 1. *Let $L = (\wp(X_L), Bel_L)$ be a g-lottery on X_L. Then for any finite $X \supseteq X_L$ there exists a unique belief function Bel'_L on $\wp(X)$ with the same focal elements of Bel_L and such that $Bel'_{L|\wp(X_L)} = Bel_L$.*

Proof. The extension Bel'_L is defined through the corresponding m'_L. For every $A \in \wp(X)$ we put $m'_L(A) = m_L(A)$ if $A \in \wp(X_L)$ and $m'_L(A) = 0$ otherwise. The function m'_L is easily seen to be a basic assignment on $\wp(X)$, moreover, the corresponding belief function Bel'_L on $\wp(X)$ is an extension of Bel_L and has the same focal elements. $\qquad\square$

Given $L_1, \ldots, L_t \in \mathcal{L}$, all rewritten on X, and a real vector $\mathbf{k} = (k_1, \ldots, k_t)$ with $k_i \geq 0$ $(i = 1, \ldots, t)$ and $\sum_{i=1}^{t} k_i = 1$, the *convex combination* of $L_1, \ldots, L_t$ according to $\mathbf{k}$ is defined as

$$\mathbf{k}(L_1, \ldots, L_t) = \left(\frac{A}{\sum_{i=1}^{t} k_i m_{L_i}(A)} \right) \quad \text{for every } A \in \wp(X) \setminus \{\emptyset\}. \quad (3)$$

Since the convex combination of belief functions (basic assignments) on $\wp(X)$ is a belief function (basic assignment) on $\wp(X)$, $\mathbf{k}(L_1, \ldots, L_t)$ is a g-lottery on X.

For every $A \in \wp(X) \setminus \{\emptyset\}$, there exists a *degenerate g-lottery* δ_A on X such that $m_{\delta_A}(A) = 1$, and, moreover, every g-lottery L with focal elements $A_1, \ldots, A_k$ can be expressed as $\mathbf{k}(\delta_{A_1}, \ldots, \delta_{A_k})$ with $\mathbf{k} = (m_L(A_1), \ldots, m_L(A_k))$.

4   Preferences over a Set of Generalized Lotteries

Consider a finite set $\mathcal{L}$ of g-lotteries with $X = \bigcup \{X_L : L \in \mathcal{L}\}$ and assume X is totally ordered by the relation $<$, which is a quite natural condition thinking at elements of X as money payoffs.

Let $\precsim$ be a *preference/indifference* relation on $\mathcal{L}$. For every $L, L' \in \mathcal{L}$ the assertion that "L is indifferent to L'", denoted by $L \sim L'$, summarizes the two assertions $L \precsim L'$ and $L' \precsim L$. Observe that not all the pairs of g-lotteries are necessarily compared. An additional strict preference relation can be elicited by assertions such as "L is strictly preferred to L'", denoted by $L \prec L'$. Let $\prec^*$ be the asymmetric relation formally deduced from $\precsim$, namely $\prec^* = \precsim \setminus \sim$. If the pair of relations $(\precsim, \prec)$ represents the opinion of the decision maker, then it is natural to have $\prec \subset \prec^*$: in fact, it is possible that, at an initial stage of judgement, the decision maker has not decided yet if $L \prec L'$ or $L \sim L'$ and he expresses his opinion only by $L \precsim L'$. Obviously if $\precsim$ is complete then $\prec = \prec^*$ and so for every pair (L, L') either $L \prec L'$ or $L \sim L'$.

We call the pair $(\precsim, \prec)$ *strengthened preference relation* if $\prec$ is not empty.

We say that a function $U : \mathcal{L} \to \mathbb{R}$ *represents* (or *agrees with*) $(\precsim, \prec)$ if, for every $L, L' \in \mathcal{L}$

$$L \precsim L' \Rightarrow U(L) \leq U(L') \text{ and } L \prec L' \Rightarrow U(L) < U(L'). \quad (4)$$

In analogy with [2], given $(\precsim, \prec)$ on $\mathcal{L}$, our aim is to find a necessary and sufficient condition for the existence of a utility function $u : X \to \mathbb{R}$ such that the Choquet expected utility of g-lotteries in $\mathcal{L}$ represents $(\precsim, \prec)$. In particular, with the money payoffs interpretation in mind we search for a *strictly increasing* u. To reach our goal we need first to define the Choquet expected utility of a g-lottery.

Definition 2. *Let* $u : X \to \mathbb{R}$ *be a utility function, then the* **Choquet expected utility**, *or* **CEU** *for short, of a g-lottery* $L = (\wp(X), Bel_L)$ *is*

$$\text{CEU}(L) = \int_{-\infty}^{0} (Bel_L(\{x : u(x) \geq t\}) - 1)\mathrm{d}t + \int_{0}^{+\infty} Bel_L(\{x : u(x) \geq t\})\mathrm{d}t. \quad (5)$$

Since $X = \{x_1, \ldots, x_n\}$ is totally ordered as $x_1 < \cdots < x_n$, we can define the *aggregated basic assigment* of a g-lottery L, for every $x_i \in X$,

$$M_L(x_i) = \sum_{x_i \in B \subseteq E_i} m_L(B), \qquad (6)$$

where $E_i = \{x_i, \ldots, x_n\}$ for $i = 1, \ldots, n$. Note that $M_L(x_i) \geq 0$ for every $x_i \in X$ and $\sum_{i=1}^n M_L(x_i) = 1$, thus M_L determines a probability distribution on X.

The next axiom requires that it is not possible to obtain two g-lotteries having the same aggregated basic assignment, by combining in the same way two groups of g-lotteries, if each g-lottery in the first group is not preferred to the corresponding one in the second group, and at least a preference is strict.

Definition 3. *A strengthened preference relation $(\precsim, \prec)$ on a set $\mathcal{L}$ of g-lotteries is said to be* **weakly rational** *if it satisfies the following condition:*

(WR) *For all $h \in \mathbb{N}$ and $L_i, L_i' \in \mathcal{L}$ with $L_i \precsim L_i'$ $(i = 1, \ldots, h)$, if*

$$\mathbf{k}(M_{L_1}, \ldots, M_{L_h}) = \mathbf{k}(M_{L_1'}, \ldots, M_{L_h'})$$

with $\mathbf{k} = (k_1, \ldots, k_h)$, $k_i > 0$ $(i = 1, \ldots, h)$ and $\sum_{i=1}^h k_i = 1$, then it can be $L_i \prec L_i'$ for no $i = 1, \ldots, h$. In particular, if $\precsim$ is complete, it must be $L_i \sim L_i'$ for every $i = 1, \ldots, h$.

Note that the convex combination referred to in condition **(WR)** is the usual one involving probability distributions on X. Moreover, it is easily proven that if $\mathbf{k}(L_1, \ldots, L_h) = \mathbf{k}(L_1', \ldots, L_h')$, then it also holds $\mathbf{k}(M_{L_1}, \ldots, M_{L_h}) = \mathbf{k}(M_{L_1'}, \ldots, M_{L_h'})$ but the converse is generally not true as shown in next example.

Example 2. Let $X = \{x_1, x_2\}$ with $x_1 < x_2$ and consider the g-lotteries

$$L_1 = \begin{pmatrix} \{x_1\} & \{x_2\} & \{x_1, x_2\} \\ \frac{1}{4} & \frac{3}{4} & 0 \end{pmatrix}, L_1' = \begin{pmatrix} \{x_1\} & \{x_2\} & \{x_1, x_2\} \\ \frac{1}{3} & \frac{2}{3} & 0 \end{pmatrix},$$

$$L_2 = \begin{pmatrix} \{x_1\} & \{x_2\} & \{x_1, x_2\} \\ 0 & \frac{2}{3} & \frac{1}{3} \end{pmatrix}, L_2' = \begin{pmatrix} \{x_1\} & \{x_2\} & \{x_1, x_2\} \\ 0 & \frac{3}{4} & \frac{1}{4} \end{pmatrix},$$

with the preferences $L_1 \precsim L_1'$ and $L_2 \precsim L_2'$. There is no $k \in [0, 1]$ such that $kL_1 + (1 - k)L_2 = kL_1' + (1 - k)L_2'$, indeed, the following system

$$\begin{cases} k\frac{1}{4} = k\frac{1}{3} \\ k\frac{3}{4} + (1 - k)\frac{2}{3} = k\frac{2}{3} + (1 - k)\frac{3}{4} \\ (1 - k)\frac{1}{3} = (1 - k)\frac{1}{4} \end{cases}$$

has not solution. Nevertheless, considering the aggregated basic assignments of L_1, L_1', L_2, L_2' we have

$$M_{L_1} = \begin{pmatrix} x_1 & x_2 \\ \frac{1}{4} & \frac{3}{4} \end{pmatrix}, M_{L_1'} = \begin{pmatrix} x_1 & x_2 \\ \frac{1}{3} & \frac{2}{3} \end{pmatrix}, M_{L_2} = \begin{pmatrix} x_1 & x_2 \\ \frac{1}{3} & \frac{2}{3} \end{pmatrix}, M_{L_2'} = \begin{pmatrix} x_1 & x_2 \\ \frac{1}{4} & \frac{3}{4} \end{pmatrix},$$

for which we have $\frac{1}{2}M_{L_1} + \frac{1}{2}M_{L_2} = \frac{1}{2}M_{L_1'} + \frac{1}{2}M_{L_2'}$.

In order to get a strictly increasing $u : X \to \mathbb{R}$ we have to require that $\mathcal{L}$ contains the set of degenerate g-lotteries on singletons $\mathcal{L}_0 = \{\delta_{\{x\}} : x \in X\}$ and that $\delta_{\{x\}} \prec \delta_{\{x'\}}$ when $x < x'$, for $x, x' \in X$. Actually, the decision maker is not asked to provide such a set of preferences, but in this case the initial partial preference $(\precsim, \prec)$ on $\mathcal{L}$ must be extended in order to reach this technical condition and, of course, the decision maker is asked to accept such an extension.

The following theorem proves that **(WR)** is a necessary and sufficient condition for the existence of a strictly increasing utility function u whose Choquet expected value on g-lotteries represents $(\precsim, \prec)$, moreover its proof provides a procedure to compute such a u.

Theorem 1. *Let $\mathcal{L}$ be a finite set of g-lotteries, $X = \bigcup\{X_L : L \in \mathcal{L}\}$ with X totally ordered by $<$, and $(\precsim, \prec)$ a strengthened preference relation on $\mathcal{L}$. Assume $\mathcal{L}_0 \subseteq \mathcal{L}$ and for every $x, x' \in X$, $x < x'$ implies $\delta_{\{x\}} \prec \delta_{\{x'\}}$. The following statements are equivalent:*

*(i) $(\precsim, \prec)$ is weakly rational (i.e., it satisfies **(WR)**);*
(ii) $(\precsim, \prec)$ is representable by a CEU of a strictly increasing function $u : X \to \mathbb{R}$ (unique up to a positive linear transformation).

Proof. Let $X = \{x_1, \ldots, x_n\}$ with $x_1 < \ldots < x_n$ and assume all g-lotteries in $\mathcal{L}$ are rewritten on X. Introduce the collections $S = \{(L_j, L_j') : L_j \prec L_j', L_j, L_j' \in \mathcal{L}\}$ and $R = \{(G_h, G_h') : G_h \precsim G_h', G_h, G_h' \in \mathcal{L}\}$ with $s = \operatorname{card} S$ and $r = \operatorname{card} R$.

(ii) $\Rightarrow$ (i). Condition *(ii)* holds if and only if there are n real numbers $w_i = u(x_i)$, with $w_1 < \ldots < w_n$, s.t. for all $(L_j, L_j') \in S$ we have $\operatorname{CEU}(L_j) < \operatorname{CEU}(L_j')$, and for all $(G_h, G_h') \in R$ we have $\operatorname{CEU}(G_h) \leq \operatorname{CEU}(G_h')$.

Setting $E_i = \{x_i, \ldots, x_n\}$ $(i = 1, \ldots, n)$ and $E_{n+1} = \emptyset$, for every g-lottery $L \in \mathcal{L}$ it holds (see [6])

$$\operatorname{CEU}(L) = \sum_{i=1}^{n} w_i \left[Bel_L(E_i) - Bel_L(E_{i+1}) \right]$$

$$= \sum_{i=1}^{n} w_i \left[\sum_{B \subseteq E_i} m_L(B) - \sum_{B \subseteq E_{i+1}} m_L(B) \right] = \sum_{i=1}^{n} w_i M_L(x_i).$$

Hence, condition *(ii)* is equivalent to the existence of a $(n \times 1)$ column vector $\mathbf{w}$ which is solution of the following system

$$\mathcal{S} : \begin{cases} A\mathbf{w} > \mathbf{0}, \\ B\mathbf{w} \geq \mathbf{0}, \end{cases}$$

where $A = (a^j)$ and $B = (b^h)$ are, respectively, $(s \times n)$ and $(r \times n)$ real matrices with rows $a^j = M_{L_j'} - M_{L_j}$ for $j = 1, \ldots, s$, and $b^h = M_{G_h'} - M_{G_h}$ for $h = 1, \ldots, r$.

Due to the homogeneity of $\mathcal{S}$ we can restrict to $\mathbf{w} \geq \mathbf{0}$, so by a known alternative theorem (see, e.g., [9]) the existence of a non-negative solution of $\mathcal{S}$ is equivalent to the non-solvability of the following system

$$\mathcal{S}' : \begin{cases} \mathbf{y}A + \mathbf{z}B \leq \mathbf{0}, \\ \mathbf{y}, \mathbf{z} \geq \mathbf{0}, \\ \mathbf{y} \neq \mathbf{0}, \end{cases}$$

where $\mathbf{y}$ and $\mathbf{z}$ are, respectively, $(1 \times s)$ and $(1 \times r)$ unknown row vectors. If $\mathbf{y}, \mathbf{z}$ is a solution of system $\mathcal{S}'$ then summing memberwise the inequalities related to $\mathbf{y}A + \mathbf{z}B \leq \mathbf{0}$ we get that the resulting inequality is verified as $0 = 0$, thus in system $\mathcal{S}'$ we can write $\mathbf{y}A + \mathbf{z}B = \mathbf{0}$. Now, let $k = \sum_{i=1}^{s} y_i + \sum_{i=1}^{r} z_i$ and $\mathbf{k}$ the $(1 \times s + r)$ row vector with $k_i = \frac{y_i}{k}$, for $i = 1, \ldots, s$, and $k_{s+i} = \frac{z_i}{k}$ for $i = 1, \ldots, r$. The solution $\mathbf{y}, \mathbf{z}$ of system $\mathcal{S}'$ implies

$$\mathbf{k}(M_{L_1}, \ldots, M_{L_s}, M_{G_1}, \ldots, M_{G_r}) = \mathbf{k}(M_{L_1'}, \ldots, M_{L_s'}, M_{G_1'}, \ldots, M_{G_r'}),$$

which can be restricted to the positive k_i's, and since $k_i > 0$ for at least one index $i \in \{1, \ldots, s\}$, this implies condition **(WR)** is violated.

$(i) \Rightarrow (ii)$. Suppose **(WR)** holds. In this case, considering a convex combination $\mathbf{k}(M_{L_1}, \ldots, M_{L_s}, M_{G_1}, \ldots, M_{G_r}) = \mathbf{k}(M_{L_1'}, \ldots, M_{L_s'}, M_{G_1'}, \ldots, M_{G_r'})$ where some of the k_i's can be 0, it must be that $k_i = 0$ for $i = 1, \ldots, s$, thus system $\mathcal{S}'$ cannot have solution, while $\mathcal{S}$ has solution $\mathbf{w}$. The hypothesis $\mathcal{L}_0 \subseteq \mathcal{L}$ and for every $x, x' \in X$, $x < x'$ implies $\delta_{\{x\}} \prec \delta_{\{x'\}}$, assures that $w_1 < \cdots < w_n$. Hence, $u(x_i) = w_i$, $i = 1, \ldots, n$, is a strictly increasing utility function on X whose CEU represents $(\precsim, \prec)$. Moreover, as any positive linear transformation of $\mathbf{w}$ produces another solution of system $\mathcal{S}$ it follows the unicity of u up to a positive linear transformation. $\square$

Example 3. Consider again the situation described in Example 1 and suppose to toss a fair coin and to choose among L_1 and G_1 depending on the face shown by the coin. In analogy, suppose to choose among L_2 and G_1 with a totally similar experiment. Let us denote with F_1 and F_2 the results of the two experiments.

We can transport the belief functions defined in Example 1 to the sets of prizes of each gamble L_i's and G_i', thus we obtain

$$L_1 = \begin{pmatrix} \{0\} & \{100\} & \{0, 100\} \\ \frac{2}{3} & \frac{1}{3} & 1 \end{pmatrix}, \quad L_2 = \begin{pmatrix} \{0\} & \{100\} & \{0, 100\} \\ \frac{1}{3} & 0 & 1 \end{pmatrix},$$

$$G_1 = \begin{pmatrix} \{10\} & \{100\} & \{10, 100\} \\ \frac{3}{4} & \frac{1}{4} & 1 \end{pmatrix}.$$

In order to express F_1 and F_2 we need to properly combine L_1, L_2 and G_1 and for this we have to rewrite them on the same set of prizes $\{0, 10, 100\}$

$$L_1 = \begin{pmatrix} \{0\} & \{10\} & \{100\} & \{0, 10\} & \{0, 100\} & \{10, 100\} & \{0, 10, 100\} \\ \frac{2}{3} & 0 & \frac{1}{3} & \frac{2}{3} & 1 & \frac{1}{3} & 1 \end{pmatrix},$$

$$L_2 = \begin{pmatrix} \{0\} & \{10\} & \{100\} & \{0, 10\} & \{0, 100\} & \{10, 100\} & \{0, 10, 100\} \\ \frac{1}{3} & 0 & 0 & \frac{1}{3} & 1 & 0 & 1 \end{pmatrix},$$

$$G_1 = \begin{pmatrix} \{0\} & \{10\} & \{100\} & \{0, 10\} & \{0, 100\} & \{10, 100\} & \{0, 10, 100\} \\ 0 & \frac{3}{4} & \frac{1}{4} & \frac{3}{4} & \frac{1}{4} & 1 & 1 \end{pmatrix}.$$

L_1, L_2 and G_1 can be simply regarded as belief functions on the same field, thus F_1 and F_2 can be defined as the convex combinations $F_1 = \frac{1}{2}L_1 + \frac{1}{2}G_1$ and $F_2 = \frac{1}{2}L_2 + \frac{1}{2}G_1$, obtaining

$$F_1 = \begin{pmatrix} \{0\} & \{10\} & \{100\} & \{0, 10\} & \{0, 100\} & \{10, 100\} & \{0, 10, 100\} \\ \frac{8}{24} & \frac{9}{24} & \frac{7}{24} & \frac{17}{24} & \frac{15}{24} & \frac{16}{24} & 1 \end{pmatrix},$$

$$F_2 = \begin{pmatrix} \{0\} & \{10\} & \{100\} & \{0,10\} & \{0,100\} & \{10,100\} & \{0,10,100\} \\ \frac{4}{24} & \frac{9}{24} & \frac{3}{24} & \frac{13}{24} & \frac{15}{24} & \frac{12}{24} & 1 \end{pmatrix}.$$

It is easily proven that for every strictly increasing $u : \{0,10,100\} \to \mathbb{R}$ the strict preferences $L_2 \prec L_1$, $L_4 \prec L_3$, $G_2 \prec G_1$, $G_4 \prec G_3$ are represented by their Choquet expected utility. Nevertheless, if we consider the further strict preference $F_1 \prec F_2$ then there is no strictly increasing $u : \{0,10,100\} \to \mathbb{R}$ whose Choquet expected utility represents our preference. Indeed, denoting $w_1 = u(0)$, $w_2 = u(10)$, $w_3 = u(100)$, one would have $\text{CEU}(F_1) = \frac{1}{3}w_1 + \frac{3}{8}w_2 + \frac{7}{24}w_3$ and $\text{CEU}(F_2) = \frac{1}{2}w_1 + \frac{3}{8}w_2 + \frac{1}{8}w_3$, thus $\text{CEU}(F_1) < \text{CEU}(F_2)$ would imply $w_3 < w_1$ and so a contradiction with the constraint $w_1 < w_2 < w_3$.

In particular, assuming the natural preferences $\delta_{\{0\}} \prec \delta_{\{10\}}$ and $\delta_{\{10\}} \prec \delta_{\{100\}}$, and introducing the aggregated basic assignments

$$M_{F_1} = \begin{pmatrix} 0 & 10 & 100 \\ \frac{1}{3} & \frac{3}{8} & \frac{7}{24} \end{pmatrix}, M_{F_2} = \begin{pmatrix} 0 & 10 & 100 \\ \frac{1}{2} & \frac{3}{8} & \frac{1}{8} \end{pmatrix}, M_{\delta_{\{0\}}} = \begin{pmatrix} 0 & 10 & 100 \\ 1 & 0 & 0 \end{pmatrix},$$

$$M_{\delta_{\{10\}}} = \begin{pmatrix} 0 & 10 & 100 \\ 0 & 1 & 0 \end{pmatrix}, M_{\delta_{\{100\}}} = \begin{pmatrix} 0 & 10 & 100 \\ 0 & 0 & 1 \end{pmatrix},$$

we get $\frac{3}{4}M_{F_1} + \frac{1}{8}M_{\delta_{\{0\}}} + \frac{1}{8}M_{\delta_{\{10\}}} = \frac{3}{4}M_{F_2} + \frac{1}{8}M_{\delta_{\{10\}}} + \frac{1}{8}M_{\delta_{\{100\}}}$, which implies that (**WR**) is violated.

5 Conclusions

We introduced a rationality principle (**WR**) for preference relations among random quantities equipped with a belief function (g-lotteries), inspired to a rationality principle introduced in [2]. We proved that (**WR**) is a necessary and sufficient condition for the existence of a strictly increasing utility function u on the prizes whose Choquet integral on g-lotteries represents the preferences. In the probabilistic framework the rationality principle assures and rules the extendibility of the relation to new lotteries, which is a very useful property for an actual use of the model. The extendibility of weakly rational preferences on generalized lotteries is one of our future aims.

References

1. Chateauneuf, A., Cohen, M.: Choquet expected utility model: a new approach to individual behavior under uncertainty and social choice welfare. In: Fuzzy Meas. and Int.: Th. and App., pp. 289–314. Physica, Heidelberg (2000)
2. Coletti, G., Regoli, G.: How can an expert system help in choosing the optimal decision? Th. and Dec. 33(3), 253–264 (1992)
3. Coletti, G., Scozzafava, R.: Toward a General Theory of Conditional Beliefs. Int. J. of Int. Sys. 21, 229–259 (2006)
4. Coletti, G., Scozzafava, R., Vantaggi, B.: Inferential processes leading to possibility and necessity. Inf. Sci. 245, 132–145 (2013)

5. Dempster, A.P.: Upper and Lower Probabilities Induced by a Multivalued Mapping. Ann. of Math. Stat. 38(2), 325–339 (1967)
6. Denneberg, D.: Non-additive Measure and Integral. Theory and Decision Library: Series B, vol. 27. Kluwer Academic, Dordrecht (1994)
7. Ellsberg, D.: Risk, Ambiguity and the Savage Axioms. Quart. Jour. of Econ. 75, 643–669 (1961)
8. Fagin, R., Halpern, J.Y.: Uncertainty, belief and probability. Comput. Int. 7(3), 160–173 (1991)
9. Gale, D.: The Theory of Linear Economic Models. McGraw Hill (1960)
10. Gilboa, I., Schmeidler, D.: Additive representations of non-additive measures and the Choquet integral. Ann. of Op. Res. 52, 43–65 (1994)
11. Jaffray, J.Y.: Linear utility theory for belief functions. Op. Res. Let. 8(2), 107–112 (1989)
12. Mc Cord, M., de Neufville, R.: Lottery Equivalents: Reduction of the Certainty Effect Problem in Utility Assessment. Man. Sci. 23(1), 56–60 (1986)
13. Miranda, E., de Cooman, G., Couso, I.: Lower previsions induced by multi-valued mappings. J. of Stat. Plan. and Inf. 133, 173–197 (2005)
14. Quiggin, J.: A Theory of Anticipated Utility. J. of Ec. Beh. and Org. 3, 323–343 (1982)
15. Savage, L.: The foundations of statistics. Wiley, New York (1954)
16. Shafer, G.: A Mathematical Theory of Evidence. Princeton University Press (1976)
17. Schmeidler, D.: Subjective probability and expected utility without additivity. Econometrica 57(3), 571–587 (1989)
18. Schmeidler, D.: Integral representation without additivity. Proc. of the Am. Math. Soc. 97(2), 255–261 (1986)
19. Smets, P.: Decision making in the tbm: the necessity of the pignistic transformation. Int. J. Approx. Reasoning 38(2), 133–147 (2005)
20. von Neumann, J., Morgenstern, O.: Theory of Games and Economic Behavior. Princeton University Press (1944)
21. Walley, P.: Statistical reasoning with imprecise probabilities. Chapman & Hall, London (1991)
22. Wakker, P.: Under stochastic dominance Choquet-expected utility and anticipated utility are identical. Th. and Dec. 29(2), 119–132 (1990)

Utility-Based Approach
to Represent Agents' Conversational Preferences

Kaouther Bouzouita[1], Wided Lejouad Chaari[1], and Moncef Tagina[2]

[1] SOIE Research Laboratory,
[2] RIADI Research Laboratory,
National School of Computer Studies, University of Manouba, Tunisia
{kaouther.bouzouita,wided.chaari,moncef.tagina}@ensi.rnu.tn

Abstract. With the growing interest in Multi-Agent Systems (MAS) based solutions, one can find multiple MAS conceptions and implementations dedicated to the same goal. Those systems with their complex behaviors are rarely predictable. They may provide different results according to agents' interactions sequences. Consequently, evaluation of the quality of MAS returned results became an urgent need. Our approach is interested in evaluating high level data by considering agent's preferences regarding performatives. By analogy with the economic field, agents may ask for services, so they are *consumers* and may receive different possible answers to their requests from other agents which are *producers*. We will then focus on the analysis of messages exchanged within standard interaction protocols and compute the utility value associated to every conversation. Then we conclude utility measures for each agent and for the whole MAS regarding some execution results.

Keywords: Multi-agent systems, rational agents, evaluation, utility, automaton, Mealy machine, interaction protocol, preferences, performatives.

1 Introduction

In this paper we intend to present an evaluation methodology of Multi-Agent Systems (MAS) based on the use of utility function and studying agents' interactions. Our approach aims at estimating the satisfaction of an agent about all undertaken conversations where she is asking for services and thereafter conclude the global agents utility in a given MAS.

Related Work. Many existing works present approaches for system low-level performances measurement like [1] where a queuing model based approach was defined to predict systems performances regarding response time. In [2] authors used the number of cycles necessary for agents to achieve a result, for time measurement. In [3], three aspects of evaluation were studied: structural, typological and statistical aspects. Nevertheless, emphasis was placed on the structural aspect of the evaluation. Such works do not consider the evaluation of the quality of the result returned by agents.

A. Laurent et al. (Eds.): IPMU 2014, Part II, CCIS 443, pp. 454–463, 2014.

The work described in [4] used a utility function to generate agents' activity plans. Such an approach is limited to agents executing daily activities, such as MATSim [10]. In addition, every agent's plan was evaluated independently from other agents' plans. Agents' preferences is an interesting field of study, since it is essential for the estimation of agents' satisfaction through exchanged messages. Thus some works have been concerned with representing agents' preferences by considering alternatives they can have during inference processes. For instance, agents interaction in automated negotiation processes were used in [5] for representing and learning preferences. Qualitative representations of preferences and reasoning were used.

Coste-Marquis et al. present in [6] a comparison between languages for representing preferences relation using propositional logic. Other works such as [7] studied numerical languages for describing utility functions using UCP-networks which consist in a directed graphical representation of utility functions combining two models: generalized additive models and CP-networks [8] which are graphical models for representing preference ordering. However, they just give a partial ordering.

Our Work Positioning. We are considering agents' interactions for the study of consumer agents' satisfaction regarding the conversations aimed at getting services or information. Our approach focuses on high level data analysis. We are considering performatives as preferences. In other words, we are trying to attribute utility values to performatives depending on their judged "usefulness" to the agent. Conversational preferences of consumer agents will be represented using finite-state automaton. By studying consumer agents' satisfaction, we evaluate the effectiveness of the multi-agent system. Indeed, we estimate that satisfied agents correspond to a system that could reach positive results, intermediate ones at least, which are a necessary requirement for system's performance validation. In this paper, we are not evaluating protocols since we work on standard ones which are widely adopted by many systems. Standards of the *Foundation for Intelligent Physical Agents (FIPA)* [11] will be used.

Overview of This Paper. The remainder of this paper is structured as follows: Section 2 presents the syntax of conversational preference representation and an example of open multi-agent system used in this paper. Section 3 is dedicated for the presentation of our representation approach. Section 4 details the evaluation methodology. Results are discussed in section 5. Sections 6 and 7 are respectively dedicated for future work and conclusion.

2   Preliminaries

We will describe the syntax of the utility function and preferences representation and we will finish by presenting the MAS used for the experimentation phase.

2.1   Utility Function and Preferences Representation Syntax

A conversation is a list of messages. It is started by an agent in order to communicate with other agents for different kinds of purposes, like asking for service or information. The agent asking for service is called a *consumer* agent. The one who has been asked is a *producer* agent. We use terminology of the economic field since an analogy can be detected between multi-agent model and economic model regarding services request and supply. We design by c_a^i the ith conversation started by the consumer agent a. C is the space of all possible conversations.

In the Mealy machine conceived for our evaluation approach, a conversation is shortened to a list of performatives leading to some states of the system. C is then the set of regular expressions recognized by the automaton. U is a utility function from C to $\mathbb{R}$. It can be deduced from the output function λ and attributes a real value to a conversation after considering the following parameters:

- $nb\_perform_a^i$: Number of messages having the performative *perform* in the considered conversation,

- $w_{perform}$: weight associated to the messages of performative *perform*.

Let $\succeq$ be a total order relation. For two conversations c_a^1 and c_a^2, $c_a^1 \succeq c_a^2$ means that c_a^1 is at least as good as c_a^2.

2.2   MAS Used for the Experimentation

An open multi-agent system of the *Java Agent DEvelopment Framework* (*JADE*) [12] which is the book-trading MAS will be used throughout this paper to illustrate our approach. That system allows final users to buy books. It uses a *FIPA* standard interaction protocol (IP), called *fipa-request*. It is composed of a buyer agent (consumer) and seller agent(s) (producer(s)). Buyer agent will requests a book from producers by sending a *CALL-FOR-PROPOSAL* (*CFP*) messages. These agents return the price if they find the book in their catalogs (*PROPOSE* message), otherwise a *REFUSE* message will be returned. Buyer agent can accept one of the proposals. Thus, the correspondent seller will retrieve the book from her catalog and inform the buyer agent of the success of purchase (*INFORM* message). If the seller does not find it anymore in her catalog , a *FAILURE* message will then be sent. The definition of automaton states and transitions will be based on the performatives mentioned above.

3   Conversational Preferences Representation

We will explain steps of our approach and present its properties.

3.1   Mealy Machine for Describing Conversations Evolution

Our Mealy machine will represent all possible states of a conversation. These states will characterize its main transformations. By "main transformations" we mean all changes that are significant in the generation of a response to the initial request started by the consumer agent. Every transition in the automaton will be adding a value to the utility measure of the conversation. That value can be positive, negative or null, depending on the corresponding performative. Let's first define regular expressions recognized by the automaton: Performatives are extracted from received messages. By considering all types of communicative acts, we will not only define positive and negative responses, but we will also consider intermediate kinds of responses so as to get a utility-based comparison between all possible conversations.

In the example of *book-trading* system, regular expressions of conversations may be represented as follows:

$$(CFP\{1, N\}REF\{N - i\}PROP\{i\}ACC\_PROP?(INF|FAIL)) + \qquad (1)$$

where N is the number of seller agents in the system and i is the number of sellers having the requested book in their catalogs and proposing to sell it ($0 \leq i \leq N$). Here we abbreviate performatives designations as follows: *CFP* for *CALL-FOR-PROPOSAL*, *REF* for *REFUSE* , *PROP* for *PROPOSE*, *ACC_PROP* for *ACCEPT-PROPOSAL*, *INF* for *INFORM* and finally *FAIL* for *FAILURE*.

Mealy machine, consisting in a tuple $(Q, \Sigma, \Omega, \delta, \lambda, q_0, F)$, is defined in our example as follows:

- Q: All possible states of a given conversation. In the *book-trading* example, states are the following:

  S_0: Initial state of the system. No CFP[1] is emitted yet, or CFP is emitted and as many refusals as the number of sellers are received. One iteration may also finish with the reception of a FAILURE message which sends the system back to the state S_0.

  $\{S_1^1, \ldots, S_1^N\}$: Service request process has been started (again) by emitting as CFP messages as the number of sellers. Every *CFP* moves the system from one S_1^i to the next state S_1^{i+1} ($1 \leq i < N$). Reception of a *REFUSE* message moves the system back to the previous state.

  S_2: Consumer agent receives one proposal from a seller agent. Eventually other requested agents may send *REFUSE* or *PROPOSE* messages (the system stays then in the same state S_2).

  S_3: The buyer agent has received all responses (that include at least one *PROPOSAL* message) and will choose the best received proposal (e.g. the lowest price). Then, she will send an *ACCEPT-PROPOSAL* message and wait for a confirmation from the seller (or a *FAILURE* message).

  S_4: A confirmation is sent to the buyer agent (*INFORM* message).

---

[1] Buyer agent sends a CFP message at the beginning of every iteration, with the filed receivers filled with seller agents' identifiers. We consider in our approach that a CFP message is sent to each seller agent, since each one of them receives a copy of that message.

- Σ: A finite (non-empty) input alphabet composed of abbreviations of all possible performatives that can be contained in ACL messages: $\Sigma = \{ACC\_PROP, CFP, FAIL, INF, PROP, REF\}$.

- Ω: is the output alphabet. In our case, it is the real space $\mathbb{R}$. Choice of output values will be presented and justified in the next paragraph.

- $\delta : Q \times \Sigma \rightarrow Q$: is the state-transition function, mapping a pair of a state of a conversation and a performative to the corresponding next state.

- $\lambda : Q \times \Sigma \rightarrow \Omega$: is the output function where $\Omega \equiv \mathbb{R}$. This function will be used to generate the utility function.

  Table 1 shows all values of state-transition and output function associated to a given pair of state and input symbol. Negative utility values were attributed to refusals and failures. Proposals add positive values since they correspond to initial positive responses. We assume that the more proposals the consumer agent gets, the better it is. Indeed, the agent will be more satisfied when receiving many proposals which, most likely, include the best one. This heuristic allows us to assess the quality of the result without necessarily going through the analysis of proposals' content. An *INFORM* message will add positive value too since it corresponds to a final positive response. *CALL-FOR-PROPOSAL*s and *ACCEPT-PROPOSAL* messages are neutral (they are emitted by consumer agent and add no supplementary service to her).

Table 1. State-transition/Output function values ($1 < i < N$)

	CFP	REF	PROP	ACC_PROP	INF	FAIL
S_0	$S_1^1/0$	-	-	-	-	-
S_1^i	$S_1^{i+1}/0$	$S_1^{i-1}/-1$	$S_2/1$	-	-	-
S_1^N		$S_1^{i-1}/-1$	$S_2/1$	-	-	-
S_2	-	$S_2/-1$	$S_2/1$	$S_3/0$	-	-
S_3	-	-	-	-	$S_4/1$	$S_0/-1$
S_4	-	-	-	-	-	-

- q_0: Initial state: S_0.

- F: the state ending with the book purchased is the final state. At this state, the buyer agent will be terminated. We have: $F = \{S_4\}$.

In the state diagram of Mealy machine, edges are labeled with an input symbol and an output symbol. Figure 1 shows the Mealy machine used for conversations' graphical representation.

3.2 Utility per Conversation

We have seen so far a graphical representation using a state diagram of one conversation of the *fipa-request* protocol. We will now deduce the utility value

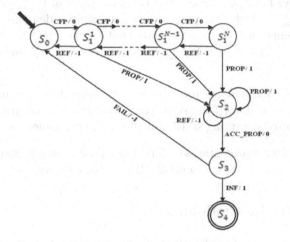

Fig. 1. Mealy machine related to a conversation in the Book-Trading System

associated to one conversation. This value will be simply defined by summing all values read on the diagram when identifying different states of that conversation. In other words, conversations will be considered as regular expressions accepted by that automaton. Utility function is:

$$U(c_a^i) = \sum_{j=1}^{nb_a^i} \lambda(S_{j-1}, p_j) \tag{2}$$

where nb_a^i is the number of messages in the conversation c_a^i, S_{j-1} the state preceding the current state and p_j is the performative of the current message. Another expression of the utility function in terms of number of performatives can be deduced from the previous expression:

$$U(c_a^i) = \sum_{performe \in \Sigma} w_{perform}.nb\_perform_a^i \tag{3}$$

where: $w_{REF} = w_{FAIL} = -1$, $w_{PROP} = w_{INF} = 1$ and $w_{CFP} = w_{ACC\_PROP} = 0$. Utility function's expression is then:

$$U(c_a^i) = w_{PROP}.nb\_PROP_a^i + w_{REF}.nb\_REF_a^i + w_{FAIL}.nb\_FAIL_a^i + w_{INF}.nb\_INF_a^i \tag{4}$$

3.3   Characteristics of the Representation

Let c_a^1 and c_a^2 be two conversations of an agent a.

- **Completeness:** Our function gives a complete preorder relation, since any two conversations c_a^1 and c_a^2 may be trivially compared using utility values: $U(c_a^1) \geq U(c_a^2) \Leftrightarrow c_a^1 \succeq c_a^2$

- **Transitivity:** Let c_a^3 be a conversation of an agent a. If $c_a^1 \succeq c_a^2$ and $c_a^2 \succeq c_a^3$ then $c_a^1 \succeq c_a^3$. In fact, this property can be simply demonstrated using to the utility function U: $c_a^1 \succeq c_a^2$ means that $U(c_a^1) \geq U(c_a^2)$, same thing for c_a^1 and c_a^3 : $U(c_a^1) \geq U(c_a^3)$. We then have $U(c_a^1) \geq U(c_a^3)$ and consequently: $c_a^1 \succeq c_a^3$.

- **Indifference:** Indifference too can be represented by our preference representation method. In fact, conversations having same utility values are considered equal regarding satisfiability for the consumer agent.

- **Efficiency:** Our representation offers a trade-off between simplicity and expressiveness by simply representing all possible conversations.

4   Evaluation Methodology

In this section, we will describe all steps in our evaluation methodology that aims at estimating consumer agents' satisfaction by all conversations.

4.1   Messages Interception

First step in our methodology consists in intercepting emitted messages during the execution of the MAS. Aspect-Oriented programming [13] was used for the implementation of two aspects: Interception of message successful emission and detection of the end of the MAS execution. After detecting one *Agent Communication Language* (ACL) message emission, that message will be stored. The aspect dedicated to the termination detection is defined according to the multi-agent system. In the example of the *book-trading* system, termination of the system corresponds to the ending of buyer agents.

4.2   Messages Semantic Analysis

Intercepted messages will be analyzed semantically to get the necessary information for the evaluation step. Every field of the message that is necessary to the evaluation such as: *performative, sender* or *receiver*, and *conversation identifier* will be extracted. Then consumers will be identified and added to a list of consumers. For this, performatives should be considered to decide whether to extract the sender field or the receiver(s) one. Performatives whose sender is the consumer: *Inform-If / Inform-Ref, Propagate / Proxy, Request / Request-When / Request-Whenever, Query-If / Query-Ref, Call-For-Proposal, Accept-Proposal, Reject-Proposal, Subscribe, Cancel.* Performatives whose sender is a producer: *Inform, Refuse, Agree, Failure, Confirm / Disconfirm, Propose.* Then we define, for each consumer, a list of conversations including all related messages (having same conversation ID).

4.3   Use of Utility Function

- **Conversation's Utility:** We apply the utility function defined above for calculating every conversation utility value. Best scenario can be defined on the automaton and by using utility function as follows: it has to include the final state (i.e. a positive response has been received) and a maximum utility.

- **Average Agent's/MAS Conversations Utility:** For every consumer agent a, an average value of all conversations' utilities will be estimated. The following function will then be used:

$$U_a = \sum_{i=1}^{nb_{conversations}} U(c_a^i)/nb_{conversations} \tag{5}$$

where $nb_{conversations}$ is the number of conversations started by a consumer agent a. After estimating every agent's satisfaction, all utility values will be considered for estimating the average utility relative to the MAS after a given execution:

$$U_{MAS} = \sum_{i=1}^{nb_{consumers}} U_i/nb_{consumers} \tag{6}$$

where $nb_{consumers}$ is the number of consumer agents.

5   Case Study

In this section, we will consider the book-trading system with multiple configuration settings. We will present two sets of results. In the first one, we used a MAS with one buyer and two sellers, and checked utility values for different numbers of iterations (see Table 2).

Table 2. Utility variation for different numbers of iterations

Number of iterations	Number of exchanged messages	Number of PROPOSE messages	Number of REFUSE messages	Number of INFORM messages	**MAS Utility**
1	5	1	1	1	1
2	8	1	3	1	−1
3	11	1	5	1	−3
4	14	1	7	1	−5
10	32	1	19	1	−17

In the second set of experiments, we considered a single buyer agent and varied the number of sellers to see the subsequent variation of the utility value. Results of Table 3 were obtained. (P.S: All sellers have the sought book in their catalogs, this is why we have one successful iteration in every execution). In both sets of experiments, number of *FAILURE* messages was equal to zero.

Table 3. Utility variation for different numbers of sellers

Number of Sellers	Number of exchanged messages	Number of PROPOSE messages	Number of REFUSE messages	Number of INFORM messages	MAS Utility
1	4	1	0	1	2
2	5	2	0	1	3
3	6	3	0	1	4
4	7	4	0	1	5
10	13	10	0	1	11

6   Results Interpretation and Discussion

In this section we will discuss the results found in the experimental section and try to interpret them in order to deduce, in the considered multi-agent system, the effect of the variation of both the number of iterations and the number of agents on the utility values.

- **First Set of Experiments Analysis:** When the number of iterations increases, this means that buyer agent did not find the requested book yet and thus, the number of refusals has increased which automatically decreases the utility of the system.

- **Second Set of Experiments Analysis:** We may clearly notice the increase of the utility value when the number of producer agents increases. In fact, in our method, the more proposals the consumer agent receives the better it is. Indeed, we estimate that it is better for the consumers to have a large set of alternatives to compare. This will lead to making a better choice.

In conclusion, best scenario corresponds to an execution with positive response, several seller agents and a minimum number of iterations.

7   Future Work

We intend to deepen our study in the field of MAS evaluation, by considering the content of positive response messages and evaluating final user's satisfaction. Weighted preferences of agents and users will then be used.

8   Conclusion

In this paper we presented an approach for the evaluation of consumer agents' satisfaction regarding services they ask from producers. Our approach uses a utilitarian preference representation that offers both a qualitative and quantitative evaluation. In fact cardinal utility was used to have an ordinal utility and give a complete preorder relation among all possible conversations. To represent those conversations we used a Mealy machine. In conclusion, we used as main criterion of evaluation the performatives of exchanged messages. Thanks to this approach, a best execution scenario can be defined as a conversation with a positive response and a maximum utility value using the automaton defined for the system's conversations recognition.

References

1. Gnanasambandam, N., Lee, S., Kumara, S.R.T., Gautam, N., Peng, W., Manikonda, V., Brinn, M., Greaves, M.: An Autonomous Performance Control Framework for Distributed Multi-Agent Systems: A Queueing Theory Based Approach. In: Proceedings of the AAMAS (2005)
2. Kaddoum, E., Gleizes, M.P., George, J.P., Glize, P., Picard, G.: Analyse des critères d'évaluation de systèmes multi-agents adaptatifs (regular paper). In: Journées Francophones Sur Les Systèmes Multi-Agents (JFSMA 2009), Lyon, Cépaduès, pp. 123–132 (2009)
3. Ben Hmida, F., Lejouad Chaari, W., Tagina, M.: Graph theory to evaluate communication in industrial multiagent systems. International Journal of Intelligent Information and Database Systems (IJIIDS) 5(4), 361–388 (2011)
4. Charypar, D., Nagel, K.: Generating complete all-day activity plans with genetic algorithms. Transportation 32(4), 369–397 (2005)
5. Aydoğan, R.: Preferences and Learning in Multi-agent Negotiation. In: AAAI-DC, Atlanta, USA, pp. 1972–1973 (2010)
6. Coste-Marquis, S., Lang, J., Liberatore, P., Marquis, P.: Expressive power and succinctness of propositional languages for preference representation. In: Proceedings of the Ninth International Conference (KR 2004), Whistler, Canada, June 2-5, pp. 203–212. AAAI Press (2004)
7. Boutilier, C., Bacchus, F., Brafman, R.: UCPnetworks: a directed graphical representation of conditional utilities. In: Proc. of UAI 2001, pp. 56–64 (2001)
8. Boutilier, C., Brafman, R.I., Domshlak, C., Hoos, H.H., Poole, D.: Cp-nets: A tool for representing and reasoning with conditional ceteris paribus preference statements. J. Artif. Intell. Res (JAIR) 21, 135–191 (2004)
9. Amato, C., Bonet, B., Zilberstein, S.: Finite-State Controllers Based on Mealy Machines for Centralized and Decentralized POMDPs. Paper presented at the Meeting of the AAAI (2010)
10. MATSim web site, http://matsim.org/ (last accessed on December 1, 2013)
11. FIPA web site, http://www.fipa.org (last accessed on October 10, 2013)
12. JADE web site, http://jade.tilab.com (last accessed on October 5, 2013)
13. AspectJ, http://eclipse.org/aspectj/ (last accessed on October 19, 2013)

Alternative Decomposition Techniques
for Label Ranking

Massimo Gurrieri*, Philippe Fortemps, and Xavier Siebert

UMons, Rue du Houdain 9, 7000 Mons, Belgium
massimo.gurrieri@umons.ac.be

Abstract. This work focuses on label ranking, a particular task of preference learning, wherein the problem is to learn a mapping from instances to rankings over a finite set of labels. This paper discusses and proposes alternative reduction techniques that decompose the original problem into binary classification related to pairs of labels and that can take into account label correlation during the learning process.

Keywords: Preference Learning, Label Ranking, Reduction Techniques, Machine Learning, Binary Classification.

1 Introduction

Preference learning [1] is gaining increasing attention in data mining and related fields. Preferences can be considered as instruments to support or identify liking or disliking of an object over others in a declarative and explicit way. In particular, the learning and modelling of preferences are being recently investigated in several fields such as knowledge discovery, machine learning, multi-criteria decision making, information retrieval, social choice theory and so on. In a general meaning, preference learning is a non trivial task consisting in inducing predictive preference models from collected empirical data. The most challenging aspect is the possibility of predicting weak or partial orderings of classes (labels), rather than single values (as in supervised classification). For this reason, preference learning can be considered as an extension of conventional supervised learning tasks, wherein the input space can be interpreted as the set of preference contexts (e.g. queries, users) while the output space consists in the preference predictions provided in the form of partial orders, linear orders, top-k lists, etc. Preference learning problems are typically distinguished in three topics [1]: object ranking, instance ranking and label ranking. **Object ranking** consists in finding a ranking function F whose input is a set X of instances characterized by attributes and whose output is a ranking of this set of instances, in the form of a weak order. Such a ranking is typically obtained by giving a score to each $x \in X$ and by ordering instances with respect to these scores. The training process takes as input either partial rankings or pairwise preferences between instances of X. In the context of **instance ranking**, the goal is to find a ranking function

---

* Corresponding author.

A. Laurent et al. (Eds.): IPMU 2014, Part II, CCIS 443, pp. 464–474, 2014.
© Springer International Publishing Switzerland 2014

F whose input is a set X of instances characterized by attributes and whose output is a ranking of this set (again a weak order on X). However, in contrast with object ranking, each instance x is associated with a class among a set of ordered classes. The output of a such a kind of problem consists in rankings wherein instances labeled with higher classes are preferred (or precede) instances labeled with lower classes. The third learning scenario concerns a set of training instances which are associated with rankings over a finite set of *labels*, i.e. **label ranking** [2, 3, 4, 5, 6, 8]. The main goal in label ranking is to predict weak or partial orderings of labels. This paper is organized as follows. In section 2, we introduce label ranking and existing approaches. In particular, we discuss learning reduction techniques that transform label ranking into binary classification. In section 3, we describe some novel reduction techniques to reduce label ranking to binary classification that are capable of taking into account correlations among labels during the learning process. Finally, in section 4 and 5, we present some experimental results, conclusions and future work, respectively.

2 Label Ranking

In label ranking, the main goal is to predict for any instance x, from an instance space X, a preference relation $\succ_x: X \to L$, where $L = \{\lambda_1; \lambda_2; ...; \lambda_k\}$ is a set of labels or alternatives, such that $\lambda_i \succ_x \lambda_j$ means that instance x prefers label λ_i to label λ_j or, equivalently, λ_i is ranked higher than λ_j. More specifically, we are interested to the case where $\succ_x$ is a total strict order over L, or equivalently, a ranking of the entire set L. This ranking can therefore be identified with a permutation $\pi_x \in \Omega$ (the permutation space of the index set of L), such that $\pi_x(i) < \pi_x(j)$ means that label λ_i is preferred to label λ_j ($\pi_x(i)$ represents the position of label λ_i in the ranking). As in classification, it is possible to associate x to an unknown *probability distribution* $\mathbb{P}(.|x)$ over the set Ω so that $\mathbb{P}(\tau|x)$ is the probability to observe the ranking τ given the instance x. Typically, the prediction quality of a label ranker M is measured by means of its *expected loss* on rankings:

$$\mathbb{E}(D(\tau_x, \tau'_x)) = \mathbb{E}(D(\tau, \tau')|x) \qquad (2.1)$$

where $D(.,.)$ is a distance function (between permutations), τ_x is the true value (ground truth) and τ'_x is the prediction made by the model M. Given such a distance metric, the best prediction is: $\tau^* = \arg\min_{\tau' \in \Omega} \sum_{\tau \in \Omega} \mathbb{P}(\tau|x)D(\tau', \tau)$. Spearman's footrule, Kendall's tau and the sum of squared distances are well-known distances between rankings [14, 15, 17]. There are two main groups of approaches to label ranking. On the one hand, decomposition (or learning reduction) methods transform label ranking problem into binary classification [2, 3, 4, 5]. On the other hand, direct methods adapt existing classification algorithms in order to deal with label ranking [6, 9, 10, 16]. This work focuses on decomposition methods because they directly learn binary preferences, i.e. simple statements like $x \succ y$ and allow to build meta-learners, i.e. rankers where any binary classifier can be used as base classifier. For example, a rule-based label ranker has been recently proposed [2, 3]. In the context of multi-label classification, it has been recently reported [7, 13] that it is crucial to take into account correlations between labels. As a consequence, it seems natural to put this issue into perspective also in label ranking. However, the standard pairwise learning reduction [5] does not take such correlations into account: a separate binary classifier is trained for each pair of labels so that each pair of labels is treated independently from the remaining pairs. In view of this, we propose in this paper alternative decomposition techniques, other than [4, 5], to take into account correlations among labels, while limiting computational complexity.

3   Reduction Framework

In the context of label ranking, the training set is $T = \{(\mathbf{x}, \pi_x)\}$, where $\mathbf{x} = (q_1, q_2, ..., q_l)$ is a vector of l attributes (the feature vector) and π_x is the corresponding target label ranking associated with the instance $\mathbf{x}$. In sections 3.1, 3.2, 3.3 we present three novel pairwise reductions techniques: Nominal decomposition (similar to the one presented in [2, 3]), Dummy Coding decomposition and Classifier Chains (based on the Classifier Chains for Multi-label Classification [7]). In section 3.4 we also discuss a method for the ranking aggregation problem. A probabilistic interpretation of the Classifier Chains for label ranking is finally presented in section 3.5.

3.1   Pairwise Decomposition: Nominal Coding

In this decomposition, each learning instance $(\mathbf{x}, \pi_x) = (q_1, q_2, ..., q_l, \pi_x)$ is transformed into a set of simpler and easier-to-learn instances $\{\mathbf{x}_{1,2}, \mathbf{x}_{1,3}, ..., \mathbf{x}_{i,j}, ...\}$, where the generic instance $\mathbf{x}_{i,j}$ is responsible to convey not only the feature vector $(q_1, q_2, ..., q_l)$ but also information about a specific pair of labels (λ_i, λ_j), according to a given decomposition of π_x into pairwise preferences. The number of pairs is at most (in case of full rankings) $k(k-1)/2$, where $k = |L|$. The learning process associated with this reduction is:

$$\mathbf{x}_{i,j} = (q_1, q_2, ..., q_l, r_{i,j}, d) \tag{3.1}$$

with $i, j \in \{1, 2, ..., k\}, i < j, d \in \{-1, +1\}$ and $r_{i,j}$ is a nominal attribute which uniquely identifies the pair (λ_i, λ_j). In this manner, each $\mathbf{x}_{i,j}$ is a learning instance responsible only for the specific pair (λ_i, λ_j). The binary attribute $d \in \{-1; +1\}$ takes into account the preference relation between the two labels (λ_i, λ_j), according to the original ranking π_x. That is, $d = +1$ when λ_i is preferred to λ_j (or ranked higher), otherwise $d = -1$, according to the provided input $(\mathbf{x}, \pi_x)$. For example, the instance $\mathbf{x} = (-1.5, 2.4, 1.6, \lambda_2 \succ \lambda_1 \succ \lambda_3)$ generates the following learning instances: $\mathbf{x}_{1,2} = (-1.5, 2.4, 1.6, r_{1,2}, -1)$, $\mathbf{x}_{1,3} = (-1.5, 2.4, 1.6, r_{1,3}, +1)$, $\mathbf{x}_{2,3} = (-1.5, 2.4, 1.6, r_{2,3}, +1)$. By using this reduction, it is possible to treat the overall pairwise preference information in a single learning set, instead of creating independent learning sets as in [5]. This allows to process the overall preference information at once in a unique learning set and therefore to learn a model M that takes into account correlations between labels, if any. The classification problem derived from the original label ranking problem can be solved by any binary classifier (e.g. Multilayer Perceptron). Assuming a given base learner whose complexity is $\Phi(l, |X|)$, the complexity of this reduction is $\Phi(l+1, |X| \times p)$, where $p = k(k-1)/2$, since the number of instances is multiplied by p (i.e. a copy of the original instance for every pair of labels). To classify a new instance x', for each pair of labels (λ_i, λ_j), with $i, j \in \{1, 2, ..., k\}, i < j$, the feature vector of the testing instance x' is augmented by adding a variable $r_{i,j}$ one at a time. This allows the model M to predict either $+1$ or -1 for the specific query pair (λ_i, λ_j).

3.2 Pairwise Decomposition: Dummy Coding

To avoid the use of nominal attributes, another reduction is based on the dummy coding so that ones and zeros are added to the feature space in order to convey the preference information about pairs of labels. Similarly as in (3.1), each learning instance $(\mathbf{x}, \pi_x) = (q_1, q_2, ..., q_l, \pi_x)$ is transformed into a set of instances $\{\mathbf{x}_{1,2}, \mathbf{x}_{1,3}, ..., \mathbf{x}_{i,j}, ...\}$. While in the previous reduction the feature space was augmented by one (a nominal attribute identifying a pair of labels), in this reduction scheme the feature space is augmented by exactly p binary attributes, where only one attribute is set to 1, the others being set to 0. The learning process associated with this reduction is:

$$\mathbf{x}_{i,j} = (q_1, q_2, ..., q_l, r_{1,2}, ..., r_{i,j}, ...r_{k-1,k}, d) \qquad (3.2)$$

with $i, j \in \{1, 2, ..., k\}, i < j$, $d \in \{-1, +1\}$ and $r_{v,z} = 1$ if $v = i \wedge z = j$, 0 otherwise. Since for every pair (i, j) only one variable $r_{i,j}$ is set to 1, the corresponding learning instance $\mathbf{x}_{i,j}$ is responsible for that specific pair. For example, the instance: $\mathbf{x} = (-1.5, 2.4, 1.6, \lambda_2 \succ \lambda_1 \succ \lambda_3)$ generates the following learning instances: $\mathbf{x}_{1,2} = (-1.5, 2.4, 1.6, 1, 0, 0, -1)$, $\mathbf{x}_{1,3} = (-1.5, 2.4, 1.6, 0, 1, 0, +1)$, $\mathbf{x}_{2,3} = (-1.5, 2.4, 1.6, 0, 0, 1, +1)$. Assuming a given base learner whose complexity is $\Phi(l, |X|)$, the complexity of this reduction is $\Phi(l + p, |X| \times p)$, where $p = k(k-1)/2$, since p binary attributes are added to each instance $\mathbf{x}_{i,j}$, while the number of instances is multiplied by p (i.e. a copy of the original instance for each pair of labels). The classification of a new instance x' is similar to the nominal decomposition.

3.3 Pairwise Decomposition: Classifier Chains

In this section we present another learning reduction technique which is based on Classifier Chains for multi-label classification [7, 13]. The proposed reduction scheme involves $p = k(k-1)/2$ binary classifiers, each binary classifier being responsible for learning and predicting the preference for a specific pair, given preference relations on previous pairs of labels, in a *chaining* scheme. In this way all previous pairs of labels are treated as additional attributes to model conditional dependence between a given pair of labels and all preceding pairs. The most interesting aspect of this reduction scheme is that it is possible to propagate preference information about pairs of labels between all classifiers throughout the chain, enabling thus to take correlations among labels into account. Moreover, there is a gain in complexity w.r.t. the previous reductions because the size of the learning set does not change at each iteration. The set of binary classifiers (the chain) $h = (h_1, h_2, ..., h_p)$ is used to model a global label ranker where each classifier h_j is trained with

$$\mathbf{x} = (q_1, q_2, ..., q_l, r_1, r_2, ..., r_{j-1}, d) \tag{3.3}$$

as a learning instance, $r_1, r_2, ..., r_{j-1}$ being the values (either $+1$ meaning $\succ$, or -1 meaning $\prec$) on the $j - 1$ previous pairs of labels provided by the ranking π_x and according to the chosen order of decomposition. The attribute $d \in \{-1; +1\}$ is the preference information about the jth pair of labels also according to π_x and to the order of decomposition. It should be noticed that a default or a random order of labels can be considered in the decomposition of the label set L. Assuming a given base learner whose complexity is $\Phi(l, |X|)$, the complexity of each single classifier h_i is $\Phi(l + c_i, |X|)$, where $1 \le c_i \le p$, since c_i attributes (binary variables) are added (at most p, in the last classifier) to each instance. For example, if $|L| = 3$ and the order decomposition is $\{(2, 1), (2, 3), (3, 1)\}$, the chain consists in (h_1, h_2, h_3) and the input instance $x = (-1.5, 2.4, 1.6, \lambda_2 \succ \lambda_1 \succ \lambda_3)$ is used as a learning instance in the following way: $h_1 \leftarrow (-1.5, 2.4, 1.6, +1)$, $h_2 \leftarrow (-1.5, 2.4, 1.6, +1, +1)$, $h_3 \leftarrow (-1.5, 2.4, 1.6, +1, +1, -1)$. The classification of a new instance x' is performed in the following way. The classifier h_1 predicts the value (either $+1$ or -1) for the first pair of labels, according to the given decomposition order. Afterwards, the feature vector of x' is augmented with the prediction on the first pair of label and the classifier h_2 predicts the value of the second pair of labels (by testing the feature vector of x' augmented by the previous prediction). In an iterative way the classifier h_j predicts the value of the jth pair using the feature vector augmented by all previous predictions provided by $(h_1, h_2, ..., h_{j-1})$. Since the order of labels could have an impact on the prediction accuracy, we also consider the ensemble scheme proposed in [7, 13]. In this manner, it is possible to avoid not only the bias due to a single (default or random) order of labels but also the effect of error propagation along the chain in case the first classifiers perform poorly. The main idea is to train T classifier chains (typically $T = 10$) where each classifier is given a random label order and moreover, each classifier is trained on a random selection of learning instances *sampled with replacement* (typically 75% of the learning set) in order to reduce time complexity without loss in prediction quality. It should be noticed that to avoid the use of an ensemble of classifier chains, some heuristics could be used to select the most appropriate order. Such heuristics are currently under study.

3.4   Ranking Generation Process

The reduction techniques (3.1), (3.2) and (3.3) require an additional step to provide a final ranking for a testing instance x'. The final ranking should be as much as possible consistent with the preference relations $\succ_{x'}$ on each pair of labels learned during the classification process. However, this is not trivial [2, 3, 4, 11, 14, 15] since the resulting preference relation is total, asymmetric, irreflexive but not transitive, in general. The underlying problem is how to find a consensus between the pairwise predictions in order to obtain a linear order? This is related to the well-known NP-hard Kemeny optimal rank aggregation problem [14, 15]. A natural choice, at least in this context, for solving the ranking aggregation problem is the *Net Flow Score* procedure [2, 3, 11] whose complexity is $O(k^2)$, where k is the number of labels. This procedure allows to obtain a ranking by ordering labels according to their net flow scores. These scores can be computed by using estimations of conditional probabilities on pairs of labels (good estimations can be provided for example by neural networks [18]) and are defined as follows. Let us define:

$$\Gamma^+_{(i,j)} = \mathbb{P}(\lambda_i \succ_{x'} \lambda_j) = \mathbb{P}(d = +1|x') \qquad (3.4)$$

as the probability that for the instance x' label λ_i is ranked higher (preferred to) than λ_j. Each label λ_i is then evaluated by means of the following score:

$$S(i) = \sum_{j \neq i}(\Gamma^+_{(i,j)} - \Gamma^+_{(j,i)}), \qquad (3.5)$$

where $\Gamma^+_{(i,j)}$ is given by (3.4). The final ranking is obtained by ordering labels according to decreasing values of (3.5), so that the higher the score, the higher the preference in the ranking: $S(i) > S(j) \Leftrightarrow \tau_i < \tau_j$. It is possible to prove, in a similar way as proved in [5], that the *Net Flow Score* procedure, as defined in (3.5), minimizes the expected loss (2.1), according to the sum of squared rank distance. This means that, if correct posterior probabilities can be obtained (or at least good estimations thereof), it is possible to find an optimal ranking by simply ordering labels according to scores (3.5). Even though the net flow score procedure does not provide in general optimal rankings w.r.t. the Kendall distance, empirically it provides good performances (see section 4).

3.5   Probabilistic Classifier Chains

In the context of multi-label classification, a Bayes-optimal probabilistic classifier chains has been recently discussed [13]. In this section, we discuss a probabilistic classifier chains for Label Ranking which relies on the same idea. Given a decomposition order of the label set into pairs, a permutation π can be identified in a unique way with a binary vector $(y^\pi_1, ..., y^\pi_p) \in \{-1, +1\}^p$ so that

$y_i^\pi = +1 \Leftrightarrow \lambda_j \succ \lambda_v$ while $y_i^\pi = -1$ otherwise. In this manner, the probability of a permutation π is equivalent to the probability of its associated vector $(y_1^\pi, ..., y_p^\pi)$ and by means of the chain rule as in a Bayesian network:

$$\mathbb{P}(\pi|x') = \mathbb{P}(y_1^\pi, ..., y_p^\pi|x') = \mathbb{P}(y_1^\pi|x') * \prod_{i=2}^{p} \mathbb{P}(y_i^\pi|x', y_1^\pi, ..., y_{i-1}^\pi). \qquad (3.6)$$

The chaining procedure (3.3) allows to learn a probabilistic classifier f_i, $i = 1, ..., p$ for each pair of labels, where $p = k * (k-1)/2$. This classifier predicts, for the ith pair of labels $y_i = (\lambda_j, \lambda_v)$, either $+1$ meaning $\lambda_j \succ \lambda_v$ or -1 meaning that $\lambda_v \succ \lambda_j$, according to the probability distribution learnt by the classifier f_i. By knowing for each pair of labels its (conditional) probability, it is possible to compute the (conditional) probability of π. By means of (3.6), it is therefore possible to rank all possible permutations for x' w.r.t. their probabilities so that an optimal prediction is given by:

$$\pi^* = \arg\max_{\pi \in S} \; [\mathbb{P}(y_1^\pi|x') * \prod_{i=2}^{p} \mathbb{P}(y_i^\pi|x', y_1^\pi, ..., y_{i-1}^\pi)] \qquad (3.7)$$

As a result, this probabilistic formulation is well-tailored for the subset $0/1$ loss function [7, 13]:

$$L(\pi, \pi') = \mathbb{1}_{\pi \neq \pi'}. \qquad (3.8)$$

Interestingly, it can easily be proved that the optimal prediction for the associated **risk** minimization problem is given by: $\pi^* = \arg\max_{\pi \in S} \mathbb{P}(\pi|x')$. Moreover, the classifier chains presented in section 3.3 can be considered as a deterministic approximation of (3.6), as similarly pointed out in [13]. While the probabilistic approach evaluates all possible permutations, the classifier chain provides, in general, a suboptimal prediction gradually obtained at each iteration of the chaining scheme by using:

$$\mathbf{h}_j(x') = \arg\max_{r_j \in \{-1,+1\}} \mathbb{P}(r_j|x', r_1, ..., r_{j-1}). \qquad (3.9)$$

As a consequence, the chaining scheme (3.9) does not provide in general neither an optimal solution w.r.t the subset $0/1$ loss function nor a linear order. Interestingly, the probabilistic approach (3.6) does not require any *ranking aggregation algorithm* since it directly evaluates permutations. However, a label ranker well-tailored for the subset $0/1$ loss is probably not reasonable in this context given that it is a quite severe loss (even a slightly different prediction gets the highest penalty). Nevertheless, a method well-tailored for the subset $0/1$ loss function should exibit good performances w.r.t. other loss functions (Kendall's tau distance, Spearman's footrule, etc.). On the other hand, the cost in terms of computational complexity is very high: in case of k labels, $k!$ permutations have to be evaluated, which imposes an upper limit of $k \approx 6$ labels. Nevertheless, it should be noticed that a stop criterion could be applied in order to reduce time complexity. If $p*$ is the probability associated with an initial permutation π_0, the evaluation of another permutation π_k can be stopped at the jth iteration as soon as:

$$\mathbb{P}(y_1^{\pi_k}|x') * \prod_{i=2}^{j} \mathbb{P}(y_i^{\pi_k}|x', y_1^{\pi_k}, ..., y_{i-1}^{\pi_k}) < p^* \tag{3.10}$$

This stop criterion allows to discard not only the permutation π_k but also all permutations wherein the first j pairs share the same values as π_k. This criterion could considerably reduce the complexity during the testing process. As in section 3.3, an ensemble of probabilistic classifier chains can be considered.

4   Experimental Setup and Results

This section is devoted to experimentations that we conducted to evaluate the performances of the proposed methods in terms of predictive accuracy. The data sets used in this paper were taken from the KEBI Data Repository[1]. The evaluation measures used in this study are the *Kendall's tau* and the *Spearman Rank Correlation coefficient* [3, 4, 16, 17]. Performance of the methods was estimated by using a cross-validation study (10-fold). We compared the standard pairwise comparison (SD) [5] (note that the Net Flow score is used for the rank aggregation issue) with the proposed reductions: the nominal decomposition (ND), the dummy coding decomposition (DD), random classifier chains (CD) and ensembled classifier chains (ECD) (the voting procedure for the final ranking is also based on the Net Flow Score procedure). In this experiment, we used Multilayer Perceptron (MLP) and Radial Basis Function (RBF) as base classifiers, both with default parameters, which generally provide good estimations of posterior probabilities [18]. All experiments were run on a 64-bit machine, allowing up to 4 GB RAM of heap memory size for larger datasets. Results w.r.t. the probabilistic classifiers chains (PCD) and ensembled probabilistic classifiers chains (EPCD) are not yet available. Tables 1 and 2 show the performances of the five classifiers in terms of Kendall's tau and Spearman's Rank correlation with MLP and RBF as base classifiers, respectively. Following the Friedman Test described in [12], we found that in both cases the null-hypothesis is rejected at a significance level of 1%. According to the post-hoc Nemenyi test [12], the significant difference in average ranks of the classifiers is 1.760 at a significant level of 5% and 1.587 at a significant level of 10%. At a significance level of 5%, ECD outperforms SD and ND when using MLP as base classifier, while the post-hoc test is not powerful enough to establish any other statistical difference. At a significance level of 10%, ECD also outperforms CC. When using RBF as base classifier, at a significant level of 5%, ECD outperforms ND and DD while SD outperforms ND and DD. Moreover, CD outperforms ND.

---

[1] See http://www.uni-marburg.de/fb12/kebi/research/repository

Table 1. Comparison of reduction techniques with MLP as base classifier

	Kendall tau				
	SD	ND	DD	CD	ECD
IRIS	.973+-.045 (4)	.964+-.055 (5)	**.991+-.017** (1)	.982+-.021 (2)	.977+-.029 (3)
GLASS	.880+-.064 (2)	.860+-.079 (5)	.865+-.062 (4)	.878+-.055 (3)	**.888+-.056** (1)
WINE	.929+-.048 (4)	.925+-.040 (5)	**.939+-.059** (1)	.936+-.036 (2.5)	.936+-.048 (2.5)
VEHICLE	.875+-.028 (4)	.877+-.023 (5)	.877+-.023 (3)	.892+-.026 (2)	**.893+-.020** (1)
VOWEL	**.910+-.014** (1.5)	.825+-.038 (5)	.861+-.022 (4)	.888+-.027 (3)	**.910+-.014** (1.5)
STOCK	.830+-.013 (5)	.874+-.010 (3)	.868+-.009 (4)	.905+-.010 (2)	**.914+-.017** (1)
CPU	.443+-.011 (4)	.472+-.015 (3)	.479+-.009 (2)	.431+-.024 (5)	**.487+-.014** (1)
BODYFAT	.229+-.054 (4)	.241+-.065 (3)	**.272+-.042** (1)	.150+-.081 (5)	.243+-.072 (2)
DDT	.062+-.040 (5)	.103+-.027 (3)	.120+-.022 (2)	.069+-.041 (4)	**.123+-.022** (1)
HOUSING	.641+-.032 (5)	.712+-.040 (2)	.699+-.032 (3)	.667+-.061 (4)	**.721+-.034** (1)
AUTORSHIP	.858+-.023 (5)	.929+-.016 (3)	.915+-.015 (4)	.937+-.010(2)	**.941+-.015** (1)
WISCONSIN	**.583+-.039** (1)	.108+-.111 (5)	.294+-.125 (4)	.451+-.014 (3)	.573+-.031 (2)
Av. Rate	3.70	3.91	2.75	3.12	1.50

	Spearman rank correlation				
	SD	ND	DD	CD	ECD
IRIS	.980+-.033 (4)	.973+-.041 (5)	**.993+-.013** (1)	.986+-.016 (2)	.983+-.022 (3)
GLASS	.908+-.059 (4)	.891+-.078 (5)	.891+-.072 (4)	.900+-.059 (2)	**.918+-.057** (1)
WINE	.944+-.042 (4)	.943+-.030 (5)	**.954+-.044** (1)	.952+-.027 (2)	.949+-.042 (3)
VEHICLE	.901+-.026 (4)	.880+-.031 (5)	.902+-.021 (3)	.913+-.025 (2)	**.916+-.018** (1)
VOWEL	**.956+-.009** (1.5)	.900+-.031 (5)	.928+-.014 (4)	.930+-.022 (3)	**.956+-.008** (1.5)
STOCK	.902+-.008 (5)	.931+-.005 (3)	.928+-.005 (4)	.947+-.007 (2)	**.953+-.011** (1)
CPU	.520+-.011 (4)	.536+-.017 (3)	.543+-.013 (2)	.475+-.028 (5)	**.547+-.017** (1)
BODYFAT	.297+-.062 (4)	.315+-.078 (3)	**.347+-.047** (1)	.196+-.098 (5)	.317+-.090 (2)
DDT	.071+-.044 (5)	.119+-.029 (3)	.135+-.028 (2)	.076+-.050 (4)	**.141+-.024** (1)
HOUSING	.751+-.028 (4)	.802+-.040 (2)	.791+-.029 (3)	.742+-.059 (5)	**.807+-.035** (1)
AUTORSHIP	.895+-.018 (5)	.954+-.011 (3)	.942+-.015 (4)	.958+-.007 (2)	**.963+-.010** (1)
WISCONSIN	**.737+-.040** (1)	.158+-.155 (5)	.400+-.161 (4)	.589+-.021 (3)	.728+-.035 (2)
Av. Rate	3.66	3.91	2.75	3.12	1.54

Table 2. Comparison of reduction techniques with RBF as base classifier

	Kendall tau				
	SD	ND	DD	CD	ECD
IRIS	.968+-.044 (3)	.808+-.063 (5)	.844+-.057 (4)	.982+-.035 (2)	**.986+-.020** (1)
GLASS	.876+-.049 (2)	.687+-.083 (5)	.707+-.053 (4)	.852+-.051 (3)	**.885+-.040** (1)
WINE	.958+-.050 (3)	.749+-.011 (5)	.828+-.061 (4)	**.977+-.033** (1)	.973+-.038 (2)
VEHICLE	.808+-.034 (3)	.544+-.041 (5)	.548+-.095 (4)	.816+-.017 (2)	**.830+-.025** (1)
VOWEL	**.815+-.015** (1)	.309+-.044 (5)	.313+-.040 (4)	.728+-.030 (3)	.786+-.019 (2)
STOCK	**.861+-.013** (1)	.527+-.072 (5)	.609+-.098 (4)	.841+-.016 (3)	.853+-.010 (2)
CPU	**.429+-.017** (1)	.245+-.034 (5)	.254+-.009 (4)	.400+-.012 (2)	.397+-.009 (3)
BODYFAT	.174+-.077 (2)	.127+-.046 (4)	.125+-.053 (5)	.143+-.038 (3)	**.179+-.053** (1)
DDT	.126+-.030 (3)	.114+-.034 (4)	.130+-.045 (2)	.112+-.027 (5)	**.144+-.018** (1)
HOUSING	.659+-.019 (2)	.441+-.111 (4)	.438+-.087 (5)	.642+-.049 (3)	**.665+-.037** (1)
AUTORSHIP	.935+-.017 (3)	.517+-.073 (5)	.584+-.103 (4)	**.936+-.013** (1.5)	**.936+-.009** (1.5)
WISCONSIN	.459+-.030 (2)	.198+-.089 (4)	.177+-.071 (5)	.427+-.047 (3)	**.465+-.037** (1)
Av. Rate	2.16	4.75	4.08	2.54	1.45

	Spearman rank correlation				
	SD	ND	DD	CD	ECD
IRIS	.976+-.033 (3)	.856+-.047 (5)	.883+-.042 (4)	.986+-.026 (2)	**.990+-.015** (1)
GLASS	.910+-.041 (2)	.739+-.084 (5)	.735+-.043 (4)	.886+-.048 (3)	**.922+-.032** (1)
WINE	.965+-.044 (3)	.803+-.085 (5)	.865+-.051 (4)	**.983+-.025** (1)	.980+-.029 (2)
VEHICLE	.842+-.029 (3)	.613+-.051 (5)	.621+-.098 (4)	.853+-.017 (2)	**.868+-.023** (1)
VOWEL	**.895+-.012** (1)	.347+-.055 (4)	.345+-.044 (5)	.808+-.030 (3)	.874+-.012 (2)
STOCK	**.923+-.009** (1)	.631+-.079 (5)	.706+-.111 (4)	.908+-.012 (3)	.918+-.006 (2)
CPU	**.480+-.019** (1)	.304+-.042 (5)	.309+-.009 (4)	.448+-.012 (2)	.442+-.012 (3)
BODYFAT	.212+-.095 (2)	.162+-.061 (5)	.166+-.074 (4)	.180+-.051 (3)	**.227+-.076** (1)
DDT	.144+-.036 (2)	.127+-.039 (5)	.149+-.045 (2)	.128+-.036 (4)	**.163+-.023** (1)
HOUSING	.760+-.022 (2)	.529+-.123 (5)	.531+-.095 (4)	.738+-.041 (3)	**.761+-.031** (1)
AUTORSHIP	.958+-.012 (3)	.631+-.059 (5)	.696+-.078 (4)	**.959+-.009** (2)	.960+-.006 (1)
WISCONSIN	.605+-.036 (2)	.281+-.126 (5)	.250+-.097 (4)	.569+-.021 (3)	**.608+-.043** (1)
Av. Rate	2.16	4.83	4	2.58	1.41

5 Conclusions

In this paper, we introduced alternative decomposition techniques for Label Ranking, closely related to the standard decomposition method [5], but that can take correlations among labels into account. We mainly investigated three decompositions that transform label ranking into binary classification and that allow to create meta learners. In particular, we adapted the classifier chains and its ensembled version for multi-label classification [7, 13] to label ranking and showed that the ensemble of classifier chains outperforms all others decomposition methods in a statistically significant way. In order to increase accuracy of classifier chains, some heuristics to determine the most appropriate label order are currently under study. Furthermore, probabilistic interpretations of the classifier chains and the ensembled version have also been introduced, though experimental results have not been provided due to their extremily high computational complexity. In particular the probabilistic classifier chains minimizes the subset 0/1 loss function in expectation. Another important result concerns the Net Flow Score procedure that provides a good approximation algorithm to the ranking aggregation problem.

References

1. Fürnkranz, J., Hüllermeier, E. (eds.): Preference Learning. Springer (2010)
2. Gurrieri, M., Siebert, X., Fortemps, P., Greco, S., Słowiński, R.: Label Ranking: A New Rule-Based Label Ranking Method. In: Greco, S., Bouchon-Meunier, B., Coletti, G., Fedrizzi, M., Matarazzo, B., Yager, R.R. (eds.) IPMU 2012, Part I. CCIS, vol. 297, pp. 613–623. Springer, Heidelberg (2012)
3. Gurrieri, M., Siebert, X., Fortemps, P., Slowinski, R., Greco, S.: Reduction from Label Ranking to Binary Classification. In: DA2PL 2012 From Multiple Criteria Decision Aid to Preference Learning, pp. 3–13. UMONS (Université de Mons), Mons (2012)
4. Har-Peled, S., Roth, D., Zimak, D.: Constraint classification for multiclass classificatin and ranking. In: Advances in Neural Information Processing Systems, pp. 785–792 (2002)
5. Hüllermeier, E., Fürnkranz, J., Cheng, W., Brinker, K.: Label Ranking by learning pairwise preference. Artif. Intell. 172(16-17), 1897–1916 (2008)
6. Cheng, W., Hühn, J., Hüllermeier, E.: Decision Tree and Instance-Based Learning for Labele Ranking. In: Proc. ICML 2009, International Conference on Machine Learning, Montreal, Canada (2009)
7. Read, J., Pfahringer, B., Holmes, G., Frank, E.: Classifier chains for multi-label classification. Machine Learning 85(3), 333–359 (2011)
8. Gärtner, T., Vembu, S.: Label Ranking Algorithms: A Survey. In: Fürnkranz, J., Hüllermeier, E. (eds.) Preference Learning. Springer (2010)
9. Dekel, O., Manning, C.D., Singer, Y.: Log-linear models for label ranking. In: Advances in Neural Information Processing Systems, vol. 16 (2003)
10. Elisseeff, A., Weston, J.: A kernel method for multi-labelled classification. In: Advances in Neural Information Processing Systems, vol. 14 (2001)
11. Bouyssou, D.: Ranking methods based on valued preference relations: A characterization of the net flow method. European Journal of Operational Research 60(1), 61–67 (1992)
12. Demšar, J.: Statistical Comparisons of Classifiers over Multiple Data Sets. Journal of Machine Learning Research 7, 1–30 (2006)
13. Cheng, W., Hüllermeier, E., Dembczynski, K.J.: Bayes optimal multilabel classification via probabilistic classifier chains. In: Proceedings of the 27th International Conference on Machine Learning (ICML 2010), pp. 279–286 (2010)

14. Dwork, C., Kumar, R., Naor, M., Sivakumar, D.: Rank aggregation revisited (2001)
15. Schalekamp, F., van Zuylen, A.: Rank Aggregation: Together We're Strong. In: ALENEX, pp. 38–51 (2009)
16. de Sá, C.R., Soares, C., Jorge, A.M., Azevedo, P., Costa, J.: Mining Association Rules for Label Ranking. In: Huang, J.Z., Cao, L., Srivastava, J. (eds.) PAKDD 2011, Part II. LNCS, vol. 6635, pp. 432–443. Springer, Heidelberg (2011)
17. Diaconis, P., Graham, R.L.: Spearman's footrule as a measure of disarray. Journal of the Royal Statistical Society, Series B (Methodological) 39, 262–268 (1977)
18. Hung, M.S., Hu, M.Y., Shanker, M.S., Patuwo, B.E.: Estimating posterior probabilities in classification problems with neural networks. International Journal of Computational Intelligence and Organizations 1(1), 49–60 (1996)

Clustering Based on a Mixture
of Fuzzy Models Approach

Miguel Pagola*, Edurne Barrenechea, Aránzazu Jurío,
Daniel Paternain, and Humberto Bustince

Universidad Publica de Navarra,
Campus Arrosadia s/n., 31006 Pamplona, Spain
miguel.pagola@unavarra.es
http://giara.unavarra.es/

Abstract. In this work we propose a clustering methodology model
named as Mixture of Fuzzy Models (MFMs). We adopt two assumptions:
the data points are generated by a membership function and the sum of
the memberships to all of the clusters must be greater or equal than zero.
The objective is to obtain a set of membership functions which represent
the data. It is formulated as a multiobjective optimization problem with
two objectives: to maximize the sum of memberships within each cluster
and to maximize the differences of memberships between clusters.

Keywords: Clustering, Membership functions, Multi-objective optimization, Outliers.

1   Introduction

Clustering an unlabeled data set $X = \{x^{(1)}, \ldots, x^{(m)}\}$ is the partitioning of X
into $1 < C < m$ subgroups such that each subgroup represents natural sub-
structure in X [10]. In statistics, a mixture model corresponds to the mixture
distribution that represents the probability distribution of observations in the
overall population. Therefore a mixture model can be regarded as providing a
statistical framework for clustering, where each cluster is modelled by a distribution, and each data point is eventually assigned to a single cluster. When dealing
with real data, Gaussian distribution (normal distribution) is the used one [9]
and this formulation is named Gaussian Mixture Models (GMMs). However any
distribution of probability can be used (Binomial, Student, Poisson, Exponential
or Log-normal).

One of the most widely used fuzzy clustering models is Fuzzy C-Means (FCM)
[1]. The FCM algorithm assigns memberships to the elements which are inversely
related to the relative distance of them to the point prototypes that are the
cluster centers.

FCM type fuzzy clustering algorithms have been shown to be closely related
to GMMs. Gan et al. [6] showed that the GMMs can be translated to an additive

---

* Corresponding author.

A. Laurent et al. (Eds.): IPMU 2014, Part II, CCIS 443, pp. 475–484, 2014.

fuzzy system. Ichihashi et al. [7] showed that the EM algorithm for the GMM can be derived from the FCM fuzzy clustering, when considering a regularized fuzzy objective function, for a proper selection of the distance metric.

Most of the existing clustering algorithms, including GMMs and FCM,impose a probabilistic constraint on the utilized membership functions. That the obtained cluster memberships of a data point must sum up to one over the derived clusters. Such constrained membership functions are not capable of distinguishing between data points which would be equally likely to belong to more than one cluster, or data points that would be unlikely to belong to anyone of the known clusters, usually referred as outliers [2].

A great deal of research work has been devoted to the attenuation of these problems, both in the case of generative mixture models, and the fuzzy clustering framework case [10].

In this work we propose a model named as Mixture of Fuzzy Models (or Mixture of Fuzzy Membership functions) in which we take two assumptions: the data points are generated by a membership function and the sum of the memberships to all of the clusters must be greater or equal to zero. Such a way a memberships can be viewed as the probability for one cluster to include one point.

The objective is to obtain a set of membership functions which represent the data. These memberships can be selected as specific functions (triangular, exponential, etc.). For example, when we create a fuzzy rule based system, we can represent the labels with triangular membership functions. Due to the relaxation of the constrain, the sum up to 1, the model obtained should not be influenced by the presence of noise or outliers.

There exist some works that present a model called Fuzzy Gaussian Mixture Models in which they used the Gaussian distribution into the calculation of the dissimilarity in the FCM calculation [8]. But this model is different from the one presented here. In this work we propose a method to fit different fuzzy membership functions to data points. We will call our model Mixture of Fuzzy Models (MFMs).

The remainder of this work is organized as follows. In Section 2, we provide an overview of the formulation of Gaussian mixture models, and a brief description of fuzzy clustering algorithm. In Section 3, the proposed model of mixture fuzzy models is formulated. In Section 4, the learning method of the MFMs is provided, giving some examples. In section 5 we evaluate the proposed model in a clustering applications with noisy data. Finally, in the concluding section, we summarize and discuss our results.

2   Preliminaries

Let an unlabeled data set $X = \{x^{(1)}, \ldots, x^{(m)}\}$ of m examples, and each example $x^{(i)} = (x_1^{(i)}, \ldots, x_n^{(i)})$ is a vector of n dimensions.

Next we describe the basis of GMMs and the FCM algorithms and we recall two common measures of cluster validity as compactness and separability.

2.1   Gaussian Mixture Models

A generative mixture model to represent this dataset is a common approach in the field of statistical pattern recognition. A C-component generative mixture model is, in essence, a superposition of weighted probability distributions of the form:

$$P(x^{(i)}) = \sum_{c=1}^{C} \alpha_c p(x^{(i)}|\theta_c) \tag{1}$$

where θ_c are the component distributions of the model, and α_c are their mixing weights (prior probabilities). The typical case of GMM is the Gaussian Mixture Model in which the probability distributions are Gaussian distributions.

The GMMS can be formulated as a maximum likelihood problem. The objective is to estimate the parameters set Θ that maximizes $\mathcal{L}(\Theta|X)$.

$$\Theta^* = \underset{\Theta}{argmax}\mathcal{L}(\Theta|X) \tag{2}$$

being $\mathcal{L}(\Theta|X)$ the likelihood function:

$$\mathcal{L}(\Theta|X) = \prod_{i=1}^{m} \left(\sum_{c=1}^{C} \alpha_c p_c(x^{(i)}|\nu_c, \Sigma_c) \right) \tag{3}$$

where the parameters of the gaussians are $(\nu_1, \ldots, \nu_C, \Sigma_1, \ldots, \Sigma_C)$ and $(\alpha_1, \ldots, \alpha_C)$ are the C mixing coefficients of the C mixed components such that: $\sum_{c=1}^{C} \alpha_c = 1$. The usual choice to obtain the maximum likelihood estimate of the GMMs parameters is the Expectation-Maximization algorithm [5].

2.2   Fuzzy Cluster Means

The aim of FCM is to find the cluster centres (centroids) ν_c that minimize a dissimilarity function. The dissimilarity function measures de distance between the point $x^{(i)}$ and the cluster prototype ν_c:

$$d_{ic}^2 = ||x^{(i)} - \nu_c||_A^2 \tag{4}$$

being $||.||_A$ an induced norm on $\mathcal{R}^n$ with A a positive definite $(n \times n)$ weight matrix. The objective function $J(U, \nu)$ is the weighted sum of dissimilarities within each cluster:

$$J(U, \nu) = \sum_{i=1}^{m} \sum_{c=1}^{C} (\mu_c(x^{(i)}))^b d_{ic}^2 \tag{5}$$

being U the fuzzy partition and b the weighting exponent called the degree of fuzziness. The processing of minimizing object function $J(U, \nu)$ depends on how centres find their ways to the best positions, as the fuzzy memberships $\mu_c(x^{(i)})$ and norm distance d_{ic}^2 would change along with the new centres' position.

2.3 Compactness and Separability

The objective of cluster validity is to find the optimal C clusters. Some validity indexes measure the degree of compactness and separation for the data structure in all of c clusters and then finds an optimal C that each one of these optimal c clusters is compact and separated from the other clusters [13].

Chen and Linkens [3] proposed the following expressions:

$$J_C = \frac{1}{m} \sum_{i=1}^{m} max(\mu_c(x^{(i)})) \tag{6}$$

$$J_S = \frac{1}{K} \sum_{c=1}^{C} \sum_{g=c+1}^{C} \left(\frac{1}{m} \sum_{i=0}^{m} min(\mu_c(x^{(i)}), \mu_g(x^{(i)})) \right) \tag{7}$$

where $K = \sum_{i=1}^{C} i$

3   Model Formulation

In our proposal, we suppose that each data point belongs to every fuzzy set, but the summation of all membership degrees must be greater or equal than zero.

The result of the clustering process in the mixture of fuzzy models is going to be a set of membership functions, and in order to obtain a cluster, we demand two conditions:

1. The sum of all the memberships to all the sets should be maximum. The membership functions obtained should cover the data points such a way that the membership degrees of the points that are clearly located inside a cluster should have large membership values to said cluster. Therefore, if the clustering process creates membership functions that represent correctly the clusters, the membership degrees of the points will be all 1 and therefore the sum of these values will be maximum.
2. The difference between the membership degrees to each cluster of any data point should be maximum. After the clustering process every data point will have a membership degree to every cluster. A desirable property is that if a point has a large membership to a cluster, then the memberships to the other clusters should be low. Therefore the differences between the membership degrees should be maximum.

Therefore we have two different objectives, the first one is to maximize the memberships:

$$J_1(\Theta) = \frac{1}{m} \sum_{c=1}^{C} \sum_{i=1}^{m} \mu_c(x^{(i)}) \tag{8}$$

and the second is to maximize the differences between membership degrees:

$$J_2(\Theta) = \frac{1}{m} \sum_{c=1}^{C-1} \sum_{g=c+1}^{C} \sum_{i=1}^{m} |\mu_c(x^{(i)}) - \mu_g(x^{(i)})| \tag{9}$$

Where Θ is the set of parameters that define the membership functions and $\bar{x}^{(c)}$ is the center of the membership function $(\mu_c(\bar{x}^{(c)}) = 1)$. We impose the following restrictions for all:

$$\bar{x}_j^{(c)} \geq min\{x_j^{(1)}, \ldots, x_j^{(m)}\} \text{ for all } j \in \{1 \ldots n\} \text{ and for all } c \in \{1 \ldots C\} \ (10)$$

$$\bar{x}_j^{(c)} \leq max\{x_j^{(1)}, \ldots, x_j^{(m)}\} \text{ for all } j \in \{1 \ldots n\} \text{ and for all } c \in \{1 \ldots C\}. \ (11)$$

These constrains mean that the center of the membership functions (clusters) must be within the maximum value and the minimum value of the data points. Depending on the membership function used we have additional constrains. For example if we use an exponential membership function:

$$\mu_c(x) = exp(-\alpha * \sum |x - \bar{x}^{(c)}|)^2 \tag{12}$$

then the value of α should be greater than 0. In case we want to use a Mahalanobis distance based membership function:

$$\mu_c(x) = \frac{1}{1 + (x - \bar{x}^{(c)})^T A(x - \bar{x}^{(c)})} \tag{13}$$

then the matrix A must be definite positive. Therefore our model is a multi-objective optimization problem (MOOP) with restrictions.

The first objective, J_1, is similar to the concept of compactness and the second objective, J_2, is similar to the concept of separability used in different cluster validity measures. Therefore we can reformulate our problem with these objective functions, i.e. maximize the compactness J_C and maximize the separability J_S.

3.1 Probability Condition or Ruspini Condition

In Mixture Models there exist a condition that all of the probabilities must sum up to one. The same restriction was inherited in the fuzzy setting by Ruspini [11] and others. As it was discussed previously, the FCM algorithm requires the sum of all of the memberships of an element to all clusters must be equal to one. In the case of a single-objective function, (where $J(\Theta) = J_1(\Theta) + J_2(\Theta)$) we can add an restriction to make the sum of the memberships of every element should be one:

$$J_r(\Theta) = J(\Theta) - K \frac{1}{m} \sum_{i=1}^{m} |1 - \sum_{c=1}^{C} \mu_c(x^{(i)})| \tag{14}$$

This term penalizes by a factor of K, the difference of the sum of the memberships and one. Therefore, the larger is K, the solution is forced to obtain a set of memberships in which the sum of the memberships in most of the training points is close to one.

4   Learning of MFMs

The optimum solution to a multi-objective optimization problem is a set of trade-off solutions known as Pareto optimal set. Each solution in Pareto optimal set is referred as Pareto optimal solution. A solution is considered Pareto optimal if no solution in entire design space is better in all objectives and constraints than it. In this section we describe how to solve the MFMs problem by means of two different ways to solve a MOOP. First, we are going to convert the original problem with multiple objectives into a single objective optimization problem. Next, by means of evolutionary algorithms we will solve the multi-objective optimization problem approximating a representative set of Pareto optimal solutions.

4.1 Weighted Sum Strategy

This is the most popular method to convert a MOOP into a single objective optimization problem (it is also known as linear scalarization). We specify weights associated with each objective and optimize the weighted sum of objectives.

$$J(\Theta) = w_1 J_1(\Theta) + w_2 J_2(\Theta) \tag{15}$$

Weights reflect the preference of each objective to be satisfied. The advantage of this method is its simplicity and ease of use. This approach works the best for problems with convex and continuous Pareto fronts.

Solving this optimization problem is a non-trivial task and may become computationally expensive. The number of parameters and constrains increase linearly with the number of clusters to be identified. We have applied the solver fmincon function implemented in the optimization toolbox of Matlab. This function provides a method for constrained non-linear optimization based on sequential quadratic programming.

We begin with a toy example on simulated data to demonstrate the performance of the MFMs in a clustering scheme. The synthetic data is generated by means of 100 samples of each of these bivariate Gaussian distribution functions:

$$f_a = \mathcal{N}\left(\begin{bmatrix} 0 \\ 2 \end{bmatrix}, \begin{bmatrix} 1 & 0.8 \\ 0.8 & 1 \end{bmatrix}\right), \; f_b = \mathcal{N}\left(\begin{bmatrix} 2 \\ 0 \end{bmatrix}, \begin{bmatrix} 1 & 0.8 \\ 0.8 & 1 \end{bmatrix}\right)$$

In Fig. 1 we show the MFMs obtained using the distance based membership functions of equation (13) for both clusters. We can obtain a similar results optimizing the following objective function based on the compactness and separability:

$$J_{CS}(\Theta) = w_1 J_C(\Theta) + w_2 J_S(\Theta) \tag{16}$$

This methodology, the same as mixture models, have problems in the initialization in sparse data sets due to flat local minima. Therefore it is useful to choose the starting points by means of the k-means algorithm or similar.

4.2 Multi-Objective Evolutionary Algorithms

Multi-objective evolutionary algorithms (MOEAs) have the potential of yielding many Pareto optimal solutions in a single simulation. Since Evolutionary Algorithms (EAs) work with a population of solutions, a simple EA can be extended to maintain a diverse set of solutions. In our case we have used the matlab function `gamultiobj` which uses a controlled elitist genetic algorithm, which is a modification of NSGA-II [4]. This algorithm uses a non-domination criterion based selection operator to handle multiple objectives. In Fig. 2(a) is depicted the Pareto front obtained for the MOOP in the simulated data set (remark: objectives have negative values due to `gamultiobj` only minimizes). The solver was applied to the problem of maximization of J_1 and J_2. Also in Fig 2(b) are shown the centers of the different solutions of the Pareto optimal subset.

5   Experimental Results

We are going to use simulated data (similar to de the one used in [4]) to illustrate the advantages of the introduced MFMs over conventional approaches.

The synthetic data is generated by means of 100 points of each of these three bivariate Gaussian distribution functions: $f_a = \mathcal{N}\left(\begin{bmatrix} 0 \\ 3 \end{bmatrix}, \begin{bmatrix} 2 & 0.5 \\ 0.5 & 0.5 \end{bmatrix}\right)$, $f_b =$ $\mathcal{N}\left(\begin{bmatrix} 3 \\ 0 \end{bmatrix}, \begin{bmatrix} 1 & 0 \\ 0 & 0.1 \end{bmatrix}\right)$, $f_c = \mathcal{N}\left(\begin{bmatrix} -3 \\ 0 \end{bmatrix}, \begin{bmatrix} 2 & -0.5 \\ -0.5 & 0.5 \end{bmatrix}\right)$

The final 100 points, outliers and noise, are generated from a bivariate Gaussian distribution with each of its components in the interval $[-10, 10]$ (see Fig. 3). We compare our result with the ones obtained by the algorithms FCM [1], GMMs [9], and PFCM [10]. All of the algorithms start with the centres obtained by the k-means algorithm. In Table 1, we provide the Rand index for every method. The values depicted are the average mean after 10 executions of every algorithm. The MFMs uses as the objective function eq. (14) with $K = 1$ and eq. (16), both with $w_1 = w_2 = 0.5$ and the membership function of eq. (13). Parameter b of FCM (see eq. 5) is equal to 2 and in PFCM $a = 2$ and $b = 4$ (see [10]).

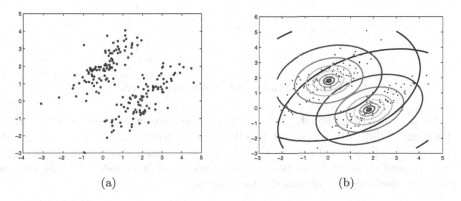

(a)   (b)

Fig. 1. (a) Simulated data (b) MFMs obtained in the simulated data

Table 1. Precision rates for each cluster

	Rand Index
FCM	0.9656
GMMs	0.7277
PFCM	0.9656
J_r	0.9610
J_{CS}	0.9347

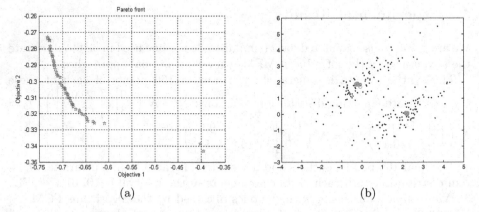

(a)   (b)

Fig. 2. (a) Pareto front obtaind by the NSGA-II algorithm (matlab version `gamultiobj` (b) Simulated data and the centers of every Pareto solution obtained by the NSGA-II algorithm (matlab version `gamultiobj`)

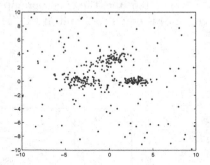

Fig. 3. Synthetic data set with outliers

As we can see although FCM have the restriction that the sum of all of the memberships of an element to all of the clusters must be one, it generate the clusters efficiently. However GMMs create a cluster that covers most of the outliers, and then it fails in most of the elements of one cluster. PFCM and the proposed methodology obtain similar results.

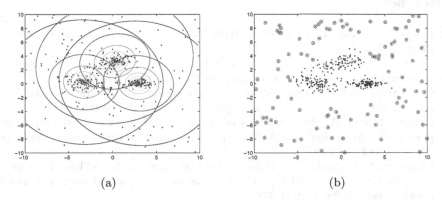

(a)   (b)

Fig. 4. (a) MFMs obtained in the synthetic data with outliers (b) Outliers detected with threshold 0.25

In Fig. 4 the MFMs obtained in the simulated data are shown. In the FCM and GMMs is not possible to identify outliers, but PFCM, through the typicalities and MFMs could detect outliers. In MFMs an outlier can be the element which all the memberships to the clusters are less than a threshold. In the PFCM an outlier can be a point such that its typicality is less than threshold to every cluster. In Fig. 4(b) are shown the points detected as outliers with the threshold equal to 0.25 in MFMs.

6   Conclusions and Future Research

In this work we have proposed a simple clustering methodology in which we fit the data to a set of membership functions relaxing the restriction of Ruspini. This a preliminary work in which we just prove in simple data sets the performance of this methodology. Future research will be to test and compare the efficiency in larger datasets and more complex problems.

Analyzing these results we plan to apply this methodology to obtain the membership functions of linguistic labels in the definition rule based systems. Also it can be easily applied to histogram fitting and thresholding. Our model has the advantage that the user can choose the membership function (if it is previously known) or to test different membership functions and select the best one. Moreover we plan to test different objective functions so the clustering structure obtained could satisfy different objectives. If the property that defines the data is the connectedness, then this approach is not useful. This type of problems are correctly solved by means of agglomerative methods like Single-link [12]. This methodology works only if the data is clustered in hyper spherical shapes.

Another problem that we plan to tackle is semi-supervised learning. The proposed model can be easily adapted to a semi-supervised learning problem adding new restrictions to the MOOP problem.

References

1. Bezdek, J.C.: Pattern Recognition With Fuzzy Objective Function Algorithms. Plenum, New York (1981)
2. Chatzis, S.P., Tsechpenakis, G.: A possibilistic clustering approach toward generative mixture models. Pattern Recognition 45, 1819–1825 (2012)
3. Chen, M.Y., Linkens, D.A.: Rule-base self-generation and simplification for data-driven fuzzy models. Fuzzy Sets and Systems 142, 243–265 (2004)
4. Deb, K., Pratap, A., Agarwal, S., Meyarivan, T.: A Fast and Elitist Multiobjective Genetic Algorithm: NSGA-II. IEEE Transactions on Evolutionary Computation 6(2), 182–197 (2002)
5. Dempster, A.P., Laird, N.M., Rubin, D.B.: Maximum Likelihood from Incomplete Data via the EM Algorithm. Journal of the Royal Statistical Society. Series B (Methodological) 39, 1–38 (1977)
6. Gan, M.T., Hanmandlu, M., Tan, A.H.: From a Gaussian mixture model to additive fuzzy systems. IEEE Transactions on Fuzzy Systems 13(3), 303–316 (2005)
7. Ichihashi, H., Honda, K., Tani, N.: Gaussian mixture pdf approximation and fuzzy c-means clustering with entropy regularization. In: Proceedings of the Fourth Asian Fuzzy System Symposium, pp. 217–221 (2000)
8. Ju, Z., Liu, H.: Fuzzy Gaussian Mixture Models. Pattern Recognition 45, 1146–1158 (2012)
9. McLachlan, G., Peel, D.: Finite Mixture Models. John Wiley & Sons, Inc. (2000)
10. Pal, N., Pal, K., Keller, J., Bezdek, J.: A possibilistic fuzzy c-means clustering algorithm. IEEE Transactions on Fuzzy Systems 13(4), 517–530 (2005)
11. Ruspini, E.H.: Numerical Methods for Fuzzy Clustering. Information Sciences (2), 319–350 (1970)
12. Sibson, R.: SLINK: an optimally efficient algorithm for the single-link cluster method. The Computer Journal (British Computer Society) 16(1), 30–34 (1973)
13. Wu, K.L., Yang, M.S.: A cluster validity index for fuzzy clustering. Pattern Recognition Letters 26, 1275–1291 (2005)

Analogical Classification: A Rule-Based View

Myriam Bounhas[1], Henri Prade[2], and Gilles Richard[2]

[1] LARODEC Laboratory, ISG de Tunis, 41 rue de la Liberté, 2000 Le Bardo, Tunisia
& Emirates College of Technology, P.O. Box: 41009, Abu Dhabi, United Arab Emirates
[2] IRIT – CNRS, 118, route de Narbonne, Toulouse, France
myriam_bounhas@yahoo.fr, {prade,richard}@irit.fr

Abstract. Analogical proportion-based classification methods have been introduced a few years ago. They look in the training set for suitable triples of examples that are in an analogical proportion with the item to be classified, on a
maximal set of attributes. This can be viewed as a lazy classification technique
since, like k-nn algorithms, there is no static model built from the set of examples.
The amazing results (at least in terms of accuracy) that have been obtained from
such techniques are not easy to justify from a theoretical viewpoint. In this paper,
we show that there exists an alternative method to build analogical proportion-
based learners by statically building a set of inference rules during a preliminary
training step. This gives birth to a new classification algorithm that deals with
pairs rather than with triples of examples. Experiments on classical benchmarks
of the UC Irvine repository are reported, showing that we get comparable results.

Introduction

Comparing objects or situations and identifying in what respects they are identical (or
similar) and in what respects they are different, is a basic type of operations at the core
of many intelligent activities. A more elaborate operation is the comparison between
pairs of objects or situations, where a comparison has already been done inside the
pairs. This corresponds to the idea of analogical proportions, i.e. statements of the form
"A is to B as C is to D", denoted $A : B :: C : D$, expressing the fact "A differs from
B as C differs from D", as well as "B differs from A as D differs from C" [5].

Analogical reasoning has been recognized for a long time as a powerful heuristic tool
for solving problems. In fact, analogical proportions are not explicitly used in general.
Compound situations identified as analogous are rather put in parallel, leading to the
plausible conclusion that what holds in one case should also hold in the other case (up
to suitable transpositions). However, analogical reasoning can also be directly based
on analogical proportions. This requires a formalized view of these proportions. Such
a modeling has been only recently developed in algebraic or logical settings [2,8,5,6].
Then analogical proportions turn to be a powerful tool in classification tasks [4].

We assume that the objects or situations A, B, C, D are represented by vectors of
attribute values, denoted a, b, c, d. The analogical proportion-based approach to classification relies on the idea that the unknown class $x = cl(d)$ of a new instance d, may
be predicted as the solution x of an equation expressing that the analogical proportion
$cl(a) : cl(b) :: cl(c) : x$ holds between the classes. This is done on the basis of triples

A. Laurent et al. (Eds.): IPMU 2014, Part II, CCIS 443, pp. 485–495, 2014.

of examples a, b and c of the training set that are such that the analogical proportion $a : b :: c : d$ holds on vector components for all, or at least on a large number of, the attributes describing the items. This approach has been tested on benchmarks [4] where results competitive with the ones of classical machine learning methods have been obtained. These good results have remained largely unexplained since it looks unclear why this analogical proportion-based approach may be so effective, beyond the general merits of analogy. In this paper, we investigate a new type of algorithm based on the induction of particular rules induced from pairs of examples, which can still be related to analogical proportions, thus providing some light on the underlying learning process.

The paper is organized as follows. First a background on analogical proportions is provided, emphasizing noticeable properties important for application to classification, before discussing how they can be applied to this task. Then the new rule-based approach is contrasted with the original triples-based approach. Algorithms are proposed and their results on machine learning benchmarks are reported and discussed.

A Short Background on Analogical Proportions

Analogical proportions are statements of the form "A is to B as C is to D", which have been supposed to continue to hold when the pairs (A, B) and (C, D) are exchanged, or when the terms B and C are permuted, just like numerical proportions, since Aristotle time; see, e.g., [7]. Thus, $A : B :: C : D$ is equivalent to $C : D :: A : B$ (symmetry), and $A : B :: C : D$ is equivalent to $A : C :: B : D$ (central permutation). By combining symmetry and permutation, this leads to 8 equivalent forms.

In this paper, A, B, C, D are represented by Boolean vectors. Let a denote a component of such a vector a. Then an analogical proportion between such vectors can be expressed componentwise, in a logical manner under various equivalent forms [5]. One remarkable expression of the analogical proportion is given by the expression $a : b :: c : d = (a \wedge \neg b \equiv c \wedge \neg d) \wedge (\neg a \wedge b \equiv \neg c \wedge d)$.

As can be seen, this expression of the analogical proportion uses only dissimilarities and could be informally read as *what is true for a and not for b is exactly what is true for c and not for d, and vice versa.* This logical expression makes clear in an analogical proportion $a : b :: c : d$ that a differs from b as c differs from d and, conversely, b differs from a as d differs from c. The 6 cases (among $2^4 = 16$ possible entries) where the above Boolean expression is *true* are given in the truth Table 1.

Table 1. When an analogical proportion is true

a	b	c	d	$a : b :: c : d$
0	0	0	0	1
1	1	1	1	1
0	0	1	1	1
1	1	0	0	1
0	1	0	1	1
1	0	1	0	1

It can be easily checked on the above truth Table 1 that the logical expression of the analogical proportion indeed satisfies symmetry and central permutation. Assuming that an analogical proportion holds between four binary items, three of them being known, then one may try to infer the value of the fourth one. The problem can be stated as follows. Given a triple (a, b, c) of Boolean values, does there exist a Boolean value x such that $a : b :: c : x = 1$, and in that case, is this value unique? It is easy to see that there are cases where the equation has no solution since the triple a, b, c may take $2^3 = 8$ values, while A is true only for 6 distinct 4-tuples. Indeed, the equations $1 : 0 :: 0 : x = 1$ and $0 : 1 :: 1 : x = 1$ have no solution. It is easy to prove that the analogical equation $a : b :: c : x = 1$ is solvable iff $(a \equiv b) \vee (a \equiv c)$ holds true. In that case, the unique solution is given by $x = a \equiv (b \equiv c)$. Note that due to symmetry and permutation properties, there is no need to consider the equations $x : b :: c : d = 1$, $a : x :: c : d = 1$, and $a : b :: x : d = 1$ that can be handled in an equivalent way.

Analogical Proportions and Classification

Numerical proportions are closely related to the ideas of extrapolation and of linear regression, i.e., to the idea of predicting a new value on the ground of existing values, and to the idea of inducing general laws from data. Analogical proportions may serve similar purposes. The equation solving property recalled above is at the root of a brute force method for classification. It is based on a kind of proportional continuity principle: if the binary-valued attributes of 4 objects are componentwise in analogical proportion, then this should still be the case for their classes. More precisely, having a 2-class classification problem, and 4 Boolean objects a, b, c, d over $\mathbb{B}^n$, 3 in the training set with known classes $cl(a), cl(b), cl(c)$, the 4th being the object to be classified in one of the 2 classes, i.e. $cl(d) \in \mathbb{B}$ is unknown, this principle can be stated as:

$$\frac{\forall i \in [1, n], a_i : b_i :: c_i : d_i = 1}{cl(a) : cl(b) :: cl(c) : cl(d) = 1}$$

Then, if the equation $cl(a) : cl(b) :: cl(c) : x = 1$ is solvable, we can allocate its solution to $cl(d)$. This principle can lead to diverse implementations; see next section. The case of attributes on discrete domains and of a number of classes larger than 2 can be handled as easily as the binary case. Indeed, consider a finite attribute domain $\{v_1, \cdots, v_m\}$. Note that the attribute may also be the *class itself*. This attribute (or the class), say $\mathcal{A}$, can be straightforwardly binarized by means of the m properties "having value v_i, or not". Consider the partial description of objects $a, b, c,$ and d wrt $\mathcal{A}$. Assume, for instance, that objects a and c have value v_1, while objects b and d have value v_2. This situation is summarized in Table 2 where the respective truth-values of the four objects wrt each binary property "having value v_i" are indicated. As can be seen on this table, an analogical proportion holds true between the four objects for each binary property, and in the example, can be more compactly encoded as an analogical proportion between the attribute values themselves, namely here: $v_1 : v_2 :: v_1 : v_2$. More generally, x and y denoting possible values of a considered attribute $\mathcal{A}$, the analogical proportion between objects $a, b, c,$ and d holds for $\mathcal{A}$ iff the 4-tuple $(\mathcal{A}(a), \mathcal{A}(b), \mathcal{A}(c), \mathcal{A}(d))$ is equal to one 4-tuple having one of the three forms (s, s, s, s), (s, t, s, t), or (s, s, t, t).

Table 2. Handling non binary attributes

	v_1	v_2	v_3	$\cdots$	v_m	
a	1	0	0	$\cdots$	0	v_1
b	0	1	0	$\cdots$	0	v_2
c	1	0	0	$\cdots$	0	v_1
d	0	1	0	$\cdots$	0	v_2

A training set TS of examples $x^k = (x^k{}_1, ..., x^k{}_i, ..., x^k{}_n)$ together with their class $cl(x^k)$, with $k = 1, t$ may also be read in an analogical proportion style: "x^1 is to $cl(x^1)$ as x^2 is to $cl(x^2)$ as $\cdots$ as x^t is to $cl(x^t)$". However note that x^k and $cl(x^t)$ are vectors of different dimensions. This may still be written (abusively) as $x^1 : cl(x^1) :: x^2 : cl(x^2) :: \cdots :: x^t : cl(x^t)$. Note that this view exactly fits with the idea that in a classification problem there exists a classification *function* that associates a unique class with each object, which is unknown, but exemplified by the training set. Indeed $x^k : cl(x^k) :: x^k : cl'(x^k)$ with $cl(x^k) \neq cl'(x^k)$ is forbidden, since it cannot hold as a generalized analogical proportion obeying to a pattern of the form (s, t, s, t) where s and t belong to different spaces.

Postulating the central permutation property, the informal analogical proportion $x^i : cl(x^i) :: x^j : cl(x^j)$ linking examples x^i and x^j can also be rewritten as $x^i : x^j :: cl(x^i) : cl(x^j)$ (still informally as we deal with vectors of different dimensions). This suggests a new reading of the training set, based on pairs. Namely, the ways vectors x^i and x^j are similar / dissimilar should be related to the identity or the difference of classes $cl(x^i)$ and $cl(x^j)$. Given a pair of vectors x^i and x^j, one can compute the set of attributes $A(x^i, x^j)$ where they agree (i.e. they are equal) and the set of attributes $D(x^i, x^j)$ where they disagree (i.e. they are not equal). Suppose, we have in the training set TS, both the pair (x^i, x^j), and the example x^k which once paired with x^0 has exactly the *same disagreement set* as $D(x^i, x^j)$ and moreover *with the changes oriented in the same way*. Note that although $A(x^i, x^j) = A(x^k, x^0)$, the 4 vectors are not everywhere equal on this subset of attributes. Then we have a perfect analogical proportion componentwise, between the 4 vectors (of the form (s, s, s, s) or (s, s, t, t) on the agreement part of the components, and of the form (s, t, s, t) on the disagreement set). Indeed, the above view straightforwardly extends from binary-valued attributes to attributes with finite domains. Thus, working with pairs, we can implicitly reconstitute 4-tuples of vectors that form an analogical proportion as in the triple-based brute force approach to classification. We now discuss the algorithmic aspects of this approach.

Analogical Classification: The Standard View

Before introducing the analogical classifiers, let us restate the classification problem. Let T be a data set where each vector $x = (x_1, ..., x_i, ..., x_n) \in T$ is a set of n feature values representing a piece of data. Each vector x is assumed to belong to a unique class $cl(x) \in C = \{c_1, ..., c_l\}$, where C is finite and covered through the data set (in the binary class case, $l = 2$). If we suppose that cl is known on a subset $TS \subset T$, given a new vector $y = (y_1, ..., y_i, ..., y_n) \notin TS$, the classification problem amounts to assign a plausible value $cl(y)$ on the basis of the examples stored in TS.

Learning by analogy, as developed in [1], is a lazy learning technique which uses a measure of *analogical dissimilarity* between 4 objects. It estimates how far 4 situations are from being in analogical proportion. Roughly speaking, the analogical dissimilarity ad between 4 Boolean values is the minimum number of bits that have to be switched to get a proper analogy. Thus $\mathrm{ad}(1,0,1,0) = 0$, $\mathrm{ad}(1,0,1,1) = 1$ and $\mathrm{ad}(1,0,0,1) = 2$. Thus, $A(a,b,c,d)$ holds if and only if $\mathrm{ad}(a,b,c,d) = 0$. Moreover ad differentiates two types of cases where analogy does not hold, namely the 8 cases with an odd number of 0 and an odd number of 1 among the 4 Boolean values, such as $\mathrm{ad}(0,0,0,1) = 1$ or $\mathrm{ad}(0,1,1,1) = 1$, and the two cases $\mathrm{ad}(0,1,1,0) = \mathrm{ad}(1,0,0,1) = 2$. When, instead of having 4 Boolean values, we deal with 4 Boolean vectors in $\mathbb{B}^n$, we add the ad evaluations componentwise to get the analogical dissimilarity between the 4 vectors, which leads to an integer belonging to the interval $[0, 2n]$. This number estimates how far the 4 vectors are from building, componentwise, a complete analogy. It is used in [1] in the implementation of a classification algorithm where the input is a set S of classified items, a new item d to be classified, and an integer k. It proceeds as follows:

Step 1: Compute the analogical dissimilarity ad between d and all the triples in S^3 that produce a solution for the class of d.

Step 2: Sort these n triples by the increasing value of ad wrt with d.

Step 3: Let p be the value of ad for the k-th triple, then find k' as being the greatest integer such that the k'-th triple has the value p.

Step 4: Solve the k' analogical equations on the label of the class. Take the winner of the k' votes and allocate this winner as the class of d.

This approach provides remarkable results and, in several cases, outperforms the best known algorithms [4]. Another equivalent approach [6] does not use a dissimilarity measure but just applies the previous continuity principle, adding flexibility by allowing to have some components where analogy does not hold. A majority vote is still applied among the candidate voters. Any triple $\mathbf{a}, \mathbf{b}, \mathbf{c}$, such that the cardinal of the set $\{i \in [1, n] | A(a_i, b_i, c_i, d_i) \text{ holds and } A(cl(\mathbf{a}), cl(\mathbf{b}), cl(\mathbf{c}), cl(\mathbf{d})) \text{ is solvable}\}$ is maximal, belongs to the candidate voters.

Analogical Classification: A Rule-Based View

We claim here that analogical classifiers behave as if a set of rules was build inductively during a pre-processing stage. To support intuition, we use an example inspired from the Golf data set (UCI repository [3]). This data set involves 4 multiple-valued attributes:

1: Outlook: sunny or overcast or rainy. ; *2: Temperature: hot or mild or cool* ;
3: Humidity: high or normal. ; *4: Windy: true or false.*

Two labels are available: 'Yes' (play) or 'No' (don't play).

Main Assumptions. Starting from a finite set of examples, 2 main assumptions are made regarding the behavior of the function cl:

- Since the target relation cl is assumed to be a function, when 2 distinct vectors x, y have different labels $(cl(x) \neq cl(y))$, the cause of the label switch is to be found in the switches of the attributes that differ. Take x and y in the Golf data set, as:
  $x = (overcast, mild, high, false)$ and $cl(x) = Yes$
  $y = (overcast, cool, normal, false)$ and $cl(y) = No$
  then the switch in attributes 2 and 3 is viewed as the cause of the 'Yes'-'No' switch.
- When 2 distinct x and y are such that $cl(x) = cl(y)$, this means that cl does not preserve distinctness, i.e. cl is not injective. We may then consider that the label stability is linked to the particular value arrangement of the attributes that differ.

Patterns. Let us now formalize these ideas. Given 2 distinct vectors x and y, they define a partition of $[1, n]$ as $A(x, y) = \{i \in [1, n] | x_i = y_i\}$ and $D(x, y) = [1, n] \setminus A(x, y) = \{i \in [1, n] | x_i \neq y_i\}$. Given $J \subseteq [1, n]$, let us denote $x|_J$ the subvector of x made of the x_j, $j \in J$. Obviously, $x|_{A(x,y)} = y|_{A(x,y)}$ and, in the binary case, when we know $x|_{D(x,y)}$, we can compute $y|_{D(x,y)}$. In the binary case, the pair (x, y) allows us to build up a *disagreement pattern* $Dis(x, y)$ as a list of pairs $(value, index)$ where the 2 vectors differ. with $n = 6$, $x = (1, 0, 1, 1, 0, 0)$, $y = (1, 1, 1, 0, 1, 0)$, $Dis(x, y) = (0_2, 1_4, 0_5)$. It is obvious that having a disagreement pattern $Dis(x, y)$ and a vector x (resp. y), we can get y (resp. x). In the same way, the disagreement pattern $Dis(y, x)$ is deducible from $Dis(x, y)$. For the previous example, $Dis(y, x) = (1_2, 0_4, 1_5)$.

In the categorical case, the disagreement pattern is a bit more sophisticated as we have to store the changing values. Then the disagreement pattern $Dis(x, y)$ becomes a list of triple $(value1, value2, index)$ where the 2 vectors differ, with $value1$ being the attribute value for x and $value2$ being the attribute value for y. For instance, with the previously described Golf dataset, for the pair of given examples x and y, $Dis(x, y)$ is $\{(mild, cool)_2, (high, normal)_3\}$. Then we have two situations:

1. x and y have different labels, i.e. $cl(x) \neq cl(y)$. Their disagreement pattern $Dis(x, y)$ is called a *change pattern*. Then $Dis(y, x)$ is also a change pattern.
2. x and y have the same label $cl(x) = cl(y)$. Their disagreement pattern $Dis(x, y)$ is called a *no-change pattern*. Then $Dis(y, x)$ is also a no-change pattern.

To build up a change (resp. no-change) pattern, we have to consider all the pairs (x, y) such that $cl(x) \neq cl(y)$ (resp. such that $cl(x) = cl(y)$). We then build 2 sets of patterns P_{ch} and P_{noch}, each time keeping only one of the 2 patterns $Dis(x, y)$ and $Dis(y, x)$ to avoid redundancy. As exemplified below, these 2 sets are not disjoint in general. Take $n = 6$, and assume we have the 4 binary vectors x, y, z, t in TS:

- $x = (1, 0, 1, 1, 0, 0)$, $y = (1, 1, 1, 0, 1, 0)$ with $cl(x) = 1$ and $cl(y) = 0$. Then, for (x, y), the disagreement pattern is a change pattern, i.e., $(0_2, 1_4, 0_5) \in P_{ch}$.
- $z = (0, 0, 1, 1, 0, 1)$, $t = (0, 1, 1, 0, 1, 1)$ with $cl(z) = cl(t)$. They have the same disagreement pattern as x and y, which is now a no-change pattern $(0_2, 1_4, 0_5) \in P_{noch}$.

Now, given an element x in TS whose label is known, and a new element to be classified y, if the disagreement pattern $Dis(x, y)$ belongs to $P_{ch} \cap P_{noch}$, we do not get any hint regarding the label of y. Then we remove the patterns in $P_{ch} \cap P_{noch}$: the remaining patterns are the *valid patterns* (still keeping the same notations for the resulting sets).

Rules. Thanks to the concept of pattern, it is an easy game to provide a formal definition of the 2 above principles. We get 2 general classification rules, corresponding to dual situations, for a new element y to be classified:

$$\text{Change Rule: } \frac{\exists x \in TS, \exists D \in P_{ch}|(Dis(x,y) = D) \vee (Dis(y,x) = D)}{cl(y) \neq cl(x)}$$

$$\text{NoChange Rule: } \frac{\exists x \in TS, \exists D \in P_{noch}|(Dis(x,y) = D) \vee (Dis(y,x) = D)}{cl(y) = cl(x)}$$

NoChange rules tell us when a new item y to be classified should get the class of its associated example(x), and Change rules tell the opposite. Let us note that if there is no valid pattern, then we cannot build up any rule, then we cannot predict anything! This has never been the case for the considered benchmarks.

Implementation

It is straightforward to implement the previous ideas.

1. Construct from TS the sets P_{ch} and P_{noch} of all disagreement patterns.
2. Remove from P_{ch} and from P_{noch} the patterns belonging to $P_{ch} \cap P_{noch}$ to get the set of valid patterns.

The remaining change patterns in P_{ch} and no-change patterns in P_{noch} are used to build up respectively the *Change Rule Set R_{ch}* and *No-Change Rule Set R_{noch}*. In this context, we have implemented two different classifiers: the *Change Rule based Classifier (CRC)* and the *No Change Rule based Classifier (NCRC)*, which have the same principles in all respect. The only difference is in the classification phase where the CRC only uses the set P_{ch} of pattern and applies the Change rules, whereas the second classifier $NCRC$ uses the no-change patterns P_{noch} and applies the No-Change rules to classify new items.

Classification. The classification process for CRC and $NCRC$ are detailed in the following algorithms 1 and 2, where the Boolean function $Analogy(x, x', y)$ is true if and only if $card(\{cl(x), cl(x'), cl(y)\}) \leq 2$. For the NCRC, the $Analogy(x, x', y)$ always has a solution since classes associated to any No-Change rule r in R_{noch} are homogeneous. In terms of complexity, the algorithms are still cubic in the size of TS since the disagreement pattern sets have a maximum of n^2 elements and we still have to check every element of TS to build up a relevant pair with y.

With our approach, contrary to k-nn approaches, we always deal with pairs of examples: i) to build up the rules, ii) to classify a new item, we just associate to this item another one to build a pair in order to trigger a rule. Moreover, the two pairs of items involved in an analogical proportion are not necessarily much similar as pairs, beyond the fact they should exhibit the same dissimilarity. An analogical view of the nearest neighbor principle could be "close/far instances are likely to have the same/possibly different class", making an assumption that the similarity of the classes is related to the similarity of the instances. This does not fit, e.g., our No-Change rules where the similarity of the classes is associated with dissimilarities of the instances. More generally, while

---

Algorithm 1. Change Rule Classifier

Given a new instance $y' \notin TS$ to be classified.
$CandidateRules(c_j) = 0$, for each $j \in [1, l]$ (in the binary class case, $l = 2$).
for each y in TS **do**
    Construct the disagreement patterns $D(y, y')$ and $D(y', y)$
    for each change rule $r \in R_{ch}$ // r has a pattern $D(x, x')$ **do**
        if $Analogy(x, x', y)$ AND $(D(y, y') = D(x, x')$ OR $D(y', y) = D(x, x'))$ **then**
            if $(cl(x) = cl(y))$ **then** $c^* = cl(x')$ **else** $c^* = cl(x)$ **end if**
            $CandidateRules(c^*) + +.$
        end if
    end for
end for
$cl(y') = arg\max_{c_j} CandidateRules(c_j)$

Algorithm 2. No Change Rule Classifier

Given a new instance $y' \notin TS$ to be classified.
$CandidateRules(c_j) = 0$, for each $j \in Dom(c_j)$.
for each y in TS **do**
    Construct the disagreement patterns $D(y, y')$ and $D(y', y)$
    for each no change rule $r \in R_{noch}$ // r has a pattern $D(x, x')$ **do**
        if $Analogy(x, x', y)$ AND $(D(y, y') = D(x, x')$ OR $D(y', y) = D(x, x'))$ **then**
            $c^* = cl(y)$
            $CandidateRules(c^*) + +.$
        end if
    end for
end for
$cl(y') = arg\max_{c_j} CandidateRules(c_j)$

---

k-nn-like classifiers focus on the neighborhood of the target item, analogical classifiers "take inspiration" of information possibly far from the immediate neighborhood.

Example. Let's continue with the previous Golf example to show the classification process in Algorithm1. Given three change rules r_1, r_2 and r_3:

$$r_{1(Yes-No)} = \{(sunny, overcast)_1, (false, true)_4\}$$
$$r_{2(No-Yes)} = \{(cool, mild)_2, (high, normal)_3\}$$
$$r_{3(No-Yes)} = \{(rainy, overcast)_1, (false, true)_4\},$$

and a new instance y' to be classified: y' : $overcast, mild, normal, true, \rightarrow$?
Assume that there are three training examples y_1, y_2 and y_3 in T_s:

y_1 : $sunny, mild, normal, false, \rightarrow Yes$
y_2 : $overcast, cool, high, true, \rightarrow No$
y_3 : $rainy, mild, normal, false, \rightarrow No$

We note that disagreement patterns p_1, p_2 and p_3 corresponding respectively to the pairs (y_1, y'), (y_2, y') and (y_3, y') match respectively the change rules r_1, r_2 and r_3. Inferring the first rule predict a first candidate class "No" for y'. In the same manner the second rule predict a class "Yes" and the third one also predict "Yes". The rule-based inference produces the following set of candidate classes for y': $Candidate = \{No, Yes, Yes\}$. So the most plausible class for y' is "Yes".

Experimental Results and Comparison

This section provides experimental results for the two analogical proportion-based classifiers. The experimental study is based on several data sets selected from the U.C.I. machine learning repository [3]. A brief description of these data sets is given in Table 3. We note that for all classification results given in the following, only half of the

Table 3. Description of datasets

Datasets	Instances	Attributes	Classes
Breast cancer	286	9	2
Balance	625	4	3
Tic tac toe	958	9	2
Car	743	7	4
Monk1	432	6	2
Monk 2	432	6	2
Monk3	432	6	2

training set is used to extract patterns. We ensured that all class labels are represented in this data set. The classification results for the CRC or NCRC are summarized in the first and second columns of Table 4. We also tested a hybrid version of these classifiers called *Hybrid Analogical Classifier (HAC)* based on the following process. Given an instance y' to classify,

1. Merge the two rule subsets R_{ch} and R_{noch} into a single rule set R_{chnoch}.
2. Assign to y' the class label with the highest number of candidate rules in R_{chnoch}.

Classification results for HAC are given in Table 4, where we also give the mean number of Change (MeanCh) and No-Change rules (MeanNoCh) generated for each data set.

In order to compare analogical classifiers with other classification approaches, Table 5 includes classification results of some machine learning algorithms (the SVM, k-nearest neighbors IBK with k=10 and the propositional rule learner JRip) obtained by using the Weka software. By analyzing classification performance in Table 4 we can see that:

• Overall, the analogical classifiers show good performance to classify test examples (at least for one of CRC and NCRC), especially NCRC.

• If we compare classification results for the two analogical classifiers, CRC and NCRC, we see that NCRC seems to be more efficient than CRC for almost all data sets, except the case of "Tic tac toe" where the two classifiers have the same accuracy.

Table 4. Classification accuracies: mean and standard deviation of 10 cross-validations

Datasets	CRC	NCRC	HAC	MeanCh	MeanNoCh
Breast cancer	50.03 ± 8.03	**74.03±7.48**	73.39±8.44	6243.4	8738.5
Balance	82.82 ±5.8	**91.02±4.44**	90.51 ± 4.27	31736.2	20805.4
Tic tac toe	**98.3±5.11**	**98.3±5.11**	**98.3±5.11**	74391.9	86394.2
Car	79.54±4.23	**95.02± 2.16**	92.6 ±2.69	36526.6	20706.1
Monk1	90.52±6.16	**100±0**	99.54 ±1.4	9001.2	8644.6
Monk2	78.02 ±4.71	**100±0**	94.68 ± 4.38	7245.9	10607.8
Monk3	91.93±7.04	**97.93±1.91**	**97.93±1.91**	10588.0	10131.7

Table 5. Classification results of some known machine learning algorithms

Datasets	SVM	IBK(k=10)	JRip
Breast cancer	69.58	73.07	70.97
Balance	90.24	83.84	71.68
Tic tac	98.32	98.64	97.80
Car	91.65	91.92	87.88
Monk1	75.0	95.60	94.44
Monk2	67.12	62.96	66.43
Monk3	100	98.37	98.61

• HAC shows good performance if compared to CRC and very close accuracies to NCRC for "Balance, Tic tac toe, Monk1 and Monk3". For the remaining datasets, the lower classification accuracy of Change rules may affect the efficiency of HAC.

• In general, analogical classifiers (especially NCRC) show very good performance when compared to some of existing algorithms. NCRC significantly outperforms all other classifiers for all tested data sets (bold results in Table 5) except to some extent for "Monk3" and SVM. We see that NCRC is largely better than other classifiers, in particular for data sets "Monk1", "Monk2" and "Car".

• The classification success of NCRC for "Monks" datasets with noisy data and "Balance" and "Car" (which have multiple classes) demonstrates its ability to deal with noisy and multiple class data sets.

• The analogy-based classifiers seem to be very efficient when classifying data sets with a limited number of attribute values and seems to have more difficulties for classifying data sets with a large number of attribute values. In order to evaluate analogical classifiers such a dataset, we tested CRC and NCRC on "Cancer" (9 attributes, each of them having 10 different labels). From this additional test, we note that analogical classifiers are significantly less efficient on "Cancer" when compared to the state of the art algorithms. By contrast, if we look at the 3 "Monks" and "Balance" data sets, we note that these data sets have a smaller number of attributes and more importantly all attributes have a reduced number of possible values (the maximum number of possible attribute values in "Balance" and "Monks" is 5, and most of attributes have only 3 possible labels). This clearly departs from the "Cancer" situation. So we may say that this latter dataset is closer to a data set with numerical rather than categorical data. The proposed classifiers are basically designed for handling categorical attributes. We plan to extend analogical rule-based classifiers in order to support numerical data in future.

• In Table 4 we see that a huge number of rules of the two kinds are generated. We may wonder if a reduced subset of rules could lead to the same accuracy. This would mean that there are some redundancy among each subset of rules, raising the question of how to detect it. We might even wonder if all the rules have the same "relevance", which may also mean that some rules have little value in terms of prediction, and should be identified and removed. This might also contribute to explain why CRC has results poorer than NCRC in most cases.

• In the case of NCRC, we come apparently close to the principle of a k-nn classifier, since we use nearest neighbors for voting, but here some nearest neighbors are disqualified because there is no NoChange rule (having the same disagreement pattern) that supports them.

Concluding Remarks

This paper has shown that analogical classification can rely on a rule-based technique, which contrasts with the existing implementations which are mainly lazy techniques. In the proposed approach, the rules are built at compile time, offline with respect to the classification process itself, where this set of rules is applied to new unclassified items in order to predict their class. This view brings new highlights in the understanding of analogical classification and may make this kind of learner more amenable to be mixed with logical ones like the ones coming from Inductive Logic Programming.

References

1. Bayoudh, S., Miclet, L., Delhay, A.: Learning by analogy: A classification rule for binary and nominal data. In: Proc. Inter. Conf. on Artificial Intelligence, IJCAI 2007, pp. 678–683 (2007)
2. Lepage, Y.: Analogy and formal languages. Electr. Notes Theor. Comput. Sci. 53 (2001)
3. Mertz, J., Murphy, P.: Uci repository of machine learning databases,
   ftp://ftp.ics.uci.edu/pub/machine-learning-databases
4. Miclet, L., Bayoudh, S., Delhay, A.: Analogical dissimilarity: definition, algorithms and two experiments in machine learning. JAIR 32, 793–824 (2008)
5. Miclet, L., Prade, H.: Handling analogical proportions in classical logic and fuzzy logics settings. In: Sossai, C., Chemello, G. (eds.) ECSQARU 2009. LNCS, vol. 5590, pp. 638–650. Springer, Heidelberg (2009)
6. Prade, H., Richard, G.: Reasoning with logical proportions. In: Lin, F.Z., Sattler, U., Truszczynski, M. (eds.) Proc. 12th Int. Conf. on Principles of Knowledge Representation and Reasoning, KR 2010, Toronto, May 9-13, pp. 545–555. AAAI Press (2010)
7. Prade, H., Richard, G.: From analogical proportion to logical proportions. Logica Universalis 7(4), 441–505 (2013)
8. Stroppa, N., Yvon, F.: Du quatrième de proportion comme principe inductif: une proposition et son application à l'apprentissage de la morphologie. Traitement Automatique des Langues 47(2), 1–27 (2006)

Multilabel Prediction with Probability Sets: The Hamming Loss Case

Sebastien Destercke

Heudiasyc, UMR 7253, Centre de recherche de royallieu, 60203 Compiegne, France
sebastien.destercke@hds.utc.fr

Abstract. In this paper, we study how multilabel predictions can be obtained when our uncertainty is described by a convex set of probabilities. Such predictions, typically consisting of a set of potentially optimal decisions, are hard to make in large decision spaces such as the one considered in multilabel problems. However, we show that when considering the Hamming loss, an approximate prediction can be efficiently computed from label-wise information, as in the precise case. We also perform some first experiments showing the interest of performing partial predictions in the multilabel case.

Keywords: Credal sets, multilabel, indeterminate classification, k-nn, binary relevance.

1 Introduction

The problem of multi-label classification, which generalizes the traditional (single label) classification setting by allowing multiple labels to belong simultaneously to an instance, has recently attracted a lot of attention. Such a setting indeed appears in a lot of cases: a film can belong to multiple categories, a music can stir multiple emotions [12], proteins can possess multiple functions [14], etc. In such problems, obtaining a complete ground truth (sets of relevant labels) for the training data and making accurate predictions is more complex than in traditional classification, in which the aim is to predict a single label.

In such a setting the appearance of incomplete observations, i.e., instances for which we do not know whether some labels are relevant, is much more likely. For example, a user may be able to tag a movie as a comedy and not as a science-fiction movie, but may hesitate about whether it should be tagged as a drama. Other examples include cases with high number of labels and where an expert cannot be expected to provide all relevant ones. Such partial labels are commonly called weak labels [9] and are common in problems such as image annotation [10] or protein function prediction [14].

Even when considering weak labels, all multilabel methods we are aware of still produce complete predictions as outputs. However, given the complexity of the predictions to make and the likely presence of missing data, it may be sensible to look for means to do cautious yet more trustful predictions. That is it may be interesting for the learner to abstain to make a prediction about a label whose relevance is too uncertain, so that the final prediction is partial but more accurate. Such an approach can be seen as an extension of the reject option implemented in learning problems [1] or of the fact of making

A. Laurent et al. (Eds.): IPMU 2014, Part II, CCIS 443, pp. 496–505, 2014.

partial predictions [4], and has been recently investigated for the related problem of label ranking [3,2].

In this paper, we consider the problem of making partial predictions in the multilabel setting using convex sets of probabilities, or credal sets [8], as our predictive model. Indeed, making partial predictions is one central feature of approaches using credal sets [4], and these approaches are also well-designed to cope with the problem of missing or incomplete data [15]. However, making partial predictions with credal sets in large decision space such as the one considered in mutlilabel is usually difficult.

In Section 3, we nevertheless demonstrate that when focusing on the Hamming loss, obtaining approximate partial predictions can be done in a quite efficient way by focusing on label-wise information. We then perform (Section 4) some experiments to demonstrate the interest of making partial predictions in the multilabel setting. Section 2 presents necessary background material.

2 Preliminary Material

In this section, we introduce the multilabel setting as well as basic notions needed to deal with sets of probabilities.

2.1 Multilabel Problem Setting

The usual goal of classification problems is to associate an instance $\mathbf{x}$ coming from an instance space $\mathscr{X}$ to a single (preferred) label of the space $\Lambda = \{\lambda_1, \ldots, \lambda_m\}$ of possible classes. In a multilabel setting, an instance $\mathbf{x}$ is associated to a subset $L_\mathbf{x} \subset \Lambda$ of labels, often called the subset of relevant labels while its complement $\Lambda \setminus L_\mathbf{x}$ is considered as irrelevant. We denote by $\mathscr{Y} = \{0,1\}^m$ the set of m-dimensional binary vectors, and identify a set L of relevant labels with a binary vector $y = (y_1, \ldots, y_m) \in \mathscr{Y}$ such that $y_i = 1$ if and only if $\lambda_i \in L$. As there is a one-to-one mapping between subsets L of Λ and $\mathscr{Y}$, we will indifferently work with one or the other.

The task in a multilabel problem is the same as in usual classification: to use the training instances $(\mathbf{x}^j, y^j)$, $j = 1, \ldots, n$ to estimate the theoretical conditional probability measure $P_\mathbf{x} : 2^{\mathscr{Y}} \to [0,1]$ associated to an instance $\mathbf{x} \in \mathscr{X}$. Ideally, observed outputs y^j should be specified vectors , however it may be the case that the value for some component y_i^j is unknown, which will be denoted by $y_i^j = *$. We will denote incomplete vectors by capital Y. Alternatively, an incomplete vector Y can be characterized by two sets $\underline{L} \subseteq \overline{L} \subseteq \Lambda$ of necessarily and possible relevant labels, defined as $\underline{L} := \{\lambda_i | y_i = 1\}$ and $\overline{L} := \{\lambda_i | y_i = 1 \vee y_i = *\}$ respectively. An incomplete vector Y describes a corresponding set of complete vectors, obtained by replacing each $y_i = *$ either by 1 or 0, or equivalently by considering any subset L such that $\underline{L} \subseteq L \subseteq \overline{L}$. To simplify notations, in the sequel we will use the same notation for an incomplete vector and its associated set of complete vectors.

Example 1. Table 1 provides an example of a multilabel data set with $\Lambda = \{\lambda_1, \lambda_2, \lambda_3\}$. $Y^3 = [* \, 1 \, 0]$ is an incomplete observed instance with $\underline{L}^3 = \{\lambda_2\}$ and $\overline{L}^3 = \{\lambda_1, \lambda_2\}$. Its corresponding set of complete vectors is $\{[0 \, 1 \, 0], [1 \, 1 \, 0]\}$

Table 1. Multilabel data set example

X_1	X_2	X_3	X_4	y_1	y_2	y_3
107.1	25	Blue	60	1	0	0
−50	10	Red	40	1	0	1
200.6	30	Blue	58	*	1	0
107.1	5	Green	33	0	1	*
...	...	...	...	...	...	...

In multilabel problems the size of the prediction space increases exponentially with m ($|\mathcal{Y}| = 32768$ for $m = 15$), meaning that estimating directly $P_\mathbf{x}$ will be intractable even for limited sizes of Λ. As a means to solve this issue, different authors have proposed so-called transformation techniques [13] that reduce the initial problem (in which 2^m parameters have to be estimated) into a set of simpler problems. For instance Binary Relevance (BR) consists in predicting relevance label-wise, solving an independent binary problem for each label. It therefore comes down to estimate m parameters $P_\mathbf{x}(y_i)$, $i = 1, \ldots, m$ and to predict $\hat{y}_i = 1$ if $P_\mathbf{x}(y_i = 1) \geq 1/2$. A common critic of the BR approach is that it does not take account of label dependencies, however it has been shown that this approach is theoretically optimal for the Hamming loss, on which this paper focuses [5]. Other reduction approaches include, for instance, Calibrated Ranking (CR) [7] that focuses on pairwise comparisons.

Another issue is that making a precise and accurate estimation of $P_\mathbf{x}$ is an extremely difficult problem given the number 2^m of alternatives and the possible presence of missing data. This problem is even more severe if little data are available, and this is why making cautious inferences (i.e., partial predictions) using as model a (convex) set $\mathscr{P}_\mathbf{x}$ of probability distributions may be interesting in the multilabel setting.

2.2 Notions about Probability Sets

We assume that our uncertainty is described by a convex set of probabilities $\mathscr{P}_\mathbf{x}$ defined over $\mathcal{Y}$ rather than by a precise probability measure $P_\mathbf{x}$. Such a set is usually defined either by a collection of linear constraints on the probability masses or by a set of extreme probabilities. Given such a set, we can define for any event $A \subseteq \mathcal{Y}$ the notions of lower and upper probabilities $\underline{P}_\mathbf{x}(A)$ and $\overline{P}_\mathbf{x}(A)$, respectively defined as

$$\underline{P}_\mathbf{x}(A) = \inf_{P_\mathbf{x} \in \mathscr{P}_\mathbf{x}} P_\mathbf{x}(A) \text{ and } \overline{P}_\mathbf{x}(A) = \sup_{P_\mathbf{x} \in \mathscr{P}_\mathbf{x}} P_\mathbf{x}(A).$$

Lower and upper probabilities are dual, in the sense that $\underline{P}(A) = 1 - \overline{P}(A^c)$. Similarly, if we consider a real-valued bounded function $f : \mathcal{Y} \to \mathbb{R}$, the lower and upper expectations $\underline{\mathbb{E}}_\mathbf{x}(f)$ and $\overline{\mathbb{E}}_\mathbf{x}(f)$ are defined as

$$\underline{\mathbb{E}}_\mathbf{x}(f) = \inf_{P_\mathbf{x} \in \mathscr{P}_\mathbf{x}} \mathbb{E}_\mathbf{x}(f) \text{ and } \overline{\mathbb{E}}_\mathbf{x}(f) = \sup_{P_\mathbf{x} \in \mathscr{P}_\mathbf{x}} \mathbb{E}_\mathbf{x}(f),$$

where $\mathbb{E}_\mathbf{x}(f)$ is the expectation of f w.r.t. P. Lower and upper expectations are also dual, in the sense that $\underline{\mathbb{E}}(f) = -\overline{\mathbb{E}}(-f)$. They are also scale and translation invariant in the sense that given two numbers $\alpha \in \mathbb{R}^+, \beta \in \mathbb{R}$, we have $\underline{\mathbb{E}}(\alpha f + \beta) = \alpha \underline{\mathbb{E}}(f) + \beta$.

In the next sections, we explore how the multilabel problem can be solved with such credal estimates. We discuss the problem, usually computationally intensive, of making partial decision and show that it can be simplified when considering the Hamming loss as our loss function. Using these results, we then perform some experiment based on label-wise decomposition and k-nn algorithms to assess the interest of making partial predictions based on credal sets.

3 Credal Multilabel Predictions with Hamming Loss

In this section, we first recall the principle of credal predictions, before proceeding to show that in the case of Hamming loss, such predictions can be efficiently approximated by an outer-approximation.

3.1 Credal Predictions

Once a space $\mathcal{Y}$ of possible observations is defined, selecting a prediction, or equivalently making a decision, requires to define:

- a space $\mathcal{A} = \{a_1, \ldots, a_d\}$ of possible actions (sometimes equal to $\mathcal{Y}$, but not necessarily);
- a loss function $\ell : \mathcal{A} \times \mathcal{Y} \to \mathbb{R}$ such that $\ell(a, y)$ is the loss associated to action a when y is the ground-truth.

Given an instance $\mathbf{x}$ and a precise estimate $\hat{P}_{\mathbf{x}}$, a decision a will be preferred to a decision a' under loss function ℓ, denote $a \succ_\ell a'$, if

$$\mathbb{E}_{\mathbf{x}} \left(\ell(a', \cdot) - \ell(a, \cdot) \right) = \sum_{y \in \mathcal{Y}} \hat{P}_{\mathbf{x}}(y) \left(\ell(a', y) - \ell(a, y) \right) > 0, \tag{1}$$

where $\mathbb{E}_{\mathbf{x}}$ is the expectation w.r.t. $\hat{P}_{\mathbf{x}}$. This equation means that exchanging a' for a would incur a positive expected loss, therefore a should be preferred to a'. In the case of a precise estimate $\hat{P}_{\mathbf{x}}$, $\succ_\ell$ is a complete pre-order and the prediction comes down to take the maximal element of this pre-order, i.e.,

$$\hat{a}_\ell = \arg\min_{a \in \mathcal{A}} \mathbb{E}_{\mathbf{x}} \left(\ell(a, \cdot) \right) = \arg\min_{a \in \mathcal{A}} \sum_{y \in \mathcal{Y}} \hat{P}_{\mathbf{x}}(y) \ell(a, y) \tag{2}$$

that is to minimize the expected loss (ties can be broken arbitrarily, as they will lead to the same expected loss). This means that finding the best action (or prediction) will therefore requires d computations of expectations.

When considering a set $\mathcal{P}_{\mathbf{x}}$ as cautious estimate, there are many ways [11] to extend Eq. (1), but the most well-founded is the maximality criterion, which states that $a \succ_\ell a'$, if

$$\underline{\mathbb{E}}_{\mathbf{x}} \left(\ell(a', \cdot) - \ell(a, \cdot) \right) > 0, \tag{3}$$

that is if exchanging a' for a is guaranteed to give a positive expected loss. In such a case, the relation $\succ_\ell$ will be a partial order, and the maximal set $\hat{A}_\ell$ of alternatives will be chosen as a prediction, that is

$$\hat{A}_\ell = \{ a \in \mathcal{A} \mid \nexists a' \in \mathcal{A} \text{ s.t. } a' \succ_\ell a \}. \tag{4}$$

Computing $\hat{A}_\ell$ requires at worst $d(d-1)$ computations, a quadratic number of comparisons with respect to the number of alternatives. Also notes that evaluating Eq. (3) usually requires solving a linear programming problem, a computationally more intensive task than evaluating Eq. (1). $\hat{A}_\ell$ is a cautious prediction, since it considers a set of potential optimal solutions.

Multilabel loss functions usually considers the set $\mathscr{A} = \mathscr{Y}$ as possible actions, or even bigger sets (for example the ranking loss considers as actions the set of complete orders over Λ). This means that getting $\hat{a}_\ell$ is already quite hard in the general case, hence computing $\hat{A}_\ell$ will be intractable in most cases, as the worst number of computation will then be 2^{2m} ($m = 15$ labels means at worst $\sim 10^9$ comparisons).

In the next subsection, we show that for the Hamming loss ℓ_H, we can get an outer approximation of $\hat{A}_\ell$ at an affordable computational cost. Offering such efficient way to make cautious predictions based on $\mathscr{P}_\mathbf{x}$ is essential to be able to use such kind of models in complex problems.

3.2   The Hamming Loss

Let the set of alternatives be $\mathscr{A} = \mathscr{Y}$. Given an observation y and a prediction $\hat{y}$, Hamming loss ℓ_H reads

$$\ell_H(\hat{y}, y) = \frac{1}{m} \sum_{i=1}^m \mathbf{1}_{(\hat{y}_i \neq y_i)} . \tag{5}$$

It counts the number of labels for which our prediction is wrong, and normalizes it. When the estimate $\hat{P}_\mathbf{x}$ is precise, it is known [5] that the optimal decision is the vector $\hat{y}$ such that $\hat{y}_j = 1$ if $\hat{P}_\mathbf{x}(y_j = 1) \geq 1/2$ and $\hat{y}_j = 0$ else. In particular, this means that the optimal decision can be derived from the sole knowledge of the marginals $\hat{P}_\mathbf{x}(y_j = 1)$, $j = 1, \ldots, n$, provided they are good estimates of $P_\mathbf{x}$.

Given a probability set $\mathscr{P}_\mathbf{x}$, let $\hat{Y}_{\ell_H}$ be the maximal set of vectors that would be obtained using Eq. (4). The next proposition shows that $\hat{Y}_{\ell_H}$ can be outer-approximated using the marginals of the cautious estimate $\mathscr{P}_\mathbf{x}$, in contrast with the precise case.

Proposition 1. *Let $\mathscr{P}_\mathbf{x}$ be our estimate, then the imprecise vector $\hat{Y}^*$ such that*

$$\hat{Y}_j^* = \begin{cases} 1 & \text{if } \underline{P}(y_j = 1) > 1/2 \\ 0 & \text{if } \underline{P}(y_j = 0) > 1/2 \qquad \text{for } j = 1, \ldots, m \\ * & \text{else, i.e. } \underline{P}(y_j = 1) \leq 1/2 \leq \overline{P}(y_j = 1) \end{cases}$$

is an outer approximation of $\hat{Y}_{\ell_H}$, in the sense that $\hat{Y}_{\ell_H} \subseteq \hat{Y}^*$.

Proof. Consider a given $j \in \{1, \ldots, m\}$ and two alternatives $\hat{y}$ and $\hat{y}'$ such that $\hat{y}_j = 1 \neq \hat{y}'_j$ and $\hat{y}_i = \hat{y}'_i$ for any $i \neq j$. Then for any y such that $y_j = 1$ we have

$$\ell_H(\hat{y}', y) - \ell_H(\hat{y}, y) = \left(\sum_{k \neq j} \mathbf{1}_{(\hat{y}'_k \neq y_k)} + \mathbf{1}_{(\hat{y}'_j \neq y_j)} \right) - \left(\sum_{k \neq j} \mathbf{1}_{(\hat{y}_k \neq y_k)} + \mathbf{1}_{(\hat{y}_j \neq y_j)} \right)$$

$$= \mathbf{1}_{(\hat{y}'_j = 0)} - \mathbf{1}_{(\hat{y}_j = 0)} = 1,$$

and for any y such that $y_j = 0$ we have

$$\ell_H(\hat{y}', y) - \ell_H(\hat{y}, y) = \left(\sum_{k \neq j} \mathbf{1}_{(\hat{y}'_k \neq y_k)} + \mathbf{1}_{(\hat{y}'_j \neq y_j)} \right) - \left(\sum_{k \neq j} \mathbf{1}_{(\hat{y}_k \neq y_k)} + \mathbf{1}_{(\hat{y}_j \neq y_j)} \right)$$

$$= \mathbf{1}_{(\hat{y}'_j = 1)} - \mathbf{1}_{(\hat{y}_j = 0)} = -1.$$

We therefore have $(\ell_H(\hat{y}', \cdot) - \ell_H(\hat{y}, \cdot) + 1)/2 = \mathbf{1}_{(y_j = 1)}$, hence

$$\underline{P}(y_j = 1) = \mathbb{E} \left(\frac{\ell_H(\hat{y}', \cdot) - \ell_H(\hat{y}, \cdot) + 1}{2} \right)$$

$$= \frac{1}{2} \mathbb{E} \left(\ell_H(\hat{y}', \cdot) - \ell_H(\hat{y}, \cdot) \right) + \frac{1}{2}$$

the last equality coming from scale and translation invariance. Hence $\mathbb{E}(\ell_H(\hat{y}', \cdot) - \ell_H(\hat{y}, \cdot)) > 0$ if and only if $\underline{P}(y_j = 1) > 1/2$. This means that, if $\underline{P}(y_j = 1) > 1/2$, any vector $\hat{y}'$ with $\hat{y}'_j = 0$ is dominated (in the sense of Eq. (3)) by the vector $\hat{y}$ where only the j-th element is modified, hence no vector with $\hat{y}'_j = 0$ is in the maximal set $\hat{Y}_{\ell_H}$. The proof showing that if $\underline{P}(y_j = 0) > 1/2$, then no vector with $\hat{y}'_j = 1$ is in the maximal set is similar. ∎

We now provide an example showing that the inclusion can be strict in general.

Example 2. Consider the 2 label case $\Lambda = \{\lambda_1, \lambda_2\}$ with the following constraints:

$$0.4 \leq P(y_1 = 1) = P(\{[1\,0]\}) + P(\{[1\,1]\}) \leq 0.6$$
$$0.9\,(P(\{[1\,0]\}) + P(\{[1\,1]\})) = P(\{[1\,0]\})$$
$$0.84\,(P(\{[0\,1]\}) + P(\{[0\,0]\})) = P(\{[0\,1]\})$$

These constraints describe a convex set $\mathscr{P}$, whose extreme points (obtained by saturating the first inequality one way or another) are summarized in Table 2. The first constraints induces that $\underline{P}(y_1 = 1) = 0.4$ and $\overline{P}(y_1 = 0) = 0.6$, while the bounds $\underline{P}(y_2 = 1) = 0.396, \overline{P}(y_2 = 1) = 0.544$, are reached by the extreme distributions $P([1\,1]) = 0.06, P([0\,1]) = 0.336$ and $P([1\,1]) = 0.04, P([0\,1]) = 0.504$, respectively. Given these bounds, we have that $\hat{Y}^* = [*\,*]$ corresponds to the complete space $\mathscr{Y}$ (i.e., the empty prediction). Yet we have that

$$\mathbb{E}\left(\ell_H([1\,1], \cdot) - \ell_H([0\,0], \cdot)\right) = 0.0008 \geq 0$$

also obtained with the distribution $P([1\,1]) = 0.06, P([0\,0]) = 0.064$. This means that the vector $[0\,0]$ is not in the maximal set $\hat{Y}_{\ell_H}$, while it is included in $\hat{Y}^*$.

Proposition 1 shows that we can rely on marginal information to provide an outer-approximation of $\hat{Y}_{\ell_H}$ that is efficient to compute, as it requires to compute $2m$ values, which are to be compared to the 2^{2m} usually required to assess $\hat{Y}_{\ell_H}$. It also indicates that extensions of the binary relevance approach are well adapted to provide partial predictions from credal sets when considering the Hamming loss, and that in this case global models integrating label dependencies are not necessary, thus saving a lot of heavy computations.

Table 2. Extreme points of $\mathscr{P}$ of Example 2

$P(\{[0\ 0]\})$	$P(\{[1\ 0]\})$	$P(\{[0\ 1]\})$	$P(\{[1\ 1]\})$
0.096	0.36	0.504	0.04
0.064	0.54	0.336	0.06

4   First Experimentations

In this section, we provide first experimentations illustrating the effect of making partial predictions with a decreasing amount of information. These experiments illustrate that such partial predictions may indeed improve the correctness of predictions.

4.1   Evaluation

Usual loss functions such as Eq. (5) are based on complete predictions. When making partial predictions, such loss functions need to be adapted. This can be done, for instance, by decomposing it into two components [3], one measuring the accuracy or correctness of the made prediction, the other measuring its completeness.

If the partial prediction is an incomplete vector such as $\hat{Y}^*$, then Hamming loss can be easily split into these two components. Given the prediction $\hat{Y}^*$ characterized by subsets $\underline{L}, \overline{L}$, let us denote $Q = \Lambda \setminus (\underline{L} \cap \overline{L})$ the set of predicted labels (i.e., labels such that $\hat{Y}^*_j = 1$ or $\hat{Y}^*_j = 0$). Then, if the observed set is y, we define correctness (CR) and completeness (CP) as

$$CR(\hat{Y}^*, y) = \frac{1}{|Q|} \sum_{\lambda_i \in Q} \mathbf{1}_{(\hat{y}_i \neq y_i)} ; \tag{6}$$

$$CP(\hat{Y}^*, y) = \frac{|Q|}{m}. \tag{7}$$

when predicting complete vectors, then $CP = 1$ and CR equals the Hamming loss (5). When predicting the empty vector, then $CP = 0$ and by convention $CR = 1$.

4.2   Method

The method we used was to apply, label-wise, the k-nn method using lower probabilities introduced in [6] (in which details can be found). This means that from an initial training data set $\mathscr{D}$, m data sets $\mathscr{D}_j$ corresponding to binary classification problems are built, this decomposition being illustrated in Figure 1. Given an instance $\mathbf{x}$, the result of the k-nn method on data set $\mathscr{D}_j$ provides an estimate of $[\underline{P}(y_j = 1), \overline{P}(y_j = 1)]$ and by duality an estimate of $\underline{P}(y_j = 0) = 1 - \overline{P}(y_j = 1)$ and $\overline{P}(y_j = 0) = 1 - \underline{P}(y_j = 1)$.

The method also automatically takes account of missing label information, and treat such missing data in a conservative way, considering them as completely vacuous information (that is, we treat them as non-MAR variables [16]).

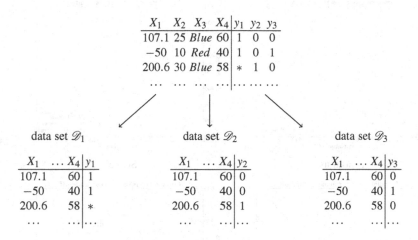

Fig. 1. Label-wise decomposition of data set $\mathscr{D}$

4.3 Results

In the experiments, the parameters of the k-nn algorithm were set to $\beta = 0.75$ and $\varepsilon_0 = 0.99$, so that results obtained when fixing the number k of neighbors to 1 display a sufficient completeness. ε_0 settles the initial imprecision, while β determines how much imprecision increases with distance (details about the role of these parameters can be found in [6]). We ran experiments on well-known multilabel data sets having real-valued features. Their characteristics are summarized in Table 3. For each of them, we ran a 10-fold cross validation with the number k of neighbors varying from 1 to 5, and with various percentages of missing labels in the training data set (0%, 20% and 40%). Varying k in the algorithm allows us to control the completeness of the prediction: the higher k is, the more imprecise become the estimations.

Table 3. Multilabel data sets summary

Name	# Features	# Labels	# Instances
emotion	72	6	593
scene	294	6	2407
yeast	103	14	2417
CAL500	68	174	502

The results of the experiment are displayed in Figure 2. From this figure, two main conclusions can be drawn: on the used data sets, allowing for partial predictions (here, by increasing the number k of neighbours) systematically improve the correctness, and missing labels only influence the completeness of the predictions, not the correctness of the results. This latter fact, however, may be due to the learning method. How fast completeness decreases with the number of neighbors, however, clearly depends on the data set.

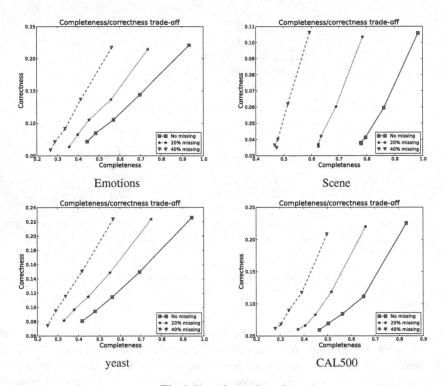

Fig. 2. Experimental results

5   Conclusions

Producing sets of optimal predictions in the multilabel setting when uncertainty is modeled by convex probability sets is computationally hard. The main contribution of this paper was to show that when using the Hamming loss, such sets can be easily outer-approximated by focusing only on the marginal probability bounds of each label being relevant. This makes both computation and learning issues easier, as one can focus on estimating such marginals (instead of the whole joint model). We can consider that as an important result, as it shows that imprecise probabilistic approaches are computationally affordable (at least under some conditions).

We also made some first preliminary experiments indicating the interest of producing such partial predictions, showing that making more cautious predictions lead to more correct predictions. In the future, we intend to make similar studies for other well-known loss functions, such as the ranking loss. We also intend to make further the experiments, i.e., to compare this approach with other methods, or to empirically assess (for small values of m) the quality of the made approximation.

Acknowledgements. Work carried out in the framework of the Labex MS2T, funded by the French Government, through the National Agency for Research (Reference ANR-11-IDEX-0004-02).

References

1. Bartlett, P., Wegkamp, M.: Classification with a reject option using a hinge loss. The Journal of Machine Learning Research 9, 1823–1840 (2008)
2. Cheng, W., Hüllermeier, E., Waegeman, W., Welker, V.: Label ranking with partial abstention based on thresholded probabilistic models. In: Advances in Neural Information Processing Systems 25 (NIPS 2012), pp. 2510–2518 (2012)
3. Cheng, W., Rademaker, M., De Baets, B., Hüllermeier, E.: Predicting partial orders: ranking with abstention. In: Balcázar, J.L., Bonchi, F., Gionis, A., Sebag, M. (eds.) ECML PKDD 2010, Part I. LNCS (LNAI), vol. 6321, pp. 215–230. Springer, Heidelberg (2010)
4. Corani, G., Antonucci, A., Zaffalon, M.: Bayesian networks with imprecise probabilities: Theory and application to classification. In: Holmes, D.E., Jain, L.C. (eds.) Data Mining: Foundations and Intelligent Paradigms. ISRL, vol. 23, pp. 49–93. Springer, Heidelberg (2012)
5. Dembczynski, K., Waegeman, W., Cheng, W., Hüllermeier, E.: On label dependence and loss minimization in multi-label classification. Machine Learning 88(1-2), 5–45 (2012)
6. Destercke, S.: A k-nearest neighbours method based on imprecise probabilities. Soft Comput. 16(5), 833–844 (2012)
7. Fürnkranz, J., Hüllermeier, E., Loza Mencía, E., Brinker, K.: Multilabel classification via calibrated label ranking. Machine Learning 73(2), 133–153 (2008)
8. Levi, I.: The Enterprise of Knowledge. MIT Press, London (1980)
9. Sun, Y.-Y., Zhang, Y., Zhou, Z.-H.: Multi-label learning with weak label. In: Twenty-Fourth AAAI Conference on Artificial Intelligence (2010)
10. Tian, F., Shen, X.: Image annotation with weak labels. In: Wang, J., Xiong, H., Ishikawa, Y., Xu, J., Zhou, J. (eds.) WAIM 2013. LNCS, vol. 7923, pp. 375–380. Springer, Heidelberg (2013)
11. Troffaes, M.: Decision making under uncertainty using imprecise probabilities. Int. J. of Approximate Reasoning 45, 17–29 (2007)
12. Trohidis, K., Tsoumakas, G., Kalliris, G., Vlahavas, I.P.: Multi-label classification of music into emotions. ISMIR 8, 325–330 (2008)
13. Tsoumakas, G., Katakis, I.: Multi-label classification: An overview. International Journal of Data Warehousing and Mining (IJDWM) 3(3), 1–13 (2007)
14. Yu, G., Domeniconi, C., Rangwala, H., Zhang, G.: Protein function prediction using dependence maximization. In: Blockeel, H., Kersting, K., Nijssen, S., Železný, F. (eds.) ECML PKDD 2013, Part I. LNCS (LNAI), vol. 8188, pp. 574–589. Springer, Heidelberg (2013)
15. Zaffalon, M.: Exact credal treatment of missing data. Journal of Statistical Planning and Inference 105(1), 105–122 (2002)
16. Zaffalon, M., Miranda, E.: Conservative inference rule for uncertain reasoning under incompleteness. Journal of Artificial Intelligence Research 34(2), 757 (2009)

Cooperative Multi-knowledge Learning Control System for Obstacle Consideration

Syafiq Fauzi Kamarulzaman[1,2] and Seiji Yasunobu[2]

[1] Faculty of Computer Systems and Software Engineering,
Universiti Malaysia Pahang, Malaysia
[2] Dept. of Intelligent Interaction Technologies, University of Tsukuba, Japan
syafiq@fz.iit.tsukuba.ac.jp,
yasunobu@iit.tsukuba.ac.jp

Abstract. A safe and reliable control operation can be difficult due to limitations in operator's skills. A self-developing control system could help assist or even replaces the operators in providing the required control operations. However, the self-developing control system is lack of flexibility in determining the necessary control option in multiple conditions where a human operator usually prevails by experiences in optimizing priority. Here, a cooperative multi-knowledge learning control system is proposed in providing flexibility for determining priority in control options, within multiple conditions by considering the required self-developing control knowledge in fulfilling these conditions. The results show that the system was able to provide consideration in prioritizing the use of the required control knowledge of the condition assigned.

Keywords: Multi-knowledge, Learning Control, Reinforcement Learning.

1 Introduction

Human operators are prone to be inefficient during any control operation due to the lack of skills in unfamiliar operation's environment and parameters. Learning Control System provides a self-developing Control Knowledge that changes according to interaction with unfamiliar environment, thus reduces the possible risk related to skills inefficiency [2]. The Control Knowledge will be continuously updated through the control operations that later provides instructions for controlling the related machine to perform at a high efficient manner. The Learning Control System reduces the dependency on human command since the Control Knowledge provides most of the required control instruction, learned from successes in previous operations [1].

However applying Learning Control operation on various conditions of state parameters is difficult and slow since the Control Knowledge is needed to be constructed for each condition. Here, issues concerning application of multiple knowledge in learning are brought to establish a way of learning an optimum action from multiple sources of individual Control Knowledge [7][9]. This study

A. Laurent et al. (Eds.): IPMU 2014, Part II, CCIS 443, pp. 506–515, 2014.

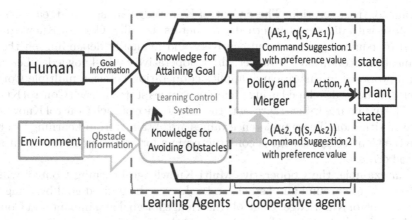

Fig. 1. The structure of cooperative multi-knowledge learning system

is driven by the need to combine multiple Control Knowledge into a single control output while putting the necessity of individual Control Knowledge into consideration. In order to combine multiple knowledge protocol, a mechanism that involves cooperation of multiple Learning Control System for producing an optimum action is presented.

In this research, the cooperation mechanism is developed by evaluating the output preference value from two sets of Control Knowledge with specific input parameters as shown in Fig. 1. Focusing on two different types of states; termed as Goal Distance and Obstacle Distance, separated learning process was conducted to obtain a Cooperative Knowledge that operates to satisfy both state condition. The first set consists of a Control Knowledge that controls the operation upon achieving a goal while the second set consists of a Control Knowledge that controls the operation upon avoiding detected obstacles. Each Control Knowledge provides the preference value of each output depending on current state. Later, a policy and merger agent evaluates the preference value from both knowledge and produce a control output.

Applying both Control Knowledge on the system provides an efficient, safe and successful operation. The success is achieved, due to the application of Cooperative Knowledge for learning and establishing a preference between two Control Knowledge specializing in two different state parameters at the same time. Learning and application of multiple Control Knowledge in a control system are simplified using the proposed method. Simulations were conducted to evaluate the proposed control technique. The results prove that the system was able to cooperate the learning process between the two Control Knowledge that results in a successful control operation.

2   Cooperative Multi-knowledge Learning Control

Multi-knowledge Learning Control is defined by having a control system that learns multiple Control Knowledge. The knowledge will then be used to perform a certain control operation. These Control Knowledge are specified, each for different task with different concerned parameters. Usually, Control Knowledge is applied by control operators consecutively and manually depending on the requirement and conditions. However, the Cooperative Multi-knowledge Learning Control proposed in this research applies cooperation between multiple Control Knowledge that is developed using preference values[stored in each Control Knowledge. The preference values represents the importance of each Control Knowledge in a given situation. These preference values are obtained from Learning Control, where the Control Knowledge is constructed in a form of value function through trial and error.

As an example, the Cooperative Multi-Knowledge Learning Control will be able to configure around an obstacle and achieved the desired goal by using two types of Control Knowledge; Control Knowledge for goal attainment and Control Knowledge for obstacle avoidance. Using "Achieving Goal" as main command, the system uses "Avoiding Obstacle" for creating a cooperative command, heading to the goal while avoiding closing obstacles. Here, Learning Control is applied twice in the protocol. Firstly, Learning Control is applied to construct two specific Control Knowledge. Then, the knowledge gained from the first iteration is combined and relearn for cooperative control operation.

2.1   Learning Control

Learning Control is a method of obtaining Control Knowledge by repeatedly construct and correct the Control Knowledge depending on the outcome of a control operation in several trials [1]. In this research, this method applies reinforcement learning where Control Knowledge is constructed in a form of value functions Q. The value will then be updated depending on the reward r that is received after performing a control operation trial [3].

The Control Knowledge is divided into state S, which defines the current situation of the control object and action A, which defines the next move of the control object. State S and Action A is divided into a set of numbers as State $S = s_1, s_2, , s_n$ and Action $A = a_1, a_2, , a_n$. During the update phase, the preference value q of the combination between state s and action a is defined through the reward obtained after performing action a. In the case successful operation, the preference value q increases, should the action contribute to a fail operation, the value decreases. A set of preference value q can be defined as value function Q, where all preference value for the combination between state S and action A is recorded. The value function Q is defined as Control Knowledge.

In this research, this Control Knowledge is updated based on Q-learning algorithm,

$$Q(S, A) = (1 - \alpha)Q(S, A) + \alpha[r + \gamma Q_{max}], \tag{1}$$

$$Q_{max} = \max_A Q(S, A) \tag{2}$$

where α is the learning rate and γ is update value discount rate [3]. The algorithm is applied in updating all the Control Knowledge constructed.

Two Control Knowledge were constructed to confirm the effectiveness of the system through a simulation. The first Control Knowledge is for Goal Attainment Control, while another is for Obstacle Avoidance Control. For each Control Knowledge, various parameters for state S were used, without imposing any changes to action A.

Goal Attainment Control. Goal Attainment Control consists Learning Control method for operating the control object towards the goal [5]. Here, the Goal Attainment Control applies goal distance $\Delta G = \{\Delta X_{goal}, \Delta Y_{goal}\}$ as state S while control output u and rotation θ as action A_{GA}. Therefore, the value function Q for Goal Attainment Control can be defined as $Q(\Delta G, A_{GA})$.

The update equation for Goal Attainment Control alone is,

$$Q_{Goal}(\Delta G, A_{GA}) = (1 - \alpha)Q_{Goal}(\Delta G, A_{GA}) + \alpha[r + \gamma Q_{max}], \qquad (3)$$

$$Q_{max} = \max_{A_{GA}} Q_{Goal}(\Delta G, A_{GA}) \qquad (4)$$

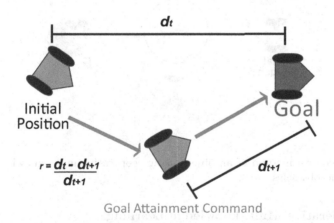

Fig. 2. Method for applying reward r to the Control Knowledge for goal attainment

Obstacle Avoidance Control. Obstacle Avoidance Control consists Learning Control method for operating the control object away from obstacles. Here, the Obstacle Avoidance Control applies obstacle distance $\Delta O = \{\Delta X_{obs}, \Delta Y_{obs}\}$ as state S while control output u and rotation θ as action A_{OA}. Therefore, the value function Q for Obstacle Avoidance Control is defined as $Q(\Delta O, A_{OA})$.

The update equation for Obstacle Avoidance Control alone is,

$$Q_{obs}(\Delta O, A_{OA}) = (1 - \alpha)Q_{obs}(\Delta O, A_{OA}) + \alpha[r + \gamma Q_{max}], \qquad (5)$$

$$Q_{max} = \max_{A_{OA}} Q_{obs}(\Delta O, A_{OA}) \qquad (6)$$

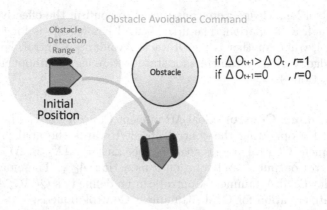

Fig. 3. Method for applying reward r to the Control Knowledge for obstacle avoidance

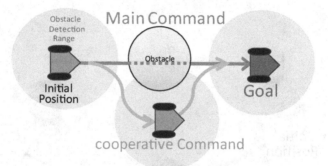

Fig. 4. The control method of an object using cooperative multi-knowledge learning control around obstacles

2.2 Cooperative Multi-knowledge Learning

Applying Learning Control for multiple Control Knowledge requires the system to analyze the value functions of both knowledge prior to the execution of the control output. The preference value of outputs supplied from each control method at a moment of state are needed to determine which Control Knowledge is more preferred at the current state. Here, update method for both Control Knowledge is modified so that the preference value is limited between 0(bad) and 1(Good) [2]. Therefore, the updated value discount rate γ in the update equation for each knowledge is applied as,

$$\gamma_{Goal} = 1 - Q_{Goal}(\Delta G, A_{GA}) \tag{7}$$

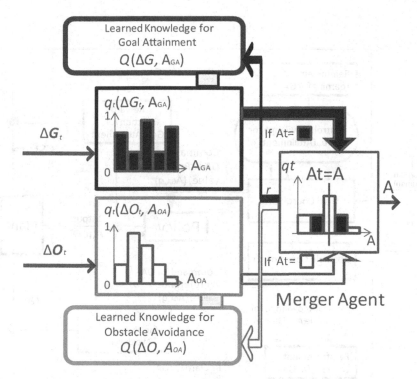

Fig. 5. The application and update process for all Control Knowledge

and

$$\gamma_{obs} = 1 - Q_{obs}(\Delta O, A_{OA}), \tag{8}$$

so that the preference value will be restricted between 0 and 1.

Fig. 5 described the method of cooperating both Control Knowledge in constructing an output. The preference values of a set of output A from both Control Knowledge are used to construct a new set of preference value for output A with the identity of the source knowledge attached; termed as Merger Output. Merger Output is constructed by selecting the minimum preference value among the two Control Knowledge for each element in the set of output A. The control output is determined from Merger Output by a *greedy* policy where the action a with the maximum preference value among the options in the set is chosen as the output at current state. The result after the output been operated will determined the rewards. Rewards will be given to the Control Knowledge based on the identity of the source knowledge in Merger output that supplies the preference value of the executed output.

The system structure involving cooperative multi-knowledge learning control is later applied in a series of simulation as validation of its effectiveness in producing an efficient, safe and reliable operation.

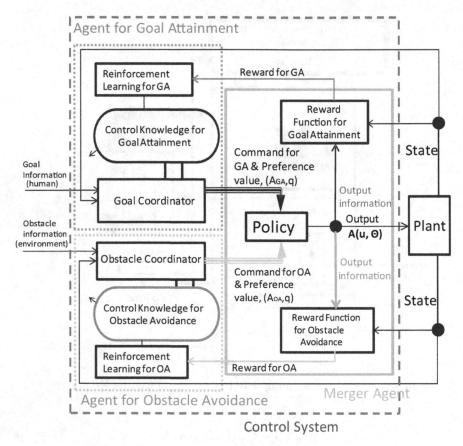

Fig. 6. The system structure for simulation experiment

3   Simulation Experiment

Simulation regarding the proposed method was constructed based on the system structure shown in Fig. 6. The simulation was conducted based on a small robot with the parameters shown in table 1 and Fig. 7b. The simulation was operated in Matlab Simulink and the result was based on operation with and without obstacles in a pre-constructed map.

The map was constructed as Fig. 7a, where 4 different goals were prepared before the simulation was done. The simulated operation results were divided into two sections for easy comparison between an operation with and without obstacles.

The simulation starts by constructing the Control Knowledge of goal attainment and obstacle avoidance separately using the Learning Control. Learning Control was done in 350 episodes in separate simulation. The simulation was continued with cooperative simulation where the constructed knowledge was integrated in the simulation. Here, the Cooperative Learning Control was done for 100 trials and the results were taken after.

Table 1. Specifications of the simulated control object

Parameters	Value
Weight	0.42 [kg]
Size:	
Length	0.53 [m]
Width	0.52 [m]
Height	0.1 [m]
Turning Radius	$-1 < \theta < 1$ [rad]
Torque Force	$-10 < V < 10$ [volt]

Table 2. Q-learning parameters

	Parameters	Range	Intervals
State (Goal)	Goal Distance, ΔG[m]	$-10 < \Delta G(x,y) < 10$	2
State (Obstacle)	Obstacle Distance, ΔO[m]	$-2 < \Delta O(x,y) < 2$	0.5
Action	Target Angle, θ[rad]	$-1 < \theta < 1$	0.5
	Torque Force, V[volt]	$-10 < V < 10$	2

Learning rate, α	0.5	Discount rate, γ	0.3

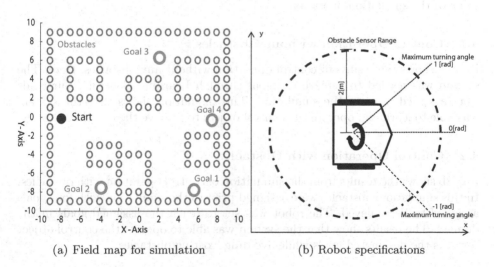

(a) Field map for simulation   (b) Robot specifications

Fig. 7. Simulation setup for field and robot

4   Experiment Results

Here, the results for control operation without obstacles and control operation with obstacles are presented. Control operation without obstacles was done to confirm the effectiveness of the Learning Control process in constructing the most

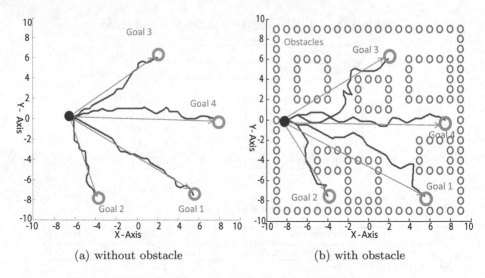

(a) without obstacle   (b) with obstacle

Fig. 8. Results of Control Operation Simulation

effective Control Knowledge for the system. Control operation with obstacles was done to confirm the effectiveness of the whole system in utilizing the cooperative multi-knowledge learning control by using two Control Knowledge constructed prior to the simulation operation.

4.1   Control Operation without Obstacles

Fig. 8a shows the results of control operation without any obstacles. Here, The system was needed to achieve the goal through Learning Control. Four goals were assigned prior to the simulation. The simulation shows that the system was able to learn and operate the control object to achieve the designated goals.

4.2   Control Operation with Obstacles

Fig. 8b shows the results from the simulation of control operation with obstacles. In this simulation, obstacles were assigned prior to the simulation with four goals assigned to be achieved. The robot was set to reach the assigned goals in the domain. The results show that the system was able to operate the control object towards the designated goals while avoiding existing obstacles.

5   Conclusion

This study presents a cooperative multi-knowledge learning control to overcome operations with obstacles. The proposed method applies learning control for multiple Control Knowledge at the same time and cooperatively shares the Control Knowledge by referring to the preference value sustained by each Control Knowledge. Control Knowledge is continuously updated through the Learning Control process. Rewards are given to the knowledge that provides the preference value of the operated control action.

The proposed method was applied in a control system for simulation of a small robot in a virtually constructed field map. The control system simulation was applied in a field, both with and without the obstacles. Results show that the system is able to utilize both Control Knowledge in performing a control operation with obstacle consideration. Therefore, Obstacle Consideration was achieved by calibrating two Control Knowledge in creating a safer cooperative command during the control operation.

Acknowledgement. This research was supported by JSPS KAKENHI Grant Number 24500272.

References

1. Schaal, S., Atkeson, C.G.: Learning Control in Robotics; Trajectory-Based Optimal Control Techniques. IEEE Robotics and Automation Magazine 7(2), 20–29 (2010)
2. Matsubara, T., Yasunobu, S.: An Intelligent Control Based on Fuzzy Target and Its Application to Car Like Vehicle. In: SICE Annual Conference (2004)
3. Xu, X., Zuo, L., Huang, Z.: Reinforcement learning algorithms with function approximation: Recent advances and applications. Journal of Information Sciences 261, 1–31 (2014)
4. Nakamura, Y., Ohnishi, S., Ohkura, K., Ueda, K.: Instance-Based Reinforcement Learning for Robot Path Finding in Continuous Space. In: IEEE SMC, pp. 1229–1234 (1997)
5. Kamarulzaman, S.F., Shibuya, T., Yasunobu, S.: A Learning-based Control System by Knowledge Acquisition within Constrained Environment. In: IFSA World Congress, vol. FC-104, pp. 1–6 (2011)
6. Chang, D., Meng, J.E.: Real-Time Dynamic Fuzzy Q-Learning and Control of Mobile Robots. In: 5th Asian Control Conference, vol. 3, pp. 1568–1576 (2004)
7. Busoniu, L., Babuska, R., Schutter, B.D.: A Comprehensive Survey of Multiagent Reinforcement Learning. IEEE Trans. Systems, Man and Cybernetics 38(2), 156–172 (2008)
8. Gullapalli, V.: Direct Associative Reinforcement Learning Methods for Dynamic Systems Control. Neurocomputing 9, 271–292 (1995)
9. Sun, R., Sessions, C.: Multi-agent reinforcement learning with bidding for automatic segmentation of action sequences. In: 4th International Conference on Multi Agent Systems, pp. 445–446 (2000)
10. Yu, J.: An Adaptive Gain Parameters Algorithm for Path Planning Based on Reinforcement Learning. In: 4th International Conference on Machine Learning and Cybernetics, pp. 3557–3562 (2005)

Building Hybrid Fuzzy Classifier Trees
by Additive/Subtractive Composition of Sets

Arne-Jens Hempel[1], Holger Hähnel[1], and Gernot Herbst[2]

[1] Technische Universität Chemnitz, Chemnitz, Germany
[2] Siemens AG, Chemnitz, Germany

Abstract. Especially for one-class classification problems, an accurate model of the class is necessary. Since the shape of a class might be arbitrarily complex, it is hard to choose an approach that is generic enough to cope with the variety of shapes, while delivering an interpretable model that remains as simple as possible and thus applicable in practice. In this article, this problem is tackled by combining convex building blocks both additively and subtractively in a tree-like structure. The convex building blocks are represented by multivariate membership functions that aggregate the respective parts of the learning data. During the learning process, proven methods from support vector machines and cluster analysis are employed in order to optimally find the structure of the tree. Several academic examples demonstrate the viability of the approach.

1 Introduction

Besides traditional classification techniques that try to distinguish between two or more classes, the demand for one-class classification methods has recently been growing [1]. In real world applications, this relevance of unary classification can be caused by two facts. First, for example, when modeling the class of "interesting products" for the customer of an online store, the learning data include no counterexamples, by which a "non-interesting" class could be determined. Secondly, the structure of counterexamples could be too extensive to be modeled appropriately, which holds e. g. for the decision of an airbag deployment in a vehicle. One approach here would be to decide on the basis of an accurately described "accident" class in a one-class classification task. This might be more apposite than discriminating between the "accident" class and a "non-accident" class, which may have a highly intricate structure making it difficult to model.

Naturally, a classification approach based on (crisp or fuzzy) sets that model a phenomenon has a "local" character and is thus especially suitable for one-class problems. This holds in contrast to discriminatory methods like linear discriminant analysis or (standard) support vector machines (SVMs), which, by construction, allow "global" decisions beneficial for two- or multi-class tasks.

The amenities of a fuzzy classification approach lie in its ability to cope with practically occuring problems such as noisy data, transitional effects, drifting and evolving classes, or overlaps in the case of two- and multi-class problems [2,3]. The high interpretability of fuzzy classification has often been stressed.

A. Laurent et al. (Eds.): IPMU 2014, Part II, CCIS 443, pp. 516–525, 2014.

It provides the opportunity to incorporate expert knowledge into the classifier [4]. Picking up the above example of an airbag deployment, interpretability might be very important when evaluating data of a car accident by an assessor.

The setup of an accurate and at the same time simple and applicable model can be challenging if the shape of the class that is to be described is non-convex or even arbitrarily complex. Such phenomena can be found in applications such as banknote authentication and machine diagnosis [5,6]. The root idea addressing this problem has been given by [7,8] and others. It revolves around the additive combination of convex fuzzy sets, such as fuzzy partitions. Besides that, a subtractive composition using so-called complementary classes has been proposed [6]. We recommend a combination of these two approaches resulting in a tree-like model, which we refer to as hybrid fuzzy classifier tree (FCT).

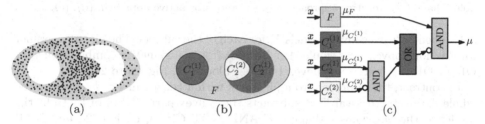

(a)   (b)   (c)

Fig. 1. Additive/subtractive classifier tree (c) built from the least possible number of convex sets (b) for two-dimensional data forming a "two-hole" shape (a)

In Fig. 1, an example leading to such a classifier tree is sketched. Aim of this article is to provide an algorithm which, when given a set of learning data as in Fig. 1a, builds up a tree as in Fig. 1c. As leaf nodes, it employs multidimensional membership functions representing convex basic building blocks similar to the sets given in Fig. 1b. This can be understood as a geometrical viewpoint for setting up the classifier. In contrast, the similar approach of fuzzy pattern trees [9] provides a rather logical description by using one-dimensional fuzzy terms as leaf nodes which form partitions of the corresponding attribute's domains.

2   Towards Hybrid Fuzzy Classifier Trees

Our aim is to build an FCT from a given set $\{x_1, \ldots, x_N\} \subset \mathbb{R}^M$ of N learning objects, each with M features. In the unary case, the classifier shall return one truth value $\mu(x) \in [0, 1]$, such that the classification of a test datum $x \in \mathbb{R}^M$ appearing in a region not supported by learning data results in a low degree of membership indicating that x does not belong to the class. Moreover, the procedure can be used for two- or multi-class problems, simply by independently building a tree for each class and comparing the truth values.

Let us assume that we possess a (fuzzy) model and learning method for sets of objects forming a convex shape in their feature space. How could we use them to cope with non-convex data? A straightforward approach would try to break the data set apart into convex subsets, e. g. using partitioning or segmentation, and

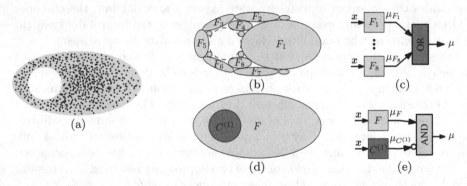

Fig. 2. Building a classifier tree for two-dimensional data forming a non-convex "one-hole" shape (a). Additive approach: (b), (c) and subtractive approach: (d), (e).

subsequently learn convex models F_i for each of the subsets. The overall classifier for this non-convex class would then be built from a disjunctive combination ("F_1 OR F_2 OR ...") of these convex building blocks, cf. Fig. 2a to 2c.

In contrast, the approach from [6] starts with a convex model F covering the whole data set. Afterwards, it subtracts the convex part $C^{(1)}$ in order to fit the model to the non-convex shape ("F AND NOT $C^{(1)}$"), cf. Fig. 2d and 2e. If a non-convex part had to be subtracted, it could be modeled recursively by a subtraction of convex elements ("$C^{(1)}$ AND NOT $C^{(2)}$ ..."). $C^{(l)}$ are called complementary classes with l indicating the level of complementation (cf. Sect. 5).

For arbitrarily complex shapes, one can expect to achieve more compact models (with a smaller number of building blocks) by combining both approaches. A classifier tree which is built up using the subtractive "AND NOT" approach as well as the additive "OR" is shown in Fig. 3. The subscripts in $C_1^{(1)}$ and $C_2^{(1)}$ enumerate the additively combined complementary classes of level 1.

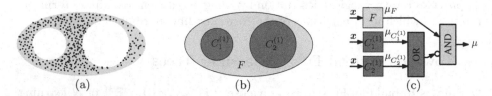

Fig. 3. Hybrid additive/subtractive classifier tree (c) for a "two-hole" shape (a)

In this paper, we start with one model F as a convex "hull" of the data and remove the parts not supported by learning data, referred to as object-unsupported class space C. C itself may be built up additively and/or subtractively from convex elements in a tree-like structure. In this way, one always gets a rough result in one step, i.e. an instance outside of F can never belong to the considered class. Subsequently, this statement is refined using the convex elements of C.

There are various options for both the structure of the tree and the type of membership functions (MFs) modelling the convex sets that are to be combined.

For the latter, we use a specific parametric MF of potential type (Sect. 3). In order to learn these models also for parts of the object-unsupported class space C, we propose to fill up C by an artificially generated set of so-called complementary objects (Sect. 4). Finally, the algorithm for setting up a model for C (Sect. 5) employs a combination of learning of MFs, cluster analysis (to find subsets for additive "OR" combinations), and the recursive "AND NOT" approach from [6].

3 Convex Building Blocks: Fuzzy Pattern Classes

Since we might need several (but as few as possible) convex building blocks to create FCTs, they should be well formalized and possess a comprehensible aggregation process when dealing with high-dimensional feature spaces. It needs to be emphasized that, in principle, any convex fuzzy set could serve as a component for building FCTs. Nevertheless, we propose the usage of a specific multivariate parametric fuzzy set. Besides the above-named properties, it features additional advantages, such as the treatment of asymmetric data distributions. We refer to this set as fuzzy pattern class (FPC), a concept which was introduced by BOCKLISCH as a generalisation of AIZERMAN's potential function [10]. It has already been applied for the modeling of traffic flows [11], medical diagnosis [12], condition monitoring [6], and time series analysis [13]. The structure of the FPC approach even suits embedded implementations in industrial applications [14].

Purpose of this section is to give a brief review of the FPC concept together with its main properties. A comprehensive description can be found in [10].

3.1 Definition and Learning of Fuzzy Pattern Classes

An FPC is a multivariate parametric fuzzy set with the membership function

$$\mu(\boldsymbol{x}) = \frac{a}{1 + \frac{1}{M}\sum_{i=1}^{M}\left(\frac{1}{b_{i,1/r}} - 1\right) \cdot \left|\frac{x_i - u_i}{c_{i,1/r}}\right|^{d_{i,1/r}}} . \tag{1}$$

It derives from the intersection of M univariate FPC basis functions of the same type, each defined by a set of well interpretable parameters.[1] These include the location parameter $u_i \in \mathbb{R}$, left and right class borders $c_{i,1/r} \in \mathbb{R}^+$ with their corresponding border memberships $b_{i,1/r} \in [0,1]$, and the specifiers for the class fuzziness $d_{i,1/r} \in [2,\infty)$. The parameters' impact on a univariate basis function can be understood by means of Fig. 4a. The quantity a can serve as a weight parameter for prioritising certain blocks of the classifier tree while decreasing the influence of others. Due to their interpretability, all parameters enable the incorporation of expert information in terms of a knowledge-based system.

An FPC is optimally fit to its supporting data by means of a translation relative to the class representative $\boldsymbol{u} = (u_1,\ldots,u_M)^\top \in \mathbb{R}^M$ and a rotation

---

[1] The intersection leading to (1) is conducted by a compensatory HAMACHER operator, preserving the function concept, parameters, and properties of the basis function [10].

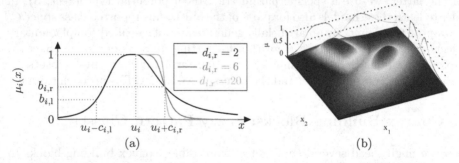

Fig. 4. 1-D (a) and 2-D (b) versions of the MF in (1) including rotation

described by the matrix $T \in \mathbb{R}^{M \times M}$ in the form $\boldsymbol{x}' = T(\boldsymbol{x} - \boldsymbol{u})$. Accordingly, the membership value for a rotated and translated FPC is given by $\mu(\boldsymbol{x}) = \mu'(\boldsymbol{x}') = \mu'(T(\boldsymbol{x} - \boldsymbol{u}))$, where μ' has the same structure as μ in (1), though with $u_i = 0$, $i = 1, \ldots, M$. Two examples of two-dimensional FPCs are illustrated in Fig. 4b along with their univariate basis functions for each dimension.

The determination of an FPC's parameters on the basis of given learning data is described in detail in [10]. In recent works, the parameterisation has been further developed, notably with regard to its robustness [15].

3.2 Fuzzy Pattern Class Properties

If a class of N learning objects is modeled by only one FPC, which is defined by $8M$ parameters, the approach will provide a data compression ratio of $\frac{N}{8}$. The ratio scales down linearly with a growing number B of building blocks in the tree. In our approach, the aim of a small value of B is achieved not only by the chosen tree structure, but also by the fact that FPCs already can be tuned quite flexibly to the properties of a data set compared to simpler choices of MFs.

4   Exploration of the Object-Unsupported Class Space

The convexity of an FPC makes it a proper description for classes with a convex data-inherent structure. In contrast, a convex set F would obviously not provide a tight description for a non-convex class like in Fig. 1. But how can we decide in practice whether a convex model is sufficiently accurate—only on the basis of a data set? An approach that has been proven to solve this problem efficiently for low- and high-dimensional data sets incorporates one-class SVMs [6]. This graph-based exploration scheme distributes complementary objects uniformly alongside the edges between "border objects", which are found in an optimal manner using SVMs. The algorithm assures that only edges within the object-unsupported class space are deployed. It also limits the number of complementary objects by $N - 1$ and thereby sets an upper limit for further computational costs.

The result of this procedure is depicted in Fig. 5 where it has been applied to an academic example featuring a shape of learning data similar to Fig. 1. Subsequently, complementary classes can be learned from the complementary objects. Since we employ the membership function proposed in Sect. 3 again, they are referred to as complementary fuzzy pattern classes (CFPCs).

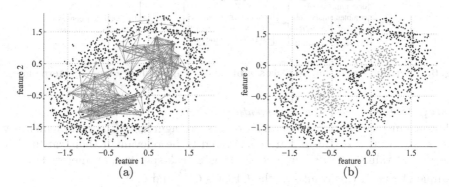

Fig. 5. (a) Learning/border objects and edges (b) learning/complementary objects

One may argue that the use of a one-class SVM sets a limit for the accuracy of the approach. Yet, we employ the SVM solely for generating complementary objects and return to the model set up based on (C)FPCs. Thus, the interpretability of the fuzzy classifier and its low complexity are retained (cf. Sect. 6). However, the accuracy can be controlled e. g. by tuning the SVM's kernel parameters.

5 Composing the Blocks and Building a Hybrid Tree

Given a set of learning data, we want to set up a shape-preserving model of a class via a fuzzy classifier tree where FPCs and CFPCs serve as basic building blocks. As mentioned, we set up an encircling fuzzy pattern class F based upon the given learning objects preliminarily (cf. Sect. 2 and Fig. 1). Obviously, F has to be combined with the "remainder" C of the classifier tree by a fuzzy-logical AND NOT. However, the locating of the CFPCs as well as their suitable interconnection for setting up C in form of a subtree is more intricate. The algorithm consists of five steps, which are processed as depicted in Fig. 6.

Step 1. *Exploration of the object-unsupported class space*
The first step is conducted by generating complementary objects according to Sect. 4. The level of complementation is set to the initial value $l = 1$.

Step 2. *Cluster analysis of complementary objects*
In order to ascertain whether there are separate groups of complementary objects, i. e. distinguishable object-unsupported partitions, we apply a density-based clustering scheme because of its property to find clusters independent of their underlying shape [16]. The clustering is governed by a distance parameter δ, whose value follows from the distribution of complementary objects in step 1.

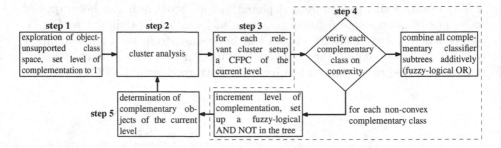

Fig. 6. Algorithm for setting up C as a subtree of complementary fuzzy pattern classes

Step 3. *Aggregation of complementary objects*
Relevant clusters of complementary objects are aggregated to complementary classes, see Sect. 3.1. Clusters are considered to be relevant if their cardinality exceeds the value $r \cdot N$, where $r \in [0,1]$ is a task-specific parameter. For the example of Fig. 1, this results in the CFPCs $C_1^{(1)}$ and $C_2^{(1)}$.

Step 4. *Verification of convexity for complementary classes*
Due to the fact that the clusters of complementary objects may also be characterized by non-convex shapes, their convex CFPC models may be inadequate as well. For $l = 1$, the suitability of each CFPC is confirmed by a classification of the learning objects x_i applying (1), where $\mu = \mu_{CFPC}$. If this classification yields low memberships $\mu_{CFPC}(x_i)$ for at least $(1 - r) \cdot N$ instances, the considered CFPC is assumed to be suitable. The description of the respective cluster is completed and the CFPC is branched off via a fuzzy-logical OR. This applies to $C_1^{(1)}$ in Fig. 1. On the contrary, if the classification of x_i results in high degrees of membership for at least $r \cdot N$ instances, the considered CFPC is expected to be inadequate (as $C_2^{(1)}$ in Fig. 1). The CFPC description itself has to be refined with one or more complementary classes of the next level, see step 5. Hence, the level of complementation is incremented. The respective CFPC is branched off via an OR and an additional AND NOT connective for its further refining. In Fig. 1, this corresponds to the combination "$C_2^{(1)}$ AND NOT $C_2^{(2)}$". However, a fuzzy description of $C_2^{(2)}$ in terms of (1) is not known at this step of the algorithm.

Generally, for $l > 1$, the verification of convexity is performed with the complementary objects of the preceding level (as defined in step 5).

Step 5. *Selection of higher-level complementary objects*
Each CFPC with an indication of non-convexity is refined separately but in the same manner, thus forming new branches in the next layer of the tree. Due to the fact that learning and complementary objects are mutually complementing each other, it follows that we can refrain from a further generation of complementary objects. For l even, the set of complementary objects is given by those learning objects with high degrees of membership to the currently refined CFPC of level $l - 1$ (cf. step 4). For l odd, this set is represented by the complementary objects, generated in step 1, exhibiting high memberships to the associated CFPC.

The selected complementary objects are fed back into the algorithm (step 2). After the cluster analysis, they form complementary classes of level l (step 3). Regarding the example from Fig. 1, the algorithm determines the description of $C_2^{(2)}$ and terminates with the verification of its convexity.

After the tree setup, all blocks (FPCs) are weighted via their parameter a in order to obtain normalized membership values $\mu(\boldsymbol{x}_i)$. For the calculation of the memberships, we apply the max operator as OR combination and the algebraic product with the natural negation for the fuzzy-logical AND NOT.

The process parameter r governs the convergence and the detailedness of the emerging classifier tree. The larger r the coarser the model (i.e. smaller B) and the faster the learning converges. The value $r \cdot N$ represents the least number of objects to form a building block. Thus, one can always choose r sufficiently large in order to achieve highly understandable and interpretable trees.[2] In the special case $r = 1$, the algorithm stops immediately resulting in the trivial tree defined upon F. Usually, r is set based on task-specific knowledge, e.g. in terms of the desired model complexity. The complexity of the algorithm itself scales quadratically with N and linearly with the dimensionality of the feature space.

6 Examples

We will now demonstrate the viability of the approach with the help of several academic examples being set up in the spirit of the introductory examples from Sect. 1 and Sect. 2. To this end, we will start with a simple distribution of learning data in a two-dimensional feature space, forming a convex shape. Subsequently, convex and non-convex parts of the shape are being taken out successively in order to resemble the shapes of Fig. 2a, Fig. 3a, and Fig. 1a.

First aim of this section is to visually confirm that the algorithm from Sect. 5 delivers a classifier tree with a minimal number B of convex sets. That is to say the resulting tree should exhibit the structure of Fig. 2c, 3c, and 1c for the examples from Fig. 2a, 3a, and 1a, respectively. It should not have any branches for the purely convex case. The second aim is to provide first findings for a comparison of the proposed tree with other one-class learners.

The learning data, the generated complementary objects, and the resulting MFs are shown in Fig. 7. The process parameter r has been set to 0.01. Obviously, the shape of each data distribution is captured very well. The resulting tree structures are not shown here since they are equivalent to Fig. 2c, 3c, and 1c.

For the example of Fig. 7d, a comparison with the purely subtractive, an additive (segmentation) tree approach, and a one-class SVM is given in Table 1. Considering only the tree approaches, the hybrid FCT is optimal w.r.t. B. Additionally, the model complexity (number of parameters) is almost one order of magnitude smaller than for a one-class SVM. In this setup, the SVM has been

---

[2] The understandability of the tree originates from a small number of blocks and their hierarchical arrangement using fuzzy-logical connectives. It is further fostered by the interpretation of a single block as a rule-based (sub)system (due to the used intersection operator) and the semantical parameters of its MF (cf. Sect. 3.1).

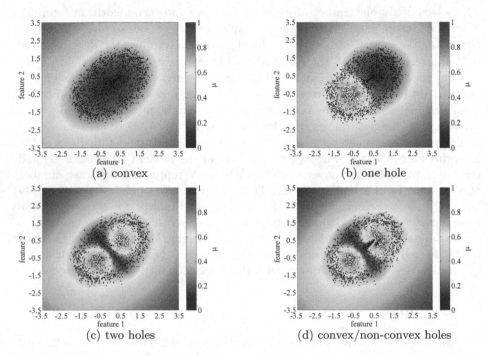

(a) convex

(b) one hole

(c) two holes

(d) convex/non-convex holes

Fig. 7. Learning data (black), complementary objects (white) and resulting membership function for examples with increasing complexity. The classifier tree structure for (b), (c) and (d) can be found in Fig. 2c, 3c and 1c, respectively.

Table 1. Comparison of hybrid fuzzy classifier trees with other one-class approaches

method	hybrid tree	subtractive tree	additive tree	one-class SVM
#convex sets (B) / #SVs	4	5	6	133
#parameters	64	80	96	400
run time	2.4s	3.1s	1.8s	0.6s

tuned in such a way that it produces a similar accuracy, which is measured by the rejection rate of the complementary objects.

7   Conclusions and Outlook

In this article, we presented the idea of a model and method for one-class learning described by a hybrid FCT. The classifier tree consists of additive and/or subtractive combinations of convex fuzzy sets learned from a given set of data with an arbitrarily complex shape. It could be demonstrated that the algorithm is able to learn and parameterize FCTs with the same minimal number and combination of convex sets as an expert would construct manually. Our future work includes a further comparison of the hybrid FCT to other unary classification methods such as Bayesian approaches, artificial neural networks, and boundary methods. This also involves an application to benchmark and real world data sets, e. g. from the field of condition monitoring.

References

1. Zhuang, L., Dai, H.: Parameter Optimization of Kernel-based One-class Classifier on Imbalance Text Learning. In: Yang, Q., Webb, G. (eds.) PRICAI 2006. LNCS (LNAI), vol. 4099, pp. 434–443. Springer, Heidelberg (2006)
2. Saez, J., Luengo, J., Herrera, F.: On the Suitability of Fuzzy Rule-based Classification Systems with Noisy Data. IEEE Transactions on Fuzzy Systems PP(99) (2012)
3. Szmidt, E., Kukier, M.: Classification of Imbalanced and Overlapping Classes Using Intuitionistic Fuzzy Sets. In: 2006 3rd International IEEE Conference on Intelligent Systems, pp. 722–727. IEEE Press, New York (2006)
4. Li, J.D., Zhang, X.J., Chen, Y.S.: Applying Expert Experience to Interpretable Fuzzy Classification System Using Genetic Algorithms. In: 4th International Conference on Fuzzy Systems and Knowledge Discovery, vol. 2, pp. 129–133 (2007)
5. Hempel, A.-J., Hähnel, H., Mönks, U., Lohweg, V.: SVM-integrated Fuzzy Pattern Classification for Nonconvex Data-inherent Structures Applied to Banknote Authentication. In: Bildverarbeitung in der Automation. inIT, Lemgo (2012)
6. Hempel, A.-J., Hähnel, H., Herbst, G.: Learning Non-convex Fuzzy Classifiers Using Single-class SVMs. In: IEEE International Conference on Fuzzy Systems, pp. 1–8. IEEE Press, New York (2013)
7. Kosko, B.: Fuzzy Systems as Universal Approximators. IEEE Transactions on Computers 43(11), 1329–1333 (1994)
8. Devillez, A.: Four Fuzzy Supervised Classification Methods for Discriminating Classes of Non-convex Shape. Fuzzy Sets and Systems 141(2), 219–240 (2004)
9. Senge, R., Hüllermeier, E.: Top-down Induction of Fuzzy Pattern Trees. IEEE Transactions on Fuzzy Systems 19(2), 241–252 (2011)
10. Hempel, A.-J., Bocklisch, S.F.: Fuzzy Pattern Modelling of Data Inherent Structures Based on Aggregation of Data with Heterogeneous Fuzziness. In: Rey, G.R., Muneta, L.M. (eds.) Modelling Simulation and Optimization, pp. 637–655. InTech (2010)
11. Päßler, M., Bocklisch, S.F.: Fuzzy Time Series Analysis. In: Hampel, R., Wagenknecht, M., Chaker, N. (eds.) Fuzzy Control: Theory and Practice, pp. 331–345. Physica-Verlag HD, Heidelberg (2000)
12. Schmidt, B., Bocklisch, S.F., Päßler, M., Czonsnyka, M., Schwarze, J.J., Klingelhöfer, J.: Fuzzy Pattern Classification of Hemodynamic Data Can Be Used to Determine Noninvasive Intracranial Pressure. Acta Neurochirurgica (suppl. 95), 345–349 (2006)
13. Herbst, G., Bocklisch, S.F.: Recognition of Fuzzy Time Series Patterns Using Evolving Classification Results. Evolving Systems 1(2), 97–110 (2010)
14. Mönks, U., Petker, D., Lohweg, V.: Fuzzy-Pattern-Classifier Training with Small Data Sets. In: Hüllermeier, E., Kruse, R., Hoffmann, F. (eds.) IPMU 2010, Part I. CCIS, vol. 80, pp. 426–435. Springer, Heidelberg (2010)
15. Hähnel, H., Hempel, A.-J., Mönks, U., Lohweg, V.: Integration of Statistical Analyses for Parameterisation of the Fuzzy Pattern Classification. In: 22. Workshop Computational Intelligence, pp. 115–131. KIT, Karlsruhe (2012)
16. Ester, M., Kriegel, H.P., Sander, J., Xu, X.: A Density-based Algorithm for Discovering Clusters in Large Spatial Databases with Noise. In: Proc. of 2nd International Conference on Knowledge Discovery and Data Mining, pp. 226–231 (1996)

Applying CHC Models to Reasoning in Fictions

Luis A. Urtubey* and Alba Massolo

Facultad de Filosofía y Humanidades, Universidad Nacional de Córdoba,
Ciudad Universitaria, 5000 Córdoba, Argentina
urtubey@ffyh.unc.edu.ar,
albamassolo@gmail.com
http://www.ffyh.unc.edu.ar

Abstract. In figuring out the complete content of a fictional story, all kinds of consequences are drawn from the explicitly given material. It may seem natural to assume a closure deductive principle for those consequences. Notwithstanding, the classical closure principle has notorious problems because of the possibility of inconsistencies. This paper aims to explore an alternative approach to reasoning with the content of fictional works, based on the application of a mathematical model for conjectures, hypotheses and consequences (abbr. CHCs), extensively developed during the last years by Enric Trillas and some collaborators, with which deduction in this setting becomes more comprehensive.

Keywords: Soft-Computing, CHC-Models, Reasoning, Fiction, Philosophy.

1 Introduction

Issues concerning fiction has increasingly attracted attention of logicians, philosophers and computer scientists during the last years. Particularly, several formal systems have been proposed and applied in order to represent the way in which a cognitive agent reasons about a work of fiction, [13], [14], [16], [15].

Talking about fictions is often restricted to conversations about literature, movies or TV-shows. Nevertheless, appealing to fiction in many areas of formal and empirical sciences has been also very fruitful. A clear example of this is the great interest in relating fiction with scientific models [11], [12]. Fiction has been applied not only to explain how a scientist builds a model but also to determine what kind of ontological entities models are. In this way, models have been understood as fictional entities; and the work scientists do while modelling different phenomena has been compared to the work of authors who create fiction. This relationship between models and fiction can also work in the opposite direction. If that were the case, it would be possible to define fiction, in its turn, as a sort of model.

During the last decade, Enric Trillas with some collaborators have worked out in [3], and more recently in [2], [7], [8], [9], a mathematical model for conjectures,

---

* Corresponding author.

A. Laurent et al. (Eds.): IPMU 2014, Part II, CCIS 443, pp. 526–535, 2014.

hypotheses and consequences (abbr. CHCs), in order to execute with this model certain mathematical and informal reasoning. These interesting mathematical models for CHCs have been established algebraically, and the statements and propositions of human thinking are represented as those elements in an ortho-complemented lattice. Additionally, several meaningful operators are defined, which act on each given set of premises, intuitively standing for the conjectures and hypotheses as well as the consequences of that set of premises. The election of Orthocomplemented lattices is justified because these are quite general algebraic structures, in order to establish a sufficiently extensive reasoning model in which CHCs can be mathematically described. Alternatively, other algebraic settings have been also studied [4], [5].

This paper will be concerned with formal reasoning applied to ordinary experience with works of fictions. It aims to explore an approach to reasoning with the content of fictional works, which especially deals with deduction in this setting, based on the application of CHC-Models. The article is organized as follows. First a short reference about current philosophical work on fiction is given. In second place, recent work on CHC-Models is addressed and it is shown how it can bear on reasoning in fictions. Finally, solutions to some problems concerning deduction in this setting are considered. The conclusion will point out some further research.

2   Philosophy and Logic on Fiction

Problems related to fiction has raised several troubles for classical conceptions in the field of philosophy of language and logic. Already starting from the work of Frege [17], the semantic role of fictional names, i.e., names of characters, creatures and places that belong to fictional works, has been far from clear. Moreover, the semantic value of fictional sentences turned out to be controversial. Notably, it is hard to establish whether a sentence like "Sherlock Holmes is a detective" is true, false or truth-valueless, but it is also embarrassing to accept that fictional sentences are never true.

From the standpoint of philosophy and logic, inference in fiction involves reasoning with incomplete information. This is due to the fact that fictional stories describe their characters, places, and events only in an incomplete way. It is not possible, for instance, to determine if Sherlock Holmes is 1.80 meters tall. Additionally, inference in fiction also involves reasoning with inconsistent information that can emerge from two sources. On the one hand, information belonging to a fiction contradicts reality in many aspects. For example, while according to Doyle's stories Sherlock Holmes used to live in London in 221B Baker Street, in the real London, there was no Sherlock Holmes who used to live there. On the other hand, some stories are based on a contradiction or contain inconsistent information. For instance, this would be the case of a story where it is said that a character x has and does not have certain property P. Specially, cases of this last type will be addressed later in this paper.

Consequently, it has turned out to be quite difficult to provide a formal account of reasoning in fiction based on classical semantics. Firstly, as it was shown, the standard approach to interpret classical languages, is objectual. According to this interpretation, the domain of discourse assumes the existence of a non-empty set of real objects. Therefore, sentences like "Sherlock Holmes is a detective" or "Sherlock Holmes is a fictional character", in this classical formal setting, must be evaluated as false. Secondly, the notion of logical consequence is defined in terms of necessary truth preservation. Clearly, it can be seen that this definition forces to take bare truth as the only semantic value acceptable in a consequence relation. In this sense, a conclusion would follow from the premises of a fictional story just in case that conclusion is as barely true as the premises. Admittedly, sentences that contain fictional names never hold in a classical interpretation. Hence, it is not possible to give a compelling formal account of reasoning in fiction inside classical semantics.

Anyway, it could be argued that it is possible to give a classical formal account of reasoning in fiction confined to propositional classical logic. In this way, it is possible to avoid speaking about objects and try to deal with fictional discourse at a propositional level. However, problems arise also in a classical propositional formal setting. On the one hand, classical propositional semantics is bivalent. The principle of bivalence states that every sentence expressing a proposition has exactly one truth value: true or false (one or zero). As a consequence, the proposition *"p or not p"* equals one. Nonetheless, as fictional works are essentially incomplete, it is impossible for some sentences to establish whether they are true or false. On the other hand, classical propositional logic obeys the principle *Ex falso quodlibet*. According to this principle, anything follows from an inconsistent set of premises. It turns out that the consequences of the proposition *"p and not p"* equals L –the entire language. But a work of fiction can contain contradictions. Hence, in a classical formal framework, dealing with inconsistent information will overgenerate propositions derived from a story. And even worse, any proposition could be drawn from a fictional work. Thence, classical propositional logic does not provide an adequate formal system for dealing with reasoning in fiction. The following sections will reaffirm this diagnosis on the basis of the relationship between classical propositional calculus and Boolean algebra.

3   Introducing CHC Models

CHCs Models have been conceived as mathematical tools for studying commonsense, everyday or ordinary reasoning. Clearly, a model of this kind is not coincidental with the reality that is modelled by it, but a simplification of the reality. Anyway, these mathematical models bring a good mean of applying formal deductive reasoning for trying to understand reality and to do more accurate and clearer philosophical reflections on it. Moreover mathematical models can be viewed as useful devices to construct new realities through computational methods.

From the perspective of CHC–Models, common-sense reasoning can be decomposed in the pair consisting of 'conjecturing+refuting', two terms that embed two different types of deducing. On the one hand there is a type of deducing, which corresponds with the informal type of deduction that people carry out in Common-sense reasoning. On the other hand, there is a more restrictive kind, which has to do with the formal or mathematical concept of deductive consequence. Admittedly this last one is the concept that Alfred Tarski [10] formulated into the well-known definition of a consequence operator.

Thus, in common-sense reasoning, deduction is an informal and weaker concept than in formal, mathematical reasoning. To better characterize what people do with common-sense deduction, a moderate dose of formalization will be introduced in the next section.

4   A Model for Reasoning in Fiction

Assuming that reasoning impose the existence of some previous information about the subject under consideration, in any reasoning task there exist, from the beginning, a body of information already available on the subject. This is usually expressed through a finite number of statements or premises in natural language, which also can include other type of expressions like numbers or functions. Thus, one can assume to start with that the information given by a fiction F is somehow stored under this form of representing knowledge. Certain other constraints will be imposed later.

As it was just said, this paper mainly aims to apply CHC Models to the formal treatment of reasoning in fiction, hence it is convenient to introduce firstly some concepts concerning CHC-Models from previous work of Enric Trillas and his collaborators. Specially paying attention to [7], [9] and to [8], which seem to be better suited to the present work.

In the setting of CHC Models knowledge is represented in an adequate algebraic structure in a set $L, (L, \leqslant, \cdot, +, ')$, containing a pre-order $\leqslant$ representing *if/then*, a unary operation $'$ representing *not*, and two binary operations representing the linguistic *and* ($\cdot$) and the linguistic *or* ($+$). Clearly the set of premises of any type of reasoning cannot be trivially inconsistent. If they were, reasoning would be absurd or utter nonsense. Thus it makes sense to assume that the set of premises must satisfy at least a somehow minimal requirement of consistency. Let be $P = \{p_1, \ldots, p_n\}$ the subset of L with these statements, which are taken as premises. It will be assumed that the element (not necessarily in P) $p_\wedge = p_1 \cdot \ldots \cdot p_n$ is not self-contradictory, i.e., $p_\wedge \not\leqslant p_\wedge'$. Let $\mathcal{F}$ be such family of subsets in L. Thus, this lack of self-contradiction means that P does not contain absurd premises.

The following example –loosely adapted from [7]– illustrate these concepts and it will help to clarify their application when reasoning with information retrieved from a fiction. The knowledge representation part revolves around the following statements, which are true about the novel "Farenheit 451" by Ray Bradbury:

Guy Montag is a fireman
A fireman burns books
Guy Montag loves books

Moreover a very clear fact also is: "Neither does Guy Montag love nor burn books".

Example 1. Let L be an ortholattice with the elements **f** for *Guy Montag is a fireman*, **b** for *A fireman burns books*, and l for *Guy Montag loves books*, and its corresponding negations, conjunctions and disjunctions. It is also the case that *Guy Montag is a fireman, Guy Montag is a fireman and does not burn books*, and *neither does Guy Montag love nor burn books*. Representing *and, or* and *not* as stated before, the set of premises is $\mathsf{P} = \{f, f \cdot b', (l \cdot b)'\}$. Thus the *core-value* of this information can be identified with $p_\wedge = f \cdot f \cdot b' \cdot (l \cdot b)' = f \cdot f \cdot b' \cdot (b' + l') = f \cdot b'$.

Some reasoning can be done in the lattice. Notably some inferences can be drawn, on the basis of the information of a fiction F, once conceded that $a \leqslant b$ means that b is a logical consequence of a. All these inferences conform the class of "conjectures" in the setting of CHC-Models. Among conjectures, consequences, hypothesis and speculations are distinguished. However to keep things as simple as possible, these classification between different types of conjectures, can be put aside and an overall distinction can be made only between "conjectures" and "consequences" for the sake of convenience. Thus, from the example it follows,

- *Guy Montag does not burn books*, b', is a consequence of P, since $p_\wedge = f \cdot b' \leqslant b'$.
- The statement *"Guy Montag is a fireman and He does not burn books and He loves books"*, $f \cdot b' \cdot l$ and *"He loves books"*, l, are conjectures of P
- The statements *"he burns books"* and *"he is not a fireman"* are refutations of P, since $f \cdot b' \leqslant b'$, and $f \cdot b' \leqslant f = (f')'$

Understandably for representing reasoning in natural language a richer and more complex framework would be highly desirable. Specially, aspects of tense and other subtleties of common language are most difficult to interpret in these more rigid algebraic structures. However, within this limited and closed framework, this example may still count as a formalization of a piece of human reasoning.

5   The Problem of Consistency

At first glance, in the previous example, a contradiction looms over the conclusions. On the one hand it holds on the story that "Guy Montag is a fireman" and, also that "a fireman burns books". On the other hand one knows that "Guy Montag does not burn books". Accordingly, it can be inferred then that "Guy Montag is not a fireman'". This is a very simple inference supported by the inference rule known as "Modus Tollens", from $a \rightarrow b$ and b', it follows a', where $\rightarrow$ is the material conditional, interpreted as $a \rightarrow b = a' + b$. Nevertheless, in spite of its obviousness, this inference presupposes some particular structural features. As argued in [1] "deduction" and "inference", must be clearly distinguished when formalizing reasoning. And one has to be aware also that the validity of formulas belonging to the language, depends on the particular deductive system at stage. In the example, for the inference to be valid, L must be endowed with the algebraic structure of a Boolean algebra, and the consequence operator should be the greatest one for such a framework. The verification of the inequality expressing Modus Tollens , i.e., $b' \cdot (a \rightarrow b) = b' \cdot (a' + b) = a' \cdot b' \leqslant a'$ in ortholattices as much as in De Morgan algebras, will cause the validity of the laws of Boolean algebras. Conversely, if the consequence operator is changed, the validity of the inference scheme can no longer be guaranteed. Arguably, the problem in this case does not have to do with the formalization of a conditional proposition as material conditional. Actually, there is no conditional statement to be formalized at all. In spite of this, according to certain naïve reading of the story, it is right to think that "either Guy Montag is not a fireman or he does burn books". A logic-minded person who sticks to classical logic, would feel that things could not be otherwise. But they are, because in the story, Guy Montag manages to do both, he remains a fireman and refrains from burning books. Unquestionably, the problem can be ascribed to the consequence operator, which is the greatest one for a Boolean algebra, and contributes to validate mentally this inference scheme.

Thus, it seems impossible to confine oneself to the premises and to ensure an overall consistency, while keeping this kind of propositional reasoning in the setting of classical logic or Boolean algebras. These are the kind of problems, which are frequent when reasoning in fictions. The closure of classical deductive consequence, together with the meaning attributed to classical connectives, implies that some undesirable conclusions must be accepted in spite of contradicting explicit information. Hence, the type of consistency concerning reasoning in fiction has to comply somehow with standards of inference and consequence other than those of classical logic. Additionally, the example about Guy Montag, helps to see that "consequences" are not the only type of deductions gained from the premises. There are also "conjectures", which are in fact a lot more useful.

The following definition from [7] specifies what is meant by a conjecture, relative to a given problem on which some information conveyed by a set $P = \{p_1, p_2, \ldots, p_n\}$ of n premises is known.

Definition 1. *q is a conjecture from* P, *provided q is not incompatible with the information on the given problem once it is conveyed throughout all p_i in* P.

The most important things in this definition are, on the one side, how to state that P is consistent; and on the other side, how to interpret the requirement that q is not incompatible with the information given by P.

Accordingly, to apply CHC-Models to reasoning in fictions, an appropriate notion of incompatibility will be needed in each case. This notion of incompatibility should be conveniently dissociated from the closure property of classical deduction.

In the development of CHC-Models during the last decade, the construction of conjecture's operators has relied persistently on standard consequences' operators. Recently, in [1] it has been shown that to keep the most typical properties of the concept of conjecture, it suffices to only consider operators that are extensive and monotonic, but without enjoying the closure property. It is worth to get a glimpse at this. Given a non empty set of sentences L let $\mathcal{F}$ be a family of subsets in L. Then a standard consequence operator, is a mapping $\mathbf{C}: \mathcal{F} \to \mathcal{F}$, such that,

- $P \subset \mathbf{C}(P)$, $\mathbf{C}$ is *extensive*
- If $P \subset Q$, then $\mathbf{C}(P) \subset \mathbf{C}(Q)$, $\mathbf{C}$ is *monotonic*,
- $\mathbf{C}(\mathbf{C}(P)) = \mathbf{C}(P)$, or $\mathbf{C}^2 = \mathbf{C}$, $\mathbf{C}$ is a *closure*.

for all P, Q in $\mathcal{F}$. In addition, *consistent* operators of consequence verify

- If $q \in \mathbf{C}(P)$, then $q' \notin \mathbf{C}(P)$

In [7] several consequence operators lacking in closure are distinguished. Especially there are three operators of consequence, which are significant to the purpose of formalizing reasoning in fictions,

- $\mathbf{C}_1(P) = \{q \in L : r(P) \cdot q' = 0\}$
- $\mathbf{C}_2(P) = \{q \in L : p_\wedge \cdot q' \leqslant (p_\wedge \cdot q')'\}$
- $\mathbf{C}_3(P) = \{q \in L : p_\wedge \leqslant q\} = \mathbf{C}_\wedge(P)$.

where $r(P)$ refers to the *core-value* of the information gathered in the premises, which could verify for example, $r(P) \leqslant p_\wedge$.

Concerning these consequence operators, for $i = \{1, 2\}$ it is $P \subset \mathbf{C}_i(P)$, and if $P \subset Q$, then $\mathbf{C}_i(P) \subset \mathbf{C}_i(Q)$. Nevertheless, $\mathbf{C}_i$ cannot be always applicable to $\mathbf{C}_i(P)$ since it easily can be $r\mathbf{C}_i(P) = 0$, due to the lack of consistency of $\mathbf{C}_i$. Furthermore, $\mathbf{C}_\wedge(P)$ is also a consistent operator of consequence.

Hence, the corresponding operator of conjectures $Conj_i$ does not come from an operator of consequences, but only from an extensive and monotonic one, for which the closure property has no sense, since $\mathbf{C}_i(P)$ cannot be taken as a body of information in the sense of [7], i.e., guaranteed free from incompatibility.

Remark 1. To have $\mathbf{C}(P) \subset Conj_C(P)$, it suffices for $\mathbf{C}$ to be a consistent operator of consequences. Hence, the consistency of $\mathbf{C}$ is what characterizes the inclusion of $\mathbf{C}(P)$ into $Conj_C(P)$, that consequences are a special type of conjectures. Thus, when the premises harbour inconsistencies, which cannot be taken away, $\mathbf{C}_1$ and $\mathbf{C}_2$ seem to be the only alternatives to take. Apparently, it is the case while reasoning in fictional stories.

6   Designing the Appropriate Framework

The problem of knowledge representation is one of the most important aspects of any formalization process. Notably, for representing ordinary reasoning, which usually involves natural language, more flexible constructions are needed. Concerning algebraic structure, it turns out that algebras of fuzzy sets enjoy such a flexibility. In [7] and [6] an abstract definition of a *Basic Flexible Algebra* is given. Latices with negations and, in particular, ortholatices and De Morgan algebras are instances of BFAs. Also the standard algebras of fuzzy sets $\left([0,1]^X, \mathsf{T}, \mathsf{S}, \mathsf{N}\right)$ are particular BFAs.

Previous to any specific application of CHC-Models, some questions must be addressed concerning the representational framework. First of all, where do the objects ('represented' statements) belong to. That is, which is L, such that $\mathsf{P} \subset \mathsf{L}$ and $q \in \mathsf{L}$? Secondly, with which algebraic structure is endowed L?. And lastly, how to state that P is consistent, and how to translate that q *is not incompatible*?

To assume that P is free of incompatible elements is to concede that there are not elements p_i, p_j in P, such that $p_i \leqslant p'_j$, or $p_i \cdot p_j = 0$. To avoid the odd case $p_i \cdot \ldots \cdot p_n = 0$ it is convinient to assume that $r(P) \neq 0$.

To keep things simpler in a first attempt, it will be convenient to pay attention only to the cases in which the BFA is an ortholatice. As a matter of fact, a De Morgan algebra offers a basic suitable structure for reasoning with the content of a fiction. Moreover, some clarification is in order concerning the formulation of the key principles of Non-contradiction and Excluded Middle. For that goal, the following distinction between the incompatibility concept of contradictory and self-contradictory elements in a BFA may be introduced.

- Two elements a, b in a BFA are said to be *contradictory* with respect to the negation $'$, if $a \leqslant b'$.
- An element a in BFA is said to be *self-contradictory* with respect to the negation $'$, if $a \leqslant a'$.

In dealing with De Morgan algebras, these principles, formulated in the way that is typical of standard modern logic $(a \cdot a' = 0; a + a' = 1)$, do not hold. Nevertheless, with an alternative formulation, De Morgan algebras also verify those principles, that is,

- **NC:** $a \cdot a' \leqslant (a \cdot a')'$
- **EM:** $(a + a')' \leqslant ((a + a')')'$

7   More on Consequence Operators and Conjectures

Arguably, it is worth to submit reasoning in fictions to the supposition that L is endowed with an ortholatice structure $\mathcal{L} = (\mathsf{L}, \cdot, +, ', 0, 1)$. Among the alternatives to express the non incompatibility between the premises and a conjecture q, there are two, which seem to be adequate in the present case: $r(P) \cdot q \neq 0$ and $r(P) \cdot q \not\leqslant (r(P) \cdot q)'$. It means that either the proposition added to the core is admissible or that the set formed this way is not auto-contradictory or impossible. The applications will make sense as much as $r(P) = p_\wedge \neq 0$.

Manifestly, it was readily seen that there is a close connection between conjectures and consequence operators. Moreover it has been also shown that there are conjecture operators $Conj_i$, which do not come from an operator of consequence. Anyway, even in the case of fictions, it seems that logical consequences are always counted as a particular case of conjectures. Then, it is natural that one wonders whether there is also a consequence operator $\mathbf{C}(P) \subset Conj_i$. Notably, it turns out that $\mathbf{C}_\wedge$ is such an operator and it is $\mathbf{C}_\wedge \subset Conj_i(P)$. Hence, $\mathbf{C}_\wedge$ turns out to be in this case the safer conjectures. The set of conjectures, which no longer include $\mathbf{C}_\wedge$, is the set

$$Conj_i(P) - \mathbf{C}_\wedge(P) = \{q \in Conj_i(P) : q < p_\wedge\} \cup \{q \in Conj_i(P) : qNCp_\wedge\} \quad (1)$$

where NC stands for non *order comparable*. Each set in (1) that remains when $\mathbf{C}_\wedge$ is taken away, has more risky deductions obtained by reasoning from the premises.

Furthermore, any operator of conjectures verify some of the following properties [7] ,

1. $Conj(P) \neq \emptyset$
2. $0 \notin Conj(P)$
3. There exist an operator $\mathbf{C}$ such that $Conj(P) = \{q \in \mathsf{L} : q' \notin \mathbf{C}(P)\}$
4. $Conj$ is expansive: $P \subset Conj(P)$
5. $Conj$ is anti-monotonic: If $P \subset Q$, then $Conj(Q) \subset Conj(P)$

Concerning this intended application to reasoning in fictions, all these features of conjecture operators are attractive. Especially, anti-monotonicity is very appealing. As noted also in [7], $Conj_\mathbf{C}$ is anti-monotonic *if and only if* $\mathbf{C}$ is monotonic. That is, conjectures and consequences are particularly linked with respect to monotony. When reasoning in fictions, it is also important to observe that a more encompassing set of premises can make showing up conjectures that one could not see before. This is an admissible interpretation of anti-monotonicity in this setting.

8   Conclusion

By using CHC-Models, deductions are treated differently in connection with reasoning. In particular, several types of deduction are distinguished. Moreover, by separating consequence and inference, a new perspective on the use of logic for reasoning is gained. It is possible now to apply more complex algebraic settings to account for different types of deductions, which taken together give a more promissory approach to the variety of human reasoning. Specifically, the application sketched in this paper, helps to show how a complex type of reasoning concerning fictional content, can be approached from the methodological perspective of CHC-Models. Deductions are no longer submitted to a classical closure principle. An alternative model, based on consequences and conjectures, can control reasoning instead. The advantages of this approach deliver promising results. Arguably, more complex information, such as inexact or fuzzy knowledge, coming from fictions, can be also represented. Even in this case, by changing the algebraic setting, this type of content shall be also accommodated. Future work on the subject can be also addressed to consistently combine suitable contextual principles into CHCs models, in order to search for conjectures and consequences thereby generated. These models can be achieved by mimicking simple fictional scenarios, on which subjects can perform specific inference tasks, whose results are somehow circumscribed by determinate constraints.

Acknowledgments. We'd like to thank the two anonymous referees for their helpful suggestions regarding this paper. First author owe special thanks to Professor Trillas for having introduce him to his appealing work on CHC models. The work on this paper has received partial financial support of SeCyT (UNC).

References

1. Trillas, E., García–Honrado, I.: Hacia un Replanteamiento del Cálculo Proposicional Clásico. Ágora–Papeles de Filosofía 32, 7–25 (2013) (in Spanish)
2. Trillas, E., García–Honrado, I., Pradera, A.: Consequences and conjectures in preordered sets. Information Sciences 180, 3573–3588 (2010)
3. Trillas, E., Cubillo, S., Castiñeira, E.: On conjectures in orthocomplemented lattices. Artificial Intelligence 117, 255–275 (2000)
4. Ying, M.S., Wang, H.: Lattice-theoretic models of conjectures, hypotheses and consequences. Artificial Intelligence 139, 253–267 (2002)
5. Qiu, D.: A note on Trillas CHC models. Artificial Intelligence 171, 239–254 (2007)
6. Trillas, E.: A Model for Crisp Reasoning with Fuzzy Sets. International Journal of Intelligent Systems 27, 859–872 (2012)
7. García-Honrado, I., Trillas, E.: On an Attempt to Formalize Guessing. In: Seising, R., Sanz González, V. (eds.) Soft Computing in Humanities and Social Sciences. STUDFUZZ, vol. 273, pp. 237–255. Springer, Heidelberg (2012)
8. Trillas, E.: Reasoning: in Black and White? In: Annual Meeting of the North American Fuzzy Information Processing Society (NAFIPS), pp. 1–4. IEEE, San Francisco (2012)
9. Trillas, E., Sánchez, D.: Conjectures in De Morgan Algebras. In: Annual Meeting of the North American Fuzzy Information Processing Society (NAFIPS), pp. 1–6. IEEE, San Francisco (2012)
10. Tarski, A.: Logic, Semantics, Metamathematics. John Corcoran (ed). Hackett, Indianapolis (1983)
11. Frigg, R., Hunter, M. (eds.): Beyond Mimesis and Convention. Representation in Art and Science. Springer, London (2010)
12. Woods, J. (ed.): Fiction and Models. New Essays. Philosophia Verlag, Munich (2010)
13. Parsons, T.: Nonexistent Objects. Yale University Press, New Haven (1980)
14. Priest, G.: Towards Non-Being: The Logic and Metaphysics of Intentionality. Oxford University Press, New York (2005)
15. Rapaport, W., Shapiro, S.: Fiction and Cognition: an Introduction. In: Ram, A., Moorman, K. (eds.) Understanding Language Understanding: Computational Models of Reading, MIT Press, Cambridge (1999)
16. Woods, J.: The Logic of Fiction. Studies in Logic, vol. 23. College Publications, London (2009)
17. Frege, G.: Posthumous Writings. Basil Blackwell, Oxford (1979)

Probabilistic Solution of Zadeh's Test Problems

Boris Kovalerchuk

Dept. of Computer Science, Central Washington University,
400 E. University Way, Ellensburg, WA 98926, USA
borisk@cwu.edu

Abstract. Zadeh posed several Computing with Words (CWW) test problems such as: "What is the probability that John is short?" These problems assume a given piece of information in the form of membership functions for linguistic terms including tall, short, young, middle-aged, and the probability density functions of age and height. This paper proposes a solution that interprets Zadeh's solution for these problems as a solution in terms of probability spaces as defined in the probability theory. This paper also discusses methodological issues of relations between concepts of probability and fuzzy sets.

1   Introduction

Zadeh's test problems include: "What is the probability that Mary is middle-aged?", "What is the probability that Mary is young?", "What is the probability that John is short?", and "What is the probability that John is tall?" [Zadeh, 2012, Belyakov et al, 2012].

These problems assume *given information I* in the form of membership functions μ_{tall}, μ_{short}, μ_{young}, $\mu_{middle\text{-}aged}$, and the probability density functions of age P_A and height P_H

Below we propose a solution that interprets Zadeh's solution [2011] for these problems as a solution in terms of probability spaces as defined in the probability theory.

First, we focus on one of the problems: "What is the probability that John is tall?" The given information *I* for this problem is: a specified membership function μ_{tall} and the probability density function of Height(John), p_H. Zadeh's solutions for other listed problems are similar.

In this notation, Zadeh defines the *probability that John is tall* as:

$$\int_R \mu_{tall}(u)\, p_H(u)\, du \tag{1}$$

where R is the real line, u belongs to R and u is the height value. Formula (1) is called the *probability measure of the fuzzy set* tall [Zadeh, 1968] and is considered as a translation of the given linguistic information *I* into a mathematical language with precisiation of meaning as Zadeh calls it. Similarly a formula for the probability that *Mary is middle-aged* is

A. Laurent et al. (Eds.): IPMU 2014, Part II, CCIS 443, pp. 536–545, 2014.

$$\int_R \mu_{middle-aged}(u)p_A(u)du \,.$$

Here $p_A(u)$ is a probability that age of Mary is u.

The solution (1) has *difficulties* to be interpreted as a probabilistic solution. First, the *probability space* associated with probability measures of the fuzzy sets (1) is unclear. Second, it is unclear if a probability measure of the fuzzy set (1) is an *additive probability measures* as required to be a probability measure in the probability theory. Third, the *robustness/invariance* of formula (1) is not clear.

The robustness/invariance issue is related to the need to clarify the *measurement scale* of membership function μ_{tall}. If μ_{tall} is not in an absolute scale then the equivalent membership function $\mu`_{tall}$ that is produced within its scale will give a different value of (1). Note that the probability is measured in the absolute scale that has no such difficultly.

To illustrate the first two difficulties consider two fuzzy sets: F = "Probably John is tall" and G = "Probably John is not tall". What is the probability measure of the disjunction of these fuzzy sets, F or G? It depends on the definition of $\mu_{F\_or\_G}(u)$. The probabilistic additive measure requires

$$\int_R \mu_{tall\ or\ not\ tall}(u)p_H(u)du = 1 \tag{2}$$

The use of a common in fuzzy logic and possibility theory max operation [Zadeh, 1968],

$$\mu_{F\_or\_G}(u) = \max(\mu_F(u),\mu_G(u))$$

does not give us (2), because

$$\max(\mu_{tall}(u),\mu_{not\ tall}(u)) < 1 \text{ when } \mu_{tall}(u) < 1 \text{ and } \mu_{not\ tall}(u) < 1.$$

2   Mathematical Models

2.1   Computing Probability P(John is Tall)

Let $\mu_{tall}(\text{Height}(X))$ be a membership function, where X is a person (e.g., John) and a *probability density function* $p_H(\text{Height}(\text{John}))$ be in the height interval $U=[u_1,u_n]$, e.g., $u_1 = 160$ cm and $u_n = 200$ cm.

The probability space for P_H contains a set of events e_i="Height of John is u_i":
e_1= "Height of John is u_1", e_2="Height of John is u_2",...,e_n="Height of John is u_n" with

$$\sum_{i=1}^{n} p_H(e_i) = 1$$

Next we build a *set of probability spaces* for the concept "tall" having a membership function $\mu_{tall}(u)$, where u=Height(X). Each of these probability spaces consists of two events g_{i1}, g_{i2}, where

   g_{i1} = "John is tall if John's height is u_i",
   g_{i2} = "John is not tall if John's height is u_i"

with probabilities defined as

$$P(g_{i1} \mid e_i) = \mu_{tall}(u_i), \; P(g_{i2} \mid e_i) = 1 - \mu_{tall}(u_i)$$

when values of $\mu_{tall}(u_i)$ are given. This requires that μ_{tall} is defined in the absolute scale by using a frequency or other approaches.

Consider a statement (event)

   $g_{i1} \& e_i$ = "(John is tall if John's height is u_i) & (height of John is u_i)".

We are interested to find the probability of this event, $P(g_{i1} \& e_i)$, in the probability space with 2 events $(g_{i1} \& e_i)$ and $(g_{i2} \& e_i)$, where

   $g_{i2} \& e_i$ = "(John is not tall if John's height is u_i) & (John's height is u_i)".

In accordance with the conditional probability rules we have

$$P(g_{i1} \& e_i) = P(g_{i1} \mid e_i)P(e_i).$$

Above $P(e_i)$ was denoted $p_H(e_i)$. Also above we have got $P(g_{1i} \mid e_i) = \mu_{tall}(u_i)$, therefore

$$P(g_{i1} \& e_i) = P(g_{i1} \mid e_i)P(e_i) = \mu_{tall}(u_i)p_H(e_i)$$

Next we are interested in getting the probability that John is tall that will include all heights. We denote this event for short as "John is tall". This means that we need to compute the probability for the disjunction (union) of all events $\{g_{i1} \& e_i\}$ that constitute "John is tall":

   John is tall $\stackrel{def}{=}$ $(g_{11} \& e_1) \vee (g_{21} \& e_2) \vee \ldots \vee (g_{n1} \& e_n) = V_{i=1}^{n}(g_{i1} \& e_i)$.

Similarly we define the event "John is not tall":

   John is not tall $\stackrel{def}{=}$ $(g_{12} \& e_1) \vee (g_{22} \& e_2) \vee \ldots \vee (g_{n2} \& e_n) = V_{i=1}^{n}(g_{i2} \& e_i)$.

These events are defined in the *probability space* that consists of all events

$$\{(g_{i1} \& e_i)\}, \{(g_{i2} \& e_i)\}, i=1{:}n.$$

Accordingly,

$$P(\text{John is tall}) = P\left(\bigvee_{i=1}^{n}(g_{i1}\&e_i)\right) =$$

$$\sum_{i=1}^{n}P(g_{i1}\&e_i) = \sum_{i=1}^{n}P_H(e_i)P(g_{i1}|e_i) = \sum_{i=1}^{n}P_H(e_i)\mu_{tall}(u_i) \tag{3}$$

For the continuous set of heights (in the interval $[u_1,u_n]$) the last formula is transformed to

$$\int_{u_1}^{u_n}\mu_{tall}P_H(u_i)du$$

This is formula (1) that has been proposed by Zadeh [2011]. Thus, we have reinterpreted Zadeh' formula in probabilistic terms.

2.2   Computing Probability P(John is Tall or Not Tall)

Next we show P(John is tall or not tall) = 1 in accordance with (2) and property

$$\mu_{tall}(u_i) + \mu_{not\ tall}(u_i) = 1:$$

$$P(\text{John is tall or John is not tall})=$$

$$P\left(\bigvee_{i=1}^{n}(g_{i1}\&e_i)\bigvee_{i=1}^{n}(g_{i2}\&e_i)\right) = \sum_{i=1}^{n}p_H(e_i)\mu_{tall}(u_i) + \sum_{i=1}^{n}p_H(e_i)\mu_{not\ tall}(u_i) =$$

$$\sum_{i=1}^{n}p_H(e_i)(\mu_{tall}(u_i) + \mu_{not\ tall}(u_i)) = \sum_{i=1}^{n}p_H(e_i) = 1$$

2.3   Computing Probability P(John is of Middle Height or Tall)

Now we show how to compute P(John is of middle height or John is tall). This requires computing $\mu_{middle\ or\ tall}(u_i)$, which in turn requires creating a probability space where it can be done.

Let the elementary events in this space be:

g_{i1} = "John is short if John's height is u_i",
g_{i2} = "John is of middle height if John's height is u_i"
g_{i3} = "John is tall if John's height is u_i"

Then the negation of g_{i1} should be equal to the disjunction of two other elementary events, $\neg g_{i1} = g_{i2} \vee g_{i3}$, and similarly $\neg g_{i2} = g_{i1} \vee g_{i3}$, $\neg g_{i3} = g_{i1} \vee g_{i2}$

We also keep the same probability space as above for P_H with a set of events

e_i="Height of John is u_i":

e_1= "Height of John is u_1",e_2="Height of John is u_2",....,e_n="Height of John is u_n"

with

$$\sum_{i=1}^{n} p_H(e_i) = 1$$

Therefore,

$$P(g_{i2} \vee g_{i3}) = P(\neg g_{i1}) = 1 - P(g_{i1}) = 1 - \mu_{short}(u_i).$$

This property requires for any u_i that

$$\mu_{short}(u_i) + \mu_{medium}(u_i) + \mu_{tall}(u_i) = 1. \tag{4}$$

This property is commonly satisfied by triangular membership functions as shown in Figure 1. We call this set of fuzzy sets (linguistic variable) an *exact complete context space* [Kovalerchuk, 1996, 2013; Kovalerchuk, Vityaev, 2000].

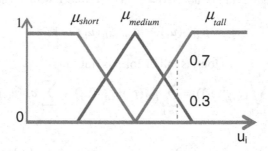

Fig. 1. Example of exact complete set of fuzzy sets (linguistic variable)

2.4 Computing Probability $P(g_{i1}\&e_i)$

Similarly to the consideration above we are interested to find the probability of event, $P(g_{i1}\&e_i)$, in the *probability space with 3 events*

$$(g_{i1}\&e_i), (g_{i2}\&e_i), (g_{i3}\&e_i),$$

where

$g_{i1}\&e_i$ = "(John is short if John's height is u_i) & (John's height is u_i)",
$g_{i2}\&e_i$ = "(John is of medium height if John's height is u_i) & (John's height is u_i)",
$g_{i3}\&e_i$ = "(John is tall if John's height is u_i) & (John's height is u_i)",

In accordance with the conditional probability rules we have $P(g_{i1}\&e_i)=P(g_{i1} \mid e_i)P(e_i)$. Above $P(e_i)$ was denoted $p_H(e_i)$. Also we define

$$P(g_{i1} \mid e_i) = \mu_{short}(u_i),\ P(g_{i2} \mid e_i) = \mu_{medium}(u_i),\quad P(g_{i3} \mid e_i) = \mu_{tall}(u_i).$$

Therefore

$$P(g_{i1}\&e_i) = P(g_{i1} \mid e_i)P(e_i) = \mu_{short}(u_i)p_H(e_i)$$
$$P(g_{i2}\&e_i) = P(g_{i2} \mid e_i)P(e_i) = \mu_{medium}(u_i)p_H(e_i) \qquad (5)$$
$$P(g_{i3}\&e_i) = P(g_{i3} \mid e_i)P(e_i) = \mu_{tall}(u_i)p_H(e_i)$$

Here (4) is assumed in formulas (5).

2.5   Probability P(John is Short) in the Space with Three Membership Functions

Next we build a *probability space* that consists of all events

$$\{(g_{i1}\&e_i)\},\ \{(g_{i2}\&e_i)\},\ \{(g_{i3}\&e_i)\},\ i=1{:}n$$

under assumption (4)

The version of (3) for $\mu_{short}(u_i)$ is

$$P(\text{John is short}) = P\left(\bigvee_{i=1}^{n}(g_{i1}\&e_i)\right) =$$

$$= \sum_{i=1}^{n} P(g_{i1}\&e_i) = \sum_{i=1}^{n} P_H(e_i)P(g_{i1}|e_i) = \sum_{i=1}^{n} P_H(e_i)\mu_{short}(u_i)$$

Similar formulas can be written place for terms medium and tall. In contrast if we use the max rule (that is common in fuzzy logic):

$$\mu_{tall\ or\ middle}(u_i) = \max\left(\mu_{middle}(u_i), \mu_{tall}(u_i)\right)$$

we will not get an additive probability space for the case shown in Figure 1, where

$$\mu_{tall\ or\ middle}(u_i) = 1,$$

but $\max\left(\mu_{middle}(u_i), \mu_{tall}(u_i)\right) < 1$ for many u_i values of u_i

3   Test Problem: What is the Probability that John is Short?

Below we propose another solution for Zadeh's task: "It is quite probable that Robert is tall. What is the probability that Robert is short?"

Model 1. An assumption is that people who know Robert answer questions if statement/event e_1 or statement/event e_2 is happened:

$$e_1: \text{Robert is tall, } e_2: \text{Robert is short.}$$

In other words we have a probability space with two events $\{e_1, e_2\}$. Let the probabilities in this space be $P(e_1) = 0.7$ and $P(e_2) = 0.3$ and computed as frequencies of respective answers. Next we construct the probability spaces to map numeric probabilities such as 0.7 and 0.3 to linguistic probabilities such as *improbable*, *unlikely, quite probable*, etc.

Consider a probability space with two elementary events: $\{a_1, a_2\}$, where

$a_1 =$ (linguistic term improbable means probability 0.3),
$a_2 =$ (linguistic term unlikely means probability 0.3).

In short we will write

$a_1 =$ (improbable=0,3), a_2=(unlikely=0.3).

Let some people be asked to select a_1 or a_2 as a preferred one with results: $P(a_1)$=0 and $P(a_2)$=1. This means that all those people prefer to associate 0.3 with *unlikely* not with *improbable*. Thus,

$$P(e_2) = P(\text{Robert is short}) = 0.3$$

is mapped to unlikely, that is the answer is: *It is unlikely that Robert is short.*

Model 2. We remove an assumption that people know Robert, and assume that his height is known, and it is 175 cm. Now people will judge 175 cm as tall or short. The rest is the same as above. Alternatively we can estimate Robert's height from the statement that it is quite probable that Robert is tall without directly assuming 175 cm.

Model 3. We change elementary events e_1 and e_2. Now

$e_1 =$ (is it quite probable that Robert is tall),
$e_2 =$ (is it quite probable that Robert is short)

asking n people which answer they prefer and getting frequencies $P(e_1)$=0.7 and $P(e_2)$=0.3. Also these can be subjective answers by a single person without computing frequencies. The rest is the same as in models 1 and 2.

For the possibility the process is the same with changing elementary events e_1 and e_2 to:

$e_1 =$ (it is quite possible that Robert is tall),
$e_2 =$ (it is quite possible that Robert is short)

asking n people which answer they prefer and getting frequencies, e.g., $P(e_1) = 0.8$, $P(e_2) = 0.2$.

We will call these numbers numeric possibility levels. For $P(e_2) = 0.2$ to map it to a linguistic possibility value we build a probability space with two elementary events: $\{a_1, a_2\}$, where

$a_1 =$ (linguistic term impossible means numeric possibility level 0.2), and
$a_2 =$ (linguistic term low possibility means numeric possibility level 0.2).

In short we can write

$$a_1 = (\text{impossible} = 0.2),\ a_2 = (\text{low possibility} = 0.2).$$

Again some people are asked to select a_1 or a_2 as a preferred answer with results $P(a_1) = 0.1$ and $P(a_2) = 0.9$, that is most of those people prefer associate 0.2 with low possibility not with impossible. Thus,

$$P(e_2) = P(\text{it is quite possible that Robert is short}) = 0.2$$

is mapped to low possibility with a high confidence (probability 0.9).

4   Probability, Possibility and Sufficiency

Zadeh formulated several test problems such as we are solving in this paper. While these tasks are challenging and this paper proposes some solutions for them, it is not obvious that solutions of these tasks will resolve the general concern about sufficiency/insufficiency of the probability theory.

There is an obvious semantic difference in the meaning of the natural language concepts of probability and possibility.

Is this fact sufficient for requiring two separate formal theories for dealing with concepts of probability and possibility?

We already have examples where the same theory can model both [Kovalerchuk, 2013] and the solutions proposed in this paper add more examples. In general, it may take time to build and refine such models. The history of the probability theory had shown that some models have been developed very quickly in a matter of days and some took much more time.

We want to emphasize that the *axiomatic probability theory*, and *probabilistic models based on it,* are not the same. The first one needs multiple additions to build rich probabilistic models with such concepts as Zadeh's very productive concept of linguistic variables.

The WCCI 2012 panel [Belyakov et al, 2012] and BISC discussion in 2013 on relations between fuzzy logic and probability theory turned to the question to clarify what is the probability theory.

We view the probability theory as consisting of several parts: (1) the *Formal Mathematical Theory of the probability spaces* that satisfy Kolmogorov's axioms, (2) multiple *Mathematical Models based on these spaces* (e.g., Markov Chains, Markov Processes, Bayesian Networks), (3) *Mathematical Statistics* as a way to link (1) with real world via its frequency interpretation, (4) *Subjective Probability* as a way to links (1) with real world via its subjective interpretation, (5) *Other* existing and future real world interpretations of (1) (e.g., using concepts of games or linguistic variables that may overlap with (3) and (4) or derived from them).

Zadeh [BISC, 12.05.2013] stated: "The problem is that computation with probability distributions is significantly more complex than computation with membership functions".

The models and formulas that we deal with in this paper are consistent with this statement.

The simplicity of fuzzy logic comes from *removing a significant part of the context* that is substituting context-dependent AND (intersection) and OR (union) operations by truth-functional operations (min/max and other T-norms/T-conorms).

In particularly, in P(A&B) = P(A | B)P(B) the conditional probability P(A | B) expresses context. In contrast, μ(A&B) = min(μ(A), μ(B)) does not express such context, because it does not involve context-dependent μ(A | B) or μ(B | A).

Zadeh responded to such type of criticism [BISC, 12.29.2013]: "With regard to truth-functionality, in fuzzy logic truth-functionality applies when there is non-interactivity. In probability theory, joint probability is the product of marginal probabilities when there is independence. Thus, in regard to truth functionality, probability theory and possibility theory are on the same footing."

This is an important statement which means that today fuzzy logic is applicable only to the tasks with non-interactivity by its design and a theory to deal with interactivity needs to be developed. In contrast in the probability theory the technique of conditional probability addresses the issue of interactivity.

Similarly Lassiter [BISC post, 01.01.2014] responded to Zadeh's statement on simplicity with a comment: "This is true, but relevant only if we have some reason to think that human language understanding relies on a cognitive system which can cope with fuzzy logic but cannot cope with probability. I do not have any reason to believe this, and in fact there is much evidence that people are able to perform complex probabilistic calculations in many cognitive domains (see, e.g., http://www.sciencemag.org/content/331/6022/1279.short and references therein). So, I do not think that relative complexity of computation is decisive about which theory is better as a model of the meanings of vague terms."

5   Conclusion

The analysis and solutions of Zadeh's test problems in this paper shows that: (1) membership functions provide a *computationally efficient way* to build a set of conditional probabilities, (2) the formula (1) proposed by Zadeh for the probability of a fuzzy set is *interpretable in rigorous probability theory terms*, and (3) the *operations* with probabilities of fuzzy sets should follow probability theory rules.

These results have been obtained by introduction of a minimal context-dependence between fuzzy sets in the linguistic variable in the form of the additive formula (4). This allows us to get a rigorous probabilistic interpretation of membership functions of fuzzy sets in accordance with the concept of the linguistic context space [Kovalerchuk, 1996, 2013; Kovalerchuk, Vityaev, 2000], which is consistent with an empirical derivation of membership functions [Hall, Szabo, Kandel, 1986] and Hisdal's probabilistic interpretation of membership functions [Hisdal, 1998].

References

1. Hall, L., Szabo, S., Kandel, A.: On the Derivation of Memberships for Fuzzy Sets in Expert Systems. Information Sciences 40, 39–52 (1986)
2. Hisdal, E.: Logical Structures for Representation of Knowledge and Uncertainty. STUDFUZZ, vol. 14. Springer, Heidelberg (1998)
3. Kovalerchuk, B.: Quest for Rigorous Combining Probabilistic and Fuzzy Logic Approaches for Computing with Words. In: Seising, R., Trillas, E., Moraga, C., Termini, S. (eds.) On Fuzziness. STUDFUZZ, vol. 298, pp. 325–336. Springer, Heidelberg (2013)
4. Kovalerchuk, B.: Context Spaces as Necessary Frames for Correct Approximate Reasoning. International Journal of General Systems 25(1), 61–80 (1996)
5. Kovalerchuk, B., Vityaev, E.: Data Mining in Finance: Advances in Relational and Hybrid Methods (ch. 7 on fuzzy systems). Kluwer, Boston (2000)
6. Lassiter, D.: BISC-group post (January 1, 2014), http://mybisc.blogspot.com/
7. Zadeh, L.: A problem in probability theory—a solution (2011); Posted in BISC in (2013)
8. Zadeh, L.: Probability measures of fuzzy events. Journal of Mathematical Analysis and Applications 23(2), 421–427 (1968)
9. Zadeh, L.: BISC-group posts (December 5, 2013) (December 29, 2013), http://mybisc.blogspot.com/
10. Zadeh, L.A.: Computing with Words. STUDFUZZ, vol. 277. Springer, Heidelberg (2012)
11. Beliakov, G., Bouchon-Meunier, B., Kacprzyk, J., Kovalerchuk, B., Kreinovich, V., Mendel, J.: Computing With Words (CWW): role of fuzzy, probability and measurement concepts, and operations. Mathware& Soft Computing Magazine 19(2), 27–45 (2012)

Some Reflections on Fuzzy Set Theory
as an Experimental Science

Marco Elio Tabacchi[1,2] and Settimo Termini[1,3]

[1] Dipartimento di Matematica ed Informatica,
Università degli Studi di Palermo, Italy
[2] Istituto Nazionale di Ricerche Demopolis, Italy
[3] ECSC, Spain
{marcoelio.tabacchi,settimo.termini}@unipa.it

Abstract. The aim of this paper is to open a critical discussion on the claim, recently presented in the community and especially heralded by Enric Trillas, that fuzzy logic should be seen as an "experimental science". The first interesting aspect of such remark is whether and in which way such position has consequences on the real development of the research, or if it is simply a (different) way of looking at the same phenomenon. As a consequence, we investigate the possible connection to Zadeh's distiction between Fuzzy logic in a restricted sense and in a general sense. We shall argue that Trillas's claim not only strongly supports the necessity for such a distinction, but provides a path of investigation which can preserve the conceptual innovativeness of the notion of fuzziness.

Keywords: Fuzzy Sets, Soft Computing, Theoretical Computer Science.

1 Introduction

The development of fuzzy set theory has been always constellated by dichotomies and contrapositions; something, perhaps, which can be considered very natural in every young field which propounds new ideas. It seems, however, that the contrapositions in such field have been particularly frequent and violent in style. A measure – in our view – of the fact that the intrinsic innovativeness of the basic ideas was so strong that a normal assimilation was not possible is given by the emergence of a few epistemological problems, which should be carefully studied and understood. We should, perhaps, summarize in a simplified way the situation through the following remark: in the first years of the development of fuzzy set theory, it was not completely clear the range of the proposed innovation.

The fact is that, if taken seriously, the fuzzy revolution shakes at its root some of the basic principles of the scientific revolution of XVII Century. This does not mean that the proposal of modeling quantities with unsharp boundaries is unscientific, but that it shakes some of the basic pillars of modern science. It is our job in the vein of our previous research [1–9] to try to preserve the innovativeness of the notion of fuzziness combining it with the unavoidable features of scientific method. Let us, just for a clear statement of the ideas involved, point

A. Laurent et al. (Eds.): IPMU 2014, Part II, CCIS 443, pp. 546–555, 2014.

in a rough way to two items which could be considered typical and paradigmatic, and postpone further details and comments:

1. Kalman's attack to fuzzy set, often quoted by Zadeh itself [10, 11].
2. The paradigmatic use of the word *perception* in some crucial papers by Lofti Zadeh (see [12]).

The classical, well known debate points are the probability vs. fuzziness conundrum, the interpretation of linguistic labels and all the cascading consequences of the assumption of this notion in a world made of numbers (of *real* numbers, measured and wrote down in their units). All considered the first, basic and founding concept – Fuzzy Sets – is the natural domain of application of fuzzy ideas and techniques, and the relationship of fuzzy sets with many valued logics brings the epistemological meaning and significance of the development of the "applications" of fuzzy techniques, especially in the setting of the so called "fuzzy control". To keep with the intended point, however, the heart of the matter is: in which direction the notion of fuzziness has been – and still is – innovative? What enduring changes does this nature have produced and in which fields and sectors of investigation? Did it integrate with the different domains of investigation with whom it had the opportunity of interacting? Will it remain as a separate field or – despite the existence of specific problems and questions will it eventually merge with all the domains in which it has shown to be useful?

Let us observe that another dichotomy exists, sometimes openly and more often surreptitiously: the one between people that substantially think that the field of "fuzzy world and research" is definitely stabilized for what regards both its aims and its relationship with other fields of research; and the alternative position of a few scientists that opine that the present "acceptance" on the part of other scientific disciplines does not correspond to a true acceptance of the most innovative aspects of the idea of fuzziness. The first position is accepted only as soon as it "conforms" to the traditional "canons" of scientific rigor, while acceptance of the other side is not so universally shared. Among the positions held in this other half of the field we shall examine and briefly take into account the ideas heralded by Enric Trillas in recent years; in particular the thesis that Fuzzy Sets theory should be approached and studied as an experimental science. Only this approach – according to Trillas – completely and fully recognizes the real and deep innovativeness of the notion of fuzziness. In the rest of the paper we will present this idea in a compact way, trying to relate it with a few of the other "founding dichotomies" of the theory of Fuzzy Sets as well as with a chain of new concepts introduced by Zadeh along the years such as linguistic labels and manipulation of perceptions. We use the term "linguistic labels", but – perhaps – the turning point was the notion of linguistic variable, something which can be represented or assumes the value of a Fuzzy Set a function assuming values in $[0,1]$ which is then suitably interpreted. The numerical assignment which we shall finally do is strictly related to the meaning associated to the linguistic parts of manipulations. If we had a way of associating numbers and numerical evaluations without referring to the meaning of the words implied and used them the same introduction of linguistic objects would be fictitious and redundant.

1.1   On the Attack on Fuzziness and All That

Now we can go back to the two questions as yet unanswered. Kalman's criticism and the implicit consequences of manipulating perceptions. Let us examine these questions now from a general epistemological point of view. Kalman's attack can be seen – from this general point of view, filtered or strained by all the personal and sociological components – as the recognition that the innovativeness implicit in taking the notion of fuzziness as a starting point for a new approach was so atypical that one could tell from the start that it could not have been correctly incorporated into the "canon" of scientific tradition that we know and accept today. So, it should be violently attacked for it role in undermining the same scientific method at its very core. In fact Zadeh [13] remembers that the attacks became stronger after the introduction of the notion of linguistic variable.

The challenges of introducing such notions as fuzzy set, fuzzy control and linguistic variable are big, but what is a bigger one is the global enterprise. This is witnessed by one of its subsequent developments, namely the idea that one should take as a starting point "the manipulation of perception" instead of "the manipulation of measurements" [12]. Such stance – in an uncritical version of the received view of scientific method - implies a violation of one of the basic pillars of scientific revolution, namely that science can develop in a secure way only when it can deal with primary qualities which we are able to afford and study with objective methods. Perceptions, instead, belong to the world of "secondary" qualities – which according to a long tradition – are not easily approachable with classically recognized scientific methodologies. Now, it is not strange that such an approach that proposes to construct something starting just from such debatable and subjective building blocks is not immediately palatable to the scientific community in its majority.

1.2   FST as an Experimental Science

The thesis that Fuzzy Sets Theory can be seen and treated as an "Experimental Science" poses some challenging and interesting questions from an epistemological point of view. To further complicate the matter, a further question arises: are all the possible declinations of Fuzzy Logic related with the distinction, periodically remarked by Zadeh, between FL in a wide sense and in a restricted sense? This network of ideas seems to be very slippery and it seems that these kind of considerations are not included in the daily routine of scientist. We hope that the bird's eye recognition we shall present in the following pages will show why these three topics are connected with the development of the same field and – far from being some additional musing about the actual scientific work without any real influence on his development, they can have a very strong influence in determining the future lines of development, and the directions along which this field will evolve in the following years. The habit of doing things following only the well established and secure paths, or duties implicitly imposed by what is considered methodologically mandatory in nearby fields of investigation are the obstacles we will find. Such rules are often more of a limit in frontier fields, and in some cases the innovativeness of the field lies exactly in rejecting them, identifying new ones in the process.

2   A Detour among a Few Quotations

In a paper now almost ten years old [14], Enric Trillas put forward some general reflections on fuzzy sets with the aim of extending "the current theories of fuzzy sets to wider areas of both language and reasoning". This path can be followed – according to him – by "rethinking fuzzy sets from their roots". This programmatic statement introduced in the abstract is expanded in the introduction of the same paper in a sort of manifesto. In fact, after a brief historical summary of the steps leading from the seminal 1965 paper [15] to the actual stance, he writes:

> Fuzzy logic not only deals with problems at the technological side of computational intelligence. Since what a fuzzy set does represent is a concrete use of a predicate (or linguistic label), and as Wittgenstein asserted, "the meaning of a word is its use in the language", fuzzy logic also deals with what is known as the Gordian Knot of computational intelligence, the problem of meaning. [...] Currently, in the thinking of some people working in fuzzy logic from almost its beginning, it is sprouting the idea that the time to rethink fuzzy sets and fuzzy logic is coming. Such rethinking is not only viewed to push ahead the knowledge of (linguistic) imprecision, and the corresponding uncertainty, but to give a better support to the more complex applications that are foreseeable in a not too long future. What this paper tries to offer from a theoretical point of view, are just some hints in the direction of extending the theories of fuzzy sets to face broader areas of language than those they can currently deal with. [...] Although the paper will refer only to mathematical models, before beginning with them it is important to declare, like in the customs, that if mathematics are really important for the understanding of the phenomena linked to imprecision, they are nothing more and nothing less than a basic tool for such a goal. What really matters is imprecision and, of course, mathematical models can help both to clarify some of its aspects and to base applications in solid grounds. But if there can be "maths for the Fuzzy" there are not "fuzzy maths". Mathematics are what they are and, as always in the history of science and technology, they are important in that they can help us is in the study of imprecision with as much precision as possible once questions on the phenomena are well posed. Only in conjunction with good questions and fine observations on them, are mathematical models interesting, and useful, for a deeper understanding of phenomena.

The introduction ends with the following passage, which can be considered a small programmatic manifesto:

Provided these ideas for rethinking fuzzy logic would be followed, what can be guessed is a ramification of current fuzzy logic in three branches: An experimental science of fuzziness, mainly dealing with imprecision in natural and specific languages; theoretical fuzzy logic, dealing with mathematical models and their linguistic counterparts as well as the necessary computing tools for their computer implementation, and a broad field of new practical applications to a multiplicity of domains, like internet, robotics, management, economy, linguistics, medicine, education, etc.

Let us now see how this manifesto is developed in a few subsequent papers. For instance, in [16] one reads:

This paper [...] is a kind of essay that tries to shed some light on those links by means of mathematical representations in algebraic frameworks as simple as possible and, hence, sufficiently general to allow the study of a wide spectrum of dynamical systems and reasonings expressed in Natural Language.

The author gives some specifications:

It should be noticed that predicates appearing in the language were usually introduced by naming a property exhibited by some elements in a 'universe of discourse'. After this, it is frequently the case that the considered predicate migrates to another universe of discourse, and that its use results in some form distorted, but showing 'family resemblance' with its former use. Hence, the use we analyze of a predicate is with reference to a given universe of discourse. The resemblance of uses is also taken into account, yet an initial study of them can be found in. This paper does not deal with the processes going from a collective towards a predicate naming it, but from a predicate on a universe towards the 'representation' in mathematical terms of the collective it can originate.

This general Weltanschaung which constructively mixes – a la Wittgenstein – meaning and use, family resemblances with predicates and usefully migrates from a universe of discourse to another cannot but escape a sort of mechanical and automatic application of a general theory, as well as the boundaries fixed by the crucial and central concepts imposed by the ortodoxy of mathematical logic as formally structured in the last decades. In fact, two leading ideas are crucial in this programme: the need for a careful design of fuzzy sets and the necessary condition for doing that, namely, to consider fuzzy logic as an experimental science. Let us see what is written in [17]:

The flexible subjects fuzzy logic deals with (that are in contraposition to the typically rigid of formal logic) force a different methodology than the one formal sciences use to approach the problem. This is like the case of physics, whose methodology is not as strictly formal as the methodology of mathematics, even though mathematical models play an important role in physics. But these models are to be experimentally tested against the world. Like it happens with the mathematical models in fuzzy logic, which are important in the amount that they allow to represent well the linguistic description of systems and/or processes.

and also

Despite its name, fuzzy logic is, in the first place, used for representing some reasonings involved with imprecision and uncertainty. Fuzzy logic is closer to an experimental science than to a formal one. In part, because of the use of real numbers and continuity properties, as well as its use in real systems.

According to Lotfi Zadeh, there are two different ways in which the definition of fuzzy logic can be understood. Since this distinction is well known in the field, we shall recollect it in a very succinct way. From a recent (January 2013) message by Zadeh to the components of the Berkeley Initiative in Soft Computing:

A major source of misunderstanding is rooted in the fact that fuzzy logic has two different meanings – fuzzy logic in a narrow sense, and fuzzy logic in a wide sense. Informally, narrow-sense fuzzy logic is a logical system which is a generalization of multivalued logic. An important example of narrow-sense fuzzy logic is fuzzy modal logic. In multivalued logic, truth is a matter of degree. A very important distinguishing feature of fuzzy logic is that in fuzzy logic everything is, or is allowed to be, a matter of degree. Furthermore, the degrees are allowed to be fuzzy. Wide-sense fuzzy logic, call it FL, is much more than a logical system. Informally, FL is a precise system of reasoning and computation in which the objects of reasoning and computation are classes with unsharp (fuzzy) boundaries. The centerpiece of fuzzy logic is the concept of a fuzzy set. More generally, FL may be a system of such systems.

We want to stress for the purposes of the present paper the statement affirming that "Wide-sense fuzzy logic is much more than a logical system"; it is "a system of reasoning and computation". The universe in which we move is not a logically-dominated world. And this notwithstanding the fact that Fuzzy Logic has become a very developed and respected brand of mathematical logic, able to develop systems in which the notion of truth (or better, the notion of truth-value) is a matter of degree.

Zadeh observes that the research in these subfields is a tiny fragment of the total quantity of papers and of all the research published. However it is clear, although not affirmed in this explicit form by Zadeh (at least in the previously mentioned article) that the main problem and question is not quantitative, but a conceptual one. The term Fuzzy Logic is generally used to indicate all the research done using the notion of Fuzzy Set and with the aim of treating in a more precise way all the situations in which "the objects of reasoning and computation are classes with unsharp (fuzzy) boundaries". But it is also clear that the approach is not proper of logic, and that the word logic is used in a sense more similar to the one used before Frege's revolution – and certainly not in a *axiomatic* sense. Maybe it could be interpreted in the sense in which Boole called his books. And, in a sense, in Zadeh's approach there is also the ambition of founding a general theory applicable to all the cases in which we are reasoning about something that has unsharp boundaries.

Let us now come back to the problem of precision: in Zadeh's words

> Fundamentally, fuzzy logic is aimed at precisiation of what is imprecise. [...] But in many of its applications fuzzy logic is used, paradoxically to imprecisiate what is precise. In such applications, there is a tolerance for imprecision, which is exploited through the use of fuzzy logic. Precisiation carries a cost. Imprecisiation reduces cost and enhances tractability. This is what I call the Fuzzy Logic Gambit. What is important to note is that precision has two different meanings: precision in value and precision in meaning. In the Fuzzy Logic Gambit what is sacrificed is precision in value, but not precision in meaning.

So, the right and correct use of FL in a specific domain is always aimed at rendering more precise what one is researching, although in some cases the "precisiation of meaning" can be strictly and inavoidably linked to an "imprecisiation in value". This is what Zadeh calls "Fuzzy Logic Gambit". In any case and without too much details, Fuzzy Logic in Zadeh's view is aimed at increasing precision, although the general and global increase can in some cases correspond to a decrease of precision in value, accepted in order to obtain a more relevant increase in meaning. If we combine this observation with his claim that FL is not an axiomatic logical system in the sense required by mathematical logic, we can certainly affirm that an enlarged setting like the one proposed by Trillas is a solution that allows preserving the scientific rigor while looking for new ways to be explored.

3   Conclusions

The aim of this paper was to present the outline of a brief analysis in order to consider and classify the proposal of considering FST as an "experimental science". First it was recognized that discussing this topic would involve to both look at the history of FS, and to indicate the most fruitful path to be followed in the future. The reason for that resides in that if by classifying a theory of a mathematical / logic and applied kind as an experimental science we are obliged to reconsider all the previous developments of the field. Not, of course, in the sense of the validation or re-evaluation of all the results obtained in the previous decades – as they remain valid for what they are – but in the sense of looking at the various, crucial results in a different way. Let us for instance have a look at the problem of extended and propositional connectives. In the previous Section we briefly remembered how Zadeh has posed the question of the difference between fuzzy logic in a strict sense and in a wide sense. Let us now see how the logician Petr Hájek [18] who has certainly worked hard toward obtaining important and deep results, but has also tried to be a good translator and ambassador between the republic of mathematical logic and the kingdom of fuzzy sets, sees the problem:

> In a broad sense, the term 'fuzzy logic' has been used as synonymous with 'fuzzy set theory and its applications'; in the emerging narrow sense, fuzzy logic is understood as a theory of approximate reasoning based on many-valued logic. Zadeh stresses that the questions of fuzzy logic in the narrow sense differ from usual questions of many-valued logic and concern more questions of approximate inferences than those of completeness, etc.; nevertheless, with full admiration to Zadeh's pioneering and extensive work a logician will first study classical logical questions on completeness, decidability, complexity, etc. of the symbolic calculi in question and then try to reduce the question of Zadeh's agenda to questions of deduction as far as possible.

And he concludes his survey paper by affirming that

> Fuzzy logic in the narrow sense is a logic, a logic with a comparative notion of truth. It is mathematically deep, inspiring and in quick development; papers on it are appearing in respected logical journals. [...] The bridge between fuzzy logic in the broad sense and pure symbolic logic is being built and the results are promising.

We think that the situation is clear enough to allow the drawing of some conclusions. Hájek attitude, which is one of the most open in the field of mathematical logic, is very optimistic. A substantial piece of the questions asked and posed by FS has been thoroughly introduced into the agenda of mathematical logic. He also recognizes that the interaction between the old program of many-valued logic and the informal questions posed by the fuzzy approach has been massively fruitful, also for mathematical many-valued logic: some new questions has been posed and answered in the affirmative way. However the agenda – as it is natural – is fixed by mathematical logic, by the questions asked and posed by its research programme. Let us observe that the attitude outlined seems more radical that the one emerging from the work done by people working in quantum logic looking for bridges with fuzzy concepts and techniques. In this context, at least following the attitude of researchers such as Pykacz [19] – who seems to be very enthusiastic towards the innovative aspects of the fuzzy approaches – an attempt can be found at considering fully classical points and requirements of the quantum logic approach. But as well a similar care in preserving all the nuances of the language of fuzzy sets theory is present. And, et *pour cause* – as he wants to recover all the innovativeness at a conceptual level of the notion of fuzziness. This can also be fully obtained by detailing a complete and total reduction of quantum logic to the language of FST. We see and frankly admit that the situation in quantum logic, and in general in mathematical logic, is different, but cannot refrain from observing that the epistemological attitude is – let us say – totalitarian. It is not a problem of passing judgements. It is exactly a question of looking for the approaches and attitudes that can be more suitable for the expansion and development of a certain field.

In a not distant future it could also happen that all the work done in fuzzy logic in an extended sense would be recognized as a province of mathematical logic (possibly in an enlarged version which beside accepting "a comparative notion of truth" as written by Hájek for what regards "Fuzzy logic in the narrow sense" will accept and assimilate many other new concepts and notions that today seem to be outside the scope and very far from mathematical logic as is intended today). It is not very likely – in our view – that such a new framework will be able to include almost all the nuances of the concepts related to language that are present in the (too) numerous quotations reported in the previous Section. However this is not the crucial and central problem at the moment.

The problem at hand today is to recognize whether many crucial questions can be better afforded in a general scheme in which the agenda is fixed by mathematical logic, or by other actors on the stage. It seems to us that due to the richness of the concepts and problems involved a good pluralism is more useful for the advancement of the field(s). This is what, with his polite and diplomatic approach, Zadeh propounds in his periodic comments on the differences between the two ways of interpreting the term "fuzzy logic". It seems that all the problems related to a treatment using approaches related or conceptually inspired to the paradigm of natural language receive their best consideration inside a general strategy like the one outlined by Trillas with his proposal inspired to the epistemology and working of experimental sciences. It does not exclude that more and more (small or big) fragments of the future technical developments of FST (or fuzzy logic) obtained by working in this direction could be embedded into the paradigm of mathematical logic. However, will this happen, the basic reason will be connected to the fact that – a posteriori – the new results will be naturally tuned with the crucial and central questions of mathematical logic, and not that – a priori – the questions to be studied were selected by their adherence to its paradigm. It is probable that if this ("experimental") approach is taken seriously, the results – as seen globally - could show that there exists a coherent *something* which has many distinguishing and specific features which prevent its total collapse into the other paradigm. So, we think that it is possible to conclude that the idea of considering fuzzy set theory as an experimental science not only is perfectly tuned with the general ideas of its founder, but strongly reinforces the possibility that the new developments preserve all the conceptual innovativeness of the original idea.

References

1. Cardaci, M., Di Gesú, V., Petrou, M., Tabacchi, M.E.: On the evaluation of images complexity: A fuzzy approach. In: Bloch, I., Petrosino, A., Tettamanzi, A.G.B. (eds.) WILF 2005. LNCS (LNAI), vol. 3849, pp. 305–311. Springer, Heidelberg (2006)
2. Cardaci, M., Di Gesu, V., Petrou, M., Tabacchi, M.E.: A fuzzy approach to the evaluation of image complexity. Fuzzy Sets Syst. 160(10), 1474–1484 (2009)
3. Petrou, M., Tabacchi, M.E., Piroddi, R.: Networks of Concepts and Ideas. The Computer Journal 53(10), 1738–1751 (2010)

4. Termini, S., Tabacchi, M.E.: Fuzzy set theory as a methodological bridge between hard science and humanities. International Journal of Intelligent Systems 29(1), 104–117 (2014)
5. Tabacchi, M.E., Termini, S.: Measures of fuzziness and information: some challenges from reflections on aesthetic experience. In: Proceedings of WConSC 2011 (2011)
6. Tabacchi, M.E., Termini, S.: Fuzziness and social life: informal notions, formal definitions. In: Proceedings of the Annual Meeting of the North American Fuzzy Information Processing Society, NAFIPS (2012)
7. Tabacchi, M.E., Termini, S.: A few remarks on the roots of fuzziness measures. In: Greco, S., Bouchon-Meunier, B., Coletti, G., Fedrizzi, M., Matarazzo, B., Yager, R.R. (eds.) IPMU 2012, Part II. CCIS, vol. 298, pp. 62–67. Springer, Heidelberg (2012)
8. Seising, R., Tabacchi, M.E.: A very brief history of soft computing. In: Pedrycz, W., Reformat, M. (eds.) 2013 Joint IFSA World Congress NAFIPS Annual Meeting. IEEE SMC (2013)
9. D'Asaro, F., Perticone, V., Tabacchi, M.E., Termini, S.: Reflections on technology and human sciences: rediscovering a common thread through the analysis of a few epistemological features of fuzziness. Archives for Philosophy and History of Soft Computing 1 (2013)
10. Zadeh, L.A.: The evolution of systems analysis and control: a personal perspective. IEEE Control Systems 16(3), 95–98 (1996)
11. Zadeh, L.A.: Is there a need for fuzzy logic? Information Sciences 178(13), 2751–2779 (2008)
12. Zadeh, L.A.: Toward a perception-based theory of probabilistic reasoning with imprecise probabilities. Journal of Statistical Planning and Inference 105(1), 233–264 (2002)
13. Zadeh, L.A.: Foreword. In: Trillas, E., Bonissone, P.P., Magdalena, L., Kacprzyk, J. (eds.) Combining Experimentation and Theory. STUDFUZZ, vol. 271, pp. IX–XI. Springer, Heidelberg (2011)
14. Trillas, E.: On the use of words and fuzzy sets. Information Sciences 176(11), 1463–1487 (2006)
15. Zadeh, L.A.: Fuzzy sets. Information and Control 8, 338–353 (1965)
16. García-Honrado, I., Trillas, E.: An essay on the linguistic roots of fuzzy sets. Information Sciences 181(19), 4061–4074 (2011)
17. Trillas, E., Guadarrama, S.: Fuzzy representations need a careful design. International Journal of General Systems 39(3), 329–346 (2010)
18. Hájek, P.: Why fuzzy logic? In: Jacquette, D. (ed.) A Companion to Philosophical Logic, pp. 595–605. Wiley (2008)
19. Pykacz, J.: Fuzzy sets in foundations of quantum mechanics. In: Seising, R., Trillas, E., Moraga, C., Termini, S. (eds.) On Fuzziness: Volume 2. STUDFUZZ, vol. 299, pp. 553–557. Springer, Heidelberg (2013)

Fuzziness and Fuzzy Concepts
and Jean Piaget's Genetic Epistemology

Rudolf Seising

European Centre for Soft Computing, Mieres, Spain
Rudolf.seising@softcomputing.es

Abstract. How do humans develop concepts? -- This paper presents a historical view on answers to this question. Psychologist Piaget was influenced by Philosopher Kant when he founded his theory of cognitive child development named "Genetic Epistemology". Biologist and historian of science Rheinberger emphasized that scientific concepts are "fluctuating objects" or "imprecise concepts" when he founded his "Historical Epistemology". In this paper we combine these approaches with that of fuzzy concepts. We give some hints to establish a new approach to extend Piaget's theory, to a so-called "Fuzzy Genetic Epistemology".

Keywords: Epistemology, Genetic Epistemology, Historical Epistemology, Fuzzy Sets, Fuzzy Concepts, Structuralism.

1   Introduction

In this paper we consider the concept of concepts. The Stanford Encyclopedia of Philosophy differentiates at least between three prevailing ways to understand what a concept is in contemporary philosophy [1]:

- Concepts as mental representations, i.e. entities that exist in the brain;
- Concepts as abilities, peculiar to cognitive agents;
- Concepts as abstract objects, where these objects are the constituents of propositions that mediate between thought, language, and referents.

In their book *Concepts and Fuzzy Logic* Belohlavek and Klir [2] use the concept of concepts from Cognitive science and they refer to Edouard Machery: "In cognitive science, concepts are the bodies of knowledge that are stored in long-term memory and are used by default in the higher cognitive processes (categorization, inductive and deductive reasoning, analogy making, language understanding, etc.)." [3]

In this paper we will argue that those concepts are unsharp or fuzzy and we will show that this fuzzy concept is also fruitful in Historical and Genetic epistemology – that are two disciplines that have been founded in the 20th century by Hans-Jörg Rheinberger and Jean Piaget, respectively.

A. Laurent et al. (Eds.): IPMU 2014, Part II, CCIS 443, pp. 556–565, 2014.

2   What Is Epistemology

2.1   Knowledge and Cognition

Epistemology is the branch of philosophy that is concerned with knowledge. The *Stanford Encyclopedia of Philosophy* defines "narrowly, epistemology is the study of knowledge and justified belief. As the study of knowledge, epistemology is concerned with the following questions: What are the necessary and sufficient conditions of knowledge? What are its sources? What is its structure, and what are its limits?" [4]

To answer these questions philosopher Immanuel Kant (1724-1804) tried to unify the two main views in former philosophy of science, empiricism and rationalism. The rationalist approach came to fundamental, logical and theoretical investigations using logics and mathematics to formulate axioms and laws, however from the empiricist point of view the source of our knowledge is sense experience and we have to use experiments to find or prove or refute natural laws. In both directions – from experimental results to theoretical laws or from theoretical laws to experimental proves or refutations – scientists have to bridge the gap that separates theory and practice in science.

In his *Critique of Pure Reason* Kant came from the view that knowledge increases through, and that cognition starts from experience. He named "concepts" the "a priori forms of cognition" and he proposed that these concepts exist within the subject of cognition. On the other hand, he said, that the object of cognition is established when the sensory content coming from the object is put in order by the subject's concepts. In Kant's epistemology the human mind provides a structure that shapes all sensory experience and thought. Perceptions and thoughts must conform into this structure in order to be representations. Kant says that humans have an active mind that produces our conception of reality by acting as a filter, and also it is organizing and enhancing. Kant says that objective reality is made possible by the form of its representation. Our mind's structure includes space, time, and causation, they are preconditions of our perceptions, and they are fundamental conditions for human experience.

2.2   Science, Nature and Concepts

All our scientific theories are abstract and formal but the systems we are intended to reconstruct by our theories are real world systems. Therefore we now draw our attention to the philosophical topic of the gap between theoretical and real entities. Again we have to mention Kant. His construction of the system of science with a progression from the formal to the most empirical phases he named an "Architectonic of Science": "By the term architectonic I mean the art of constructing a system. Without systematic unity, our knowledge cannot become science; it will be an aggregate, and not a system. Thus architectonic is the doctrine of the scientific in cognition, and therefore necessarily forms part of our methodology." [5]

In this *Critic of Pure Reason* he also developed a system of physical nature starting with "the most formal act of human cognition, called by him the transcendental unity of apperception, and its various aspects, called the logical functions of judgment. He then proceeds to the pure categories of the understanding, and then to the schematized categories, and finally to the transcendental principles of nature in general." [6]

It is this concept of a "schema" that Kant used to both the structure of human knowledge itself and the procedure by which the human mind produces and uses such structures, that Jean Piaget adopted to his epistemology, the Genetic epistemology.

2.3   Genetic Epistemology

In the history of Developmental Psychology where some psychologists became interested in the field of child development in the early 20th-century, mainly two doctrines are significant to date:

- In his Analytic Psychology (*Psychoanalysis*) Sigmund Freud (1856-1939) stressed the importance of childhood events and experiences. In his *Three Essays on the Theory of Sexuality* he described child development as a series of 'psychosexual stages' or phases: oral (ages 0–2), anal (2–4), phallic (3–6), latency (6-puberty), and mature genital (puberty-onward). Each phase involves the satisfaction of a libidinal desire and can later play a role in adult personality. [7]
- When he was a young scientist, Jean Piaget was engaged in Psychoanalysis, but later he established a very different stage theory of children's cognitive development: *Genetic Epistemology*. In his theory it is assumed that children think differently than adults and they play an active role in gaining knowledge of the world. They actively construct their knowledge!

2.4   Historical Epistemology

The Swiss historian of science and biologist Hans-Jörg Rheinberger proposed in the last decade of the 20th century the program of "historical epistemology" as a turnaround in philosophy and history of science. He made allowance for the situation that historians and philosophers of science in the 1980s and 1990s concentrated their attention to experiments in science. Historical epistemology deals with the concept of so-called "experimental systems". What is an experimental system? – Rheinberger illustrated: "This notion is firmly entrenched in the everyday practice and vernacular of the twentieth-century life scientists, especially of biochemists and molecular biologists. Scientists use the term to characterize the scope, as well as the limits and the constraints, of their research activities. Ask a laboratory scientist what he is doing, and he will speak to you about his "system". Experimental systems constitute integral, locally manageable, functional units of scientific research." ([8], p. 246)

In Rheinberger's epistemology we find again two parts of scientific research, that he named epistemic and technical, respectively, but he emphasizes that there is no sharp boundary between, moreover this boundary is vague or fuzzy. Here is a brief sketch of his approach: "If there are concepts endowed with organizing power in a research field, they are embedded in experimental operations. The practices in which the sciences are grounded engender epistemic objects, epistemic things as I call them, as targets of research." We follow Rheinberger's arguments to show that theses "epistemic things" are vague or fuzzy and that they "move the world of science" ([9], p. 220). He considered these "fluctuating objects" and "imprecise concepts" – as he also called them - in detail in his historical work.

3   Inexact or Fuzzy Concepts

3.1   Fuzzy Sets and Fuzzy Systems

In the early 1960s Lotfi A. Zadeh, a professor of Electrical Engineering at the University of California at Berkeley "began to feel that complex systems cannot be dealt with effectively by the use of conventional approaches largely because the description languages based on classical mathematics are not sufficiently expressive to serve as a means of characterization of input-output relations in an environment of imprecision, uncertainty and incompleteness of information." [10] In the year 1964 he discovered how he could describe real systems as they appeared to people. "I'm always sort of gravitated toward something that would be closer to the real world" [11]. In order to provide a mathematically exact expression of experimental research with real systems, it was necessary to employ meticulous case differentiations, differentiated terminology and definitions that were extremely specific to the actual circumstances, a feat for which the language normally used in mathematics could not provide well. The circumstances observed in reality could no longer simply be described using the available mathematical means. These thoughts indicate the beginning of the genesis of Fuzzy Set Theory." ([12], p. 7) In his first article "Fuzzy Sets" he launched new mathematical entities as classes or sets that "are not classes or sets in the usual sense of these terms, since they do not dichotomize all objects into those that belong to the class and those that do not." He introduced "the concept of a fuzzy set, that is a class in which there may be a continuous infinity of grades of membership, with the grade of membership of an object x in a fuzzy set A represented by a number $f_A(x)$ in the interval $[0,1]$." [13, 14]

Some years later Zadeh compared the strategies of problem solving by computers on the one hand and by humans on the other hand. He called it a paradox that the human brain is always solving problems by manipulating "fuzzy concepts" and "multidimensional fuzzy sensory inputs" whereas "the computing power of the most powerful, the most sophisticated digital computer in existence" is not able to do this. Therefore, he stated that "in many instances, the solution to a problem need not be exact", so that a considerable measure of fuzziness in its formulation and results may be tolerable. The human brain is designed to take advantage of this tolerance for imprecision whereas a digital computer, with its need for precise data and instructions, is not." ([15], p. 132)

3.2   Fuzzy Concepts

Zadeh had served as first reviewer and his Berkeley-colleague and mathematician Hans-Joachim Bremermann (1926–1996), as second for the Ph.D. thesis of mathematician Joseph A. Goguen's (1941-2006) entitled *Categories of Fuzzy Sets*. [16] Here, Goguen generalized the fuzzy sets to so-called "L-sets". An L-set is a function that maps the fuzzy set carrier X into a partially ordered set L. The partially ordered set L Goguen called the "truth set" of A. The elements of L can thus be interpreted as "truth values"; in this respect, Goguen then also referred to a *Logic of Inexact Concepts* [17].

Zadeh's efforts to use his fuzzy sets in linguistics led to an interdisciplinary scientific exchange between him and Goguen on the one hand and between Berkeley-psychologist Eleanor Rosch and the Berkeley-linguist George Lakoff on the other.

In her psychological experiments Rosch could show that concept categories are graded. Consequently she argued that concepts are not adequately represented by classical sets. Rosch developed her prototype theory on the basis of these empirical studies. This theory assumes that people perceive objects in the real world by comparing them to prototypes and then ordering them accordingly. In this way, according to Rosch, word meanings are formed from prototypical details and scenes and then incorporated into lexical contexts depending on the context or situation. It could therefore be assumed that different societies process perceptions differently depending on how they go about solving problems. [18]

Lakoff referred to the fact that also statements in natural language are graded, "that sentences of natural languages (at least declarative sentences) are either true or false or, at worst, lack a truth value, or have a third value." He argued "that natural language concepts have vague boundaries and fuzzy edges and that, consequently, natural language sentences will very often be neither true, nor false, nor nonsensical, but rather true to a certain extent and false to a certain extent, true in certain respects and false in other respects. (19, p. 458) In this paper Lakoff wrote that Zadeh's fuzzy set theory is an appropriate tool of dealing with degrees of membership, and with (concept) categories that have unsharp boundaries. Because he used the term "fuzzy logic" he deserves the credit for first introducing this expression in the scientific literature but based on his later research, however, he came to find that fuzzy logic is not an appropriate logic for linguistics [20]. Now, we move to fuzzy concepts in psychology!

4   Cognitive Development

4.1   A Brief Sketch of a Biography on Jean Piaget

Jean Piaget (1896-1980) was born in Neuchâtel, in Switzerland. After his graduation in biology his interest turned to psychoanalysis and he moved to Paris where he became a teacher in Alfred Binet's (1857-1911) school for boys. Here he helped to mark intelligence and during that time he noticed that young children consistently gave wrong answers to certain questions. Piaget did not focus so much on the fact of the children's answers being wrong, but that young children consistently made types of mistakes that older children and adults did not. He established the theory that young children's cognitive processes are inherently different from those of adults and finally he proposed a global theory of cognitive developmental stages in which individuals exhibit certain common patterns of cognition in each period of development [21, 22]. In 1921, Piaget returned to Switzerland as director of the *Rousseau Institute* in Geneva. In 1923, Piaget married his co-worker Valentine Châtenay. The couple had three children, whom the parents studied from infancy. Piaget was professor of psychology, sociology, and philosophy of science at the University of Neuchatel (1925-1929). In 1929 he became Director of the *International Bureau of Education* (IBE) and he remained the head of this international organization until 1968. Piaget died in 1980.

4.2   Structuralism

How is the making of ideas, concepts, and structures (or their "preforms") in the development of human beings' minds, i.e. in the minds of children? An answer to this question was given by Piaget with his "Genetic epistemology" that he also named "mental embryology" or "embryology of intelligence". In this approach the concept of structures is most significant. In his early years Piaget noticed a tendency to consider structures in various academic disciplines, as in psychology (Gestalt psychology, e.g. Kurt Koffka, Max Wertheimer, and Wolfgang Köhler), in sociology (Karl Marx) and particularly in linguistics (Roman O. Jakobson and Ferdinand de Saussure), in physics (e.g. field theory of James C. Maxwel)l and in mathematics (Nicolas Bourbaki). Last mentioned "Nicolas Bourbaki" was a pseudonym under which a group of French mathematicians wrote a series of books. The group members' view of modern advanced mathematics was to found all of mathematics on set theory. With this goal the group strove for rigour and generality in modern mathematics [23].

This programme is the root of an approach in the 1950s to determine the mathematical structure of a theory in a precise way by use of informal set theory without recourse to formal languages, created by the American mathematician and philosopher Patrick Suppes (born 1922) [24-26]. In the 1970s, Suppes' Ph D.-student and physicist Joseph D. Sneed (born 1938) developed informal semantics meant to include not only mathematical aspects, but also application subjects of scientific theories in the framework, based on this method. In *The Logical Structure of Mathematical Physics* [27] he presented the view that all empirical claims of physical theories have the form "x is an S", where "is an S" is a set-theoretical predicate (e.g., "x is a classical particle mechanics").

In the last third of the 20th century this structuralist approach has been elaborated by Sneed, W. Stegmüller, C. U. Moulines, W. Balzer and others. Their developments act as a bridge between philosophy and history of science to describe logical structures and dynamics of scientific theories. These authors have published their results as *An Architectonic for Science* as [28].

To adapt this structuralist metatheory to fuzzy set theory the physician and philosopher Kazem Sadegh-Zadeh requires in his *Handbook of Analytical Philosophy of Medicine* "to render the metatheory applicable to real world scientific theories, it needs to be fuzzified because like everything else in science, scientific theories are vague entities and implicitly or explicitly fuzzy. To explicitly fuzzify scientific theories he claims the "Introduction of the theory's set-theoretical predicate as a fuzzy predicate ("x is a fuzzy S" instead of "x is an $\underline{S}$")" and may bc "also any other component of the theory appearing in the structure that defines the predicate may be fuzzified." He concludes with the following outlook: "Fuzzifications of both types will impact the application and applicability of theories as well as the nature of the knowledge produced by using them" [29, p. 439f]. For further analyses and assessments we refer to [30, 31]

The early structuralist approaches encouraged Piaget to create his new epistemological approach with a genetic view and with the concept of abstract structures as central entities. In his view humans have adaptive mental structures. These mental structures assimilate external events; humans convert these external events to fit their mental structures. On the simplest level of these mental structures Piaget introduced so-called "schemata" as categories of knowledge to describe mental or physical actions.

4.3   Genetic Epistemology

As a biologist by training Piaget was skilled to observe that organism adapted to their environment. He then applied this model to cognitive development and in his theory the mind organizes internalized regularities or operations into dynamic cognitive structures. Piaget named these structures "schema". Schemata are structured clusters of concepts that can be used to represent objects, scenarios or sequences of events or relations between concepts. As we mentioned already, the original idea was proposed by philosopher Kant as innate structures used to help us perceive the world. Piaget's concept of a schema covers category of knowledge and a process to receive that knowledge. When we obtain new knowledge then our schema may be modified or changed. In his view, the cognitive development (the adaption to the environment) is a process of four aspects: schema, assimilation, accommodation, and equilibrium.

- *Schema*: Humans develop "cognitive structures" or "mental categories" that Piaget named schemata in order to name and organize, and to make sense of life and reality. [32, p. 10]
- *Assimilation*: An individual uses its existing schemata to make sense of a new event. This process involves trying to understand something new by fitting it into what we already know.
- *Accommodation*: Also existing schemata can change to respond to a new situation. If new information cannot be made to fit into existing schemata, a new, more appropriate structure must be developed. There are also instances when an individual encounters new information that is too unfamiliar that neither assimilation nor accommodation will occur because the individual may choose to ignore it. Equilibration: This is the complex act of searching for the balance in organizing, assimilating, and accommodating.

On the other hand, it is the state of Disequilibrium that motivates us to search for a solution through assimilation or accommodation.

In *Psychogenesis and the History of Science* – a coauthored and posthumous published book of Piaget and physicist Rolando Garcia, Piaget referred to an evolutionary view of theory change in science as MacIsaac [33] quoted: "Our notion of epistemic framework ...", which describes "...an explanatory schema for the interpretation of the evolution of knowledge, both at the level of the individual and that of social evolution". Piaget further recognizes the role of social construction in schemata, stating "... when language becomes the dominating means of communication ... what we might call direct experience of objects becomes subordinated ... to the system of interpretations attributed to it by the social environment." He claimed that there is a connection of the psychogenesis of logico-mathematical thinking and historical development, he distinguished between "prescientific" (other authors name it "immature") and "scientific" periods in each field, he mentions "analogies" between ontogenesis and historical development in both periods [34, p. 63].

Quoting Wadsworth, "...the child's active assimilation of objects and events results in the development of structures (schemata) that reflect the child's concepts of the world or reality. As the child develops these structures, reality or his knowledge of the world changes." [35]

4.4   Unsharp Concepts in Piaget's Genetic Epistemology

Piaget defined four stages of cognitive development:

- *Sensorimotor stage: Birth through ages 18-24 months.* – "In this stage, infants construct an understanding of the world by coordinating experiences (such as seeing and hearing) with physical, motoric actions. Infants gain knowledge of the world from the physical actions they perform on it. An infant progresses from reflexive, instinctual action at birth to the beginning of symbolic thought toward the end of the stage."[36]
- *Preoperational stage: 18-24 months – 7 years.* – The children do not yet understand concrete logic and cannot mentally manipulate information. They increase in playing but this is mainly categorized by symbolic play and manipulating symbols. They still have trouble seeing things from different points of view. The children lack basic logic. An example of transitive inference: when a child is presented with the information " *a* is greater than *b* and *b* is greater than *c*." This child may have difficulty here understanding that *a* is also greater than *c*. [37, ch. 3] [38, ch. 8]
- *Concrete operational stage: Ages 7 – 12 years.* – In this stage children use logic appropriate. They start solving problems in a more logical fashion but abstract, hypothetical thinking has not yet developed. The children can only solve problems that apply to concrete events or objects. They can draw inferences from observations in order to make a generalization (inductive reasoning) but they struggle with deductive reasoning, which involves using a generalized principle in order to try to predict the outcome of an event. The Children cannot figure out logic, e.g., a child will understand $A>B$ and $B>C$, however when asked is $A>C$, the child might not be able to logically figure the question out in their heads."
- *Formal operational stage: Adolescence – adulthood.* – Now, the person has the ability to distinguish between their own thoughts and the thoughts of others. He or she recognizes that their thoughts and perceptions may be different from those around them. The children are now able to classify objects by their number, mass, and weight, they can think logically about objects and events and they have the ability to fluently perform mathematical problems in both addition and subtraction.

However, it was quite plain to Piaget that not all children may pass through the stages at the exactly the same age. Also he acknowledged that children could show at a given time the characteristics of more than one of the stages above. Nevertheless he emphasized that cognitive development always follows the sequence of the enumeration above. No stage can be skipped, and every stage prepares the child with new intellectual abilities and a more complex understanding of the world. Therefore, Piaget's four stages of cognitive development are unsharp defined stages. Also the analysis of Piaget's model by MacIsaac' leads to the assumption that this model comprises non-crisp concepts, e.g. Piagetian assimilation "is a much more complex and fluid activity. There is a great deal of flexibility displayed within assimilation [...]." [28]

5   Outlook on Fuzzy Concepts in Genetic Epistemology

In our view the concepts that Piaget named "schema" are fuzzy concepts, Piaget's processes of assimilation and accommodation are fuzzy relations, and the stages in Piaget's constitution of the series of the cognitive development of children have no sharp borders but are fuzzy stages. A fuzzy structuralist approach to model Piaget's theory, i.e. a "Fuzzy Genetic Epistemology" could be established by using fuzzy sets and fuzzy relations instead of usual sets and relations. With fuzzy structures we will create fuzzy models for unsharp concepts ("fuzzy schemata") that children use to represent objects, scenarios or sequences of or relations between events. Fuzzy relations will model changes of fuzzy schemata like Piaget's "assimilation" and "accommodation".

Acknowledgements. Work leading to this paper was partially supported by the *Foundation for the Advancement of Soft Computing* Mieres, Asturias (Spain).

References

1. Margolis, E., Lawrence, S.: Concept. Stanford Encyclopedia of Philosophy. Metaphysics Research Lab at Stanford University (retrieved November 6, 2012)
2. Belohlavek, R., Klir, G.J. (eds.): Concepts and Fuzzy Logic. The MIT Press, Cambridge (2011)
3. Machery, E.: Concepts are not a Natural Kind. Phil. of Science 72, 444–467 (2005)
4. Steup, M.: Epistemology. The Stanford Encyclopedia of Philosophy (Winter 2013 Edition), http://plato.stanford.edu/archives/win2013/entries/epistemology
5. Kant, I.: Critic of Pure Reason, ch. III. The Architectonic of Pure Reason, The Cambridge Edition of the Works of Immanuel Kant. Cambridge University Press (1998), http://www.philosophy-index.com/kant/critique_pure_reason/ii_iii.php
6. Ellington, J.W.: The Unity of Kant's Thought in his Philosophy of Corporeal Nature. Philosophy of Material Nature. Hackett Publishing Company, Indianapolis (1985)
7. Freud, S.: Three Essays on the Theory of Sexuality, VII, 1905, 2nd edn. Hogarth Press (1955)
8. Rheinberger, H.-J.: Experimental complexity in biology: Some Epistemological and historical remarks. Philosophy of Science, Proc. of the 1996 Biennial Meeting of the Philosophy of Science Association 64(suppl. 4), 245–254 (1997)
9. Rheinberger, H.-J.: Gene Concepts. Fragments from the Perspective of Molecular Biology. In: Beurton, P., Falk, R., Rheinberger, H.-J. (eds.) The Concept of the Gene in Development and Evolution, pp. 219–239. Cambridge University Press, Cambridge (2000)
10. Zadeh, L.A.: Autobiographical Note, undated type-written manuscript (after 1978)
11. Seising, R.: Interview with L. A. Zadeh, UC Berkeley, Soda Hall, see [14] (July 26, 2000)
12. Zadeh, L.A.: My Life and Work – A Retrospective View. Applied and Computational Mathematics 10(1), 4–9 (2011)
13. Zadeh, L.A.: Fuzzy Sets. Information and Control 8, 338–353 (1965)
14. Seising, R.: The Fuzzification of Systems. The Genesis of Fuzzy Set Theory and Its Initial Applications – Developments up to the 1970s. STUDFUZZ, vol. 216. Springer, Heidelberg (2007)

15. Zadeh, L.A.: Fuzzy Languages and their Relation to Human and Machine Intelligence. In: Marois, M. (ed.) Man and Computer. Proc. of the First International Conference on Man and Computer, Bordeaux, June 22-26, 1970, pp. 130–165. S. Karger, Basel (1972)
16. Goguen, J.A.: Categories of Fuzzy Sets: Applications of a Non-Cantorian Set Theory. Ph.D. Thesis. University of California at Berkeley (June 1968)
17. Goguen, J.A.: The Logic of Inexact Concepts. Synthese 19, 325–373 (1969)
18. Rosch, E.: Natural Categories. Cognitive Psychology 4, 328–350 (1973)
19. Lakoff, G.: Hedges: A study in meaning criteria and the logic of fuzzy concepts. Journal of Philosophical Logic 2, 458–508 (1972)
20. Lakoff, G.: R. S. Interview, UC Berkeley, Dwinell Hall (August 6, 2002)
21. Piaget, J.: The Child's Conception of Causality, transl. by M. Gabain, London (1930)
22. Piaget, J.: Les notions de mouvement et de vitesse chez l'enfant, Paris (1946)
23. For more information to "Nicolas Bourbaki", http://www.bourbaki.ens.fr/
24. Suppes, P.: A set of independent axioms for extensive quantities (1951), http://suppescorpus.stanford.edu/browse.html?c=mpm&d=1950
25. Suppes, P.: Some remarks on problems and methods in the philosophy of science (1954), http://suppescorpus.stanford.edu/browse.html?c=mpm&d=1950
26. Suppes, P.: Introduction to Logic. Van Nostrand, New York (1957)
27. Sneed, J.D.: The Logical Structure of Mathematical Physics. Reidel, Dordrecht (1971)
28. Balzer, W., Moulines, C.U., Sneed, J.D.: An Architectonic for Science. The Structuralist Program. Reidel, Dordrecht (1987)
29. Sadegh-Zadeh, K.: Handbook of Analytical Philosophy of Medicine, Dordrecht (2012)
30. Seising, R.: Fuzzy Sets and Systems and Philosophy of Science. In: Seising, R. (ed.) Views on Fuzzy Sets and Systems. STUDFUZZ, vol. 243, pp. 1–35. Springer, Heidelberg (2009)
31. Seising, R.: A "Goodbye to the Aristotelian Weltanschauung" and a Handbook of Analytical Philosophy of Medicine. In: Seising, R., Tabacchi, M. (eds.) Fuzziness and Medicine. STUDFUZZ, vol. 302, pp. 19–76. Springer, Heidelberg (2013)
32. Wadsworth, B.: Piaget's Theory of Cognitive Development: An Introduction for Students of Psychology and Education. David McKay & Comp., NY (1971)
33. MacIsaac, D.: The Pedagogical Implications of Parallels between Kuhn's Philosophy of Science and Piagets' Model of Cognitive Development, http://physicsed.buffalostate.edu/danowner/kuhnpiaget/KP1.html
34. Piaget, J., Garcia, R.: Psychogenesis and the history of science (H. Fieder, Trans.). Columbia University Press, New York (1988) (Original published in 1983)
35. Wadsworth, B.: Piaget for the classroom teacher. Longman, New York (1978)
36. Santrock, W.: A Topical Approach To Life-Span Development, NY, pp. 211–216 (2008)
37. Loftus, G., et al.: Introduction to Psychology, 15th edn. Cengage, London (2009)
38. Santrock, J.W.: Life-Span Development, 9th edn. McGraw-Hill College, Boston (2004)

Paired Structures in Logical and Semiotic Models of Natural Language

J. Tinguaro Rodríguez[1], Camilo Franco De Los Ríos[2], Javier Montero[1], and Jie Lu[3]

[1] Faculty of Mathematics, Complutense University of Madrid, Madrid, Spain
{jtrodrig,javier_montero}@mat.ucm.es
[2] Department of Food and Resource Economics, Faculty of Science,
University of Copenhagen, Frederiksberg, Denmark
cf@ifro.ku.dk
[3] Faculty of Engineering and Information Technology,
University of Technology, Sydney, Australia
jie.lu@uts.edu.au

Abstract. The evidence coming from cognitive psychology and linguistics shows that pairs of reference concepts (as e.g. *good/bad*, *tall/short*, *nice/ugly*, etc.) play a crucial role in the way we everyday use and understand natural languages in order to analyze reality and make decisions. Different situations and problems require different pairs of landmark concepts, since they provide the referential semantics in which the available information is understood accordingly to our goals in each context. In this way, a semantic valuation structure or system *emerges* from a pair of reference concepts and the way they oppose each other. Such structures allow representing the logic of new concepts according to the semantics of the references. We will refer to these semantic valuation structures as *paired structures*. Our point is that the semantic features of a paired structure could essentially depend on the semantic relationships holding between the pair of reference concepts from which the valuation structure emerges. Different relationships may enable the representation of different types of *neutrality*, understood here as an epistemic hesitation regarding the references. However, the standard approach to natural languages through logical models usually assumes that reference concepts are just each other complement. In this paper, we informally discuss more deeply about these issues, claiming in a positional manner that an adequate logical study and representation of the features and complexity of natural languages requires to consider more general semantic relationships between references.

Keywords: knowledge representation, natural languages, logic, fuzzy sets.

1   Introduction

In many situations of our everyday life, whenever we analyze the available information in order to understand a certain reality and make decisions in view of our knowledge, we elect and use pairs of references to assess such information and the available options in adequate terms. In this way, for instance, the information we collect may be

A. Laurent et al. (Eds.): IPMU 2014, Part II, CCIS 443, pp. 566–575, 2014.

relevant or *irrelevant* in terms of our knowledge; the possible decisions we can make are usually regarded in terms of their *appropriateness* or *inappropriateness* in order to achieve our goals; and the possible scenarios that can arise as a consequence of our decisions may be judged in terms of whether they are *positive* or *negative* for our interests. Or one can think about whether some previous actions have been *good* or *bad* in order to act in a better way next time. Or whether a person is being *honest* or *fallacious* in its words, or whether a situation is *fair* or *unjust* for a person or a community. Or whether some music, movies or news makes you feel *happy* or *sad*. And many different things, or even the same in different contexts, can be judged in terms of different pairs as *cold/warm, night/day, dry/raining, fast/slow, black/white, cheap/expensive, big/small, sure/impossible victory/defeat, success/failure, entertaining/boring, legal/illegal, safe/dangerous, quiet/loud, tall/short, rich/poor, nice/ugly, coward/temerarious, smart/dumb, weak/strong, young/old, love/hate, life/death* and so on.

Thus, different problems or situations, related to different contexts, objects of study and pieces of information, require different pairs of references providing an adequate contextual meaning (or *semantics*) to objects and information. That is, the concepts in a pair provide the semantic references in terms of which objects and information are assessed regarding a specific problem or context. In this way, for instance we can regard a *day* or a particular *moment of the day* as e.g. *definitely cold*, or *slightly warm*, or *not-warm* at all. Similarly, regarding its *speed*, a *car* could be e.g. *very fast*, or *quite slow*, while simultaneously, in relation to its *price*, it could be e.g. *neither cheap nor expensive*, or *both not-cheap and not too expensive*. And though we can judge a person in terms of *tall/short*, perhaps a specific problem, e.g. choosing a peer to climb a dangerous mountain, would rather require judging persons in terms of *coward/temerarious*, or *weak/strong*. These instances show how a pair of concepts provides a referential context for assessing objects and information accordingly to the semantics of the concepts in the pair. Moreover, an object or piece of information can be simultaneously assessed through different pairs, each referring to a different characteristic or *criterion*, whenever an object possesses different features to be analyzed. In this sense, a specific problem or situation could require to elect a particular pair (or set of pairs) of references among all those pairs that may be applied to a certain object, since some of these references may define the adequate semantic context for a given problem, while some others may not provide any relevant information for the current analysis.

It is important to note that the two concepts making up any of the mentioned pairs are related in a very general sense: both refer to the same underlying characteristic or criterion, but at the same time they represent *opposite* references, acting on the same information while simultaneously defining somehow *extreme* and *antagonistic* semantic landmarks for the evaluation of such criterion. That is, it is easy to devise a situation in which objects are assessed through e.g. the pairs *young/old* and *smart/not-smart* (for instance, the evaluation of candidates for a post-graduate research grant), but it is rather weird (if not impossible) to analyze objects in terms of e.g. the pairs *young/day* and/or *smart/raining*. Somehow, these last pairs cannot serve for the purpose of reference since they are made up of concepts that hardly refer to the same characteristic (though some poets can made surprising associations). Similarly, note that e.g. the pairs *old/very old* or *smart/very smart* does not provide adequate references for assessing *age* or *intelligence* in a general way, and could only provide meaningful landmarks once they are considered as extreme, somehow opposite references on a constrained subset of the range of these characteristics.

Another important remark concerns the linguistic character of the references as well as of the assessments to be made in terms of the former. That is, these pairs are all made of terms or concepts of a natural language, providing a referential semantics in terms of which further linguistic evaluations (i.e. further terms or concepts) can be obtained. In other words, it is the semantics of the references and the relationships between them and the new concepts what provide the semantics (i.e. the meaning) of the new concepts. Thus, pairs of reference concepts provide the semantics for a linguistic evaluation of reality [13]. Furthermore, note that even when we explicitly take only a single concept as reference, e.g. let us say *tall*, allowing us to express *degrees* of verification as *certainly tall, quite tall* or *little tall*, we are implicitly (even perhaps unconsciously) taking the lack of verification of such reference, i.e. *not-tall*, as a second reference.

2 Logical Models of Language

On the other hand, different formal tools exist for the representation and the treatment of natural language and the semantics of its concepts (see for instance [14], [21], [30]). Let us focus on logical models of language. Logic, through set theory, provide a classical representation of the semantics of a concept (or predicate) P in terms of the collection or set of objects x of a universe of discourse X (i.e. the whole collection of objects under consideration) for which the assertion "the object x is (or verifies) P", denoted as $P(x)$, is *true*. In formal terms, using (classical) logic we can represent (by extension) the semantics of P as the set

$$P = \{x \in X \mid P(x)\}. \tag{1}$$

This representation is equivalent to that provided (by assignation) through an indicator function $1_P : X \rightarrow \{0,1\}$ such that for any object $x \in X$, it is $1_P(x) = 1$ if and only if $P(x)$ holds, i.e. is *true*. And note that in this classical model the semantics of the predicate *not-P* can be obtained from that of P, since it is $1_{not-P}(x) = 1$ if and only if $1_P(x) = 0$ if and only if $P(x)$ does not hold, i.e. is *false*.

In this sense, classical logic can be regarded as a formal discourse built upon the semantics of the pair *true/false*, in fact a discourse based upon a particular semantics of this pair of notions, in which both of them are regarded as the only possible evaluations, being these each other complements. And note that through this discourse we can model and describe the semantics of other concepts, in such a way that particularly we can identify *precise*, crisp concepts to those concepts having its semantics defined through classical logic (e.g. all the usual mathematical notions).

In this way, fuzzy logic and fuzzy set theory [30], as a generalization of classical logic and set theory, provides a discourse based upon the same pair of references *true/false*, but now allowing these notions to be a matter of degree, i.e. notions with an *imprecise* semantics, and therefore allowing more than just two opposite, extreme evaluations 0 and 1 in order to estimate the veracity of a predicate P. Let us recall that, in the context of fuzzy logic, the indicator function 1_P is replaced by a *member-ship* function $\mu_P : X \rightarrow [0,1]$, in such a way that $\mu_P(x) \in [0,1]$ represents the degree (or the intensity) up to which the object x verifies P, i.e. up to which $P(x)$ is *true*. Note also that fuzzy logic still considers *true* and *false* as complementary notions, since any degree of verification or *truth* $\mu_P(x)$ is always associated with an inverse degree of lack of verification or *falsehood* $\mu_{not-P}(x) = n(\mu_P(x))$, obtained from the former through a *negation* $n : [0,1] \rightarrow [0,1]$, usually defined by $n(v) = 1 - v$ for any $v \in [0,1]$. Thus, in consonance with its status of generalization of classical logic, note again that fuzzy logic provides a formal tool through which we can model and de-scribe the semantics of *imprecise* predicates, as many of the concepts we use in our everyday life, i.e. the mathematically ill-defined concepts of a natural language.

   Then, our first observation is that, though the modeling of the semantics of con-cepts of a natural language through classical and fuzzy logic is a quite extended prac-tice (see for instance [31]), much less attention has been paid through these tools to the correlative semantics that a *pair* of concepts should maintain in order to be re-garded as appropriate references for linguistic evaluation (see [24]). And even less attention (if actually any) has received the possibility that the features of the semantic evaluation *structure* enabled by a pair of references (i.e. *emerging*, see [3,4], from these references) could depend in a drastic manner on the semantic relationship hold-ing between the references (and see [15]). That is, our point is that different semantic relationships between the concepts making up a pair of references could determine semantically-different structures for the evaluation and the interpretation of reality in terms of a natural language.

   To some extent, this previous observation should not be surprising, since classical logic has somehow introduced the bias of thinking in terms of complementary no-tions, in order to analyze the world *logically* (see [6], [1], [28,29]). This bias has also profound philosophical roots in the Western cultural tradition, historically attracted by *dualism*, in the form of pairs of notions regarded as each other complements, as e.g. *being/not-being, material/immaterial, body/mind* (or *body/soul*), *God/Devil*, etc.

   Particularly, fuzzy representations suffer from this bias, since, as discussed above, we can only consider degrees of verification of a concept P when we use the implicit complementary reference *not-P*. In this sense, as also pointed above, a concept P and its complement *not-P* maintain a very specific semantic relationship in terms of being *opposite* and *extreme* notions, allowing us to construct a pair of references from which semantically analyze reality in terms of just the semantics of a single concept P. And observe that, emerging from such complementarity relationship between a concept P and its negation or complement *not-P*, enabling to form the reference pair *P/not-P*, we in fact obtain a semantic structure for evaluating a certain reality (e.g. the *height* of a person): particularly, the semantic structure of precise, binary notions allowed by classical logic (it is either $1 \equiv tall$ or $0 \equiv not\text{-}tall$), or, in a more general framework, the semantic structure of imprecise, gradable concepts allowed by grada-tion procedures and particularly fuzzy logic (it could then be e.g. $1 \equiv certainly\ tall$,

0.75 ≡ *quite tall*, 0.5 ≡ *half tall*, 0.25 ≡ *not so tall* and 0 ≡ *not tall at all*). In fact, we can regard fuzzy logic, in its usual formulation, as the study of the class of gradable and imprecise semantics representable through the assumption of implicit complementary references.

3   Paired Concepts and Structures

Therefore, the semantic evaluation structure arising from a pair of references has been widely studied by means of logical tools, but mainly under the assumption that paired references are each other complements, due to the binary bias long time ago attached to our scientific conception of reality. For this reason, little attention has been paid to the possibility (indeed the key idea of this paper) that different semantic relationships between the two concepts in a reference pair could in fact define essentially different semantic structures for a linguistic evaluation of reality.

Let us stress that two natural language concepts P, Q making up a reference pair P/Q could be far away from being regarded as each other complements, i.e. we could understand them in such a way that $P \neq not\text{-}Q$ and $Q \neq not\text{-}P$. For instance, a person not being *tall* could be also *not-short*, and a *not-fast* car is not necessarily *slow*, in the same manner as it could also be simultaneously *not-cheap* and *not-expensive*. Or, e.g. a *not-happy* song could be far away from being *sad* (people can even sing *sad* songs and feel *happy* for that, or reciprocally). And a day could be forecasted as having both *cold* and *warm* moments, making us hesitate about the adequate clothes to wear. Also, a certain thing could be *not-good* and at the same time *not-bad*, and even simultaneously *good* and *bad*: in fact, the evidence coming from neuroscience [5] points out that our brain has different channels to process *negative* emotions as pain, sadness or loneliness than to process *positive* feelings as happiness, joy or euphoria. And precisely for this reason, because they are evaluated separately, *positive* and *negative* emotions are not each other complement, but rather they harvest independent families of arguments, and thus both can appear together (conforming different mixed, conflictive or contradictory emotions as thrill or tornness) or not appear at all (producing neutral, flat emotions as indifference or serenity). In fact, every day we find that many things can have simultaneous *good* and *bad* sides, and in certain regions of the world this is indeed a form of ancestral wisdom (e.g. Taoism and its *yin/yang* principle).

These ideas suggest that the semantic structure arising from a pair of non-complementary references may in fact enable the emergence of new linguistic concepts as categories for valuation, exhibiting a somehow neutral or mixed character (or semantics) with respect to the landmark concepts, and which cannot be properly regarded as degrees of verification of a single reference concept correlative to its complement. For instance, when somebody is neither *coward* nor *temerarious* we could rely on or define an intermediate category as *courageous*, with a proper semantics of its own, to refer to such neutral assessment in terms of the considered reference pair (and note that such neutrality could represent the best alternative for a given problem, as in the mountain-climbing example). Similarly, the hesitation we feel when we find something being simultaneously *good* and *bad* (e.g. the pleasure of smoking even knowing it makes you ill), or both *desirable* and *unattainable* (e.g. a *nice, fast* car too *expensive* to afford it) is hardly interpretable as a degree expressed in terms of a reference and its complement (i.e. in terms of the semantics of a single concept). Rather, it provides a differentiated semantics, better fitting to the idea of a new concept available for valuation, emerging as a consequence of considering non-complementary references.

Thus, the second key idea we propose is that different *types of neutrality* can arise in a semantic structure as a consequence of the different relationships holding between the pair of references. That is, non-complementary references could define new concepts or categories with a differentiated semantics with respect to the references, not showing the particular semantics of degrees either. Particularly, these neutral categories could not allow an ordering, or not refer to the same notion of order as degrees (and see [19] for an instance in paraconsistent logics). In other words, there could be many ways of being *in between* a pair of references (see [15]).

Let us insist that apparently complementary notions could be in fact regarded as non-complementary and define new categories (rather than degrees) for valuation. For instance, the references *true* and *false*, even in a rigorous mathematical setting, can be understood as allowing a neutral category, let us call it *undecidability*, applicable to those statements, as e.g. the twin prime or Hardy–Littlewood conjectures, for which either a proof or a counterexample has been not yet found. In this sense, if mathematical *truth* is identified with *provability* from a set of axioms, the Second Gödel's Theorem [10] could be interpreted as stating that for certain propositions P (e.g. the logical consistency of arithmetic), both P and *not-P* are (and will forever be) *undecidable* once we choose a set of axioms to fundament our logical system.

However, let us also note that choosing two antonym words (e.g. *tall/short*) as references does not necessarily entail considering them as non-complementary concepts. For instance, in a binary setting we could understand that $1 \equiv tall \equiv not\text{-}short$ and $0 \equiv short \equiv not\text{-}tall$, and even allow a linear or fuzzy gradation between these references, which are in this way regarded as each other complements. And conversely, in practice we could use the terms P and *not-P* to refer to non-complementary references, e.g. when there is not an antonym word for P [23]. For instance, there is not an antonym word for *yellow*, so we could be forced to use the term *not-yellow* to refer to an opposite idea or concept to *yellow*, even when we understand that a color not being *yellow*, as e.g. *white*, *light green* or *beige*, could not fulfill such (artificially defined) opposite notion of *not-yellow*, instead verified by other colors as e.g. *blue*, *black* or *violet*. Thus, it is the semantics of the reference concepts, and not the labels or words that represent them, what determines the semantic features of the linguistic valuation structure arising from such a pair of references. And, in this sense, perhaps a more general notion of opposition than antonymy, let us call it *duality* or *antagonism*, should be devised to represent in a general way the semantic relationships holding between pairs of references. Therefore, in principle there could be duality relationships different from antonymy and complementation, and again, our basic observation is that different duality relationships between reference concepts could determine different semantic structures for valuation.

A final remark concerns the notions of *decomposable* and *primitive* references. Many (if not all) concepts can be explained in terms of other concepts, as we usually do to explain the meaning of a word of our mother language to a child or a non-native speaker, or as it is done in a dictionary. In this sense, many reference pairs could be *decomposable*, in the sense of being high-level references explainable in terms of, or obtainable from, a set of lower-level references, then in charge of defining the semantic features of the high-level valuation structure (possibly through a hierarchy of concepts, see [25]) But, as discussed for instance in [20], by starting from explaining a single concept in terms of other concepts and successively, recursively explaining these lower-level concepts in terms of further concepts, we would end obtaining the *whole* language in the process, in such a way that a circular reference (returning to the starting concept) is obtained sooner or later. That is, there are not *lowest-level* concepts, i.e. absolute linguistic references (but note that some basic physiological emotions could in fact provide absolute *pre-linguistic* semantic references, as suggested in [9], [17]).

To some extent, at this respect language is similar to the Universe: it constitutes the whole system in which we (our minds) move, in which no absolute references can be laid down as it has not a *center* or starting point. For this reason, in any problem or situation we face, at some point we have to regard some concepts as *primitive* references, i.e. not admitting decomposition in terms of further references and thus providing the basic, primitive semantics for building the meaning of other concepts. As mentioned before, that is in fact the essence of reference pairs, and the reason why we organize our thinking in such terms. Moreover, this is also related to the way we learn a language when we are children [9], [17]: from a small set of linguistic references (e.g. *mummy/daddy*, or even *mummy/not-mummy*) learned through both mimicry and the basic, non-verbal semantics of physiological emotions, we are able to successively add further levels of concepts into our vocabulary by relating the semantics of the new concepts to previous semantic references, in a dynamical, always-changing (learning) process. And to some extent, this is also the reason why mathematics and logic have to be built from primitive axioms, the truth of which is supposed in order to develop further theories (and see [22] for a discussion on how axioms can be related to the empirical knowledge the human kind has obtained throughout its cultural, linguistic evolution).

Thus, in a few words, this work proposes, in a positional manner, to join together the three following ideas:

1. Pairs of references, and the semantic structures arising from them, constitute a key aspect of the manner we use, organize and understand a natural language in order to evaluate reality and make decisions.
2. The semantics of concepts, and even the semantic relationships between reference concepts, can be modeled through logical reasoning tools.
3. The semantic features of the linguistic evaluation structure arising from a pair of references (particularly the types of neutrality being allowed) could depend on the semantic relationships holding between these references.

Then, departing from these ideas, our objective will be to study the semantic structures arising from pairs of reference concepts by means of logical and semantic tools, and particularly focusing on how the expressive (or representative) power of these structures could in turn depend on how the references are semantically related. For simplicity, let us refer to these semantic structures as *paired structures*. Particularly, let us remark that our position/proposal is not about formal logic or its interpretation, but rather it deals with knowledge and natural language representation by means of logical tools, under the assumption that the semantic relationships between references may determine the expressive power of our representations.

Therefore, paired structures could constitute a relevant step towards a more general and useful model of language, allowing for several kinds of neutrality and linguistic uncertainty to be addressed and represented. In the long term, our vision aims a more general and ambitious objective: enabling us to speak in logical terms (but perhaps through a discourse not developed in terms of complementary references) of those things about which the *first* Wittgenstein [26] thought we should stay in silence, but about which the *second* Wittgenstein [27] referred to as being "the really important things" [7]: our everyday feelings, emotions and intentions.

4   Final Remarks

As a preliminary and positional work, this paper has focused on presenting our argument and its potential relevance, rather than on providing a strictly formal approach addressing all possible technical difficulties. A more formal paper will follow, focused on addressing these aspects in more rigorous terms, although some first steps have been already presented in [18].

Let us anticipate that in a general sense, besides containing the references themselves, paired structures have to be regarded as valuation structures providing a finite (possibly void) set of interrelated categories (with a differentiated semantics) as primary values, together with appropriate scales in which a secondary value, representing the correlative verification status of each primary category, is measured. In this sense, as each neutral category allowed in a paired structure can be associated with a different type of neutrality –in turn associated to a different kind of epistemic hesitation provoking difficulties to understand information, decide or act–, paired structures may provide a general framework in which several features of our everyday reasoning process (as e.g. different types of uncertainty and hesitation) can be represented through logical terms.

Then, in practice, paired structures may be used and applied in two complementary ways: either a) the allowed categories emerging from the considered references and their verification degrees are estimated directly from the objects; or b) the relevant categories and their verification are obtained through aggregation from more primitive information, as the verification degrees of the references or even from those of the concepts in lower-level structures. In either way, however, some semantic constraints should be imposed (on either the estimations or the aggregation operators) in order to relate the different categories and their verification accordingly to the semantics they are intended to exhibit. Thus, different types of scales or valuation spaces (from univariate lattices to complex multidimensional polyhedrons) may be obtained depending on the semantic assumptions of the model. These two complementary methodologies should then provide a manner of empirically testing whether these semantic assumptions are appropriate or not in a given context and situation.

Another issue that will be addressed in forthcoming works is the relationship between paired structures and the existing models for knowledge and language representation. We conjecture that many of the existing models and approaches (as e.g. type-n fuzzy sets [31], computing with words [32], [11], bipolarity [8], *intuitionistic* fuzzy sets [1], probability [12], etc.) can be understood and represented as special kinds of paired structures. In this way, paired structures would then provide a general, unifying notion allowing a deeper understanding of the relationships holding between many of the formalisms commonly used in knowledge representation.

Acknowledgements. This research has been partially supported by the Government of Spain, grant TIN2012-32482.

References

1. Atanassov, K.T.: Intuitionistic Fuzzy-Sets. Fuzzy Sets and Systems 20(1), 87–96 (1986)
2. Barnes, B.: On the conventional character of knowledge and cognition. Philosophy of the Social Sciences 11(3), 303–333 (1981)
3. Bunge, M.: Emergence and the mind. Neuroscience 2(4), 501–509 (1977)
4. Bunge, M.: Emergence and convergence: Qualitative novelty and the unity of knowledge. University of Toronto Press (2003)
5. Cacioppo, J.T., Berntson, G.G.: The affect system, architecture and operating characteristics - Current directions. Psycological Science 8, 133–137 (1999)
6. Collins, R.: The sociology of philosophies: A global theory of intellectual change. Harvard University Press (1998)
7. Doxiádis, A.K.: Logicomix. Bloomsbury Publishing (2009)
8. Dubois, D., Prade, H.: An introduction to bipolar representations of information and preference. International Journal of Intelligent Systems 23(8), 866–877 (2008)
9. Edelman, G.M.: Bright air, brilliant fire: On the matter of the mind. The Penguin Press, London (1992)
10. Gödel, K.: On Formally Undecidable Propositions Of Principia Mathematica And Related Systems. Monatshefte für Mathematik und Physik 38, 173–198 (1931) (in German)
11. Herrera, F., Martínez, L.: A 2-tuple fuzzy linguistic representation model for computing with words. IEEE Transactions on Fuzzy Systems 8(6), 746–752 (2000)
12. Kolmogorov, A.: Foundations of the Theory of Probability. Julius Springer, Berlin (1933) (in German)
13. Lindsay, P.H., Norman, D.A.: Human information processing: An introduction to psychology. Academic Press, New York (1972)
14. Manning, C.D., Schütze, H.: Foundations of statistical natural language processing, vol. 999. MIT Press, Cambridge (1999)
15. Montero, J., Gómez, D., Bustince, H.: On the relevance of some families of fuzzy sets. Fuzzy Sets and Systems 158(22), 2429–2442 (2007)
16. Osgood, C.E., Suci, G.J., Tannenbaum, P.H.: The measurement of meaning. University of Illinois Press, Urbana (1957)

17. Piaget, J.: The language and thought of the child, vol. 5. Psychology Press (1959)
18. Rodríguez, J.T., Franco, C.A., Montero, J.: On the semantics of bipolarity and fuzziness. In: Melo-Pinto, P., Couto, P., Serôdio, C., Fodor, J., De Baets, B., et al. (eds.) Eurofuse 2011. AISC, vol. 107, pp. 193–205. Springer, Heidelberg (2011)
19. Rodríguez, J.T., Turunen, E., Ruan, D., Montero, J.: Another paraconsistent algebraic semantics for Lukasiewicz–Pavelka logic. Fuzzy Sets and Systems (June 25, 2013), http://dx.doi.org/10.1016/j.fss.2013.06.011 ISSN 0165-0114
20. Rosch, E.: Linguistic relativity. In: Silverstein (ed.) Human Communication: Theoretical Perspectives (1974)
21. Schank, R.C.: Conceptual dependency: A theory of natural language understanding. Cognitive Psychology 3(4), 552–631 (1972)
22. Searle, J.R.: Mind, language and society: Philosophy in the real world, vol. 157. Basic books, New York (1999)
23. de Soto, A.R., Trillas, E.: On antonym and negate in fuzzy logic. International Journal of Intelligent Systems 14(3), 295–303 (1999)
24. Trillas, E., et al.: Computing with antonyms. In: Nikravesh, M., Kacprzyk, J., Zadeh, L.A. (eds.) Forging New Frontiers: Fuzzy Pioneers I. STUDFUZZ, vol. 217, pp. 133–153. Springer, Heidelberg (2007)
25. Willie, R.: Concept lattice and conceptual knowledge systems. Computers and Mathematics with Applications 23, 493–515 (1992)
26. Wittgenstein, L.: Tractatus logico-philosophicus. Routledge & Kegan Paul, Rotterdam (1921)
27. Wittgenstein, L.: Philosophical investigations (1953)
28. Woolgar, S.: Configuring The User: the case of usability trials. In: Law, J. (ed.) A Sociology of Monsters: Essays on Power, Technology and Domination. Routledge, London (1991)
29. Woolgar, S., Hamilton, P.: Science, the very idea. Ellis Horwood, Chichester (1988)
30. Zadeh, L.A.: Fuzzy sets. Information and control 8(3), 338–353 (1965)
31. Zadeh, L.A.: The concept of a linguistic variable and its application to approximate reasoning—I. Information sciences 8(3), 199–249 (1975)
32. Zadeh, L.A.: Fuzzy logic=computing with words. IEEE Transactions on Fuzzy Systems 4(2), 103–111 (1996)

A Fuzzy Rule-Based Haptic Perception Model for Automotive Vibrotactile Display*

Liviu-Cristian Duţu[1], Gilles Mauris[1], Philippe Bolon[1],
Stéphanie Dabic[2], and Jean-Marc Tissot[2]

[1] Univ. Savoie, LISTIC, F-74000 Annecy, France
{liviu-cristian.dutu,gilles.mauris,philippe.bolon}@univ-savoie.fr
[2] Valeo Interior Controls, Annemasse, France
{stephanie.dabic,jean-marc.tissot}@valeo.com

Abstract. Currently, tactile surfaces implemented in automobiles are passive, i.e., *feedbackless*, thus forcing the user to visually check the device. To improve drivers' interaction, surface vibrations can be used to deliver feedback to the finger when touched, and an associated perception model is required. Hence, this paper introduces a fuzzy model for the comfort degree of vibrotactile signals. System input variables are chosen from the physical characteristics of the signals, and are validated on a dissimilarity judgment task. The system achieves an error of 9% and correctly classifies 17 out of 18 signals within a reasonable interval. A graphical user interface to interact with the system is also presented.

Keywords: haptic interfaces, sensory evaluation, fuzzy system, wavelets.

1 Introduction

Mechanical interfaces installed in car cockpits are evolving and are currently being replaced by tactile surfaces. Many of the functionalities of an automobile, such as choosing a radio channel or controlling the CD player, have already migrated on tactile surfaces, and it is expected that many others will soon be migrated. This change forced car manufacturers to investigate the impact of this new technology on passengers and on the security in a driving situation.

Its important deficiency is that it relies *only* on the visual sense for feedback, forcing the driver to visually check the device after an interaction, thus distracting the attention away from the road and increasing the risk of car accident [1].

To counterbalance this side effect, feedback might be delivered directly to the tactile sense using vibrations or *vibrotactile signals*. This is consistent with the *Multiple Resources Theory* [2], stating that humans are able to process information from different senses without any significant deterioration in the cognitive performance (e.g., drive a car and listen to the radio). Thus, activating the tactile sense through vibrotactile signals as a way to deliver the feedback of touch interfaces, will allow the driver to focus his/her visual resources on the road.

---

* Developed under the FUI-MISAC project, approved by the French Government with the contract number F-11-06-048-V.

A. Laurent et al. (Eds.): IPMU 2014, Part II, CCIS 443, pp. 576–585, 2014.
© Springer International Publishing Switzerland 2014

This calls for a deep understanding of the human tactile sense and its interactions with the vibrotactile signal's physical characteristics. In order to investigate these interactions, we propose to define a perception model associating users' subjective evaluations to measured properties of the vibrotactile signals.

Section 2 describes the device and the experiments conducted to collect users subjective evaluations. Using these evaluations as ground truth, we confirmed the existence of a link between users' perception and the measured physical characteristics of the signals. Section 3 then proposes an intuitive 2-variable fuzzy model associating the characteristics of the acceleration and velocity of the vibrating tactile surface to the perceived comfort degree. The model is customizable through a graphical user interface.

2 Signal Characteristics and Perceptual Distance

2.1 Apparatus and Psychophysical Procedures

Two experiments were first conducted to collect users evaluations of vibrotactile signals. The apparatus used to produce the signals was developed by *Valeo Annemasse France*. It includes a voice coil actuator encapsulated under a resistive screen layer which the users can touch to perceive the vibrotactile signals. They are produced by the moving actuator receiving electrical stimuli with varying frequency, duration and waveform parameters, while the amplitude was constant.

To filter out the low frequency mechanical noise produced by the device while vibrating, participants wear a pair of noise-canceling headphones.

The first experiment consists in evaluating the perceived dissimilarity between 18 vibrotactile signals. Their corresponding electrical stimuli differ in terms of frequency, duration and waveform. Using the fingertip, 24 subjects evaluated the dissimilarity between each pair of signals on a 1 to 7 Likert scale, where 1 was equivalent to *the signals are very similar* and 7 to *the signals are very different*. The order within a pair and the pairs occurrence order was counterbalanced between subjects. A complete description of the experiment is available in [3]. An 18×18 perceptual dissimilarity matrix D_P was created in [3] based on the values collected for each pair of signals. Later, D_P will be used as ground truth to find the variables accounting for signals dissimilarity.

The second experiment, conducted on the same set of signals, aimed at obtaining an absolute appreciation degree (i.e., comfort) for each vibrotactile signal. Ten subjects rated the appreciation of each signal, presented in a random order, on a 1 to 7 Likert scale with 1 being the lowest appreciation degree and 7 the highest. These values are used in Section 3 to define a fuzzy model of perception.

Due to the multiple layers between the actuator and the tactile surface, electrical stimuli might be altered before they reach the tactile layer. Thus, we consider that they can not fully account for human perception, and we have focused on the acceleration and displacement data measured on the device output layer, where users *perceive* the signals. To extract a pair of variables from these data, we first need to investigate the psychophysical properties of the tactile sense and its relationship to the vibrational signals' physical characteristics.

2.2   Psychophysical Overview of the Tactile Sense

According to the *four channels theory* [4], the tactile sense is mediated by four
psychophysical channels: the Pacinian (P) channel and three Non-Pacinian channels (NP I, II, III). Each channel is activated by a duo of vibration frequency
and contact area, but with the current experimental setup, the NP II and NP
III channels are unreachable. Thus, we can assume that the tactile sensations
elicited by the signals used in section 2.1 mainly originate from the activation of
the P and NP I channels. Next, we will focus on these channels to find the variables that simulate their neural activation, based on their inherent properties.

The Pacinian channel is the most sensitive among the four and operates in
range [40 Hz, 800 Hz] [4], with a sensitivity peak around 200Hz. It is the only
channel to exhibit temporal summation capabilities [6] and according to [5] it
shows *"near perfect integration of stimulus energy"* over time, which is compatible with the neural-integration theory of temporal summation for the auditory
sense elaborated by Zwislocki in [7]. In [8], the authors found that stimuli in the
range of the P channel are discriminable mainly based on their differences in
power, and that their waveform differences are not perceived.

These suggest that the P channel is activated by a variable akin to the energy
gathered within its frequency range in the vibrating tactile surface acceleration.

The NP I channel is responsible for the perception of tactile stimuli in the
range [10 Hz, 100 Hz] [4]. It is optimally tuned at 30-50 Hz [6]. Unlike the P
channel, it does not exhibit temporal summation. Although for the NP I channel
there seems not to be a consensus in the literature on which variables better
simulate its activation, one of the early references [9] suggests that it might be
sensitive to the dynamic stimulus' velocity. Therefore, in the next sections we
will use the mean positive velocity of the tactile surface to reflect its activation.

2.3   Pacinian Channel Activation and Wavelet Analysis

To obtain the energy delivered to the P channel, we inspected the measured
acceleration signal of the surface using the time-frequency technique of the continuous wavelet transform (CWT), since both the channel *frequency* selectivity
and its *temporal* summation capacity dictate its neural activity, as shown above.

Wavelets are finite-energy functions which can be shifted across the signal and
scaled by dilation or compression. In wavelet analysis, the notion of *frequency* is
replaced by that of *scale*. The CWT coefficient of $f(t)$ using wavelet ψ at scale
σ and time-shift b is defined below, where ψ^* is the complex conjugate of ψ.

$$C_f(\sigma, b) = \int\limits_{-\infty}^{\infty} f(t) \frac{1}{\sqrt{\sigma}} \psi^* \left(\frac{t-b}{\sigma} \right) dt \ . \tag{1}$$

The matrix C_f of CWT coefficients is obtained by iterating (1) for a finite
number of scales σ_i and time shifts b_j. It can be seen as a 2D representation of
the similarity between $f(t)$ and the wavelet's shifted and scaled versions.

The appropriate way to implement the CWT is to choose a wavelet which most resembles the signal to be analyzed. In this paper we have chosen the real-valued *Morlet* wavelet $(\psi(t) = e^{\frac{-t^2}{2}} cos(5t))$, but exploratory studies realized in [10] show that relatively similar results can be obtained with different wavelets.

As wavelets are practically band-limited, $|C_f(\sigma, b)|^2$ is the energy of $f(t)$ at time b and in a frequency band around $1/\sigma$. Generalizing, we obtain the signal energy distribution in the time-scale plane, called *scalogram* [11]: $E_s(\sigma, b) \equiv |C_f(\sigma, b)|^2$, $\forall \sigma, \forall b$. Fig. 1 illustrates the scalograms, using the Morlet wavelet, of two of the most perceptually dissimilar acceleration signals. We can clearly see the differences between their energy patterns in the time-scale plane.

To switch from the intrinsic time-scale plane of the wavelet transform to a time-frequency plane, the scale-frequency relationship is $F_\sigma = F_c/\sigma\delta$, where F_c is the central frequency of the wavelet (e.g., $F_c = 0.81$ Hz for the Morlet wavelet), F_σ is the equivalent frequency of scale σ, and δ is the sampling period.

Linear interpolation of $|C_f|^2$ coefficients was employed to get the scalograms to a common duration N_A. Energy is preserved by multiplying the interpolated coefficients $|C_f^I|^2$ with the ratio between the original duration and N_A.

Relying on section 2.2, we set the threshold between the P and NP I channel at 70 Hz, i.e., scale $\sigma = 59$ for the Morlet wavelet and a sampling frequency of 5 KHz. The P channel upper limit of 800 Hz is equivalent to scale $\sigma = 5$.

The neural activity of the P channel is defined as the sum of the squared modulus of the acceleration scalogram coefficients within $\sigma = [5, 59]$ normalized by the P channel frequency sensitivity $(T(\sigma))$, as shown in (2).

$$E_{P_i} = \sum_{\sigma=5}^{59} \sum_{b=0}^{N_A} |C_i^I(\sigma, b)|^2 T(\sigma)^2, \ \forall i = 1, 2, ..., 18 \tag{2}$$

with $T = 1$ for frequencies above 175 Hz $(\sigma < 23)$, and decays linearly towards $T = 0.5$ as we reach 70 Hz $(\sigma = 59)$. Fine tuning of $T(\sigma)$ needs to be investigated.

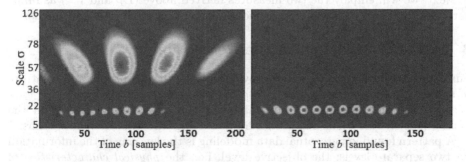

Fig. 1. Scalograms of two vibrating surface acceleration signals for scales $\sigma \in [1, 126]$. The visual difference between the two is compatible with the high perceptual distance of their corresponding tactile signals as evaluated by the users in section 2.1.

Based on the E_{P_i}, a scalogram dissimilarity matrix, D_{En}, holding the pairwise P channel energy-related dissimilarities between the signals was defined with the same structure as the D_P matrix obtained in section 2.1. The details of the method can be found in [10], where a constant $T = 1, \forall \sigma$, was assumed. The mean column Pearson's correlation between the two matrices was around *88%*. This suggests that the P channel dominates the perception for the 18 tactile signals and that the above measure is adequate to simulate its neural activation.

2.4   Non-Pacinian I Channel Activation

As mentioned in section 2.2, the debate in the literature on which variables better reflect the activation of the NP I channel is still open. One of the variables used for this purpose in [9], and which can be acquired, is the stimulus velocity.

The velocity data were obtained through numerical differentiation of the measured displacement data of the tactile surface. With the velocity data for all 18 signals obtained, we computed the signal *mean positive velocity* (V^+), which was then weighted by the percentage of energy delivered in the range of the NP I channel [10 Hz, 70 Hz], to obtain the *normalized mean positive velocity* $(\overline{V}^+)$.

$$V_i^+ = \frac{\sum_{t=0}^{N_V} v_i(t)}{L}; \quad \overline{V}_i^+ = V_i^+ \cdot \frac{\sum_{f=10Hz}^{70Hz} |X_i(f)|^2}{\sum_{f=10Hz}^{800Hz} |X_i(f)|^2}; \quad \forall i = 1, 2, ..., 18; \quad (3)$$

$\forall t$ such that $v_i(t) > 0$; where N_V is the length of $v_i(t)$, L is the number of points for which $v_i(t) > 0$, and X_i is the Fourier Transform of the ith acceleration signal.

With the $\overline{V}_i^+$ defined, we have constructed a velocity dissimilarity matrix $D_{\overline{V}^+}$, where $D_{\overline{V}^+}(i, j) = |\overline{V}_i^+ - \overline{V}_j^+|/(\overline{V}_i^+ + \overline{V}_j^+)$. As the D_{En} matrix, $D_{\overline{V}^+}$ too, has the same structure as the perceptual matrix D_P. The mean column correlation between the two matrices was *66%*, implying that a certain link between the normalized mean positive velocity of the signals and their perception exists. It might also indicate that the NP I channel accounted for a small fraction of perception in the signals, being dominated by the P channel, or that velocity alone is not enough to fully characterize its activation.

Next, we will employ the two measures derived above (E_P and $\overline{V}^+$) as input variables for a fuzzy system used to model perception of vibrotactile signals.

3   A Fuzzy Model of Tactile Perception

Since its early days, fuzzy logic was proven to assess the complexity and volatility of subjective information properly [13]. Sensory information, a particular type of subjective information, deals mainly with the quality evaluations and the perceptions elicited by an object in a human being through one or more senses.

A pattern found in perceptual data modeling is the layering of the information on two separate levels: the objective level, i.e., the *physical characteristics* of an object, and the subjective level, i.e., the *perceptions* induced by the object. According to [13] fuzzy logic is particularly adapted to find relations between the two levels and therefore it might be suited to define a tactile perception model.

3.1   Fuzzy Inference System (FIS) Architecture

In order to model the perceived comfort of the tactile signals, we considered a fuzzy rule-based symbolic system, endowed with signals physical character-istics as input variables. Our proposed system uses two input fuzzy variables (E_P and $\overline{V}^+$), whose universes are divided into three fuzzy sets labeled Low_{E_P}, $Medium_{E_P}$, $High_{E_P}$ and Low_V, $Medium_V$, $High_V$, respectively. Each set has a corresponding membership function (MF) μ associating it a degree of belonging for every numerical value. To maximize both system intuitiveness and results intelligibility, the sum of the input degrees were set to one for both variables.

$$\mu_{(x_{E_P})}(Low_{E_P}) + \mu_{(x_{E_P})}(Medium_{E_P}) + \mu_{(x_{E_P})}(High_{E_P}) = 1, \forall x_{E_P}$$
$$\mu_{(x_V)}(Low_V) + \mu_{(x_V)}(Medium_V) + \mu_{(x_V)}(High_V) = 1, \quad \forall x_V \tag{4}$$

Hence, a signal is initially characterized by its *crisp* numerical input $\{E_P, \overline{V}^+\}$, computed using (2) and (3). After fuzzification, they are converted into a fuzzy linguistic description (e.g., $0.3/Low_{E_P} + 0.7/Medium_{E_P} + 0.0/High_{E_P}$).

The system's rule base is initially constructed by the *a priori* knowledge that the neural activity of the P channel is inversely proportional to the perceived comfort [10]. The actual rules used are shown in Table 1 where the × marks denote areas where none of the 18 signals physical characteristics fall in.

The system output classes correspond to a degree of appreciation for the signals belonging to them. Thus, for enhanced intuitiveness we can associate a label to each class (e.g., *Excellent, Good, Acceptable, Intolerable*).

With the set of rules from Table 1, the inference from the input space $E_P \times \overline{V}^+$ described by trapezoidal MFs, to the output classes is made using Zadeh's com-positional rule of inference [12]. The choice of the T-norm combination operator and of the T-conorm projection operator is subjugated to the limitation that the sum of the output degrees is required to equal one. This is related to human rea-soning, since a signal described as $0.8/Excellent + 0.5/Good + 0.8/Acceptable + 0.7/Intolerable$ lacks intuition. In [15] it is shown that if the sum of the input degrees equals one (eq. (4)), then, choosing the arithmetic product as combi-nation operator, $\top(x, y) = x * y$, and the bounded sum as projection operator, $\bot(x, y) = min(x+y, 1)$, satisfies this limitation even for a many-to-one mapping.

Table 1. Fuzzy Symbolic Rules of Haptic Perception

E_P ＼ $\overline{V}^+$	Low_V	$Medium_V$	$High_V$
Low_{E_P}	Class 1 (*Excellent*)	Class 3 (*Acceptable*)	Class 2 (*Good*)
$Medium_{E_P}$	Class 3 (*Acceptable*)	Class 3 (*Acceptable*)	Class 4 (*Intolerable*)
$High_{E_P}$	Class 4 (*Intolerable*)	×	×

Further, we need to assign a modal value m_k for each output class k, and using the height method of defuzzification [16] we retrieve the numerical value of the predicted preference for a signal i, whose input crisp values are $(E_{P_i}, \overline{V}_i^+)$:

$$y_i = \frac{\sum_k [\mu_{(E_{P_i}, \overline{V}_i^+)}(Class_k) m_k]}{\sum_k \mu_{(E_{P_i}, \overline{V}_i^+)}(Class_k)}, \forall i,\, k = 1, 2, 3, 4 \tag{5}$$

where $\mu_{(E_{P_i}, \overline{V}_i^+)}(Class_k)$ is the belonging degree of signal i to the output class k, computed by the compositional rule of inference using $\top$ and $\bot$ defined above.

3.2   FIS Performance

Though the main goal of a fuzzy symbolic system is to produce a *qualitative* intelligible model of human perception, we cannot ignore its *quantitative* aspects given by the system performance indicators.

The 18 signals input values $(E_{P_i}, \overline{V}_i^+)$ are normalized to $[0, 1]$. Their appreciation degrees (section 2.1), are averaged across subjects and standardized by the *z-score* procedure to obtain a sample of mean 0 and variance 1. After that, their variation interval is $I_A = [-2.1, 2]$. Three thresholds $U_{X\%}, X \in \{5, 10, 20\}$, accounting for the evaluations uncertainty, were defined to assess the quality of the predictions. Their values correspond to 5%, 10% and 20% of the length of I_A. If the absolute difference between the actual and predicted degrees of a signal is greater than $U_{X\%}$, it will be labeled as misclassified w.r.t. $U_{X\%}$. The global error (Δ) is defined in (6), where M is the number of signals, A and P the arrays of actual and predicted comfort degrees, $MaxErr_i$ the maximal error the system can make when estimating the degree of signal i, and k is the midpoint of I_A.

$$\Delta = \frac{\sum_{i=1}^{M} |P(i) - A(i)|}{\sum_{i=1}^{M} MaxErr_i}, \text{ with } MaxErr_i = \begin{cases} A(i) - min(I_A), & \text{if } A(i) \geq k \\ max(I_A) - A(i), & \text{otherwise} \end{cases} \tag{6}$$

To find the MFs for the input variables and the modal values for the output classes (m_k) we have used Δ as a performance indicator and minimized it by *simulated annealing* (SA) [17]. Using the optimized set of parameters, the system error was $\Delta = 9\%$. Fig. 2(a) shows the plot of the actual comfort degrees and of those predicted by the fuzzy system after the SA optimization, for the set of 18 signals from section 2.1. The uncertainties $U_{X\%}$ are represented at a $1 : 1$ scale. The thick vertical lines mark misclassified signals w.r.t. a certain $U_{X\%}$. There were four misclassified signals when matched against the $U_{5\%}$ threshold, three for the $U_{10\%}$ threshold and only one for the $U_{20\%}$ threshold. Fig. 2(b) shows the SA-optimized MFs for the fuzzy variables and also the final rule base. Due to the equation (4), the intersection between two adjacent symbols is at 0.5.

The optimized system was then validated on a new set of 10 vibrational signals whose electrical stimuli are sine waves with different frequencies and durations, produced by a novel haptic bench developed within the FUI-MISAC project.

After a *post factum* analysis of the second experiment from section 2.1, we have decided to measure the appreciation degrees for this new set of signals on

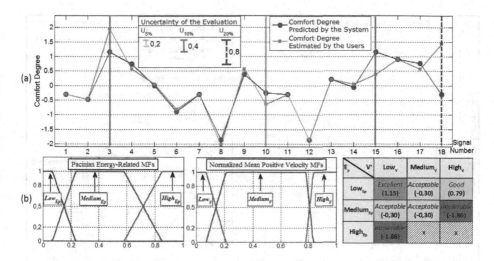

Fig. 2. (a) Actual (green) and predicted (blue) comfort degrees for the set of 18 signals. The thick vertical lines mark discrepancies greater than the different values of $U_{X\%}$; (b) SA-tuned MFs for the fuzzy variables and the final rule base with modal values.

a bipolar scale with only 5 levels. Thus, thirteen subjects rated the appreciation for the 10 signals, randomly presented to them, on a -2 to $+2$ scale, where -2 was the lowest degree of appreciation and $+2$ the highest. We sustain that this new scale leads to a higher reliability of the responses, as the number of levels is lowered by two, and that, thanks to its bipolarity and to its 0 neutral value, it better represents human reasoning on hedonic evaluations. That is why all our future experiments will be carried on this bipolar, 5 levels, -2 to $+2$ scale.

The 10 signals input values $(E_{P_i}, \overline{V_i^+})$ were normalized to $[0,1]$, and their appreciation degrees standardized by the z-score procedure. As these degrees were acquired on a different scale than the one employed for the 18 signals used to adjust the model, we limited the validation to the Pearson's correlation ($\rho = 88\%$) and the Spearman's rank correlation ($\rho_S = 85\%$) between the vectors of predicted and actual appreciation degrees of the set of 10 signals.

3.3 FIS Customization through Graphical User Interface (GUI)

Even though the optimized set (not to be confused with the *optimal set*) of parameters provided by the simulated annealing can boost system performances, the rules derived might be somehow counterintuitive for the human user. Therefore, we risks of losing the system structure comprehensibility. Comprehensibility along with readability accounts for the fuzzy system interpretability [18].

This is why we created a customizable GUI, where the expert can see the signals distribution in the space of the normalized input variables and the performance indicators. At first, the optimized set of parameters is loaded, but the operator can alter the modal values m_k with a set of sliders, and also the limits of the MFs by changing with the mouse the positions of the lines on the 2D grid.

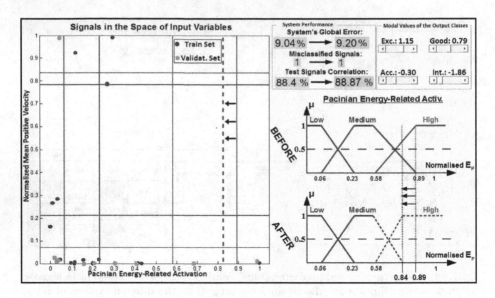

Fig. 3. Qualitative illustration of the GUI, where the effect of altering the lower limit of $High_{E_P}$ can be seen on the system indicators and on the membership functions of the E_P variable. Same procedure goes for all fuzzy sets and variables of the system.

Fig. 3 reveals the two above described sets of signals dispatched in the 2D space of the input variables $(E_P, \overline{V}^+)$. The customizable limits of the trapezoidal MFs are depicted by the series of vertical lines for E_P and horizontal lines for $\overline{V}^+$. Their position and the four modal values of output classes, as illustrated above, represent the optimized set of parameters found by SA to best minimize Δ.

The use of trapezoidal membership functions, as presented in Fig. 3, justifies the choice of both the lower and upper limit of the sets as degrees of freedom for the system. For triangular fuzzy partitions, the choice of the 0.5-intersection of two partitions can be used as a degree of freedom [14]. As we can see in Fig. 3, every change is reflected on the system's performance indicators, Therefore, the user can decide to trade-off some performance for a more comprehensible system.

4 Conclusions and Perspectives

An intuitive two variables fuzzy symbolic system was proposed to model the sensory information of the tactile sense. Initially, a psychophysical study of the haptic sense guided us towards the appropriate two measures that simulate the neural activation of two psychophysical channels of the skin. We then analytically validated their choice using a perceptual dissimilarity matrix as ground truth.

These two measures were later used as input variables for the fuzzy symbolic system, which under the supervision of the simulated annealing managed to obtain an error of 9% and to correctly classify 17 out of 18 signals. The correlation between the predicted and actual values for the validation set was 88%.

To increase the interpretability of the model, a GUI that easily customizes the system parameters is available to the user. The intuitiveness of the GUI and the optimization strength of the simulated annealing can be seen as a step toward a *semi-automatic* fuzzy symbolic system of haptic perception. In the future, loop interactions between the two might be implemented.

References

1. Van Erp, J.B., Van Veen, H.A.H.C.: Vibro-tactile Information Presentation in Automobiles. In: Proceedings of Eurohaptics, pp. 99–104 (2001)
2. Wickens, C.D.: Multiple Resources and Performance Prediction. Theoretical Issues in Ergonomics Science 3, 159–177 (2002)
3. Dabic, S., Tissot, J.-M., Navarro, J., Versace, R.: User Perceptions and Evaluations of Short Vibrotactile Feedback. J. of Cognitive Psychology 25, 299–308 (2013)
4. Bolanowski, S.J., Gescheider, G.A., Verrillo, R.T., Checkosky, C.M.: Four Channels Mediate the Mechanical Aspects of Touch. J. Acoust. Soc. Am. 84, 1680–1694 (1988)
5. Gescheider, G.A., Berryhill, M.E., Verrillo, R.T., Bolanowski, S.J.: Vibrotactile Temporal Summation: Probability Summation or Neural Integration? Somatosensory and Motor Research 16, 229–242 (1999)
6. Gescheider, G.A., Bolanowski, S.J., Verrillo, R.T.: Some Characteristics of Tactile Channels. Behavioural Brain Research 148, 35–40 (2004)
7. Zwislocki, J.: Theory of Temporal Auditory Summation. J. Acoust. Soc. Am. 32, 1046–1060 (1960)
8. Bensmaia, S.J., Hollins, M.: Complex Tactile Waveform Discrimination. J. Acoust. Soc. Am. 108, 1236–1245 (2000)
9. Looft, F.J.: Response of Monkey Glabrous Skin Mechanoreceptors to Random Noise Sequences: II. Dynamic Stimulus State Analysis. Somatosensory and Motor Research 13, 11–28 (1996)
10. Duțu, L.-C., Mauris, G., Bolon, P., Dabic, S., Tissot, J.-M.: A Fuzzy Model Relating Vibrotactile Signal Characteristics to Haptic Sensory Evaluations. In: IEEE International Conference on Computational Intelligence and Virtual Environments for Measurement Systems and Applications (CIVEMSA), pp. 49–54 (2013)
11. Rioul, O., Flandrin, P.: Time-Scale Energy Distributions: A General Class Extending Wavelet Transforms. IEEE Trans. on Signal Processing 40, 1746–1757 (1992)
12. Zadeh, L.A.: The Concept of a Linguistic Variable and Its Application to Approximate Reasoning - Part 1. Information Sciences 8, 199–249 (1975)
13. Bouchon-Meunier, B., Lesot, M.-J., Marsala, C.: Modeling and Management of Subjective Information in a Fuzzy Setting. Int. J. Gen. Syst. 42, 3–19 (2013)
14. Valet, L., Mauris, G., Bolon, P., Keskes, N.: A Fuzzy Rule-Based Interactive Fusion System for Seismic Data Analysis. Information Fusion 4, 123–133 (2003)
15. Mauris, G., Benoit, E., Foulloy, L.: The Aggregation of Complementary Information via Fuzzy Sensors. Measurement 17, 235–249 (1996)
16. Bouchon-Meunier, B., Yager, R.R., Zadeh, L.A.: Fuzzy Logic and Soft Computing. World Scientific (1995)
17. Kirkpatrick, S., Gelatt, C.D., Vecchi, M.P.: Optimization by Simulated Annealing. Science 220, 671–680 (1983)
18. Alonso, J.M., Magdalena, L.: Special Issue on Interpretable Fuzzy Systems. Information Sciences 181, 4331–4339 (2011)

A Linguistic Approach to Multi-criteria and Multi-expert Sensory Analysis

José Luis García-Lapresta[1], Cristina Aldavero[2], and Santiago de Castro[2]

[1] PRESAD Research Group, IMUVA, Dept. de Economía Aplicada,
Universidad de Valladolid, Spain
[2] Academia Castellana y Leonesa de Gastronomía y Alimentación, Valladolid, Spain
lapresta@eco.uva.es, {acaldavero,casalfsa12}@gmail.com

Abstract. In this paper, we introduce a multi-criteria and multi-expert decision-making procedure for dealing with sensory analysis. Experts evaluate each product, taking different criteria into account. If they are confident in their opinions, the evaluation is presented using specific linguistic terms. If not, linguistic expressions generated from several consecutive linguistic terms are used. Products are ranked according to the average distance between the obtained ratings and the highest possible assessment. The procedure is applied to a field experiment in which six trained sensory panelists assessed a variety of wines and wild mushrooms.

Keywords: sensory analysis, imprecision, linguistic assessments, distances.

1 Introduction

According to ISO 5492 [6], the sensory analysis is defined as "the examination of the organoleptic properties of a product using the human senses". Sensory analysis includes a variety of tools and techniques for evaluating consumer products taking into account human sensations provided by trained experts and/or potential consumers (see Stone and Sidel [15], Meilgaard et al. [11] and Lawless and Heymann [7], among others).

In all evaluations, a panel of tasters use their senses to perceive qualities such as color, size, shape, smell, flavor, texture, malleability and sound. These sensations are translated into graphical, verbal or numerical values of previously defined quality descriptors for each type of product (some analyses regarding the use of different kinds of hedonic scales in sensory analysis can be found in Lim et al. [8]).

Clearly, judges who analyze sensorial aspects of food, beverages, cosmetics, textile materials, etc. have to deal with imprecision. Although the use of exact numerical values could be inappropriate for assessing vague attributes such as appearance, smell and taste, in tasting procedures judges usually have to fill in some forms using finite numerical scales. Once the ratings have been provided by the judges, an aggregation procedure (most often using the arithmetic mean) generates a ranking of the products.

A. Laurent et al. (Eds.): IPMU 2014, Part II, CCIS 443, pp. 586–595, 2014.

Due to the vagueness and imprecision involved in sensory evaluation, the use of finite scales formed by linguistic terms is clearly more suitable than using numerical scales.

In this paper, we propose a multi-criteria and multi-expert decision-making procedure for dealing with sensory analysis from a linguistic approach[1]. To this end, a finite linguistic scale is fixed. When experts are confident in their opinions, they assign a linguistic term to each alternative in each criterion.

It is important to note that there is empirical evidence showing that experts may hesitate when they assess alternatives through linguistic terms[2]. For this reason, we have considered linguistic expressions generated by consecutive linguistic terms. Thus, experts are allowed to assign linguistic expressions to each alternative in each criterion.

Once experts have provided their assessments, we calculate the distances between the linguistic ratings and the highest possible assessment for each alternative and each criterion. These distances are obtained by adding the geodesic distances in the graph associated with the linguistic expressions set and the penalization values which are dependent upon the increase of imprecision.

Taking into account a weighting vector that reflects the different level of importance attributed to each criterion, for each alternative we calculate the weighted average distance between the obtained ratings and the highest possible assessment. Then, the alternatives are ranked according to the above mentioned weighted average distances (the less, the better), similarly to Falcó [3, Chapter 4].

In order to show how the proposed decision-making process works, we have considered the data obtained in a field experiment in which six trained experts gave their opinions on different wines and wild mushrooms through linguistic terms and linguistic expressions.

The paper is organized as follows. Section 2 includes the basic notation and some concepts that are necessary to develop the proposal. Section 3 is devoted to introduce the proposal of multi-criteria and multi-expert sensory analysis. Section 4 contains the description of the field experiment and the results. Finally, Section 5 includes some concluding remarks.

2 Preliminaries

Let $A = \{1, \ldots, m\}$, with $m \geq 2$, be a set of agents and let $X = \{x_1, \ldots, x_n\}$, with $n \geq 2$, be the set of alternatives which have to be evaluated. Under total certainty, each agent assigns a linguistic term to every alternative within a linguistic ordered scale $L = \{l_1, \ldots, l_g\}$, with $l_1 < l_2 < \cdots < l_g$. It is assumed that the linguistic scale is balanced and consecutive terms are equispaced.

---

[1] Some fuzzy linguistic approaches to sensory analysis can be found in Davidson and Sun [2], Martínez [9], Martínez et al. [10] and Agell et al. [1], among others.

[2] For instance, in the tasting described in Agell et al. [1], 40% of the assessments were linguistic expressions with two or more linguistic terms.

Taking into account the *absolute order of magnitude spaces* introduced by Travé-Massuyès and Piera [16], the *set of linguistic expressions* is defined as

$$\mathbb{L} = \{[l_h, l_k] \mid l_h, l_k \in L, \, 1 \le h \le k \le g\},$$

where $[l_h, l_k] = \{l_h, l_{h+1}, \ldots, l_k\}$. Given that $[l_h, l_h] = \{l_h\}$, this linguistic expression can be replaced by the linguistic term l_h. In this way, $L \subset \mathbb{L}$.

Example 1. Consider the set of linguistic terms $L = \{l_1, l_2, l_3, l_4, l_5\}$ with the meanings given in Table 1.

Table 1. Meaning of the linguistic terms

l_1	l_2	l_3	l_4	l_5
very bad	bad	acceptable	good	very good

Since linguistic expressions are intervals of linguistic terms, their meanings are straightforward. For instance, $[l_2, l_3]$ means 'between bad and acceptable', $[l_3, l_5]$ means 'between acceptable and very good', or 'at least acceptable', etc.

Taking into account the approach introduced in Roselló *et al.* [14], the set of linguistic expressions can be represented by a graph $G_{\mathbb{L}}$. In the graph, the lowest layer represents the linguistic terms $l_h \in L \subset \mathbb{L}$, the second layer represents the linguistic expressions created by two consecutive linguistic terms $[l_h, l_{h+1}]$, the third layer represents the linguistic expressions generated by three consecutive linguistic terms $[l_h, l_{h+2}]$, and so on up to last layer where we represent the linguistic expression $[l_1, l_g]$. As a result, the higher an element is, the more imprecise it becomes.

The vertices in $G_{\mathbb{L}}$ are the elements of $\mathbb{L}$ and the edges $\mathcal{E} - \mathcal{F}$, where $\mathcal{E} = [l_h, l_k]$ and $\mathcal{F} = [l_h, l_{k+1}]$, or $\mathcal{E} = [l_h, l_k]$ and $\mathcal{F} = [l_{h+1}, l_k]$. Fig. 1 shows the graph representation of Example 1. When an agent is confident about his opinion on an alternative, he can assign a linguistic term $l_h \in L$ to this alternative. However, if he is unconfident about his opinion, he might use a linguistic expression $[l_h, l_k] \in \mathbb{L}$, with $h < k$. For more details, see Roselló *et al.* [12], [13], [14] and Falcó *et al.* [5].

A binary relation $\succcurlyeq$ on a set $Z \ne \emptyset$ is a *weak order* if it is complete ($x \succcurlyeq y$ or $y \succcurlyeq x$, for all $x, y \in Z$) and transitive (if $x \succcurlyeq y$ and $y \succcurlyeq z$, then $x \succcurlyeq z$, for all $x, y, z \in Z$). On the other hand, a *linear order* on $Z \ne \emptyset$ is an antisymmetric[3] weak order on Z. Given a weak or linear order $\succcurlyeq$ on $Z \ne \emptyset$, the asymmetric part of $\succcurlyeq$ is denoted by $\succ$; in other words, $x \succ y$ if not $y \succcurlyeq x$.

3   The Multi-criteria and Multi-expert Decision Process

In order to introduce our proposal, we first consider a family of parameterized metrics in the set of linguistic expressions. They will be used for comparing the alternatives and for generating a weak order over them.

---

[3] $\succcurlyeq$ is antisymmetric if for all $x, y \in Z$ ($x \succcurlyeq y$ and $y \succcurlyeq x$) implies $x = y$.

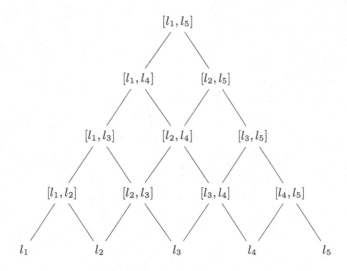

Fig. 1. Graph representation of the linguistic expressions for $g = 5$

3.1   Comparing Linguistic Expressions

The set of linguistic expressions can be ranked in different ways. We now introduce a linear order on this set that will be compatible with the ordering induced by the distance with respect to the highest possible assessment l_g.

Proposition 1. *The binary relation* $\succeq_L$ *on* $\mathbb{L}$, *defined as*

$$[l_h, l_k] \succeq_L [l_{h'}, l_{k'}] \Leftrightarrow \begin{cases} h + k > h' + k' \\ or \\ h + k = h' + k' \text{ and } k - h \leq k' - h', \end{cases}$$

is a linear order, and it is called the canonical order *on* $\mathbb{L}$.

For $g = 5$, the elements of $\mathbb{L}$ are ordered as follows (see also Fig. 2):

$$l_5 \succ_L [l_4, l_5] \succ_L l_4 \succ_L [l_3, l_5] \succ_L [l_3, l_4] \succ_L [l_2, l_5] \succ_L l_3 \succ_L [l_2, l_4] \succ_L$$

$$\succ_L [l_1, l_5] \succ_L [l_2, l_3] \succ_L [l_1, l_4] \succ_L l_2 \succ_L [l_1, l_3] \succ_L [l_1, l_2] \succ_L l_1.$$

The geodesic distance[4] between two linguistic expressions $\mathcal{E}, \mathcal{F} \in \mathbb{L}$ is defined as the distance in the graph $G_{\mathbb{L}}$ between their associated vertices:

$$d_G([l_h, l_k], [l_{h'}, l_{k'}]) = |h - h'| + |k - k'|.$$

In Falcó *et al.* [4] the geodesic distance is modified for penalizing the imprecision by means two parameters, α and β. For simplicity, in this paper we only consider the first penalization through the parameter α.

---

[4] The geodesic distance between two vertices in a graph is the number of edges in one of the shortest paths connecting them.

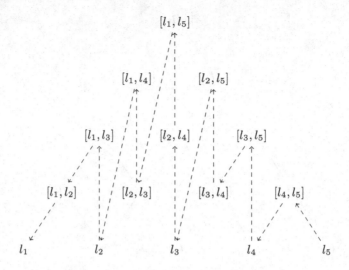

Fig. 2. Canonical order in $\mathbb{L}$ for $g = 5$

Given $\mathcal{E} = [l_h, l_k] \in \mathbb{L}$, with $\#\mathcal{E}$ we denote the cardinality of $\mathcal{E}$, i.e., the number of linguistic terms in the interval $[l_h, l_k]$: $\#\mathcal{E} = k + 1 - h$.

Proposition 2. *For every* $\alpha \geq 0$, *the function* $d : \mathbb{L} \times \mathbb{L} \longrightarrow \mathbb{R}$, *defined as*

$$d(\mathcal{E}, \mathcal{F}) = d_G(\mathcal{E}, \mathcal{F}) + \alpha \, |\#\mathcal{E} - \#\mathcal{F}|,$$

is a metric, and it is called the metric associated with α.

The following result is a direct consequence of Falcó *et al.* [4, Prop. 3].

Proposition 3. *Let* $d_\alpha : \mathbb{L}^2 \longrightarrow \mathbb{R}$ *be the metric associated with* $\alpha \geq 0$. *The following statements are equivalent:*

1. $\forall \mathcal{E}, \mathcal{F} \in \mathbb{L} \quad \left(\mathcal{E} \succ \mathcal{F} \Leftrightarrow d_\alpha(\mathcal{E}, l_g) < d_\alpha(\mathcal{F}, l_g) \right)$.

2. $\alpha \in T_g$, *where* $T_g = \left(0, \frac{1}{g-2} \right)$, *if* g *is odd, and* $T_g = \left(0, \frac{1}{g-1} \right)$, *if* g *is even.*

Example 2. Following Example 1, where $g = 5$, Proposition 3 shows that $\alpha < \frac{1}{3}$ should be satisfied for avoiding inconsistencies. Then,

$d(l_5, l_5) = 0 < d([l_4, l_5], l_5) = 1 + \alpha < d(l_4, l_5) = 2 < d([l_3, l_5], l_5) = 2 + 2\alpha <$
$d([l_3, l_4], l_5) = 3 + \alpha < d([l_2, l_5], l_5) = 3 + 3\alpha < d(l_3, l_5) = 4 <$
$d([l_2, l_4], l_5) = 4 + 2\alpha < d([l_1, l_5], l_5) = 4 + 4\alpha < d([l_2, l_3], l_5) = 5 + \alpha <$
$d([l_1, l_4], l_5) = 5 + 3\alpha < d(l_2, l_5) = 6 < d([l_1, l_3], l_5) = 6 + 2\alpha <$
$d([l_1, l_2], l_5) = 7 + \alpha < d(l_1, l_5) = 8 \,.$

3.2 From the Individual Assessments to a Social Ranking

A *profile* V is a matrix (v_i^a) consisting of m rows and n columns of linguistic expressions, where the element $v_i^a \in \mathbb{L}$ represents the linguistic assessment given by the agent $a \in A$ to the alternative $x_i \in X$. Then,

$$V = \begin{pmatrix} v_1^1 & \cdots & v_i^1 & \cdots & v_n^1 \\ \cdots & \cdots & \cdots & \cdots & \cdots \\ v_1^a & \cdots & v_i^a & \cdots & v_n^a \\ \cdots & \cdots & \cdots & \cdots & \cdots \\ v_1^m & \cdots & v_i^m & \cdots & v_n^m \end{pmatrix} = (v_i^a).$$

Given a profile $V = (v_i^a)$,

$$D(x_i) = \frac{1}{m} \sum_{a=1}^{m} d\left(v_i^a, l_g\right)$$

is the average distance between the ratings of x_i and the highest possible assessment l_g.

Proposition 4. *Given $\alpha \geq 0$ satisfying the condition of Proposition 3, let d be the metric associated with α. The binary relation $\succcurlyeq$ on X defined as*

$$x_i \succcurlyeq x_j \iff D(x_i) \leq D(x_j)$$

is a weak order on X.

Consider now that agents have to assess each alternative from different criteria: $c_1, \ldots, c_r$. Then we have r profiles $V^1 = (v_i^{a,1}), \ldots, V^r = (v_i^{a,r})$, where $v_i^{a,k} \in \mathbb{L}$ is the rating given by the agent a to the alternative x_i with respect to the criterion k.

Since each criterion may have different importance in the decision, consider a weighting vector $\boldsymbol{w} = (w_1, \ldots, w_r) \in [0,1]^r$, with $w_1 + \cdots + w_r = 1$.

Given a profile $V = (v_i^a)$ and a weighting vector $\boldsymbol{w} = (w_1, \ldots, w_r)$,

$$D_{\boldsymbol{w}}(x_i) = \sum_{k=1}^{r} w_p \frac{1}{m} \sum_{a=1}^{m} d\left(v_i^{a,k}, l_g\right)$$

is the weighted average distance between the ratings of x_i and the highest possible assessment l_g.

Proposition 5. *Given $\alpha \geq 0$ satisfying the condition of Proposition 3, let d be the metric associated with α. The binary relation $\succcurlyeq_{\boldsymbol{w}}$ on X defined as*

$$x_i \succcurlyeq_{\boldsymbol{w}} x_j \iff D_{\boldsymbol{w}}(x_i) \leq D_{\boldsymbol{w}}(x_j)$$

is a weak order on X.

4   A Field Experiment

The proposed multi-criteria and multi-expert sensory analysis was demonstrated in a field experiment carried out in *Trigo* restaurant in Valladolid (November 30th, 2013), under appropriate conditions of temperature, light and service.

A total of six judges (one female and five males between the ages of 30 and 50) trained in the sensory analysis of wine and wild mushrooms were recruited through the *Gastronomy and Food Academy of Castilla y León*. When the test was being carried out there was no communication between judges. The samples were given without any identification. The tasting was divided into two parts.

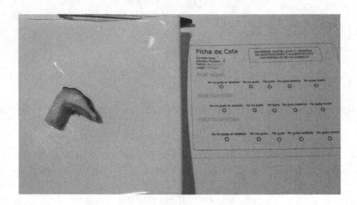

Fig. 3. Cooked *Boletus pinophilus* and the linguistic hedonic tasting sheet

Firstly, the experts gave their evaluations using a classical numerical hedonic tasting sheet on a scale of 1 to 10. Secondly, they repeated the tasting –in a different order– assessing the same products through linguistic terms from a linguistic hedonic scale and the corresponding linguistic expressions.

After the tasting, the six experts were asked if they had difficulties filling in linguistic hedonic tasting sheet and whether they were more comfortable using numerical or linguistic assessments. All of them declared that they did not have difficulties filling in the linguistic hedonic tasting sheet. Four out of the six experts preferred the linguistic hedonic tasting sheet over the numerical one; one was indifferent regarding the two styles; and another preferred the numerical hedonic tasting sheet over the linguistic one.

We now describe the tasting in its linguistic part. The six experts assessed the products included in Table 2 through the five linguistic terms of Table 3 (or the corresponding linguistic expressions, when they hesitated) under three criteria: appearance, smell and taste[5].

Five of the six experts sometimes hesitate about their opinions and they assigned linguistic expressions with two linguistic terms. This happened in 19 of the 108 ratings, i.e., 17.59% of the cases.

---

[5] We have presented the results all together. Clearly, wines, raw wild mushrooms and cooked wild mushrooms have to be analyzed separately.

Table 2. Alternatives

x_1	White wine 'Rueda Zascandil 2012'
x_2	Red wine 'Toro Valdelacasa 2007'
x_3	Raw *Boletus pinophilus*
x_4	Raw *Tricholoma pertentosum*
x_5	Cooked *Boletus pinophilus*
x_6	Cooked *Tricholoma pertentosum*

Table 3. Linguistic terms

l_1	I don't like it at all
l_2	I don't like it
l_3	I like it
l_4	I rather like it
l_5	I like it so much

Taking into account the weights[6] $w_1 = 0.2$ for appearance, $w_2 = 0.3$ for smell and $w_3 = 0.5$ for taste, i.e., $\boldsymbol{w} = (0.2, 0.3, 0.5)$, and the parameter $\alpha = 0.3$, we obtain the following weighted average distances from the alternatives ratings to the highest possible assessment l_5:

$$D_{\boldsymbol{w}}(x_1) = 2.965 \qquad D_{\boldsymbol{w}}(x_2) = 2.081 \qquad D_{\boldsymbol{w}}(x_3) = 1.246$$

$$D_{\boldsymbol{w}}(x_4) = 3.515 \qquad D_{\boldsymbol{w}}(x_5) = 1.148 \qquad D_{\boldsymbol{w}}(x_6) = 2.193.$$

Then, $x_5 \succ_{\boldsymbol{w}} x_3 \succ_{\boldsymbol{w}} x_2 \succ_{\boldsymbol{w}} x_6 \succ_{\boldsymbol{w}} x_1 \succ_{\boldsymbol{w}} x_4$.

We note that the alternatives were similarly ordered when using the numerical assessments with the same weights, after calculating the average of the individual ratings in each phase: $x_5 \succ_{\boldsymbol{w}} x_3 \succ_{\boldsymbol{w}} x_6 \succ_{\boldsymbol{w}} x_2 \succ_{\boldsymbol{w}} x_1 \succ_{\boldsymbol{w}} x_4$, with x_2 and x_6 being the only alternatives where the ranking is different.

Although all six products are presented in the ranking, they can be divided into three parts: wines (alternatives x_1 and x_2), raw wild mushrooms (alternatives x_3 and x_4) and cooked wild mushrooms (alternatives x_5 and x_6).

Sensory Description. According to one of the judges who participated in the tasting, the most aromatic and tasty wild mushroom was *Boletus pinophilus*, the low temperature of the cooking process enhances volatilization of its essential oils, flavor development and improvement of texture, making it a favorite of the panel of tasters in both the numerical rating and verbal evaluation methods. Raw *Tricholoma portentosum* had a medium flavor intensity and did not have a very strong flavor, this mushroom improves with the cooking process, which enhances all of its qualities outstandingly. The red wine presented medium-high primary aromas, a good structure and a good balance of acidity, alcohol and tannins, while the white wine had a lower aromatic intensity and a slight imbalance of acidity.

---

[6] These weights are usual in this kind of tasting.

Table 4. Ratings

		Judge 1	Judge 2	Judge 3	Judge 4	Judge 5	Judge 6
x_1	Appearance	l_3	l_4	l_4	l_5	l_3	l_3
	Smell	$[l_3, l_4]$	l_3	l_4	l_4	l_4	l_4
	Taste	l_3	l_3	l_4	l_4	l_3	l_3
x_2	Appearance	l_4	l_4	l_2	l_5	l_4	l_4
	Smell	$[l_4, l_5]$	l_5	l_2	l_5	l_3	l_5
	Taste	$[l_4, l_5]$	l_5	l_2	l_5	l_3	$[l_4, l_5]$
x_3	Appearance	l_4	l_4	l_5	l_5	l_3	l_4
	Smell	$[l_4, l_5]$	l_3	$[l_4, l_5]$	l_5	l_3	l_5
	Taste	l_5	l_5	$[l_4, l_5]$	l_5	l_4	$[l_4, l_5]$
x_4	Appearance	l_4	l_3	l_4	l_5	l_3	$[l_2, l_3]$
	Smell	l_3	$[l_2, l_3]$	l_4	l_3	l_2	$[l_2, l_3]$
	Taste	l_3	l_3	$[l_4, l_5]$	l_4	l_2	l_4
x_5	Appearance	l_5	l_4	l_4	l_4	l_3	l_4
	Smell	l_5	l_5	l_5	l_4	l_3	$[l_4, l_5]$
	Taste	$[l_4, l_5]$	l_5	l_5	l_5	l_4	$[l_4, l_5]$
x_6	Appearance	l_5	l_3	$[l_4, l_5]$	l_5	$[l_2, l_3]$	l_4
	Smell	l_5	l_3	l_5	l_3	$[l_3, l_4]$	l_4
	Taste	l_4	l_4	l_5	l_3	l_4	$[l_3, l_4]$

5   Concluding Remarks

A generalization of our proposal consists of using appropriate aggregation functions for defining the aggregated distance between the ratings of an alternative and the highest possible assessment. In this way, the arithmetic mean can be replaced by an OWA operator (see Yager [17]).

Acknowledgments. The authors are grateful to the participants in the tasting, as well as Víctor Martín and Noemí Martínez (*Trigo* restaurant in Valladolid). The financial support of the Spanish *Ministerio de Economía y Competitividad* (project ECO2012-32178) and *Consejería de Educación de la Junta de Castilla y León* (project VA066U13) are also acknowledged.

References

1. Agell, N., Sánchez, G., Sánchez, M., Ruiz, F.J.: Selecting the best taste: a group decision-making application to chocolates design. In: Proceedings of the 2013 IFSA-NAFIPS Joint Congress, Edmonton, pp. 939–943 (2013)
2. Davidson, V.J., Sun, W.: A linguistic method for sensory assessment. Journal of Sensory Studies 13, 315–330 (1998)
3. Falcó, E.: Voting Systems with Linguistic Assessments and their Application to the Allocation of Tenders. PhD Dissertation, University of Valladolid (2013)

4. Falcó, E., García-Lapresta, J.L., Roselló, L.: Aggregating imprecise linguistic expressions. In: Guo, P., Pedrycz, W. (eds.) Human-Centric Decision-Making Models for Social Sciences. SCI, vol. 502, pp. 97–113. Springer, Heidelberg (2014)
5. Falcó, E., García-Lapresta, J.L., Roselló, L.: Allowing agents to be imprecise: A proposal using multiple linguistic terms. Information Sciences 258, 249–265 (2014)
6. ISO 5492 (1992)
7. Lawless, H.T., Heymann, H.: Sensory Evaluation of Food: Principles and Practices. Springer (2010)
8. Lim, J., Wood, A., Green, B.G.: Derivation and evaluation of a labeled hedonic scale. Chemical Sense 34(9), 739–751 (2009)
9. Martínez, L.: Sensory evaluation based on linguistic decision analysis. International Journal of Approximate Reasoning 44, 148–164 (2007)
10. Martínez, L., Espinilla, M., Liu, J., Pérez, L.G., Sánchez, P.J.: An evaluation model with unbalanced linguistic information applied to olive oil sensory evaluation. Journal of Multiple-Valued Logic and Soft Computing 15, 229–251 (2009)
11. Meilgaard, M.C., Carr, T., Civille, G.V.: Sensory Evaluation Techniques, 4th edn. CRC Press (2006)
12. Roselló, L., Prats, F., Agell, N., Sánchez, M.: Measuring consensus in group decisions by means of qualitative reasoning. International Journal of Approximate Reasoning 51, 441–452 (2010)
13. Roselló, L., Prats, F., Agell, N., Sánchez, M.: A qualitative reasoning approach to measure consensus. In: Herrera-Viedma, E., García-Lapresta, J.L., Kacprzyk, J., Fedrizzi, M., Nurmi, H., Zadrożny, S. (eds.) Consensual Processes. STUDFUZZ, vol. 267, pp. 235–261. Springer, Heidelberg (2011)
14. Roselló, L., Sánchez, M., Agell, N., Prats, F., Mazaira, F.A.: Using consensus and distances between generalized multi-attribute linguistic assessments for group decision-making. Information Fusion 17, 83–92 (2014)
15. Stone, H., Sidel, J.L.: Sensory Evaluation Practices. Academic Press (2004)
16. Travé-Massuyès, L., Piera, N.: The orders of magnitude models as qualitative algebras. In: Proceedings of the 11th International Joint Conference on Artificial Intelligence, Detroit, pp. 1261–1266 (1989)
17. Yager, R.R.: Ordered weighted averaging operators in multicriteria decision making. IEEE Transactions on Systems, Man and Cybernetics 8, 183–190 (1988)

Using Fuzzy Logic to Enhance Classification of Human Motion Primitives

Barbara Bruno[1], Fulvio Mastrogiovanni[1],
Alessandro Saffiotti[2], and Antonio Sgorbissa[1]

[1] University of Genova, Dept. DIBRIS,
via Opera Pia 13, 16145 Genova, Italy
{barbara.bruno,fulvio.mastrogiovanni,antonio.sgorbissa}@unige.it
[2] Örebro University, AASS Cognitive Robotic Systems Lab.,
Fakultetsgatan 1, S-70182 Örebro, Sweden
asaffio@aass.oru.se

Abstract. The design of automated systems for the recognition of specific human activities is among the most promising research activities in Ambient Intelligence. The literature suggests the adoption of wearable devices, relying on acceleration information to model the activities of interest and distance metrics for the comparison of such models with the run-time data. Most current solutions do not explicitly model the uncertainty associated with the recognition, but rely on crisp thresholds and comparisons which introduce brittleness and inaccuracy in the system. We propose a framework for the recognition of simple activities in which recognition uncertainty is modelled using possibility distributions. We show that reasoning about this explicitly modelled uncertainty leads to a system with enhanced recognition accuracy and precision.

Keywords: Activity recognition, Activities of Daily Living, wearable sensors, possibility measures.

1 Introduction

Automatic recognition of human activities is a vivid area of research, whose impact ranges from smart homes to future factory automation and to social behavioural studies. One of the most timely application is the process of determining the level of autonomy of an elderly person. Ever since the publication of the *Index of Activities of Daily Living* (ADL) by Katz and colleagues [1], this process is usually accomplished by analysing the person's ability to carry out a set of daily activities, each one involving the use of different motor and cognitive capabilities. Unfortunately, the most commonly adopted indexes and sets of ADL have been defined assuming that a caregiver examines the person's performance on a *qualitative* basis; this makes the design of automated systems for the monitoring, recognition and classification of ADL particularly challenging.

Existing solutions take two different approaches: *smart environments* rely on heterogeneous sensors distributed in the environment [2–4]; *wearable sensing systems* rely on sensors located on the person body [5, 6].

A. Laurent et al. (Eds.): IPMU 2014, Part II, CCIS 443, pp. 596–605, 2014.

Wearable sensing is quickly becoming the preferred approach to monitor either body gestures or bio signals. Most systems aim at engineering a single sensing device, either based on the integration of different sensors [7] or on the use of a single sensing mode. Among the latter, systems based on a single tri-axial accelerometer are the most common [8, 9]. As outlined in [5], most wearable sensing systems have a similar architecture, whose main tasks are: (i) to extract relevant *features* from the available sensory data; (ii) to create *representations* of the target activities in terms of the features; and (iii) to *classify* the run-time sensory data according to the representations. Features are chosen so that they discriminate well different activities while being invariant across different executions of the same activity: for wearable sensors, gravity and body acceleration are the most commonly adopted features [8],[10]. The representation of the target activities, and the classification of run-time sensor data, are typically based on the definition of rules and the adoption of decision trees [10]; or on the creation of models (e.g., Hidden Markov Models or Gaussian Mixture Models) and the definition of adequate distance measures [8],[11, 12]. In both cases, the system labels the run-time data as an occurrence of the activity that most closely represents them.

The above approaches produce a crisp decision on the recognized activity, and do not model the *uncertainty* associated with this decision. Suppose that a person executes an activity A which is not modelled in the system; and suppose that, although none of the existing models fully fits the observed data, B is the closest one. A crisp system would label this activity as B without providing any warning of the *unreliability* of this decision. As another example, suppose that the observed data fit both the model of A and, to a less degree, the one of B. A crisp system would label the activity as A without providing any warning of the *ambiguity* of this decision. Failure to model the uncertainty associated with a decision limits the possibility to integrate the decision with complementary context-assessing systems [2],[13], and may reduce the applicability of the system to real-life scenarios [14].

We propose a framework for the recognition of ADL that takes recognition uncertainty into account. Building upon the work in [12],[15], we use information from a single wrist-placed tri-axial accelerometer, and rely on Gaussian Mixture Modelling and Gaussian Mixture Regression to create models of ADL. We extend that work by building a possibility distribution over the space of ADL [16], where degrees of possibilities are computed from the Mahalanobis distance between the observed data and the model. The use of possibility measures is justified by the semantics of our models, which is inherently similarity-based [17]. We then show how to leverage possibility values to improve the accuracy and precision of recognition, by reasoning about the evolution of these values in time.

This article is organized as follows. Section 2 outlines the architecture of the system. Section 3 describes the classification and analysis procedures that we propose. Section 4 reports experimental results. Finally, conclusions follow.

2   System Architecture

In this paper, we reserve the term ADL for a whole complex activity, and define as *human motion primitives* (HMP) those stereotyped motions that can uniquely identify an ADL. For example, the ADL *Feeding* defined in Katz Index comprises the HMP *pick up a glass* and *put down a glass*, while the ADL *Mobility* comprises the HMP *walk* and *climb the stairs*. Typically, HMP can be monitored with wearable sensing systems while ADL may require a whole smart environment.

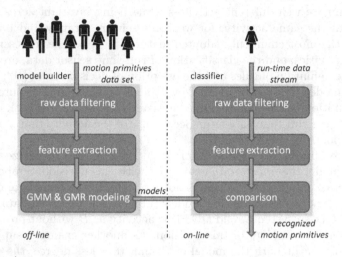

Fig. 1. System architecture for HMP classification

The classical part of our system for the recognition of HMP builds upon the techniques described in [12],[15], and it is composed of two distinct subsystems. During the off-line phase (Figure1, left) one or more individuals are provided with a wrist-mounted tri-axial accelerometer and asked to perform each HMP multiple times. The acceleration data are recorded and each occurrence of any HMP is tagged by a human observer. The result is a dataset which, for each HMP, contains a large number of examples in the form of sequences of acceleration data along the three axes x, y and z. Once the training set is available, the *model builder* module performs a number of steps: *raw data filtering* is aimed at reducing high frequency noise through a median filter; *feature extraction* isolates the gravitational components by applying a low-pass filter to the acceleration signal and then obtains the body acceleration components by subtraction; *GMM and GMR modelling* generate a probabilistic model of each HMP in terms of its features. Since keeping the axes correlation into account allows for an increase in the recognition accuracy [12], both the features and their models are defined in the $4D$ space of time and tri-axial acceleration. A description of the *model builder* block can be found in [12]. During the on-line phase (Figure 1, right), the monitored person wears a wrist-mounted device providing acceleration data.

Analogously to the off-line phase, a number of steps are sequentially executed: raw data filtering and feature extraction execute the very same algorithms of the off-line phase; then the comparison step performs the classification by considering a temporal window moving over the run-time data stream and comparing the features extracted from acceleration data therein with all the stored models. A description of the *classifier* block can be found in [15].

It is important to notice that the presented system is independent from the sensing device used for the acquisition of acceleration data. In particular, it is compatible with already available smart watches, i.e., wrist-worn devices equipped with various sensors.[1] The positioning of the device, however, is crucial for the reliability and the effectiveness of the monitoring system, and it depends on the target set of HMPs: e.g., hand gestures require a wrist placement, while lower limbs movements are better monitored with waist-placed sensors.

3   Representing Uncertainty in the HMP Classifier

The basic approach presented in the previous section is typically applied without taking uncertainty into account: the best matching model is simply returned as the classification result. We now see how we can model classification uncertainty in a possibilistic way, and how we can reason about this uncertainty to enhance the classification performance.

3.1   Basic Concepts and Definitions

Let us denote with M the number of models available for the classification, created by the *model builder* in Figure 1, and with K_m the number of data points used to model the HMP m. Let us also denote with ξ the generic feature of interest, i.e. ξ can either correspond to gravity g or body acceleration b. The following definitions are in order.

- $\hat{\xi}_{m,k}$ is the data point k of feature ξ of model m, defined as:

$$\hat{\xi}_{m,k} = (t^{\xi}_{m,k}, \hat{a}^{\xi}_{m,k}, \Sigma^{\xi}_{m,k}),\tag{1}$$

  where $t^{\xi}_{m,k} \in \mathbb{R}$ is the time information, $\hat{a}^{\xi}_{m,k} \in \mathbb{R}^3$ contains the expected x, y and z acceleration components, and $\Sigma^{\xi}_{m,k} \in \mathbb{R}^{3\times3}$ is the conditional covariance associated to $\hat{a}^{\xi}_{m,k}$.
- $\hat{\Xi}^{\xi}_m$ is the generalized version of feature ξ for HMP m, defined as:

$$\hat{\Xi}^{\xi}_m = \{\hat{\xi}_{m,1}, \dots, \hat{\xi}_{m,K_m}\}.\tag{2}$$

- $\hat{\Xi}_m = (\hat{\Xi}^g_m, \hat{\Xi}^b_m)$ is the model of HMP m.

---

[1] Popular smart watches are Pebble SmartWatch (https://getpebble.com/) or Sony SmartWatch (http://store.sony.com/smartwatch/cat-27-catid-Smart-Watch)

In a similar way, given a window $\Xi_w = (\Xi_w^g, \Xi_w^b)$ of size N moving over the run-time acceleration data stream, we define the feature ξ extracted from the window as:

$$\Xi_w^\xi = \{\xi_{w,1}, \dots, \xi_{w,N}\}, \tag{3}$$

where $\xi_{w,n} = (t_{w,n}^\xi, a_{w,n}^\xi)$ is the data point at position n inside the window. The task of the classifier is to compare Ξ_w with the M models $\{\hat{\Xi}_1, \dots, \hat{\Xi}_M\}$ in order to identify the model $\hat{\Xi}_m$ that most likely represents the run-time data.

3.2 Modelling Uncertainty by Possibility Distributions

The "classical", crisp approach to human activities recognition would select, at each time instant, the activity, among those modelled by the system, that most closely matches the acceleration data inside the window. In our case this is done by first computing Mahalanobis distance between the window of run-time data Ξ_w and each of the models $\hat{\Xi}_m$, then labelling the window as an occurrence of the HMP with minimum distance. Figure 2 (a) reports the output of such classifier when the executed motion is the HMP *standing up from a chair*. It is immediate to see that this classic, crisp classifier: (i) does not allow for a correct definition of the starting and ending moments of the recognized motion; and (ii) has a high rate of false positive recognitions. In order to overcome these limitations, we have designed a possibility based classifier which, at each time instant, returns the degree of possibility π_m^w that the modelled HMP m is the one representing the run-time data inside the current window w, computed as:

$$\pi_m^w = \begin{cases} 1 - D_m^w/\tau_m & D_m^w < \tau_m \\ 0 & D_m^w \geq \tau_m \end{cases}. \tag{4}$$

D_m^w is the distance between model m and the run-time data, defined as:

$$D_m^w = \alpha \cdot d(\hat{\Xi}_m^g, \Xi_w^g) + \beta \cdot d(\hat{\Xi}_m^b, \Xi_w^b), \tag{5}$$

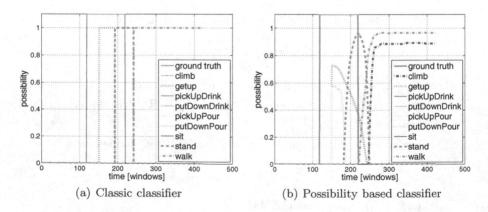

(a) Classic classifier   (b) Possibility based classifier

Fig. 2. Output of a (a) classic classifier and (b) possibility based classifier when tested with an execution of the HMP *stand up from a chair* (within orange solid lines)

where d is the Mahalanobis distance between the samples in the off-line model $\hat{\Xi}_m^\xi$ and the on-line window Ξ_w^ξ, and α and β are the relative weight of the gravity and the body accelerations. The normalization factor τ_m is the distance between model m and the farthest curve which is likely to be generated by the model itself:

$$\tau_m = \alpha \cdot d(\hat{\Xi}_m^g, \bar{\Xi}_m^g) + \beta \cdot d(\hat{\Xi}_m^b, \bar{\Xi}_m^b), \tag{6}$$

where the "farthest curve" $\bar{\Xi}_m^\xi$ generated by model m for feature ξ is defined as:

$$\bar{\Xi}_m^\xi = \{\bar{\xi}_{m,1}, \dots, \bar{\xi}_{m,K_m}\},$$
$$\bar{\xi}_{m,k} = \left(t_{m,k}^\xi, \hat{a}_{m,k}^\xi + \gamma \cdot \mathrm{diag}(\Sigma_{m,k}^\xi)\right). \tag{7}$$

As it can be seen in Figure 2 (b), the possibilities computed by (4) allow an immediate qualitative evaluation of the consistency of the classifier output.

3.3   Reasoning about Uncertainty

Consider again Figure 2 (b), and assume that the monitored person starts to execute the motion *stand up from a chair*. At the beginning no sample referring to *stand up* is inside the window w of run-time data, and $\pi_{stand}^w = 0$. While the person executes the motion, the samples referring to *stand up* fill the window w and thus the value of π_{stand}^w steadily increases up to its maximum value, which corresponds to the ending moment of the motion, when all the samples in w refer to *stand up*. As the person moves on to execute another activity, the samples referring to *stand up* leave the window, thus making the value of π_{stand}^w steadily decrease. In case of a false positive, since there is no correlation between the samples filling the window and the model, the possibility does not follow the outlined *rise-fall* pattern. This argument suggests that a simple temporal analysis of the possibility values can lead to: (i) the identification of the starting and ending moments of each execution of a modelled HMP and (ii) a significant reduction in the number of false positive recognitions.

On such basis, we define the interval I_m of recognition of HMP m as:

$$I_m = \langle t_s, t_e, open, \pi_s, \pi_{max} \rangle, \tag{8}$$

where:

– t_s and t_e denote, respectively, the starting and ending time of the interval;
– *open* is a Boolean value which indicates if the interval is currently open;
– π_s stores the possibility π_m^w recorded at time t_s;
– π_{max} stores the maximum possibility π_m^w recorded until time t_e.

The recognition procedure based on the temporal analysis of the possibility values π_m is outlined in Algorithm 1 (we omit the w superscript for simplicity). Parameter T is introduced to check the symmetry of the rise-fall temporal pattern of the possibility. Lines 4, 5 open a new potential interval, when the possibility $\pi_m > 0$. Procedure **init** initializes template interval I_m as

---

Algorithm 1. Identify intervals I_m of recognition

Require: the possibilities π_m, π_m^{old} computed at time instants t and $t-1$ respectively.
Require: the template intervals $\{I_1, \ldots, I_M\}$.

```
 1: for all m ∈ {1, ..., M} do
 2:     if π_m > 0 then
 3:         if !open then
 4:             init(t, π_m, π_m^old, I_m)
 5:             T ← 1
 6:         else if π_m ≥ π_max then
 7:             update(t, π_m, I_m)
 8:             T ← T + 1
 9:         else
10:             T ← T - 1
11:             if (π_m > π_m^old) ∨ (T = 0 ∧ π_m < π_s) then
12:                 open ← false
13:             else if T ≥ 0 ∧ π_m = π_s ∧ t_s ≠ t_e then
14:                 publish(I_m)
15:             end if
16:         end if
17:     else if open = true ∧ π_s = 0 ∧ t_s ≠ t_e then
18:         publish(I_m)
19:     end if
20: end for
```

$I_m = \langle t, t, true, \pi_m, \pi_m^{old} \rangle$. Lines 7, 8 correspond to the steady increase in the possibility while the person executes the motion. Procedure **update** updates template interval I_m as $I_m = \langle t_s, t, true, \pi_s, \pi_m \rangle$. Lines 10 onward cover the situation in which the person has just finished executing the motion and check whether the possibility follows the expected pattern. More specifically, line 12 resets the interval in case of a false positive recognition, while line 14 corresponds to the recognition of a rise-fall pattern. Procedure **publish** makes the recognition information available to other procedures and resets interval I_m by setting $open \leftarrow false$. Line 18 is the analogous of Line 14 for the case $\pi_s = 0$.

4   Experimental Results

To validate the system, we collected 243 executions of 9 HMP (listed in Fig. 2) and split them into a modelling dataset (37%) and a validation dataset (63%).[2]

We chose the 9 HMP under the assumption of bounded rationality of the user, to be representative of commonly considered ADL. We applied the model builder described in Section 2 to the modelling dataset to build the 9 models.

Table 1 reports the results of the experiment we have designed to estimate, for each motion primitive: the number of true positive recognitions corresponding to

---

[2] We have collected a larger dataset, available at: http://archive.ics.uci.edu/ml/datasets/Dataset+for+ADL+Recognition+with+Wrist-worn+Accelerometer

Table 1. Experimental validation of the rise-fall pattern of the possibility values

HMP	TRF	TnRF	FRF	FnRF
Get up from the bed	73.33%	26.67%	0%	100%
Pick up a glass (drink)	100%	0%	–	–
Put down a glass (drink)	100%	0%	12.5%	87.5%
Pick up a bottle (pour)	93.75%	6.25%	45.45%	54.55%
Put down a bottle (pour)	76.92%	23.08%	2.78%	97.22%
Sit down on a chair	76.92%	23.08%	0%	100%
Stand up from a chair	100%	0%	46.67%	53.33%

a rise-fall possibility pattern (TRF); the number of true positives corresponding to a non rise-fall pattern (TnRF); the number of false positives corresponding to a rise-fall pattern (FRF); and the number of false positives corresponding to a non rise-fall pattern (FnRF). In all cases we have provided the classifier with the trials of the validation dataset and recorded the corresponding possibilities. The motion *pick up a glass (drink)* was never misclassified as a false positive and thus has no entries for the columns FRF and FnRF. The high percentages in columns TRF and FnRF of Table 1 suggests that the rise-fall pattern can be used to reduce the number of false positive recognitions of the classifier.

Table 2. Accuracy, Precision and Recall of the possibility based classifier

HMP	A	P	R
Get up from the bed	92.94%	90%	60%
Pick up a glass (drink)	96.47%	100%	82.35%
Put down a glass (drink)	96.47%	100%	82.35%
Pick up a bottle (pour)	91.76%	77.78%	82.35%
Put down a bottle (pour)	88.24%	76.92%	62.50%
Sit down on a chair	92.94%	100%	64.7%
Stand up from a chair	98.82%	100%	94.74%
AVERAGE	93.95%	92.1%	75.57%

Table 2 reports the accuracy (A), precision (P) and recall (R) of the recognition system implementing Algorithm 1. As expected, motions with higher values of TRF in Table 1 also have higher recall values in Table 2, and vice versa. A comparison with [15] shows that explicit reasoning about uncertainty results in a significant increase in both the recognition accuracy and precision.

Finally, Figure 3 reports the results of a real-time test in which a user performs a full sequence of motion activities. The figure shows the possibility values for each model computed by the possibility based classifier at each time instant. The results indicate that the correlation between the rise-fall pattern and true positive recognitions holds valid also in the case of sequences of motion.

It should be noted that all the motions considered in our experiments are one-shot. For cyclic motions, like *climb the stairs* and *walk*, the simple rise-fall pattern would not apply. We expect that a similar analysis on the temporal

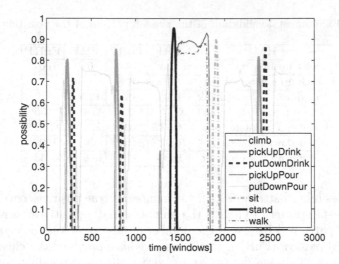

Fig. 3. Output of the possibility based classifier during an execution of the sequence of actions: *pick up a glass (drink), put down a glass (drink), pick up a glass (drink), put down a glass (drink), stand up from a chair, sit down on a chair, pick up a glass (drink), put down a glass (drink)*

evolution of the possibility values can still be made, though, based on their periodic behavior. The validation of this hypothesis is part of our current work.

5   Conclusions

We have described a framework for the recognition of simple human activities based on accelerometer data. Our framework extends classical approaches by modelling uncertainty in recognition through possibility values, and by reasoning about the temporal patterns of these values. Experimental results show that our approach can lead to a significant increase in recognition accuracy and precision.

One of the tenets of our approach is that we can add reasoning about uncertainty on top of existing tools. Specifically, we did not replace the existing classifier, e.g., by designing a classical fuzzy classifier. Rather, we modified it in a minimal way in order to compute possibility values associated to its decisions. Once we do that, we can then design a downstream reasoner module that generates the final decisions based on the analysis of these values and of their temporal evolution. We expect that this modular approach may be profitably applied to other domains as well.

References

1. Katz, S., Chinn, A., Cordrey, L.: Multidisciplinary studies of illness in aged persons: a new classification of functional status in activities of daily living. J. Chron. Dis. 9(1), 55–62 (1959)

2. Mastrogiovanni, F., Scalmato, A., Sgorbissa, A., Zaccaria, R.: An integrated approach to context specification and recognition in smart homes. In: Helal, S., Mitra, S., Wong, J., Chang, C.K., Mokhtari, M. (eds.) ICOST 2008. LNCS, vol. 5120, pp. 26–33. Springer, Heidelberg (2008)
3. Aggarwal, J., Ryoo, M.: Human activity analysis: a review. ACM Comput. Surv. 43(3), 16:1–16:43 (2011)
4. Cirillo, M., Pecora, F., Saffiotti, A.: Proactive Assistance in Ecologies of Physically Embedded Intelligent Systems – A Constraint-Based Approach. In: Mastrogiovanni, F., Chong, N.Y. (eds.) Handbook of Research on Ambient Intelligence and Smart Environments. Information Science Reference, pp. 534–557. IGI Global, Hershey (2011)
5. Lara, O.D., Labrador, M.A.: A survey on human activity recognition using wearable sensors. IEEE Communications Surveys & Tutorials 15(3), 1192–1209 (2012)
6. Bao, L., Intille, S.S.: Activity recognition from user-annotated acceleration data. In: Ferscha, A., Mattern, F. (eds.) PERVASIVE 2004. LNCS, vol. 3001, pp. 1–17. Springer, Heidelberg (2004)
7. Maurer, U., Rowe, A., Smailagic, A., Siewiorek, D.: Location and activity recognition using eWatch: A wearable sensor platform. In: Cai, Y., Abascal, J. (eds.) Ambient Intelligence in Everyday Life. LNCS (LNAI), vol. 3864, pp. 86–102. Springer, Heidelberg (2006)
8. Allen, F., Ambikairajah, E., Lovell, N., Celler, B.: Classification of a known sequence of motions and postures from accelerometry data using adapted gaussian mixture models. Physiol. Meas. 27(10), 935–953 (2006)
9. Lee, M., Khan, A., Kim, T.: A single tri-axial accelerometer-based real-time personal life log system capable of human activity recognition and exercise information generation. Personal and Ubiquitous Computing 15(8), 887–898 (2011)
10. Krassnig, G., Tantinger, D., Hofmann, C., Wittenberg, T., Struck, M.: User-friendly system for recognition of activities with an accelerometer. In: Int. Conf. on Pervasive Computing Technologies for Healthcare, pp. 1–8. IEEE Press, New York (2010)
11. Minnen, D., Starner, T.: Recognizing and discovering human actions from on-body sensor data. In: IEEE Int. Conf. on Multimedia and Expo, pp. 1545–1548. IEEE Press, New York (2005)
12. Bruno, B., Mastrogiovanni, F., Sgorbissa, A., Vernazza, T., Zaccaria, R.: Human motion modelling and recognition: A computational approach. In: IEEE Int. Conf. on Automation Science and Engineering (CASE), pp. 156–161. IEEE Press, New York (2012)
13. Pecora, F., Cirillo, M., Dell'Osa, F., Ullberg, J., Saffiotti, A.: A constraint-based approach for proactive, context-aware human support. Journal of Ambient Intelligence and Smart Environments 4, 347–367 (2012)
14. Bloch, I., Hunter, A.: Fusion: general concepts and characteristics. Int. J. of Intelligent Systems 16(10), 1107–1134 (2001)
15. Bruno, B., Mastrogiovanni, F., Sgorbissa, A., Vernazza, T., Zaccaria, R.: Analysis of human behavior recognition algorithms based on acceleration data. In: IEEE Int. Conf. on Robotics and Automation (ICRA), pp. 1602–1607. IEEE Press, New York (2013)
16. Zadeh, L.A.: Fuzzy Sets as a Basis for a Theory of Possibility. Fuzzy Sets and Systems 1, 3–28 (1978)
17. Ruspini, E.H.: On the semantics of fuzzy logic. Int. J. of Approximate Reasoning 5, 45–88 (1991)

Optimization of Human Perception on Virtual Garments by Modeling the Relation between Fabric Properties and Sensory Descriptors Using Intelligent Techniques

Xiaon Chen[1,2], Xianyi Zeng[1,2], Ludovic Koehl[1,2],
Xuyuan Tao[1,2], and Julie Boulenguez-Phippen[1,2,3]

[1] Université Lille 1 Sciences et Technologies, 59655 Lille, France
[2] GEMTEX, ENSAIT, 2 allée Louise et Victor Champier, 59056 Roubaix Cedex 1, France
[3] GEMTEX, HEI, 13 rue de Toul, 59046 Lille cedex, France

Abstract. 3D virtual garment design using specific computer-aided-design software has attracted a great attention of textile/garment companies. However, there generally exists a perceptual gap between virtual and real products for both designers and consumers. This paper aims at quantitatively charactering human perception on virtual fabrics and its relation with the technical parameters of real fabrics. For this purpose, two sensory experiments are carried out on a small number of fabric samples. By learning from the identified input (technical parameters of the software) and output (sensory descriptors) data, we set up a series of models using different techniques. The fuzzy ID3 decision tree model has shown better performance than the other ones.

Keywords: virtual garment, human perception, CAD, sensory evaluation, fuzzy ID3.

1 Introduction

Under the worldwide economic pressure, there is a strong need for industrial enterprises to quickly design new various products with short life cycles and low costs meeting personalized requirements of consumers in terms of functionalities, comfort and fashion style [1]. Currently, 3D virtual garment design using specific computer-aided-design (CAD) software has attracted a great attention of textile/garment companies. Virtual garment design can be considered as an optimal combination of designers, computer technology and animation technology, permitting to realize and validate design ideas and principles within a very short time [2]. New knowledge and design elements on garments can be obtained from human-machine interactions. The application of virtual technology in garment industry can effectively accelerate new product development and reduce design and production cost in order to enhance the competitiveness of garment companies in the worldwide market [3].

Several popular garment CAD software systems supporting virtual garment collaborative design have been developed by Lectra Company in France (Modaris 3D Fit) , OptiTex, Clo3d and others. In garment industry, the most appreciated CAD systems

A. Laurent et al. (Eds.): IPMU 2014, Part II, CCIS 443, pp. 606–615, 2014.

are based on mechanical models, built according to the mechanical properties of real cloth measured on devices such as KES (Kawabata evaluation system) and FAST (Fabric Assurance by Simple Testing). These models can effectively simulate fabric deformable structures and be accurate enough to deal with nonlinearities and deformations occurring in cloth, such as folds and wrinkles. However, as these models are generated from mechanical laws such as the finite element method and particle system, there generally exists a perceptual gap between virtual and real products for both designers and consumers. In this context, the characterization of human perception on virtual garments with different styles and different fabric materials has become a very important element for the success of new garment design. In practice, the perception of virtual garments can be modified by adjusting the technical parameters of the corresponding garment CAD software so that their perceptual effects are as close as possible to those of real garments. In this way, designers can control some sensory criteria such as softness, smoothness and draping effects in virtual products according to their requirements. The perceptual quality of textile materials and garments is mainly related to two components, i.e. tactile and visual properties. Visual properties essentially include fabric appearance, color, fashion style and garment fitting effects while tactile properties (hand feeling) are generally related to the nature of materials. Researches have shown that a significant portion of tactile properties can be perceived by human's eyes. Thus, creating a quantitative model characterizing the relation between technical parameters and human perception on virtual garments is a very significant approach for designing new user-oriented products.

Many researches have studied the relation between technical parameters and sensory descriptors on real fabrics. The classical analysis methods such as linear regression, multiple factor analysis, PCA (principal component analysis) are usually used tools for that. Also, intelligent techniques, including artificial neural network [4], genetic algorithm [5], fuzzy inference systems, decision tree (such as ID3 [6], CART) and their combinations are applied and found efficient. Recently, a fuzzy decision tree like Fuzzy-ID3 algorithm [7] as an extension of ID3 has been studied, which supports not only symbolic or discrete data but also linguistic data (e.g. small, warm, low) and numerical data represented by fuzzy sets.

This paper aims at charactering human perception on virtual products and its relation with the technical parameters of real fabrics. For this purpose, two sensory experiments are carried out on a small number of fabric samples displayed on a cylinder drapemeter. The target of the first experiment is to find the most appropriate combination of the parameters of the CAD software permitting to minimize the overall perceptual difference between real and virtual fabric draping. The second sensory experiment is aimed to extract normalized tactile and visual sensory descriptors characterizing human perception on the concerned fabric samples. In product design, sensory descriptors can be considered as one part of fabric features and used for communications between designers and consumers. By learning from the identified inputs (technical parameters of the software) and outputs (sensory descriptors) data, a series of models using different techniques, including multi-linear regression, neural network and fuzzy ID3 decision tree are setup. A comparative analysis of the 3 methods is given at the end of this paper.

2   Virtual Fabric and Sensory Tests

2.1   Software for Virtual Fabric Realization

Of all the well-known commercialized 3D garment CAD software systems supporting virtual collaborative design, Clo3d, developed by South Korean company CLO Virtual Fashion, is used in our study for creating virtual fabrics. This selection is not only related to its powerful simulation speed and high garment rendering quality, but also due to its capacity of supporting both static and dynamic effects of the 3D garment simulation. For virtual fabrics, the working scheme of Clo3d is described in Fig.1. Mechanical and optical parameters are required as inputs to generate the behavior of output virtual fabric.

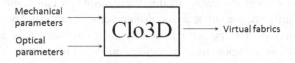

Fig. 1. A scheme of the Clo3D software for generating virtual fabrics

In order to realize virtual fabrics using the Clo3D software, two categories of parameters, i.e. optical and mechanical parameters, are required as inputs. The optical parameters include texture picture, color, brilliance, and transparence while the mechanical ones are composed of fabric stretch resistance, shear resistance, basic mass and others. Those parameters are combined and integrated into the nonlinear models of the software in order to generate fabric deformations, folds and wrinkles.

Different from some classical garment CAD software systems which require precise values of mechanical parameters as input variables, most of the parameters of Clo3D only require relative values from 0 to 99 with respect to a great number of standard references. In this context, the precise values measured on KES or FAST devices are not necessary and the software is then more adapted to industrial companies in terms of cost and time. For determining relative values of the optical and mechanical parameters of each fabric sample, we introduce a design of sensory experiments in order to minimize static and dynamic perceptions between virtual fabrics generated by different values of the parameters and the real ones.

2.2   Virtual Fabric Realization (Sensory Experiment I)

The proposed design of experiments is a qualitative sensory test equivalent to the comparison of a set of products with a standard reference in order to select the best one. The real fabric draping is considered as standard reference and the virtual ones, generated from an orthogonal table defining different combinations of parameters of software, are taken as products. In the qualitative sensory test, each evaluator only needs to compare the real fabric and each virtual product by assigning it a similarity degree (linguistic score).

2.2.1   Fabric Real and Virtual Representation

The static and dynamic behavior of a real fabric draping is realized on a drapemeter. The fabric is let dropped down freely over the drapemeter disc until it finally reaches a balance. The final shape of the fabric is characterized through several images taken by a camera from different angles. For showing optimized effects of the virtual fabric draping, we first extract numerical textures of the real fabric using a calibrated scanner and then integrate them into the software in order to restore original fabric optical rendering.

Of the 9 mechanical parameters required by the software, basic mass and thickness are 2 commonly used parameters in textile industry and they can be easily measured or provided by fabric suppliers. The other 7 parameters are determined from a series of human observations conducted by a design of sensory experiments. An orthogonal table of all the seven parameters with two levels (L_{16}, 2^7) is built for setting two levels (minimal and maximal values of the range) of each mechanical parameter in order to keep the number of sensory experiments as small as possible. For simplicity, the processing of the optical parameters is not discussed in this paper.

From the orthogonal table, we obtain 16 combinations of the mechanical parameters for each fabric. Each combination permits to generate one virtual fabric draping (static and dynamic effects) using the garment CAD software. The experimental environment (e.g. size of the drapemeter and fabric samples, gravity) is the same as that of the real fabric draping. As a result, for each fabric we obtain 16 virtual fabric drapings. They are captured from different angles.

2.2.2   Sensory Experiment I

A sensory experiment is carried out in order to find the optimal combination of the mechanical inputs related to the real fabric draping. A panel of 10 members (4 males and 6 females) is recruited for evaluating the total 19 fabric samples. For each fabric, 16 numerical slides representing 16 combinations of mechanical parameters are diffused to the panelists independently, in which the photos of both real and virtual fabric drapings at the view of 0°, 45°, 90°, 135° and top are put together (Fig.2).

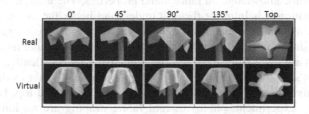

Fig. 2. One example for comparing the real and virtual draping

The panelists are first invited to have a quick overview of all the 16 slides in order to obtain a general impression of the fabrics. Then they are required to evaluate the differences between the real and virtual fabric draping for each combination of the parameters in the previous orthogonal table by selecting a score from 0 to 3. These scores are interpreted as follows: 0 - identical, 1 - close, 2 - different and 3 - very different. The above procedure is repeated for all the 19 fabrics. Finally, all the scores are collected for each fabric. The combination of parameters corresponding to the minimal averaged evaluation score for all the panelists is considered as the optimized inputs to the software, leading the virtual fabric draping very close to the real one.

2.3   Tactile Perceptions Evaluation on Virtual Fabrics (Sensory Experiment II)

In this section, the tactile perceptions are evaluated on the 19 virtual fabrics obtained from sensory experiment I. 6 pairs of fabric hand descriptors (Table 1) are chosen by experts for the evaluation experiment in brain-storm. They represent different features of the fabric in the field of bending, surface, stretch or assemblage of them. The selection of sensory descriptors is carried out according the following three principles. First, the descriptors representing the hand touch proprieties should be attractive for fabric developers, designers and consumers. Second, the tactile descriptors can really be perceived through eyes without direct touch on the real fabrics. Third, descriptors with ambiguity and uncertainty should be avoided.

Table 1. 6 pairs of tactile descriptors

Nm	Descriptor pair	Nm	Descriptor pair	Nm	Descriptor pair
D^1	Pliable – Stiff	D^2	Draped – non draped	D^3	Soft – harsh
D^4	Smooth – rough	D^5	Thin – thick	D^6	Light – Heavy

For each descriptor pair, a scale of 11 scores is used. For example, the scores of descriptor pair "pliable - stiff" is interpreted by: 0 - extremely pliable, 1 - very pliable, 2 - quite pliable, 3 - fairly pliable, 4 - less than medium, 5 - medium, 6 - more than medium, 7 - fairly stiff, 8 - quite stiff, 9 - very stiff, 10 - extremely stiff.

In this scenario, a panel of 12 members (7 males and 5 females) is recruited. About half of them are professionals in textile industry with more than 10 years experiences in the textile product design, development or inspection fields. The other half are researchers including of professors, lecturers or PhD. students in textile universities who have mastered some knowledge on fabric hand properties. The training of the panelists before the evaluation is helpful for them to understand better the meaning of each descriptor and strengthen their evaluation-related knowledge. In this stage, a number of real fabrics are shown to panelists, which are different from the final fabrics to be evaluated. Panelists are free to touch and feel the fabric hand independently.

The panelists start by giving an overall perception of all 19 fabrics draping photos in the virtual environment with a view of 0°, 45°, 90°, 135° and top. Different from the first sensory experiment, photos of real fabric draping are no longer shown to them. This step is important because it can help panelists not only to have a general idea about the fabric features but also 'pre-position' each virtual fabric in the 'scale' for each descriptor. Next, during the evaluation, there is no time restriction for the panelists to evaluate the virtual fabric photos. Each person is free to compare, weigh and judge differences between any pair of the 19 fabrics. The scores of the evaluation result are filled on an answer sheet, and finally collected and treated using the statistical method for each fabric. The averaged value given by all the panelists is the final result for each specific fabric and each descriptor.

3   Characterizing the Relation between Fabric Parameters and Sensory Descriptors

The virtual garment software can provide a platform permitting to simulate fabric static and dynamic behaviors according to its real technique parameters. Perception on virtual fabrics as well as related virtual ambiance can be easily modified according to the preference of consumers by adjusting the previously identified optical and mechanical parameters of the 3D CAD software. In this context, a consumer-oriented new fabric product can be realized in this platform through interactions between fashion designers, material developers and consumers. In this platform, a mathematical model characterizing the relation between technical parameters of fabrics and identified sensory descriptors will be helpful for optimizing these interactions in order to generate the most appropriate virtual fabric parameters. Their relation can be considered as a complex system in which the physically measured parameters and the sensory descriptors are taken as input and output variables respectively (See Fig.3).

Fig. 3. The complex relation between the technical parameters and sensory descriptors

After the previous sensory test (Experiment II), we finally obtain a (19×6)-dimensional matrix, representing all the averaged evaluation data for the 6 pairs of descriptors on 19 fabric samples. They are considered as outputs of the model. The input data constitute a (19×9)-dimensional matrix, corresponding to the 9 identified technical parameters and 19 fabrics.

For setting up an appropriate model for characterizing the relation between the 9 technical parameters of fabrics and each sensory descriptor from the previous learning data, many methods can be available. Before selecting an appropriate model, a pre-processing of the existing data is performed for checking the validity of the input variables, their internal correlation, and predicting the type of the modeling problem (linear or nonlinear) . In our study, linear regression method is first employed.

3.1   Linear Regression

In our case, the 9 inputs (mechanical parameters) are denoted as A^1, A^2...A^9 and 6 outputs (sensory descriptor pairs) as D^1, D^2...D^6 respectively. The method of Leave-One-Out is used for detecting the performance of the model. Namely, of the 19 available input-output data pairs, 18 pairs are taken as the training data to build a regression model. The remaining one is used for testing the performance of the model, expressed by the Root Mean Squared Error (RMSE). This procedure repeats 19 times by testing with each of the data pairs. Before the regression procedure, each of the input and output variables is normalized into the interval [0, 1] in order to remove the effects of different scales.

Fig. 4 shows the performance of the linear regression model characterizing the relation between the physical parameters and the descriptor 'pliable-stiff'. We can see that most RMSEs are around 0.2-0.4. The accuracy of the model is rather low because the normalized evaluation scores are included in [0, 1]. For T13 and T18, the values of RMSE even reach 0.8 and 0.5 respectively, meaning that there is a very large gap between the predicted output and the real output.

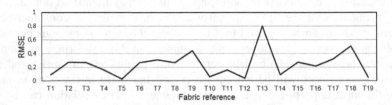

Fig. 4. Performance of the linear regression model between physical parameters and descriptor 'pliable-stiff'

The previous result shows that the linear regression model cannot effectively process nonlinear relations existing between the technical parameters of fabrics and the sensory descriptors.

3.2   Artificial Neural Networks (ANN)

ANN is considered as a dynamic, flexible and adaptive method with high precision for solving complex nonlinear prediction problems. However, it usually requires a great quantity of learning data for determining the appropriate parameters of an ANN network, including the number of layers and the number of neurons inside each layer. A too few number of learning data will cause "over fitting", leading to a very instable performance of the model output.

The characterization of the relation between the technical parameters of fabrics and one sensory descriptor is a problem of modeling with very few learning data. We need to set up the model of 9 input variables from the available 19 learning data. In this situation, we have to reduce the number of inputs in order to conform to the number of learning data. Principal Component Analysis (PCA) is an efficient method for reducing a space of high dimensions into a low dimensional subspace while minimizing information loss of the original data. In our study, PCA is used to the original 9 dimensional input data and we find that the first 6 principle components (independent with each other) represent more than 95% of the original data. Next, a 3-2-1 network structure (3 inputs, 1 hidden layer with 2 neurons, 1 output) is built up (see Fig.5). We also use 'Leave One Out' method to test the performance of the ANN model. Its performance is shown in Fig.6.

The result shows that the combination of the ANN network and PCA cannot give efficient prediction results. Like the results of the linear regression, the values of RMSE are still around 0.2-0.4. The low accuracy is caused by 1) the information lost related to the application of PCA (the explanation rate of the three first principal components is only 60%), 2) there exist a number of local optima, preventing the ANN learning algorithm from converging to the global optimum.

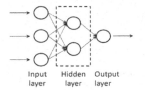

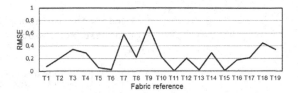

Fig. 5. ANN structure of 3-2-1

Fig. 6. Performance of ANN model between physical parameters and descriptor 'pliable-stiff'

In our experiments, the 3-2-1 structure of the ANN model is very stable because the variance of the testing results for 10 times is very small. However, when we increase the number of neurons in the hidden layer, the results become more and more unstable due to the over fitting problem.

3.3 Fuzzy-ID3 Decision Tree

Let S be a training set of examples dealing with 9 attributes or input variables (9 technical parameters) A^{k}'s $(k=1...9)$, and D one pair of sensory descriptors (output variable). Each variable of A^{k}'s and D is normalized into the interval [0, 1] before its discretization and generation of the corresponding membership functions. After the discretization, we have $A^1, ..., A^7 \in \{0,1\}$ because their corresponding values take two levels only. Also, we have $A^8, A^9, D \in \{0, 0.1, 0.2, 0.3, 0.4, 0.5, 0.6, 0.7, 0.8, 0.9, 1\}$ because their values are continuously distributed in [0, 1]. The membership functions are defined in Fig.7(a) for $A^1, ..., A^7$ while in Fig.7(b) for A^8, A^9 and D.

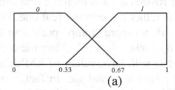

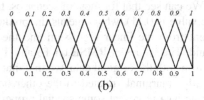

(a)   (b)

Fig. 7. Membership functions for different variables (a) $A^1...A^7$ (b) A^8, A^9 and D

Fig.8 shows an example of the fuzzy decision tree obtained by the FUZZY-ID3 algorithm. It is built for the 9 input attributes and 1 output attribute (pliable-stiff). From this example, we can find that A^8 ("basic mass") is first selected. This selection accords with the knowledge of 'stiffness' because a heavy fabric with big basic mass is usually composed of the gross yarns or weaved by a high density structure. Next, for each branch of A^8, the selection of further attributes is performed according to the computed values of Entropy $E(A,S)$ and Gain $G(A,S)$. Moreover, we define a threshold for them, i.e. $\tau E(A,S)=0.2$ or $\tau G(A,S)=0.1$, in order to stop the generation of the decision tree. For example, the decision tree continues to be split when $A^8=0.2$ and it selects A^4 as its successive node. The further splitting node falls in A^6 with 2 leaves when $A^4=0$ and $A^8=0.2$. Each final classification result (a leaf) is expressed in the form of membership function, which can be defuzzified for prediction.

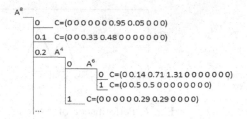

Fig. 8. Fuzzy decision tree characterizing the relation between the physical parameters and the descriptor 'pliable-stiff'

Considering the fact that the number of learning data is too limited, we still use the method of Leave-One-Out to test the effectiveness of the model. The values of RMSE are calculated for each experiment. As an example, the performance of the Fuzzy ID3 model characterizing the relation between the physical properties of fabrics and the descriptor 'pliable-stiff' is shown in Fig.9.

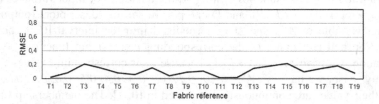

Fig. 9. The performance of Fuzzy ID3 model in the prediction of the descriptor 'pliable-stiff'

We can see that for most of scenarios, the values of RMSE are less than 0.2 (the average value is 0.11), which means the predicted output is rather close to the real one. Different from the linear regression and the ANN method, no more 'jump' point can be found. The values of RMSE on T3 and T15 are slightly higher than 0.2. After the comparison with the original data, we find that the testing data with large values of RMSE are usually "marginal" related to the other data left in the learning database. In fact, these testing data represent some special properties quite different from the others. This problem can be improved by increasing the capacity and quality of the learning database. The more the training data is uniformly distributed, the more the model is efficient.

The fuzzy decision tree with FUZZY-ID3 algorithm is finally proved to be an efficient method for modeling with our dataset. It can model complex nonlinear relations of multiple inputs and single output with small quantity of learning data. Moreover, the fuzzy decision tree is composed of a set of IF…THEN rules, which is more interpretable, flexible and robust than the other models. The problem of over-fitting can be effectively solved.

4   Conclusion

This paper presents the method of realizing virtual fabric using a 3D garment CAD software. For obtaining perceptual effects of a virtual fabric very close to the corresponding real fabric, we carry out a sensory experiment in order to select the most appropriate technical parameters, which constitute the input data to the software. Next, another sensory experiment is carried out to quantitatively characterize the human perception on the obtained virtual fabrics using a set of normalized sensory descriptors.

Based on the learning data of the fabric technical parameters and the sensory data describing virtual fabrics, we try to find the most appropriate model in order to predict and control human perception on virtual fabrics by adjusting the technical parameters of the software. Compare with the other modeling methods, the fuzzy decision tree with Fuzzy-ID3 algorithm has shown its effectiveness. Robustness, flexibility and capacities of interpretation and processing very few data are the main advantages of this method. The proposed model can generate a decision tree for garment designers, which could effectively help them to identify technical criteria of required fabric materials satisfying a specific perceptual preference of consumers in terms of color, texture, style and even fabric hand feeling. In this context, garment consumers can be strongly involved in the product design process and quickly identify satisfying fabrics through a series of interactions with fashion designers and material developers.

References

1. Zeng, X., Li, Y., Ruan, D., Koehl, L.: Computational Textile. SCI, vol. 55. Springer, Heidelberg (2007)
2. Volino, P., Thalmann, N.M.: Virtual clothing theory and practices. Springer (2000)
3. Fontana, M., Rizzi, C., Cugini, U.: 3D virtual apparel design for industrial applications. Computer-Aided Design 37(6), 609–622 (2005)
4. Haykin, S.O.: Neural Networks and Learning Machines, 3rd edn. Pearson (2008)
5. Ruan, D.: Intelligent hybrid systems: fuzzy logic neural networks, and genetic algorithms. Kluwer, Boston (1997)
6. Quinlan, J.R.: Induction on Decision Trees. Machine Learning 1, 81–106 (1986)
7. Bartczuk, Ł., Rutkowska, D.: A new version of the fuzzy-ID3 Algorithm. In: Rutkowski, L., Tadeusiewicz, R., Zadeh, L.A., Żurada, J.M. (eds.) ICAISC 2006. LNCS (LNAI), vol. 4029, pp. 1060–1070. Springer, Heidelberg (2006)

Customization of Products Assisted by Kansei Engineering, Sensory Analysis and Soft Computing

Jose M. Alonso, David P. Pancho, and Luis Magdalena

European Centre for Soft Computing, 33600 Mieres, Asturias, Spain
{jose.alonso,david.perez,luis.magdalena}@softcomputing.es

Abstract. This paper presents a new methodology aimed at making simpler the product/market fit process. We propose a user-centered approach inspired on the Oriental philosophy that is behind Kansei Engineering. In essence, we advocate for customization of products guided by users' expectations. Our proposal combines Sensory Analysis and Soft Computing techniques in order to uncover what users think but also what they feel and desire when facing new products. That is elicitation of the so-called *kanseis* or "psychological feelings". Then, we can design new prototypes that truly matter to people because they fit the deepest users' demands. Thus, improving innovation and marketing success rate. We have illustrated the details of our proposal in a case study related to gin packaging.

Keywords: Fuzzy Logic, Sensory Analysis, Consumer Analysis, Kansei, Iterative Design, Marketing, Gin Packaging.

1 Introduction

Nine out of ten startups fail [11]. The success of business model innovation comes up with one of the following strategies [18]: (1) *to satisfy existing but unanswered market needs*; (2) *to bring new technologies, products, or services to market*; (3) *to improve, disrupt, or transform an existing market with a better business model*; or (4) *to create an entirely new market*.

This paper tries to yield some light regarding the first two strategies. Unfortunately, many new products fail mainly because consumers' desires and needs are not clearly identified on advance [8]: *Marketing research should be employed as a tool to qualitatively and quantitatively specify the product's role among consumers.*

Hence, market research becomes crucial in the product/market fit process [20]. However, eliciting consumers' expectations is not straightforward. According to Maslow, human motivations generally move up through a hierarchy of needs that can be represented by a pyramid with the most basic needs at the bottom [10]. They are physiological needs such as breathing, food, sleep, and so on. Once these basic needs are satisfied then more and more complex needs turn up (safety, belongingness and love, esteem, self-actualization and self-transcendence).

A. Laurent et al. (Eds.): IPMU 2014, Part II, CCIS 443, pp. 616–625, 2014.

Consumers look for products satisfying their own needs. Therefore, consumers' expectations about new products can be seen in a pyramid inspired on Maslow's hierarchy of needs. The basic expectations appear at the bottom (security, healthy, etc.) but more and more demanding expectations (functionality, usability, and so on) turn up while reaching the top (full affective satisfaction).

This paper presents a user-centered methodology for increasing the success rate of introducing new products into market. As a side effect, the consumer relationship management is also improved because we offer to consumers what they actually need. The proposed methodology combines tools provided by Kansei Engineering, Sensory Analysis, and Soft Computing.

The rest of the manuscript is organized as follows. Section 2 presents some preliminaries. Section 3 introduces the proposed framework for customizing products. Section 4 describes a case study which illustrates some of the main benefits coming out from our proposal. Finally, some conclusions and future works are pointed out in Section 5.

2   Preliminaries

2.1   Kansei Engineering

Kansei Engineering establishes a framework to formalize not only what consumers verbalize but what they actually sense when exposed to new products [7,13,14]. The translation of Japanese word *kansei* is not straightforward. It represents a kind of psychological thinking/feeling with deep roots in the Japanese culture. Thus, *kansei* corresponds to affection, feeling and/or emotion [21]. Nagamachi, who is world-wide recognized as the father of Kansei Engineering, started with the so-called *Jocho* technology which focuses on dealing with emotions [13]. However, he understood quickly that customized product design is a multidisciplinary challenge in which considering only emotions is not enough. That is why Nagamachi has actively promoted Kansei Engineering since the 1970's [14]. Kansei Engineering is aimed at designing products that consumers will enjoy and be satisfied with because human feelings and emotions (subjective consumer insights) are driving the design process [15]. Nowadays, Kansei Engineering is still growing all around the world but the emphasis is on the interaction with complementary methodologies and technologies [16].

2.2   Sensory and Consumer Analysis

Sensory analysis is the science of sensory measurements where different techniques have tackled for years with consumers' expectations, likes and dislikes [6, 12], yielding powerful consumer analysis techniques [12]. Sensory sciences [24] have deep roots into physiology, psychology and psychophysics. They apply principles of experimental design and statistical analysis to evaluate consumer products [12]. Moreover, it is important to remark that consumer analysis studies are not reproducible in nature contrary to sensory measurements.

Sensory data coming from human senses (sight, smell, taste, touch and hearing) can be translated into valuable information for companies in order to reinforce/discard what they believe to know about consumer behavior [5]. Therefore, most large companies have departments devoted to sensory analysis what is becoming a key part of their business strategy. They are actually multidisciplinary teams which include sensory professionals, brand managers, marketing researchers, engineers, psychologists, etc. The goal is to increase their knowledge about consumers' needs and expectations with the aim of predicting consumers' responses. To do so, sensory evaluation is carried out by trained and/or untrained panels of human assessors who, for instance, are in charge of testing the new products before they are introduced into market [3].

2.3   Soft Computing

The term Soft Computing was coined in the 1990's by L. A. Zadeh [28] who was already world-wide recognized as father of Fuzzy Logic [26]. It raised rapidly the interest of many other researchers [1]. There are several definitions but probably the most popular ones describe Soft Computing in terms of its essential properties, as a complement of hard computing, as a tool for coping with imprecision and uncertainty, and as a family of complementary techniques (Fuzzy Logic, Neuro-computing, Probabilistic Reasoning, Evolutionary Computation, etc.) which are able to solve lots of complex real-world problems for which classical techniques cannot give accurate results [9]. The constituent technologies are well-known because of their complementary and cooperative nature, i.e., because of their ability to work in a cooperative way, taking profit from the main advantages (and overcoming the main drawbacks) of each other.

3   Proposal

Qualitative and quantitative assessment of consumers' insights into products is inherently subjective. It becomes a key task for designing products fitting the market but it is not straightforward. Soft Computing techniques are well-known because of their ability to tackle with imprecision and uncertainty in system identification [27] but also due to their suitability for computing with perceptions [29, 30]. Vague concepts (such as *tasteful, elegant, exclusive,* and so on) can be formalized, in an approximate but even precise way, thanks to the tools provided by Soft Computing. In consequence, Soft Computing techniques can assist both Sensory Analysis and Kansei Engineering in the design and customization of products.

This section presents a novel framework (Fig. 1) aimed at simplifying both product/market fit and customer relationship management processes.

First of all, a product specification is given to the designer (that is, the person in charge of designing the new product). The specification normally includes technical requirements (regarding security, usability, etc.) but also a set of desired attributes or insights (usually defined by vague concepts like *originality,*

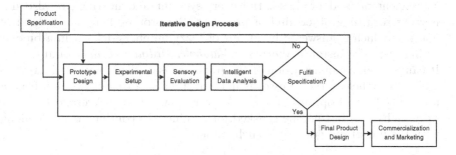

Fig. 1. Proposed framework

exclusiveness, and so on). The product is expected to satisfy consumers' desires and needs identified on advance, in a careful market study. The selected insights correspond to the so-called *kanseis* and are usually characterized by the well-known Osgood's semantic differential space [17,23]. Notice that, the selection of the right *kanseis* (which are going to guide the synthesis of the new product) is a key task deserving careful attention [16].

Then, the designer has to look for the best combination of design parameters (that we call "prototype labels" and can be defined either numerically or linguistically depending on their nature) to satisfy the product specification. The right labels can be uncovered in an iterative design process (the core of Fig. 1) which is made up of four main stages:

1. **Prototype Design.** The given set of labels (such as *height, weight, color,* etc.) is translated into a set of prototypes to be validated. Each label is defined in a range of values, thus several alternative prototypes are built (all of them defined by selecting random values inside the range of interest of each label, for instance *height* between 10 and 20 cm in case of a bottle).
2. **Experimental Setup.** This stage requires a strong statistical background with the aim of setting up the right experiment. Firstly, we must establish the set of *kanseis* to evaluate. Then, we have to select a panel of assessors. The selection of a good panel becomes essential and it depends on the kind of test to perform but also on the target market. In addition, a minimum number of assessors are required in order to provide results with statistical significance. Finally, we should collect a set of samples that is representative enough. Samples can include products provided by rival companies and/or alternative prototypes. Notice that we have to set the right number of samples but also to select their most suitable way of representation. Each sample can consist on a physical object but also on an image/audio/video describing it. In the case of opinion polls they are usually carried out by means of web questionnaires. They can include a set of questions with different kinds of possible answers: yes/no, multiple options, checkbox, open text, ranking, rating scale, etc.
3. **Sensory Evaluation.** The type of evaluation depends on the product (along with the set of *kanseis*) under consideration. An example of pure sensory evaluation (involving human senses like smelling, tasting or touching) is the

assessment of food products. However, eyesight and hearing are also important regarding other kind of products. We propose here a new kind of question which is answered by means of fuzzy numbers defined in a bipolar rating scale (for instance *Worthless-Valuable*, *Ordinary-Unique*, and so on). It can be seen as an extension of the popular Osgood's semantic differential scale. Nevertheless, the assessor is not asked for choosing one option from a pre-defined set of options (1 / 2 / 3 ; Very low / Low / Average / High / Very high; etc.) but he/she is asked for a value in a continuous range along with the uncertainty degree of such value.

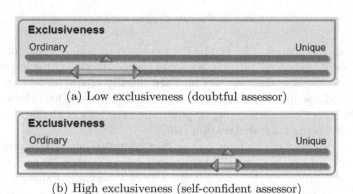

(a) Low exclusiveness (doubtful assessor)

(b) High exclusiveness (self-confident assessor)

Fig. 2. Interpretation of fuzzy answers

Fig. 2 shows an illustrative example. *Exclusiveness* is assessed in a continuous range from *Ordinary* to *Unique*. In the case of the picture at the top (Fig. 2(a)) the assessor gives a small value to *Exclusiveness*. That is, the upper pointer is closer to *Ordinary*. Moreover, the assessor's answer is not very trustful as it can be appreciated by the fact that the interval defined by the lower pointers (which express the uncertainty degree of the assessors' answer) is quite large. The larger the interval the higher imprecision and uncertainty are attached to the given answer which is very likely to be determined by the upper pointer but assuming that in practice it can take whatever value in the whole interval. Notice that, the three pointers define a fuzzy number. The picture at the bottom (Fig. 2(b)) shows the answer provided by a self-confident assessor because the related uncertainty degree is small. Furthermore, that assessor regards the sample under consideration as somehow exclusive (the upper pointer is closer to *Unique*).

4. **Intelligent Data Analysis.** Here, the focus is at processing and analyzing all the answers collected in the previous stage. To do so, we propose the use of Soft Computing techniques, mainly fuzzy logic [26], in order to tackle with the imprecision and uncertainty inherent to the assessors' evaluations which are clearly subjective. To start with, we have to check the goodness and reliability of the available data. Outliers, rare answers very far from the rest, must be identified and discarded, in the search for consensus. In addition, assessor

profiles can be compared and grouped according to different socio-economic criteria. Afterwards, all collected answers for each sample should be properly aggregated [2]. Finally, we can consider learning algorithms, mainly fuzzy modeling [19, 25] and fuzzy association rules [4], in order to match labels (defining the product) and *kanseis* (characterizing the customers' expectations).

The iterative design process ends when at least one prototype fulfills the product specification or no further improvement is achieved in comparison with the previous iteration. The definition of labels is refined, thus their range of possible values is reduced, at the end of each iteration. Therefore, prototypes are expected to be closer to consumers' demands in each new iteration. Once we have a prototype satisfying the given product specification it is time to transform it into a final product ready for the market.

4   An Illustrative Case Study

This section presents a case study aimed at illustrating how the framework previously introduced works in practice. For the sake of clarity, we concentrate just in the last iteration of the iterative design process. Thus, we compare several rival commercial products against the final prototype. Moreover, notice that product specification is given on advance while details related to the final product design, commercialization and marketing remain secret[1]. The four stages in the proposed iterative design process are detailed below:

1. **Prototype Design.** Our case study deals with gin packaging design. We analyzed a representative sample of well-known gin bottles (Fig. 3). Nowadays, choosing the right packaging has a great impact for selling products: *With packaging being now a representation of the product, the development of effective and attractive packaging is critical to the ability to distribute and deliver food that satisfies consumers* [8].
2. **Experimental Setup.** We considered a panel made up of 50 untrained assessors (62% men and 38% women). They were asked to assess 24 gin bottles according to formal and aesthetic criteria. Notice that they had to give their opinion about the bottle design, no matter their drinking preferences.
3. **Sensory Evaluation.** Each assessor had to evaluate five *kanseis* (*Glamour, British Personality, Gender, Exclusiveness*, and *Originality*) for each bottle through an opinion web poll. Fig. 4 shows an example of answers related to the design prototype.
4. **Intelligent Data Analysis.** We first checked all the collected answers looking for outliers. In search of consensus, we discarded the most doubtful answers, i.e. those above the third quartile regarding the related uncertainty degree. Then, we grouped answers according to seven usual linguistic terms (from VL="Very Low" to VH="Very High", through A="Average") represented by a uniform strong fuzzy partition (Fig. 5(a)).

---

[1] This study is part of a real project subject to preserve some confidential information. That is why we cannot reveal prototype design details.

Fig. 3. Bottles of gin under consideration

Fig. 4. Screenshot of the webpoll

In order to provide a linguistic interpretation of the collected answers, we computed the similarity [22] of each fuzzy answer (A) with respect to each fuzzy set (B) in Fig. 5(a):

$$S(A, B) = 1 - \frac{|A \cap B|}{|A \cup B|} \tag{1}$$

where A and B are the two fuzzy sets to be compared, $|\cdot|$ denotes the cardinality of a set, and the $\cap$ and $\cup$ operators represent the intersection and union, respectively. Notice that, the linguistic term related to the fuzzy set B yielding the highest similarity to A is selected for each answer.

Fig. 5(b) shows a histogram with the percentage of answers assigned to each linguistic term (regarding our prototype design in Fig. 4). Most assessors

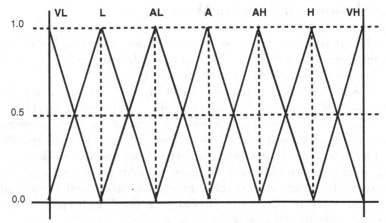

(a) Strong fuzzy partition with seven labels

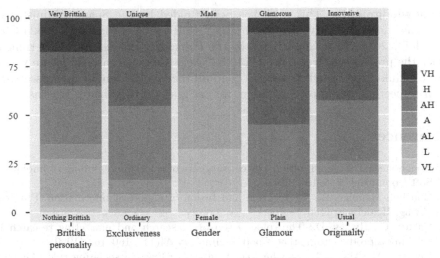

(b) Collected answers after grouping by linguistic terms

Fig. 5. Linguistic interpretation of collected answers

gave high score (between AH and VH) to all *kanseis* except to *Gender*. Comparing all 24 bottles under consideration, the prototype was ranked as the second more *Glamorous*, the ninth more *British*, the second more *Female*, the sixth more *Innovative*, and the fourth more *Unique*. It is worthy to remark that product specification demanded a gin packaging expected to be *Innovative*, *Glamorous*, and *Unique*, but also to be considered as quite *British* and somehow *Male*. Each *kansei* is given by an expert panel and has attached a degree of importance expressed in [0,1]: *Glamour*=0.9; *British Personality*=0.6; *Gender*=0.3; *Exclusiveness*=0.7; and *Originality*=0.8. Our prototype design was the first ranked when considering all five *kanseis* together (in a weighted linear combination). Hence, product specification was successfully fulfilled.

5   Conclusions and Future Work

We have introduced a new framework for human-centered design of products that is supported by three well-known and established research fields: Kansei Engineering, Sensory Analysis and Soft Computing.

It has already been applied in a real case study about gin packaging. We have reported some preliminary results which only illustrates a small part of the entire framework that is only the top of the iceberg. A lot of work is in progress.

In the near future we will extend this preliminary study by exploring different groups of users with clustering techniques, trying several aggregation operators, exploring alternate graphical representations, looking for correlated answers, learning rules (fuzzy association rules, fuzzy rule bases, and so on) supporting the design of new prototypes, statistical tests, etc. We also plan to apply our framework in other related case studies.

Acknowledgments. This work has been funded by the Spanish Ministry of Economy and Competitiveness under ABSYNTHE project (TIN2011-29824-C02-01 and TIN2011-29824-C02-02) and by the Principality of Asturias Government under the project with reference CT13-53. Regarding the case study, we would like to thank "Rød brand consultans" and all the involved anonymous assessors. With respect to the prototype design, we are grateful to Valentín Iglesias.

References

1. Bonissone, P.: Soft computing: The convergence of emerging reasoning technologies. Soft Computing 1, 6–18 (1997)
2. Bouchon-Meunier, B.: Aggregation and fusion of imperfect information. Physica-Verlag, Heidelberg (1997)
3. Carter, C., Riskey, D.: The roles of sensory research and marketing research in bringing a product to market. Food Technology 44(11), 160–162 (1990)
4. Delgado, M., Marín, N., Sánchez, D., Vila, M.A.: Fuzzy association rules: General model and applications. IEEE Trans. on Fuzzy Systems 11(2), 214–225 (2003)
5. Douglas, S.P., Craig, C.S.: The changing dynamic of consumer behavior. Implications for cross-cultural research. International Journal of Research in Marketing 14, 379–395 (1997)
6. Galmarini, M.V., Symoneaux, R., Chollet, S., Zamora, M.C.: Understanding apple consumers' expectations in terms of likes and dislikes. Use of comment analysis in a cross-cultural study. Appetite 62, 27–36 (2013)
7. Ikeda, G., Nagai, H., Sagara, Y.: Development of food kansei model and its application for designing tastes and flavors of green tea beverage. Food Science and Technology Research 10(4), 396–404 (2004)
8. Lord, J.B.: New product failure and success. In: Developing New Food Products for a Changing Marketplace, pp. 4.1–4.32. CRC Press (1999)
9. Magdalena, L.: What is soft computing? revisiting possible answers. International Journal of Computational Intelligence Systems 3(2), 148–159 (2010)
10. Maslow, A.H.: A theory of human motivation. Psychological Review 50(4), 370–396 (1943)

11. Maurya, A.: Running Lean: Iterate from plan A to a plan that works. O'Reilly Vlg. Gmbh & Co. (2010)
12. Naes, T., Brockhoff, P.B., Tomic, O.: Statistics for sensory and consumer science. Wiley (2010)
13. Nagamachi, M.: A study of emotional technology. Japanese Journal of Ergonomics 10(2), 121–130 (1974)
14. Nagamachi, M.: Kansei Engineering. Kaibundou, Tokyo (1989)
15. Nagamachi, M.: Kansei engineering: The implication and applications to product development. In: IEEE International Conference on Systems, Man, and Cybernetics, pp. 273–278 (1999)
16. Nagamachi, M.: Perspectives and the new trend of kansei/affective engineering. TQM 20(4), 290–298 (2008)
17. Osgood, C.E., Suci, G., Tannenbaum, P.: The measurement of meaning. University of Illinois Press (1957)
18. Osterwalder, A., Pigneur, Y.: Business model generation. John Wiley & Sons, Inc. (2010)
19. Pedrycz, W., Gomide, F.: Fuzzy modeling: Principles and methodology. In: Fuzzy Systems Engineering: Toward Human-Centric Computing (2007)
20. Schneider, J., Hall, J.: The new launch plan: 152 tips, tactics and trends from the most memorable new products. BNP Publisher (2010)
21. Schütte, S.: Engineering emotional values in product design. PhD thesis, Linköpings Universitet, Kansei Engineering Group (2005)
22. Setnes, M., Babuska, R., Kaymak, U., Van Nauta Lemke, H.R.: Similarity measures in fuzzy rule base simplification. IEEE Trans. on Systems, Man and Cybernetics, Part B 28(3), 376–386 (1998)
23. Snider, J.G., Osgood, C.E.: Semantic differential technique: A sourcebook. Aldine, Chicago (1969)
24. Stone, H., Bleibaum, R.N., Thomas, H.A.: Sensory evaluation practices. Food Science and Technology, 4th edn. International Series. Academic Press (2012)
25. Yager, R.R., Filev, D.P.: Essentials of fuzzy modeling and control. John Wiley, New York (1994)
26. Zadeh, L.A.: Fuzzy sets. Information and Control 8, 338–353 (1965)
27. Zadeh, L.A.: Outline of a new approach to the analysis of complex systems and decision processes. IEEE Trans. on Systems, Man, and Cybernetics 3, 28–44 (1973)
28. Zadeh, L.A.: Soft computing and fuzzy logic. IEEE Software 11(6), 48–56 (1994)
29. Zadeh, L.A.: From computing with numbers to computing with words - from manipulation of measurements to manipulation of perceptions. IEEE Trans. on Circuits and Systems - I: Fundamental theory and applications 45(1), 105–119 (1999)
30. Zadeh, L.A.: A new direction in AI toward a computational theory of perceptions. AI Magazine, American Association for Artificial Intelligence 1, 73–84 (2001)

Author Index